GENERAL THEORY OF HIGH SPEED AERODYNAMICS

VOLUME VI

HIGH SPEED AERODYNAMICS

AND JET PROPULSION

GENERAL THEORY OF HIGH SPEED AERODYNAMICS

EDITOR: W. R. SEARS

PRINCETON, NEW JERSEY

PRINCETON UNIVERSITY PRESS

1954

London: Geoffrey Cumberlege, Oxford University Press

L. C. Card 52-8779

Printed in the United States of America by
The Maple Press Inc., York, Pennsylvania

FOREWORD

On behalf of the Editorial Board, I would like to make an acknowledgement to those branches of our military establishment whose interest and whose financial support were instrumental in the initiation of this publication program. It is noteworthy that this assistance has included all three branches of our Services. The Department of the Air Force through the Air Research and Development Command, the Department of the Army through the Office of the Chief of Ordnance, and the Department of the Navy through the Bureau of Aeronautics, Bureau of Ships, Bureau of Ordnance, and the Office of Naval Research made significant contributions. In particular, the Power Branch of the Office of Naval Research has carried the burden of responsibilities of the contractual administration and processing of all manuscripts from a security standpoint. The administration, operation, and editorial functions of the program have been centered at Princeton University. In addition, the University has contributed financially to the support of the undertaking. It is appropriate that special appreciation be expressed to Princeton University for its important over-all role in this effort.

The Editorial Board is confident that the present series which this support has made possible will have far-reaching beneficial effects on the further development of the aeronautical sciences.

Theodore von Kármán

PREFACE

Rapid advances made during the past decade on problems associated with high speed flight have brought into ever sharper focus the need for a comprehensive and competent treatment of the fundamental aspects of the aerodynamic and propulsion problems of high speed flight, together with a survey of those aspects of the underlying basic sciences cognate to such problems. The need for a treatment of this type has been long felt in research institutions, universities, and private industry and its potential reflected importance in the advanced training of nascent aeronautical scientists has also been an important motivation in this undertaking.

The entire program is the cumulative work of over one hundred scientists and engineers, representing many different branches of engineering and fields of science both in this country and abroad.

The work consists of twelve volumes treating in sequence elements of the properties of gases, liquids, and solids; combustion processes and chemical kinetics; fundamentals of gas dynamics; viscous phenomena; turbulence; heat transfer; theoretical methods in high speed aerodynamics; applications to wings, bodies and complete aircraft; nonsteady aerodynamics; principles of physical measurements; experimental methods in high speed aerodynamics and combustion; aerodynamic problems of turbomachines; the combination of aerodynamic and combustion principles in combustor design; and finally, problems of complete power plants. The intent has been to emphasize the fundamental aspects of jet propulsion and high speed aerodynamics, to develop the theoretical tools for attack on these problems, and to seek to highlight the directions in which research may be potentially most fruitful.

Preliminary discussions, which ultimately led to the foundation of the present program, were held in 1947 and 1948 and, in large measure, by virtue of the enthusiasm, inspiration, and encouragement of Dr. Theodore von Kármán and later the invaluable assistance of Dr. Hugh L. Dryden and Dean Hugh Taylor as members of an Editorial Board, these discussions ultimately saw their fruition in the formal establishment of the Aeronautics Publication Program at Princeton University in the fall of 1949.

The contributing authors and, in particular, the volume editors, have sacrificed generously of their spare time under present-day emergency conditions where continuing demands on their energies have been great. The program is also indebted to the work of Dr. Martin Summerfield who guided the planning work as General Editor from 1949–1952. The cooperation and assistance of the personnel of the Princeton University Press and of the staff of this office has been noteworthy. In particular, Mr. H. S. Bailey, Jr., the Director of the Press, and Mr. R. S. Snedeker, who has supervised the project at the Press and drawn all the figures, have been of great help. Special mention is also due to Mrs. H. E. H. Lewis of this office who has handled the bulk of the detailed editorial work for the program from its inception.

Joseph V. Charyk
General Editor

PREFACE TO VOLUME VI

This volume summarizes present-day knowledge in the theory of high speed aerodynamics. It begins with an over-all sketch of the subject by Theodore von Kármán. Largely nonmathematical, Professor von Kármán's section touches on the major features of subsonic, transonic, supersonic, and hypersonic flows and emphasizes both important recent developments and some problems still unsolved. K. O. Friedrichs presents a new study of the mathematical nature of hyperbolic flows, i.e. supersonic or unsteady compressible flows.

Following these introductory sections, attention is directed especially toward the theories based on the small disturbance hypothesis, which have proved so useful. W. R. Sears presents and summarizes subsonic small perturbation theory, with emphasis on recent studies that unify and clarify this subject, and on rotational and unsteady flows. The largest section of the volume is by Max. A. Heaslet and Harvard Lomax, who present in a unified manner the whole subject of linearized supersonic theory, especially with application to wings. Their section also includes the transonic and hypersonic nonlinear approximations. M. J. Lighthill's contribution concerns the methods of higher approximation which are used to improve the first order theories and to explain and eliminate their shortcomings. This piece of work is unique in aeronautical literature and should be an especially valuable feature of the volume.

Departing now from the ideas of perturbation theories, Y. H. Kuo and W. R. Sears collaborate in a section presenting a review of a highly complicated and confused subject: the theory of potential subsonic and mixed supersonic-subsonic plane flows based on the hodograph method. Finally, in two sections, Antonio Ferri writes on the method of characteristics applied to steady two- and three-dimensional flows and flow patterns characterized by the presence of shock waves. These sections include Professor Ferri's recent work on the linearized methods of characteristics.

The seven authors of this volume have cooperated admirably with the editors throughout the preparation of the work. To them, to the General Editor and his staff, and to Princeton University Press the volume editor expresses his sincere thanks.

W. R. Sears
Volume Editor

CONTENTS

Y. H. Kuo, Graduate School of Aeronautical Engineering, Cornell University, Ithaca, New York
W. R. Sears, Graduate School of Aeronautical Engineering, Cornell University, Ithaca, New York

Antonio Ferri, Department of Aeronautical Engineering, Polytechnic Institute of Brooklyn, Brooklyn, New York

GENERAL THEORY OF HIGH SPEED AERODYNAMICS

SECTION A

ON THE FOUNDATION OF HIGH SPEED AERODYNAMICS

TH. VON KÁRMÁN

A,1. Incompressible Flow. Looking back on fifty years of aerodynamics research during the first half of this century, it appears to me most remarkable that the crude approximation which considers the air as an incompressible nonviscous fluid has proved itself so valuable in solving many practical problems of aircraft design.

In certain cases, for example in the performance calculations of airplanes and propellers, useful results were obtained even by a further simplification: by approximating the actual flow by infinitely small perturbations of a uniform and parallel airflow. Of course it was recognized at the time that certain phenomena, especially those connected with drag and stalling, require the consideration of viscosity. However, even for such problems Prandtl's classical idea of restricting the influence of viscosity to the neighborhood of solid walls—i.e. the concept of the boundary layer—proved to be sufficiently exact for the description, and in some cases for the prediction, of the phenomena.

This state of affairs was fundamentally changed by the advent of high speed aircraft. First it seemed that compressibility troubles could be met by "compressibility corrections." However, very soon it became clear that the engineer needed a full grasp and knowledge of fluid mechanics over the entire speed range, extending from incompressible flow to flows with velocities large in comparison to the velocity of sound, and over a density range extending almost to a complete vacuum.

The concept of an incompressible fluid ignores the fact that pressure variations are propagated in a fluid with a finite velocity. The assumption of instantaneous propagation of pressure introduces essential mathematical simplifications which make possible the application of Laplace's equation and the techniques of conformal transformation, i.e. the simplest and most popular methods of mathematical physics. It also justifies the concept of an *apparent mass* in the study of nonsteady phenomena. It is one of the fundamental theorems of the mechanics of incompressible nonviscous fluids that an arbitrary continuous sequence of irrotational flow patterns always represents a dynamically correct transient flow. It is evident that this means an enormous reduction of difficulties in dealing with nonsteady phenomena.

A,2. Propagation of Pressure. I believe the first calculation of the propagation of a pressure wave through air was made by Sir Isaac Newton. Since he could not know the difference between isothermal and isentropic compression, his formula for the velocity of sound has a wrong numerical factor. He says quite clearly, however, in Theorem XXXVIII of his second book [*1*]: "The velocities of pulses propagated in an elastic fluid are in a ratio compounded of the square root of the ratio of the elastic force directly, and the square root of the ratio of the density inversely; supposing the elastic force of the fluid to be proportional to its condensation."

In the same book he gives the following description of the mechanism of the resistance of a solid body moving in an elastic fluid: "Projectiles excite a motion in fluids as they pass through them, and this motion arises from the excess of the pressure of the fluid at the fore parts of the projectile above the pressure of the same at the hinder parts; and cannot be less in mediums infinitely fluid than it is in air, water, and quicksilver, in proportion to the density of matter in each. Now this excess of pressure does, in proportion to its quantity, not only excite a motion in the fluid, but also acts upon the projectile so as to retard its motion; and therefore the resistance in every fluid is as the motion excited by the projectile in the fluid; and cannot be less in the most subtle ether in proportion to the density of that ether, than it is in air, water, and quicksilver, in proportion to the densities of those fluids." We shall see later that this concept of the drag was correct for the case of the supersonic motion of projectiles, whereas in the subsonic case—at least for rounded bodies—his description is at variance with d'Alembert's theorem.

A,3. Supersonic vs. Subsonic Flow. From Newton's concept of the propagation of pressure (Fig. A,3a) in an elastic fluid, one arrives directly at the well-known picture of subsonic and supersonic flows. For the following considerations we restrict ourselves to flows produced by small disturbances and neglect viscosity, i.e. the absorption of energy in the air.

Since a slight pressure change is propagated at sound velocity, it is evident that the effect of pressure changes produced in the air by a body moving faster than sound cannot reach points ahead of the body. It may be said that the body is unable to send signals ahead. It is seen that there is a fundamental difference between subsonic and supersonic motion.

Consider the case of subsonic steady motion—for example, the uniform level flight of an airplane. Here a pressure signal travels ahead at sound velocity minus flight velocity, relative to the airplane, whereas a signal

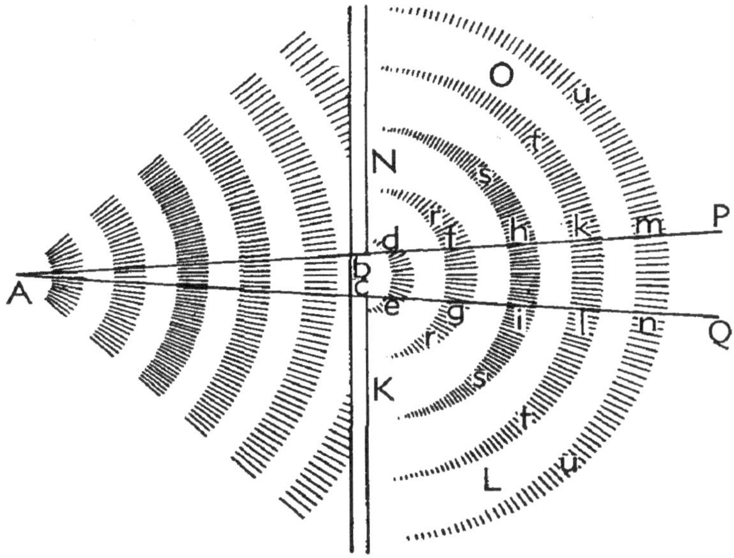

Fig. A,3a. Newton's illustration of sound propagation in the case of diffraction through an opening. This illustration is used here merely to show his concept of the propagation of pressure disturbances in the shape of spherical waves.

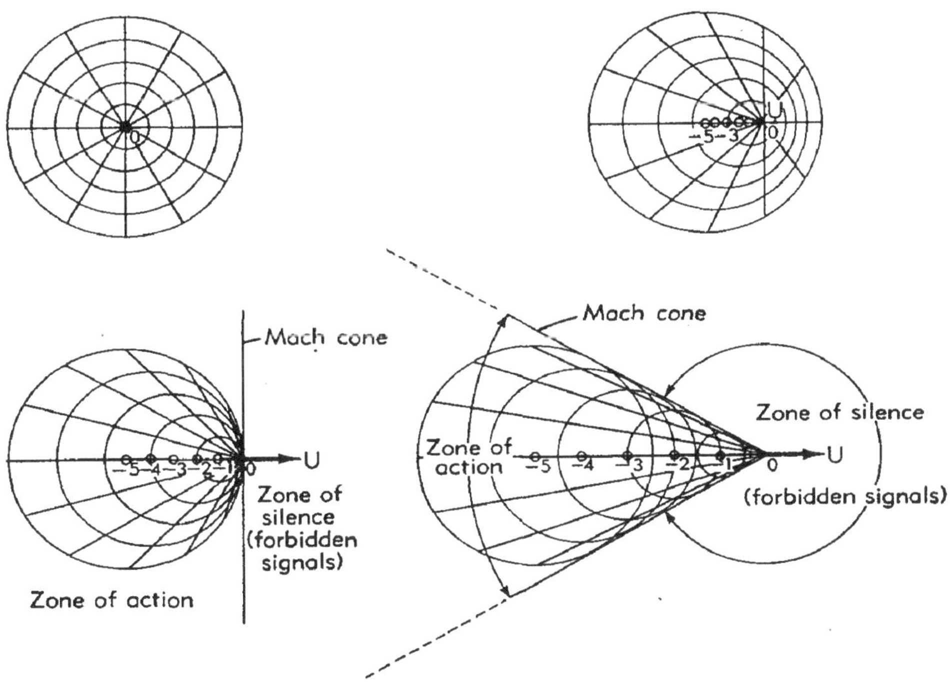

Fig. A,3b. Propagation of sound signals from stationary and moving sources.

travels backward at a speed equal to the sum of flight and sound velocities. So, although the mechanism of pressure propagation is no longer symmetric, every point in space is reached by a signal, provided the flight started from an infinitely remote point. As can easily be seen, this is not the case in supersonic flight.

Take the simplest case of a point source (Fig. A,3b). The upper left diagram shows the spherical surfaces reached by the pressure effect in equal time intervals in the case of a point source at rest. The upper right shows the same surfaces relative to the point source moving with a speed less than that of the sound. The lower left represents the case of a point source moving with sonic velocity, and the lower right, the case of a source moving faster than sound. It is seen that in the last case all action is restricted to the interior of a cone that includes all the spheres emitted by the source before the instant considered. The region outside of this cone can be called the *zone of silence.* It is easily seen that the sine of the half vertex angle of the cone is equal to the reciprocal of the Mach number. This angle is called the Mach angle. The cone that separates the zone of action from the zone of silence is called the *Mach cone.*

The points plotted in Fig. A,3b show the location of mass points that are supposed to be emitted from the source and to move at sound velocity in all directions. They illustrate qualitatively the distribution of the density of action in the various cases. In the subsonic case one finds that the pressure effect not only decreases with increasing distance from the source, but is also dispersed in all directions. In the case of a body moving at supersonic velocity, the bulk of the effect is concentrated in the neighborhood of the Mach cone that forms the outer limit of the zone of action.

A,4. Linearized Theory of Supersonic Flow. The so-called linearized theory of supersonic flow builds up the flow produced by the motion of a body by superposition of small disturbances such as considered in the last paragraph. One can develop in this way relatively simple methods for the computation of velocity and pressure distributions in the field and also for the computation of the forces, lift and drag, acting on moving bodies. In the last decade very extensive analytical work has been done using the linearized theory of supersonic flow. This grew out of such modest beginnings as Ackeret's work on the lift of a two-dimensional thin airfoil, published in 1925 [*2*], the computation of the drag of slender bodies, published by N. B. Moore and the writer in 1932 [*3*], the article on "Problems of resistance in compressible fluids," presented by the writer at the Volta Congress for high speed in 1935 in Rome [*4*], and Busemann's work of 1935 [*5*].

In the case of irrotational flow, the equations of motion can be reduced to equations analogous to the wave equation. The coordinate parallel to the direction of the main flow or the motion of the missile, wing, or body plays the role of the time coordinate. Hence, the well-established methods of finding solutions of the wave equation can be used in a great number of problems of practical importance, similarly to the capabilities of the methods for the solution of Laplace's equation in the case of incompressible flow. Such methods constitute the subject of Sec. D of the present volume.

Wave drag. The concept of the linearized supersonic flow theory also reveals the existence of a novel kind of drag, which we do not encounter in subsonic motion and which we designate as "wave drag."

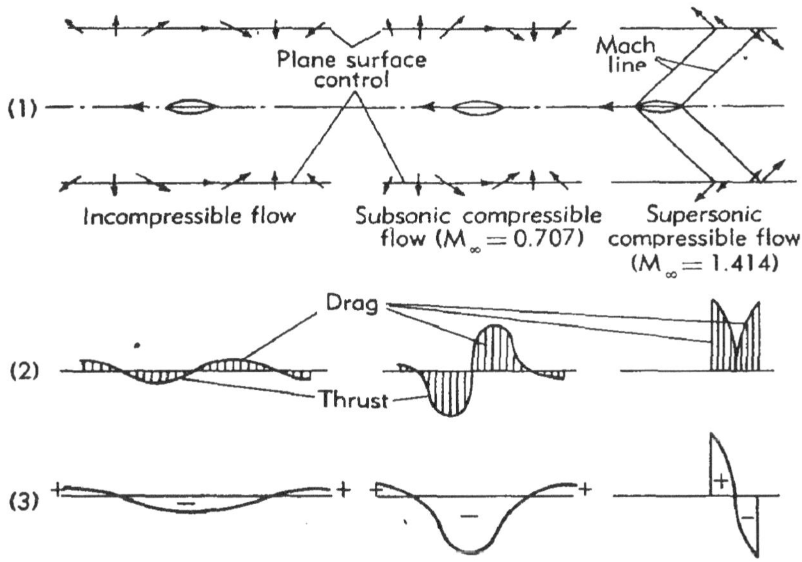

Fig. A,4. Comparisons of disturbance fields of airfoils in various speed regimes.

As before, we neglect viscosity and assume that the motion of the body produces disturbances that can be considered small. At a certain distance from the moving body this second assumption will, in general, be satisfied. Now consider the body and the surrounding air inside a cylindrical control surface as one mechanical system. Then one finds that, because of the concentrated action that characterizes the propagation of pressure from a source moving at supersonic velocity, the total flux of momentum of the air masses entering and leaving the cylindrical boundary remains finite even when the boundary is removed to an arbitrarily large distance.

Fig. A,4 refers to the case of a two-dimensional symmetric airfoil, with sharp leading edge, moving through air initially at rest. Let us consider the flow through a plane parallel to the plane of symmetry at a certain distance from the body. The diagram first shows the distribution of induced velocities (1) and the horizontal component of momentum transfer (2) along this plane for three cases. It is evident that the reaction of outgoing flow having a horizontal component opposite to the flight direction and incoming flow with a horizontal component in the flight direction is equivalent to a propulsive thrust acting on the body. Conversely, outgoing flow with a component in the flight direction and incoming flow directed opposite to the flight direction give rise to a drag. In the two subsonic cases ($M_\infty \to 0$ and $M_\infty = 0.707$), thrust and drag contributions are balanced, and the total horizontal momentum transfer is equal to zero. This is in accordance with d'Alembert's theorem.

The influence of increasing Mach number is essentially to increase the magnitude of the induced velocities and to increase the concentration of the disturbance in the region extending laterally outward from the body. The increase in the concentration of action is also illustrated by the pressure distribution (3) on the control surface. In the supersonic case ($M_\infty = 1.414$), the disturbance is restricted to two strips bounded by two Mach lines. These lines are the intersections of planes that are envelopes of the Mach cones starting from points of the leading and trailing edges of the airfoil. The horizontal component of the outward flow is in the flight direction; that of the inward flow is the opposite. Hence, both represent drag on the body; this is called *wave drag.*

Linearized theory limitations. The linearized theory, however, has serious limitations. First, it gives only a first approximation, since all deviations from the uniform parallel flow are considered infinitely small and therefore additive. If the disturbance caused by the body cannot be considered small, the linearized theory needs important corrections. To this end considerable progress has recently been made in the development of methods of higher approximation (see Sec E). Such methods permit taking into account the existence of discontinuities (shock waves) in supersonic flow.

Second, there are speed ranges in which the linearization of the equation of motion, even for small disturbances, is not justified. The conditions for the validity of the linearized theory of steady flow are twofold:

1. The perturbation velocities must be small in comparison with both the main-stream velocity and the velocity of sound.
2. The perturbation velocities must be small in comparison with the difference between the mainstream velocity and the sound velocity.

Condition (1) excludes the case of very high velocities; evidently, if the main-stream velocity is several times larger than the sound velocity, disturbances which are small relative to the main-stream velocity may be of the same order of magnitude as the sound velocity. This speed range is called the *hypersonic range.*

On the other hand, condition (2) excludes the range near $M_\infty = 1$ from the scope of the linearized theory. We call this range the *transonic range.* In these two cases, even the study of small perturbation flows requires solution of nonlinear equations (D,32-D,34).

A,5. Finite Disturbances in Compressible Flow. Shock Waves. The solution of the equations of motion in the case of finite disturbances of a uniform parallel stream of an ideal compressible fluid is, in general, a cumbersome mathematical problem. The problems of proceeding beyond the successive approximation techniques treated in Sec. E, are somewhat simpler for a supersonic than for a subsonic stream. In the former, at least in the domains where the flow remains supersonic, the method of characteristics (Sec. G) can be applied with good results. The transfer of the computation from the physical plane to the so-called hodograph plane, i.e. the replacement of space coordinates by velocity coordinates as independent variables (Sec. F), is helpful in many problems, since the equations in the hodograph plane become linear. As was mentioned on the preceding page, a new phenomenon has to be taken into account: from the mathematical point of view, the possibility of discontinuous solutions and, from the physical point of view, the existence of shock waves.

Shock waves. In subsonic flow the only possible discontinuous change of the velocity is tangential to a surface consisting of streamlines. Such a discontinuity does not violate any mechanical law because of the non-existence of shear stresses in the fluid. In supersonic flow, however, a discontinuous change of velocity along streamlines is also possible. Such a discontinuous change is called a *shock*. In fact, the laws of mechanics are satisfied provided the fluid element conserves its mass, momentum, and energy.

The physical reason why a discontinuous change is possible only in supersonic flow can easily be seen. The theorem of conservation of mass calls for the equality of the so-called Fano number (product of density and normal velocity) on both sides of the surface. The question arises: Can this product have the same value for different individual values of velocity and density while at the same time the momentum and energy of the fluid element remain unchanged? Now, if we consider the expansion or compression process of a gas, we find that the Fano number has a maximum when the velocity of the gas is equal to the velocity of sound. Consequently, in the case of a shock, the velocity normal to the discontinuity surface has to be subsonic on one side and supersonic on the other side. Therefore no discontinuous change can occur in a purely subsonic steady flow. The rule, however, does not exclude a discontinuous change in a purely supersonic flow, since it refers to the components of the velocity normal to the discontinuity surface. Thus an *oblique shock* is possible also in a purely supersonic flow.

One gains a certain insight into the physical relationships by the consideration of the following problem. Assume a gas flow in a cylindrical pipe of constant cross section starting with a given pressure p_0, velocity V_0, temperature T_0, and stagnation enthalpy h^0. Compute the variation of these quantities when heat in the amount Q is introduced into the gas. One finds that, if we start with an arbitrary subsonic velocity V_0, the velocity can be increased only to the sonic limit by introduction of heat. If one computes the velocity as function of the introduced heat beyond the sonic limit, one finds that heat has to be taken out of the gas. The curve showing the velocity and the temperature as functions of the introduced heat is represented in Fig. A,5. Evidently there are two states for every value of the total enthalpy, or the sum $h^0 + Q$, such that both states correspond to the same values of mass flow and momentum.

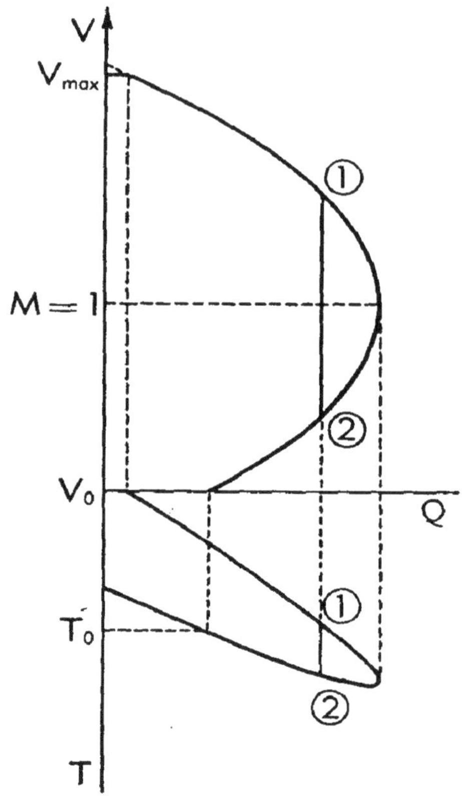

Fig. A,5. Velocity and temperature produced in a gas flowing in a cylindrical pipe by addition of heat.

Mathematically, a sudden transition would be possible either from subsonic to supersonic velocity or from supersonic to subsonic velocity. However, if we consider the transition between the points ① and ②, i.e. from high to low velocity along the continuous curve shown in Fig. A,5, we find that, in order to realize this transition, heat is introduced at a low temperature level and taken out at a high temperature level; whereas, if the process is carried out in the opposite sense, from low to high velocity, heat is taken out at a low temperature and introduced at a high temperature. Evidently, according to the second law of thermodynamics, the second process is impossible without applying mechanical work. Hence, the only possible transition is the compression shock, i.e. the transition from supersonic to subsonic flow. Of course, the same answer can be obtained by computing the entropy corresponding to the two states; then we find that the compression shock involves increase of entropy whereas the expansion wave would mean decrease of entropy. No discontinuous change of state can occur without change of entropy, but in so-called weak shocks the entropy rise is small. This explains why the linearized theory gives reasonable results, even in cases in which weak shocks are involved.

The concept of shock waves goes back almost a century to Riemann [*6*], who, I believe, was the first to recognize the essential difference between the propagation of infinitely small and of finite pressure variations. The theory of shock waves was developed by Rankine, Hugoniot, Jouguet, Hadamard, and many others.

In the last ten or fifteen years very intensive research activities were carried out in several countries to clear up questions related both to stationary and nonstationary shock waves and shock-wave combinations. The problem of shock waves has a bearing on many problems outside of supersonic aeronautics—for example detonation waves—but also has great importance for several practical aeronautical problems. In fact shock waves may cause sudden change in the aerodynamic behavior of high speed aircraft, affecting their balance, stability, and control and producing undesirable vibrations. Of course, the appearance of shock waves is also closely correlated with the increase of drag experienced when we approach sonic velocity.

Drag. In the case of purely subsonic flow of a nonviscous fluid—disregarding wakes produced by flow separation—one obtains zero drag because of the lack of any surplus in the momentum transfer through a far-removed control surface. In the supersonic case, according to the linearized theory, the momentum transfer through the sides of the control surface remains finite even at infinity and the result of this effect is the wave drag. On the other hand, if there are shock waves present, either attached to or detached from the body, a certain wake necessarily follows the body. The reason is that, even if the pressure becomes uniform far behind the body and equal to the ambient pressure, the velocity relative to the body cannot everywhere reach the magnitude of the velocity in front of the body because of the change in stagnation pressure, or change in entropy, through the shock wave.

It appears at first sight that the "wave drag" and the "wake drag" are additive components of the total drag. However, on closer consideration it becomes difficult to discriminate between wave drag as defined in the linearized theory and the wake drag caused by the "shadow" of the shock wave or waves. As a matter of fact, probably even in the case of a nonviscous fluid, if the linearized solution for the flow is replaced by an exact solution of the equations of aerothermodynamics, the concept of wave drag fades away. In other words, the exact solution probably does not show momentum transfer sideways to infinity; the shock waves are "attenuated," so that the entire drag is due to the difference of momentum in front of the body and in the wake. There is no general proof for this statement, however; the interesting point is that the wave drag computed by means of the linearized theory gives quite a fair approximation to the actual drag. The situation may be similar to the case of the linearized theory of incompressible fluids, which leads to the formula for the induced drag. Also in this case, one calculates the momentum at infinity, in the so-called Trefftz plane, although one realizes that the idealized downwash system at infinity has to be quite different from the real one because of Helmholtz's law of the conservation of vorticity. However, if we consider the flow in a plane sufficiently far behind the wing, but not so far that the rolling up of the vortex system has already essentially changed the velocity distribution, the linearized theory of the induced drag furnishes a good approximation.

A,6. The Hypersonic Range. As was mentioned in Art. 4 above, the linearized theory does not apply to very high stream velocities (hypersonic flow).

The main practical application of the aerodynamics of hypersonic flow is the study of the motion of projectiles and similar bodies with speeds at velocities large in comparison to the velocity of sound. The linearization of the equations of motion requires, first, that the ratio u^2/a^2 (u being the velocity relative to the body in the direction of motion of the body, a the local sound velocity) be replaced by U^2/a_∞^2, where U is the velocity of the body and a_∞ is the sound velocity corresponding to the temperature of the undisturbed flow; and, secondly, that Uv/a^2, Uw/a^2, v^2/a^2, and w^2/a^2 be small in comparison with unity.

Consider, for example, the motion of a slender body of revolution. Assume that the body has a sharp ogival nose. The linearized theory cannot lead to a shock wave of finite amplitude. According to the exact theory one obtains a shock wave attached to the ogive, at least above a limiting Mach number that for sharp ogives is only slightly larger than unity.

We obtain a very clear idea of the difference between the results of the linearized theory and the actual conditions in the hypersonic range by computing the coefficient of the pressure acting on an inclined plane surface in two-dimensional flow, as, for example, on the front part of a wedge airfoil.

The angle of inclination α of the shock wave relative to the main-stream direction is a function of the vertex angle of the wedge θ and the Mach number of the stream, M_∞. For small vertex angles one obtains from the general shock laws

$$\alpha = \frac{\gamma + 1}{4}\,\theta + \sqrt{\left(\frac{\gamma + 1}{4}\right)^2 \theta^2 + \frac{1}{M_\infty^2}} \tag{6-1}$$

It is seen that for small values of M_∞ the Mach number has an essential influence on the ratio α/θ, whereas for large values of M_∞, the ratio between shock angle and vertex angle becomes independent of M_∞ and approaches the constant value $(\gamma + 1)/2$. Since this number is only slightly larger than 1 for air, the conditions on the forward part of the body are very similar to the picture conceived by Newton: the air particles hit the body and continue their paths along the surface. The difference is that the shock does not occur at the surface of the body, but at a plane surface in the air that has a slightly larger vertex angle than the wedge itself.

Since it follows from Eq. 6-1 that

$$\frac{\alpha}{\theta} = \frac{\gamma + 1}{4}\left[1 + \sqrt{1 + \frac{1}{\left(\frac{4}{\gamma + 1}\right)^2 (M_\infty\theta)^2}}\right] \tag{6-2}$$

one concludes that the governing parameter for the ratio between shock angle and vertex angle is the product of Mach number and vertex angle. We shall show later that this is in accordance with the general similarity law of hypersonic flows.

The linearized theory shows only an infinitely weak, straight shock wave whose inclination α is given by the relation $\sin \alpha = 1/M_\infty$; the pressure coefficient is equal to $2\theta/\sqrt{M_\infty^2 - 1}$. Hence the shock angle is independent of the angle of inclination of the surface and depends only on the Mach number, whereas the pressure coefficient is proportional to the surface angle and decreases with increasing Mach number.

For the hypersonic case, $M_\infty \gg 1$, the shock angle is actually proportional to the surface angle and becomes independent of Mach number. The pressure coefficient is given by

$$C_p = \frac{4}{\gamma + 1}\left(\alpha^2 - \frac{1}{M_\infty^2}\right) \tag{6-3}$$

i.e. for very high Mach numbers it is proportional to the square of the surface angle θ and has the value $(\gamma + 1)\theta^2$. The value of C_p according to Newton's theory would be $2\theta^2$; the two values are equal if $\gamma = 1$. One may remember that the same discrepancy exists between the values of the sound velocity computed by Newton and by the modern theory, if $\gamma \neq 1$.

Fig. A,6 shows the values of C_p according to the linear and the hypersonic approximations. Points computed from the exact solution are also included. One sees that the linearized theory always holds for very small values of the angle of inclination θ, but the limit of its validity shrinks to zero as the Mach number increases to infinity. For $M_\infty = 5$ the hypersonic approximation is almost correct, and for $M_\infty = 10$ it is very good. Similar comparisons have been made by Linnell [23].

The assumption that the main-stream velocity is large in comparison with the sound velocity leads to a simplification of the general equations

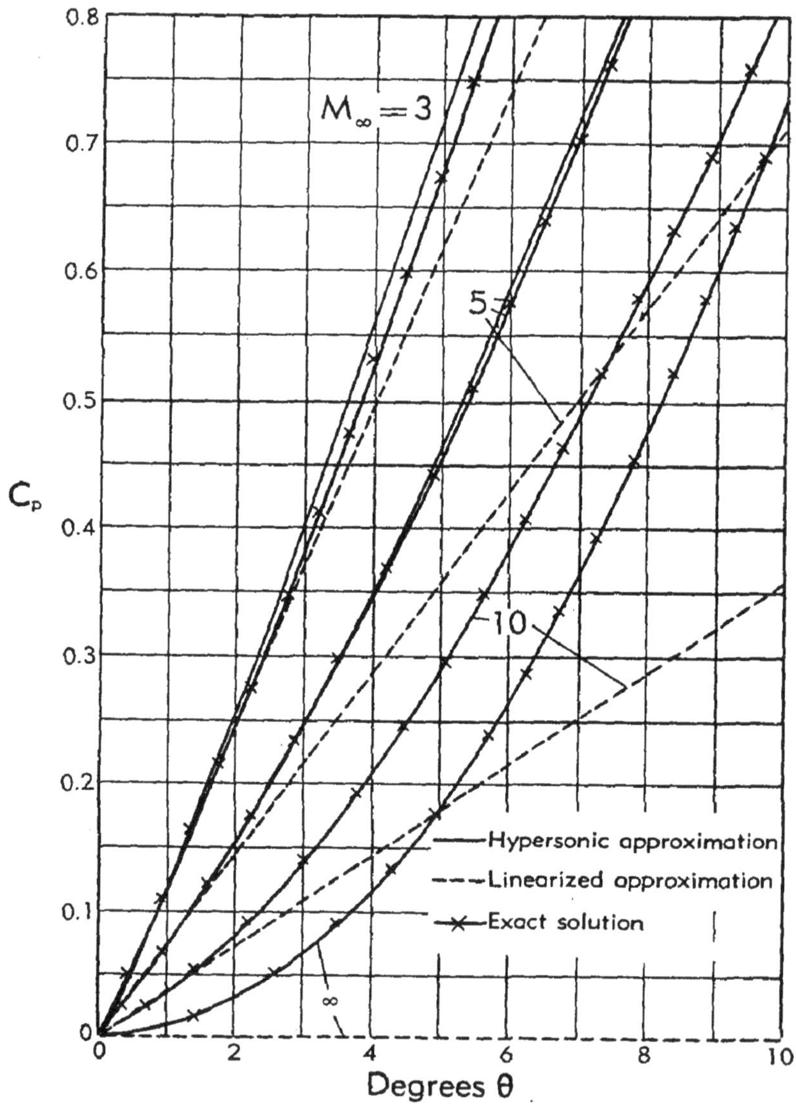

Fig. A,6. Comparison of exact and approximate pressure coefficients for two-dimensional supersonic flow past a wedge of semiangle θ.

of the flow of compressible fluids ([7]; also D,34). From these simplified equations one obtains a similarity law which allows the transfer of results obtained for a given geometrical arrangement to arrangements which are similar in the sense of an affine transformation (i.e. corresponding to a change of the ratio between the scale in the flow direction and the scale in directions normal to the flow), provided the Mach number is changed in an appropriate way. We restrict ourselves to two-dimensional isentropic motion. The equation for the perturbation velocities u and v can be written in the form:

$$[a^2 - (U + u)^2] \frac{\partial u}{\partial x} - (U + u)v \left(\frac{\partial u}{\partial y} + \frac{\partial v}{\partial x}\right) + (a^2 - v^2) \frac{\partial v}{\partial y} = 0 \quad (6\text{-}4)$$

We assume that u/U, v/U, and a/U are small (U = the main-stream velocity, a = the local sound velocity). Then $a^2 - (U + u)^2$ will be replaced by $-U^2$, $(U + u)v$ by Uv, and $a^2 - v^2$ will be replaced by $a_\infty^2 - (\gamma - 1)Uu - \frac{1}{2}(\gamma + 1)v^2$. In order to obtain these results we have to take into account that the difference between the local velocity of sound a and the velocity of sound in the main stream a_∞ is relatively large. In fact,

$$a_\infty^2 + \tfrac{1}{2}(\gamma - 1)U^2 = a^2 + \tfrac{1}{2}(\gamma - 1)[(U + u)^2 + v^2]$$

and therefore

$$a^2 \cong a_\infty^2 - (\gamma - 1)Uu - \frac{\gamma - 1}{2} v^2$$

Hence we obtain as the differential equation for hypersonic approximation:

$$U^2 \frac{\partial u}{\partial x} + Uv \left(\frac{\partial u}{\partial y} + \frac{\partial v}{\partial x}\right) - \left[\frac{U^2}{M_\infty^2} - (\gamma - 1)Uu - \frac{\gamma + 1}{2} v^2\right] \frac{\partial v}{\partial y} = 0 \quad (6\text{-}5)$$

We want the answer to the following question: How are the values of u, v, and M_∞^2 to be changed, so that the equation remains satisfied provided we keep U and the x scale unchanged but increase n times the y coordinates of the geometrical arrangement? One sees immediately that by increasing the v components in proportion to n and the u components in proportion to n^2, all terms with the exception of the term containing the Mach number remain proportional and become n^2 times larger. Consequently M_∞^2 has to be reduced in proportion to $1/n^2$, in order to have all forces balanced. In other words, the similarity law can be expressed in this form: The flow remains similar for an affine transformation of the geometrical configuration if the quantity $M_\infty \delta/c$ remains constant, where δ/c is a characteristic thickness parameter of the body. This law was discovered by H. S. Tsien [7].

It is seen that this rule is in accordance with the rule we found for the shock wave formation at an ogive. Our discussion, based on Eq. 6-4, involves the assumption that the flow is irrotational and isentropic. Since, as we shall see later, irrotational flow may become rotational after the fluid passes through a shock wave, the exactitude of the similarity law for cases in which shock waves occur has to be critically examined. This has been done by Hayes [8], Goldsworthy [9], and Eggers and Syvertson [24], and also in D,34 of the present volume.

A,7. The Transonic Range. When the Mach number of the flow approaches unity, we encounter another case in which the linearized theory does not apply. The physical reason can be seen in the following process. Consider two-dimensional supersonic flow against an inclined surface. According to the linearized theory, the u and v components of the perturbation velocity are in the ratio $u/v = 1/\sqrt{M_\infty^2 - 1}$ (which is equal to the tangent of the Mach angle). Consequently, for $M_\infty = 1$, u/v would be infinitely large. The same result is obtained when one approaches $M_\infty = 1$ from the low speed side, i.e. from subsonic flow. According to the linearized theory known as the theory of Prandtl and Glauert (Sec. C), the same ratio would be of the order of $1/\sqrt{1 - M_\infty^2}$. One can express this conclusion from the linearized theories also in the form of a similarity law, saying that (at least for the two-dimensional case) the flow remains similar if the thickness ratio is changed in proportion to $\sqrt{M_\infty^2 - 1}$ or $\sqrt{1 - M_\infty^2}$, respectively.

We may ask now: What is the similarity rule and the appropriate approximative equation in the transonic range?

If we consider Eq. 6-4, the main difficulty is that $a^2 - (U + u)^2$ cannot legitimately be replaced by $a_\infty^2 - U^2$, since a and U are of the same order of magnitude. It appears that the best method is to assume that the difference between local velocity $U + u$ and an appropriately chosen value of the sound velocity, for example the critical sound velocity a^*, is small. Also v/a has to be taken as small.

Then, denoting $U + u - a^*$ as u', and neglecting u'/a^* and v/a^* in comparison to unity, one can easily show that $a^2 - (U + u)^2$ can be replaced by $-(\gamma + 1)a^*u'$. The computation is analogous to the hypersonic case. The term $(U + u)v$ will be replaced by a^*v and $a^2 - v^2$ by a^{*2}.

Hence one obtains

$$(\gamma + 1)u' \frac{\partial u'}{\partial x} + v\left(\frac{\partial u'}{\partial y} + \frac{\partial v}{\partial x}\right) - a^* \frac{\partial v}{\partial y} = 0 \qquad (7\text{-}1)$$

In addition we assume that

$$\frac{\partial u'}{\partial y} - \frac{\partial v}{\partial x} = 0 \qquad (7\text{-}2)$$

Eq. 7-2 expresses the absence of vorticity.

One sees, first, that $\partial v/\partial y$ must be of the order of u'^2 if a^* is considered of the order of unity. Let us assume that u' is of the order ϵ, v of the order ϵ^α; then the operation $\partial/\partial y$ must be of the order $\epsilon^{2-\alpha}$ and from the condition of vanishing vorticity it follows that $3 - \alpha = \alpha$ or $\alpha = \frac{3}{2}$. Hence we conclude that the first and the last terms of Eq. 7-1 are of the order ϵ^2, whereas the second term is of the order ϵ^3 and consequently is negligible. One arrives at the equations

$$(\gamma + 1)u' \frac{\partial u'}{\partial x} = a^* \frac{\partial v}{\partial y}$$
$$\frac{\partial u'}{\partial y} = \frac{\partial v}{\partial x} \tag{7-3}$$

One also concludes that, in order that the flow remain similar when the body thickness is increased n times, u' has to be increased in proportion to $n^{\frac{2}{3}}$. Since the quantity $U = a^* + u'_\infty = M_\infty a^*$ represents the velocity of the stream ahead of the body, it follows that $u'_\infty = (M_\infty - 1)a^*$, so that the solution of Eq. 7-3 has to satisfy the condition $u' = (M_\infty - 1)a^*$ for $x = -\infty$. Therefore, in order to satisfy Eq. 7-3 and the boundary condition for $x = -\infty$, we have to assume that for similar flows the quantity $M_\infty = 1$ is also increased in proportion to $n^{\frac{2}{3}}$. The forces remain balanced when the ratio

$$\frac{M_\infty - 1}{\left[(\gamma + 1)\frac{\delta}{c}\right]^{\frac{2}{3}}} \tag{7-4}$$

remains constant. In this expression δ/c is an appropriately chosen thickness ratio. We can see that the similarity law also allows a comparison of flows of gases with different values of $\gamma = c_p/c_v$. This result is of importance when wind tunnel experiments have to be compared with flow phenomena in a gas with a different value of γ.

Pressure coefficient. The coefficient of the pressure acting on an inclined surface appears in the form

$$C_p = \frac{\theta^{\frac{2}{3}}}{(\gamma + 1)^{\frac{1}{3}}} f\left(\frac{M_\infty - 1}{(\gamma + 1)^{\frac{2}{3}}\theta^{\frac{2}{3}}}\right) \tag{7-5}$$

where θ is the inclination of the surface. We assume that C_p has a finite value for $M_\infty = 1$. Then $C_p \doteq \theta^{\frac{2}{3}}(\gamma + 1)^{-\frac{1}{3}}f(0)$ for $M_\infty = 1$. We see that, whereas the theoretical value of C_p for the linearized theory is proportional to θ, and for very high Mach numbers proportional to θ^2, for $M_\infty = 1$ it is proportional to $\theta^{\frac{2}{3}}$.

In many configurations a weak detached shock wave appears ahead of a body exposed to a stream of a velocity slightly higher than sound velocity. In this case the Mach number of the subsonic flow behind the shock wave and before the body is approximately equal to $1 - m$, if $1 + m$ is the Mach number of the flow upstream of the shock wave. Thus one may assume that the local Mach number M at a given point of the surface will be the same for the case of a subsonic flow at a Mach number equal to $1 - m$ and for the case of a supersonic flow, with detached shock wave, at a Mach number equal to $1 + m$, the flow pattern near the body being essentially the same.

Hence we can assume that for $M_\infty = 1$, $dM/dM_\infty = 0$.

Denoting the pressure at the point on the surface considered by p, and that at $x = -\infty$ by p_∞, the pressure coefficient is defined as $C_p = 2(p - p_\infty)/\rho_\infty U^2$, where U and ρ_∞ are the speed and the density at $x = -\infty$ respectively. Now, $\rho_\infty U^2/p_\infty = \gamma M_\infty^2$; on the other hand, one finds that for isentropic flow, if M_∞ and M are both near unity, $p/p_\infty \cong 1 + [2\gamma/(\gamma + 1)](M_\infty - M)$. Hence, with this approximation,

$$C_p \cong \frac{4}{\gamma + 1}\left(1 - \frac{M}{M_\infty}\right) \tag{7-6}$$

and

$$\left(\frac{dC_p}{dM_\infty}\right)_{M_\infty=1} \cong \frac{4}{\gamma + 1} \tag{7-7}$$

It is seen that we can obtain much information about conditions near the sonic velocity by rather simple considerations. If we measure the pressure coefficient at a point on the surface for $M_\infty = 1$, the pressure coefficient for a similar configuration with increased angle of inclination will be proportional to the two-thirds power of the angle of inclination θ. This gives us the intersections of the $C_p(M_\infty)$ curves with the vertical line $M_\infty = 1$. The inclination of the curves at $M_\infty = 1$ will be independent of θ; all curves pass through $M_\infty = 1$ with the same gradient.

The almost constant value of the Mach number at any point through a velocity range between $M_\infty = 1 - m$ and $M_\infty = 1 + m$ is shown in Fig. A,7a, which represents the local Mach number distribution along the surface of a wedge for different free stream Mach numbers.[1] There are no published data for pressure coefficient curves quite near to $M_\infty = 1$. Fig. A,7b shows the drag coefficients for three wedges,[1] reduced according to the formula

$$\tilde{C}_D = \frac{(\gamma + 1)^{\frac{1}{3}}}{\theta^{\frac{5}{3}}} C_D = f\left(\frac{M_\infty - 1}{(\gamma + 1)^{\frac{1}{3}}\theta^{\frac{2}{3}}}\right) \tag{7-8}$$

which is in line with our Eq. 7-5 for the pressure coefficient.

According to the similarity law, $\tilde{C}_D$ should be a function of the parameter $\zeta = (M_\infty - 1)/(\gamma + 1)^{\frac{1}{3}}\theta^{\frac{2}{3}}$, as in Eq. 7-8. This is fairly well substantiated by the data of Fig. A,7b. It must be mentioned, however, that the similarity law is deduced for the potential flow of a nonviscous fluid. One can show that it remains approximately valid for the case of weak shock waves, because in the case of transonic flow the shock waves are, to a first approximation, normal to the direction of the mean stream, and the vorticity behind the shock waves is small.

[1] These data are from the measurements of Liepmann and Bryson [*10*].

From the mathematical point of view, Eq. 7-3 can be reduced to one equation for the perturbation potential $\varphi(x, y)$ [11]:

$$\frac{\partial\varphi}{\partial x}\frac{\partial^2\varphi}{\partial x^2} = \frac{a^*}{\gamma + 1}\frac{\partial^2\varphi}{\partial y^2} \tag{7-9}$$

This partial differential equation is elliptic when $\partial\varphi/\partial x < 0$ and hyper-

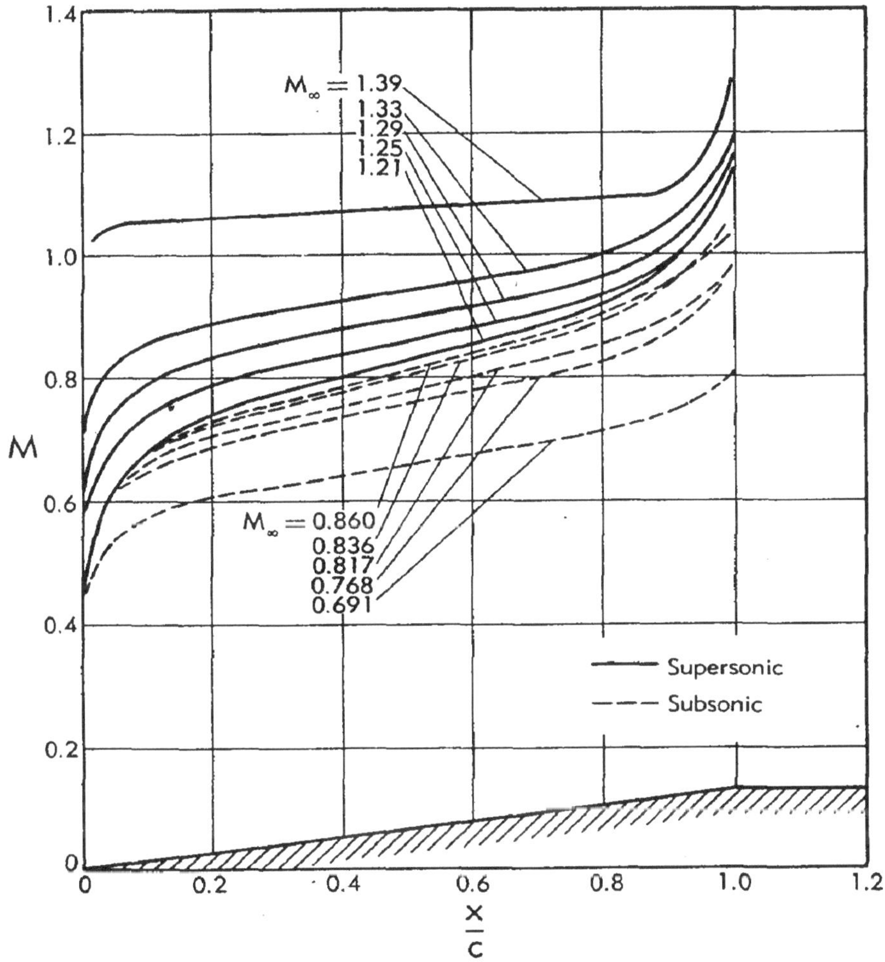

Fig. A,7a. Local Mach number distribution on surface of wedge for various free-stream Mach numbers M_∞ near $M_\infty = 1$. (From [10].)

bolic when $\partial\varphi/\partial x > 0$. Thus in general we obtain solutions consisting of subsonic and supersonic flow domains side by side.

The Eq. 7-3 can also be transformed into two linear equations for x and y by choosing quantities proportional to the velocity components u' and v as independent variables (the hodograph method; cf. Sec. F). We obtain

$$\frac{\partial^2 x}{\partial \eta^2} = \frac{1}{\xi}\frac{\partial^2 x}{\partial \xi^2} - \frac{1}{\xi^2}\frac{\partial x}{\partial \xi}$$
$$\frac{\partial^2 y}{\partial \xi^2} = \xi \frac{\partial^2 y}{\partial \eta^2} \tag{7-10}$$

with $\xi = (\gamma + 1)u'/a^*$, $\eta = (\gamma + 1)v/a^*$.

Tricomi's equation. The second of Eq. 7-10 is called Tricomi's equation. Many authors have given so-called fundamental solutions of this equation. In connection with these solutions, the following interesting

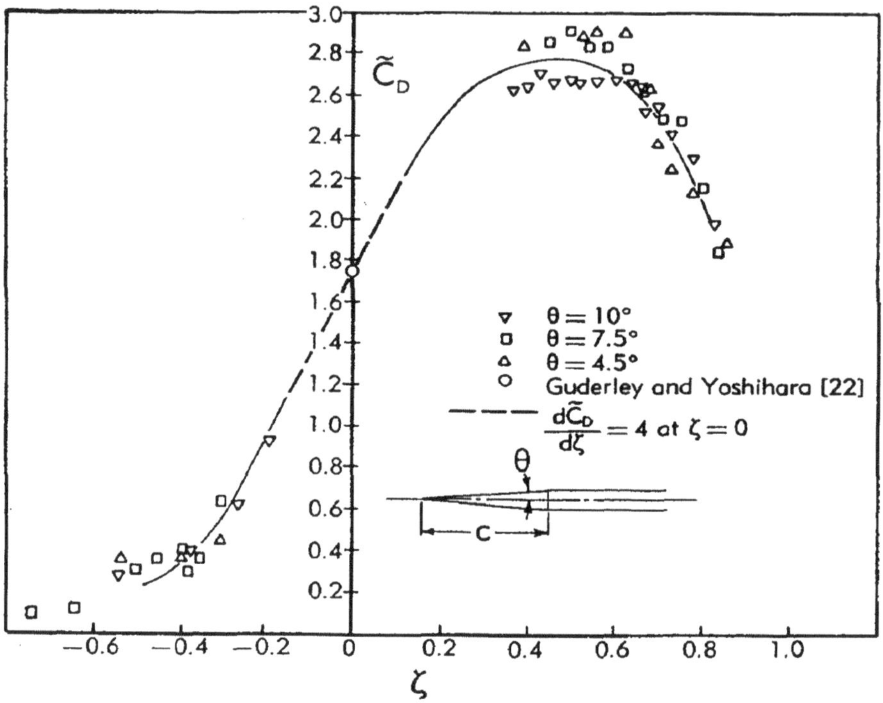

Fig. A,7b. Reduced drag coefficient $\tilde{C}_D = (\gamma + 1)^{\frac{1}{3}}\theta^{-\frac{5}{3}}C_D$ plotted against the parameter $\zeta = (\gamma + 1)^{-\frac{2}{3}}\theta^{-\frac{2}{3}}(M_\infty - 1)$ for three wedge semiangles θ in the transonic range. (Experimental data taken from [*31*], and the theoretical value at $\zeta = 0$ from [*22*].)

question arises. Undoubtedly, by means of such solutions, one can construct flow fields such that the flow is subsonic everywhere, with the exception of closed domains near the body surface, in which the flow is supersonic—why are such flows not observed in experiment?

For example, J. Fabri and the writer [*12*] gave simple analytic solutions of Eq. 7-10, which represent a flow along a wavy wall. These solutions are continuous also in the supersonic domain near the crests of the wall. In other words, there are no shock waves. The resistance force acting on the wall is theoretically zero until the velocity at infinity becomes supersonic. Actual observation reveals, however, that in such a case shock waves practically always appear; i.e. the return from supersonic to subsonic speed is discontinuous. Also, the drag is found to increase appreciably in the high subsonic range, which indicates that there must be shock waves.

There are three possible explanations for this contradiction:

1. Some authors, for example Busemann and Guderley [*13,25,26*] believe that continuous solutions exist only for specially selected wall configurations. For example, if we change the Mach number of the undisturbed flow in the problem of the wavy wall but keep the same analytic solution, the shape of the wall will change. It has not been proved that one can obtain a sequence of continuous solutions by changing the Mach number while keeping the shape unchanged.

2. A continuous flow of the type considered is unstable and therefore in all practical cases it goes over into a flow with shock.

3. The presence of viscosity produces a boundary layer which causes shock waves.

These possibilities are discussed at greater length in Sec. F. It is probable that hypothesis (1) is correct. It would be useful, however, if a convincing proof of its correctness were available.

A,8. Entropy and Vorticity. Let us return to the problem of shock waves. We noted in Art. 5 that the appearance of a shock wave involves an entropy change. On the other hand, when we considered the drag problem we found that the flow at infinity behind the body is nonuniform, i.e. it is not irrotational and this necessarily involves the existence of drag. This suggests that there ought to be general relations between entropy change and vorticity, and between entropy change and drag. In the next two articles we indicate the nature of these correlations.

It is known that in a nonviscous fluid the vorticity remains zero if at the beginning it was equal to zero and if the fluid is subjected only to forces which have a potential. Therefore, if the circulation integral of the velocity along a closed line within the fluid is zero at a certain instant, it remains zero.

V. Bjerknes [*14*] has shown that the time derivative of the circulation integral of the velocity is equal to the line integral of the acceleration. In a nonviscous fluid the acceleration vector is equal to $-(1/\rho)\nabla p$, where p is the pressure and ρ the density. Hence the time derivative of the circulation integral is equal to the line integral

$$-\oint(1/\rho)(dp/ds)ds$$

This is Bjerknes' theorem.

From this theorem it is possible to obtain an expression which allows a direct calculation of the vorticity vector $\boldsymbol{\omega}$. We shall restrict ourselves to the case of a steady flow. The acceleration in a steady flow of a fluid is equal to the sum of two parts. The first is $\nabla(q^2/2)$, the gradient of the kinetic energy per unit mass; this term is evidently a generalization of the linear acceleration along a streamline, $q(dq/ds)$. The other part is a generalization of the centripetal acceleration and is equal to $-(\mathbf{q} \times \boldsymbol{\omega})$. One has $\mathbf{q} \times \boldsymbol{\omega} = (1/\rho)\nabla p + \nabla(q^2/2)$, since the acceleration is equal to $-(1/\rho)\nabla p$. On the other hand, due to the conservation of the energy in

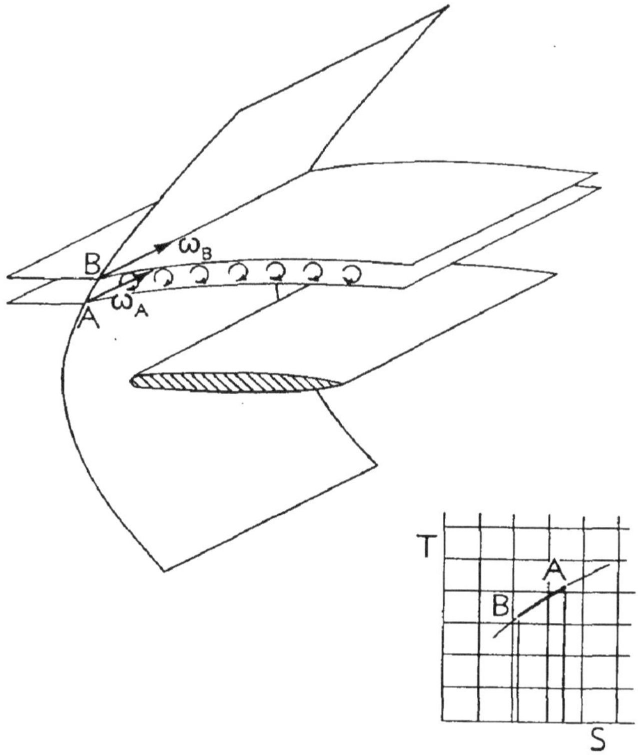

Fig. A,8. Vorticity flux and entropy rise in a flow through a curved shock wave.

the fluid, $\nabla(h + q^2/2) = 0$ (where the symbol h stands for the enthalpy) and therefore

$$\mathbf{q} \times \boldsymbol{\omega} = \frac{1}{\rho}\nabla p - \nabla h \tag{8-1}$$

According to the definition of the entropy s, the right side of the last equation is equal to $-T\nabla s$, so that

$$\mathbf{q} \times \boldsymbol{\omega} = -T\nabla s \tag{8-2}$$

This result was obtained by L. Crocco [*15*]. It shows that, for example, if the fluid passes through a stationary shock wave, the flow can conserve its irrotational character only if the entropy rise is uniform along the shock wave. Otherwise, as is the case in general for curved shock waves, vorticity is created in proportion to the derivative of the entropy along the shock wave. Consider, for example, two points A and B at unit distance on a shock wave (Fig. A,8). Then one can say that the component of the vector product of velocity and vorticity is equal to the heat input necessary to transform the state of the gas from the state A to the state B. Or, at least in the case of plane motion, one can interpret Crocco's equation by stating that the vorticity flux $q_n\omega$ through $\overline{AB}$ is numerically equal to the heat input corresponding to the change of state from A to B.

A,9. Entropy Rise and Drag. We mentioned before that, according to d'Alembert's theorem, a body moving uniformly in a nonviscous fluid does not encounter resistance, provided that no wake formation or vortex shedding occur. Since wakes and vortices are either produced by viscous effects or by sharp corners or edges, d'Alembert's theorem is correct for all bodies whose surfaces have no points or lines where the curvature is infinite. The theorem holds for incompressible and compressible fluids as long as no shock waves occur. Hence it appears that the existence of drag must be in correlation with entropy rise.

The proof of d'Alembert's theorem is based on the statement that the force acting on the moving body or on the body at rest in a fluid stream is equal to the resultant of the pressure acting on a suitably chosen control surface and of the momentum transferred in unit time through the same surface. If there are no shock waves, one can show that this resultant converges to zero when the control surface is extended to infinity. If we take, for example, a sphere of radius R as control surface, the pressure difference $p - p_\infty$ and the components of the momentum transfer converge to zero faster than R^{-2}, since for $R = \infty$, the resultant force converges to zero. For the case of a flow with shock waves it is more convenient to take as control surface a cylinder of radius R and length L. One finds in this case that the contributions of the cylindrical surface converge to zero because of the attenuation of the shock waves, but there is a difference between the momentum entering at $x = -L/2$ and leaving the cylinder at $x = L/2$, which remains finite when $L \to \infty$. This difference of momentum furnishes a drag of the amount:

$$D = \int_A \rho u(U - u)dA \tag{9-1}$$

where the integral is to be extended over the end surface of the cylinder at $x = \infty$; U is the velocity at $x = -\infty$ whereas u is the velocity at an arbitrary point at $x = \infty$. The components v and w are negligible at $x = \infty$.

Evidently

$$U - u = -\int_\infty^P \left(\frac{\partial u}{\partial y} dy + \frac{\partial u}{\partial z} dz\right)$$

where the integral is taken along an arbitrary curve in the end surface of the cylinder starting at a point far from the body's wake and ending at the arbitrary point P. Now according to Crocco's theorem

$$-\left(\frac{\partial u}{\partial y}\,dy + \frac{\partial u}{\partial z}\,dz\right) = \frac{T}{u}\left(\frac{\partial s}{\partial y}\,dy + \frac{\partial s}{\partial z}\,dz\right)$$

and therefore

$$D = \int_A \rho u dA \int_\infty^P \frac{T}{u}\left(\frac{\partial s}{\partial y}\,dy + \frac{\partial s}{\partial z}\,dz\right) \tag{9-2}$$

Let us denote $\rho u dA$ (the mass flow through the cross-sectional area dA) by dm and replace T/u by T_∞/U, which is justified if we neglect terms of second order. Then

$$D = \frac{T_\infty}{U}\int_A dm\,(s - s_\infty) \tag{9-3}$$

or

$$DU = T_\infty \int_A dm\,(s - s_\infty) \tag{9-4}$$

The physical meaning of Eq. 9-4 is the following: The work per unit time required to move a slender body with the uniform velocity U through a nonviscous compressible fluid is equal to the product of the temperature of the fluid and the entropy rise of the fluid per unit time. This statement is correct to the first approximation, i.e. for a small increase of entropy by weak shocks. This subject is discussed in greater detail in E,4 of this volume, where the contributions of higher order terms are investigated.

Oswatitsch [*16*] has shown that the Eq. 9-4 is correct for viscous fluids provided the viscous stresses converge to zero at infinity. H. S. Tsien pointed out that in the case that the body is not thermally insulated one has also to take into account the heat transferred from the body to the fluid or from the fluid to the body.

A,10. The Role of Viscosity. In classical, low speed aerodynamics, viscous effects are studied essentially for two reasons: first, because viscous forces contribute to drag (in some cases they are responsible for the major part of it), and second, because viscosity has a great influence on flow separation, which causes loss of lift and increase of drag by wake formation and vortex shedding. In high speed aerodynamics the same problems arise. There is, however, one aspect of the phenomena in real (viscous) fluids that is secondary in low speed aerodynamics and very important in high speed aerodynamics, namely, the heat balance connected with the viscous flow. At low speeds the heat produced by friction does not appreciably affect the velocity distribution and frictional resistance. Therefore—at least in cases in which the temperature differences in the flow are not very large—the heat transfer problem can be treated separately, or rather in parallel with, the flow problem. In high speed phenomena the two problems are intimately interconnected.

Boundary layers. As in low speed aerodynamics, the concept of the boundary layer is of great importance for the study of frictional drag and heat transfer at high speeds. The theory of the laminar boundary layer in compressible fluids, initiated by Busemann, Crocco, Tsien, and the writer, has been considerably developed by many authors in recent years (see Vol. IV). As perhaps the most advanced from the analytic viewpoint, Crocco's work [*17*] may be mentioned.

As we stated above, the balance between heat production by dissipation and heat conduction across the boundary layer is an important item in the theory of compressible boundary layers. In the laminar case it has been found that for a particular value of the ratio between the coefficients of heat conduction and viscosity, corresponding to a Prandtl number equal to 1, the balance is complete in the sense that the sum of the enthalpy c_pT and the kinetic energy $u^2/2$ of the unit mass is constant across the boundary layer. (The Prandtl number is $Pr = \mu c_p/k$ where μ is the viscosity coefficient, k the coefficient of the heat conduction, and c_p the specific heat at constant pressure.) This is a remarkable result, since the same law holds for isentropic compression in a compressible gas along a streamline. In fact, at the stagnation point the temperature reaches the value of the so-called stagnation temperature, which is higher by the amount $U^2/2c_p$ than the temperature in the undisturbed stream of velocity U. Now we find that in the viscous and heat-conducting fluid (at least in the specific case of $Pr = 1$) the temperature at the wall rises to the same value that it would reach by isentropic compression. Since the Prandtl number is near to unity for practically every gas, there will always be a very essential temperature rise at the wall. This effect has great practical importance for the design of high speed missiles, fast-moving turbine blades, etc.

The theory of the turbulent boundary layer in compressible fluids is in a much less satisfactory state than the theory of the laminar boundary layer. This is easily understood if one remembers that even the theory of the incompressible turbulent boundary layer is built up on a partly empirical basis. It rests on assumptions like the similarity of the turbulent flow pattern, in the present author's theory, or a rather arbitrary law for the "mixing length," corresponding to Prandtl's concept. Nevertheless, for incompressible fluids the semiempirical logarithmic law for the velocity distribution gives a powerful means for calculation of frictional drag and heat transfer, especially if we take into account the existence of a laminar sublayer and a buffer layer between the latter and a fully turbulent domain where the logarithmic law holds. Unfortunately there is no corresponding universal law for the velocity distribution in the case of compressible fluids. Many attempts have been made to adapt the theories valid for incompressible fluids to the mechanics of compressible fluids. These attempts necessarily contain much guesswork. It is the author's belief that real progress will be made only after the basic laws of the exchange mechanisms in turbulent flow are better understood.

The experimental investigations show a significant "enthalpy recovery" at the wall similar to that computed for and measured in laminar boundary layers. One finds in general that the "recovery factor" is almost equal to unity.

Shock wave boundary layer interaction. Flow separation effects also have a new, interesting aspect, especially in transonic and supersonic flow, because of the presence of shock waves. Since, in a viscous fluid, the velocity relative to the wall is equal to zero at the wall, the flow in the inner portion of the boundary layer is necessarily subsonic. Consequently, a shock wave cannot reach the wall. When, however, a shock wave exists outside of the boundary layer, reaching into the outer portion of it, the pressure rise along the wall will be rather steep, and is likely to produce separation. Hence flow separation behind a shock wave will be a more frequent case than flow without at least temporary separation.

Also, it was noticed by Ackeret, Feldmann, and Rott [*18*] and later by Liepmann [*19*] that the formation and configuration of shock waves is greatly affected by the laminar or turbulent character of the boundary layer. Especially, the reflection of shock waves at a wall, both in steady and unsteady flow, depends appreciably on the type and size of the boundary layer. In fact, the interference between shock wave and boundary layer became an important and highly interesting chapter of theoretical and experimental high speed aerodynamics. So far, however, there are few results that are directly applicable to design problems.

Another sort of shock wave boundary layer interaction occurs in hypersonic flow. The hypersonic flow of an ideal compressible fluid was discussed in Art. 6 above, especially the case of a wedge of small vertex angle. Such a flow is characterized by the fact that the angle of inclination of the shock wave is very small, of the same order as the half vertex angle (Eq. 6-1). If one desires to account for the influence of viscosity, it does not appear permissible to calculate the flow for an ideal fluid and then to employ the concept of the boundary layer without considering the modification of the outside flow by the viscous layer. Even in the case of a flat plate parallel to the main flow, the deceleration of the flow inside the boundary layer produces a kind of detached shock wave ahead of the leading edge of the plate, and there occurs a mutual interference between this shock and the boundary layer. At present there is no satisfactory theory for the description of this complex phenomenon and for the computation of the skin friction for hypersonic speeds. Some studies of the subject have recently been made by Shen [27,28], Lees [29], and Li and Nagamatsu [30].

Shock wave structure. In connection with shock waves, we have to note that no mathematical discontinuity can exist in a viscous and heat-conducting fluid. This applies also to low speed flow. For example, when two fluid masses glide on each other, the viscosity smoothes out the abrupt change of velocity which would be possible in a nonviscous fluid. Similarly, in a viscous and heat-conducting fluid, shock waves have a finite thickness.

The mathematical discontinuity is replaced by a narrow strip, in which viscous dissipation and heat exchange take place.

The thickness of the shock wave and the distribution of temperature, pressure, and density across it can be approximately calculated by solving a one-dimensional flow problem. The values of these parameters before and after the shock are known by the conservation equations and appear as boundary values at $\mp\infty$. Since the process is considered adiabatic, the sums of the enthalpy and kinetic energy ahead and behind the shock are equal. Inside the shock, work done by the viscous forces is transformed into heat and heat is transferred by conduction in the direction opposite to the flow direction. Becker [20], in 1922, found that when the Prandtl number has the particular value $\frac{3}{4}$, a balance between viscous and heat-conduction effects exists, so that the sum of the enthalpy and the kinetic energy is constant at every point across the entire shock wave. This represents an interesting analogy to the result found in the theory of a compressible boundary layer.

One finds that the thickness of the shock wave is of the order of magnitude of ν/a (where ν is the coefficient of kinematic viscosity, and a the velocity of sound), i.e. of the order of the mean free path of the molecules in the gas. For this reason it is somewhat doubtful whether the dynamics of continuous fluids can be applied without a correction to account for the molecular structure of the gas. It has been proposed to attack the problem directly from the viewpoint of the kinetic gas theory by trying to determine the most probable velocity distribution function of the molecules for each point within the shock wave. No significantly different result, however, has yet been obtained by this refined method.

A,11. Limitations of the Theory of Continuous Fluids. Up to now, with the exception of the last remark in the preceding article, we have not left the domain of the mechanics of continuous media and phenomenological thermodynamics. There are, however, problems in high speed flow which definitely require the consideration of the molecular or atomic structure of matter. The concept of the fluid as a continuous medium with definite thermodynamic properties supposes that there exists at least approximately a statistical equilibrium for the distribution of the energy between the various degrees of freedom of the molecules at every point.

One interesting case is the adiabatic expansion of a gas at such a high time rate that there is not sufficient time to establish the distribution of energy between the various degrees of freedom corresponding to an isentropic process. In such a fast process carbon dioxide (CO_2), for example, does not behave as a three-atomic gas, because the degree of freedom corresponding to the intramolecular vibration has a relatively large "relaxation time" [*21*].

Another, perhaps more important group of flow phenomena which especially requires a modification of the theory of viscosity has been designated as *superaerodynamics*. In the theory of viscous fluids the stress components are expressed as products of a viscosity coefficient and certain combinations of the derivatives of the velocity components with respect to the space coordinates. In the simplest case of a parallel flow the viscous shear stress is equal to the product of the viscosity coefficient μ and the velocity gradient du/dn across the flow. The viscosity coefficient is proportional to the density, the mean free path l_m, and the mean velocity c of the molecules. The physical idea behind this definition is the following: Assume that a molecule is moving with the velocity c across the shear flow; then it will transfer an amount of momentum per unit mass equal to the difference in flow velocity between the point whence the molecule started and the point where it was stopped by the next collision. This physical picture requires, however, that the amount of velocity differential $l_m(du/dn)$, be small in comparison with the molecular velocity c. Otherwise, the momentum transfer cannot be considered as a transfer process between fluid masses in a state of statistical equilibrium established by the collisions. Now, if we consider, for example, a solid body moving in a fluid or a flow problem with a certain given wall configuration, the derivative du/dn will be of the order of magnitude of U/L, where U is the velocity of the flow relative to the body or the wall and L is a characteristic length of the body or the flow problem. Taking into account that the kinematic viscosity coefficient $\nu = \mu/\rho \sim l_m c$, one obtains from $(l_m/c)(du/dn) \ll 1$,

$$\frac{\nu}{c^2}\frac{U}{L} = \frac{\nu U^2}{ULc^2} \ll 1 \tag{11-1}$$

For this consideration we can replace the molecular velocity c by the sound velocity a, thus $(\nu/UL)(U/a)^2 \ll 1$. On the other hand, $UL/\nu = Re$, the Reynolds number of the flow configuration. Therefore, the condition for the validity of the theory of continuous viscous flow can be expressed in the form

$$\frac{M^2}{Re} \ll 1 \tag{11-2}$$

It is to be expected that for missiles moving at high Mach numbers in a very thin atmosphere, this condition is not satisfied. It is worthwhile to mention that, if we postulate that the ratio between mean free path and boundary layer thickness is small, we arrive at the same criterion for the limitation of the mechanics of continuous fluids. Also, the postulate that the number of collisions in the gas along a body of the length L is large leads to the same criterion.

If M^2/Re is not a small quantity, there is no justification for computing the frictional drag as a resultant of viscous stresses acting on a body; there is also no justification for assuming that the relative velocity of the fluid at the wall vanishes. One has to consider the individual impacts of molecules and their reflections from the surface. Thus one of the main questions is the probable distribution of the direction and magnitude of the velocity of the reflected molecules when the distribution of the incoming molecules is known.

Because the condition of zero relative mean velocity at the wall does not hold, such flow is sometimes called a *slip flow*. Also, the temperature of the fluid at the surface is no longer necessarily equal to the temperature of the surface. Slip flow and "temperature jump" were investigated theoretically and experimentally for slow motion of highly rarefied gases many decades ago, for example by Knudsen and Smoluchowski. Superaerodynamics, a new branch of aerodynamics, introduces as a new element the combination of high velocity and low density.

Since the ambient pressure is evidently equal to $\rho a^2/\gamma$, we can say that the ratio between the viscous stresses (of the order of magnitude $\mu U/L$) and the ambient pressure is proportional to the quantity M^2/Re. In other words the characteristic feature of the slip flow is that the viscous stress is no longer small in comparison with the ambient pressure. On the other hand, the transonic and supersonic speed range is characterized by the fact that the momentum per unit area and unit time, ρU^2, is not small in comparison with the ambient pressure, since evidently $\rho U^2/p = \gamma M^2$. In the most interesting problems of superaerodynamics both M^2 and M^2/Re are large. In other words both the momentum ρU^2 (or the dynamic pressure $\frac{1}{2}\rho U^2$) and the viscous stresses are large in comparison with the ambient pressure.

A,12. Cited References.

1. Newton, Sir Isaac. *Principia,* Book II. Univ. Calif. Press, 1947. p. 378.
2. Ackeret, J. Luftkräfte auf Flügel, die mit grösserer als Schallgeschwindigkeit bewegt werden. *Z. Flugtech. u. Motorluftschiffahrt 16,* 72–74 (1925). Translated in *NACA Technical Memorandum 317.* 1925.
3. von Kármán, Th., and Moore, N. B. Resistance of slender bodies moving with supersonic velocities, with special reference to projectiles. *Trans. ASME (Appl. Mech.) 54,* 303-310 (1932).
4. de Kármán, Th. The problem of resistance in compressible fluids. *Atti del Convegno della Fondazione Alessandro Volta 1935,* 222–276 (1936). Rome.

5. Busemann, A. Aerodynamischer Auftrieb bei Überschallgeschwindigkeit. *Atti del Convegno della Fondazione Alessandro Volta 1935*, 328–360 (1936). Rome.
6. Riemann, B. Über die Fortpflanzung ebener Luftwellen von endlicher Schwingungsweite. *Abhandlungen der Gesellschaft der Wissenschaften zu Göttingen, Mathematisch-physikalische Klasse 8*, 43 (1860), or *Gesammelte Werke*, 1876, p. 144.
7. Tsien, H. S. Similarity laws of hypersonic flows. *J. Math. and Phys. 25*, 247–251 (1946).
8. Hayes, W. D. On hypersonic similitude. *Quart. Appl. Math. 5*, 105–106 (1947).
9. Goldsworthy, F. A. Two-dimensional rotational flow at high Mach number past thin airfoils. *Quart. J. Mech. and Appl. Math. 5*, 54–63 (1952).
10. Liepmann, H. W., and Bryson, A. E., Jr. Transonic flow past wedge sections. *J. Aero. Sci. 17*, 745–755 (1950).
11. von Kármán, Th. The similarity law of transonic flow. *J. Math. and Phys. 26*, 182–190 (1947).
12. von Kármán, Th., and Fabri, J. Écoulement transsonique à deux dimensions le long d'une paroi ondulée, *C.R. Acad. Sci. Paris 231*, 1271–1274 (1950).
13. Busemann, A. The drag problem at high subsonic speeds. *J. Aero. Sci. 16*, 337–344 (1949).
14. Bjerknes, V. *Vid.-Selsk. Skrifter*, Kristiania, 1918. See also Lamb, H. *Hydrodynamics*, 6th ed. Cambridge, 1932. p. 247.
15. Crocco, L. Eine neue Stromfunktion für die Erforschung der Bewegung der Gase mit Rotation. *ZAMM 17*, 1–7 (1937).
16. Oswatitsch, K. Der Verdichtungsstoss bei der stationären Umströmung flacher Profile. *ZAMM 29*, 129–141 (1949).
17. Crocco, L. Lo strato limite laminare nei gas. *Monografie Scientifiche di Aeronautica 3*, 1946; also *North American Aviation Trans. AL-684*, 1948. Los Angeles.
18. Ackeret, J., Feldmann, F., and Rott, N. Untersuchungen an Verdichtungsstössen und Grenzschichten in schnell bewegten Gasen. *Mitt. No. 10, Inst. für Aerodynamik, Eidg. Tech. Hochschule.* Zürich, 1946.
19. Liepmann, H. W. The interaction between boundary layer and shock waves in transonic flow. *J. Aero. Sci. 13*, 623–637 (1946).
20. Becker, R. Stosswelle und Detonation. *Zeits. f. Physik 8*, 321–362 (1922).
21. Kantrowitz, A. Heat capacity lag in gas dynamics. *J. Chem. Phys. 14*, 150–164 (1946); see also *NACA Wartime Report L-457* (formerly *Advance Restricted Report 4A22*). 1944.
22. Guderley, G., and Yoshihara, H. The flow over a wedge profile at a Mach number one. *Air Force Technical Report 5783*. 1949.
23. Linnell, R. D. Two-dimensional airfoils in hypersonic flows. *J. Aero. Sci. 16*, 22–30 (1949).
24. Eggers, A. J., Jr., and Syvertson, C. A. Inviscid flow about airfoils at high supersonic speeds. *NACA Technical Note 2646*. 1952.
25. Busemann, A., and Guderley, G. The problem of drag at high subsonic speeds. *MOS Reports and Translations 184*. 1947.
26. Guderley, G. On the transition from a transonic potential flow to a flow with shocks. *Wright Field Report F-TR-2160-ND*. 1947.
27. Shen, S. F. On the boundary layer equations in hypersonic flow. *J. Aero. Sci. 19*, 500–501 (1952).
28. Shen, S. F. Hypersonic flow over an insulated wedge with viscosity effects. *Part I Sc.D. Thesis, Department of Aeronautical Engineering, Mass. Inst. Technol.* 1949.
29. Lees, L. On the boundary layer equations in hypersonic flow and their approximate solutions. *J. Aero. Sci. 20*, 143–145 (1953).
30. Li, T. Y., and Nagamatsu, H. T. Shock wave effects on the laminar skin friction of an insulated flat plate at hypersonic speeds. *J. Aero. Sci. 20*, 345–355 (1953).

SECTION B

MATHEMATICAL ASPECTS OF FLOW PROBLEMS OF HYPERBOLIC TYPE

K. O. FRIEDRICHS

To solve a mathematical problem originally meant to find its complete numerical solution. Gradually it became clear that such explicit solutions are possible only in exceptional cases, that in general one must be satisfied with a scheme by which the solution may be determined approximately, though with any desired accuracy. Something quite different is very frequently offered as the solution of a mathematical problem, namely, a representation of the solution in terms of the data of the problem; although it is in principle possible to devise a scheme for numerical calculation from such a representation, the question remains: What actually is the solution? Mathematicians, in their search for representations of solutions, have often modified the meaning of "solution" even further; to solve a problem has become simply to prove the unique existence of a solution.

Clearly, if a mathematical problem is the correct expression of a physical one, it has a unique solution, for the physical situation to be determined from given data does actually occur. Thus to know that certain mathematical problems have unique solutions might seem to have no significance in mathematical physics. It would perhaps have none if it were obvious that the physical problem is correctly expressed in mathematical terms. But frequently this is not obvious at all. The statement that a unique solution exists may then serve as a partial verification of the correctness of the mathematical expression of the problem. If the solution is not unique the data given are not sufficient; if the solution does not exist, the data are incompatible and fewer should have been given.

Following Hadamard [*1,4*], a mathematical problem which possesses a unique solution is said to be *correctly posed* or formulated. For a large class of mathematical problems, the way in which they are posed has never been questioned, just because of their physical significance. These problems are mostly of a standard, rather regular, type. Doubts arise, however, when, for simplicity, the actual physical problem has been replaced by an idealized problem. Such idealized problems may be considered as limiting cases of actual problems, arising when, for example, the domain is extended to infinity, forces are concentrated on surfaces, lines, or points, or terms in the equations are simply omitted as insignificantly small. To the understanding of such idealized problems, purely mathematical existence and uniqueness considerations have made, and may still make, valuable contributions.

As Hadamard emphasized, not only existence and uniqueness, but also a third abstract property of the solution should be required of the problem if it is to be called correctly posed: the property of continuous dependence on the data. Since physical data are not given with absolute precision, the mathematical problem is certainly not the appropriate expression of an actual physical situation if an arbitrarily small variation of the data may have a finite effect on the solution, or even destroy its existence or uniqueness. If the solution does not depend continuously on the data, it might be called unstable. It should, however, be noted that in the customary sense the term instability refers to problems in which the continuous dependence on the data breaks down only for exceptional values of the data.

On the other hand, there are important problems, problems in transonic flow, for example, which possess solutions only for exceptional values of the data; thus the solutions do not depend continuously on the data even when they exist.

The mathematical methods by which the unique existence of a solution and its continuous dependence on the data are established will, in general, furnish information as to the sensitivity of the solution against variation of the data. Roughly speaking, this information will be simply a measure of the rate of change of the solution for a given small change of the data. Naturally, such information may be very valuable if one is faced with the task of selecting an approximation scheme for numerical computation and estimating the chances of its rapid convergence. Here, therefore, is another aspect under which abstract mathematical considerations appear helpful.

Whether or not the problems in the field of gas dynamics have been correctly posed, has been, or should have been, doubtful; and many doubts still exist. In this section an account is presented of some essential points in the mathematical arguments that have been used to answer questions about the uniqueness, existence, and continuous dependence of the solutions of mathematical problems by which problems in gas dynamics have been expressed.

We shall discuss the questions raised in connection with nonsteady flow problems in the development of the gas flow in the course of time. It is true that problems of steady flow are at present considered more important for high speed aerodynamics than problems of unsteady flow. Questions of unique existence can, however, be discussed much more easily and more completely in connection with nonsteady problems. In this section we therefore, confine ourselves to treating one-dimensional nonsteady problems.

The facts described in the considerations of this section are typical for all problems governed by hyperbolic differential equations for functions of two independent variables. This applies in particular to steady two-dimensional supersonic flow problems. The insight gained in studying one-dimensional unsteady flow can be used to solve questions of existence, uniqueness, and continuous dependence in connection with steady two-dimensional flows. There are, however, additional phenomena arising in such flows which have no counterpart in one-dimensional unsteady flow. A few remarks about these phenomena will be made at the end of this section.

It should be said once more that the considerations of this section will be concerned merely with the general mathematical structure of flow problems governed by hyperbolic differential equations. Methods of actual calculation will be treated elsewhere (see III,C and Sec. D, E, F, and G of the present volume).

B,1. One-dimensional Flow. We shall first be concerned with questions arising in connection with flow in a straight cylindrical tube, assuming that all pertinent quantities are constant across the tube (cf. III,B and III,C). The pertinent quantities are the component u of the velocity in the axial direction, the density ρ, and the force per unit volume F. These quantities depend on the abscissa x in the axial direction and the time t. The Eulerian equations of motions are then

$$\rho_t + (\rho u)_x = 0 \tag{1-1}$$

$$\rho u_t + \rho u u_x = F \tag{1-2}$$

where partial differentiation is indicated by subscripts. We consider first the case of a perfect gas in which the force per unit volume F is the pressure gradient

$$F = -p_x \tag{1-3}$$

We assume the gas to be isentropic so that the pressure p is a given function of the density

$$p = f(\rho) \tag{1-4}$$

with a positive derivative

$$f'(\rho) = a^2 \tag{1-5}$$

a being the speed of sound.

Solutions $u = u(x, t)$ and $\rho = \rho(x, t)$ of the two differential equations (1-1) and (1-2) are to be found and must satisfy appropriate conditions. From the mathematical point of view the most natural such conditions are the Cauchy initial conditions, prescribing the distribution of velocity

u and density ρ (or, what is equivalent, pressure p) as a function of the abscissa x at a certain initial time $t = t_0$,

$$u = u_0(x), \qquad \rho = \rho_0(x) \quad \text{for } t = t_0 \tag{1-6}$$

The question is whether or not a unique solution of this problem exists. This question cannot be answered in the affirmative without qualification. It is true that a unique solution exists for a certain time $t \geqq t_0$ provided the functions $u_0(x)$ and $\rho_0(x)$ possess continuous derivatives. In general, however, at a certain time $t_1 > t_0$ and a certain place x_1, a discontinuity in the derivatives of $u(x, t)$ and $\rho(x, t)$ may occur. This discontinuity is such that continuation for times $t > t_1$ in the neighborhood of the place $x = x_1$ is not possible. This fact implies that, from the time $t = t_1$, description of the gas as *ideal* is no longer sufficiently accurate, i.e. we can no longer suppose that the only force F acting results from the gradient of the pressure assumed to be a given function of the density. More, however, can be stated. Not only is it true that gases as they actually occur in nature can no longer be described as ideal under these circumstances, but it is also impossible in principle that an ideal gas could exist under these circumstances.

It is interesting that Lord Rayleigh assumed the converse of the latter fact. He writes [*2*]:

"When discontinuity sets in, a state of things exists to which the usual differential equations are inapplicable; and the subsequent progress of the motion has not been determined. It is probable, as suggested by Stokes, that some sort of reflection would ensue. In regard to this matter we must be careful to keep purely mathematical questions distinct from physical ones. We shall see later how the tendency to discontinuity may be held in check by forces of a dissipative character. But this has nothing directly to do with the mathematical problem of determining what would happen to waves of finite amplitude in a medium, free from viscosity, whose pressure is under all circumstances proportional to the density. To suppose that the problem has no solution would seem to be tantamount to admitting an inherent contradiction in the assumption, usually made in hydrodynamics, of a continuous fluid subject to Boyle's law. It would be strange if the necessity of a molecular constitution for gases could be established by such an argument" (*Theory of Sound*, 1878, §251).

This quotation is given to show that at times, though perhaps rarely, a natural conception about a physical situation does not lead to the right assumption about the existence of a solution of an associated mathematical problem.

Cauchy's data, the initial velocity and the initial density, are the most natural ones mathematically, but in most physical problems data are prescribed differently. Initial data are combined with boundary data.

If, for example, a piston is pushed into the gas or retracted, the velocity of the gas on the piston path is prescribed to be equal to the piston velocity

$$u(x, t) = X_t(t) \quad \text{for} \quad x = X(t) \tag{1-7}$$

when $x = X(t)$ represents the motion, or "path," of the piston. The gas may then be confined to one side of the piston, $x \geqq X(t)$, for example. More generally, one may prescribe the gas velocity u at the cross section at $x = X(t)$ as any given function $U(t)$ of the time,

$$u(x, t) = U(t) \quad \text{for} \quad x = X(t) \tag{1-8}$$

implying that the appropriate flow through the cross section $x = X(t)$ is supplied. Of course, this problem is rather artificial; nevertheless, in the discussion of the propagation of shocks, it will prove helpful to have investigated it.

Data may just as well be prescribed on two paths $x = X_1(t)$, $x = X_2(t)$ which enclose the gas under consideration. Also, the density or, what is equivalent, the pressure may be prescribed instead of the velocity. We then speak of "boundary data." Of course, Cauchy data should be prescribed in addition at an initial time $t = t_0$ on the section

$$X_1(t_0) \leqq x \leqq X_2(t_0)$$

In any case it should be stated that for such data a unique solution does not exist unless the flow velocity u differs from the path velocity X_t by less than the speed of sound. If, however,

$$|u(x,t) - X_t| < a \quad \text{for} \quad x = X(t) \tag{1-9}$$

the solution exists at least for some time $t \geqq t_0$. Since the value of the sound speed depends on the solution, it is not always a priori clear that this condition is satisfied except at the initial time $t = t_0$.

B,2. Domain of Dependence, Range of Influence. The role of the speed of sound as the speed of propagation of disturbances can best be explained in connection with a uniqueness question. To this end we consider Cauchy's problem in which velocity $u = u_0(x)$ and density $\rho = \rho_0(x)$ are given everywhere at the time $t = t_0$. Furthermore we consider another set of initial data $u = u_0^{(1)}(x)$ and $\rho = \rho_0^{(1)}(x)$ which agree with the data $u_0(x)$ and $\rho_0(x)$ in an interval $\mathfrak{D}_0$ of the x axis,

$$u_0^{(1)}(x) = u_0(x), \qquad \rho_0^{(1)}(x) = \rho_0(x) \quad \text{on } \mathfrak{D}_0 \quad x_+ \leqq x \leqq x_- \tag{2-1}$$

Then the corresponding solutions agree with each other at a later time $t > t_0$ in a certain interval $\mathfrak{D}_t$, called the "domain of determinacy" of the interval $\mathfrak{D}$ at the time t (Fig. B,2a).

The left and right points $x_+(t)$ and $x_-(t)$ of this interval $\mathfrak{D}_t$ are just those points which can be reached from the end points x_+ and x_- of the interval $\mathfrak{D}_0$ by propagation with speed of sound forward and backward respectively. More precisely, let $u(x, y)$, $\rho(x, t)$ be the solution associated with the data $u_0(x)$, $\rho_0(x)$ at the time $t = 0$. Then $x = x_+(t)$ and $x = x_-(t)$ are the solutions of the differential equation

$$\frac{dx}{dt} = u(x, t) \pm a(x, t) \tag{2-2}$$

satisfying the initial condition $x_+(0) = x_+$ and $x_-(0) = x_-$ respectively. Assuming that the functions $u(x, t)$, $a(x, t)$ have continuous derivatives,

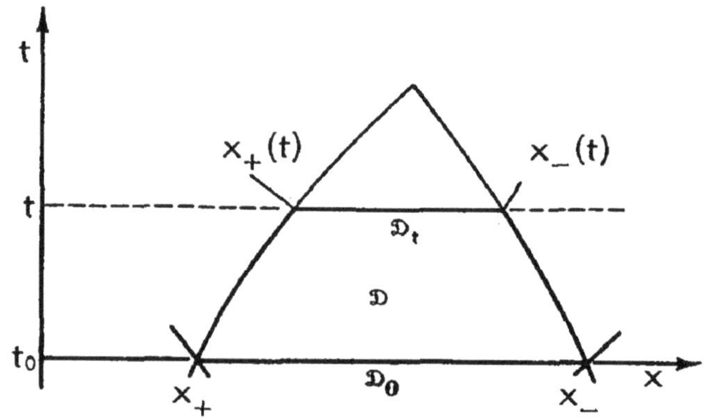

Fig. B,2a. Domain of determinacy $\mathfrak{D}_t$ of the interval $\mathfrak{D}_0$.

we know that these solutions $x_1(t)$, $x_2(t)$ exist. Accordingly, there is a determinacy domain $\mathfrak{D}_t$ as long as $x_+(t) \leqq x_-(t)$, as in Fig. B,2a. If this condition is violated at a certain time t_1, i.e. if $x_+(t_1) = x_-(t_1)$, there is no determinacy domain for $t > t_2$. The uniqueness statement is now that the solution $u^{(1)}(x, t)$, $\rho^{(1)}(x, t)$ agrees with $u(x, t)$, $\rho(x, t)$ in every interval $\mathfrak{D}_t$,

$$u^{(1)}(x, t) = u(x, t), \qquad u^{(2)}(x, t) = u(x, t) \text{ in } \mathfrak{D}_t, \qquad 0 \leqq t \leqq t_1 \tag{2-3}$$

We again consider an interval $\mathfrak{R}_0$ at the time $t = 0$ and again a set of initial data $u_0(x)$, $\rho_0(x)$ and $u_0^{(1)}(x)$, $\rho_0^{(1)}(x)$, which are now supposed to agree with the data $u_0(x)$, $\rho_0(x)$ outside of the interval $\mathfrak{R}_0$ but may differ from them in $\mathfrak{R}_0$. In other words, differences of the data $u_0^{(1)}(x) - u_0(x)$, $\rho_0^{(1)}(x) - \rho_0(x)$ represent "disturbances" of the original data in the interval $\mathfrak{R}_0$. We then ask which part of the x axis may be influenced by the disturbance at later times. This part of the x axis is an interval $\mathfrak{R}_t$, the "range of influence" of the interval $\mathfrak{R}_0$. For a discussion of this and similar notions see [3]. In order to determine this range of influence we need only determine those parts of the x axis in which the two solutions agree. Then $\mathfrak{R}_t$ is evidently the remaining part of the x axis. According to the definition given above, the part of the x axis in which the two solutions agree is just the determinacy domain $\mathfrak{R}_t^*$ of the complement

$\mathcal{R}_0^*$ of the interval $\mathcal{R}_0$, i.e. of all points on the x axis not in $\mathcal{R}_0$. Thus $\mathcal{R}_t$ is seen to be the complement of $\mathcal{R}_t^*$. Consequently, the end points of $\mathcal{R}_t$ satisfy again the differential equation (2-2), if we now denote by x_- the left end point and by x_+ the right end point as in Fig. B,2b. Accordingly, the end points of the range of influence move with the speed of sound relative to the gas.

The deviation of the disturbed from the undisturbed gas motion is conveniently called a "sound signal," provided that the initial data are disturbed only on a small interval.

In the discussion of determinacy and influence, disturbed gas motions were considered which initially differed from the undisturbed motion

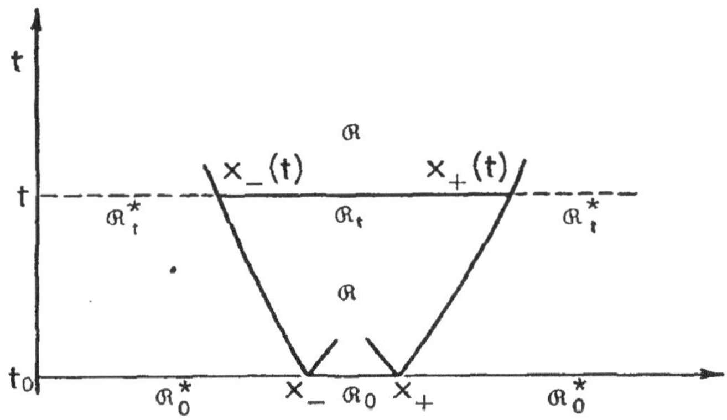

Fig. B,2b. Range of influence $\mathcal{R}_t$ of the interval $\mathcal{R}_0$.

in only a finite segment. The difference between disturbed and undisturbed data cannot be given in terms of a regular analytic function, since such a function cannot vanish along a segment without vanishing identically. Thus it is seen that it would be impossible to understand the most prominent features of the differential equations governing a propagation of a disturbance if the mathematical tools were restricted to regular analytic functions.

Another closely related question can be answered immediately. Let x be any place and $t > t_0$ any time later than the initial time t_0. The solution u,ρ at the place x and the time t does not depend on the initial data on the whole x axis, but only on part of them; the question is, on which part? The answer is that the solution at x, t depends only on the initial data on a certain interval $\mathcal{P} = \mathcal{P}(x, t)$ of the x axis, called the "domain of dependence" of the place x at the time t; the end points are characterized by the condition that "sound signals" sent out from them at the time $t = t_0$ meet precisely at the time t at the place x (Fig. B,2c). Indeed, the place x at the time t does not then belong to the range of influence of any parts of the x axis outside of $\mathcal{P}$, and consequently the solution at x, t is not influenced by parts outside of $\mathcal{P}$. In fact, the determinacy domain of the interval $\mathcal{P}$ has just shrunk to the point x at the time t.

We shall indicate below how, and under which conditions, one can prove the uniqueness assertions on which these considerations are based. Before doing so we must introduce a few closely connected notions.

For the mathematical discussion it is convenient not to keep a strict distinction between space and time. Instead, it is advantageous to introduce an x, t plane and consider u and ρ as functions of the "point" x, t. Instead of motions described by functions $x = x(t)$ we then consider curves in the x, t plane, called "paths," described by such functions. Instead of intervals on the x axis which vary in time, we shall consider regions in the x, t plane swept out by the moving intervals.

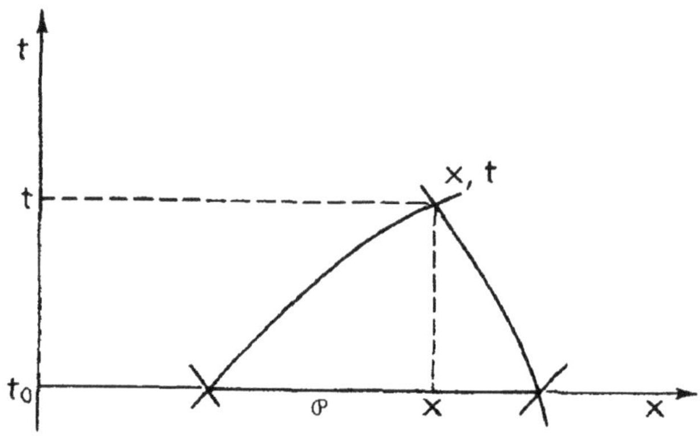

Fig. B,2c. Domain of dependence $\mathcal{P}$ of point x, t.

The curves representing the paths of "sound signals," given by the solutions of the differential equations (2-2), are the "characteristics" of the system of differential equations (1-1) and (1-2). We distinguish between "forward" and "backward" characteristics corresponding to forward and backward motion of a sound signal (see A,3; III,C, and [3]).

As the determinacy domain $\mathfrak{D}$ of an interval $\mathfrak{D}_0$ on the x axis we shall then denote the triangle-like domain in the x, t plane which is bounded by $\mathfrak{D}_0$, the forward characteristic through the left end point, and the backward characteristic through the right end point of $\mathfrak{D}_0$. The range of influence $\mathcal{R}$ of an interval $\mathcal{R}_0$ on the x axis is bounded by $\mathcal{R}_0$, the backward characteristic through the left end point, and the forward characteristic through the right end point of $\mathcal{R}_0$. The domain of dependence of a point x, t is cut out of the x axis by the two characteristics through the point x, t.

Consider a point x, t in the x, t plane and the set of all directions through it, given by vectors with the components dx, dt. We assume $dt \geqq 0$, i.e. we assume that the directions correspond to nondecreasing time. A direction is called characteristic if

$$dx = (u + a)dt \quad \text{or} \quad dx = (u - a)dt \tag{2-4}$$

It is called "spacelike" if

$$|dx - udt| > adt \tag{2-5}$$

because the pure space direction $(dx, 0)$ satisfies that condition. It is called "timelike" if

$$|dx - udt| < adt \tag{2-6}$$

because the pure time direction $(0, dt)$ corresponding to rest satisfies this condition. A curve is called spacelike or timelike if its direction is everywhere spacelike or timelike respectively. Of course, these qualities are defined only with reference to a given flow, i.e. a given solution $u(x, t)$, $\rho(x, t)$.

These notions are helpful in describing possible data through which a solution is uniquely determined. For example, Cauchy's initial value

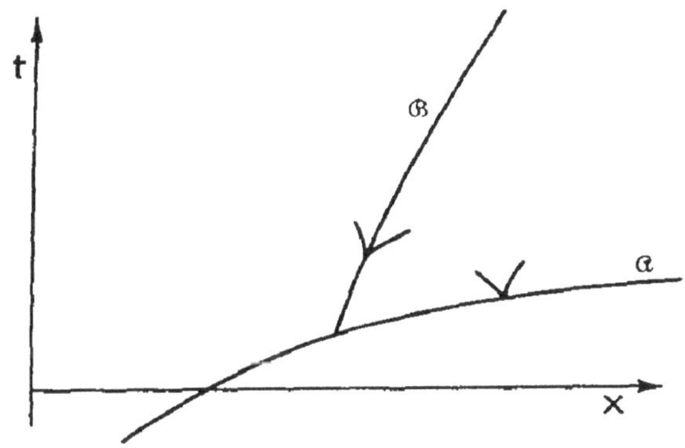

Fig. B,2d. **Spacelike curve 𝒜 and timelike curve ℬ defined with reference to characteristic directions.**

problem may be generalized by prescribing initial data, $u_0(x)$, $\rho_0(x)$ on any spacelike curve $\mathcal{A}$, described by an equation $t = t_{\mathcal{A}}(x)$. (The function $t_{\mathcal{A}}(x)$ will be assumed to have continuous derivatives.) The condition that this curve be spacelike depends on the data prescribed on it. Therefore, one should rather say: if the (continuously differentiable) data $u_0(x)$, $\rho_0(x)$ on the curve $t = t_{\mathcal{A}}(x)$ are such that the curve becomes spacelike through these data, then there exists a unique solution of the differential equations (1-1) and (1-2), at least in a neighborhood of $\mathcal{A}$, which assumes the prescribed values of $\mathcal{A}$.

On the other hand, data may be prescribed on a curve which consists of a spacelike section $\mathcal{A}$ and a timelike section $\mathcal{B}$ (Fig. B,2d). The condition (1-9), which was formulated earlier, did exactly express the timelike character of the curve $x = X(t)$. Two data may be prescribed on $\mathcal{A}$, provided the data make the curve $\mathcal{A}$ spacelike, but only one quantity, u or ρ, may be prescribed on $\mathcal{B}$ as a function of t. Whether or not the curve $\mathcal{B}$ is timelike depends on both quantities ρ and u there, but only one quantity is known on $\mathcal{B}$ beforehand. Therefore, it may not be known beforehand whether the curve $\mathcal{B}$ is timelike for the resulting solution. In any case, it should be required that $\mathcal{B}$ be timelike at the point of intersection with $\mathcal{A}$; this requirement can be met since two quantities are known at this point. We assume that the data on $\mathcal{B}$ have a continuous derivative with respect to t and agree at the intersection with $\mathcal{A}$ with the value there prescribed on $\mathcal{A}$. Then a unique solution exists at least in a neighborhood of the curve $\mathcal{A}$. Certainly, the curve $\mathcal{B}$ will remain timelike at least in a neighborhood of the curve $\mathcal{A}$. A breakdown of the existence of the solution may be due to the causes alluded to in Art. 1, or it may result if the curve $\mathcal{B}$ ceases to be timelike.

It should be mentioned that there is one case in which the timelike character of the curve $\mathcal{B}$ is assured beforehand, namely, if on $\mathcal{B}$ the flow velocity u is prescribed to be precisely equal to the rate of change dx/dt on the curve $\mathcal{B}$. For then $|dx - udt| = 0$ and condition (2-6) is always satisfied. In fact, prescribing the velocity u in this way corresponds to the problem in which a piston moves into the gas-filled tube with a given velocity. This was the situation expressed by condition (1-8).

B,3. Uniqueness Proofs. In this article we intend to prove the uniqueness of the solution of a few simple problems and indicate how such proofs can be carried out in more complicated cases.

We first consider the problem of sound motion which results when the velocity is small and the density differs little from a constant. Accordingly, in this case, we may replace u, ρ, p, and a by δu, $\rho + \delta\rho$, $a + \delta a$, $p = p + \delta p = p + a^2\delta\rho$, assuming ρ, p, and $a^2 = p'(\rho)$ to be constant. When only terms of first order in the differential equations (1-1) and (1-2) are kept, they reduce to the linear equations

$$\delta\rho_t + \rho\delta u_x = 0, \qquad \rho\delta u_t + a^2(\rho)\delta\rho_x = 0 \tag{3-1}$$

We prescribe initial data $\delta u_0(x)$, $\delta\rho_0(x)$ on the initial curve $t = 0$ and proceed to prove that there is at most one solution δu, $\delta\rho$ which possesses continuous derivatives. A pair of functions $\delta u(x, t)$, $\delta\rho(x, t)$ satisfying Eq. 3-1 and the initial conditions is said to be one solution here; the pair of equations is accordingly said to be *one* equation.

Of course, Eq. 3-1 can be solved explicitly. If δu, $\delta\rho$ is a solution, then, as is well known, there are two functions $f(\alpha)$, $g(\beta)$ with continuous derivatives such that

$$\begin{aligned}\delta u &= af(x - at) - ag(x + at)\\ \delta\rho &= \rho f(x - at) + \rho g(x + at)\end{aligned}$$

Setting $t = 0$ we find

$$\begin{aligned}f(x) &= (2a)^{-1}\delta u_0(x) + (2\rho)^{-1}\delta\rho_0(x)\\ g(x) &= -(2a)^{-1}\delta u_0(x) + (2\rho)^{-1}\delta\rho_0(x)\end{aligned}$$

so that indeed δu and $\delta \rho$ are determined uniquely through the initial data. Moreover, we observe that the value of u, ρ at a point x_0, t_0 depends on the initial data only at those points that are cut out of the initial line by the two characteristics, $x \pm at = \text{const}$, through the point x_0, t_0. From this fact one deduces easily the statements made in Art. 2 about determinacy domains, domains of dependence, and ranges of influence.

In spite of the simplicity of this argument we shall present another argument which has the advantage that the underlying idea can be generalized to problems which cannot be solved explicitly.

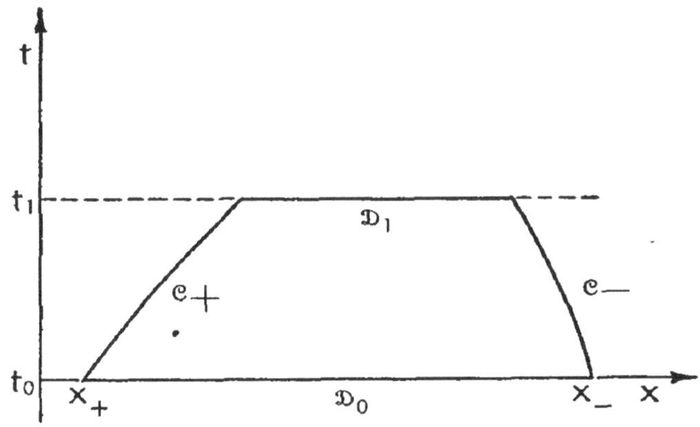

Fig. B,3a. Domain $\mathfrak{D}_1$ in which the uniqueness of the solution is proved in terms of the data on the interval $\mathfrak{D}_0$.

This second argument can be derived from mathematical considerations related to those that express the energy balance. We multiply the second equation of Eq. 3-1 by δu, the first one by $a^2\rho^{-1}\delta\rho$ and add; the result can be written in the form

$$[\tfrac{1}{2}\rho\delta u^2 + \tfrac{1}{2}a^2\rho^{-1}\delta\rho^2]_t + [a^2\delta\rho\delta u]_x = 0 \tag{3-2}$$

Here $\delta u^2 = (\delta u)^2$ and $\delta\rho^2 = (\delta\rho)^2$. The term $[\frac{1}{2}\rho\delta u^2]_t$ is evidently the time rate of change of the kinetic energy of the gas per unit volume. The term $[a^2\delta\rho\delta u]_x$ may be interpreted as the second order contribution to the work per unit volume per unit time $[(p + a^2\delta\rho)\delta u]_x$ done by the pressure on the gas passing the cross section x; this term may also be called the flux of energy per unit time through that cross section. The term $[\frac{1}{2}a^2\rho^{-1}\delta\rho^2]_t$ would accordingly be a second order contribution to the time rate of change of intrinsic energy. The details of this interpretation are irrelevant in the present context.

We now consider a domain bounded by the section $\mathfrak{D}_0$, $x_+ \leqq x \leqq x_-$, on the x axis, the two sections $\mathfrak{C}_+$ and $\mathfrak{C}_-$ of characteristics $x = x_+ + at$, $x = x_- - at$ for $0 \leqq t \leqq t_1$ and the section $\mathfrak{D}_1$, $x_+ + at_1 \leqq x \leqq x_- - at_1$ on the line $t = t_1$, assuming $x_+ + at_1 \leqq x_- - at_1$ (Fig. B,3a). Integrating relation (3-2) over this domain we obtain

$$\int_{\mathfrak{D}_1} [\tfrac{1}{2}\rho\delta u^2 + \tfrac{1}{2}a^2\rho^{-1}\delta\rho^2]dx + \int_{\mathfrak{C}_+} [\tfrac{1}{2}\rho\delta u^2 - a\delta\rho\delta u + \tfrac{1}{2}a^2\rho^{-1}\delta\rho^2]dx + \int_{\mathfrak{C}_-} [\tfrac{1}{2}\rho\delta u^2 + a\delta\rho\delta u + \tfrac{1}{2}a^2\rho^{-1}\delta\rho^2]dx = \int_{\mathfrak{D}_0} [\tfrac{1}{2}\rho\delta u^2 + \tfrac{1}{2}a^2\rho^{-1}\delta\rho^2]dx \quad (3\text{-}3)$$

We note that the integrands of the integrals over $\mathfrak{C}_+$ and $\mathfrak{C}_-$ can be written in the form $\tfrac{1}{2}\rho^{-1}[\rho\delta u \mp a\delta u]^2$; these integrands are therefore non-negative. Consequently, we can deduce from Eq. 3-3 the inequality

$$\int_{\mathfrak{D}_1} [\rho^2\delta u^2 + a^2\delta\rho^2]dx \leqq \int_{\mathfrak{D}_0} [\rho^2\delta u^2 + a^2\delta\rho^2]dx \quad (3\text{-}4)$$

Suppose now two solutions δu, $\delta\rho$ are given which on $t = 0$ assume the same initial values δu_0, $\delta\rho_0$. Then we consider the difference of these two solutions; since this difference, which for the moment we simply denote by δu, $\delta\rho$, also satisfies the linear equation (3-1), the inequality (3-4) holds for it. Since the difference δu, $\delta\rho$ vanishes on the initial line $t = 0$, the right member of inequality (3-4) vanishes. Since the left member of this inequality is non-negative, it also vanishes. Consequently, $\delta u = \delta\rho = 0$ on $\mathfrak{D}_1$. Since the time t_1 was arbitrary so long as $x_+ + at_1 \leqq x_1 - at_1$, we may conclude that $\delta u = \delta\rho = 0$ in the whole region $x_+ + at \leqq x \leqq x_- - at$ for $0 \leqq t \leqq t_2$, the time t_2 being defined by $x_+ + at_2 = x_- - at_2$. Thus it is seen that the difference of the two solutions vanishes in the determinacy domain of the section $\mathfrak{D}_0$. In other words, there is at most one solution in this determinacy domain assuming prescribed values on the section $\mathfrak{D}_0$. (Uniqueness proofs of this type were first given by Zaremba [*1,4*].)

Another conclusion can be drawn from the inequality (3-4). Let u, ρ be the difference of two solutions which do not necessarily agree on $\mathfrak{D}_0$. Then the right member of inequality (3-4) may be considered as, in a sense, a measure of the deviation of the two solutions from each other on $\mathfrak{D}_0$. The statement of inequality (3-4) then is that the deviation of the two solutions in this sense on the segment $\mathfrak{D}_1$ is not greater than that on the segment $\mathfrak{D}_0$. This statement implies, in a sense, that the solution on $\mathfrak{D}_1$ depends continuously on the initial data. Statements about the continuous dependence of the solution on the initial data in the customary sense can be deduced from inequalities which can be derived in a manner similar to that in which inequality (3-4) was derived. In general, one may say that every proof of the continuous dependence of the solution on the data implies a uniqueness proof.

Next we shall consider Cauchy's problem for the nonlinear equations (1-1) and (1-2), i.e. for the equation

$$\begin{aligned} \rho u_t + \rho u u_x + a^2(\rho)\rho_x &= 0 \\ \rho_t + u\rho_x + \rho u_x &= 0 \end{aligned} \tag{3-5}$$

Let $u(x, t)$ and $\rho(x, t)$ be a solution of this equation which assumes initial values $u_0(x)$, $\rho_0(x)$ on an interval $\mathfrak{D}_0$ of the x axis. We assume that this solution is given in a neighborhood of $\mathfrak{D}_0$ which contains the determinancy domain $\mathfrak{D}$ of $\mathfrak{D}_0$ and possesses continuous second derivatives there. The determinacy $\mathfrak{D}$ domain is bounded by $\mathfrak{D}_0$ and the two sections $\mathfrak{C}_+$, $\mathfrak{C}_-$ of the forward and backward characteristics from the left and right end points of $\mathfrak{D}_0$, respectively, to their intersection. Note that these characteristics, and hence also the domain $\mathfrak{D}$, depend on the solution u, ρ. Suppose now that another solution $u + \delta u$, $\rho + \delta\rho$ of Eq. 3-5 is given in the domain $\mathfrak{D}$, which assumes the same initial values on $\mathfrak{D}_0$ as u, ρ, i.e.

$$\delta u = \delta\rho = 0 \text{ on } \mathfrak{D}_0 \tag{3-6}$$

(We also assume that $u + \delta u$, $\rho + \delta\rho$ has continuous second derivatives in $\mathfrak{D}$.) We then state that $\delta u = \delta\rho = 0$ throughout in $\mathfrak{D}$. In other words, the solution u, ρ is uniquely determined in $\mathfrak{D}$ through the initial value on $\mathfrak{D}_0$.

The proof of this statement is based on the fact that the difference δu, $\delta\rho$ of the two solutions itself satisfies a linear differential equation with coefficients which depend on the two solutions u, ρ and $u + \delta u$, $\rho + \delta\rho$. Since these two solutions are given, the coefficients of the linear differential equation are given. This linear differential equation is obtained by subtracting Eq. 3-5 taken for u, ρ from the same equation taken for $u + \delta u$, $\rho + \delta\rho$. The result can be written in the form

$$\begin{aligned} \rho\delta u_t + \rho u\delta u_x + a^2(\rho)\delta\rho_x &= A\delta u + B\delta\rho \\ \delta\rho_t + u\delta\rho_x + \rho\delta u_x &= C\delta u + D\delta\rho \end{aligned} \tag{3-7}$$

The functions A, B, C, and D of x and t are easily expressed in terms of the given functions u, ρ, δu, and $\delta\rho$ and their first derivatives; accordingly, they themselves have continuous first derivatives. In deriving this fact one should observe that the difference $a^2(\rho + \delta\rho) - a^2(\rho)$ can be written in the form ξdp, in which the function $\xi(x, t)$ is a difference quotient when $\delta\rho \neq 0$, a differential quotient when $\delta\rho = 0$.

Next one adds the first equation (3-7) multiplied by δu and the second one multiplied by $a^2\rho^{-1}\delta\rho$. For reasons that will become apparent later on, the result is multiplied by the exponential function $e^{-\lambda t}$, the constant λ being kept at our disposal. The result can then be written in the form

$$\frac{1}{2}\frac{\partial}{\partial t} e^{-\lambda t}[\rho \delta u^2 + a^2(\rho)\rho^{-1}\delta\rho^2]$$
$$+ \frac{1}{2}\frac{\partial}{\partial x} e^{-\lambda t}[\rho u \delta u^2 + 2a^2(\rho)\delta u \delta\rho + a^2(\rho)\rho^{-1}u\delta\rho^2]$$
$$+ \frac{1}{2}\lambda e^{-\lambda t}[\rho\delta u^2 + a^2(\rho)\rho^{-1}\delta\rho^2] + \frac{1}{2}e^{-\lambda t}[P\delta u^2 + 2Q\delta u\delta\rho + R\delta\rho^2] = 0 \tag{3-8}$$

in which P, Q, and R are given functions of x and t. We now integrate this identity over the domain $\mathfrak{D}$. The contributions resulting from the first two terms become integrals over the boundary $\mathfrak{D}_0 + \mathfrak{C}_+ + \mathfrak{C}_-$ of $\mathfrak{D}$. The integral over $\mathfrak{D}_0$ vanishes since $\delta u = \delta\rho = 0$ on $\mathfrak{D}_0$. We take relations (2-2) into account, which hold along the characteristics $\mathfrak{C}_-$, $\mathfrak{C}_+$ respectively. Then the integrals over $\mathfrak{C}_+ + \mathfrak{C}_-$ reduce to

$$\frac{1}{2}\int_{\mathfrak{C}_+ + \mathfrak{C}_-} e^{-\lambda t}[\rho\delta u^2 + \rho^{-1}a^2(\rho)\delta\rho^2 \mp 2a(\rho)\delta u\delta\rho]a(\rho)dt$$

Since the expression in the bracket is evidently a pure square, the integral is non-negative. Consequently, integration of identity (3-8) over the domain $\mathfrak{D}$ yields the inequality

$$\iint_{\mathfrak{D}} e^{-\lambda t}[(\lambda\rho + P)\delta u^2 + 2Q\delta u\delta\rho + (\lambda a^2(\rho)\rho^{-1} + R)\delta\rho^2]dxdt \leqq 0 \tag{3-9}$$

Evidently, the constant λ can be chosen so large that the quadratic form in the bracket is here positive definite. Inequality (3-9) can therefore hold only if $\delta u = \delta\rho = 0$ in $\mathfrak{D}$. Thus the uniqueness statement is proved.

The remarks about the continuous dependence of the solution on the initial data made above in connection with the linear problem apply also to the nonlinear problem just considered.

Although it is true that the basic idea underlying the uniqueness proof discussed originated from a consideration of the energy balance, it would be pointless to try to interpret the form the proof has now acquired in terms which have a familiar physical significance.

Finally we shall prove the uniqueness of the solution of a problem in which boundary data are prescribed on timelike lines. For simplicity we assume that both timelike lines issue from one point. The proof to be presented could be carried through in essentially the same manner if the two timelike lines did not meet at the initial line, but cut a spacelike segment out of it. Of course, the proof could also be carried through if a datum were prescribed on only one spacelike curve. We restrict ourselves to giving the uniqueness proof for the linear equation (3-1). The proof could be carried over to the case of the nonlinear equation in the way it was carried over for Cauchy's initial problem.

Let $\mathcal{B}^{(1)}$, $\mathcal{B}^{(2)}$ be two timelike or characteristic curve segments which issue from the point $O = 0, 0$ and do not cross each other. We describe the two curves by two functions $x = X^{(1)}(t)$ and $x = X^{(2)}(t)$ defined for $0 \leqq t \leqq t_1$ and assumed to have continuous second derivatives. We then have $X^{(1)}(0) = X^{(2)}(0) = 0$ and $X^{(1)}(t) \leqq X^{(2)}(t)$. The timelike, or characteristic, nature of the two curves, with reference to the undis-

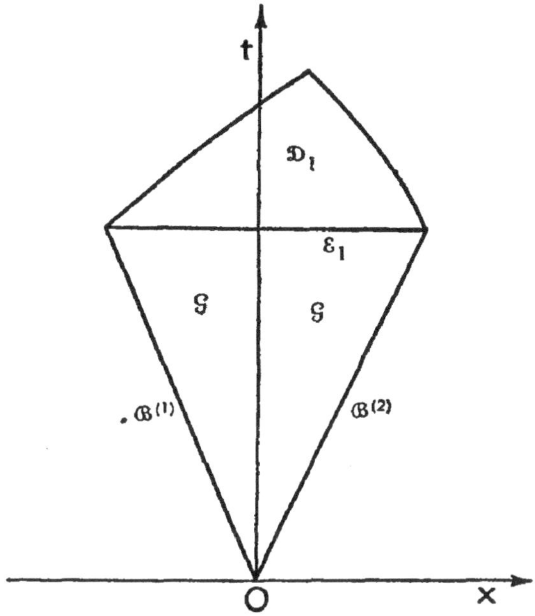

Fig. B,3b. Domain $\mathcal{G}$ in which the uniqueness of the solution is proved in terms of the data on the timelike segments $\mathcal{B}^{(1)}$ and $\mathcal{B}^{(2)}$.

turbed flow $u = 0$, $a = \text{const}$, is expressed by the conditions

$$|X_t^{(1)}(t)| \leqq a, \qquad |X_t^{(2)}(t)| \leqq a, \quad \text{for } 0 \leqq t \leqq t_1 \tag{3-10}$$

In addition we assume

$$X_t^{(1)}(t) < 0, \qquad X_t^{(2)}(t) > 0 \tag{3-11}$$

without serious restriction.

We shall establish the uniqueness of the solution of the linear equation (3-1) in the domain $\mathcal{G}$, which is bounded by the two curve segments $\mathcal{B}^{(1)}$, $\mathcal{B}^{(2)}$, and the segment $\mathcal{E}_1$: $X^{(1)}(t_1) \leqq x \leqq X^{(2)}(t_1)$, $t = t_1$, between them on the line $t = t_1$ (Fig. B,3b). (When the two curves $\mathcal{B}^{(1)}$, $\mathcal{B}^{(2)}$ are characteristic so that equality holds in Eq. 3-10 the problem reduces to "Goursat's problem.") This domain $\mathcal{G}$ is not the determinacy domain of the two curve segments $\mathcal{B}^{(1)}$, $\mathcal{B}^{(2)}$; this determinacy domain is obtained from the domain $\mathcal{G}$ by joining to it the determinacy domain $\mathfrak{D}_1$ of the segment $\mathcal{E}_1$. Certainly, it is sufficient to prove the uniqueness of the solution in the domain $\mathcal{G}$; for then the uniqueness in the total determinacy domain $\mathcal{G} + \mathfrak{D}_1$ follows from the previous uniqueness theorem applied to the initial domain $\mathcal{E}_1$, and its determinacy domain $\mathfrak{D}_1$.

We now assume δu, $\delta\rho$ to be a solution of differential equation (3-1) (with continuous derivatives) defined in the domain $\mathcal{G}$. We maintain that there is no other solution for which δu on the segments $\mathcal{B}^{(1)}$ and $\mathcal{B}^{(2)}$ and $\delta\rho$ at the point O assume the same values as for the given solution.

Clearly, it is sufficient to prove that $\delta u = \delta\rho = 0$ in $\mathcal{G}$ if

$$\delta u = 0 \quad \text{on } \mathcal{B}^{(1)} \text{ and } \mathcal{B}^{(2)} \tag{3-12}$$

$$\delta\rho = 0 \quad \text{at } O \tag{3-13}$$

The statement would not be true if the condition $\delta\rho = 0$ at O were omitted; for the functions $\delta u = 0$, $\delta\rho = \text{const} \neq 0$ would form a counterexample.

In order to prove the statement we employ an identity similar to Eq. 3-3, but involving the derivatives of the function δu instead of δu and $\delta\rho$.

From Eq. 3-1 we derive the relation

$$[\rho^2\delta u_x^2 + a^2\delta\rho_x^2]_t + 2[\rho\delta\rho_t\delta u_t]_x = 0$$

or

$$[\delta u_x^2 + a^{-2}\delta u_t^2]_t - 2[\delta u_x\delta u_t]_x = 0$$

Note that the coefficients in Eq. 3-1 are constant. After integrating this relation over the domain $\mathcal{G}$ we obtain

$$\int_{\mathcal{E}_1} [\delta u_x^2 + a^{-2}\delta u_t^2]dx + \int_{\mathcal{B}^{(1)}} \{[\delta u_x^2 + a^{-2}\delta u_t^2]X_t^{(1)} + 2\delta u_x\delta u_t\}dt$$
$$- \int_{\mathcal{B}^{(2)}} \{[\delta u_x^2 + a^{-2}\delta u_t^2]X_t^{(2)} + 2\delta u_x\delta u_t\}dt = 0$$

Making use of relation

$$\delta u_t + X_t^{(1)}\delta u_x = 0 \quad \text{on } \mathcal{B}^{(1)}$$

$$\delta u_t + X_t^{(2)}\delta u_x = 0 \quad \text{on } \mathcal{B}^{(2)}$$

we obtain

$$\int_{\mathcal{E}_1} [\delta u_x^2 + a^{-2}\delta u_t^2]dx - \int_{\mathcal{B}^{(1)}} [1 - a^{-2}(X_t^{(1)})^2]\delta u_x^2 X_t^{(1)}dt$$
$$+ \int_{\mathcal{B}^{(2)}} [1 - a^{-2}(X_t^{(2)})^2]\delta u_x^2 X_t^{(2)}dt = 0 \tag{3-14}$$

The last two terms are non-negative by virtue of assumptions (3-10) and (3-11). Consequently, relation (3-14) implies that $\delta u_x = \delta u_t = 0$ on $\mathcal{E}_1$.

Instead of the domain $\mathcal{G}$ we could have chosen the domain $\mathcal{G}'$ cut out of $\mathcal{G}$ by the condition $t \leqq t'$ with any positive value $t' \leqq t_1$. We would then have found $\delta u_x = \delta u_t = 0$ on $\mathcal{E}_1'$. Since all the segments $\mathcal{E}_1'$

exhaust domain $\mathcal{G}$, we may conclude that $\delta u_x = \delta u_t = 0$ in $\mathcal{G}$. In other words, δu is constant in $\mathcal{G}$ and hence vanishes since it vanishes on $\mathcal{B}^{(1)}$ and $\mathcal{B}^{(2)}$ by Eq. 3-12. From Eq. 3-1 we infer that $\delta\rho$ is constant in $\mathcal{G}$ and from condition (3-13) we finally conclude that also $\delta\rho = 0$ in $\mathcal{G}$. Thus the uniqueness statement has been proved.

By similar arguments we can prove the uniqueness of the solution if $\delta\rho$ instead of δu is prescribed on one or the other segment $\mathcal{B}$. Also, it is possible to prove the uniqueness of the solution of the nonlinear differential equations (3-5) by extending the argument in the way it was extended for Cauchy's initial value problem.

B,4. The Characteristic Equations. Existence. Let $x = x_+(\alpha)$, $t = t_+(\alpha)$ be a parametric representation of a forward characteristic curve $\mathcal{C}_+$; then the two functions $x(\alpha)$, $t(\alpha)$ satisfy the equation

$$x_\alpha - (u + a)t_\alpha = 0 \qquad (4\text{-}1)_\alpha$$

The characteristic feature of a characteristic curve is that the values of u and ρ derived by solving Eq. 3-5 and considered as functions $u = u_+(\alpha)$, $\rho = \rho_+(\alpha)$ along the curve $\mathcal{C}_+$, also satisfy an equation of the same type as Eq. 4-1, viz.

$$\rho u_\alpha + a\rho_\alpha = 0 \qquad (4\text{-}2)_\alpha$$

which can be derived as an appropriate linear combination of Eq. 3-5. Similarly, we let $x = x_-(\beta)$, $t = t_-(\beta)$ be a parametric representation of a backward characteristic $\mathcal{C}_-$, and consider u and ρ as functions $u = u_-(\beta)$, $\rho = \rho_-(\beta)$ along $\mathcal{C}_-$. Then the relations

$$x_\beta - (u - a)t_\beta = 0 \qquad (4\text{-}1)_\beta$$

$$\rho u_\beta - a\rho_\beta = 0 \qquad (4\text{-}2)_\beta$$

hold. Relations (4-2) illuminate the role of the characteristic curves as boundaries of determinacy and influence domains; for they imply that the values of u and ρ on a characteristic curve depend on each other.

In 1927, H. Lewy [5] observed that the characteristic equations can be made the basis of a general existence proof, which had up until then been limited to special classes of equations. (Lewy considered the general hyperbolic equation of second order for one function of two variables; his treatment is equally applicable to a hyperbolic system of two equations of first order, of which Eq. 3-5 is a special case.) Lewy's idea was to consider the characteristic parameters α, β as new independent variables and to consider the former dependent and independent variables as functions of α, β. The characteristic equations are then considered as a system of partial differential equations to be satisfied by these functions of α and β.

In our case the four functions $x(\alpha, \beta)$, $t(\alpha, \beta)$, $u(\alpha, \beta)$, and $\rho(\alpha, \beta)$ are to be determined from the four equations (4-1) and (4-2). Thus a system of two partial differential equations has been replaced by a system of four such equations. However, the new equations are of a simpler type than the original ones, since in each equation differentiation with respect to only one variable occurs. Also, the equations no longer depend explicitly on the independent variables, so that the domain in which the new functions are to be determined is known beforehand.

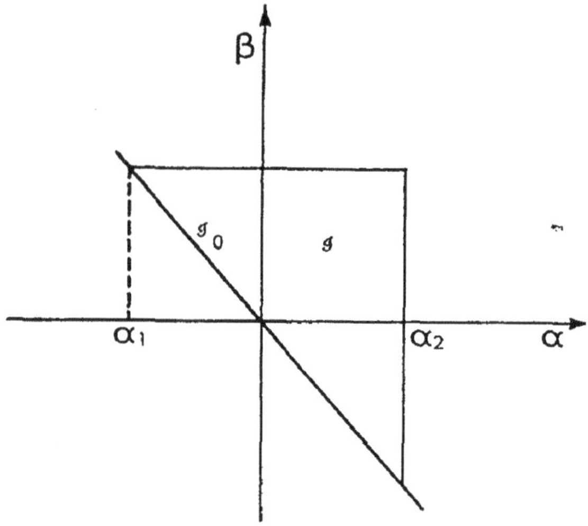

Fig. B,4a. Images $\mathfrak{J}_0$ and $\mathfrak{J}$ of the initial segment $\mathfrak{D}_0$ and its determinacy domain $\mathfrak{D}$ in the plane of the characteristic parameters α, β.

Let $\mathfrak{J}_0$ be the image of the initial segment $\mathfrak{D}_0$ in the x, t plane (Fig. B,4a). For simplicity we assume it is given in the form

$$\mathfrak{J}_0\colon\ \beta = -\alpha, \qquad \alpha_1 \leqq \alpha \leqq \alpha_2 \tag{4-3}$$

This restriction is appropriate since $\mathfrak{D}_0$ was assumed to be spacelike, and since the parameter α could be replaced by an arbitrary function of $\mathfrak{D}_0$. The image $\mathfrak{J}$ of the determinacy domain $\mathfrak{D}$ is then found to be given by

$$\mathfrak{J}\colon\ -\alpha \leqq \beta \leqq -\alpha_1, \qquad \alpha_1 \leqq \alpha \leqq \alpha_2 \tag{4-4}$$

Denote the four functions x, t, u, ρ of α, β summarily by $x^{(k)}(\alpha, \beta)$, $k = 1, 2, 3, 4$. Then the initial data supply four continuously differentiable functions $x_0^{(k)}(\alpha)$, and the initial condition is

$$x^{(k)}(\alpha, -\alpha) = x_0^{(k)}(\alpha) \tag{4-5}$$

The four equations (4-1) and (4-2) can be reduced to a manageable form by differentiating Eq. $(4\text{-}1)_\alpha$ and $(4\text{-}2)_\alpha$ with respect to β and Eq. $(4\text{-}1)_\beta$ and $(4\text{-}2)_\beta$ with respect to α. Four equations result which, by solving for the four second derivatives $x_{\alpha\beta}^{(k)}$, can be rewritten in the form

$$x_{\alpha\beta}^{(k)} = f(x_\alpha, x_\beta, x), \qquad k = 1, \ldots, 4 \tag{4-6}$$

in which x in the argument of f refers to all four functions. In addition to the initial values, Eq. 4-5, the values of $x_\alpha^{(k)}$ and $x_\beta^{(k)}$ on the initial line $\mathcal{I}_0$ are known. These eight initial values can be determined from the eight equations which consist of the derivatives of the four equations (4-5) with respect to α and the four equations (4-1) and (4-2) specialized to $\beta = -\alpha$.

Thus the problem has been reduced to the initial value problem for Eq. 4-6. The existence of solutions of problems of this type was established by Picard as early as 1890 [4]. Picard used his famous method of itera-

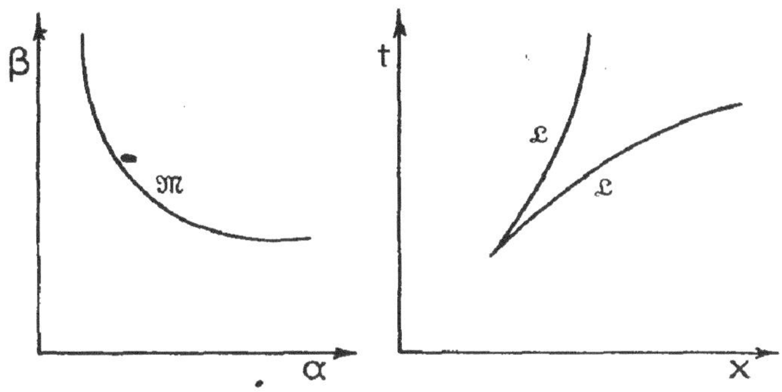

Fig. B,4b. A curve $\mathfrak{M}$ in the α, β plane along which the Jacobian J vanishes corresponds to a limiting line $\mathfrak{L}$ in the x, t plane.

tions (or successive approximations), applying it to a system of integral equations into which Eq. 4-6, together with the initial conditions, is easily transformed.

It is easily seen that those solutions of Eq. 4-6 which satisfy the initial conditions also satisfy Eq. 4-1 and 4-2, for relation (4-6) implies that the left members of Eq. 4-1 and 4-2 are constant, and the initial relations then imply that these members are zero.

Another question arises: Does the solution of Eq. 4-1 and 4-2 furnish a solution of the original equation (3-5)? This would be the case in a (sufficiently small) region in which the Jacobian

$$J = x_\alpha t_\beta - x_\beta t_\alpha = -2at_\alpha t_\beta \tag{4-7}$$

does not vanish; for x and t can then be reintroduced as independent variables instead of α and β. From the assumption that the initial line is spacelike it follows easily that $J \neq 0$ on it. The same is then true in a neighborhood $\mathcal{I}$, so that the existence of the solution is assured at least for a small neighborhood.

However, it may very well happen that the Jacobian J does vanish along a certain curve $\mathfrak{M}$ in $\mathcal{I}$ in such a way that the mapping of the α, β domain $\mathcal{I}$ into the x, y plane forms a fold, the image $\mathfrak{L}$ of the curve $\mathfrak{M}$, and covers part of the x, t plane two or more times (Fig. B,4b). Then the solution of our problem just does not exist.[1] The smallest value t^* of t that occurs on the curve $\mathfrak{M}$ gives the time up to which the solution of the original problem exists. The solution $u(x, t)$, $\rho(x, t)$ cannot, however, be continued beyond the time t^*.

The fold $\mathfrak{L}$ which here is responsible for nonexistence of the solution is called a "limiting line" (see F,12).

Suppose now that the Jacobian J does not vanish in the domain $\mathfrak{g}$. Then the solution of the original problem exists in the full determinacy domain $\mathfrak{D}$ of the initial segment $\mathfrak{D}_0$. The uniqueness of the solution in $\mathfrak{D}$ and its continuous dependence on the data is a by-product of this existence proof.

It would seem likely that one could, in a similar fashion, design a proof of the existence of the solution of a problem in which boundary data are prescribed on curve segments $\mathfrak{B}^{(1)}$, $\mathfrak{B}^{(2)}$ which are timelike, at least at their juncture with the initial curve. The images $\mathfrak{J}^{(1)}$, $\mathfrak{J}^{(2)}$ in the α, β plane may be taken rather arbitrarily, for example, as straight segments $\alpha - \beta = \text{const}$. The main difficulty to be overcome stems from the fact that the boundary conditions to be imposed on these segments are not of a standard type and depend nonlinearly on the functions to be determined. It seems that such a general existence proof for this case has not yet been carried out.

B,5. Method of Finite Differences. The iteration process referred to in Art. 4, which is used for an existence proof, can in principle be used for numerical determination of the solution. Experience, however, has shown that in complicated problems only a few steps in such an iteration scheme are suitable for numerical computations, since eventually the evaluation of the integrals involved becomes too cumbersome.

In general, another approximation procedure is more suitable, the "method of finite differences," which is very powerful and easily used because it is primitive.

After having chosen two mesh widths Δx and Δt, one may place in the x, t plane a net consisting of the points $x = l\Delta x$, $t = m\Delta t$ with integral values of l and m. The values of any quantity or function q at these net points may be denoted by $q_{l,m}$. The time derivative q_t of the quantity q may be replaced by the difference quotient $(\Delta t)^{-1}[q_{l,m+1} - q_{l,m}]$. This difference quotient will be assigned to the point $(l\Delta x, m\Delta t)$. The differential quotient q_x may be replaced by the central difference quotient $(2\Delta x)^{-1}[q_{l+1,m} - q_{l-1,m}]$. (Of course, q_t could also be replaced by a central difference quotient.) Correspondingly, the differential equations are replaced by difference equations. The two equations which thus result

[1] In order to make this statement precise enough to refute Rayleigh's conjecture (see Art. 1), a somewhat more delicate argument would be needed.

from Eq. 3-5 enable one to express the values of $u_{l,m}$ and $\rho_{l,m}$ by the values of u and ρ at the three points $(l \pm 1, m - 1)$, $(l, m - 1)$ of the preceding row. If initial data on the line $t = 0$ are prescribed, one can successively express $u_{l,m}$ and $\rho_{l,m}$ by means of the initial data at the points on the line $t = 0$.

The question arises whether or not a good approximation of the actual solution is obtained in this way if the mesh widths are taken fine enough. Mathematically expressed, the question is: Do the solutions of the difference equation problems converge to the solution of the differential equation problems if one lets the mesh widths Δx and Δt approach zero? As a partial answer one can state that, except for special initial data, there is no convergence if the ratio of the time mesh width to the space mesh width exceeds a definite value.

The reason for this phenomenon is rather obvious [*6*]. The value of the solution $u_{l,m}$, $\rho_{l,m}$ of the difference equation does not depend on the initial data on the whole initial line, but only on part of it. This part is cut out from the initial line by the two lines $x\Delta t \pm t\Delta x = \text{const}$ through the point $l\Delta x$, $m\Delta t$. If in the limit, this segment remained in the interior of the actual domain of dependence of the point $l\Delta x$, $m\Delta t$, convergence of the solutions of the difference equation to that of the differential equation would lead to the contradictory conclusion that the solution of the differential equation depends on the initial values on a section of the initial line contained in its actual domain of dependence.

It has been observed [*7*] that the restriction on the ratio of the mesh widths can be weakened by modifying the setup of the difference equations. Suppose q_t is replaced by $(\Delta t)^{-1}[q_{l,m+1} - q_{l,m}]$; then q_x is replaced by $(4\Delta x)^{-1}[q_{l+1,m+1} - q_{l-1,m+1} + q_{l,m+1} - q_{l,m-1}]$. Now the difference equations no longer allow us to express the value of the quantities at the point l, $m + 1$ in terms of those at the points $l \pm 1$, m and l, m. Instead, a system of linear equations has to be solved in order to express the values of the quantities in the row $(m + 1)$ in terms of those in the row (m); consequently, the values at one point of the row $(m + 1)$ will, in general, depend on the values at all points of the row (m). This approach appears valuable for the solution of linear differential equations of a type similar to Eq. 3-5.

For actual numerical computation of the solution of the nonlinear equation (3-5) one need not use the finite differences here explained (incidentally, for this reason we do not discuss the question of convergence for sufficiently small $\Delta t/\Delta x$). Preferably one introduces the net in the α, β plane, introducing the net points $\alpha = k\Delta$, $\beta = n\Delta$ with the same mesh width Δ. One will then replace the derivatives q_α and q_β by $\Delta^{-1}(q_{k,n} - q_{k-1,n})$ and $\Delta^{-1}(q_{k,n} - q_{k,n-1})$ respectively. From Eq. 4-1 and 4-2 so modified, one can express the values of x, t, u, ρ at the point k, n through the values of these quantities at the points $k - 1$, n and

k, $n - 1$. The values of the quantities at the point k, n through the prescribed values of these quantities at the initial line $\beta = -\alpha$ can thus be expressed successively.

Clearly, the setup of the difference scheme in this case guarantees that the domain of dependence of the solutions of the difference equations be the same as that of the solution of the differential equation. This is the main reason why the present setup is preferable to the one discussed previously.

The difficulties that arise when boundary data are imposed are of a more special nature and are not discussed here.

Various refinements in the setup of finite difference equations have been worked out for practical numerical calculations and have been found useful in particular classes of problems. They are discussed in Sec. G and H and in III,C.

B,6. Nonisentropic Gas Motion. In a nonisentropic gas the force per unit volume is still the pressure gradient (see Eq. 1-3) but the pressure p is a given function of the entropy s (per unit mass) in addition to the density ρ,

$$p = f(\rho, s) \tag{6-1}$$

In the absence of heat conduction the entropy s of a particle remains unchanged. Together with Eq. 1-1 and 1-2, this condition leads to the three differential equations

$$\begin{aligned} \rho_t + (\rho u)_x &= 0 \\ \rho u_t + \rho u u_x + f_\rho \rho_x + f_s s_x &= 0 \\ s_t + u s_x &= 0 \end{aligned} \tag{6-2}$$

for the three functions u, ρ, and s of x and t.

The question arises: What kind of data may properly be prescribed, and what are the determinacy domains and ranges of influence?

Since the system (6-2) is of the third order, there are three characteristic directions; they are given by $dx = (u \pm a)dt$ and $dx = udt$. Here $a^2 = f_\rho(\rho, s)$. The first two directions correspond to those of isentropic flow, the third is that of the particle paths.

A basic fact in the theory of a hyperbolic differential equation of nth order, of which Eq. 6-2 is one example, is that the dependency and influence properties are governed by the two outer characteristics. These are the first two in the present case. Spacelike and timelike directions will then be defined with respect to them as in Art. 2.

As initial data on a spacelike arc the values of u, ρ, and s should be prescribed. Then the statements about the dependency domain and the range of influence made in Art. 2 remain valid.

The basic fact is again that the solution in the dependency domain

$\mathfrak{D}$ of a segment $\mathfrak{D}_0$ is uniquely determined by the data given on $\mathfrak{D}_0$. This statement can easily be proved by an "energy integral" argument of the type described in Art. 3. (In doing this it is convenient to introduce p instead of ρ as one of the unknowns.)

In prescribing boundary data on timelike arcs a new situation arises. Whether or not one or two quantities may be prescribed on such a timelike arc depends on the direction of the middle characteristic with respect to it. If this direction, with $dt > 0$, points into the interior of the domain in which the solution is sought, two quantities must be prescribed there, such as u and s or p and s. If, on the other hand, this direction points into the exterior of this domain, only one quantity, such as u or p, should be prescribed. That is just what might be expected. In the first case gas is forced to enter through the end cross section whose motion is described by the curve $\mathfrak{B}$; clearly, the entropy of the entering gas may be prescribed. In the second case gas is forced to leave through the cross section; it carries its entropy with it and this entropy may therefore not be prescribed.

It is not quite arbitrary which quantities may be prescribed at the timelike arc. There is a definite rule that should be followed. To explain the rule it is necessary to introduce the characteristic equations. In the present case, in which the differential equations are Eq. 6-2, they are

$$\begin{aligned} p_\alpha - \rho a u_\alpha &= 0 \\ p_\beta + \rho a u_\beta &= 0 \end{aligned} \tag{6-3}$$

$$s_\gamma = 0 \tag{6-4}$$

when differentiation with respect to α, β, γ is defined by

$$\begin{aligned} x_\alpha - (u + a)t_\alpha &= 0 \\ x_\beta - (u - a)t_\beta &= 0 \end{aligned} \tag{6-5}$$

$$x_\gamma - ut_\gamma = 0 \tag{6-6}$$

The rule then is this: If both the α and γ characteristic directions at the arc $\mathfrak{B}$ point into the interior of the domain (Fig. B,6), one should prescribe on $\mathfrak{B}$ two of the quantities s, p, u in terms of the third such that the relation

$$dp + \rho a du = 0$$

does not hold along $\mathfrak{B}$ unless $ds = dp = du = 0$. This condition implies, for example, that the entropy s is prescribed at each point of $\mathfrak{B}$, either directly or as a function of p or u. If the α characteristic alone points into the interior, one should prescribe on $\mathfrak{B}$ one of the quantities s, p, u in terms of the two others in such a way that the relations

$$dp + \rho a du = 0, \qquad ds = 0$$

do not hold simultaneously along $\mathfrak{B}$ unless $ds = dp = du = 0$.

The statement that the solutions are uniquely determined through these data in the appropriate determinacy domains could be proved with the aid of the energy integral method as explained in Art. 3.

While the uniqueness proof with the aid of the energy integral method can be carried over from problems of second to those of higher order, a completely new difficulty arises if one attempts to carry over the existence proof which for a problem of second order was explained in Art. 4.

Although three families of characteristic curves are now available, only two of them can be introduced as new coordinate lines. It is natural

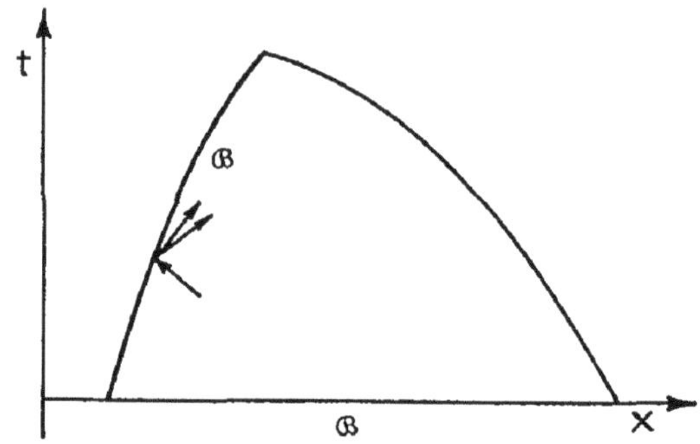

Fig. B,6. Boundary arc 𝔅 of a region at which two characteristic directions are "outgoing," while one is "incoming."

to choose the outer characteristics for this purpose. Accordingly, we consider x, t, u, p, s as functions of α and β. These functions are required to satisfy Eq. 6-3 and 6-5. In addition, we require that Eq. 6-4 is satisfied provided that this equation is interpreted appropriately.

Using Eq. 6-6, Eq. 6-4 may first be written in the original form

$$s_t + us_x = 0$$

(see Eq. 6-2). Then s_t and s_x are expressed in terms of s_α and s_β with the aid of Eq. 6-5. The result is

$$s_\alpha t_\beta - s_\beta t_\alpha = 0$$

It is remarkable that this equation is not linear in the derivatives. For this reason the idea of the existence proof of Art. 4 cannot be carried over without considerable modifications.

Various papers have been written [*8*] in order to establish rigorously the unique existence of solutions of systems of hyperbolic differential equations, i.e. equations which can be brought into the characteristic form of which Eq. 6-3 to 6-6 are typical examples.

Particular attention may be called to a paper by Courant and Lax [*9*], because the procedure used there comes closest to an approach that might be used for numerical computation. First the equations are differentiated with respect to the two variables, and the derivatives are introduced as new quantities. The number of relations between these quantities which are available is greater than the number of quantities. Hence a selection is necessary. This is done in such a way that a system of equations results which is linear in the derivatives and permits a treatment similar to that described in Art. 4. The main difficulty lies in proving that all original differential equations are satisfied. It is shown that this difficulty can be overcome by proceeding in a rather unsymmetrical manner; specifically, by treating the derivatives with respect to t and those with respect to x quite differently.

In setting up a finite difference scheme for numerical computation of the solution one might do well to follow one of these procedures which have been successful in an existence proof. It is possible to replace Eq. 6-3, 6-5, and 6-7 by difference equations in a simple manner, and the equations so obtained are rather simple in form. It is, however, doubtful whether or not the solutions of these difference equations give a good approximation to the solutions of the differential equations, since it is doubtful whether an approximation scheme based on these equations converges.

B,7. Gas Flow Involving Shock and Reaction Fronts. As has been mentioned several times, the solutions of the differential equations which govern the motion of a perfect gas may cease to exist from a certain time on. This fact indicates that the idealization of the gas as perfect cannot be maintained. In addition to the pressure gradient, one must consider forces which result from molecular exchange of momentum, such as viscous forces; also one must take into account changes of entropy which result from molecular exchange of energy, such as heat transfer. It is remarkable that a very severe idealization leads to an excellent description of actual occurrences under wide classes of circumstances. This severe idealization consists in assuming that the gas is nonperfect only at certain surfaces, at which discontinuities may occur. If this surface is not crossed by the gas it is called a "contact" surface; all quantities except the pressure and the velocity may be different on both sides. If the gas crosses the surface but does not undergo a chemical change, the surface is a "shock front" (III,E). If a chemical reaction takes place, the surface is called a "reaction front" (III,G). The well-known three discontinuity or jump conditions then take the place of the three differential equations.

There are several questions concerning the unique determinacy of flows involving discontinuity fronts. They have been answered completely in the case of constant flow quantities (such as velocity, pressure, and so on) between the discontinuity surfaces. Only a few questions have been completely answered for a general flow.

One question that has been answered completely is this: Suppose that at the time $t = 0$ functions $u(x)$, $p(x)$, and $s(x)$ are given which possess a discontinuity at a particular point $x = x_0$, but satisfy the jump conditions which characterize the discontinuity surface as a shock or reaction front; then the question is whether or not a solution of the flow differential equations exists which possesses discontinuities at a certain curve F through the point $x = x_0$, $t = 0$ at which the appropriate jump conditions are satisfied. The answer is affirmative for shocks, but for reaction fronts it is affirmative only under conditions which can be specified.

The situation may perhaps become clear if the facts are explained in connection with a general system of n differential equations for n functions $p^{(k)}(x, t)$. We assume that differential equations are written in characteristic form

$$\sum_{l=1}^{n} b_{kl} \left(\frac{\partial}{\partial t} + a^{(k)} \frac{\partial}{\partial x} \right) p^{(l)} = f^{(k)}, \qquad k = 1, \ldots, n \tag{7-1}$$

in which b_{kl}, $a^{(k)}$, and $f^{(k)}$ are given functions of x, t and $p^{(1)}, \ldots, p^{(n)}$. The n characteristic directions are given by

$$dx = a^{(k)} dt \tag{7-2}$$

All quantities may have different values at the two sides of a discontinuity line F given by $x = X(t)$. The values at the positive and negative sides, $x > X(t)$ and $x < X(t)$, will be distinguished by superscripts or subscripts $(+)$ and $(-)$ respectively.

With reference to the line F, we distinguish between incoming and outgoing characteristic directions on each side of F. A direction with $dt > 0$ is said to be "incoming" on one side if it points toward the curve F from this side; it is said to be "outgoing" on this side if it points away from this curve toward this side. Since the quantities $a^{(k)}$ may have different values on both sides of F, the number of incoming and outgoing characteristics on each side may be different from each other. There may be r^+ incoming and $n - r^+$ outgoing characteristics on the positive side, $x > X(t)$, r^- incoming, and $n - r^-$ outgoing characteristics on the negative side $x < X(t)$. Here $r^\pm$ may be any number $0, 1, \ldots, n$. The subscripts k associated with incoming characteristics will be denoted by k_ρ^+, $\rho = 1, \ldots, r^+$, and k_ρ^-, $\rho = 1, \ldots, r^-$ respectively.

The jump conditions to be imposed at each point x, t of the curve F should consist in prescribing the velocity X_t of the front and $2n - r^+ - r^-$ of the quantities $p_+^{(k)}$ and $p_-^{(k)}$, $k = 1, \ldots, n$, as functions of x, t arbitrarily, and the remaining $r^+ + r^-$ among the quantities $p_\pm^{(k)}$ in such a way that the relations

$$\sum_{l=1}^{n} b_{k_\rho^+ l} \, dp_+^{(l)} = 0, \qquad \rho = 1, \ldots, r^+$$

$$\sum_{l=1}^{n} b_{k_\rho^- l} \, dp_-^{(l)} = 0, \qquad \rho = 1, \ldots, r^-$$

at a point of F are not satisfied unless

$$dp_+^{(l)} = 0,\ dp_-^{(l)} = 0, \qquad l = 1, \ldots, n$$

If this condition is satisfied, the unique continuation of the solution is assured (provided appropriate differentiality requirements are met). This fact, in a somewhat different formulation, was proved by Göldner [*10*].

It is easily seen that the indicated rule is satisfied if the discontinuity surface is a shock front. As is well known (e.g. [*3*]) the path of the shock front is spacelike when observed from the gas ahead of it, so that all three characteristic directions on that side are incoming. On the other hand, the path of the shock front is timelike when observed from behind. More specifically, there are two outgoing characteristics on the back side; they are associated with the differentials ds and $dp + \rho a du$ (if the back side is to the left) as seen from Eq. 6-3. With the aid of the three familiar shock jump conditions, the shock velocity X_t together with s and p or u on the back side can be expressed in terms of the other quantities. One immediately verifies that $du = dp = ds = 0$ on both sides if $dp \pm \rho a du = 0$, $ds = 0$ on the front side and $dp - \rho a du = 0$ on the back side. Therefore, the rule formulated above is applicable. This assures unique continuation of the solution of the flow differential equation involving a discontinuity at which the shock conditions are satisfied.

The situation is quite different if the jump conditions allow a chemical reaction to take place. The difference of the reaction jump conditions from the shock jump conditions consists in the presence of a term in the energy balance condition expressing the given energy released in the reaction process. This small difference is strong enough to modify completely the determinacy of the flow through differential equations and jump conditions. Two types of reaction processes are admitted by these jump conditions, "detonations," in which the pressure rises, and "deflagrations" in which the pressure drops. According to "Jouguet's rule," a detonation front when observed from the gas ahead is spacelike and a deflagration front is timelike, e.g. [*3*]. In the latter case there is just one outgoing characteristic ahead of the front. One may distinguish between strong and weak detonations by noting whether the front is timelike or spacelike when observed from behind, and between strong and weak deflagrations by noting whether the front is spacelike or timelike when observed from behind. Processes at the transition between strong and weak, are called Chapman-Jouguet detonations or deflagrations.

At Chapman-Jouguet and strong detonation fronts, two of the three characteristics are outgoing just as at shock fronts, and the unique continuation of a flow involving such a front is assured.

Three outgoing characteristics occur at weak detonation and deflagration fronts, and even four at strong deflagration fronts. Therefore, the continuation of flows involving such processes is not uniquely determined by differential equations and jump conditions. The idealizing assumption that the chemical reaction takes place instantaneously when the gas crosses a mathematical surface breaks down. Additional information must be secured by an analysis of the reaction process assumed to take place in a narrow zone and not instantaneously. The result is that strong deflagrations and weak detonations are altogether impossible, while weak deflagrations are possible only at a definite speed which depends on the heat conductivity of the gas (in addition to the other flow quantities). If this condition on the burning speed is added to the three other jump conditions, the unique continuation of a flow involving a deflagration (or flame) front is again assured.

B,8. Remarks About Two-dimensional Steady Flow. The mathematical structure of two-dimensional flow problems is essentially the same as that of one-dimensional unsteady flow problems, provided the steady flow is supersonic throughout. This fact was alluded to at the end of the introduction. We add a few amplifying remarks.

Consider a flow in a duct which begins with an "entrance" section that has straight parallel walls and then in some way extends to infinity. We assume that a supersonic flow with constant velocity arrives in the entrance section. The question is: How will the flow continue? We assume the flow to take place in the x, y plane; for simplicity we assume that the walls of the duct are given by two curves $y = Y_{\pm}(x)$, $Y_{\pm}(x) = \text{const}$ for $x < 0$, so that the duct covers the region $Y_{-}(x) < y < Y_{+}(x)$.

In the entrance, the flow is isentropic and irrotational, but it will lose this character behind shocks. For the description of the flow we use the speed q and the angle of flow direction θ; to describe the state of the fluid we use the entropy s per unit mass and, most conveniently, the pressure p. The density ρ and the speed of sound $a = [\partial\rho/\partial p]^{-\frac{1}{2}}$ are then functions of s and p.

For the formulation of the flow equation we may introduce the differentiation

$$\nabla_\sigma = \cos\theta\frac{\partial}{\partial x} + \sin\theta\frac{\partial}{\partial y}, \qquad \nabla_n = -\sin\theta\frac{\partial}{\partial x} + \cos\theta\frac{\partial}{\partial y} \tag{8-1}$$

in the direction of the flow and perpendicular to it. Then the continuity equation assumes the form

$$\nabla_\sigma(\rho q) + \rho q \nabla_n \theta = 0 \tag{8-2}$$

The force equations are

$$\rho q \nabla_\sigma q + \nabla_\sigma p = 0 \tag{8-3}$$

$$\rho q^2 \nabla_\sigma \theta + \nabla_n p = 0 \tag{8-4}$$

while the adiabatic character of thermodynamic changes is expressed by

$$\nabla_\sigma s = 0 \tag{8-5}$$

These equations can be easily brought into characteristic form. In fact, Eq. 8-5 and 8-3 are already in characteristic form. The corresponding characteristic direction is evidently the flow direction; this characteristic direction is thus to be counted twice. Eliminating $\nabla_\sigma q$ from Eq. 8-2 and 8-3 we obtain the equations

$$\nabla_n p + \rho q^2 \nabla_s \theta = 0$$
$$\nabla_\sigma p + \rho q^2 \tan^2 \beta \, \nabla_n \theta = 0$$

in which we have employed the Mach angle β defined by $\sin \beta = a/q$, provided the flow is supersonic, $q > a$. Introducing the differentiation

$$\nabla^{\pm} = \cos \beta \, \nabla_\sigma \pm \sin \beta \, \nabla_n \tag{8-6}$$

in the directions of the Mach lines, i.e. in directions making angles $\pm\beta$ with the flow direction, we find

$$\nabla^{\pm} p \pm \rho q^2 \tan \beta \, \nabla^{\pm} \theta = 0 \tag{8-7}$$

These are the characteristic equations associated with the Mach directions as characteristic directions. They, together with Eq. 8-5, are the analogues of the characteristic equations (6-3) and (6-4) of one-dimensional unsteady motion.

Suppose the entrance flow is supersonic. Then the continued flow will be uniquely determined by the entrance flow as long as it remains supersonic. Just as for one-dimensional flow the existence of continuous flow may break down; in general the flow will then pass through a shock front. In contrast to one-dimensional flow, however, there are cases in which no such continuation of the flow through a shock is possible; namely, when the wall suddenly turns through an angle exceeding a definite limit (see III, E). Nevertheless it may be said that as long as a continuation of the flow exists which is supersonic throughout, this continuation is unique.

There are, however, situations in which a continued flow exists, but not without being partly subsonic. That is in particular the case if "regular reflection" of a shock front at the wall or "regular intersection" of two shock points in the interior is not possible and instead "Mach reflection" or "Mach intersection" takes place (see III,E). The important point to be made here is this: If the flow is partly subsonic, it is not uniquely determined by the entrance flow. One quantity at the exit of the duct must be prescribed in order to make the flow determined. If the flow approached a parallel flow with a definite velocity on each streamline, it would be natural to consider the pressure at the exit as the quantity which should be prescribed to make the flow determined. In general, though, the flow will show a complicated shock pattern and not approach a parallel flow, provided the influence of viscosity can be disregarded. In such a case it would seem impossible to formulate a condition on the behavior of the flow at the exit infinity that would make it uniquely determined. Actually, of course, the influence of viscosity will eventually become predominant.

Thus it appears that it is impossible to characterize steady flows by conditions at the downstream end without taking effects of viscosity into account.

B,9. Cited References.

1. Hadamard, J. *Le problème de Cauchy et les équations aux derivées partielles linéaires hyperboliques*, Herman, Paris, 1932.
2. Rayleigh, Lord. Aerial plane waves of finite amplitude. *Proc. Roy. Soc. A84*, 47–84 (1910).
3. Courant, R., and Friedrichs, K. O. *Supersonic Flow and Shock Waves*. Interscience, 1948.
4. Courant, R., and Hilbert, D. *Methoden der Mathematischen Physik*, Vol. II. Springer, Berlin, 1937.
5. Courant, R., Friedrichs, K., and Lewy, H. Über die partiellen Differenzengleichungen der mathematischen Physik. *Math. Annalen 100*, 32–74 (1928).
6. O'Brien, G., Hyman, M. A., and Kaplan, S. A study of the numerical solution of partial differential equations. *J. Math. and Phys. 29*, 223–251 (1951).
7. Friedrichs, K. O. Nonlinear hyperbolic differential equations for functions of two independent variables. *Am. J. Math. 70*, 555–589 (1948), and papers quoted there.
8. Douglis, A. Some existence theorems for hyperbolic systems of partial differential equations in two independent variables, *Comm. on Pure and Appl. Math. 5*, 119–154 (1952).
9. Courant, R., and Lax, P. On nonlinear partial differential equations with two independent variables. *Comm. on Pure and Appl. Math. 2*, 255–273 (1949).
10. Göldner, S. R. Existence and uniqueness theorems for two systems of nonlinear hyperbolic differential equations for functions of two independent variables. *New York Univ. Ph.D. Thesis*, 1949.

SECTION C

SMALL PERTURBATION THEORY

W. R. SEARS

C,1. Introduction. There is an important class of problems in aerodynamics in which the flow pattern consists of a small perturbation of a simple and well-understood flow. The most obvious case is the classical one of acoustics, where the unperturbed situation is simply a gas at rest. A more useful case in aeronautical applications, perhaps, is that of a slightly perturbed parallel uniform gas stream. In this section we consider these cases and some others. We shall find, as has already been suggested in Sec. A, that the mathematical problem of compressible fluid flow is greatly simplified by the small perturbation approximations, and that the resulting equations still exhibit some of the most significant features of the complete equations and are adequate to describe some of the important phenomena of high speed aerodynamics.

The idea of attacking nonlinear problems of mechanics by the small perturbation method has been used many times in the history of the subject and in applications much different from gas flows. If we try to trace the history of this technique in gas dynamics, we must first recognize that the basic concept is implicit in any estimation of the velocity of sound. Sir Isaac Newton (1642–1727) probably made the first such estimation, and in doing so made a famous error, for he assumed that a small disturbance in a gas would compress the gas isothermally. His error was rectified by Laplace, who calculated the sound speed for the correct adiabatic process. In the eighteenth century, problems of acoustics were treated by Euler, D. Bernoulli, and Lagrange. The second order differential equation of sound propagation was known to them and to d'Alembert. The subject reached a state of great development in the nineteenth century, in the work of Lord Rayleigh, Helmholtz, Stokes, Kirchhoff, and others.

In his famous treatise on *The Theory of Sound* [*1*] Rayleigh proceeded from a small perturbation hypothesis to the partial differential equation of sound propagation, in a manner substantially the same as we shall do later in this section.

Turning now to the application of the same basic idea in the realm of aerodynamics, we find that apparently this did not evolve directly from the science of acoustics. Ludwig Prandtl, who had already made a series of contributions of essential importance in fluid mechanics, introduced the small perturbation theory of steady flows in his lectures at Göttingen in 1922. It was applied to both subsonic and supersonic cases by his student, J. Ackeret, in a publication of 1928 [*2*]. Independently, the same method was proposed and applied to the subsonic case by H. Glauert in 1928 [*3*]. Apparently all these writers began their deductions having in mind the typical aeronautical case mentioned above, i.e. the slightly perturbed parallel stream. This is also true of an earlier work of Ackeret [*4*], where he considered small perturbations of a supersonic flow on the basis of the exact solution, already known, for supersonic flow around a corner. In fact, the identity of these methods with those of classical acoustics—to which they are reduced by a simple translation of coordinate axes—seems not to have been noticed by aerodynamicists until somewhat later.

It is not difficult to see how this would occur, first because of the aeronautical preoccupation with parallel streams and also because, in reality, the idea of *small perturbations* of a parallel stream was already well known, having been exploited for somewhat different reasons in the literature of airfoil theory. Prandtl's famous lifting-line wing theory (1918), although it treated wings in incompressible flow, was a small perturbation theory, as were the thin airfoil theories of Munk (1922), Birnbaum, Glauert, and others. Consequently, it was more natural to proceed from this background toward an approximate theory of compressible fluids than to return to the equations of acoustics.

To be sure, there are undoubtedly prior investigations in the field of acoustics that anticipated salient features of the Prandtl-Glauert or Ackeret theories, and could be thought of as linking the acoustic and aeronautical versions of small perturbation theory. One of these is a 1907 paper by von Kármán [*5*] concerning standing waves in supersonic jets. Nevertheless, the main course of the idea, so far as aeronautical engineers are concerned, has clearly been through airfoil theory in the work of Prandtl and Glauert, rather than from Rayleigh directly.

In spite of this, our treatment of small perturbation flows in the present section will begin with the acoustical situation of a gas at rest and proceed to Prandtl's and Ackeret's problems as subcases. This procedure simplifies and clarifies certain matters, especially pertaining to unsteady flows.

After introducing the basic ideas in this way, and obtaining the fundamental equations in several different forms, we shall devote ourselves in this section principally to their exploitation in subsonic flow situations. In Sec. D the applications to supersonic flight conditions are treated. This constitutes a report on the vast and rapidly growing body of literature on the aerodynamics of thin wings at supersonic speeds, and on some important aspects of the mathematical techniques found useful for these applications. Sec. D closes with a consideration of those special situations (transonic and hypersonic) where more careful analysis is required and nonlinear terms remain, even in the small perturbation equations.

Finally, in Sec. E, we complete our study of small perturbation methods by considering their shortcomings and methods that have been proposed to improve them. Actually, these considerations will bring us to a much clearer understanding of the first order perturbation theories themselves; consequently the authors of Sec. C and D present these theories without complete justification of the approximations involved, depending upon Sec. E to make up these deficiencies.

C,2. The Equation of Sound Propagation. As in most of the sections of this volume, the equations of inviscid fluid flow provide the basis for our considerations. These are satisfactory to describe the flow of real gases in regions where neither friction nor heat transfer are important: namely, as we have seen (III,A,6), where velocity and temperature gradients are not large.

It has already been noted (III,A,7) that, in the absence of friction and heat transfer, each particle of the fluid undergoes reversible adiabatic changes of state. Consequently, if the entropy of the gas was initially uniform, as in the case of a homogeneous gas at rest or flowing in a uniform stream, it must remain constant and uniformly distributed throughout the flow. The equations of the flow are:

$$\frac{D\mathbf{q}}{Dt} = -\frac{1}{\rho}\operatorname{grad} p; \qquad p\rho^{-\gamma} = \text{const}$$
$$\frac{D\rho}{Dt} + \rho \operatorname{div} \mathbf{q} = \frac{1}{a^2}\frac{Dp}{Dt} + \rho \operatorname{div} \mathbf{q} = 0 \tag{2-1}$$

where D/Dt is the usual "convective" or "substantive" derivative that measures a rate of change for an individual particle of gas, and a^2 denotes the quantity $dp/d\rho$ for reversible adiabatic compression, i.e. the square of the speed of sound. Clearly, the replacement of $D\rho/Dt$ by $a^{-2}Dp/Dt$ can only be made on the basis stated above, that each particle moves about under adiabatic conditions. Both of the statements in Eq. 2-1 are nonlinear, and consequently have been solved only for a few special cases and under rather severe limitations.

In the particular case of sound propagation through a body of gas uniformly at rest, we proceed to make the small perturbation approximation in Eq. 2-1. We assume the velocity and its derivatives to be small, and the corresponding density and pressure to differ only slightly from the values ρ_∞, p_∞, which pertain to the undisturbed gas. Denoting these small departures from the steady values by ρ', p', and neglecting second order terms in these and their derivatives, we have

$$\mathbf{q}_t = -\operatorname{grad}\left(\frac{p'}{\rho_\infty}\right)$$
$$\frac{1}{a_\infty^2}p'_t + \rho_\infty \operatorname{div}\mathbf{q} = 0 \tag{2-2}$$

where a_∞^2 denotes $\gamma p_\infty/\rho_\infty$, the square of the speed of sound in the undisturbed gas.

Eq. 2-2 are linear. Moreover, in view of the initial condition of uniform rest, the first of Eq. 2-2 implies that $\mathbf{q}$ is the gradient of a scalar function, i.e. that a potential function ϕ exists. Eq. 2-2 can then be written as follows:

$$\rho_\infty\phi_t = -p' \tag{2-3}$$

$$\phi_{tt} = a_\infty^2\nabla^2\phi \tag{2-4}$$

These are the classical equations of sound propagation [*1*, p. 15]. (Eq. 2-4 is known as the wave equation.) They are useful in aerodynamics to describe a great variety of small perturbation situations outside of boundary layers and where shock waves either do not exist or are so weak as to justify our approximations. They can sometimes be used to describe the flow behind stronger shock waves, which are nearly uniform in strength, but this case requires special care, as is made clear below (Art. 10). An essential assumption in the derivation of Eq. 2-3 and 2-4 is that the initial condition is one of uniform rest, which has been used to obtain $\mathbf{q}$ as the gradient of a potential function.

Clearly, these equations must be discarded whenever there is evidence that the small disturbance hypothesis is violated; we shall see (Art. 4) that this sometimes occurs even when the flow is due to the motion of a slender body.

C,3. The Prandtl-Glauert Equation. Suppose now that the undisturbed flow is the familiar one of aerodynamics, namely, a uniform steady stream. This is reduced to the preceding case by steady translation of the reference axes. Let the x axis be chosen in the stream direction and the stream speed be called U. The coordinate transformation between the moving (x', y', z', t') system and the stationary (x, y, z, t) one is, of course,

$$x = x' - Ut', \qquad y = y', \qquad z = z', \qquad t = t' \tag{3-1}$$

Thus, Eq. 2-3 and 2-4 become

$$\rho_\infty(\phi_{t'} + U\phi_{x'}) = -p' = p_\infty - p \tag{3-2}$$

$$\phi_{t't'} + 2U\phi_{x't'} + U^2\phi_{x'x'} = a_\infty^2\nabla'^2\phi \tag{3-3}$$

where again ρ_∞, p_∞, a_∞ are values pertaining to the undisturbed stream, and ∇'^2 involves derivatives with respect to x', y', z'.

The equations for the special case where the flow is steady when viewed from the moving coordinate system are called the Prandtl-Glauert equations [2,3]:

$$p - p_\infty = -\rho_\infty U \phi_{x'} \tag{3-4}$$

$$\left(1 - \frac{U^2}{a_\infty^2}\right) \phi_{x'x'} + \phi_{y'y'} + \phi_{z'z'} = 0 \tag{3-5}$$

Eq. 3-5 illustrates clearly the change of mathematical character of gas flow in the three major speed regimes, subsonic, transonic, and supersonic. This equation is elliptic, parabolic, or hyperbolic when the coefficient of the first term is positive, zero, or negative, i.e. when the stream speed falls into the three regimes mentioned. It is a result of our small perturbation hypothesis, of course, that the character of the equation depends only on the Mach number of the undisturbed stream; this will unavoidably restrict the usefulness of this approximate theory, especially in the transonic regime.

Eq. 3-5 is also of value because it establishes the significance of incompressible flow theory in relation to the aerodynamics of compressible gases. It is apparent that there will be a range of Mach numbers near zero, at least for slender bodies, for which the compressible and incompressible flow patterns will be nearly alike. There is another way to establish this fact, which is better because it is not restricted to slender bodies, namely, the expansion method of Janzen and Rayleigh, which is treated in E,2. The Janzen-Rayleigh theory confirms that the first departures from incompressible fluid theory, as the Mach number of steady flow, M_∞, is increased from a small value, appear in terms of order M_∞^2, as might be inferred from Eq. 3-5.

This equation is put into the form of Laplace's celebrated equation by a simple affine transformation. Dropping the primes in Eq. 3-4 and 3-5 and writing β^2 for $(1 - U^2/a_\infty^2)$, which is a positive number for subsonic stream flow, we put

$$x = m\xi, \qquad y = n\eta, \qquad z = n\zeta \tag{3-6}$$

with $m/n = \beta$. Eq. 3-4 and 3-5 are immediately transformed into

$$p - p_\infty = -\frac{1}{m} \rho_\infty U \varphi_\xi \tag{3-7}$$

and

$$\varphi_{\xi\xi} + \varphi_{\eta\eta} + \varphi_{\zeta\zeta} = 0 \tag{3-8}$$

where $\varphi(\xi, \eta, \zeta)$ has been written for the function obtained by substituting Eq. 3-6 in $\phi(x, y, z)$; that is, $\varphi(\xi, \eta, \zeta) \equiv \phi(m\xi, n\eta, n\zeta)$.

Whatever boundary conditions are available can also be expressed in the new variables ξ, η, ζ, whereupon any steady, subsonic, small perturbation problem is brought into the scope of the powerful and well-known techniques of potential theory.

C,4. Application to Wing Theory. To illustrate the situation just described we shall now consider in some detail certain typical examples from the realm of practical aerodynamics. The first of these is the application of the Prandtl-Glauert theory to a wing problem. (Other examples follow in Art. 5, including one that provides a unified general method for a wide class of small perturbation flows.)

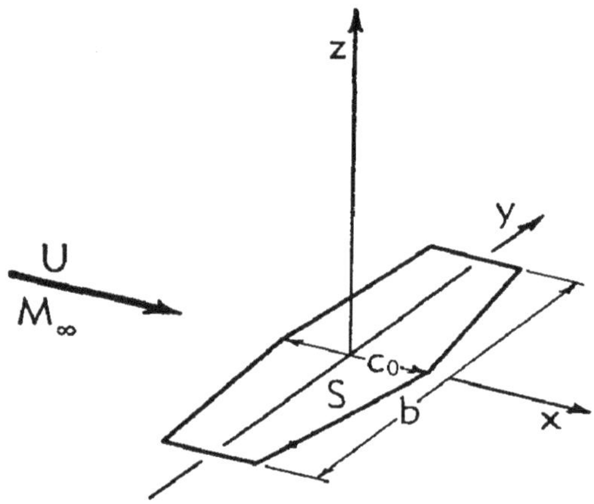

Fig. C,4a. Thin wing in subsonic stream.

Consider a thin wing lying approximately in the x,y plane (Fig. C,4a) whose upper and lower surface contours are defined by

$$\begin{array}{ll} \text{Upper:} & z = Z^{(u)}(x, y) \\ \text{Lower:} & z = Z^{(l)}(x, y) \end{array} \Bigg\} \; x, y \text{ in } S \tag{4-1}$$

The boundary condition at the wing surface is that the normal component of the velocity vanishes there. We assume for the present that the wing is in an unlimited body of fluid, so that the only additional condition is the vanishing of perturbations far from the wing.

In the approximations of thin wing theory, it is sufficient to replace the boundary condition at the wing by the following:

$$\begin{aligned} \phi_z(x, y, +0) &= UZ_x^{(u)}(x, y) \equiv U\sigma_u(x, y) \\ \phi_z(x, y, -0) &= UZ_x^{(l)}(x, y) \equiv U\sigma_l(x, y) \end{aligned} \tag{4-2}$$

It will be seen that σ_u and σ_l denote slopes of the wing contours in the x direction.

The differential equation to be satisfied is, of course, Eq. 3-5, which is brought into the form of Laplace's equation by means of the affine transformation (3-6). In particular, for the present example, we choose

$m = \beta$ and $n = 1$, so that the transformation is $x = \beta\xi$, $y = \eta$, $z = \zeta$. Let us write down the boundary conditions in terms of the transformed coordinates ξ, η, ζ, again writing $\varphi(\xi, \eta, \zeta)$ for $\phi(\beta\xi, \eta, \zeta)$:

$$\left.\begin{aligned} \varphi_\zeta(\xi, \eta, +0) &= U\sigma_u(\beta\xi, \eta) \\ \varphi_\zeta(\xi, \eta, -0) &= U\sigma_l(\beta\xi, \eta) \end{aligned}\right\} \beta\xi, \eta \text{ in } S \tag{4-3}$$

The boundary condition at infinity is unchanged in the transformation.

It is clear that a new thin wing problem in the ξ, η, ζ space has been defined, and that it involves a different wing in an incompressible fluid stream of speed U. Since $\beta < 1$, the planform of the new wing of Eq. 4-3 is one of reduced aspect ratio (Fig. C,4b). Its aspect ratio, defined as

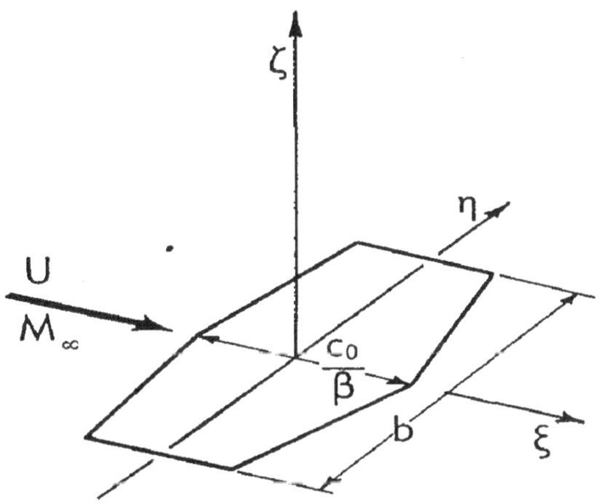

Fig. C,4b. Fictitious related thin wing in incompressible stream.

(span)²/area, is, in fact, $\beta Ꭿ$, if the aspect ratio of the original wing is Ꭿ. The slopes of its profile are equal to those of the original; in other words the thickness ratios and incidences of its profiles are the same.

One now proceeds to solve this fictitious incompressible flow problem by whatever technique of wing theory is appropriate for its geometry. It may be pointed out that the reduction of aspect ratio noted above may necessitate the use of a more elaborate method than if the original aspect ratio had been carried over into the incompressible fluid. It should be noted that in this particular case the transformation of the boundary-value problem has led to another, recognizable *wing* problem, and that this is advantageous when we proceed to carry out the calculations for the incompressible flow.

Suppose that this calculation has been carried out, so that the incompressible flow perturbation potential $\varphi(\xi, \eta, \zeta)$ can be considered known. The velocity components in the compressible fluid are then calculated immediately by differentiation:

$$
\begin{aligned}
\phi_x(x, y, z) &= \frac{1}{\beta}\varphi_\xi(\xi, \eta, \zeta) \\
\phi_y(x, y, z) &= \varphi_\eta(\xi, \eta, \zeta) \\
\phi_z(x, y, z) &= \varphi_\zeta(\xi, \eta, \zeta) \\
x = \beta\xi, \qquad y &= \eta, \qquad z = \zeta
\end{aligned} \tag{4-4}
$$

The pressure in the compressible fluid is also immediately computed according to Eq. 3-7; it can be expressed as

$$
C_p(x, y, z) = \frac{1}{\beta} C_{p_{\text{inc}}}(\xi, \eta, \zeta) = \frac{1}{\beta} C_{p_{\text{inc}}}\left(\frac{x}{\beta}, y, z\right) \tag{4-5}
$$

where the pressure coefficient C_p is defined as $(p - p_\infty)/(\frac{1}{2}\rho_\infty U^2)$, and $C_{p_{\text{inc}}}$ denotes the corresponding function for the fictitious incompressible flow. Such useful quantities as total lift, moments, and span load distribution can now be evaluated by integration. In terms of the usual dimensionless coefficients, it is easily verified that

$$
\begin{aligned}
&\text{Total lift:} && C_L = \frac{1}{\beta} C_{L_{\text{inc}}} \\
&\text{Lift distribution:} && c_l(y) = \frac{1}{\beta} c_{l_{\text{inc}}}(y)
\end{aligned} \tag{4-6}
$$

etc.

where in each case the "incompressible" coefficient is defined in terms of the stream dynamic pressure and the dimensions of the fictitious airfoil in the ξ, η, ζ space.

If the fluid is not unlimited, i.e. if there are other bodies or boundaries to be accounted for, the treatment does not require any essential change and the results are the same as Eq. 4-4 to 4-6. The additional boundary conditions must be transformed to the ξ, η, ζ space and there accounted for in the incompressible flow problem. If our wing is being tested in a wind tunnel, for example, the fictitious one of increased chord is also in a wind tunnel of the same diameter. Similarly, a given biplane cellule is related to one of greater chord and stagger but equal gap. Thus, a more complete statement of Eq. 4-4 to 4-6 would define the "inc" quantities as including the effects of any additional boundary conditions properly transformed according to the definition $\varphi(\xi, \eta, \zeta) \equiv \phi(\beta\xi, \eta, \zeta)$. The effects of Mach number on wind tunnel boundary corrections according to this theory have been considered by several authors [7,8].

The function $\beta^{-1}(M_\infty)$, which plays so important a role in these formulas, is plotted in Fig. C,4c. It becomes infinite at $M_\infty = 1$, and therefore, according to Eq. 4-4, so do the perturbation velocity components in general. This would certainly constitute a breakdown of our perturbation method, for its basic assumption would be violated, as the stream Mach number increased toward 1. We must exercise some restraint in drawing this conclusion at this point, however, because the derivatives of $\varphi(\xi, \eta, \zeta)$ into which β^{-1} is multiplied in Eq. 4-4 are functions of β themselves. As M_∞ approaches 1, the affine transformation stretches the wing farther and farther in the ξ direction, and at $M_\infty = 1$ it has zero aspect ratio. There are special cases, as we shall see directly below, and again in Art. 9 and in Sec. D, where the small perturbation

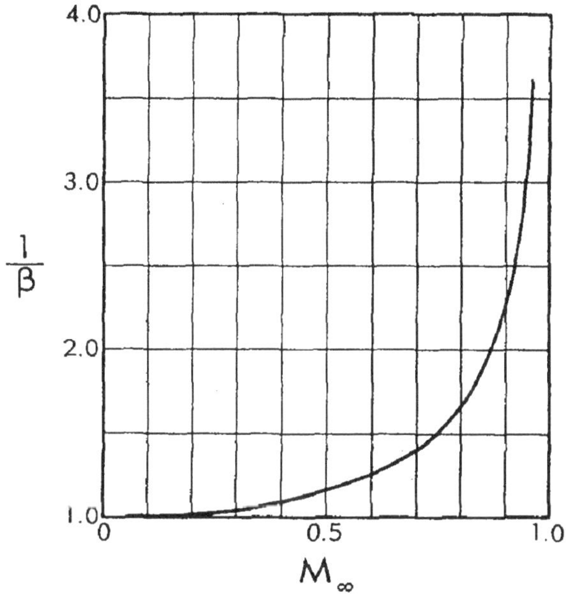

Fig. C,4c. Variation of the factor $1/\beta$ with stream Mach number M_∞.

theory remains self-consistent at transonic speeds and above (cf. [*32*]). When this is not the case, the present methods have to be discarded in favor of a different type of approximation (D,30).

Lifting line theory. If the aspect ratio $\beta \mathcal{R}$ is sufficiently great and the wing lies nearly straight across the stream (small sweepback), the approximations of Prandtl's lifting line theory may be applicable to the incompressible problem encountered above. The equation satisfied by the circulation $\Gamma_{\text{inc}}(\eta)$ for a wing in an unbounded stream, according to this theory, is

$$\Gamma_{\text{inc}}(\eta) = \frac{1}{2} U c_{\text{inc}} \kappa_{\text{inc}} \left(\alpha_{\text{inc}} - \frac{w_{\text{inc}}}{U} \right) \tag{4-7}$$

where c, κ, α, and w denote respectively the chord length, profile constant, incidence, and downwash component, all functions of η, and the subscript "inc" denotes incompressible flow values as before. The profile constant κ_{inc} is the slope of the lift coefficient curve for the profile of the fictitious wing; to the first approximation it depends only on the Reynolds number of the flow and the character of the boundary layer on the wing. We assume $\kappa_{ino}(\eta)$ to be known.

The circulation $\Gamma_{ino}(\eta)$ is easily calculated in terms of φ_ξ by an integration from leading edge (l.e.) to trailing edge (t.e.): it is

$$\Gamma_{ino}(\eta) = \int_{\text{l.e.}}^{\text{t.e.}} [\varphi_\xi(\xi, \eta, +0) - \varphi_\xi(\xi, \eta, -0)]d\xi \tag{4-8}$$

Its relation to the corresponding quantity $\Gamma(y)$ in the real flow follows immediately, upon substitution for φ_ξ and $d\xi$ in terms of ϕ_x and dx:

$$\Gamma_{ino}(\eta) = \Gamma(y) \tag{4-9}$$

It is also clear that $c_{ino}(\eta)$ is just $c(y)/\beta$. The evaluation of α_{ino} follows from Eq. 4-3, which defines the fictitious wing; we see that $\alpha_{ino}(\eta) = \alpha(y)$. Finally, according to Prandtl's theory, w_{ino} is given by an integration of $\Gamma'_{ino}(\eta)$, which sums up the effects of the sheet of trailing vortices behind the wing:

$$\begin{aligned} w_{ino}(\eta) &= \frac{1}{4\pi}\,\mathcal{P}\int_{-b/2}^{b/2} \frac{\Gamma'_{ino}(\bar{\eta})d\bar{\eta}}{\eta - \bar{\eta}} \\ &= \frac{1}{4\pi}\,\mathcal{P}\int_{-b/2}^{b/2} \frac{\Gamma'(\bar{y})d\bar{y}}{y - \bar{y}} \end{aligned} \tag{4-10}$$

(The symbol $\mathcal{P}$ indicates that the Cauchy principal value is to be taken.) Eq. 4-10 is an interesting result in itself; it states that the downwash distribution produced by any given circulation distribution is unaffected by compressibility, in our first order theory. We are now able to write the Prandtl equation (4-7) in terms of given quantities:

$$\Gamma(y) = \frac{1}{2\beta}\,Uc\kappa_{ino}\left[\alpha - \frac{1}{4\pi U}\,\mathcal{P}\int_{-b/2}^{b/2} \frac{\Gamma'(\bar{y})d\bar{y}}{y - \bar{y}}\right] \tag{4-11}$$

In this form, the fictitious wing problem has been completely eliminated, except for the appearance of the function $\kappa_{ino}(y)$, whose evaluation has already been mentioned. Obviously one can consider $\beta^{-1}\kappa_{ino}$ as a compressible flow profile constant; then Prandtl's equation has the same form for either compressible or incompressible flow.

We can proceed to generalize certain familiar results of the lifting line theory to account for compressibility. Some of these results pertain to the wing of elliptic circulation distribution, for which Eq. 4-10 reduces to

$$w_{ino} = \text{const} = U\,\frac{C_{L_{ino}}}{\pi \text{Æ}_{ino}} \tag{4-12}$$

and for the flat wing of elliptic planform, in particular,

$$C_{L_{ino}} = \frac{\kappa_{ino}}{1 + \dfrac{\kappa_{ino}}{\pi \text{Æ}_{ino}}}\,\alpha_{ino} \tag{4-13}$$

According to our present results, these become

$$w = U \frac{C_L}{\pi A\!\!R} \tag{4-14}$$

and

$$C_L = \frac{\kappa_{\text{inc}}}{\beta + \dfrac{\kappa_{\text{inc}}}{\pi A\!\!R}} \alpha \tag{4-15}$$

This interesting formula, which might be used as a "rule-of-thumb" for wings in subsonic flow, was first given by Göthert [6].

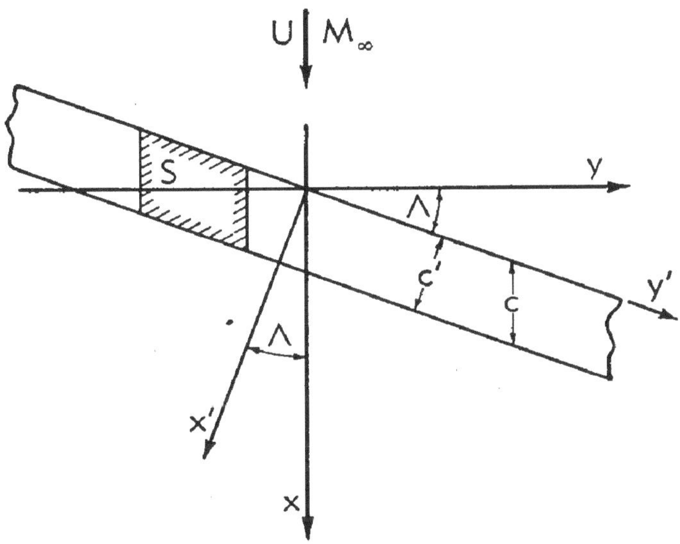

Fig. C,4d. Yawed or sweptback infinite wing.

Yawed infinite wing. We consider now a cylindrical wing of infinite extent, yawed or swept back in the x, y plane, as sketched in Fig. C,4d. The boundary conditions of the problem are most conveniently expressed in terms of an x', y', z' coordinate system, which differs from the x, y, z only by being rotated around the z axis so that the y' axis coincides with the leading edge of our sweptback wing. The wing surface condition is then (cf. Eq. 4-2)

$$\left.\begin{aligned} \phi_{z'}(x', +0) &= U \frac{\partial Z^{(u)}}{\partial x} = U\sigma_u(x') \\ \phi_{z'}(x', -0) &= U \frac{\partial Z^{(l)}}{\partial x} = U\sigma_l(x') \end{aligned}\right\} 0 \leqq x' \leqq c' \tag{4-16}$$

The cylindrical character of the wing is expressed by the fact that $Z^{(u)}$, $Z^{(l)}$, σ_u, and σ_l are all functions of x' only; nevertheless, σ_u and σ_l are the slopes measured in the stream (x) direction. The relation between these and the slopes measured normal to the leading edge, which we denote by

τ_u and τ_l, is simply $\sigma = \tau \cos \Lambda$; hence, Eq. 4-16 can be expressed as

$$\left.\begin{aligned} \phi_{z'}(x', +0) &= U \cos \Lambda\, \tau_u(x') \\ \phi_{z'}(x', -0) &= U \cos \Lambda\, \tau_l(x') \end{aligned}\right\} 0 \leqq x' \leqq c' \tag{4-17}$$

Let us now express our Prandtl-Glauert equation (3-5) in terms of the x', y', z' coordinates. The calculation is a simple one; in view of the infinite cylindrical character of the wing, we know that all velocity components are independent of y', which eliminates a number of terms in the equation. The result is

$$(\beta^2 \cos^2 \Lambda + \sin^2 \Lambda)\phi_{x'x'} + \phi_{z'z'} = 0 \tag{4-18}$$

or, because β^2 has been defined as $1 - M_\infty^2$,

$$[1 - (M_\infty \cos \Lambda)^2]\phi_{x'x'} + \phi_{z'z'} = 0 \tag{4-19}$$

Taken together with conditions (4-16) and the condition of vanishing perturbations far from the wing, this constitutes precisely the same boundary-value problem as for the same infinite airfoil at zero angle of yaw, in a stream whose Mach number is $M_\infty \cos \Lambda$. Thus it appears that conditions of flow about the wing, expressed as perturbations of the stream, are completely defined by the normal (to the leading edge) stream component and the geometry of the normal cross section of the wing. To be sure, this result can be explained more simply, for a yawed infinite cylindrical wing is equivalent to the same wing, without yaw, being translated steadily in its axial direction, and surely such axial translation cannot have any effect in a frictionless fluid.

We proceed now to apply the Prandtl-Glauert affine transformation to Eq. 4-19. For this case the transformation to be used is $x' = \beta'\xi$, $z' = \zeta$, where $\beta'^2 \equiv 1 - (M_\infty \cos \Lambda)^2$. The process is completely analogous to what we have done above, and the results are the following:

$$\begin{aligned} \phi_{x'}(x', z') &= \frac{1}{\beta'} \varphi_\xi(\xi, \zeta) \\ \phi_{z'}(x', z') &= \varphi_\zeta(\xi, \zeta) \end{aligned} \tag{4-20}$$

where $\varphi(\xi, \zeta)$ is the perturbation potential of the plane, incompressible flow, at stream speed $U \cos \Lambda$, about the airfoil whose profile is defined by the slopes $\tau_u(\beta'\xi)$ and $\tau_l(\beta'\xi)$. This is an airfoil of increased chord and equal slopes, compared to the original, viewed in normal cross section; thus its thickness ratio and incidence are the same as those of the original normal cross section. Furthermore, the pressure coefficient $C_p(x', z')$, based of course on the free stream dynamic pressure $\frac{1}{2}\rho_\infty U_\infty^2$, is still approximately $-2\phi_x/U$; hence

$$C_p(x', z') = -\frac{2}{U}\phi_{x'} \cos \Lambda = -\frac{2}{\beta' U}\varphi_\xi(\xi, \zeta) \cos \Lambda \tag{4-21}$$

In Eq. 4-21, we have succeeded in relating the pressure coefficient in the compressible flow about the yawed airfoil to the perturbation potential of plane incompressible flow about a related profile. The relation can be expressed still more conveniently by introducing here the pressure coefficient $C^{(2)}_{p_{\text{inc}}}(\xi, \zeta)$ of this plane flow; i.e.

$$\varphi_\xi(\xi, \zeta) = -\tfrac{1}{2}U \cos \Lambda\, C^{(2)}_{p_{\text{inc}}}(\xi, \zeta)$$

Then Eq. 4-21 becomes a relation between the pressure coefficients of compressible three-dimensional flow and the related plane incompressible flow, viz.

$$C_p(x', z') = \frac{1}{\beta'} C^{(2)}_{p_{\text{inc}}}(\xi, \zeta) \cos^2 \Lambda$$

or

$$C_p(x', z') = \frac{1}{\beta'} C^{(2)}_{p_{\text{inc}}}\left(\frac{x'}{\beta'}, z'\right) \cos^2 \Lambda \tag{4-22}$$

These expressions show clearly the effects of both Mach number and yaw: for example, with $\beta' = 1$, we have in Eq. 4-22 the formula for the effect of yaw of an infinite wing in incompressible flow:

$$C_{p_{\text{inc}}}(x', z') = C^{(2)}_{p_{\text{inc}}}(x', z') \cos^2 \Lambda \tag{4-22a}$$

Now, formula 4-22 must surely be included in our more general result for thin wings, Eq. 4-5, as a special case. The difference of appearance of these two formulas stems from the fact that $C^{(2)}_{p_{\text{inc}}}$ in Eq. 4-22 refers to plane flow and, moreover, the fictitious wings in the two formulations are not quite the same. With a little care regarding this point, the reader should have no difficulty in verifying that Eq. 4-5 and 4-22 are in agreement. The latter is certainly a much more convenient formula for yawed infinite wings.

We can proceed without difficulty to define force and moment coefficients and obtain formulas analogous to Eq. 4-6. For example, the coefficient of lift on any panel of the wing, such as S in Fig. C,4d, based on the area of this panel and the stream dynamic head is

$$C_L = \frac{1}{\beta'} c_{l_{\text{inc}}} \cos^2 \Lambda \tag{4-23}$$

where $c_{l_{\text{inc}}}$ is the lift coefficient of the fictitious airfoil in plane incompressible flow in the ξ, ζ plane.

The most interesting result expressed by Eq. 4-20 to 4-23, however, is that the significant Mach number is not M_∞ but $M_\infty \cos \Lambda$. Thus the inconsistency of the small perturbation theory noticed previously, due to the vanishing of β' with increasing stream Mach number, does not occur until the component Mach number $M_\infty \cos \Lambda$ approaches 1. In the present application, then, the Prandtl-Glauert theory is self-consistent at transonic speeds, provided the sweepback is great enough. This result is confirmed by experiment, as is mentioned again below.

These important properties of yawed or sweptback wings were apparently discovered independently by A. Busemann and R. T. Jones.

C,5. Application to Other Bodies. In this article we shall present additional examples of the application of the Prandtl-Glauert method. The first two of these, which pertain to flow past a corrugated circular cylinder of infinite length and to flow about slender bodies of revolution,

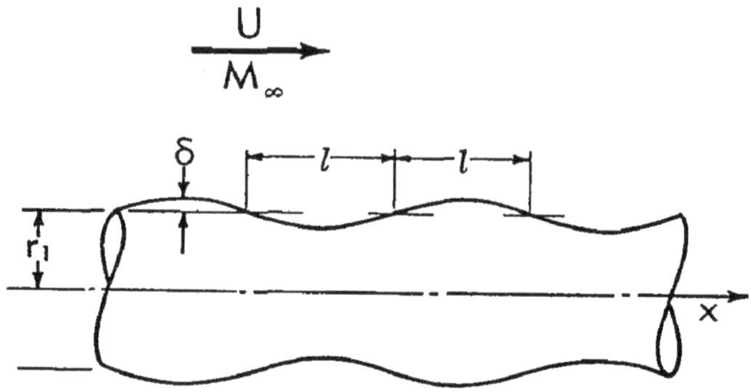

Fig. C,5. Corrugated infinite cylinder in subsonic axial flow.

illustrate the straightforward technique of transformation of boundary conditions and solution of Laplace's equation. The last example, which will be shown to include the preceding examples as special cases, will show how a wide class of flows can be related in an interesting way to incompressible flows about altered bodies.

Corrugated cylinder. Consider an infinitely long corrugated circular cylinder in symmetrical compressible flow, as sketched in Fig. C,5. This case has been treated by E. Reissner [9]. We introduce cylindrical coordinates x, r in place of x, y, z used above, and ξ, ϖ in place of ξ, η, ζ.

Let the surface of the cylinder be given by

$$r = R(x) = r_1 + \delta \cos\left(\pi \frac{x}{l}\right) \tag{5-1}$$

Let us assume, furthermore, that $\delta \ll l$, so that the small perturbation type of flow may be expected. The differential equation to be satisfied by the perturbation velocity potential is the cylindrical form of Eq. 3-5, viz.

$$\beta^2 \phi_{xx} + \frac{1}{r} \phi_r + \phi_{rr} = 0 \tag{5-2}$$

and the boundary condition at the cylinder surface is, to the first order,

$$\phi_r(x, r_1) = UR'(x) = -\pi U \frac{\delta}{l} \sin\left(\pi \frac{x}{l}\right) \tag{5-3}$$

An additional boundary condition is the vanishing of the perturbation far from the axis.

As before, Eq. 5-2 is transformed into Laplace's equation by use of Eq. 3-6; in the present example we select $m = 1$ and $n = 1/\beta$, so that $x = \xi$ while $r = \varpi/\beta$. Then, using the notation $\phi(x, r) = \varphi(\xi, \varpi)$, boundary condition (5-3) becomes

$$\beta\varphi_{\varpi}(\xi, \beta r_1) = -\pi U \frac{\delta}{l} \sin\left(\pi \frac{\xi}{l}\right) \tag{5-4}$$

It is well known [*10,* p. 216] that a solution of Laplace's equation having cylindrical symmetry and sinusoidal variation with ξ is

$$\varphi(\xi, \varpi) = \left[AI_0\left(\pi \frac{\varpi}{l}\right) + BK_0\left(\pi \frac{\varpi}{l}\right)\right] \sin\left(\pi \frac{\xi}{l}\right) \tag{5-5}$$

where I_0 and K_0 are modified Bessel functions of the first and second kinds of zero order, and A and B are arbitrary constants. It is necessary to take A equal to zero to satisfy our boundary condition of vanishing perturbation far from the axis. Comparison of Eq. 5-4 and 5-5 then leads to

$$B = \frac{U\delta}{\beta K_1\left(\pi\beta \frac{r_1}{l}\right)}$$

and therefore

$$\varphi(\xi, \varpi) = \frac{U\delta}{\beta} \frac{K_0\left(\pi \frac{\varpi}{l}\right)}{K_1\left(\pi\beta \frac{r_1}{l}\right)} \sin\left(\pi \frac{\xi}{l}\right) \tag{5-6}$$

and

$$\phi(x, r) = \frac{U\delta}{\beta} \frac{K_0\left(\pi\beta \frac{r}{l}\right)}{K_1\left(\pi\beta \frac{r_1}{l}\right)} \sin\left(\pi \frac{x}{l}\right) \tag{5-7}$$

From this result the velocity components and pressure can be obtained immediately by differentiation. In particular, the coefficient of pressure at the body surface is given by

$$\begin{aligned} C_p(x, r_1) &\cong -\frac{2}{U}\phi_x(x, r_1) \\ &\cong -\frac{2\pi}{\beta}\frac{\delta}{l} \frac{K_0\left(\pi\beta \frac{r_1}{l}\right)}{K_1\left(\pi\beta \frac{r_1}{l}\right)} \cos\left(\pi \frac{x}{l}\right) \end{aligned} \tag{5-8}$$

If the cylinder is very slender, so that r_1/l is a small number, we may substitute in Eq. 5-8 the appropriate asymptotic forms of the Bessel functions, which are

$$K_0(\mu) \cong -\ln \mu; \qquad K_1(\mu) \cong \mu^{-1} \tag{5-9}$$

The surface pressure coefficient (5-8) then becomes, approximately,

$$C_p(x, r_1) \cong 2\pi^2 \frac{\delta r_1}{l^2} \ln\left(\pi\beta \frac{r_1}{l}\right) \cos\left(\pi \frac{x}{l}\right) \tag{5-10}$$

At the other extreme, when r_1 is large compared to l the asymptotic forms for large argument are appropriate; these are

$$K_0(\mu) \cong K_1(\mu) \cong \sqrt{\frac{\pi}{2\mu}}\, e^{-\mu} \tag{5-11}$$

The approximate expression for the surface pressure coefficient is then

$$C_p(x, r_1) \cong -\frac{2\pi}{\beta} \frac{\delta}{l} \cos\left(\pi \frac{x}{l}\right) \tag{5-12}$$

Reissner's paper also includes the case of the corrugated cylinder in a circular wind tunnel. The additional boundary condition, if the tunnel has a closed throat of radius b, is

$$\phi_r(x, b) = 0 \tag{5-13a}$$

and if the tunnel has an open throat

$$\phi_x(x, b) = 0 \tag{5-13b}$$

After transformation to the ξ, ϖ space, these appear, respectively, as

$$\varphi_\varpi(\xi, \beta b) = 0 \quad \text{or} \quad \varphi_\xi(\xi, \beta b) = 0 \tag{5-14}$$

It is not difficult to choose the constants in Eq. 5-5 so as to satisfy simultaneously Eq. 5-4 and one or the other of Eq. 5-14. It may be noted that, since the tunnel boundary condition is satisfied at $\varpi = \beta b$, this boundary effect becomes more severe as the stream Mach number is increased.

Slender bodies of revolution at zero incidence. For very slender bodies of revolution there are simple approximate methods using source-sink and doublet distributions, which can be used to solve the incompressible flow problem. As an example of these procedures, we shall consider here the flow past such a body in an unlimited stream at zero incidence. Let us again adopt cylindrical coordinates ξ and ϖ in the transformed space, corresponding to x and r in the original space. Suppose the body of revolution is defined by $r = R(x)$, which we shall assume is a smooth continuous curve.

In the incompressible fluid theory applicable to this case, the flow perturbations are imagined to be due to a distribution of sources and sinks of strength $f(\xi)$ along the ξ axis. The resulting velocity potential is

$$U\xi - \frac{1}{4\pi}\int_a^b \frac{f(\bar{\xi})d\bar{\xi}}{\sqrt{(\xi-\bar{\xi})^2+\varpi^2}} + \text{const} \tag{5-15}$$

so that the disturbance potential $\varphi(\xi, \varpi)$ of our Prandtl-Glauert theory is given by the integral in Eq. 5-15. Here, $\xi = a$ represents the nose, and $\xi = b$ the tail, of the fictitious body. The corresponding perturbation velocities are

$$\varphi_\xi(\xi, \varpi) = \frac{1}{4\pi}\int_a^b \frac{(\xi-\bar{\xi})f(\bar{\xi})d\bar{\xi}}{[(\xi-\bar{\xi})^2+\varpi^2]^{\frac{3}{2}}} \tag{5-16}$$

and

$$\varphi_\varpi(\xi, \varpi) = \frac{\varpi}{4\pi}\int_a^b \frac{f(\bar{\xi})d\bar{\xi}}{[(\xi-\bar{\xi})^2+\varpi^2]^{\frac{3}{2}}} \tag{5-17}$$

Clearly, these formulas are adequate to calculate the velocities throughout the flow, and in the real compressible flow by simple transformation, if the appropriate function $f(\xi)$ can be determined for the boundary conditions under consideration.

The first order approximation appropriate to very slender bodies consists in replacing Eq. 5-17, for points near the body, with its limiting form for $\varpi \to 0$; it can be verified that this is

$$\varphi_\varpi(\xi, \varpi) \cong \frac{f(\xi)}{2\pi\varpi} \quad \text{as } \varpi \to 0 \tag{5-18}$$

The boundary condition in the present problem is

$$\phi_r(x, R) = UR'(x) \tag{5-19}$$

or, with $x = \xi$, $r = \varpi/\beta$, and $\phi(x, r) = \varphi(\xi, \varpi)$,

$$\beta\varphi_\varpi(\xi, \beta R) = UR'(\xi) \tag{5-20}$$

It may be remarked here that Eq. 5-20 need not be recognized as the boundary condition for incompressible flow about any particular body; nevertheless, it is perfectly satisfactory to determine the source distribution $f(\xi)$. Combining Eq. 5-18 and 5-20, which means selecting the source distribution so as to produce the required radial velocity at the location specified in Eq. 5-20, we have

$$\varphi_\varpi(\xi, \beta R) = \frac{1}{\beta}UR'(\xi) \cong \frac{f(\xi)}{2\pi\beta R} \tag{5-21}$$

Thus, the approximate source distribution function is

$$f(\xi) \cong 2\pi UR(\xi)R'(\xi) = US'(\xi) \tag{5-22}$$

where $S(x)$ denotes the cross-sectional area distribution of the original body. It is notable that this is independent of β.

Substitution of this function for $f(\xi)$ in the integral of Eq. 5-15 gives us $\varphi(\xi, \varpi)$, and the perturbation potential of the compressible flow pattern is

$$\phi(x, r) = -\frac{U}{4\pi}\int_a^b \frac{S'(\xi)d\xi}{\sqrt{(x-\xi)^2+\beta^2r^2}} \tag{5-23}$$

The velocity components are found immediately by differentiation. It is interesting that the axial component ϕ_x at points on the axis before or behind the body is independent of β.

The axial component ϕ_x at the body surface can now be calculated. It would seem consistent to use again the limiting value for $r \to 0$, in view of the similar approximation already made in calculating φ_ϖ. The leading term of this value is found to be

$$\phi_x(x, r) \cong \frac{U}{2\pi} S''(x) \ln\left(\beta\frac{r}{l}\right) \quad \text{as } r \to 0 \tag{5-24}$$

The reader should be warned, however, that this formula (5-24) is principally of theoretical interest and is not generally useful for numerical computation, because for practical bodies the logarithm does not increase so rapidly for small r as to make the neglected terms unimportant. In particular, the formula leaves us with an arbitrariness in the choice of the normalizing length l, since any change of this reference quantity introduces only a term of the order already neglected. Naturally, this arbitrariness can be eliminated by repeating the reduction of Eq. 5-23 and carrying all the terms of the next order of magnitude; the result is

$$\phi_x(x, r) \cong \frac{U}{4\pi}\left[S''(x)\ln\left(\frac{\beta^2r^2}{4(b-x)(x-a)}\right) + \int_a^b \frac{S''(x)-S''(\xi)}{|x-\xi|}d\xi + \frac{S'(b)}{b-x} - \frac{S'(a)}{x-a}\right] \quad \text{as } r \to 0 \tag{5-24a}$$

Improvement of pressure formula. Before proceeding to our next example, we digress here slightly to consider an important improvement which can be made in our first order formula for the pressure coefficient, $C_p \cong -2\phi_x/U$, for points near the surface of a slender body of revolution. The improved expression is (E,7; also [*47*, p. 75])

$$C_p = -2\frac{\phi_x}{U} - \left(\frac{\phi_r}{U}\right)^2 \tag{5-25}$$

It has frequently been observed that Eq. 5-25 affords an improvement in the approximation to C_p on the surface of a slender body in cases where a comparison can be made with more exact solutions or with experimental values. On a body of infinitesimal thickness ratio τ, the two terms of Eq.

5-25 are generally of orders $\tau^2 \ln \tau$ and τ^2 respectively[1] so that the second term is of higher order than the first. Moreover, the inclusion of a quadratic term is not obviously consistent with the approximations already made in deriving the Prandtl-Glauert differential equation (3-5). Nevertheless, it can be deduced by more careful analysis that Eq. 5-25 is the correct second approximation[2] to C_p, where ϕ is the solution of our first order equation (3-5), and of course terms of order τ^2 in ϕ_x must be retained, as in Eq. 5-24a. This is fortunate because, on bodies of practical interest, the two terms of Eq. 5-25 are likely to be of about the same numerical size, as has already been suggested above.

For points farther away from the axis, the term $(\phi_r/U)^2$ is comparatively smaller, and our original approximation $C_p \cong -2\phi_x/U$ is again appropriate, where ϕ_x is to be calculated from Eq. 5-23.

Flow about three-dimensional bodies in general. We have now seen, in a series of examples, how the Prandtl-Glauert theory permits the translation of first order boundary value problems into problems involving Laplace's equation ("potential problems"), which can, in a rather general way, be considered solvable. In these examples, we have sometimes (Art. 4) recognized the transformed problem as one of flow about a related solid body and sometimes, as in the case directly above, we have not.[3] There is sometimes an important advantage to being able to make such a recognition, because it may bring us into a class of potential problems whose solutions we know (as in Art. 4). In other cases, the recognition may be valuable because the related incompressible flow problem can then be set up experimentally, in flight or in a wind tunnel, to provide the solution.

We shall now show, as our final example, how, for all bodies for which the first order theory is applicable, the Prandtl-Glauert method can be used in such a way as to make the related flow a recognizable one involving a solid body. To do this, we adopt a simple scheme proposed by Göthert [*6*]. The presentation here is substantially that of Hess and Gardner [*11*]

We consider any body whose surface F is given by the equation

$$F(x, y, z) = 0 \tag{5-26}$$

[1] Compare our results in Eq. 5-10 and 5-24.

[2] Our nomenclature differs here from that of Sec. E, where this subject is investigated in detail for the analogous supersonic case. There Eq. 5-25 is called the "first approximation," the name "second approximation" being reserved for a formula containing all the terms arising from the solution of a more exact differential equation.

[3] When the transformed problem is not a "body problem," Göthert [*6*] says that the body surface and the stream surfaces have been distorted differently in the transformation. Actually, this language seems a little harsh, for there is no a priori reason why the transformed boundary value problem must be a recognizable "body problem," as is discussed here. Nevertheless, Göthert's observation led him to the discovery and correction of a widespread error in the application of the Prandtl-Glauert theory (see p. 81.).

We assume, furthermore, that it is a "slender body," in the sense that the inclination of the surface F to the stream direction is everywhere small, so that the perturbations that it introduces into an initially uniform steady stream can be expected to be small, except, of course, in the neighborhood of stagnation points. Then, if the stream speed is U and its Mach number M_∞, the differential equation to be satisfied by the perturbation potential ϕ is the Prandtl-Glauert equation (3-5)

$$\beta^2\phi_{xx} + \phi_{yy} + \phi_{zz} = 0; \qquad \beta^2 \equiv 1 - M_\infty^2$$

The boundary condition at the body surface is expressed by the statement that the normal velocity component vanishes there. Since the normal direction to the surface F is (F_x, F_y, F_z), the statement of this boundary condition is

$$(U + \phi_x)F_x + \phi_y F_y + \phi_z F_z = 0 \quad \text{on } F(x, y, z) = 0 \tag{5-27}$$

Now, an equivalent statement of the slender body hypothesis to the one stated above is $|F_x| \ll \sqrt{F_y^2 + F_z^2}$: i.e. the normal to the surface F is inclined only slightly in the x direction. But since we are assuming small velocity perturbations, this implies that the term $\phi_x F_x$ in Eq. 5-27 is of second order compared to the other terms; consequently, we neglect it and write the approximate boundary condition as

$$UF_x + \phi_y F_y + \phi_z F_z = 0 \quad \text{on } F(x, y, z) = 0 \tag{5-28}$$

Let us now apply the affine transformation (3-6), say in the form $x = \beta\xi$, $y = \eta$, $z = \zeta$. Eq. 3-5 then becomes Laplace's equation in the ξ, η, ζ space, while Eq. 5-28 is transformed into

$$\frac{1}{\beta} U\mathfrak{F}_\xi + \varphi_\eta \mathfrak{F}_\eta + \varphi_\zeta \mathfrak{F}_\zeta = 0 \quad \text{on } \mathfrak{F}(\xi, \eta, \zeta) = 0 \tag{5-29}$$

where $\varphi(\xi, \eta, \zeta)$ is written for $\phi(\beta\xi, \eta, \zeta)$, and $\mathfrak{F}(\xi, \eta, \zeta)$ for $F(\beta\xi, \eta, \zeta)$.

The important point to notice here is that boundary condition (5-29), if interpreted correctly, is the boundary condition for a fictitious, incompressible fluid, slender body problem in the ξ, η, ζ space. For example, after multiplication by β, it is clearly the correct condition for flow of a stream at speed U about the surface $\mathfrak{F}(\xi, \eta, \zeta) = 0$, provided that $\beta\varphi(\xi, \eta, \zeta)$ is taken to be the perturbation potential, rather than φ itself.[4] The surface $\mathfrak{F}$ differs from the original surface F in that its dimensions in the stream direction are stretched in the ratio $1:\beta$.

We can now express the velocity components of the real flow in terms of those of the fictitious one, which we shall call u', v', and w':

[4] Obviously, another interpretation of Eq. 5-29 is that it states the condition for the surface $\mathfrak{F}$ in a stream of speed U/β, where φ is the perturbation potential. This choice (like any number of others that can be made at this point) leads to results exactly equivalent to those obtained here. To agree with Hess and Gardner, and also to simplify the identification of these general results with our earlier ones, we prefer to retain the stream speed U in the fictitious flow.

$$\phi_x(x, y, z) = \frac{1}{\beta}\varphi_\xi(\xi, \eta, \zeta) = \frac{1}{\beta^2}\, u'(\xi, \eta, \zeta)$$

$$\phi_y(x, y, z) = \varphi_\eta(\xi, \eta, \zeta) = \frac{1}{\beta} v'(\xi, \eta, \zeta) \qquad (5\text{-}30)$$

$$\phi_z(x, y, z) = \varphi_\zeta(\xi, \eta, \zeta) = \frac{1}{\beta} w'(\xi, \eta, \zeta)$$

$$x = \beta\xi, \qquad y = \eta, \qquad z = \zeta$$

Let us state this result verbally: to calculate the velocity components in the compressible flow about any slender body, first calculate the components u', v', and w' in the incompressible flow, at the same stream speed, about a stretched body whose streamwise dimensions are β^{-1} times as great. The desired velocity components are then $\beta^{-2}u'$, $\beta^{-1}v'$, and $\beta^{-1}w'$ at corresponding points of the affine transformation which stretches the space in the same fashion.

This is Göthert's "extended Prandtl rule" [*6*]. Previously, the application of the Prandtl-Glauert theory to axially symmetric and three-dimensional bodies had been performed erroneously by a number of writers [*7,8,12–15*]. The correct application has now been given in several places [*16–21*].

The identity of our general equations (5-30) with our earlier results for wings (Eq. 4-4) is not difficult to establish. The missing factor β^{-1} in Eq. 4-4 is provided by decreasing the ordinates of the fictitious wing in accordance with our general rule. That is, in our general method we should have stretched the wing in the stream direction without simultaneously increasing its ordinates so as to maintain its shape and incidence. This would, of course, require the additional factor β^{-1} to make the results the same.

It can also be established that our general rule coincides with Reissner's results for the corrugated cylinder and ours for slender bodies of revolution. In each of these cases we have preferred to use a slightly different affine transformation, but of course, the flow patterns, described in terms of velocity ratio for example, depend only on the *shape* of a body and not on the size. The identity of the flows, in this sense, can be worked out by means of a little algebra.

Finally, it should be pointed out explicitly that our general rule holds equally well when the slender body involved is immersed in a bounded stream or when it consists of an arrangement of individual bodies. In fact, the surface $F(x, y, z) = 0$ need not be a simple closed surface, but may have various unconnected parts, and the same deduction still holds. Thus, the fictitious incompressible flow is one in which the body or bodies and the boundaries of the stream are all stretched in the stream direction.

C,6. Pressure-Correction Formulas for Steady Flow. Although steady, subsonic, small perturbation flows have already been solved, in principle, either by straightforward transformation of the boundary-value problem or by adoption of Göthert's "extended Prandtl rule" (which is not essentially different), it is convenient to carry the matter one step further and to devise "pressure-correction formulas" that display explicitly the variation of pressure with Mach number (at least for a certain family of bodies) in the form

$$C_p = C_p(C_{p_{\text{inc}}}, M_\infty) \tag{6-1}$$

This can be accomplished only for special families, however, as we shall see.

Plane flow. The pressure-correction formula for thin cylinders in unconfined plane flow follows immediately from our treatment of thin wings in Art. 4. Since the cylindrical wing is independent of the reduction of aspect ratio involved there, Eq. 4-4 to 4-6 are explicit statements regarding the relationship of compressible and incompressible flows about identically the same cylinders (except for a change of scale which, as has already been mentioned, is of no consequence here). Thus, Eq. 4-5 is a pressure-correction formula for plane flow about thin cylinders in unbounded steady flow. Since the pressure coefficient is really a function of dimensionless coordinates, say x/c, y/c, etc., and the airfoil chord in the ξ, η plane is c/β, this rule might best be written as

$$C_p\left(\frac{x}{c}, \frac{y}{c}\right) = \frac{1}{\beta} C_{p_{\text{inc}}}\left(\frac{x}{c}, \beta\frac{y}{c}\right) \tag{6-2}$$

and for points on the wing surface,

$$C_p\left(\frac{x}{c}, \pm 0\right) = \frac{1}{\beta} C_{p_{\text{inc}}}\left(\frac{x}{c}, \pm 0\right) \tag{6-3}$$

Thus, the pressure coefficient is proportional to the function $\beta^{-1}(M_\infty)$ plotted in Fig. C,4c.

Similarly, formulas (4-6) for force and moment coefficients now become explicit compressibility-correction formulas for any given cylinder, and state that these coefficients vary as β^{-1}.

The same rules, of course, can be obtained from our general rule in the form worked out in Art. 5, above. Taking $C_p \cong -2\phi_x/U$, we have, from Eq. 5-30,

$$C_p\left(\frac{x}{c}, \frac{y}{c}\right) = \frac{1}{\beta^2} \frac{u'\left(\frac{x}{c}, \beta\frac{y}{c}\right)}{U} \tag{6-4}$$

which appears, at first glance, to disagree with Eq. 6-2 by a factor β^{-1}. However, according to our general rule, the incompressible flow component u' pertains to a cylinder of reduced thickness ratio and incidence. In plane, unbounded, incompressible flow about a thin cylinder, in the first approximation, all velocity components are proportional to a typical body ordinate, i.e. to thickness ratio and incidence. Thus, one of the factors β^{-1} in Eq. 6-4 is required to bring the fictitious cylinder back to the original thickness and incidence, and we find that the two expressions are in complete agreement.

Slender bodies of revolution at zero incidence. An analogous pressure-correction formula cannot be obtained for bodies of revolution, even for the very slender ones that correspond to our approximate result, Eq. 5-24. This result gives us, for points on the body $r = R$,

$$C_p(x, R) \cong -2\frac{\phi_x}{U} \cong -(R^2)'' \ln\left(\beta\frac{R}{l}\right) \tag{6-5}$$

This is an explicit formula for the variation of the pressure coefficient at the surface of a very slender body, with varying Mach number, but clearly cannot be put into the form $C_p = C_p(C_{p_{\text{ino}}}, M_\infty)$. On the contrary Eq. 6-5 gives the compressibility correction as an additive effect:

$$C_p(x, R) \cong C_{p_{\text{ino}}}(x, R) - (R^2)'' \ln \beta \tag{6-6}$$

Furthermore, this convenient formula also results if our improved pressure coefficient approximation (5-25) is used, together with the complete second order expression for ϕ_x, Eq. 5-24a. This is because β occurs only in the logarithmic term in Eq. 5-24a and does not occur at all in the term $(\phi_r/U)^2$, which is simply R'^2 at the body surface. Thus, to the second approximation,[5] which means $O(\tau^2)$ for bodies of thickness ratio τ, the formula for the surface pressure coefficient is Eq. 6-6 (cf. [*20*]).

Ellipsoids of revolution and streamline bodies. Rough pressure-correction formulas of a somewhat different character have been proposed by some authors, who have simply applied the general results of our Prandtl-Glauert method to representative pressures on convenient typical families of bodies and have suggested that the results be adopted as approximations for wider application. For example, Sears [*16*], Young, and Kirkby [*18*], and others have plotted typical curves of C_p against M_∞ for slender ellipsoids of revolution (prolate spheroids) in unbounded stream flow. Curves of this type are shown in Fig. C,6a. Although the curves have been computed for the minimum pressure point at the midsection of the ellipsoid, it is proposed to use these same curves for various locations on various slender bodies, as indicating roughly the relationship between C_p and $C_{p_{\text{ino}}}$. Comparison with Fig. C,4c will disclose that the increase of $(-C_p/C_{p_{\text{ino}}})$ suggested here is considerably slower with increasing M_∞ than for cylinders in plane flow.

[5] See footnote 2.

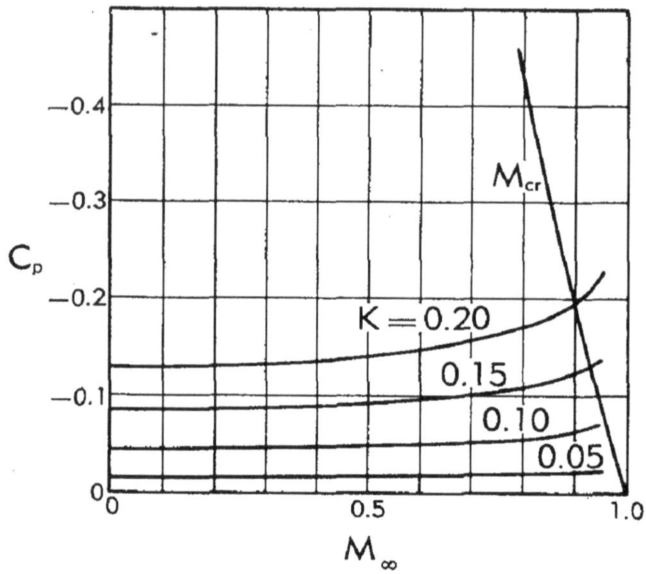

Fig. C,6a. Variation of minimum pressure coefficient C_p with stream Mach number M_∞ for ellipsoids of revolution (prolate spheroids) of thickness ratio K, in unbounded stream flow.

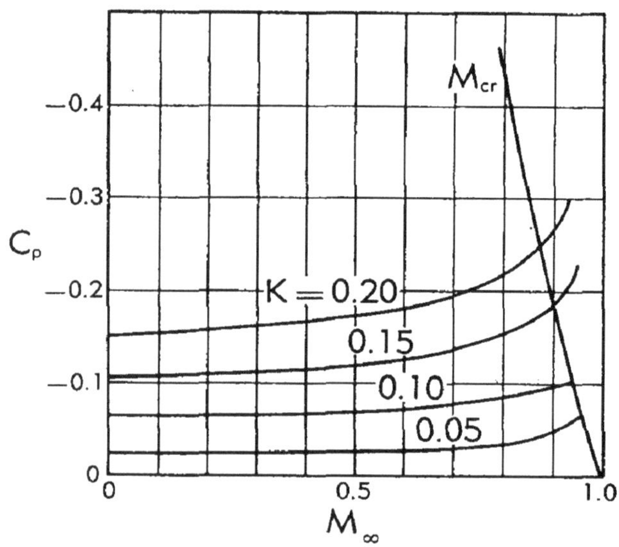

Fig. C,6b. Variation of minimum pressure coefficient C_p with stream Mach number M_∞ for the Young-Kirkby family of streamline bodies of revolution [*18*] of thickness ratio K, in unbounded stream flow.

Analogous curves are presented in Fig. C,6b, taken from the report of Young and Kirkby [*18*]. These pertain to the minimum pressures on a special family of streamline bodies of revolution (developed by Young and Young in an earlier paper [*22*]) having moderate velocity gradients ahead of the location of minimum pressure, which occurs about 37 per cent of the length back from the nose. These shapes are probably closer to the shapes of practical fuselages, nacelles, and external fuel tanks than the ellipsoids, and it is interesting that for them $(-C_p/C_{p_{\text{ino}}})$ increases somewhat more rapidly with increasing M_∞. Again it is proposed that these curves be adopted as rough pressure correction diagrams in the form $C_p = C_p(C_{p_{\text{ino}}}, M_\infty)$, regardless of thickness ratio or body abscissa, for a variety of practical streamline bodies in unconfined stream flow.

Corrugated cylinder. Reissner's investigation [9] of the flow past an infinite corrugated circular cylinder, which has already been discussed in detail in Art. 5, throws some light on the question of pressure-correction formulas.

For a cylinder that is neither very slender nor very large compared to the dimensions of the corrugation, the ratio of compressible to incompressible flow pressure coefficients at the body surface is, from Eq. 5-8,

$$\frac{C_p}{C_{p_{\text{ino}}}} = \frac{1}{\beta}\,\frac{K_0\left(\pi\beta\frac{r_1}{l}\right)}{K_0\left(\pi\frac{r_1}{l}\right)}\,\frac{K_1\left(\pi\frac{r_1}{l}\right)}{K_1\left(\pi\beta\frac{r_1}{l}\right)} \tag{6-7}$$

This cannot, of course, be reduced to the form $C_p = C_p(C_{p_{\text{ino}}}, M_\infty)$. If the cylinder is very slender $(r_1/l \ll 1)$, as we have assumed in Eq. 5-10, this reduces to a form analogous to our result (6-6) for slender bodies of revolution (except, of course, that only the first approximation $C_p \cong -2\phi_x/U$ has been used—we could improve it by including the ϕ_r^2 term, but this would not affect our conclusions here).

It is only where the cylinder radius is large compared to the corrugation, so that formula (5-12) is appropriate, that we have a pressure-correction formula:

$$\frac{C_p}{C_{p_{\text{ino}}}} \cong \frac{1}{\beta} \tag{6-8}$$

This, of course, is the formula for plane flow, for the corrugation becomes practically a cylindrical surface when the basic cylinder is very large.

C,7. The Critical Mach Number. It has become common to designate as "critical Mach number" the lowest value of the stream Mach number M_∞ for which a local Mach number of unity occurs in a flow. The terminology is based on the reasoning, first suggested by G. I. Taylor, that this Mach number at least provides a lower limit for the onset of local shock waves and related phenomena, which are usually observed in transonic flow and are characteristic of local supersonic flow conditions. H. S. Tsien has suggested the name "lower critical Mach number" for this value.

There is some logical doubt whether the steady flow small perturbation theory is generally valid for Mach numbers as great as the critical. On one hand, if the body is not-very slender, it cannot fall within the scope of this theory at any Mach number. On the other hand, if it is very slender, its critical Mach number will occur only when M_∞ is close to 1, and there, as we have seen, our theory is likely to become self-inconsistent. Nevertheless, it is generally assumed in practical work that the estimation of critical Mach number on the basis of this theory may correctly indicate certain trends and thus be helpful in comparing various shapes proposed for use in high speed flow.

The relation between minimum C_p and the critical Mach number M_{cr} follows immediately from the relations (III,A,8)

$$\frac{q^2}{2} + \frac{\gamma}{\gamma - 1}\frac{p}{\rho} = \text{const}, \qquad p\rho^{-\gamma} = \text{const} \tag{7-1}$$

which hold throughout any steady flow of the type we are considering. It might be noted that the only approximations involved here are the neglect of viscosity and heat conduction; there can be no shock waves. We obtain a solution of these equations for C_p, which is again defined as $(p - p_\infty)/\frac{1}{2}\rho_\infty U_\infty^2$; viz. (cf. III,B)

$$C_p = \frac{2}{\gamma M_\infty^2}\left[\left(\frac{M_\infty^2 + \dfrac{2}{\gamma - 1}}{M^2 + \dfrac{2}{\gamma - 1}}\right)^{\frac{\gamma}{\gamma - 1}} - 1\right] \tag{7-2}$$

Hence, the relation between minimum C_p and M_{cr} for any body is

$$C_{p_{\min}} = C_{p_{\text{sonic}}} = \frac{2}{\gamma M_{cr}^2}\left[\left(\frac{2}{\gamma + 1} + \frac{\gamma - 1}{\gamma + 1} M_{cr}^2\right)^{\frac{\gamma}{\gamma - 1}} - 1\right] \tag{7-3}$$

This function is plotted in Fig. C,6a and C,6b (at the right-hand side of each Fig.).

For small perturbation flow, it should be legitimate to approximate formula (7-3) by assuming that M_{cr} will be close to unity. If this is done and only the leading terms involving $1 - M_{cr}$ are retained, a very simple formula results:

$$M_{cr}^2 = \left(1 - \frac{\gamma + 1}{2} C_{p_{\min}}\right)^{-1} \tag{7-4}$$

A numerical check will confirm that this is a good approximation to Eq. 7-3 for small $(-C_{p_{\min}})$.

Now, in the case of plane flow in an unlimited stream, where $C_{p_{\min}}$ is related to minimum $C_{p_{\text{inc}}}$ by a universal pressure-correction formula (6-3), it is easy to calculate M_{cr} directly in terms of the minimum $C_{p_{\text{inc}}}$:

$$M_{cr}^2 = 1 - \left[\frac{\gamma + 1}{2}\left(-C_{p_{\text{inc}}}\right)\right]^{\frac{2}{3}} \tag{7-5}$$

In the same way, if Young and Kirkby's results are adopted as a first approximation for a variety of streamline bodies of revolution, an analogous formula for M_{cr} can be obtained for these bodies. In Fig. C,7 are plotted both of these relations for M_{cr}, as well as the relation appropriate for ellipsoids of revolution, which were also discussed in Art. 6.

The reader will notice that in the last two paragraphs above we have tacitly assumed that our formulas for pressure coefficients at the body surface are applicable to the calculation of M_{cr}, i.e. that the minimum

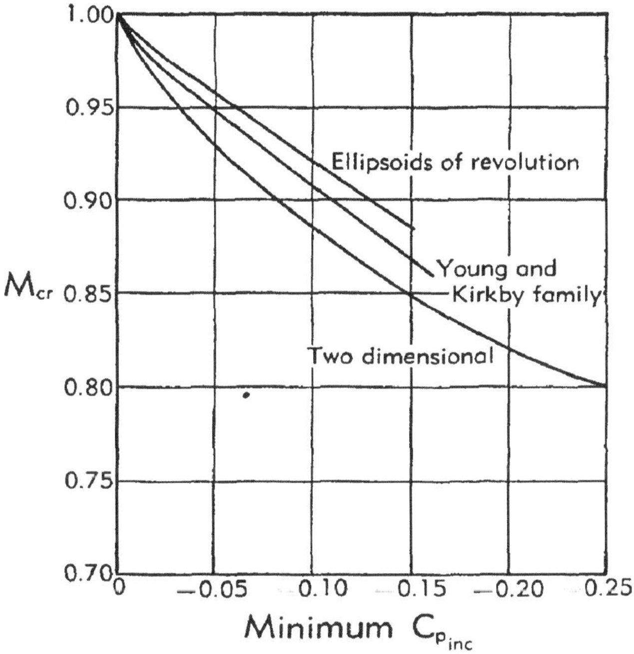

Fig. C,7. Critical Mach number as a function of minimum pressure coefficient in incompressible flow, for two families of slender bodies of revolution and for thin cylinders in plane flow.

pressure always occurs at the body surface. This is, in fact, generally true, for there is a general theorem to this effect in incompressible flow, and the Prandtl-Glauert theory tells us that the same must be true for steady, small perturbation, compressible flow as well.

Yawed and sweptback cylinders. We have already found (Art. 4) that the significant Mach number in the case of flow past a yawed or sweptback infinite cylinder is that of the stream component normal to the cylinder axis. We have also argued that this must be so, in a frictionless fluid by transporting our frame of reference along the cylinder. These considerations hold equally well for the concept of the critical Mach number; i.e., it is a critical value of the component Mach number.

To see this, we need only return to Taylor's original suggestion, mentioned above. He proposed M_{cr} as a lower limit below which it was impossible for a shock wave to form. But a shock wave produced by a yawed infinite cylinder is an oblique shock wave, and its appearance is determined by conditions of the normal component flow. In particular, it cannot occur unless the normal velocity component locally, i.e. everywhere along one of the cylinder's generators, exceeds the sonic speed.

Therefore, let us repeat our calculation of critical Mach numbers for this case, where the angle of yaw is Λ. Eq. 7-3 and 7-4 become relations between $C_{p_{\min}}/\cos^2 \Lambda$ and $M_{\mathrm{cr}}^2 \cos^2 \Lambda$; in particular Eq. 7-4 is

$$M_{\mathrm{cr}}^2 \cos^2 \Lambda = \left(1 - \frac{\gamma + 1}{2} \frac{C_{p_{\min}}}{\cos^2 \Lambda}\right)^{-1} \tag{7-6}$$

Combining this with Eq. 4-22, we obtain, approximately,

$$M_{\mathrm{cr}}^2 \cos^2 \Lambda = 1 - \left[\frac{\gamma + 1}{2} \left(-C_{p_{\mathrm{inc}}}^{(2)}\right)\right]^{\frac{2}{3}} \tag{7-7}$$

where $C_{p_{\mathrm{inc}}}^{(2)}$ is the coefficient defined just before Eq. 4-22, namely, the pressure coefficient of two-dimensional incompressible flow about the given cylinder. Thus, the "two-dimensional" curve of Fig. C,7 is applicable to yawed or sweptback thin cylinders if the ordinate and abscissa are interpreted as $M_{\mathrm{cr}} \cos \Lambda$ and $C_{p_{\mathrm{inc}}}^{(2)}$.

The significance of this result is very great, especially since there are usually some parts of a finite span yawed cylinder where the conditions of flow can be expected to resemble closely the flow considered here. The result emphasizes the significance of the usual critical Mach number as a lower limit only. For the special case of infinite yawed cylinders, knowing the angle at which a shock wave must stand relative to the stream, we have been able to raise the lower limit. In general, in three-dimensional cases, it is not so easy to predict the shock-wave configuration, and the lower critical Mach number may well be exceeded before a shock wave can form.

C,8. Experimental Confirmation. In this article we shall present certain experimental data measured in steady subsonic flow past slender objects, in comparison with the predictions of the small perturbation theory discussed above. It will not be practical to reproduce here, or even to describe and discuss in detail, all of the extensive literature on this subject. Instead, we shall give only a few typical results selected from this literature, and also give references to some other data on the subject.

Plane steady flow. In Fig. C,8a are shown the pressure coefficient distributions measured on an airfoil profile in plane flow, in comparison with the distributions predicted by our results of Art. 4. The experimental values are taken from a report by Amick [*23*]. They were measured in a 4-inch by 18-inch rectangular closed throat wind tunnel at the Langley Laboratory, on a 5-inch chord airfoil of NACA 0012 profile at zero incidence.

For this comparison, the theoretical distributions have been calculated by applying formula (6-3) to the values measured at $M_\infty = 0.40$. Since the model was confined within wind tunnel walls, attention had to be given, in using Eq. 6-3, to the change of effective tunnel dimensions

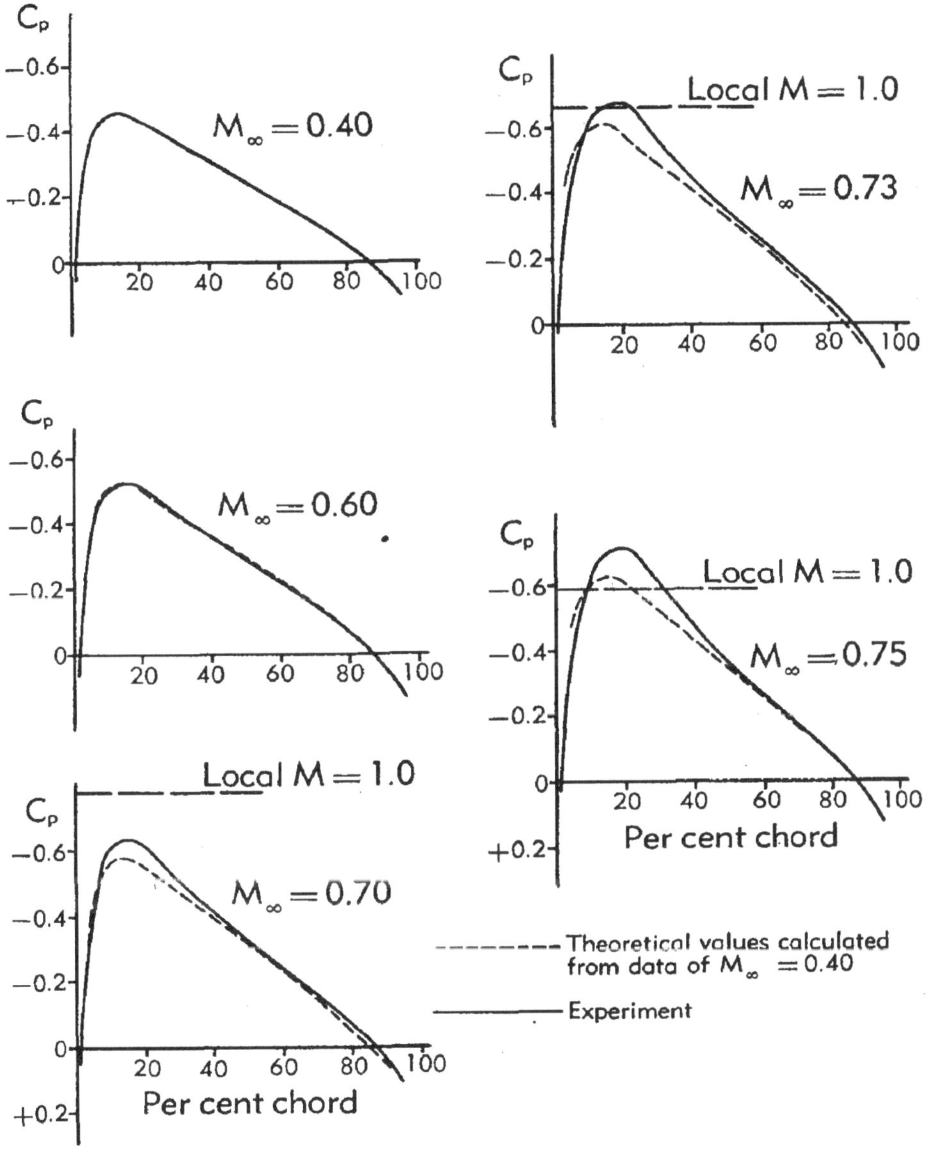

Fig. C,8a. Comparison of experimental pressure distributions of [23] with predictions of Prandtl-Glauert theory. NACA 0012 profile in closed-threat wind tunnel.

with varying M_∞. For this purpose, the wall interference has been estimated from the results given by Tsien and Lees [8], and the increase of this interference with increasing M_∞ has been included in the theoretical pressure distributions. The interference effects are small for the particular geometry involved here.

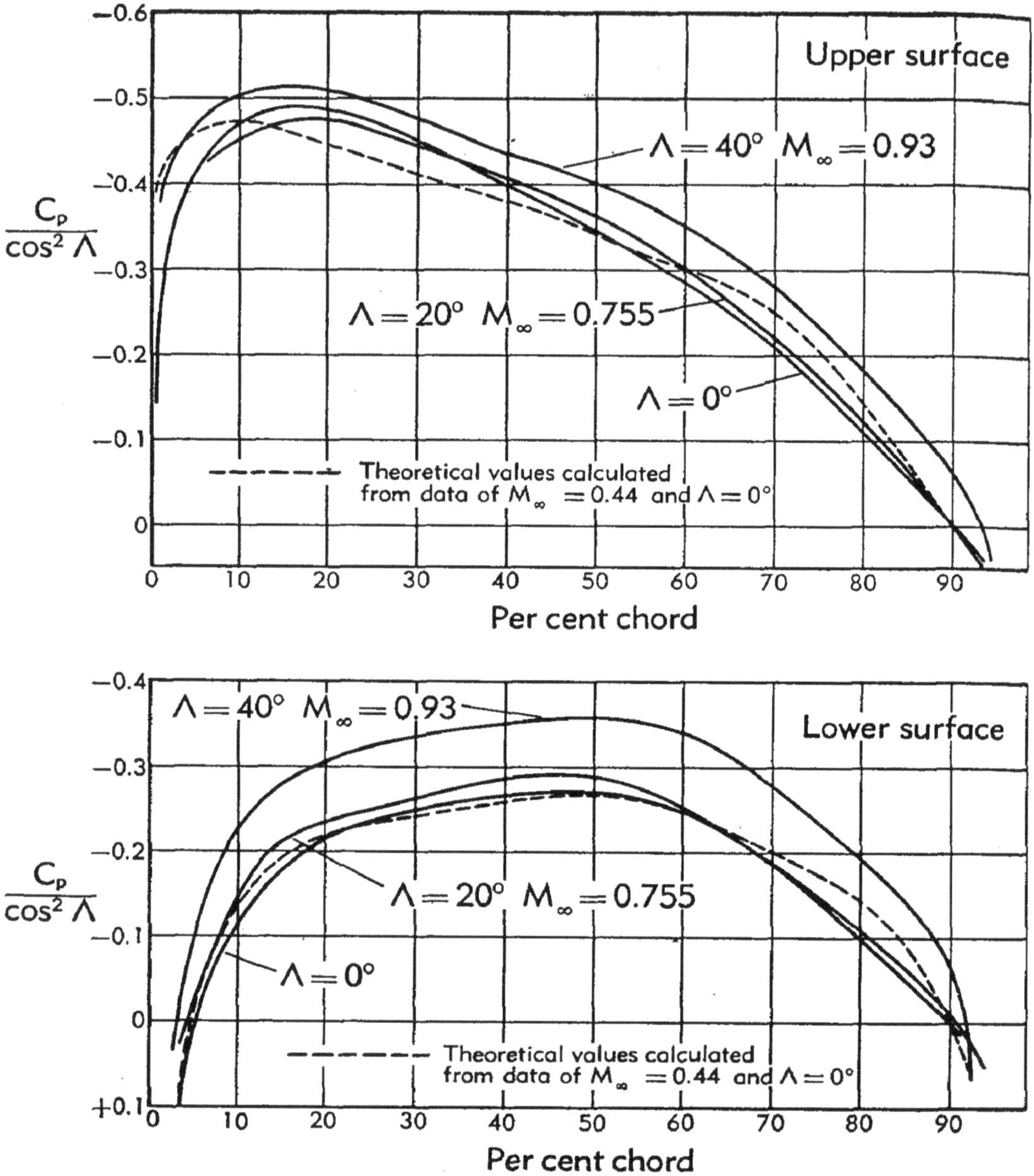

Fig. C,8b. Comparison of experimental pressure distributions of [24] with predictions of Prandtl-Glauert theory. Symmetrical 9 per cent thick airfoil at 1° incidence in closed-throat tunnel at three angles of yaw Λ.

The comparison shows that the pressure curves based on the small perturbation theory fall appreciably below the measured values as the critical Mach number is approached, and that the shape of the distribution curve begins to deviate from the predicted one (which does not change with M_∞) as M_{cr} is exceeded. The good agreement seen in Fig. C,8a for moderate Mach numbers, as well as the type of deviation encountered at the above M_{cr}, are typical. There are available in published literature a considerable number of such comparisons [*13*, *28–31*] and nearly all exhibit the same characteristics—although in many of these the experimental data are not as reliable as those used here, and sometimes no account has been taken of wind tunnel boundary effects.

Yawed airfoils. Another useful set of experimental data has been given by Lippisch and Beuschausen [*24*] in connection with an investigation of yawed airfoils. Pressure distributions were measured on a 9 per

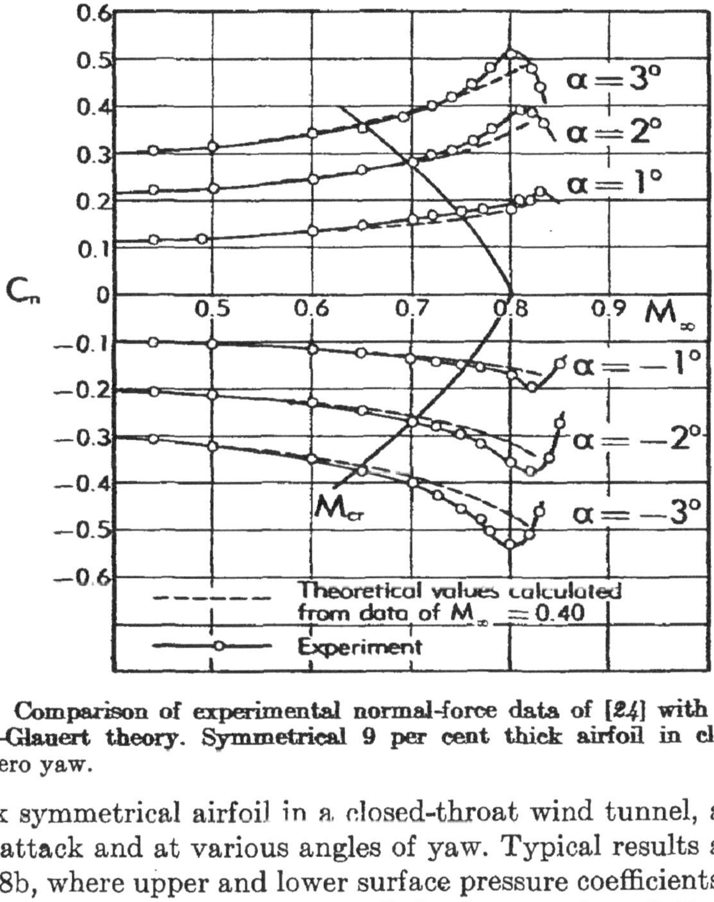

Fig. C,8c. Comparison of experimental normal-force data of [*24*] with predictions of Prandtl-Glauert theory. Symmetrical 9 per cent thick airfoil in closed-throat tunnel at zero yaw.

cent thick symmetrical airfoil in a closed-throat wind tunnel, at various angles of attack and at various angles of yaw. Typical results are shown in Fig. C,8b, where upper and lower surface pressure coefficients for three different angles of yaw Λ, in the form $C_p/\cos^2 \Lambda$, are plotted. Two of these sets of data have been obtained from [*24*] by interpolation, so that all correspond to the same value of $M_\infty \cos \Lambda$. For comparison, a fourth set of curves has been obtained by correcting the measured data for zero yaw and $M_\infty = 0.44$ to this same value in accordance with Eq. 4-22.

According to Eq. 4-22, these three sets of data should fall together. It is seen that the agreement for $\Lambda = 0°$ and 20° is reasonably good, and that even the curves for $\Lambda = 40°$ are in rough agreement.

Fig. C,8c and C,8d are also based on the data of [*24*]. In these figures C_n denotes the coefficient of normal force, and is approximately equal to

C_L. Fig. C,8c is believed to be self-explanatory; the agreement with first order theory seems quite-satisfactory, even for Mach numbers above the critical, which is roughly indicated on this figure. The comparison with the predictions of the Prandtl-Glauert theory is also typical, the predicted increments due to increasing M_∞ being somewhat too small at high Mach numbers. At still higher values of M_∞, the experimental curves break sharply toward $C_n = 0$. This effect is due to boundary layer separation,

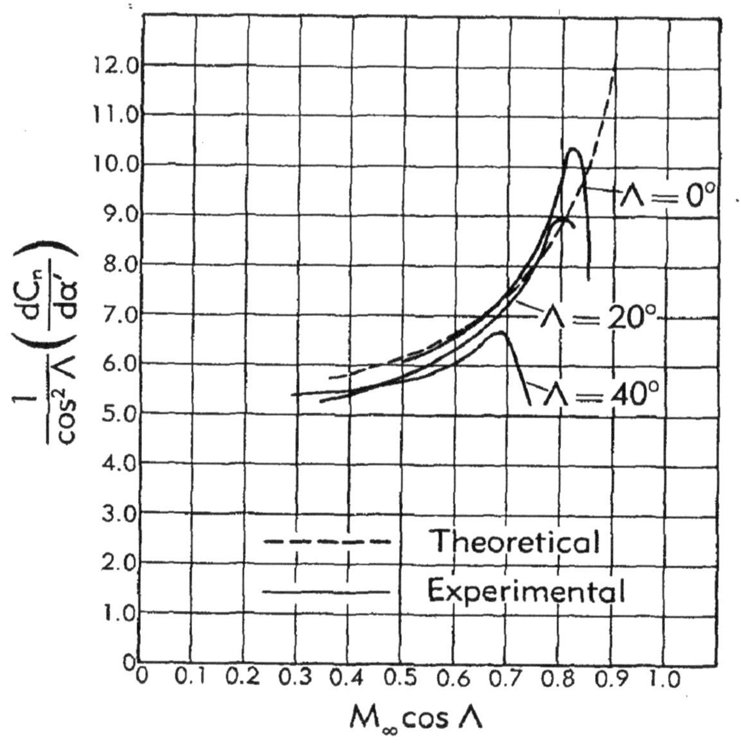

Fig. C,8d. Comparison of experimental normal-force data of [24] with predictions of Prandtl-Glauert theory. Symmetrical 9 per cent thick airfoil at three angles of yaw Λ.

which in turn results from the presence of a shock wave near the upper surface of the airfoil.

Fig. C,8d presents another check on the theory for yawed airfoils. According to Eq. 4-23, the slope $dC_n/d\alpha'$ should be proportional to $(1/\beta') \cos^2 \Lambda$, where α' is defined as the incidence measured in the normal (to the leading edge) cross section. The results in Fig. C,8d show that this is approximately true over a restricted range of the parameter $M_\infty \cos \Lambda$. In evaluating this result and the results of Fig. C,8b, the reader must remember that Lippisch and Beuschausen's airfoils were not of infinite span, but were tested spanning a straight-walled tunnel of width only five times the normal chord length. The pressure distributions were measured at the midspan section, and were integrated to give C_n. Undoubtedly the presence of the tunnel side walls had considerable effect on the results, especially at large angles of yaw and high Mach numbers. Moreover, the straight-walled tunnel would be expected to choke near $M_\infty = 1$, in spite of the sweepback of the airfoil. Thus the premature breaks in the curves for $\Lambda = 20°$ and $40°$ in Fig. C,8d, which actually occurred at $M_\infty = 0.85$ and 0.9, should be considered of no significance in the comparison. Similarly, the deviation of the pressure curves for $\Lambda = 40°$ in Fig. C,8b, which occurred at a tunnel Mach number 0.93, should not be alarming.

Bodies of revolution. Pressure distributions on several slender bodies of revolution at various angles of attack have been measured by Matthews [*21*] in the Langley 8-foot high speed wind tunnel. Typical results from this investigation provide a valuable check on the validity of the first order theory for such bodies. In Fig. C,8e and C,8f are plotted experimental and theoretical pressure distributions for prolate spheroids (ellipsoids of revolution) of thickness ratios 1/6 and 1/10. The experimental values were corrected for wind tunnel boundary effects by Matthews, using the method of [*25*].

The theoretical values shown in these figures were calculated by Matthews by direct application of the "extended Prandtl rule" of Art. 5, above. For this purpose he employed the rather complicated formula giving $C_{p_{\text{inc}}}$ on the surface of a prolate spheroid. The agreement between experiment and theory seems to be fairly satisfactory. The critical Mach numbers of these ellipsoids are estimated in [*21*] to be 0.916 and 0.947 respectively.

It is also interesting to compare these experimental and theoretical results with the somewhat simpler theories given in our preceding articles. First, we have applied to Matthews' two spheroids, at zero yaw, the slender body, source-sink theory given in Eq. 5-19, 5-23, and 5-25, above. For the special case of the spheroids, the integral in Eq. 5-23 can be evaluated analytically and an explicit formula written out for the pressure coefficient at any subsonic Mach number. The results lie very close to Matthews' theoretical curves plotted in Fig. C,8e and C,8f, except for points closer to the nose than 5 per cent of the body length. (In the interests of clarity, these curves are not reproduced in the figures.) These results for the source-sink method are not in agreement with Matthews' results for the same method, because he used the rough formula $C_p = -2\phi_x/U$, instead of the improved pressure formula (5-25), in his source-sink method. Inclusion of the $(\phi_r/U)^2$ term brings his curves for the source-sink method into good agreement with his results for the more elaborate one mentioned above.

Secondly, we have tried the extremely simple additive correction rule (6-6) in these two cases. We find that, starting with either the exact-incompressible or the source-sink curve for $M_\infty = 0$, the change of

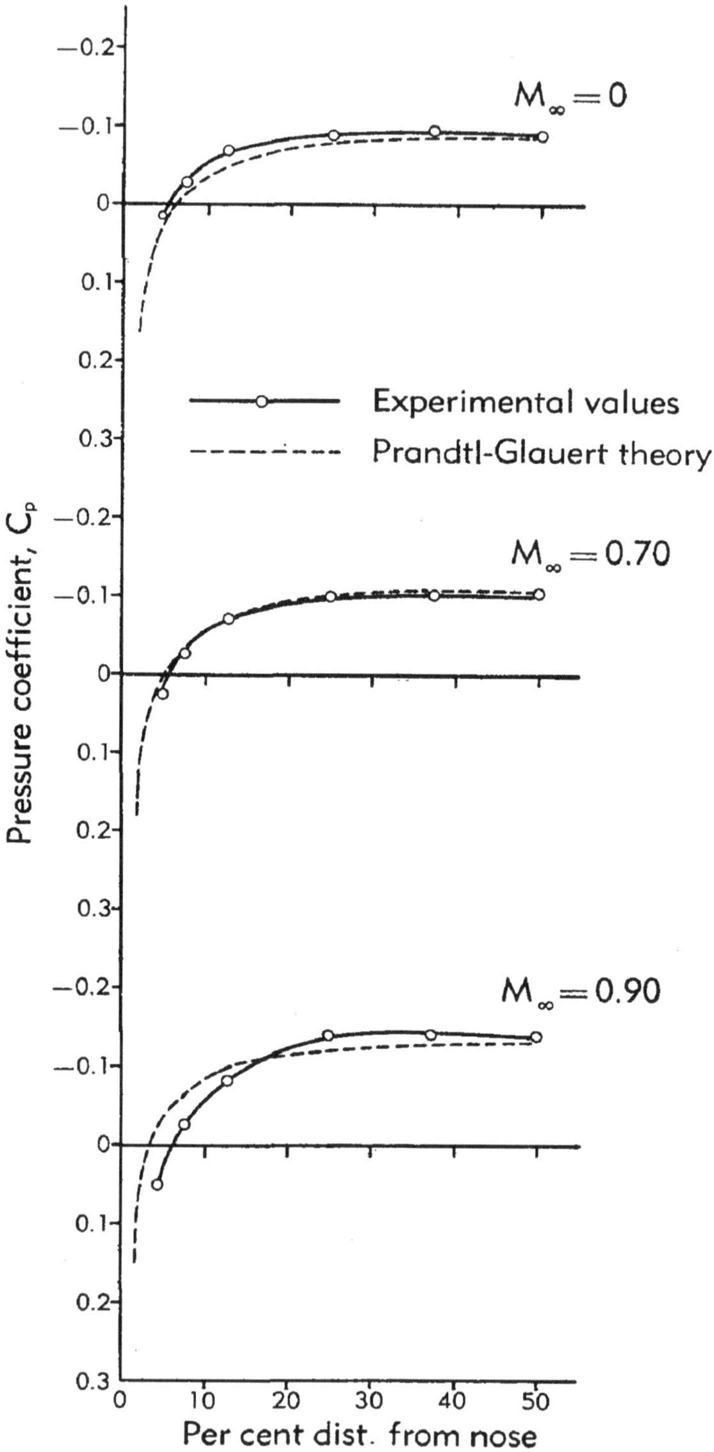

Fig. C,8e. Comparison of experimental pressure distributions with predictions of the Prandtl-Glauert theory for an ellipsoid of revolution (prolate spheroid) in a closed-throat circular tunnel at zero yaw. Thickness ratio 1/6. (From [21].)

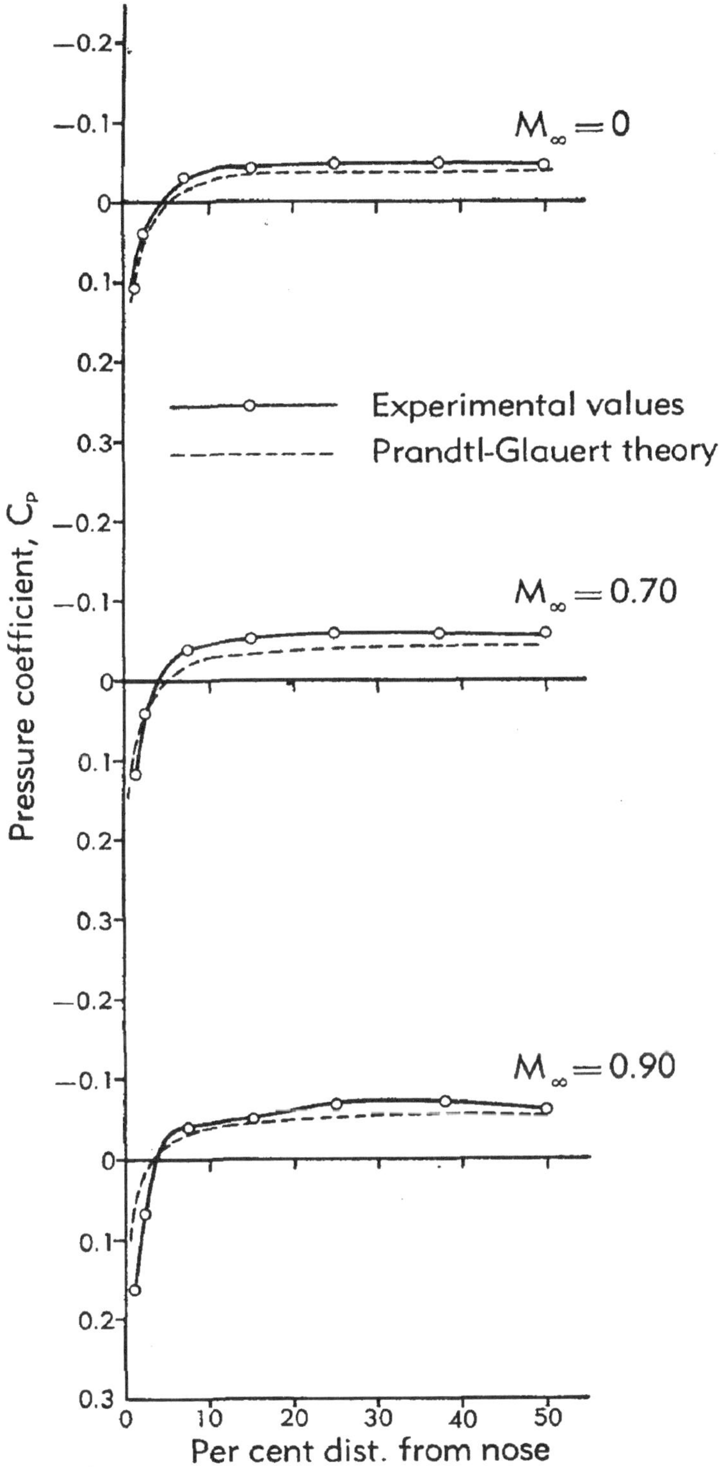

Fig. C,8f. Comparison of experimental pressure distributions with predictions of the Prandtl-Glauert theory for an ellipsoid of revolution (prolate spheroid) in a closed-throat circular tunnel at zero yaw. Thickness ratio 1/10. (From [21].)

theoretical pressure coefficient with Mach number is very well predicted by this simple rule. We might conclude, therefore, that more elaborate theoretical methods are of doubtful value in predicting compressibility effects on slender bodies if the incompressible flow pressures are known.

This conclusion seems inescapable in view of the comparison between any of these theoretical results and the *measured* ones; as suggested previously, this agreement can only be considered fair. Whereas, as has just been mentioned, the theoretical compressibility correction is additive, and in fact is constant for each M_∞ for a spheroid, the experimental curves for $M_\infty = 0.90$ do not exhibit this behavior at all.

It should be mentioned explicitly that [*21*] includes considerably more experimental and theoretical data than have been specifically referred to here. Matthews gives both measured and calculated results for ellipsoids of revolution at angles of attack, and also considers two nonelliptic bodies, one being a "typical transonic body" and the other an ellipsoid with an annular bump. Two other experimental investigations of slender bodies of revolution in subsonic flow should be mentioned; these were both carried out at the Technische Hochschule in Zürich, and are reported in [*26,27*].

C,9. The Slender Body Theory of Munk and Jones. It has been known for some time that there are certain incompressible flows about elongated bodies in which the conditions near the body can be approximated by neglecting the first term in the potential equation

$$\phi_{xx} + \phi_{yy} + \phi_{zz} = 0 \tag{9-1}$$

where the x direction is the stream direction and the direction of elongation of the body. This was first discovered by Munk and given in his famous paper on the cross-force distribution on airships at yaw [*33*]. It has been extended to compressible flows and to the calculation of pressure distributions more recently, especially by R. T. Jones [*34*]. Naturally, this approximation depends primarily on the assumption that, for a very long slender body whose cross section varies slowly with x, the first term in Eq. 9-1 is negligible compared to the others.

Let us first assume that the Munk approximation is valid for a certain body in incompressible flow, and demonstrate, on the basis of the Prandtl-Glauert theory, that the analogous approximation is valid for compressible flow. Let the body surface be given by $F(x, y, z) = 0$; it is our hypothesis that, for incompressible flow about this body,

$$\phi_{xx} \ll \phi_{yy} + \phi_{zz} \tag{9-2}$$

and that the flow near the body is adequately described by

$$\phi_{yy} + \phi_{zz} \cong 0 \tag{9-3}$$

Now, in the compressible flow, the perturbation potential satisfies

$$\beta^2\phi_{xx} + \phi_{yy} + \phi_{zz} = 0; \qquad \beta^2 \equiv 1 - M_\infty^2 \tag{9-4}$$

According to our extended Prandtl rule (Art. 5), a determination of ϕ can be made by calculating the flow about an elongated body whose surface is $F(\beta\xi, \eta, \zeta) \equiv \mathfrak{F}(\xi, \eta, \zeta) = 0$, in incompressible flow in the ξ, η, ζ space, and then putting ϕ equal to $1/\beta$ times the perturbation potential of that flow. Let this potential be called $\chi(\xi, \eta, \zeta)$; then $\phi(x, y, z) = \beta^{-1}\chi(\xi, \eta, \zeta)$.

But the body $\mathfrak{F}(\xi, \eta, \zeta) = 0$ is even more slender than the original one, and since it is considered for incompressible flow, there is no doubt that the Munk-Jones theory applies to it as well as to the original, i.e.

$$\chi_{\xi\xi} \ll \chi_{\eta\eta} + \chi_{\zeta\zeta} \quad \text{and} \quad \chi_{\eta\eta} + \chi_{\zeta\zeta} \cong 0 \tag{9-5}$$

Transforming these back to the original variables, we have,

$$\beta^3\phi_{xx} \ll \beta\phi_{yy} + \beta\phi_{zz} \quad \text{and} \quad \phi_{yy} + \phi_{zz} \cong 0 \tag{9-6}$$

In other words, according to the Prandtl-Glauert theory, the increase of ϕ_x with M_∞ for a given body is compensated by the increased slenderness of the reference body, so that the Munk-Jones theory remains valid as the Mach number increases. Consequently, for this class of problems, the calculation of the flow is independent of Mach number, and there is obviously no need for transformation to a related incompressible flow. It is clear that this is another striking case of the validity of small perturbation methods over a range of Mach numbers including the transonic.

Let us proceed now to carry out the calculation of the flow near the body according to the Munk-Jones approximation. For any slender body, the surface boundary condition is, approximately, a specification of v_n, the velocity component normal to the transverse section of the body. Together with the abbreviated differential equation (9-6), this poses a two-dimensional potential problem in the transverse plane; say,

$$\begin{gathered} \varphi_{yy} + \varphi_{zz} = 0 \\ \varphi_n \text{ specified on body cross section} \end{gathered} \tag{9-6a}$$

Since the boundary condition in Eq. 9-6a varies from section to section, i.e. with x, let us write the solution of Eq. 9-6a as $\varphi(x; y, z)$.

Clearly, we can identify these two-dimensional potentials with $\phi(x, y, z)$ only up to an undetermined function of x:

$$\phi(x, y, z) = \varphi(x; y, z) + g(x)$$

If the function $g(x)$ can be identified, the calculation is complete. When $g(x)$ cannot be identified, the Munk-Jones theory fails and recourse must be made to a more complete theory.[6]

[6] The relationship between the Munk-Jones theory and the more complete small perturbation theory in which all terms of Eq. 9-4 are carried becomes very clear in

Fortunately, there is an important class of problems for which $g(x)$ can be shown to be at most a constant, namely, those in which ϕ_x must be antisymmetric with respect to a longitudinal plane. For example, suppose $\phi(x, y, z)$ is the perturbation potential for a flat wing at incidence. For such a case, ϕ_x must be antisymmetric with respect to the wing plane, and this can be true only if $g'(x) = 0$.

It is interesting to note that, since the present theory applies only in the vicinity of the body, we cannot make use of boundary conditions on ϕ or its derivatives at large distances from the body. This point has sometimes been overlooked in applications of the Munk-Jones theory.

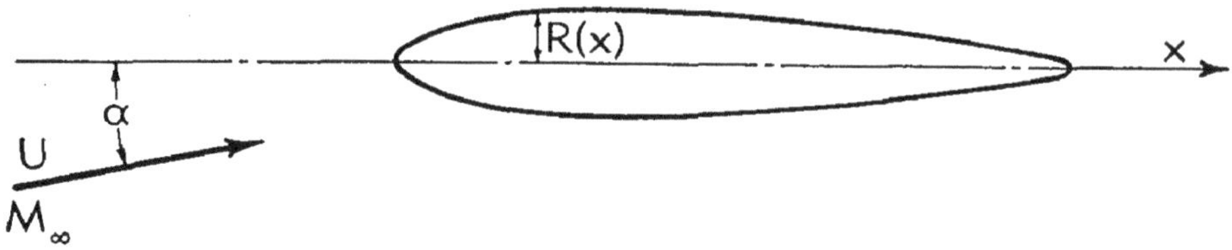

Fig. C,9. Slender body of revolution at incidence or yaw.

We shall now illustrate the Munk-Jones theory by applying it to a typical aerodynamic problem.

Example: Body of revolution at yaw. Consider the body of revolution sketched in Fig. C,9 in an unconfined stream at a small angle of attack, α. The boundary condition at the body surface is approximately

$$\phi_r(x, R, \omega) = U \cos \alpha R'(x) + U\alpha \cos \omega \tag{9-7}$$

Here we have adopted a cylindrical coordinate system x, r, ω with ω measured from the plane containing the stream velocity vector.

The first term in condition (9-7) is the same as for the body in symmetric flow, and, by superposition, our previous solution of this problem, e.g. Eq. 5-23, can be carried over to satisfy this part of the boundary condition. We therefore need only consider the new problem whose boundary condition is

$$\phi_r^{(2)}(x, R, \omega) = U\alpha \cos \omega \tag{9-8}$$

At any value of x, the solution of the plane potential problem stated in Eq. 9-6 and 9-8 is well known. We can immediately write

$$\phi^{(2)}(x, r, \omega) = -U\alpha \frac{R^2(x)}{r} \cos \omega + g(x) \tag{9-9}$$

As stated above, the requirement that $\phi_x^{(2)}$ be antisymmetric with respect to the horizontal plane tells us that $g(x)$ is at most a constant.

---

the work of Ward [*48*], at least in the supersonic case. For example, for symmetric flow problems Ward is able to identify $g(x)$ by comparing $\varphi(x; y, z)$ with an appropriate solution of Eq. 9-4. Ward's work is discussed at some length in E,7 and [*49*].

Thus, the velocities due to the angle of yaw are

$$\begin{aligned}\phi_x^{(2)} &= -U\alpha \frac{dR^2(x)}{dx}\frac{\cos\omega}{r}\\ \phi_r^{(2)} &= U\alpha R^2(x)\frac{\cos\omega}{r^2}\\ \frac{1}{r}\phi_\omega^{(2)} &= U\alpha R^2(x)\frac{\sin\omega}{r^2}\end{aligned} \tag{9-10}$$

Finally, the two terms given by Eq. 5-23 and 9-9 are to be combined, giving the total perturbation potential for the yawed body. Let u, v, and w denote the total fluid velocity components in the x, r, and ω directions, and let $\phi^{(1)}(x, r)$ denote the symmetric flow perturbation potential of Eq. 5-23; then the formulas for the velocity components are

$$\begin{aligned}u(x, r, \omega) &= U\cos\alpha + \phi_x^{(1)}(x, r) - U\alpha\frac{dR^2(x)}{dx}\frac{\cos\omega}{r}\\ v(x, r, \omega) &= \phi_r^{(1)}(x, r) + U\alpha\left[\frac{R^2(x)}{r^2} - 1\right]\cos\omega\\ w(x, r, \omega) &= U\alpha\left[1 + \frac{R^2(x)}{r^2}\right]\sin\omega\end{aligned} \tag{9-11}$$

In particular, at the body surface, these become

$$\begin{aligned}u(x, R, \omega) &= U\cos\alpha + \phi_x^{(1)}(x, R) - 2U\alpha R'(x)\cos\omega\\ v(x, R, \omega) &= UR'(x)\\ w(x, R, \omega) &= 2U\alpha\sin\omega\end{aligned} \tag{9-12}$$

We recall (cf. Eq. 5-24) that $\phi_x^{(1)}(x, R)$ is of order $\tau^2 \ln \tau$, for a slender body whose thickness ratio is τ, while the other perturbation term in $u(x, R, \omega)$ is of order $\alpha\tau$, which we shall assume to be the same order as α^2 or τ^2. Clearly, v and w in Eq. 9-12 are of the first order, i.e. α or τ. As before (Art. 5), it is advisable to carry certain second order terms in calculating the surface pressure coefficient, viz.

$$C_p(x, R) = \alpha^2 - 2\frac{\phi_x^{(1)}(x, R)}{U} + 4\alpha R'(x)\cos\omega - [R'(x)]^2 - 4\alpha^2\sin^2\omega \tag{9-13}$$

The cross-force loading at any station is found by integration of $C_p(x, R)(\rho U^2/2)\cos\omega$ around the body contour at that station; it is evident that the only contribution comes from the third term of Eq. 9-13. The result is $(\rho U^2/2)2\pi[R^2(x)]'\alpha$, in agreement with Munk's classical result [33], which was calculated directly from the rate of change of cross-wind momentum.

This slender body theory of Munk and Jones has been applied with considerable success to highly sweptback wings, wing-body combinations, and the like, as discussed in Sec. D and in VII,A, B, and C. There is considerable experimental verification, as is mentioned there; in particular, regarding the pressure distribution on yawed bodies of revolution, good agreement was found by Ackeret, Degen, and Rott [*27*].

C,10. Rotational Small Perturbation Flow. In our derivation (Art. 2) of the equation of sound propagation, we assumed explicitly that the entropy of the initial state was uniform. Nevertheless, the sound propagation equation (2-4) derived there and exploited in the succeeding articles requires no modification if the undisturbed state of the gas is one of rest involving small variations of entropy. This might be the case, for example, in a room filled with air at rest but with small differences between densities at various points, due to variations of temperature. The last of Eq. 2-1 is still correct and the small perturbation approximations lead again to Eq. 2-2, 2-3, and 2-4, where now ρ_∞ and a_∞ denote average values for the nonuniform initial state.

Our conclusion is that any small perturbation flow produced from a state of initial rest must be irrotational, to the first order, regardless of first order nonuniformities of the undisturbed gas. Obviously, by the same coordinate transformation as in Art. 3, the equation of sound propagation leads again to the equations of that article for unsteady and steady perturbations of a stream, even though the undisturbed stream carries first order variations of density and entropy, provided its velocity is uniform.

Now, although the class of flows described in the preceding paragraph is of importance in acoustics and also includes interesting aeronautical cases of bodies moving through a stratified atmosphere, it does not include steady flows with uniform stagnation enthalpy and small variations of entropy and vorticity. These are especially important, since they are the steady flows downstream of shock waves. Although the changes of velocity, density, and the other properties of the gas through a stationary shock all depend upon the strength of the wave, the stagnation enthalpy is always conserved through the shock (cf. III,B). Consequently, a steady flow that is produced from a uniform stream and contains shock waves is generally one of uniform stagnation enthalpy (i.e. isoenergic) and nonuniform entropy and vorticity (III,A,11), at least outside of boundary layers and similar viscous regions. In particular, if the shock waves are weak or are nearly straight, the variations of entropy and vorticity will be small; this type of flow occurs frequently in practical applications (cf. [*44*]). Our derivation in Art. 2 fails for these cases because, relative to the acoustical frame of reference, the initial state is not one of uniform rest, but involves small rotational velocities.

We are led, therefore, to consider flows having, initially, first order vorticity. For maximum generality it is easiest to begin with flow nearly at rest relative to our coordinate axes. Repeating the analysis of Art. 2, we arrive again at the approximate relations

$$\mathbf{q}_t = -\operatorname{grad}\left(\frac{p'}{\rho_\infty}\right)$$
$$\frac{1}{a_\infty^2} p'_t + \rho_\infty \operatorname{div} \mathbf{q} = 0 \tag{10-1}$$

where ρ_∞, a_∞ are average values, say, for the initial flow, and p', $\mathbf{q}$ include both perturbations and small initial departures from uniformity. We cannot, in this case, conclude that $\mathbf{q}$ is the gradient of a function, but only that it has the form

$$\mathbf{q}(x, y, z, t) = \operatorname{grad} \phi(x, y, z, t) + \mathbf{q}_1(x, y, z) \tag{10-2}$$

where ϕ denotes $-\int (p'/\rho_\infty)dt$ and $\mathbf{q}_1(x, y, z)$ represents the initial, first order, velocity distribution, which is, in general, rotational. Eq. 10-2 states that the velocity at any time is made up of an irrotational perturbation plus the initial velocity at the same point. Thus, it can be said that an initial small vorticity is not disturbed by acoustic perturbations, but remains fixed in the space. The differential equation for the perturbation potential ϕ follows immediately from Eq. 10-1:

$$\nabla^2\phi - \frac{1}{a_\infty^2}\phi_{tt} = -\operatorname{div} \mathbf{q}_1$$
$$p' = -\rho_\infty \phi_t \tag{10-3}$$

The corresponding steady flow equations are

$$(1 - M_\infty^2)\phi_{x'x'} + \phi_{y'y'} + \phi_{z'z'} = -(v_1)_{y'} - (w_1)_{z'}$$
$$p - p_\infty = p' = -\rho_\infty U \phi_{x'} \tag{10-4}$$

The right-hand side of the first of these equations has the form assumed by $-\operatorname{div} \mathbf{q}_1(x, y, z)$ if the flow is steady with respect to the x', y', z', t' system of coordinates defined in Eq. 3-1, as is postulated here.

Here the velocity components at any point are $U + u_1(y', z') + \phi_{x'}$, $v_1(y', z') + \phi_{y'}$, and $w_1(y', z') + \phi_{z'}$. The irrotational perturbations $\phi_{x'}$, $\phi_{y'}$, and $\phi_{z'}$, vanish far upstream. In this case we say that first order vorticity is transported by the main parallel stream. The Prandtl-Glauert affine transformation puts Eq. 10-4 into the form of Poisson's equation, familiar in potential theory. Its solution can be carried out by standard techniques.

Equations for pressure perturbation. An equivalent method of handling these problems is to return to Eq. 10-1, differentiate the second of these with respect to t, and substitute for $\mathbf{q}_t$; the result is

$$\frac{1}{a_\infty^2} p'_{tt} - \nabla^2 p' = 0 \tag{10-5}$$

Thus the variable part of the pressure (which includes the initial variations as well as time-dependent perturbations) satisfies the sound propagation equation whether there is first order vorticity or not. From the solution of Eq. 10-5 one would determine velocities by integration of Eq. 10-1, if they were required. Essentially this method has been used by Lighthill [*35*] to study the rotational unsteady flow behind a diffracting shock wave (see below).

If our frame of reference is transported with a uniform speed U, Eq. 10-5 becomes

$$\frac{1}{a_\infty^2}(p'_{t't'} + 2Up'_{x't'} + U^2 p'_{x'x'}) - \nabla'^2 p' = 0 \tag{10-6}$$

The corresponding steady flow equation would be

$$(1 - M_\infty^2)p'_{x'x'} + p'_{y'y'} + p'_{z'z'} = 0 \tag{10-7}$$

In a subsequent paper [*36*] on shock-wave boundary layer interaction, Lighthill generalized this subject, working out the equations for the steady flow case of small perturbations of an initially parallel stream having large vorticity. Later, the same generalization was made by Chang and Chu [*37*].

Let the undisturbed parallel stream have the speed $U(y, z)$, density $\rho_\infty(y, z)$, speed of sound $a_\infty(y, z)$, and pressure p_∞, a constant. We suppose this stream to be perturbed by small steady disturbances u', v', w', ρ', a', and p'. Our general equations (2-1) are, for steady motion,

$$\rho \mathbf{q} \cdot \nabla \mathbf{q} + \operatorname{grad} p = 0 \tag{10-8}$$

$$\mathbf{q} \cdot \nabla p + a^2 \rho \operatorname{div} \mathbf{q} = 0 \tag{10-9}$$

Upon substituting, and by neglecting second order quantities, we have

$$\rho_\infty(Uu'_x + v'U_y + w'U_z) + p'_x = 0 \tag{10-10}$$

$$\rho_\infty Uv'_x + p'_y = 0, \qquad \rho_\infty Uw'_x + p'_z = 0 \tag{10-11}$$

$$Up'_x + a_\infty^2 \rho_\infty(u'_x + v'_y + w'_z) = 0 \tag{10-12}$$

Now u'_x can be eliminated between Eq. 10-10 and 10-12; also it is convenient to notice that $U^2\rho_\infty = \gamma p_\infty M_\infty^2(y, z)$. The equations then appear in the forms

$$[1 - M_\infty^2(y, z)]\frac{p'_x}{p_\infty} = \gamma M_\infty^2(y, z)\left[\left(\frac{v'}{U}\right)_y + \left(\frac{w'}{U}\right)_z\right] \tag{10-13}$$

and

$$\left(\frac{v'}{U}\right)_x = -\frac{1}{\gamma M_\infty^2}\frac{p'_y}{p_\infty}, \qquad \left(\frac{w'}{U}\right)_x = -\frac{1}{\gamma M_\infty^2}\frac{p'_z}{p_\infty} \tag{10-14}$$

Further, eliminating v' and w' between these equations by differentiation, we have the equation satisfied by the pressure perturbation p' in this case:

$$[1 - M_\infty^2(y, z)]p'_{xx} - 2\frac{M_{\infty y}(y, z)}{M_\infty(y, z)}p'_y - 2\frac{M_{\infty z}(y, z)}{M_\infty(y, z)}p'_z + p'_{yy} + p'_{zz} = 0 \quad (10\text{-}15)$$

It is easily seen how Eq. 10-15 reduces to Eq. 10-7 in the special case when M_∞ is constant except for first order variations.

Lighthill obtains solutions of Eq. 10-15, in its two-dimensional form, and applies them to represent the case of a compressible boundary layer perturbed by an incident shock wave. In general, the flow at every point depends upon the whole profile of the undisturbed stream, and not just on local values as in the small vorticity case.

Lighthill's potential function. Lighthill [38] has suggested another technique which, at least formally, makes certain steady, rotational, small perturbation cases equivalent to the irrotational. We return to the consideration of nearly uniform steady streams; for these cases the equations of motion are simply

$$\frac{U}{a_\infty^2}p'_x + \rho_\infty \operatorname{div} \mathbf{q} = 0 \quad (10\text{-}16)$$

and

$$U\mathbf{q}_x = -\operatorname{grad}\left(\frac{p'}{\rho_\infty}\right) \quad (10\text{-}17)$$

or

$$Uu'_x = -\frac{p'_x}{\rho_\infty}, \qquad Uv'_x = -\frac{p'_y}{\rho_\infty}, \qquad Uw'_x = -\frac{p'_z}{\rho_\infty} \quad (10\text{-}18)$$

where $U + u'$, v', w' are the components of the velocity $\mathbf{q}$, and now U is constant.

Now, it is easily seen from Eq. 10-18 that the vector $(-p'/\rho_\infty U, v', w')$ is irrotational provided that $w'_y = v'_z$, as is always the case in plane or axially symmetric flow, for example. Let us fix our attention on these two cases, letting x, y now denote either plane-Cartesian or cylindrical coordinates and u', v' the corresponding velocity components. The third of Eq. 10-18 is not needed; the second is identically satisfied if we write

$$\left(\frac{-p'}{\rho_\infty U}, v'\right) = \operatorname{grad} \chi(x, y) \quad (10\text{-}19)$$

Combining Eq. 10-16, 10-19, and the first of 10-18, we find the differential equation satisfied by χ. For plane flow,

$$(1 - M_\infty^2)\chi_{xx} + \chi_{yy} = 0 \quad (10\text{-}20)$$

and for axially symmetric flow

$$(1 - M_\infty^2)\chi_{xx} + \frac{1}{y}(y\chi_y)_y = 0 \quad (10\text{-}21)$$

Thus χ satisfies the Prandtl-Glauert equation and the pressure is found immediately as one of its derivatives. Lighthill has employed this technique in the paper referred to above [*38*].

Example: Shock diffraction. As an example of a practical case of rotational small perturbation flow, we consider here in brief fashion the case of the diffraction of a plane shock wave at a small corner. This problem has been treated by Lighthill [*35*], whose method will be used here. A previous treatment by Bargmann [*39*] was confined to diffraction of weak shocks, where the flow could be considered everywhere irrotational.

The physical situation is shown in Fig. C,10a. An originally plane shock S, propagating with speed U_S into still air at density ρ_0 and pres-

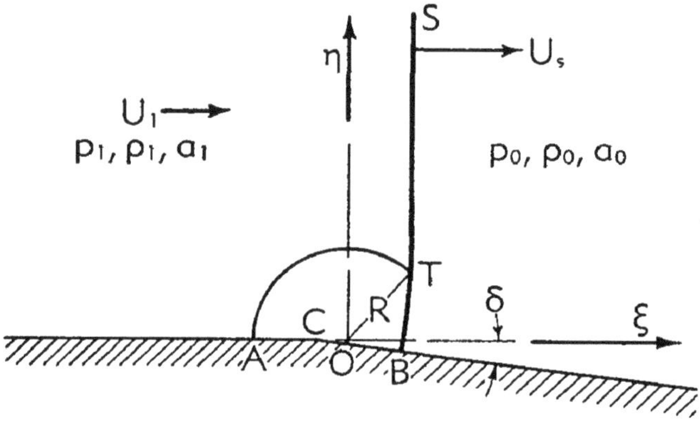

Fig. C,10a. Diffraction of plane shock wave.

sure p_0, passes over a corner whose edge is parallel to the shock front at C. In passing C it is partially diffracted, as sketched. The conditions behind the original shock will be designated U_1, ρ_1, p_1, and a_1. The Mach number U_1/a_1 may be subsonic or supersonic, depending on the strength of S We consider here only the subsonic case. If the corner angle δ is small, conditions in region R will differ from U_1, ρ_1, p_1, and a_1, by small perturbations, of order δ, which we shall designate u', v', ρ', p', and a'. The portion of the shock S between T and B is known as the diffracted shock or the "Mach shock." This part of the shock wave is slightly distorted from its original plane form; thus, the flow behind it must be expected to involve first order variations in entropy and vorticity. On the other hand, the "reflected wave" AT is a weak wave propagating into a uniform region, so that the flow directly behind it, according to the deductions at the beginning of this article, is irrotational to the first order.

Now, the "triple point" T has progressed from the corner C to its present location, and in doing so has traced the boundary between the air affected directly by the Mach shock BT and that affected by the reflected shock AT. Due to the motion at substantially the speed U_1, this boundary now appears in some such position as line OT in Fig. C,10a. In other words, we have two adjoining regions, $ACOT$ of irrotational perturbation flow and OBT of rotational perturbation flow. To treat this problem, Lighthill used the method of the pressure perturbation, for which a single equation (10-5) or (10-6) suffices for the entire disturbed region R. In our present notation, choosing a ξ, η, t coordinate system moving with speed U_1 in the $+\xi$ direction, we have the case of small perturbations of a body of air nearly at rest. The appropriate equation is (10-5), i.e.

$$\frac{1}{a_1^2} p'_{tt} = p'_{\xi\xi} + p'_{\eta\eta} \tag{10-22}$$

The boundary conditions are: (1) the vanishing of p' along the arc AT (cf. [*35*, p. 458]); (2) the conditions imposed by the impermeable wall along AC and CB respectively; and (3) the shock-wave conditions along BT. Consistent with a first order theory, it is permissible to apply conditions 2 and 3 along approximate locations, viz.

$$\begin{aligned} v'(\xi, 0) &= 0 \text{ for } \xi < -U_1 t \\ &= -U_1\delta \text{ for } -U_1 t < \xi < (U_S - U_1)t \end{aligned} \tag{10-23}$$

and the shock-wave conditions along the straight line $\xi = (U_S - U_1)t$. We shall consider the latter below.

The problem can be simplified substantially by observing that it has a kind of conical symmetry in a ξ, η, t space. In its simplest terms, it follows from the absence of any characteristic length that the configuration sketched in Fig. C,10a at time t can differ from that at any other time only in absolute scale. This means that all the physical variables of the problem must actually be functions of two variables, ξ/t and η/t, or, in dimensionless form, ξ/a_1t and $\eta/a_1 t$. Let these dimensionless variables be called x and y, as in Lighthill's paper [*35*]; their introduction into Eq. 10-22 leads to

$$a_1^2(p'_{xx} + p'_{yy}) = \left(x\frac{\partial}{\partial x} + y\frac{\partial}{\partial y} + 1\right)(xp'_x + yp'_y) \tag{10-24}$$

Finally, this equation is transformed into Laplace's equation in two dimensions by means of a transformation used by A. Busemann and others in conical flow problems.

The boundary condition (10-23) gives us a condition on p', by virtue of the first of Eq. 10-1, i.e. $\rho_1 v'_t = -p'_\eta$. When expressed in the x, y plane and compared with Eq. 10-23, this states that $p'_y(x, 0) = 0$ for $x < -U_1/a_1$ and for $-U_1/a_1 < x < (U_S - U_1)/a_1$. At $x = -U_1/a_1$ there is a singularity, and Lighthill introduces another condition there to assure the proper jump in the value of v'.

In order to apply the shock-wave conditions along the line $x = (U_S - U_1)/a_1 = k$, Lighthill calculates the normal velocity of the shock BT in terms of its position $x = k + f(y)$, making use again of the conical property. Since $f(y)$ can be presumed to be small for small δ, he is able to approximate the shock conditions at $x = k$, obtaining finally[7]

$$k^2 p'_x + kyp'_y = p'_x - Ayp'_y + Bky^{-1}p'_y \text{ on } x = k \qquad (10\text{-}25)$$

where A and B are functions of U_S/a_0, the shock Mach number.

Having transformed the problem to one involving Laplace's equation, Lighthill solves it by complex variable methods. His results include pres-

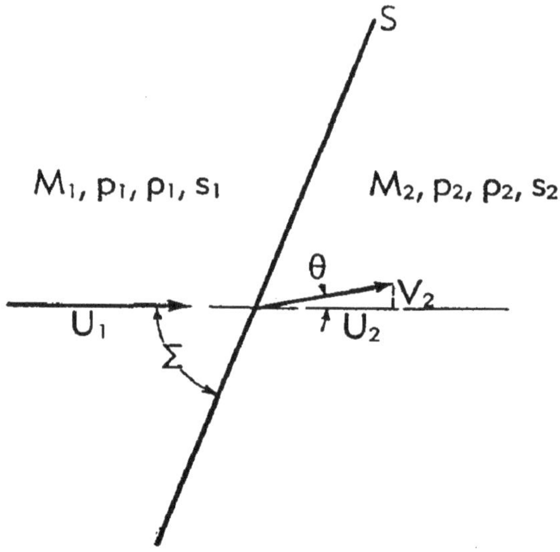

Fig. C,10b. Oblique shock wave.

sure distributions along the wall and the configurations of the diffracted and reflected shocks.

More recently the same problem has been studied by Ting and Ludloff [*40*], who have attacked Eq. 10-22 directly, i.e. without transformation to Laplace's equation. They calculate the field of pressures p' and, from this and knowledge of the entropy distribution in OBT, the field of densities. The entropy distribution follows from the knowledge of shock strength (pressure rise) along BT, the fact that this shock has expanded "conically" from C, and from our principle (Eq. 10-2) that first order vorticity remains fixed.

In a recent report [*41*], White presents the results of interferometric experiments on shock-wave diffraction. He shows that the agreement with the first order theories of Lighthill and Ting-Ludloff is good, for angles δ up to the order of 10°.

Perturbations of shock waves. The shock-diffraction problem involves, along the Mach shock BT, small perturbations of a strong shock wave.

[7] This process is analogous to our treatment of perturbed shock waves below in this article.

This problem is typical of a number of problems that have arisen, involving slightly perturbed strong shocks. Let us consider briefly how these may be attacked.

The Rankine-Hugoniot shock wave equations (III,D,1) are statements of the conservation of mass, momentum, and energy across the wave. Consider the basic case of the oblique shock shown in Fig. C,10b. S denotes the shock wave, making an angle Σ with the upstream flow direction. On passing through the shock, the flow is turned through an angle θ and its speed is reduced, with accompanying increases of pressure, temperature, and entropy. For our purposes it is convenient to calculate the velocity components parallel and perpendicular to the direction of upstream flow. There is a form of the shock wave equations that is expressed in these terms—often called the equation of the shock polar (H,7, Eq. 7-1; also III,E,2)—viz.

$$V_2^2 = \frac{(U_1 - U_2)^2(U_1U_2 - a^{*2})}{\frac{2}{\gamma + 1} U_1^2 - U_1U_2 + a^{*2}} \equiv F_1(U_1, U_2) \tag{10-26}$$

where a^* denotes the speed of sound for unit Mach number, a parameter characterizing the total enthalpy of the flow, which is conserved through the shock. Together with Eq. 10-26, we can write other expressions relating conditions before and after the oblique shock; for example, the pressure p_2 is given by

$$p_2 = p_1 + \rho_1 U_1(U_1 - U_2) \tag{10-27}$$

which can be put into the form

$$p_2 = F_2(U_1, U_2) \tag{10-28}$$

and the shock-wave angle Σ is given by

$$\tan \Sigma = \frac{U_1 - U_2}{V_2} = F_3(U_1, U_2) \tag{10-29}$$

The entropy behind the wave can be similarly expressed:

$$s_2 - s_1 = F_4(U_1, U_2) \tag{10-30}$$

It is clear that the conditions behind the shock, like the shock angle, can be determined only when one of the variables U_2, V_2, p_2, s_2, . . . , or Σ, is specified. In the formulas (10-26) to (10-30), we have selected U_2 as the downstream parameter identifying the solution; naturally this choice is quite arbitrary.

Now, the first order effects of a small change of U_1, say u_1, are most easily evaluated by differentiation of these functions. Thus, the increments resulting from this u_1 are related as follows:

$$
\begin{aligned}
v_2' &= \frac{1}{2V_2}\left(\frac{\partial F_1}{\partial U_1} u_1 + \frac{\partial F_1}{\partial U_2} u_2'\right) \\
p_2' &= \left(\frac{\partial F_2}{\partial U_1} u_1 + \frac{\partial F_2}{\partial U_2} u_2'\right) \\
\sigma' &= \cos^2 \Sigma \left(\frac{\partial F_3}{\partial U_1} u_1 + \frac{\partial F_3}{\partial U_2} u_2'\right) \\
s_2' &= \frac{\partial F_4}{\partial U_1} u_1 + \frac{\partial F_4}{\partial U_2} u_2'
\end{aligned}
\tag{10-31}
$$

etc.

In general, the increment u_1 is accompanied by an increment v_1, as in Fig. C,10c. Then our formulas (10-31) are still correct, but they refer

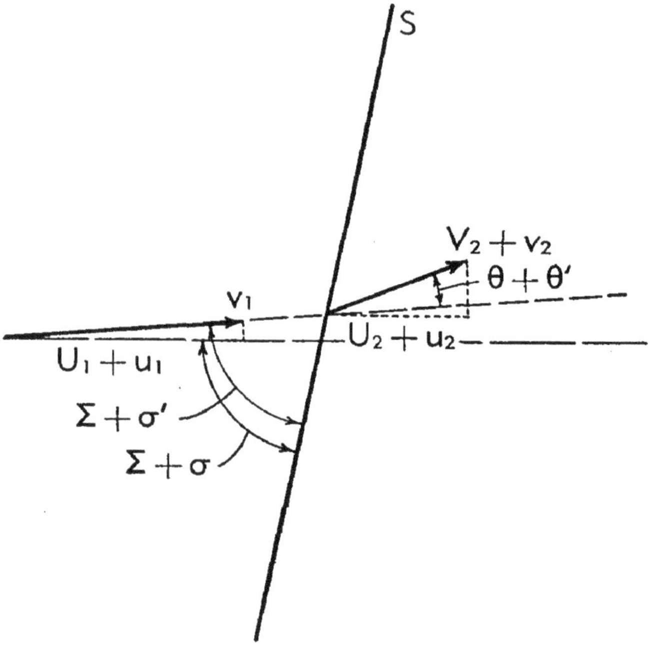

Fig. C,10c. Perturbed oblique shock wave.

to increments u_1, u_2', and v_2' measured parallel and perpendicular to the new upstream velocity vector. To compute components aligned with our original undisturbed stream direction, we must add certain terms, viz.

$$
\begin{aligned}
v_2 &= v_2' + \frac{U_2}{U_1} v_1, \qquad u_2 = u_2' - \frac{V_2}{U_1} v_1 \\
\sigma &= \sigma' + \frac{v_1}{U_1}
\end{aligned}
\tag{10-32}
$$

where the quantities ( )′ are those calculated in Eq. 10-31, and those without primes are defined with relation to the undisturbed upstream

Table C,10. Coefficients of Eq. 10-33. $f_i = A_i u_1 + B_i v_1 + C_i u_2$.

i	f_i	A_i	B_i	C_i
1	v_2	$\frac{\cot^3 \Sigma}{2}\left[\frac{\frac{2}{\gamma+1}U_1U_2}{U_1U_2 - a^{*2}} \times \left(1 + \frac{U_1^2 - a^{*2}}{U_1U_2 - a^{*2}}\right) - 2\right]$	$\frac{V_2}{U_1}\frac{\cot^3 \Sigma}{2}\left[\frac{\frac{2}{\gamma+1}U_1^2}{U_1U_2 - a^{*2}} \times \left(\frac{U_1^2 - U_1U_2}{U_1U_2 - a^{*2}} - 2\right) + 2\right] + \frac{U_2}{U_1}$	$\frac{U_1}{V_2}B_1 - \frac{U_2}{V_2}$
2	σ	$\frac{\cos^2 \Sigma}{V_2}(1 - A_1 \tan \Sigma)$	$\frac{\cos^2 \Sigma \tan \Sigma}{V_2}\left(\frac{U_2}{U_1} - B_1\right) + \frac{\sin^2 \Sigma}{U_1}$	$-\frac{\cos^2 \Sigma}{V_2}(1 + C_1 \tan \Sigma)$
3	p_2'	$\frac{p_1 U_1}{a_1^2}\frac{1 - M_1^2}{M_1^2}\left(\frac{p_2}{p_1} - 1\right)$	$-\frac{\gamma p_1}{a_1^2}V_2$	$-\frac{\gamma p_1}{a_1^2}U_1$
4	s_2'	$\frac{\gamma+1}{2\gamma}\mathcal{R}\frac{p_1}{p_2}\frac{\left(\frac{p_2}{p_1} - 1\right)^2}{1 + \frac{\gamma-1}{2}M_1^2 \sin^2 \Sigma} \times \left(\frac{\gamma+1}{2U_1}\frac{a^{*2}}{a_1^2} + A_2 \cot \Sigma\right)$	$\frac{\gamma+1}{2\gamma}\mathcal{R}\frac{p_1}{p_2}\frac{\left(\frac{p_2}{p_1} - 1\right)^2}{1 + \frac{\gamma-1}{2}M_1^2 \sin^2 \Sigma} \times \cot \Sigma \left(B_2 - \frac{1}{U_1}\right)$	$\frac{\gamma+1}{2\gamma}\mathcal{R}\frac{p_1}{p_2}\frac{\left(\frac{p_2}{p_1} - 1\right)^2}{1 + \frac{\gamma-1}{2}M_1^2 \sin^2 \Sigma} \times C_2 \cot \Sigma$

vector, as shown in Fig. C,10c. Naturally, the expressions for p_2', s_2', and similar scalar quantities in terms of u_1 and u_2' remain unchanged, but they can be expressed instead in terms of u_1 and u_2. Combining the equations above, we arrive at the forms:

$$\begin{aligned} v_2 &= A_1u_1 + B_1v_1 + C_1u_2 \\ \sigma &= A_2u_1 + B_2v_1 + C_2u_2 \\ p_2' &= A_3u_1 + B_3v_1 + C_3u_2 \\ s_2' &= A_4u_1 + B_4v_1 + C_4u_2 \end{aligned} \tag{10-33}$$

The A_i, B_i, and C_i of Eq. 10-33 have been calculated from the shock-wave equations and are tabulated in Table C,10. It is clear that the selection of u_2 as the defining downstream parameter has been arbitrary, and any of the others can be used equally well.

Generally, conditions downstream of the shock will be determined by a differential equation, such as Eq. 10-3, 10-5, or one of the other forms of this article or of Art. 11 below, and by boundary conditions on the velocity components or pressure at the boundaries of the downstream flow region, as in the shock-diffraction problem above. In such problems, Eq. 10-33 are to be used to formulate additional boundary conditions at the approximate shock-wave location. This process of solution is analogous to that used in Sec. H of this volume, but is greatly simplified by the small perturbation approximation.

C,11. The Stream Function in Rotational Steady Flow. In plane or axially symmetric steady flow, the equation of continuity

$$\operatorname{div}(\rho\mathbf{q}) = 0 \tag{11-1}$$

is identically satisfied by writing

$$\frac{\rho}{\rho_1}u = y^{-\epsilon}\Psi_y, \qquad \frac{\rho}{\rho_1}v = -y^{-\epsilon}\Psi_x \tag{11-2}$$

where ρ_1 is a reference density, a constant, and ϵ is defined as having the value 0 for plane flow and 1 for axially symmetric flow. To represent small deviations from a uniform steady stream, we write

$$\Psi = \frac{1}{1+\epsilon}Uy^{1+\epsilon} + \psi \tag{11-3}$$

Thus ψ is the stream function of the perturbations and the initial first order nonuniformities of the stream. In this case Eq. 11-2 reads

$$\begin{aligned} \left(1 + \frac{\rho'}{\rho_1}\right)(U + u') &= U + y^{-\epsilon}\psi_y \\ \left(1 + \frac{\rho'}{\rho_1}\right)v' &= -y^{-\epsilon}\psi_x \end{aligned} \tag{11-4}$$

or, to the first order,

$$u' + \frac{\rho'}{\rho_1} U = y^{-\epsilon}\psi_y$$
$$v' = -y^{-\epsilon}\psi_x \tag{11-5}$$

We must now express ρ'/ρ_1 for this type of flow. For this purpose we make use of the equation of state:

$$\frac{p}{p_1} = \left(\frac{\rho}{\rho_1}\right)^{\gamma} e^{\frac{s-s_1}{c_v}} \tag{11-6}$$

where s denotes the specific entropy. We also have the energy equation for steady isoenergic flow:

$$\frac{1}{2}(u^2 + v^2) + \frac{\gamma}{\gamma - 1}\frac{p}{\rho} = h^0 = \text{const}$$
$$= \frac{1}{2}U^2 + \frac{\gamma}{\gamma - 1}\frac{p_1}{\rho_1} \tag{11-7}$$

It will be noted that at this point we have specifically assumed that the stream is one of uniform stagnation enthalpy. It should also be emphasized that the uniform reference conditions U, p_1, ρ_1, and s_1 may be fictitious; nevertheless, they are convenient reference conditions, and are completely defined, for given h^0, by the equations above.

When Eq. 11-6 and 11-7 are combined and second order terms are neglected, the result is

$$\frac{\rho}{\rho_1} = 1 + \frac{\rho'}{\rho_1} = 1 - M_1\frac{u'}{a_1} - \frac{s - s_1}{\mathcal{R}} \tag{11-8}$$

where M_1 denotes $U\sqrt{\rho_1/\gamma p_1}$, the Mach number corresponding to the reference conditions, and $\mathcal{R}$ is the gas constant $c_p - c_v$. Thus, after substitution of Eq. 11-8 in Eq. 11-5, we have

$$(1 - M_1^2)u' - \frac{U}{\mathcal{R}}(s - s_1) = y^{-\epsilon}\psi_y$$
$$v' = -y^{-\epsilon}\psi_x \tag{11-9}$$

Moreover, the entropy term can be expressed in terms of velocity components, for Crocco's theorem (III,A,12) for this type of flow is

$$\Omega = v'_x - u'_y = \frac{p}{\rho_1\mathcal{R}} y^{\epsilon} \frac{ds}{d\Psi} \tag{11-10}$$

Thus

$$s - s_1 = \rho_1\mathcal{R}\int (v'_x - v'_y)y^{-\epsilon}\frac{d\Psi}{p}$$
$$= \mathcal{R}\int (v'_x - u'_y)(udy - vdx)\frac{\rho}{p} \tag{11-11}$$

To the first order, this is

$$s - s_1 = \mathcal{R}U \frac{\rho_1}{p_1} \int (v'_x - u'_y)dy \tag{11-12}$$

The differential equation satisfied by ψ is now obtained by substituting Eq. 11-12 in Eq. 11-9, differentiating to find u'_y and v'_x, and calculating the vorticity Ω:

$$\begin{aligned} (1 - M_1^2)u'_y - \gamma M_1^2(v'_x - u'_y) &= \psi_{yy}y^{-\epsilon} - \epsilon\psi_y y^{-1-\epsilon} \\ v'_x &= -\psi_{xx}y^{-\epsilon} \end{aligned} \tag{11-13}$$

from which

$$u'_y = \frac{\psi_{yy} - \epsilon y^{-1}\psi_y - \gamma M_1^2 \psi_{xx}}{1 + (\gamma - 1)M_1^2} y^{-\epsilon} \tag{11-14}$$

and finally

$$(1 - M_1^2)\psi_{xx} + \psi_{yy} - \frac{\epsilon}{y}\psi_y = -\{1 + (\gamma - 1)M_1^2\}y^{\epsilon}\Omega \tag{11-15}$$

This is the equation for the stream function $\psi(x, y)$, defined in Eq. 11-3, in the case of rotational flow. As has already been pointed out in Art. 10, the vorticity is to be considered as being carried with the main stream; thus, Ω in Eq. 11-15 is to be taken as a function of y alone. Eq. 11-15 can be solved (when $M_1^2 < 1$) by applying the Prandtl-Glauert affine transformation, which puts it into the form of Poisson's equation (see below). This can be attacked by standard methods of potential theory, and the velocity components and pressure can be calculated without difficulty from Eq. 11-5, 11-6, and 11-8, since in this approximation s is a function of y alone. This method has successfully been applied to practical flow problems by Adams [*42*] and Pai [*43*].

Naturally, Eq. 11-15 is also useful in certain irrotational flow problems, for which, of course, the right-hand side is replaced by zero. The small perturbation stream function was used in this way by Goldstein and Young [*7*].

It might be mentioned that, except in the small perturbation case considered here, the conventional stream function defined in Eq. 11-2 is not a particularly useful device in rotational compressible flow problems. This has led Crocco to define a different stream function, which differs from Ψ by a factor involving the entropy (III,A,12). Crocco worked out the differential equation satisfied by his new function, and it can be verified that Eq. 11-15 is consistent with it, in the small perturbation case.

The choice of the method used for small perturbation rotational flows will be governed by the type of boundary conditions available. When these are prescribed for velocity components, the present method or that of Eq. 10-3 may be most convenient. When pressure conditions are known, one of the other procedures of Art. 10 will probably be best.

Sears [45] has shown how, for subsonic flow, Eq. 11-15 can be transformed by the usual Prandtl-Glauert affine transformation, Eq. 3-6, to give a related, incompressible, rotational flow. His results are as follows: the stream function of the compressible flow is

$$\Psi(x, y) = \frac{1}{1+\epsilon} Uy^{1+\epsilon} + \psi(x, y) = \frac{1}{1+\epsilon} Uy^{1+\epsilon} + \beta_1^{-1-\epsilon} f(x, \beta_1 y) \quad (11\text{-}16)$$

where β_1^2 denotes $1 - M_1^2$, and the stream function of the related incompressible flow is

$$\Psi_{\text{inc}}(\xi, \eta) = \frac{1}{1+\epsilon} U\eta^{1+\epsilon} + f(\xi, \eta) \quad (11\text{-}17)$$

The equation satisfied by $f(\xi, \eta)$ is

$$f_{\xi\xi} + f_{\eta\eta} - \frac{\epsilon}{\eta} f_\eta = -\frac{1}{\beta_1} [1 + (\gamma - 1)M_1^2]\eta^\epsilon \Omega\left(\frac{\eta}{\beta_1}\right) \quad (11\text{-}18)$$

If $\Psi(x, y)$ describes the flow past a slender body $y = Y(x)$, the related flow described by $\Psi_{\text{inc}}(\xi, \eta)$ is past a more slender body, $\eta = \beta_1 Y(\xi)$, immersed in an incompressible fluid stream whose vorticity is defined by Eq. 11-18.

Perturbation of a normal shock wave. As a simple example of the utility of the stream function in rotational flow, and also to illustrate the use of our shock-perturbation relations (Art. 10), we consider here the two-dimensional case of a nominally normal shock wave in a channel, disturbed by small velocities originating in the supersonic region (Fig. C,11). These disturbances are described by a two-dimensional perturbation potential $\phi(x, y)$. As is well known, such perturbations are constant along the Mach lines, as sketched. Where these lines intersect the shock, the shock will be distorted slightly, in accord with Eq. 10-26, 10-27, and 10-29 above. We assume that the boundaries of the flow are such as to fix the location of the normal shock wave and that they are far away from the limited region of disturbance of the shock. We can assume, then, that the effects of the disturbance are local. In particular, the pressure perturbations in the subsonic downstream region must die out at large distances from the region sketched in Fig. C,11.

For this case, for which $\Sigma = 90°$ in the notation of Art. 10 and Fig. C,10b, the shock-perturbation formulas are particularly simple. For example (cf. Eq. 10-31 to 10-33),

$$u_2 = u_2' = -\frac{U_2}{U_1} u_1$$

$$\sigma = \frac{v_1 - v_2}{U_1 - U_2} \quad (11\text{-}19)$$

$$s_2' = \frac{2\gamma\mathcal{R}(M_1^2 - 1)^2}{2\gamma M_1^2 - (\gamma - 1)} \frac{u_1}{U_1}$$

It will be noticed that both u_2 and s'_2 are dependent upon u_1 only. This permits considerable simplification of the calculation, as will be seen below.[8]

Let us adopt the stream function method to solve this problem. The differential equation to be solved is then Eq. 11-15 above, with $\epsilon = 0$ for the two-dimensional case. In this problem, the quantities with subscripts 1 in Eq. 11-1 to 11-8 are to be identified with those carrying the subscript 2 in Art. 10. Let us adopt a coordinate system whose y

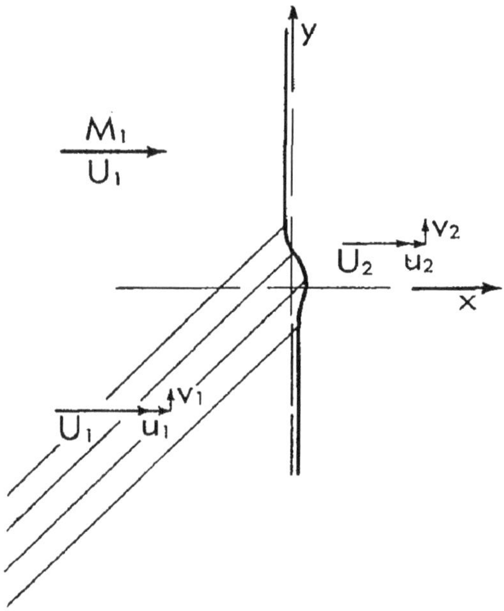

Fig. C,11. "Normal" shock wave perturbed by small, steady effects originating in supersonic stream.

axis coincides with the undisturbed normal shock. We adopt the notation of Art. 10, so that the subscript 2 denotes quantities downstream of the shock and u, v and the primed quantities denote perturbations. The values given by Eq. 10-33 or 11-19 are then boundary values to be satisfied at $x = 0$. Thus, we write Eq. 11-15 as

$$\beta_2^2\psi_{xx} + \psi_{yy} = -[1 + (\gamma - 1)M_2^2]\Omega(y) = C\Omega(y) \tag{11-20}$$

and Eq. 11-5 as

$$\begin{aligned} u_2 + \frac{\rho'_2}{\rho_2} U_2 &= \psi_y \\ v_2 &= -\psi_x \end{aligned} \tag{11-21}$$

An appropriate solution of Eq. 11-20, obtained from classical potential theory with the aid of the Prandtl-Glauert transformation as de-

[8] This simplification occurs not only at $\Sigma = 90°$ but also over a range of Σ near 90° where U_2 is of the same order of magnitude as the perturbation velocities.

scribed above, is

$$\psi(x, y) = -\frac{C}{4\pi\beta_2}\int_{-\infty}^{\infty}\int_{-\infty}^{\infty}\Omega(\bar{y})\ln\left[(y-\bar{y})^2+\frac{(x-\bar{x})^2}{\beta_2^2}\right]d\bar{x}d\bar{y}$$
$$+\frac{1}{2\pi}\int_{-\infty}^{\infty}g(\bar{y})\tan^{-1}\left(\beta_2\frac{y-\bar{y}}{x}\right)d\bar{y} \quad (11\text{-}22)$$

where $g(\bar{y})$ is a function to be determined.

Thus, by differentiation according to Eq. 11-9,

$$\beta_2^2u_2 - \frac{U_2}{\mathcal{R}}s_2' = -\frac{C\beta_2}{2\pi}\int_{-\infty}^{\infty}\int_{-\infty}^{\infty}\Omega(\bar{y})\frac{(y-\bar{y})d\bar{x}d\bar{y}}{\beta_2^2(y-\bar{y})^2+(x-\bar{x})^2}$$
$$+\frac{\beta_2 x}{2\pi}\int_{-\infty}^{\infty}g(\bar{y})\frac{d\bar{y}}{\beta_2^2(y-\bar{y})^2+x^2} \quad (11\text{-}23)$$

$$v_2 = \frac{C}{2\pi\beta_2}\int_{-\infty}^{\infty}\int_{-\infty}^{\infty}\Omega(\bar{y})\frac{(x-\bar{x})d\bar{x}d\bar{y}}{\beta_2^2(y-\bar{y})^2+(x-\bar{x})^2}$$
$$+\frac{\beta_2}{2\pi}\int_{-\infty}^{\infty}g(\bar{y})\frac{(y-\bar{y})d\bar{y}}{\beta_2^2(y-\bar{y})^2+x^2} \quad (11\text{-}24)$$

In the limit $x \to 0$, the second integral in Eq. 11-23 becomes $\frac{1}{2}g(y)$ and that in Eq. 11-24 becomes

$$-\frac{1}{2\pi\beta_2}\mathcal{P}\int_{-\infty}^{\infty}g(\bar{y})\frac{d\bar{y}}{y-\bar{y}} \quad (11\text{-}25)$$

where the symbol $\mathcal{P}$ denotes Cauchy's principal value.

Crocco's theorem, Eq. 11-10, gives us the vorticity in terms of the entropy, $s_2(y)$, of which the variable part is tabulated in Eq. 11-19; we find

$$\Omega(y) = \frac{p_2}{\rho_2\mathcal{R}}\frac{1}{U_2}\frac{ds_2'}{dy} = \frac{p_2}{\rho_2U_1U_2}\frac{2\gamma(M_1^2-1)^2}{2\gamma M_1^2-(\gamma-1)}(u_1)_y \quad (11\text{-}26)$$

where $(u_1)_y$ is the y derivative of the function u_1 immediately before the shock.

Combining Eq. 11-23, 11-24, and 11-26 with the boundary values of Eq. 11-19, we find our problem solved, for Eq. 11-23, in the limit $x \to 0$, has the form

$$\tfrac{1}{2}g(y) = \text{const } u_1 + [\text{integral involving } (u_1)_y]$$

Thus $g(y)$ can be considered known in terms of the given upstream disturbance, and consequently u_2 and v_2 can be calculated anywhere in the downstream region. In particular, all of the interesting quantities at the shock, including the shock angle σ', can be evaluated.

Adams [*42*] has used this procedure to determine the configurations assumed by a "normal" shock wave under the influence of typical sinusoidal and "roof-top" disturbances upstream.

C,12. Unsteady and Periodic Flow Problems. Fundamental Solutions. The fundamental equations of small perturbation flow in a gas at rest are Eq. 2-3 and 2-4, i.e.

$$\phi_{tt} = a_\infty^2 \nabla^2 \phi \tag{12-1}$$

$$\rho_\infty \phi_t = -p' \equiv p - p_\infty \tag{12-2}$$

It is easily verified that Eq. 12-1 is satisfied by functions of the form

$$\phi(r, t) = \frac{1}{r}[f(r - a_\infty t) + g(r + a_\infty t)] \tag{12-3}$$

where r denotes the spherical radius from the origin, $\sqrt{x^2 + y^2 + z^2}$. These solutions are of the form representing wavelike disturbances propagating outward or inward with speed a_∞, their amplitudes varying as $1/r$ as they progress.

Küssner [46] noted that Eq. 12-1 is invariant with respect to the Lorentz transformation, and that this fact leads immediately to the solution appropriate for disturbances moving with respect to the undisturbed fluid with velocity U, say, provided $U < a_\infty$. Let β^2 denote $1 - U^2/a_\infty^2$ as before; then the Lorentz transformation is

$$x = \frac{X + UT}{\beta}, \qquad y = Y,$$

$$z = Z, \qquad t = \frac{T + \dfrac{UX}{a_\infty^2}}{\beta} \tag{12-4}$$

The differential equation, as has been mentioned, is invariant:

$$\phi_{TT} = a_\infty^2(\phi_{XX} + \phi_{YY} + \phi_{ZZ}) \tag{12-5}$$

while the solutions given in Eq. 12-3 are immediately transformed to the new coordinate system by noting that

$$r = \left[\left(\frac{X + UT}{\beta}\right)^2 + Y^2 + Z^2\right]^{\frac{1}{2}} \tag{12-6}$$

Thus the singularity of these solutions appears at the point

$$X = -UT, \qquad Y = 0, \qquad Z = 0.$$

The general solution Eq. 12-3, in terms of the new variables, assumes the following form, where we write ϖ^2 for $Y^2 + Z^2$ and M_∞ for U/a_∞:

$$\begin{aligned}\phi = [(X + UT)^2 + \beta^2\varpi^2]^{-\frac{1}{2}}[f(\sqrt{(X + UT)^2 + \beta^2\varpi^2} - a_\infty T - M_\infty X) \\ + g(\sqrt{(X + UT)^2 + \beta^2\varpi^2} + a_\infty T + M_\infty X)]\end{aligned} \tag{12-7}$$

This solution is particularly useful to describe the unsteady flow due to moving wings or other bodies. Let us interpret the coordinates X, Y, Z, T as ordinary Cartesian distances and time, stationary with respect to a body of gas nominally at rest. This we can do in view of the invariance of the differential equation (12-5). We then see that Küssner's device has succeeded in giving us a class of solutions (12-7) characterized by singularities that move with constant speed U in the negative X direction. These are also conveniently described in an x', y', z', t' coordinate system which is translated with respect to the X, Y, Z, T system with speed $-U$. Let $X = x' - Ut'$, $Y = y'$, $Z = z'$, $T = t'$; then the differential equation is the same as Eq. 3-3, viz. dropping the primes,

$$\phi_{tt} + 2U\phi_{xt} + U^2\phi_{xx} = a_\infty^2\nabla^2\phi \tag{12-8}$$

and the solutions Eq. 12-7 become

$$\phi = \frac{1}{\sigma}\left[f(\sigma - a_\infty\beta^2 t - M_\infty x) + g(\sigma + a_\infty\beta^2 t + M_\infty x)\right] \tag{12-9}$$

where σ denotes $\sqrt{x^2 + \beta^2(y^2 + z^2)}$.

These solutions have singularities of the type $1/\sigma$, i.e. they may be thought of as sources and sinks of variable strength. The potentials for moving doublets are simply obtained by differentiation.

Sinusoidal oscillations. Let us now assume that the motion to be described is that produced by a body vibrating harmonically while moving through the gas at constant speed U. The potential ϕ must have the form

$$\phi(x, y, z, t) = e^{i\omega t}F(x, y, z) \tag{12-10}$$

According to Eq. 12-8, the function F must satisfy the equation

$$-\omega^2 F + 2i\omega U F_x + U F_{xx} = a_\infty^2\nabla^2 F \tag{12-11}$$

or, writing G for $Fe^{-i\omega Ux/a_\infty^2\beta^2}$

$$\beta^2 G_{xx} + G_{yy} + G_{zz} + \frac{1}{\beta^2}\frac{\omega^2}{a_\infty^2}G = 0 \tag{12-12}$$

This equation is obviously amenable to treatment by the familiar Prandtl-Glauert transformation, which reduces it to the form of the equation for harmonic sound vibrations in a gas at rest, viz., from Eq. 12-1,

$$\nabla^2 f + kf = 0, \qquad k = \text{const} \tag{12-13}$$

It admits solutions of the form

$$G(x, y, z) = \frac{\text{const}}{\sigma}\, e^{\pm i\omega\sigma/a_\infty\beta^2} \tag{12-14}$$

where σ is the same quantity introduced in Eq. 12-9. This solution could also have been deduced immediately by comparison of Eq. 12-9 and Eq. 12-10.

Plane flow. The plane flow counterpart of Eq. 12-1 may be written in the form

$$\frac{1}{a_\infty^2}\phi_{tt} - \phi_{xx} - \phi_{yy} = 0 \tag{12-15}$$

An important elementary solution is

$$\begin{aligned}\phi &= \frac{\text{const}}{\sqrt{a_\infty^2 t^2 - x^2 - y^2}} \qquad \text{for } a_\infty^2 t^2 > x^2 + y^2 \\ &= 0 \qquad\qquad\qquad\qquad\quad \text{for } a_\infty^2 t^2 < x^2 + y^2\end{aligned} \tag{12-16}$$

This represents the potential due to an impulse which occurs at the origin at $t = 0$ and whose effects spread out in a widening circle about the origin. A distribution of these singularities over an interval of the time variable can be used to represent the effects of time-dependent small disturbances in a fluid otherwise at rest, i.e.

$$\phi(r, t) = \int_0^{t - r/a_\infty} \frac{f(\bar{t})\, d\bar{t}}{\sqrt{a_\infty^2 (t - \bar{t})^2 - r^2}} \tag{12-17}$$

where $r^2 = x^2 + y^2$. Solution (12-16) is called an acoustic two-dimensional source. This is one of the solutions of the two-dimensional wave equation that will be found of great usefulness in supersonic small perturbation theory in Sec. D. Its exploitation will be postponed to that section. Other solutions with higher degree singularities are easily found by differentiation of it.

Other useful solutions of Eq. 12-15 are obtained from the following, which is appropriate for oscillating flows:

$$\phi = \left[A J_0\left(\frac{\omega r}{a_\infty}\right) + B Y_0\left(\frac{\omega r}{a_\infty}\right)\right] e^{i\omega t} \tag{12-18}$$

Here A and B are arbitrary constants, and J_0 and Y_0 are Bessel functions of the first and second kinds of zero order. Some idea of the character of this flow can be obtained from the asymptotic behaviors of these Bessel functions for large arguments, which are

$$J_0(\mu) \sim \frac{\cos \mu}{\sqrt{\mu}}, \qquad Y_0(\mu) \sim \frac{\sin \mu}{\sqrt{\mu}}, \qquad \mu \to \infty \tag{12-19}$$

When these asymptotic forms are valid, the solution (12-18) has the form

$$\phi \cong r^{-\frac{1}{2}} \left[C e^{i\omega\left(t + \frac{r}{a_\infty}\right)} + D e^{i\omega\left(t - \frac{r}{a_\infty}\right)} \right] \tag{12-20}$$

where C and D are complex constants. In this limiting form, we see for the first time a solution of the two-dimensional equation (12-15) analogous to the forms (12-3) in three dimensions and

$$\phi(x, t) = f(x - a_\infty t) + g(x + a_\infty t) \tag{12-21}$$

which is the well-known solution of the one-dimensional equation

$$\phi_{tt} = a_\infty^2 \phi_{xx} \tag{12-22}$$

In general, the two-dimensional solutions do not have this form. The result is that two-dimensional sound waves do not preserve their shape, and the wave produced by a square impulse propagates with a "tail" of gradually diminishing strength.

Unsteady flow problems in both subsonic and supersonic flow are treated further in Sec. D of this volume, where solutions are constructed by use of the elementary solutions derived here and others. Practical problems of airfoils in unsteady motion are treated in more detail in VII,F.

C,13. Cited References.

1. Rayleigh, Lord. *The Theory of Sound* (1894). Dover, 1945.
2. Ackeret, J. Über Luftkräfte bei sehr grossen Geschwindigkeiten insbesonders bei ebenen Strömungen. *Helv. Phys. Acta 1*, 301–322 (1928).
3. Glauert, H. The effect of compressibility on the lift of an aerofoil. *Proc. Roy. Soc. A118*, 113–119 (1928).
4. Ackeret, J. Luftkräfte auf Flügel, die mit grösserer als Schallgeschwindigkeit bewegt werden, *Z. Flugtech. u. Motorluftschiffahrt 16*, 72–74 (1925). Translated in *NACA Technical Memorandum 317*. 1925.
5. von Kármán, Th. Über stationäre Wellen in Gasstrahlen. *Phys. Zeits. 8*, 209–211 (1907).
6. Göthert, B. Ebene und räumliche Strömung bei hohen Unterschallgeschwindigkeiten. *Jahrb. der deut. Luftfahrtforschung 1941*, I156–I157 (1941). Also *Bericht 127 der Lilienthal-Gesellschaft für Luftfahrtforschung*, 97–101 (1940).
7. Goldstein, S., and Young, A. D. The linear perturbation theory of compressible flow, with applications to wind-tunnel interference. *ARC Reports and Memoranda 1909*. 1943.
8. Tsien, H. S., and Lees, L. The Glauert-Prandtl approximation for subsonic flows of a compressible fluid. *J. Aero. Sci. 12*, 173–187, 202 (1945).
9. Reissner, E. On compressibility corrections for subsonic flow over bodies of revolution. *NACA Technical Note 1815*. 1949.
10. Margenau, H., and Murphy, G. M. *The Mathematics of Physics and Chemistry*. Van Nostrand, 1943.
11. Hess, R. V., and Gardner, C. S. Study by the Prandtl-Glauert method of compressibility effects and critical Mach number for ellipsoids of various aspect ratios and thickness ratios. *NACA Technical Note 1792*. 1949.
12. Prandtl, L. Allgemeine Überlegungen über die Strömung zusammendrückbarer Flüssigkeiten. *Atti del Convegno della Fondazione Alessandro Volta 1935*, 169–197 (1936). Rome. Translated in *NACA Technical Memorandum 805*. 1936.
13. von Kármán, Th. Compressibility effects in aerodynamics. *J. Aero. Sci. 8*, 337–356 (1941).
14. Sauer, R. *Theoretische Einführung in die Gasdynamik*. Springer, Berlin, 1943. Also Edwards Bros., Ann Arbor, 1945 (in English, 1947).
15. Sears, W. R. On compressible flow about bodies of revolution. *Quart. Appl. Math. 4*, 191–193 (1946).
16. Sears, W. R. A second note on compressible flow about bodies of revolution. *Quart. Appl. Math. 5*, 89–91 (1947).

17. Lees, L. A discussion of the application of the Prandtl-Glauert method to subsonic compressible flow over a slender body of revolution. *NACA Technical Note 1127.* 1946.
18. Young, A. D., and Kirkby, S. Application of the linear perturbation theory to compressible flow about bodies of revolution. *The College of Aeronautics, Cranfield, Report 11.* 1947.
19. Schmieden, C., and Kawalki, K. H. Beiträge zum Umströmungsproblem bei hohen Geschwindigkeiten. *Lilienthal-Gesellschaft für Luftfahrtforschung Bericht S 13/1,* 40–68 (1942). Translated in *NACA Technical Memorandum 1233.* 1949.
20. Laitone, E. V. The extension of the Prandtl-Glauert rule. *J. Aero. Sci. 17,* 250–251 (1950).
21. Matthews, C. W. A comparison of the experimental subsonic pressure distributions about several bodies of revolution with pressure distributions computed by means of the linearized theory. *NACA Technical Note 2519.* 1952.
22. Young, A. D., and Young, E. A family of streamline bodies of revolution suitable for high-speed or low-drag requirements. *ARC Reports and Memoranda 2204.* 1945.
23. Amick, J. L. Comparison of the experimental pressure distribution on an NACA 0012 profile at high speeds with that calculated by the relaxation method. *NACA Technical Note 2174.* 1950.
24. Lippisch, A., and Beuschausen, W. Druckverteilungsmessungen bei Hochgeschwindigkeit und Schräganblasung. *Deut. Luftfahrtforschung, Forschungsbericht 1669.* 1942. Translated in *NACA Technical Memorandum 1115.* 1947.
25. Herriot, J. G. Blockage corrections for three-dimensional-flow closed-throat wind tunnels, with consideration of the effect of compressibility. *NACA Research Memorandum A7B28.* 1947.
26. Van Driest, E. R. Die linearisierte Theorie der dreidimensionalen kompressiblen Unterschallströmung und die experimentelle Untersuchung von Rotationskörpern in einem geschlossenen Windkanal. *Mitt. No. 16, Inst. für Aerodynamik, Eidg. Tech. Hochschule.* Zürich, 1949.
27. Ackeret, J., Degen, M., and Rott, N. Über die Druckverteilung an schräg angestromten Rotationskörpern bei Unterschallgeschwindigkeit. *L'Aerotecnica 31,* 11–19 (1951).
28. Feldmann, F. K. Untersuchung von symmetrischen Tragflügelprofilen bei hohen Unterschallgeschwindigkeiten in einem geschlossenen Windkanal. *Mitt. No. 14, Inst. für Aerodynamik, Eidg. Tech. Hochschule.* Zürich, 1948.
29. Murray, H. E. Comparison with experiment of several methods of predicting the lift of wings in subsonic compressible flow. *NACA Technical Note 1739.* 1948.
30. Ferri, A. Investigations and experiments in the Guidonia supersonic wind tunnel, *NACA Technical Memorandum 901.* 1939; Experimental results with airfoils tested in the high speed tunnel at Guidonia. *NACA Technical Memorandum 946.* 1940.
31. Stack, J. Compressibility effects in aeronautical engineering. *NACA Advance Confidential Report.* 1941.
32. Robinson, A., and Young, A. D. Note on the application of the linearised theory for compressible flow to transonic speeds. *The College of Aeronautics, Cranfield, Report 2.* 1947.
33. Munk, M. M. Elements of the wing section theory and of the wing theory. *NACA Report 191.* 1924.
34. Jones, R. T. Properties of low-aspect-ratio pointed wings at speeds below and above the speed of sound. *NACA Technical Note 1032.* 1946. Also *NACA Report 835.* 1946.
35. Lighthill, M. J. The diffraction of blast I. *Proc. Roy. Soc. A198,* 454–470 (1949).
36. Lighthill, M. J. Reflection at a laminar boundary layer of a weak steady disturbance to a supersonic stream, neglecting viscosity and heat conduction. *Quart. J. Mech. and Appl. Math. 3,* 303–325 (1950).
37. Chang, C. C., and Chu, B. T. *The General Theory of Linearized Rotational Compressible Flow and Applications.* The Johns Hopkins University, 1951. (Presented at 1951 annual meeting of the American Physical Society, New York City, February 2, 1951.)

38. Lighthill, M. J. The flow behind a stationary shock. *Phil. Mag. 40*, 214–220 (1949).
39. Bargmann, V. On nearly glancing reflection of shocks. *Applied Mathematics Panel Report 108.2R* (*Applied Mathematics Group—Institute for Advanced Study*). 1945.
40. Ting, L., and Ludloff, H. F. Aerodynamics of blast. *J. Aero. Sci. 18*, 143–144 (1951).
41. White, D. R. An experimental survey of the Mach reflection of shock waves. *Princeton Univ. Dept. of Phys. Technical Report II-10*. August 21, 1951.
42. Adams, Mac C. On shock waves in inhomogeneous flow. *J. Aero. Sci. 16*, 685–690 (1949).
43. Pai, S. I. On the flow behind an attached curved shock. *J. Aero. Sci. 19*, 734–742 (1952).
44. Pai, S. I. On the vorticity of the flow behind curved shock. *J. Aero. Sci. 17*, 188–189 (1950).
45. Sears, W. R. The linear-perturbation theory for rotational flow. *J. Math. and Phys. 28*, 268–271 (1950).
46. Küssner, H. G. Allgemeine Tragflächentheorie. *Luftfahrtforschung 17*, 370–378 (1940).
47. Lighthill, M. J. Methods for predicting phenomena in the high-speed flow of gases, *J. Aero. Sci. 16*, 69–83 (1949).
48. Ward, G. N. Supersonic flow past slender pointed bodies. *Quart. J. Mech. and Appl. Math. 2*, 75–97 (1949).
49. Adams, Mac C., and Sears, W. R. Slender-body theory—review and extensions. *J. Aero. Sci. 20*, 85–98 (1953).

SECTION D

SUPERSONIC AND TRANSONIC SMALL PERTURBATION THEORY

MAX. A. HEASLET
HARVARD LOMAX

CHAPTER 1. THE WAVE EQUATION AND METHODS OF SOLUTION

D,1. Introductory Remarks. The previous sections of this volume provide the basic equations of high speed gas flow; and, in particular, the development of small perturbation theory is given in Sec. C with principal attention to application to stream or flight velocities less than the speed of sound. In this section we extend the range of investigation beyond the subsonic regime and thus are essentially concerned with flight or stream velocities comparable to or greater than that of sound.

The feasibility of constructing efficient supersonic aircraft is dependent upon the determination of aerodynamic shapes inducing small perturbations in the surrounding air, since large disturbances are generally associated with the formation of strong shock waves and thus involve nonisentropic flow which in turn produces excessive values of drag. It has been found by experience that a large field of practical application does exist within the framework of small perturbation theory and that, furthermore, conventional linear analysis suffices for many engineering purposes.

In all analytic investigations of physical problems the theorist must strive to prescribe the properties of his idealized model in such a manner that, first, the properties are reducible to mathematical form; second, the subsequent analysis does not impose excessive demands in the application of the mathematical techniques; and, finally, that reasonable agreement with observed conditions can be obtained. For high speed aerodynamics the resolution of these somewhat opposing requirements is considered in detail in the earlier sections, and reference can be made to these discussions. Throughout the major part of this section we confine ourselves to an investigation of phenomena which can be characterized by means of the wave equation, and the work consequently falls within the broad outlines of acoustic theory. The following assumptions are made:

1. The viscosity of the air is neglected.
2. The heat conductivity of the air is neglected.
3. The flow field is free of vorticity (and thus possesses a velocity potential) except for certain prescribed regions downstream of a body.
4. All external force fields are excluded.
5. The equation of state for perfect gases holds.

The main objective before us is to study the motion of an aerodynamic configuration in both steady and unsteady high speed flight, and to provide mathematical equipment that will determine to the fullest extent possible the lift, resistance, and the moment of the forces experienced by the specified configuration. In order to attain a degree of success in this task, one must first impose some limitations on the geometry of the body to be considered and also define the nature of its motion. The guide to such an approach is, however, provided by the historical development of low speed theory, and it therefore appears natural to subdivide the theory into such categories as profile theory, wing theory, lifting and thickness effects, steady and unsteady motion, etc. These topics are discussed in detail in the following chapters. Before taking up such subdivisions, however, it is convenient to present first the different forms of the wave equation and to outline the manner in which solutions are to be obtained.

D,2. Partial Differential Equation of Fluid Flow. The basic partial differential equation satisfied by the velocity potential in three-dimensional fluid flow is derived from the Eulerian momentum equations, the continuity condition, and the relation for the speed of pressure propagation.[1] In order to introduce the assumptions of small perturbation theory into the derivation, it is possible to choose either of two avenues of attack. In the first place we can maintain as our immediate objective the linearization of the fundamental equation. This is not difficult since groups of terms in the momentum and continuity equations may be disregarded if we utilize directly the fact that the perturbation velocity components are to be small. If the velocity of the medium at the point (x', y', z') and at time t' is $\mathbf{q}$, the velocity potential $\varphi(x', y', z', t')$ satisfies the relation grad $\varphi = \mathbf{q}$. Further, if the axes are assumed fixed relative to the fluid at infinity so that $\mathbf{q} = 0$ for $x'^2 + y'^2 + z'^2 = \infty$, if the system of equations is modified to take account of the fact that $\mathbf{q}$ and its gradients are everywhere small, and if pressure p and density ρ

[1] These relations are all given analytically in Art. 32 of this section, as well as in C,2 and, in a very general discussion, in III,A.

are eliminated from the modified system, we find the desired differential equation is

$$\varphi_{x'x'} + \varphi_{y'y'} + \varphi_{z'z'} - \frac{1}{a_\infty^2}\varphi_{t't'} = 0 \tag{2-1}$$

where $a_\infty^2 = (dp/d\rho)_\infty$ is the constant speed of pressure propagation measured in the undisturbed region of the fluid. Eq. 2-1, from which the analysis in the first seven chapters of this section evolves, is the well-known wave equation (cf. C,2; Eq. 2-4) and, because of its intimate connection with acoustic and electromagnetic theory, has well-established methods of solution.

The alternate method of investigation is to retain first the most general form of the kinematic and dynamic laws and to impose the assumption that velocity components and gradients are small without attempting to force the analysis into a linearized theory. Such a procedure leads, of course, to a higher degree of exactness in specifying the physical properties of a flow field but, in practice, requires inevitably more complex methods of investigation. Since linear theory proves to be inadequate for the complete development of transonic and hypersonic flow fields, we return to this method of attack in Art. 32-34. It must suffice to state here that the most general potential equation is

$$\begin{aligned}\varphi_{t't'} + 2\varphi_{x'}\varphi_{x't'} + 2\varphi_{y'}\varphi_{y't'} + 2\varphi_{z'}\varphi_{z't'} &= \varphi_{x'x'}(a^2 - \varphi_{x'}^2) + \varphi_{y'y'}(a^2 - \varphi_{y'}^2) \\ &+ \varphi_{z'z'}(a^2 - \varphi_{z'}^2) - 2\varphi_{y'z'}\varphi_{y'}\varphi_{z'} - 2\varphi_{x'z'}\varphi_{x'}\varphi_{z'} - 2\varphi_{x'y'}\varphi_{x'}\varphi_{y'}\end{aligned} \tag{2-2}$$

where a, the speed of pressure propagation, is not a constant but is a point function given by $(dp/d\rho)^{\frac{1}{2}}$, where p and ρ are local pressure and density respectively.

Eq. 2-1 is valid for a flow field produced by a body which we may assume is traveling in the direction of the negative x' axis at a velocity U and moving away from the fixed coordinate system. Before proceeding further, it is pertinent to consider the possibility of (1) finding a transformation fixing the axial system on the wing and (2), at the same time, retaining the wave equation as the governing equation of the flow. These conditions can indeed be fulfilled [*1*] by means of the Lorentz transformation (see also C,12)

$$x_1 = \frac{x' + Ut'}{[1 - (U/a_\infty)^2]^{\frac{1}{2}}}, \qquad y_1 = y', \qquad z_1 = z', \qquad t_1 = \frac{t'a_\infty + Ux'}{[1 - (U/a_\infty)^2]^{\frac{1}{2}}}$$

provided U is less than the speed of sound a_∞. In the supersonic case, however, it can be shown that no transformation exists which satisfies the above conditions. Although we shall use Eq. 2-1 for unsteady motion (see Art. 25-28), it should be pointed out that in a large number of theoretical investigations the axes are fixed on the wing by means of the transformation

$$x = x' + Ut', \qquad y = y', \qquad z = z', \qquad t' = t' \tag{2-3}$$

The potential equation then becomes

$$\varphi_{xx}(1 - M_\infty^2) + \varphi_{yy} + \varphi_{zz} - 2\frac{M_\infty}{a_\infty}\varphi_{xt'} - \frac{1}{a_\infty^2}\varphi_{t't'} = 0 \qquad (2\text{-}4)$$

where we have introduced Mach number M_∞ defined as the ratio of U to a_∞.

In Eq. 2-4 we may now think of the wing and axes being fixed in a uniform free stream of velocity U directed parallel to the positive x axis. In the steady state case, where there is no variation with time, Eq. 2-4 reduces to the familiar Prandtl-Glauert equation ([*2,3*]; also Art. C,3)

$$(1 - M_\infty^2)\varphi_{xx} + \varphi_{yy} + \varphi_{zz} = 0 \qquad (2\text{-}5)$$

which is valid for both subsonic and supersonic flows.

The expression for the static pressure at any point in the flow field is conveniently related to the free stream static and dynamic pressures so as to form a dimensionless pressure coefficient C_p. Setting $C_p = (p - p_\infty)/\frac{1}{2}\rho_\infty U^2$, and imposing conditions consistent with the assumptions of small perturbation theory, it is easy to show from Euler's equation that, *provided* the perturbation velocity components are of the *same order of magnitude*, the various relations for pressure coefficient are

$$-\frac{2}{U}\varphi_x, \qquad -\frac{2}{U^2}(\varphi_{t'} + U\varphi_x), \qquad -\frac{2}{U^2}\varphi_{t'} \qquad (2\text{-}6)$$

for Eq. 2-5, 2-4, and 2-1 respectively. When the velocity components differ in order of magnitude, a modification of these expressions is necessary (see, e.g., Art. 15-24).

D,3. Steady State Boundary Conditions and Methods of Solution. Eq. 2-5 is the basic potential equation of linearized compressible flow, steady state theory. In application to wings it leads directly to the consideration of thin airfoil problems, that is, problems for which the conditions that force the flow to conform to the wing surface may be specified in the plane of the wing. Thin airfoil theory, as introduced by Munk [*4*], Jeffreys [*5*], and others, was originally developed for incompressible flow and applied to wings inducing small perturbations in the free stream and obviously, in that case, it is merely a subdivision of the classical linear analysis associated with Laplace's equation. We have found, on the other hand, that the compressible flow potential equation is linearized only after making strong restrictions in the nature of the flow so that, insofar as the study of wings is concerned, linear theory and thin airfoil theory are now virtually coextensive for both subsonic and supersonic flows.

For supersonic flow, Eq. 2-5 is a hyperbolic equation and has associated with it certain characteristic surfaces; in the present case these surfaces are right circular cones having their axes of rotation parallel to the x axis or free stream direction and with semivertex angles equal to $\csc^{-1} M_\infty$. The family of characteristics is expressed analytically as

$$(x - x_1)^2 - (M_\infty^2 - 1)[(y - y_1)^2 + (z - z_1)^2] = 0$$

where the point P with coordinates x, y, z is the vertex and x_1, y_1, z_1 are running coordinates. Physically, the surface of the cone extending downstream from its apex represents the outermost region of the flow field which can be influenced by a disturbance at the apex and is called the Mach cone. The surface of the cone extending upstream is the boundary of the region containing all points from which a disturbance will affect the flow at the cone apex and is called the forecone of the point P.

A distinctive feature of linearized thin airfoil theory is the fact that the geometry of a wing can be assumed to be composed of two additive elements: the nonlifting and the lifting portions. Suppose a thin wing at small angle of attack has the ordinates of its upper and lower surfaces given by $z = z_u(x, y)$ and $z = z_l(x, y)$ respectively and let us introduce the variables z_t and z_c defined by the relations

$$z_t = \tfrac{1}{2}(z_u - z_l), \qquad z_c = \tfrac{1}{2}(z_u + z_l)$$

Since the basic potential equation is linear and since the boundary conditions are specified in the plane of the wing (we arbitrarily take this plane to be $z = 0$), we may, by virtue of the fact that

$$z_u = z_c + z_t, \qquad z_l = z_c - z_t$$

study separately the flow over a symmetrical wing with a thickness distribution given by $\pm z_t$ and the flow over a cambered and twisted surface given by z_c. The flow field around the wing can then be obtained by combining the solutions for the symmetrical or nonlifting portion with those for the lifting surface.

The detailed study of wings in two- and three-dimensional supersonic flow is reserved for Art. 5–14. At the present time we confine ourselves to an outline of a general solution of the Prandtl-Glauert equation for the case when Mach number is greater than 1. To this end consider first the solution employed for incompressible flow where Laplace's equation

$$\nabla^2 \Omega \equiv \Omega_{\bar{x}\bar{x}} + \Omega_{\bar{y}\bar{y}} + \Omega_{\bar{z}\bar{z}} = 0 \tag{3-1}$$

applies. (Eq. 3-1 has been written in terms of Ω rather than the velocity potential since occasion will arise to use other dependent variables.) The method is provided by Green's theorem which relates a volume integral over a region R to a surface integral over the surface S enclosing R. If σ and Ω are any two functions which, together with their first and second derivatives, are finite and single-valued throughout R, then

$$\iint_S \left(\sigma \frac{\partial \Omega}{\partial n} - \Omega \frac{\partial \sigma}{\partial n}\right) dS = -\iiint_R (\sigma \nabla^2 \Omega - \Omega \nabla^2 \sigma) dR \qquad (3\text{-}2)$$

where the directional derivatives in the left member are taken along the normal n drawn into the region R. Identify now the function σ with the fundamental solution

$$\frac{1}{\bar{r}} = \frac{1}{[(\bar{x} - \bar{x}_1)^2 + (\bar{y} - \bar{y}_1)^2 + (\bar{z} - \bar{z}_1)^2]^{\frac{1}{2}}}$$

If Ω satisfies Laplace's equation and S_0 is a small sphere enclosing the point P at $\bar{x}$, $\bar{y}$, $\bar{z}$, then if R is the region bounded externally by some closed surface S_1 and internally by S_0, we have, from Eq. 3-2,

$$\iint_{S_1+S_0} \left[\frac{1}{\bar{r}} \frac{\partial \Omega}{\partial n} - \Omega \frac{\partial}{\partial n}\left(\frac{1}{\bar{r}}\right)\right] dS = 0.$$

If the radius of S_0 tends to zero, the fundamental relation follows:

$$\Omega(\bar{x}, \bar{y}, \bar{z}) = -\frac{1}{4\pi} \iint_{S_1} \left[\frac{1}{\bar{r}} \frac{\partial \Omega}{\partial n} - \Omega \frac{\partial}{\partial n}\left(\frac{1}{\bar{r}}\right)\right] dS \qquad (3\text{-}3)$$

The physical significance of this last relation follows immediately: the term $-1/4\pi\bar{r}$ represents a fluid source and the term $-\partial(1/4\pi\bar{r})/\partial n$ represents a doublet with its axis lying along the normal to S_1, both source and doublet being situated at the surface point $\bar{x}_1$, $\bar{y}_1$, $\bar{z}_1$. The value of the function Ω at the point $\bar{x}$, $\bar{y}$, $\bar{z}$ is therefore given as an integral of a source and doublet distribution, the strengths of the two being determined by the respective boundary values of $\partial\Omega/\partial n$ and Ω.

If in Eq. 3-2 the surface S_1 is again the external boundary of R and the point P is chosen outside R, the integral in the right member of Eq. 3-3 is equal to zero.

Problems in wing theory usually involve boundary conditions in which Ω or $\partial\Omega/\partial n$ have prescribed discontinuities over a portion of the plane occupied by the wing. Consider now a region R bounded by the x, y plane and a hemispherical dome of infinite radius lying above this plane. For all problems to which the results are applied, the values of Ω and $\partial\Omega/\partial n$ vary inversely as the first and second powers of the radius.[2] For an infinite radius, the contribution of the surface integral over the hemisphere is

[2] It is sufficient to assume that Ω and $\partial\Omega/\partial n$ vary as stated on the sphere at all points having radius vectors which make finite (nonzero) angles with the positive x axis while Ω and $\partial\Omega/\partial\bar{x}$ are merely bounded at all points infinitely distant from the wing and at a finite distance from the positive x axis. Thus the results of the analysis can be applied to lifting surface problems with trailing vortex sheets for the case when $\Omega = \varphi$.

zero, and from Eq. 3-3

$$\Omega(\bar{x}, \bar{y}, \bar{z}) = -\frac{1}{4\pi} \iint \left[\left(\frac{1}{\bar{r}} \right)_s \frac{\partial \Omega_u}{\partial \bar{z}_1} - \Omega_u \left(\frac{\partial}{\partial \bar{z}_1} \frac{1}{\bar{r}} \right)_s \right] d\bar{x}_1 d\bar{y}_1$$

where the integration extends over the entire plane and the subscript s indicates the function is to be calculated at $\bar{z}_1 = 0$. The directional derivatives are necessarily in the direction of the positive $\bar{z}$ axis and the subscript u denotes conditions on the upper side of the plane. Keeping P fixed and integrating over the lower side of the $\bar{x}$, $\bar{y}$ plane, it follows that

$$0 = -\frac{1}{4\pi} \iint \left[\left(\frac{1}{\bar{r}} \right)_s \frac{\partial \Omega_l}{\partial \bar{z}_1} - \Omega_l \left(\frac{\partial}{\partial \bar{z}_1} \frac{1}{\bar{r}} \right)_s \right] d\bar{x}_1 d\bar{y}_1$$

where the negative direction of the normal may be ignored since the integral is equal to zero. Introducing the notation

$$\Delta\Omega = \Omega_u - \Omega_l, \qquad \Delta \frac{\partial \Omega}{\partial \bar{z}_1} = \frac{\partial \Omega_u}{\partial \bar{z}_1} - \frac{\partial \Omega_l}{\partial \bar{z}_1}$$

and subtracting the above two equations, we have

$$\Omega(\bar{x}, \bar{y}, \bar{z}) = -\frac{1}{4\pi} \iint_\tau \left[\left(\frac{1}{\bar{r}} \right)_s \Delta \frac{\partial \Omega}{\partial \bar{z}_1} - \Delta\Omega \left(\frac{\partial}{\partial \bar{z}_1} \frac{1}{\bar{r}} \right)_s \right] d\bar{x}_1 d\bar{y}_1 \qquad (3\text{-}4)$$

the integral extending only over the area τ for which the integrand does not vanish.

Eq. 3-4 is the fundamental equation of low speed wing theory and holds for any point $\bar{x}$, $\bar{y}$, $\bar{z}$ either above or below the $\bar{x}$, $\bar{y}$ plane.

Setting $\beta^2 = |1 - M_\infty^2|$, the Prandtl-Glauert transformation

$$\bar{x} = x, \qquad \bar{y} = \beta y, \qquad \bar{z} = \beta z \qquad (3\text{-}5)$$

furthermore transforms Eq. 3-1 to the subsonic form of Eq. 2-5, and Eq. 3-4 is correspondingly extended to include the entire subsonic range. In this case, the modification of the fundamental solution leads to the expression $-1/4\pi[(x - x_1)^2 + \beta^2(y - y_1)^2 + \beta^2(z - z_1)^2]^{\frac{1}{2}}$ for the subsonic source.

It remains now to develop for supersonic flow a general solution analogous to Eq. 3-4. We can assume, by virtue of the Prandtl-Glauert transformation, that the partial differential equation is

$$\Omega_{\bar{x}\bar{x}} - \Omega_{\bar{y}\bar{y}} - \Omega_{\bar{z}\bar{z}} = 0 \qquad (3\text{-}6)$$

since retransformation to the original space variables can be carried out later. The first step in the derivation is, once more, the introduction of Green's theorem for Eq. 3-6. Employing the operator

$$\Box^2 = \frac{\partial^2}{\partial \bar{x}^2} - \frac{\partial^2}{\partial \bar{y}^2} - \frac{\partial^2}{\partial \bar{z}^2}$$

we have

$$\iint_S \left(\sigma \frac{\partial \Omega}{\partial \nu} - \Omega \frac{\partial \sigma}{\partial \nu}\right) dS = - \iiint_R (\Omega \Box^2 \sigma - \sigma \Box^2 \Omega) dR \tag{3-7}$$

where ν is the so-called conormal to the surface S. By definition, ν has direction cosines equal to ν_1, ν_2, ν_3 such that $\nu_1 = -n_1$, $\nu_2 = n_2$, $\nu_3 = n_3$ where n_1, n_2, n_3 are the direction cosines of the normal to the surface S. (The conormal at any point $\bar{x}_1$, $\bar{y}_1$, $\bar{z}_1$ of a surface is the mirror image in the plane $\bar{x} = \bar{x}_1$ of the normal through the same point.)

The formal appearance of Eq. 3-7 is quite similar to that of Eq. 3-2 and in supersonic theory relations are obtained that parallel closely the source, doublet, and vortex formulas of subsonic flow. As a start toward this objective, consider the differences that arise in the physical nature of the flow field. The essential difference is produced by the existence of the Mach cones. Thus, a given wing produces perturbations that lie wholly within the Mach cones with vertices along the leading edge of the wing. The envelope of these foremost Mach cones is a surface we denote by γ. A point P with coordinates $\bar{x}$, $\bar{y}$, $\bar{z}$ is affected by all disturbances lying within its forecone C and at the same time behind the forward surface γ, the nature of γ being dictated by the leading edge of the wing. In the applications of Eq. 3-7 the volume integral is limited to the portion of space common to the surfaces C, γ, and the plane of the wing and the surface integral involves a discussion of conditions on these three surfaces.

Since there is no way of determining Ω and $\partial\Omega/\partial\nu$ along C, the attempted solution is not easy unless the particular solution σ and its derivative along the conormal vanish everywhere on C. It was with this idea in mind that Volterra [6] suggested the choice

$$\sigma = \cosh^{-1} \frac{\bar{x} - \bar{x}_1}{[(\bar{y} - \bar{y}_1)^2 + (\bar{z} - \bar{z}_1)^2]^{\frac{1}{2}}}$$

The value of σ is equal to zero on C since the equation of the forecone in the present system of coordinates is

$$(\bar{x} - \bar{x}_1)^2 - (\bar{y} - \bar{y}_1)^2 - (\bar{z} - \bar{z}_1)^2 = 0$$

and further, since the conormal is always directed along the forecone, $\partial\sigma/\partial\nu$ is the gradient of σ along C and is also zero.

The particular solution σ is infinite on the line given by the equation $(\bar{y} - \bar{y}_1)^2 + (\bar{z} - \bar{z}_1)^2 = 0$, and in the application of Green's theorem it is necessary to exclude this line by means of a cylinder κ of infinitesimal radius. Assuming Ω satisfies Eq. 3-6 and that $z > 0$, we apply Eq. 3-7 to two regions: first, the part above the $\bar{x}$, $\bar{y}$ plane, within C and γ, and external to κ; second, the part below the $\bar{x}$, $\bar{y}$ plane and at the same time bounded by C and γ. Eq. 3-7 then yields, respectively, the expressions

$$\iint_{\tau+\kappa+\gamma_1} \left(\Omega \frac{\partial \sigma}{\partial \nu} - \sigma \frac{\partial \Omega}{\partial \nu}\right) = 0 \tag{3-8}$$

and

$$\iint_{\tau+\gamma_2} \left(\Omega \frac{\partial \sigma}{\partial \nu} - \sigma \frac{\partial \Omega}{\partial \nu}\right) = 0 \tag{3-9}$$

where τ is the portion of the x, y plane bounding the region of integration, and γ_1 and γ_2 denote the portions of γ above and below the plane. In the limit, as the radius of κ approaches zero, Eq. 3-8 yields

$$\Omega(\bar{x}, \bar{y}, \bar{z}) = \frac{1}{2\pi} \frac{\partial}{\partial \bar{x}} \iint_{\tau+\gamma_1} \left(\Omega \frac{\partial \sigma}{\partial \nu} - \sigma \frac{\partial \Omega}{\partial \nu}\right) dS \tag{3-10}$$

Subtracting the derivative of Eq. 3-9 from Eq. 3-10, we have [7]

$$\Omega(\bar{x}, \bar{y}, \bar{z}) = -\frac{1}{2\pi} \frac{\partial}{\partial \bar{x}} \iint_{\tau} \left(\sigma \Delta \frac{\partial \Omega}{\partial \bar{z}_1} - \Delta \Omega \frac{\partial \sigma}{\partial \bar{z}_1}\right) d\bar{x}_1 d\bar{y}_1$$
$$+ \frac{1}{2\pi} \frac{\partial}{\partial \bar{x}} \iint_{\gamma_1} \left(\Omega \frac{\partial \sigma}{\partial \nu} - \sigma \frac{\partial \Omega}{\partial \nu}\right) dS + \frac{1}{2\pi} \frac{\partial}{\partial \bar{x}} \iint_{\gamma_2} \left(\Omega \frac{\partial \sigma}{\partial \nu} - \sigma \frac{\partial \Omega}{\partial \nu}\right) dS \tag{3-11}$$

The complete discussion of the integrations over γ_1 and γ_2 is possible only after identifying more specifically the nature of Ω. When Ω is the perturbation velocity potential φ, its value on γ_1 and γ_2 can be shown to be zero; but if, for example, Ω is equal to u, the x component of perturbation velocity, it is not zero on γ_1, γ_2 for cases when the leading edge of the wing is inclined to the x axis at an angle greater than $\csc^{-1} \sqrt{2}$. Analysis reveals, however, that in this case the integration over γ_1 just cancels the integration over γ_2 [8]. We limit ourselves to these two forms of Ω and the fundamental equation may therefore be written from Eq. 3-11,

$$\Omega(\bar{x}, \bar{y}, \bar{z}) = -\frac{1}{2\pi} \frac{\partial}{\partial \bar{x}} \iint_{\tau} \left(\sigma \Delta \frac{\partial \Omega}{\partial \bar{z}_1} - \Delta \Omega \frac{\partial \sigma}{\partial \bar{z}_1}\right) d\bar{x}_1 d\bar{y}_1$$

or, after using the transformations (3-5),

$$\Omega(x, y, z) = -\frac{1}{2\pi} \frac{\partial}{\partial x} \iint_{\tau} \left\{\Delta \frac{\partial \Omega}{\partial z} \cosh^{-1} \frac{x - x_1}{\beta \sqrt{(y - y_1)^2 + z^2}}\right.$$
$$\left. - \Delta \Omega \frac{(x - x_1) z}{[(y - y_1)^2 + z^2] \sqrt{(x - x_1)^2 - \beta^2 (y - y_1)^2 - \beta^2 z^2}}\right\} dx_1 dy_1 \tag{3-12}$$

Eq. 3-12 holds, as in the case of Eq. 3-4, for x, y, z either above or below the x, y plane.

D,4. Concluding Remarks. No attempt is made to trace here the historical development of linearized supersonic aerodynamic theory, although the various chapters are intended to contain adequate references to the research literature. The underlying mathematical and physical principles associated with the wave equation and acoustic phenomena have long been well established. From the standpoint of the aerodynamicist, however, supersonic wing theory did not evolve until the decade 1925–1935 and is connected principally with the names Ackeret [*9*], Busemann [*10*], Prandtl [*2*], and von Kármán [*11*]. In particular, the series of lectures given by the last three men at the Fifth Volta Congress in Rome (1935) provided, in a general way, the groundwork for all subsequent applications. Since that time the attention paid to the field and the number of published results have increased notably. As a result of war conditions and classification restrictions together with the rapid growth of the theory, considerable duplication of effort has inevitably occurred. It is of interest to observe, however, that aside from certain deviations in method, the investigations have followed a logical and almost parallel growth in all countries engaged actively in research.

The arrangement of the following chapters is based on the attempt to study problems of increasing geometric and theoretical complexity although these dual objectives are, of course, not always compatible. Chap. 2 (Art. 5–9) considers steady state supersonic profile theory and is confined essentially to well-established results. In Chap. 3 (Art. 10–14) steady state wing theory is treated at considerable length and in as unified a manner as was found possible. The development of supersonic wing theory has freed the analyst of many of the difficulties inherent in subsonic theory and of the onerous details of lifting line and discrete vortex methods. As a result, a wealth of information is now at hand, particularly in lifting surface theory. Despite this fact much work remains to be done, for example, in the design of optimum planforms and lifting surfaces, and a sizeable portion of the chapter deals with methods rather than results. Chap. 4 (Art. 15–18) is confined to a short discussion of bodies of revolution in supersonic flow and of the solutions obtained by means of source and doublet distributions. Since this subject is to be treated elsewhere in this series and at greater length (E,G,H, and VII,B), the applications have been limited to the study of slender bodies, where the boundary conditions can be treated with relative ease. In Chap. 5 (Art. 19–24) the concepts of slender wing theory for lifting wings and wing-body combinations are applied to a number of examples. The simplicity of the basic assumptions permits the study of relatively complicated configurations and, in particular, gives first order approximations to the pressures on sweptback wings and of sweptback wings mounted on cylindrical fuselages. The analysis of such problems by other existing methods is usually lengthy and involved. Slender wing theory becomes doubly important by virtue of the fact that it can also be thought of as a first order approximation to lifting surface problems at sonic flight speeds. (See also VII,A,B, and C.) Chap. 6 (Art. 25–28) is principally concerned with the application of Kirchhoff's solution of the wave equation to problems in unsteady motion. The repeated use of acoustic concepts and of analogues taken from steady state lifting surface theory characterizes the theoretical approach. Chap. 7 (Art. 29–31) introduces reciprocity relations for wing theory and gives an indication of their extreme usefulness in specific calculations of wing characteristics in steady and unsteady motion. Chap. 8 (Art. 32–34) treats in some detail the derivation of the basic potential equations of compressible flow and also gives derivations of similarity rules in transonic, supersonic, and hypersonic flow.

Before proceeding to the study of wings in a supersonic stream, it is pertinent to introduce at this point a terminology that is particularly useful in classifying edge conditions. We can assume, for all practical purposes, that the planform of a wing is a simple closed figure with the property that lines parallel to the stream direction cut the boundary of the planform in two points. In the majority of cases the bounding curves are straight lines and the planform is a rectilinear figure. Now let lines be drawn parallel to the stream direction and tangent to the lateral boundaries of the planform. The portion of the boundary upstream of the points of tangency is the *leading edge* of the wing, the portion downstream of the points of tangency is the *trailing edge,* and any portion of the boundary coincident with or parallel to the tangent lines is a *side edge.* These distinctions are, of course, a familiar part of all wing theory. In supersonic flow, however, it is customary to make a further subdivision and to refer to any segment of the leading or trailing edges as supersonic or subsonic provided the component of the stream velocity normal to the edge is greater than or less than the speed of sound, that is, provided the Mach number normal to the edge is greater than or less than one. It is obvious that any point on a supersonic leading edge is not influenced by any other point on the wing and that, moreover, the upper and lower surfaces do not influence each other on wings whose edges are entirely supersonic. Two-dimensional profiles obviously possess supersonic leading and trailing edges and the purely supersonic-type flow that results is always associated with such edges. Subsonic leading and trailing edges, on the other hand, display many of the characteristics of edges in subsonic analysis. In Art. 10–14, when the loadings on wings are calculated, it is necessary to impose the Kutta condition at the trailing edge in order to establish uniqueness of flow, and square-root singularities in the loading at the leading edge of a flat plate result exactly as in subsonic and low speed theory. Similar considerations arise in the discussion of symmetrical nonlifting wings with prescribed pressure distributions when the wings have subsonic edges.

It can perhaps be stated that no other single concept has played such an important role in the modern development of wings for high speed flight as that of the effect of leading- and trailing-edge sweep on the wing's characteristics. As pointed out by R. T. Jones [*12*], wings with subsonic leading and trailing edges permit the designer to obtain more favorable drag characteristics than those associated with wings having supersonic-type edges. Resultant improvements in lift-drag ratio and airplane performance are therefore possible. Attention to sweptback planforms has thus dominated both theoretical and experimental investigations during recent years.

CHAPTER 2. TWO-DIMENSIONAL SUPERSONIC STEADY STATE FLOW

D,5. Basic Potential Equation and Solutions. The perturbation velocity potential is known, from Art. 2, to satisfy the partial differential equation

$$\beta^2\varphi_{xx} - \varphi_{zz} = 0 \tag{5-1}$$

where $\beta^2 = M_\infty^2 - 1$. Utilizing the Prandtl-Glauert transformation $\bar{x} = x$, $\bar{z} = \beta z$, we may reexpress Eq. 5-1 in the canonical form

$$\varphi_{\bar{x}\bar{x}} - \varphi_{\bar{z}\bar{z}} = 0 \tag{5-2}$$

This latter equation plays the same role in supersonic theory as Laplace's two-dimensional equation in subsonic investigations. Thus, just as solutions at an arbitrary free stream Mach number less than 1 can be related to solutions at a zero Mach number, so also can the solutions at arbitrary supersonic Mach numbers be related to solutions at a Mach number equal to $\sqrt{2}$. This convenient change of variables has already been used in Art. 3.

The mathematical literature presents a variety of methods of solving Eq. 5-2 for the type of boundary conditions encountered in wing section theory. The most straightforward attack, however, stems from the use of the expression

$$\varphi(\bar{x}, \bar{z}) = f(\bar{x} - \bar{z}) + g(\bar{x} + \bar{z}) \tag{5-3}$$

where the functions f and g are to be determined from the geometry of the problem. Transforming to the original space variables, we therefore have for the general solution the expression

$$\varphi(x, z) = f(x - \beta z) + g(x + \beta z) \tag{5-4}$$

Because of the wide applicability of source and sink distributions to wing theory in general, it is of interest at this point to see how such concepts arise in the present case. To this end we first determine the perturbation potential for a unit two-dimensional supersonic source. If the unit source is located at the point x_1, 0, z_1 its potential in the x, z plane is that induced by three-dimensional sources distributed uniformly along a line parallel to the y axis and passing through x_1, 0, z_1. The perturbation potential of a three-dimensional source is

$$\varphi(x, y, z) = -\frac{1}{2\pi[(x - x_1)^2 - \beta^2(y - y_1)^2 - \beta^2(z - z_1)^2]^{\frac{1}{2}}}$$

where x_1, y_1, z_1 is the source position and x, y, z is the field point, and it follows that the potential at the point x, 0, z of a source at x_1, y_1, z_1 is

$$\varphi(x, 0, z) = -\frac{1}{2\pi[(x - x_1)^2 - \beta^2 y_1^2 - \beta^2(z - z_1)^2]^{\frac{1}{2}}}$$

The potential at the point x, 0, z can only be affected by disturbances within the forecone from the point; hence it follows that, setting

$\beta Y = [(x - x_1)^2 - \beta^2(z - z_1)^2]^{\frac{1}{2}}$, we have, for $(x - x_1)^2 \geqq \beta^2(z - z_1)^2$

$$\begin{aligned}\varphi(x, z) &= -\frac{2}{2\pi}\int_0^Y \frac{dy_1}{[(x - x_1)^2 - \beta^2 y_1^2 - \beta^2(z - z_1)^2]^{\frac{1}{2}}} \\ &= -\frac{1}{\beta\pi}\sin^{-1}\frac{\beta y_1}{[(x - x_1)^2 - \beta^2(z - z_1)^2]^{\frac{1}{2}}}\Bigg]_0^Y = \frac{-1}{2\beta}\end{aligned}$$

In general, the potential of a two-dimensional source of strength m located at the point x_1, z_1 is

$$\begin{aligned}\varphi(x, z) &= -\frac{m}{2\beta}, \qquad (x - x_1)^2 \geqq \beta^2(z - z_1)^2 \\ &= 0, \qquad (x - x_1)^2 < \beta^2(z - z_1)^2\end{aligned} \tag{5-5}$$

This gives the interesting result that the perturbation potential in two dimensions is a constant within the region downstream of the source and bounded by the Mach lines passing through the source position.

A sink is by definition a source of negative strength. The physical connotation of the terms is the same here as in subsonic flow since the introduction of a source into a stream has the effect of parting the streamlines in the field, while a sink acts effectively as a repository of the fluid and thus causes the streamlines to draw closer together.

The properties of a single source are indicated in Fig. D,5a. If the physical plane is cut by lines $z - z_1$ = const, the value of φ along these lines is zero ahead of the Mach lines $x - x_1 = \pm\beta(z - z_1)$, and for all points behind the Mach lines the value is a constant depending on the free stream Mach number. Because of this constancy in φ, induced velocities $u = \varphi_x$ and $w = \varphi_z$ are zero everywhere except at the Mach lines. The potential is thus a step function and both of its derivatives are pulse functions or, as they are usually termed, Dirac δ functions. It is apparent that *a two-dimensional source influences the flow field only along the Mach lines emanating from it, and induced velocities remain constant along these same lines.* Infinitely far out conditions are the same as in the vicinity of the source itself.

Suppose now that a two-dimensional airfoil has its leading edge at the origin, has a chord length equal to c, is situated in a free stream of Mach number equal to $\sqrt{2}$, and can be represented by boundary conditions specified at $z = 0$. Our problem is to calculate the value of the perturbation potential $\varphi(\bar{x}, \bar{z})$ at the point P with coordinates $\bar{x}$, $\bar{z}$ (see Fig. D,5b).

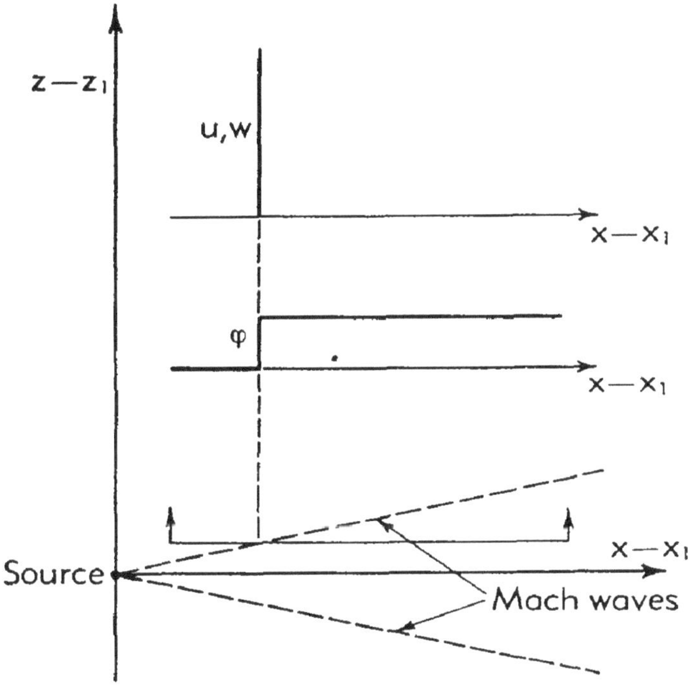

Fig. D,5a. Velocity and potential fields induced by a two-dimensional supersonic source.

Green's theorem for Eq. 5-2 may be expressed in a form relating a surface integral over the surface S to a line integral around the contour C enclosing S. If σ and φ are any two functions which, together with their first and second derivatives, are finite and single-valued throughout the region S, we have

$$\iint_S (\sigma \Box^2 \varphi - \varphi \Box^2 \sigma) dS = \int_C \left(\sigma \frac{\partial \varphi}{\partial \nu} - \varphi \frac{\partial \sigma}{\partial \nu} \right) ds \tag{5-6}$$

where ν is the conormal to the contour C and is obtained by simply reversing the $\bar{x}$ component of the normal through the same point. If σ is made a constant and φ is the desired solution of the differential equation, it follows that

$$\int_C \frac{\partial \varphi}{\partial \nu}\, ds = 0 \tag{5-7}$$

where C is an arbitrary closed curve. We choose C so that it is composed of line segments formed by portions of the $\bar{x}$ axis, the Mach lines extending upstream from P, and the Mach line extending downstream from the leading edge; that is, the segments OA, AP, PB, and BO of Fig. D,5b.

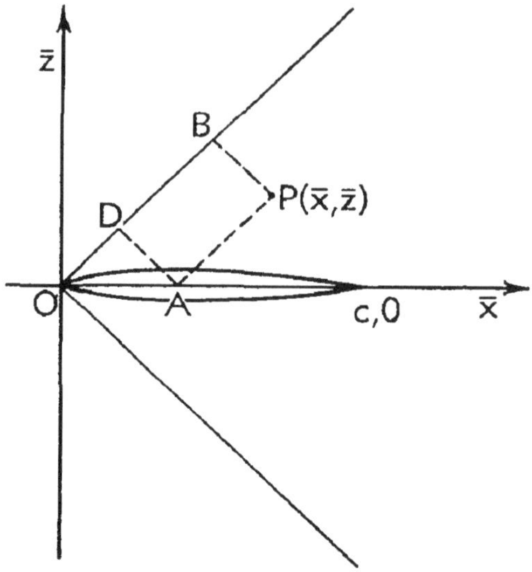

Fig. D,5b. Integration regions used in developing solutions to two-dimensional supersonic wing problems by Green's theorem.

Application of Eq. 5-7 yields the equation

$$\int_0^{\bar{x}-\bar{z}} \left(\frac{\partial \varphi}{\partial \bar{z}}\right)_u d\bar{x}_1 + \int_{\overline{AP}} \frac{\partial \varphi}{\partial s}\, ds + \int_{\overline{BP}} \frac{\partial \varphi}{\partial s}\, ds + \int_{\overline{BO}} \frac{\partial \varphi}{\partial s}\, ds = 0$$

and direct integration of the last three terms leads to the form

$$2\varphi(\bar{x}, \bar{z}) = \varphi_A - \int_0^{\bar{x}-\bar{z}} \left(\frac{\partial \varphi}{\partial \bar{z}}\right)_u d\bar{x}_1 \tag{5-8}$$

where use has been made of the fact that the potential vanishes on the foremost Mach lines and where the subscript u denotes conditions on the upper side of the $\bar{x}$ axis.

Consider now the contour C' composed of the line segments OA, OD, and AD. Eq. 5-7 then gives

$$0 = \varphi_A + \int_0^{\bar{x}-\bar{z}} \left(\frac{\partial \varphi}{\partial \bar{z}}\right)_u d\bar{x}_1 \tag{5-9}$$

If Eq. 5-9 is subtracted from Eq. 5-8, we get the desired expression

$$\varphi(\bar{x}, \bar{z}) = -\int_0^{\bar{x}-\bar{z}} \left(\frac{\partial\varphi}{\partial\bar{z}}\right)_u d\bar{x}_1 \tag{5-10}$$

This result holds for positive values of z and a similar derivation yields, for negative $\bar{z}$,

$$\varphi(\bar{x}, \bar{z}) = \int^{\bar{x}+\bar{z}} \left(\frac{\partial\varphi}{\partial\bar{z}}\right)_l d\bar{x}_1 \tag{5-11}$$

Denoting by $\lambda_u(x)$ and $\lambda_l(x)$ the slopes of the profile on the upper and lower surfaces, one has

$$\frac{w_u}{U} = \lambda_u(x), \qquad \frac{w_l}{U} = \lambda_l(x)$$

and if the retransformation of Eq. 5-10 and 5-11 to the variables x and z is carried out, the expressions for the perturbation velocity potential at an arbitrary supersonic Mach number are

$$\varphi(x, z) = -\frac{1}{\beta}\int_0^{x-\beta z} \left(\frac{\partial\varphi}{\partial z}\right)_u dx_1 = -\frac{U}{\beta}\int_0^{x-\beta z} \lambda_u(x_1)dx_1, \qquad z \geqq 0 \tag{5-12}$$

and

$$\varphi(x, z) = \frac{1}{\beta}\int_0^{x+\beta z} \left(\frac{\partial\varphi}{\partial z}\right)_l dx_1 = \frac{U}{\beta}\int_0^{x+\beta z} \lambda_l(x_1)dx_1, \qquad z \leqq 0 \tag{5-13}$$

In both cases, the value of φ can be interpreted by means of Eq. 5-5 as the integral of a distribution of sources with strengths per unit of length equal respectively to $2(\partial\varphi/\partial z)_u = 2w_u$ and $-2(\partial\varphi/\partial z)_l = -2w_l$, where w_u and w_l are the prescribed values of vertical induced velocity on the upper and lower surfaces. In the case of a symmetrical profile, these values are equal in magnitude and sign and the result may be stated as follows: *The perturbation velocity potential of a symmetric airfoil is obtained from a distribution of sources along the chord line, with the source strength per unit of length proportional to twice the slope of the profile.* The same statement is also known to apply to source distributions in subsonic thin airfoil theory.

A corollary of the results in Eq. 5-12 and 5-13 is the fact that conditions on either side of a two-dimensional supersonic wing are independent of conditions on the other side. The incapability of the sides to signal to each other their existence is the reason sources alone are sufficient to treat all cases. Similar considerations arise in portions of three-dimensional wing theory.

D,6. Airfoil Pressure Distributions. From Eq. 5-4 or 5-12 and 5-13 we may proceed immediately to the calculation of the perturbation field of a given airfoil. In the case of the former equation it follows from physical considerations that the orientation of the Mach lines is such that they slope downstream from the body and that the solution must be of the form

$$\varphi(x, z) = f(x - \beta z) \tag{6-1}$$

in the upper half-plane, and

$$\varphi(x, z) = g(x + \beta z) \tag{6-2}$$

in the lower half-plane. From the form of these solutions, the value of the potential is obviously constant along the Mach lines in both cases. In order to calculate φ from a knowledge of the airfoil shape we have, after taking derivatives with respect to z in Eq. 6-1 and 6-2 and letting z approach zero,

$$w_u = -\beta f'(x) \quad \text{and} \quad w_l = \beta g'(x)$$

where the primes denote differentiation. It follows that f and g can be expressed as integrals of the vertical induced velocity on the two surfaces and the expressions for φ are in exact agreement with Eq. 5-12 and 5-13.

Airfoil pressure distributions or load distributions follow directly from a knowledge of the streamwise component of perturbation velocity on the airfoil surface. Differentiating Eq. 5-12 and 5-13 with respect to x and setting z equal to zero, we have

$$\frac{u_u(x)}{U} = -\frac{\lambda_u(x)}{\beta} \quad \text{and} \quad \frac{u_l(x)}{U} = \frac{\lambda_l(x)}{\beta} \tag{6-3}$$

These simple relations between local velocity and local slope of an arbitrary profile, together with the fact that conditions remain the same along Mach lines, introduce an unusual degree of simplicity in the study of two-dimensional supersonic airfoil theory. Two- and three-dimensional subsonic analysis and three-dimensional supersonic studies lead invariably to integral relations between the velocity components, and as a result the way in which boundary conditions are specified has an important bearing on the method whereby the solution is effected. In the present case, regardless of whether the geometry of the wing or the pressure distribution is prescribed first, the solution follows from the use of Eq. 6-3.

Two-dimensional supersonic airfoil theory was first investigated by Ackeret [9], and velocity distributions that are directly proportional to wing slope are referred to, in general, as being of the Ackeret type.

For a symmetrical airfoil at zero angle of attack, the pressure distribution on either the upper or lower surface can be written in coefficient form as

$$C_p = -\frac{2u}{U} = \frac{2\lambda_u(x)}{\beta} = -\frac{2\lambda_l(x)}{\beta} \tag{6-4}$$

and a camber line with a slope distribution given by $\lambda_c(x)$ has the load distribution

$$\frac{\Delta p}{\frac{1}{2}\rho_\infty U^2} = (C_p)_l - (C_p)_u = -\frac{4\lambda_c(x)}{\beta} \tag{6-5}$$

For example, a thin biconvex airfoil of thickness t and chord c with an upper surface given by the equation $c^2z = 2t(cx - x^2)$ has a pressure distribution on the wing of the form

$$C_p = \frac{4t}{\beta c^2}(c - 2x) \tag{6-6}$$

A flat plate at a small angle of attack α radians is expressible in the form $z = -\alpha x$ and the load distribution is

$$\frac{\Delta p}{\frac{1}{2}\rho_\infty U^2} = \frac{4\alpha}{\beta} \tag{6-7}$$

D,7. Airfoil Characteristics. If

$$z_c = z_c(x), \qquad z_t = z_t(x)$$

denote respectively the ordinates of the camber line and thickness distribution of a profile, the ordinates of the upper and lower surfaces are found from the relations

$$z_u = z_c(x) + z_t(x), \qquad z_l = z_c(x) - z_t(x)$$

A knowledge of these functions together with Eq. 6-3 is sufficient to permit the calculation of lift, drag, and pitching moment of an airfoil. Thus, since lift coefficient is by definition the z component of force (normal to the free stream) given by the expression

$$c_l = \frac{1}{c}\int_0^c \frac{\Delta p}{\frac{1}{2}\rho_\infty U^2}\,dx = \frac{2}{c}\int_0^c \frac{u_u - u_l}{U}\,dx$$

it follows that

$$c_l = -\frac{4}{\beta c}\int_0^c \frac{dz_c}{dx}\,dx = -\frac{4}{\beta c}[z_c(c) - z_c(0)] \tag{7-1}$$

We therefore find that lift coefficient is independent of the camber line shape and is a function only of the angle of attack of the line connecting the leading and trailing edges of the airfoil.

In a similar fashion, drag coefficient is by definition the x component of force (parallel to the free stream) given by the expression

$$\begin{aligned} c_d &= \frac{1}{c}\int_0^c \left[(C_p)_u \frac{dz_u}{dx} - (C_p)_l \frac{dz_l}{dx}\right] dx \\ &= \frac{2}{\beta c}\int_0^c [\lambda_u^2 + \lambda_l^2]dx = \frac{2}{\beta c}\int_0^c [\lambda_c^2 + \lambda_t^2]dx \end{aligned} \tag{7-2}$$

The appearance, even in a perfect fluid, of drag at supersonic speeds is in striking contrast to its nonexistence in subsonic flow (d'Alembert's paradox). For reasons that will become apparent in the two following articles this retarding force is referred to as *wave drag* (cf. A,4).

Finally, moment coefficient about the leading edge (positive when the trailing edge is forced down) is

$$\begin{aligned} c_m &= -\frac{1}{c^2}\int_0^c x \frac{\Delta p}{\frac{1}{2}\rho_\infty U^2}\,dx \\ &= \frac{4}{\beta c^2}\int_0^c x \frac{dz_c}{dx}\,dx \end{aligned} \tag{7-3}$$

D,8. Plane Waves and Applications. The general solution

$$\varphi(x, z) = f(x - \beta z) + g(x + \beta z)$$

of the basic potential equation (5-1) represents two standing waves of small magnitude; and if the airfoil in supersonic flight is thought of as a

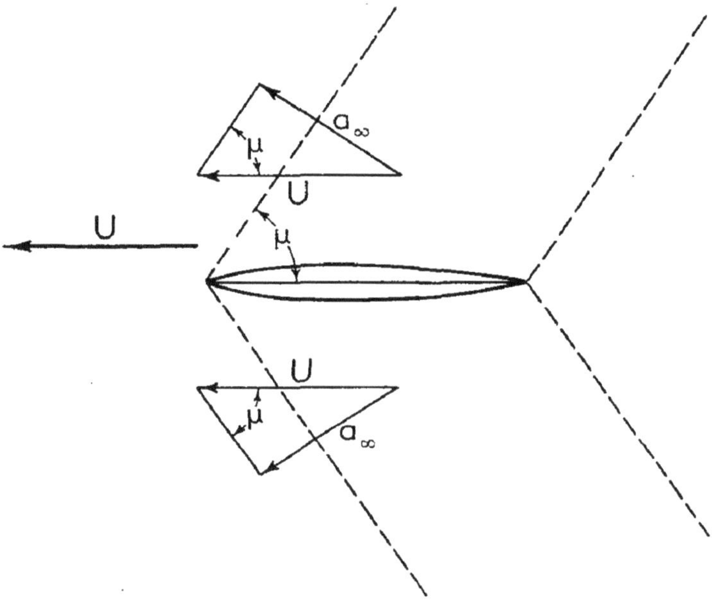

Fig. D,8a. Airfoil in steady supersonic flight into undisturbed medium.

generator of such waves, some added insight is gained as to the nature of the two-dimensional disturbance field [*13*]. The waves are inclined at the angles $\pm\mu = \pm \tan^{-1}(1/\beta) = \pm \sin^{-1}(1/M_\infty)$ to the direction of flight where μ is the Mach angle. In Fig. D,8a an airfoil is shown moving toward the left with velocity U greater than a_∞, the velocity of sound in the undisturbed medium. It is obvious from the geometry that the velocity of the wave front normal to itself is equal to a_∞ and the wave is therefore subject to laws of acoustic disturbances. The fore and aft limits of the waves are determined by the Mach lines from the leading and trailing edges of the airfoil.

In order to calculate the velocity profile of the wave we note first that, if $\delta \cong \tan \delta = \lambda_u(x)$ is the angle between the upper surface at a point and the direction of motion, then the velocity component of the upper surface normal to itself is $U \sin \delta$. Moreover, this expression must equal the velocity component of the air normal to the surface. Hence

$$U \sin \delta = u_n \cos (\mu - \delta) \tag{8-1}$$

where u_n is the component of velocity normal to the Mach wave. For small values of δ, Eq. 8-1 becomes

$$u_n = U\lambda_u \sec \mu = \frac{U^2\lambda_u}{(U^2 - a_\infty^2)^{\frac{1}{2}}} \tag{8-2}$$

Since $u = -u_n \sin \mu$, it follows that

$$C_p = \frac{2\lambda_u}{\beta}$$

and we have thus provided an alternate method of deriving the results given in Eq. 6-4.

In the development of classical acoustic theory, Rayleigh [*14*, Vol. 2, p. 17] and Lamb [*15*, p. 479] divided the energy in a wave into two parts: kinetic and intrinsic energy. The kinetic energy K can be expressed in the form

$$K = \tfrac{1}{2}\rho_\infty \iiint u_n^2 dxdydz \tag{8-3}$$

where the volume integral encompasses the disturbed region. The intrinsic energy I, on the other hand, measures the work done by the fluid against external pressure in expanding from its actual volume to its volume in the undisturbed condition. Applying this calculation first to a unit mass of fluid of volume v where $v = 1/\rho$, we have

$$I_{\text{unit mass}} = -\int_{v_\infty}^{v} p dv$$

Since the variation in the variables is assumed small, the integral may be approximated by evaluating p at the mid-point of the interval and then multiplying by the variation in v. Thus, using a series expansion for p, we have

$$I_{\text{unit mass}} \cong -\left[p_\infty + \left(\frac{dp}{dv}\right)_\infty \frac{dv}{2}\right] dv$$

If this expression is integrated over all the mass elements of the system, the contribution of the term $p_\infty dv$ will vanish whenever the total change of volume is zero. Moreover, the second term becomes

$$-\frac{dpdv}{2} = \frac{(dp)^2}{2\rho^2}\frac{d\rho}{dp}$$

so that, after setting $dp = -\rho_\infty U\, u = \rho_\infty u_n a_\infty$ and $dp/d\rho = a_\infty^2$, we have $I_{\text{unit mass}} = u_n^2/2$. It follows that

$$I = \tfrac{1}{2}\rho_\infty \iiint u_n^2 dx dy dz \tag{8-4}$$

and *the kinetic and the intrinsic energies of the wave system are equal.*

Since the motion of an airfoil produces plane waves and since the energy in the waves must equal the work done by the airfoil in moving through the air, that is, the sum of the kinetic and intrinsic energies must

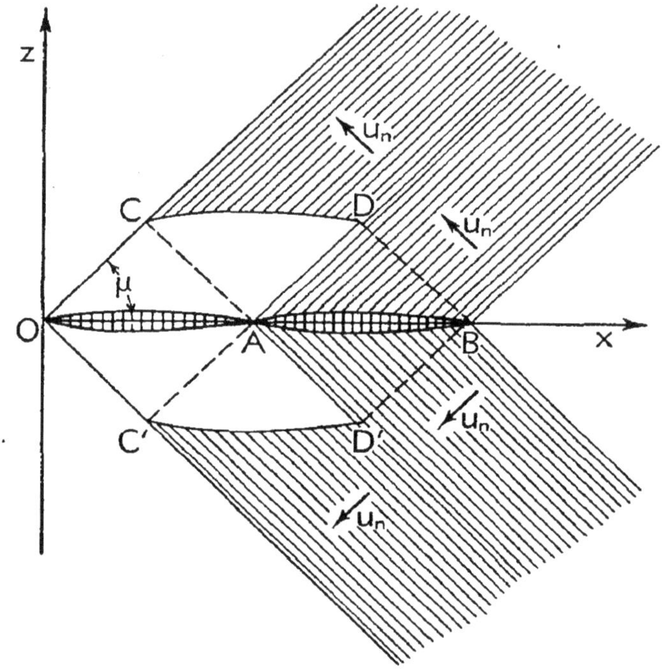

Fig. D,8b. Areas used to calculate airfoil drag by energy methods.

equal the product of drag and distance traveled, it follows that the calculation of drag by means of energy methods is feasible.

The total energy in the waves associated with the airfoil is infinite since in this steady state case the waves are infinite in extent and the airfoil has supposedly traveled far enough to reach asymptotic conditions. Consider, however, the energy added to the wave system while the airfoil moves one chord length; this must equal Dc where D is drag and c is chord length.

In Fig. D,8b the airfoil is shown as originally at AB and in its later position at OA. The two waves move obliquely on the plane, that is, normal to the Mach lines so that in the later position CD and $C'D'$ fix the lower limits of the waves abutting first on the airfoil. Thus, the regions $OADC$ and $OAD'C'$ contain the newly induced field and from Eq. 8-2, u_n is known in terms of airfoil slope which is in turn a function of x.

The fact that no interaction exists between the two wave systems permits us to calculate drag on the upper surface by evaluating the energy in $OADC$ and drag on the lower surface from the energy in $OAD'C'$. Moreover, the total energy has been shown to equal twice the kinetic energy. Consider first the upper region and, in order to utilize the fact that conditions are constant along Mach lines, introduce the coordinate system $\xi = x$, $\eta = z \csc \mu$. Denoting section drag on the upper surface by D_u, we have, from Eq. 8-3 and 8-4 applied to a strip of unit width,

$$\begin{aligned} D_u c &= \rho_\infty \int_0^c d\xi \int_0^{c \cos \mu} u_n^2 \sin \mu d\eta \\ &= \rho_\infty \frac{U^4}{U^2 - a_\infty^2} \int_0^c \lambda_u^2 c \sin \mu \cos \mu d\xi \end{aligned}$$

It follows that

$$D_u = \frac{\rho_\infty U^2}{\beta} \int_0^c \lambda_u^2 d\xi \tag{8-5}$$

and by exactly the same type of analysis, the drag on the lower surface is

$$D_l = \frac{\rho_\infty U^2}{\beta} \int_0^c \lambda_l^2 d\xi \tag{8-6}$$

When combined, these equations give a result in agreement with Eq. 7-2.

D,9. Momentum Relations and Applications. Although the generation of plane waves by a supersonic airfoil permits us to use energy methods in the calculation of drag, the most successful applications to the prediction of force characteristics for all classes of wings are based on momentum considerations. This type of analysis always depends on the calculation of contour integrals around a control surface C enclosing the body and, as in the case of contour integrations in general, the choice of the proper control surface sometimes enables one to avoid involved integrations through the use of some ingenuity. In this article we confine ourselves to two-dimensional flow and in subsequent chapters return to the three-dimensional aspects of momentum theory.

Consider, as in Fig. D,9a, an airfoil immersed in a stream of velocity U and let the airfoil be contained within a simple closed curve C. If the components of the resultant force on the wing are F_x and F_z, the components of pressure forces on the body and the curve will just balance the rate of increase of momentum of the fluid leaving the given region and we may consequently write

$$F_x = -\int_C p dz - \int_C \rho (U + u)[(U + u)dz - w dx] \tag{9-1}$$

$$F_z = \int_C p dx + \int_C \rho w[w dx - (U + u)dz] \tag{9-2}$$

Since

$$\int_C p_\infty dz = 0 \quad \text{and} \quad U \int_C \rho[(U+u)dz - wdx] = mU$$

where m is the net mass of fluid generated within the region, Eq. 9-1 and 9-2 become

$$F_x = -\int_C (p - p_\infty)dz - \int_C \rho[(uU + u^2)dz - uwdx] - mU \quad (9\text{-}3)$$

$$F_z = \int_C (p - p_\infty)dx + \int_C \rho w[wdx - (U+u)dz] \quad (9\text{-}4)$$

We assume now that m is always zero and proceed to modify these equations under the assumptions of small perturbation theory. The nature

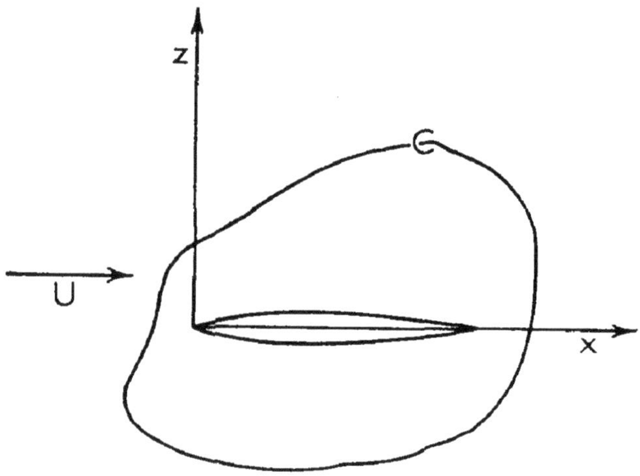

Fig. D,9a. Airfoil and enclosing contour, C.

of the expression for F_x is such that the first order terms exactly balance and it is therefore necessary to retain in that equation the second order. Using the relations

$$\rho = \rho_\infty\left(1 - M_\infty^2 \frac{u}{U}\right),$$

$$p - p_\infty = -\rho_\infty\left[Uu + \frac{1}{2}(u^2 + w^2)\right] + \frac{1}{2}\rho_\infty M_\infty^2 u^2 \quad (9\text{-}5)$$

we have

$$F_x = \tfrac{1}{2}\rho_\infty \int_C \{2uwdx + [(M_\infty^2 - 1)u^2 + w^2]dz\} \quad (9\text{-}6)$$

and

$$F_z = -\rho_\infty U \int_C [udx + wdz] \quad (9\text{-}7)$$

or, in coefficient form,

$$c_d = \frac{1}{c} \int_C \left\{ 2 \frac{u}{U} \frac{w}{U} dx + \left[(M_\infty^2 - 1) \frac{u^2}{U^2} + \frac{w^2}{U^2} \right] dz \right\} \tag{9-8}$$

and

$$c_l = \frac{2}{c} \int_C \left[\frac{u}{U} dx + \frac{w}{U} dz \right] \tag{9-9}$$

These equations hold for both subsonic and supersonic linearized theory.

Eq. 9-7 may also be written

$$L = -\rho_\infty U \int_C V_s ds \tag{9-10}$$

where V_s is the tangential component of perturbation velocity enclosing the airfoil and L is the lift. By definition, the circulation Γ around a simple closed curve C in a flow field is

$$\Gamma = - \int_C V_s ds = - \int_C (udx + wdz) \tag{9-11}$$

and, from Eq. 9-10 and 9-11, it follows that the Kutta theorem for lift

$$L = \rho_\infty U \Gamma \tag{9-12}$$

holds in linear theory for subsonic and supersonic Mach numbers.

Eq. 9-8 also provides an easy method of calculating drag for supersonic airfoils. In Fig. D,9b an arbitrary profile is shown with one example of an integration contour. Since along AD, BA, and BC the perturbation velocities vanish, we have

$$c_d = \frac{1}{c} \int_{CE} \left[\beta^2 \left(\frac{u}{U}\right)^2 + \left(\frac{w}{U}\right)^2 \right] dz + \frac{1}{c} \int_{ED} \left[\beta^2 \left(\frac{u}{U}\right)^2 + \left(\frac{w}{U}\right)^2 \right] dz$$

Denoting by $\lambda_u(x)$ and $\lambda_l(x)$ the slopes of the upper and lower surfaces, it follows that on ED we have

$$\frac{u}{U} = \frac{-\lambda_u(c - \beta z)}{\beta}, \qquad \frac{w}{U} = \lambda_u(c - \beta z)$$

and on EC

$$\frac{u}{U} = \frac{\lambda_l(c + \beta z)}{\beta}, \qquad \frac{w}{U} = \lambda_l(c + \beta z)$$

Hence, the drag coefficient is

$$c_d = \frac{1}{c} \int_0^{c/\beta} 2\lambda_u^2(c - \beta z) dz + \frac{1}{c} \int_{-c/\beta}^0 2\lambda_l^2(c + \beta z) dz$$

and the substitutions $c \mp \beta z = \xi$ lead to previously obtained results.

It is of interest to consider the determination of drag coefficient by means of Eq. 9-8 in the case when the contour C is shrunk so as to approach the surface of the airfoil as a limit. Provided no singularities in the velocity components appear on the profile, the resulting expression is

$$c_d = \frac{1}{c} \int_C 2 \frac{u}{U} \frac{w}{U} \, dx \tag{9-13}$$

Remembering that line integrals are always evaluated in a counterclockwise direction, and noting that either u or w merely changes sign in

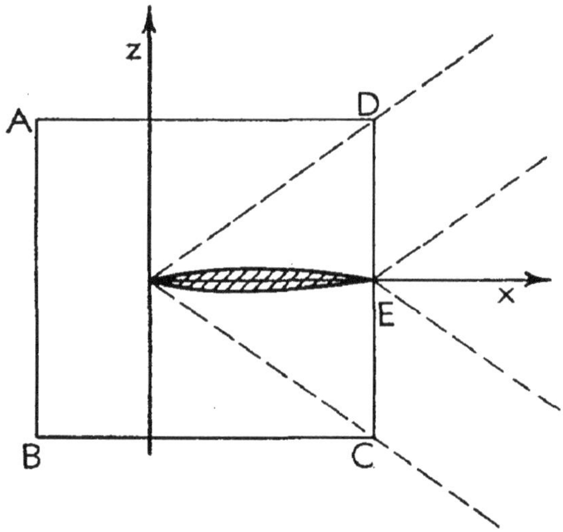

Fig. D, 9b. Airfoil with particular contour.

going from the lower to the upper surface of a camber line or a symmetrical airfoil, we get for these particular cases

$$c_d = \frac{2}{c} \int_0^c -2 \frac{u_u}{U} \frac{w_u}{U} \, dx \tag{9-14}$$

The term $-2u_u/U$ in this latter integral is, of course, the coefficient of pressure normal to the upper surface while the factor w_u/U is surface slope and, in thin airfoil theory, the product is the force component measured nondimensionally in the free stream direction.

In the application of Eq. 9-14 to a flat plate at an angle of attack $\alpha = -w_u/U$, we have $c_d/c_l = \alpha$; and as a consequence the resultant force vector is normal to the airfoil surface in conformity with the fact that it is the integral of the normal pressure forces. Fig. D,9c shows the distribution of lift on the flat plate in supersonic flow. From Eq. 6-7 we have $\Delta p/\frac{1}{2}\rho_\infty U^2 = 4\alpha/\beta$ and therefore $c_l = 4\alpha/\beta$, $c_d = 4\alpha^2/\beta$.

The general momentum relation of Eq. 9-6 is applicable to both subsonic and supersonic flows so it would appear that Eq. 9-13 and 9-14 are also applicable for the entire Mach number range. For example, a flat plate at an angle of attack α in a subsonic stream has a load distribution given by the expression

$$\frac{\Delta p}{\frac{1}{2}\rho_\infty U^2} = \frac{4\alpha}{\beta}\sqrt{\frac{c-x}{x}}$$

so that presumably

$$c_l = \frac{2\pi\alpha}{\beta}, \qquad c_d = \frac{2\pi\alpha^2}{\beta}$$

It is well known, however, that the resultant force vector is normal to the free stream for subsonic flow and that drag is zero. The existence of this

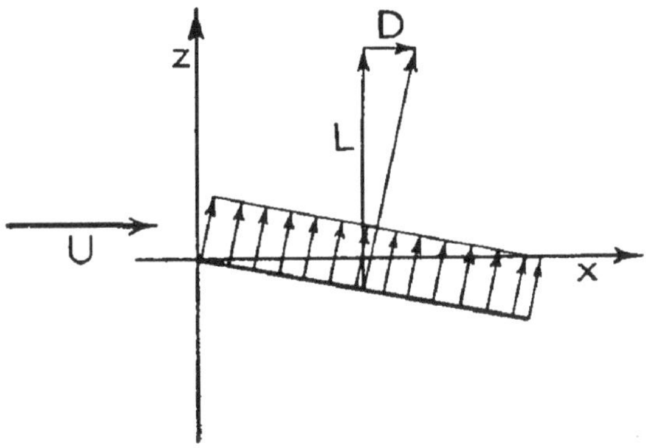

Fig. D,9c. Lift and drag vector orientations.

paradoxical result is brought about by the fact that the velocity components do not remain finite in the latter case and thus the assumptions underlying the derivation of Eq. 9-13 and 9-14 no longer apply. Actually, the infinite value of u at the leading edge of the airfoil produces a finite thrust or suction force on the airfoil even though applied over an infinitesimal area.

In order to predict quantitatively the thrust on an arbitrary airfoil it is sufficient to apply the momentum integral of Eq. 9-6 to a contour enclosing the leading edge and to evaluate the integral in the limit as the contour shrinks about the point. If the singularity in the x component of the perturbation velocity is expressible in the form

$$\frac{u}{U} = \frac{G}{\beta\sqrt{x}} \tag{9-15}$$

the contribution to the drag coefficient of the pressure at the point is

$$c_d = -\frac{2\pi G^2}{\beta c} \tag{9-16}$$

For a flat plate, G is equal to $\alpha\sqrt{c}$ and the thrust is just sufficient to overcome the drag calculated above. The concept of leading edge thrust plays an important part in the next chapter in the calculation of drag for supersonic wings with leading edges swept behind the foremost Mach cone.

Eq. 9-16 is based on the existence of infinite values of u. A companion result can be derived for airfoils with blunt ends ([16]; also VII,A). Thin airfoil theory introduces infinite values of vertical induced velocity in this case and, again, the integration of Eq. 9-13 gives incorrect results. If the vertical induced velocity at the nose of the airfoil satisfies the relation

$$\frac{w}{U} = \frac{G}{\sqrt{x}} \tag{9-17}$$

analysis of the problem reveals that the drag coefficient contributed at that point is

$$c_d = \frac{2\pi G^2}{\beta c} \tag{9-18}$$

The infinite positive pressure at the nose thus produces a leading edge drag.

If the drag is calculated, for an arbitrary subsonic airfoil with a blunt leading edge, by using Eq. 9-13, the result will in general indicate that the airfoil has thrust. The drag force calculated by Eq. 9-18 is just sufficient to destroy this thrust and produce a resultant zero drag. In the case of three-dimensional sweptback supersonic wings at zero angle of attack, however, the resultant drag is not zero and specific knowledge of the leading edge drag in terms of the nose shape is important. In this nonlifting case we have considered only singularities in velocity distribution at the leading edge. In the case of the Joukowski-type airfoil the trailing edge is cusped and the flow leaves the airfoil smoothly. When the trailing edge is rounded, Eq. 9-18 still suffices in the treatment of such problems provided G is chosen to conform with the geometry of the airfoil and the direction of the force is considered.

CHAPTER 3. THREE-DIMENSIONAL SUPERSONIC STEADY STATE FLOW

D,10. Basic Potential Equation and Solutions. The linearized partial differential equation governing three-dimensional supersonic steady state flow is known to be

$$\beta^2\varphi_{xx} - \varphi_{yy} - \varphi_{zz} = 0 \tag{10-1}$$

where $\beta^2 = M_\infty^2 - 1$. The present chapter is devoted to the study of problems in wing theory and, as a consequence, is repeatedly concerned with obtaining solutions of Eq. 10-1 subject always to certain physical boundary conditions. The problems before us are of a dual character:

first, the prediction of the pressure on the wing surface resulting from either the thickness of a symmetrical wing or the twist and camber of a lifting surface; second, the prediction of the wing shape corresponding to a specified pressure or load distribution. Our first concern is the development, from the basic solution given in Eq. 3-12, of certain fundamental formulas that apply directly to the known problems. These formulas are subsequently used to obtain solutions of several types of wing problems. In all cases the plane of the wing is conveniently situated in the x, y plane.

Volterra's solution of Eq. 10-1, as given in Art. 3, depended chiefly on the use of an elementary solution that vanished on the characteristic surface or Mach forecone with vertex at the field point x, y, z. Experience in wing theory for subsonic speeds has shown, however, that use of particular elementary solutions provides a natural development of the analysis. With the idea of establishing a parallelism between subsonic and supersonic studies, we attempt to express our fundamental formulas in terms of sources, doublets, and elementary vortices or lifting elements. The perturbation potentials of each of these elements of strength m are, respectively,

Source
$$-\frac{m}{2\pi\sqrt{(x-x_1)^2-\beta^2(y-y_1)^2-\beta^2(z-z_1)^2}}$$

Doublet with vertical axis
$$\frac{m\beta^2(z-z_1)}{2\pi[(x-x_1)^2-\beta^2(y-y_1)^2-\beta^2(z-z_1)^2]^{3/2}} \tag{10-2}$$

Elementary vortex
$$-\frac{m(z-z_1)(x-x_1)}{2\pi[(y-y_1)^2+(z-z_1)^2]\sqrt{(x-x_1)^2-\beta^2(y-y_1)^2-\beta^2(z-z_1)^2}}$$

In each case the position of the element is at the point with coordinates x_1, y_1, z_1 and is defined only for points within the Mach cone with vertex at that point. We have chosen to identify the analytic expressions with the perturbation potential $\varphi(x, y, z)$ but it should be borne in mind that they can be equated to any function satisfying the basic differential equation. Thus, since $\varphi_x = u$ satisfies Eq. 10-1, and since u is directly proportional to the pressure coefficient, in some investigations it is convenient to set u equal to one of the above expressions and such concepts as "pressure source" then arise.

The three fundamental formulas in supersonic wing theory. Before deriving the three fundamental formulas to be used, we first recall the meaning of the above elementary solutions in terms of their effect on the flow field. A source involves the concept of introducing fluid into the free stream and thus causes a splitting of the stream lines. On the other hand, doublets and vortices introduce no net amount of fluid into the stream, but they do generate circulation and thus can produce a jump in potential or pressure respectively in their plane of distribution. For planar distributions of such elements we can say specifically that a source distribution is one that relates the velocity potential to a discontinuity, in the plane, of the vertical induced velocity Δw; a doublet distribution is one that relates φ to a jump in the potential $\Delta\varphi$; and an elementary vortex or lifting element distribution is one that relates φ to a jump in the pressure Δp, or, in other words, to a loading.

With these ideas in mind it follows that the velocity potential induced at the point x, y, z by a distribution of sources over the $z_1 = 0$ plane is evidently given by the part of Eq. 3-12 which can be written, for $\Omega = \varphi$,

$$\varphi(x, y, z) = -\frac{1}{2\pi}\frac{\partial}{\partial x}\iint_{\tau} \Delta w \cosh^{-1}\frac{x - x_1}{\beta\sqrt{(y - y_1)^2 + z^2}}\,dx_1dy_1 \tag{10-3}$$

In this form the interpretation of Eq. 10-3 in terms of sources is not apparent. If, however, the partial derivative with respect to x is taken through the double integral sign,[3] the expression

$$\varphi = -\frac{1}{\pi}\iint_{\tau}\frac{w_u(x_1, y_1)dx_1dy_1}{\sqrt{(x - x_1)^2 - \beta^2(y - y_1)^2 - \beta^2z^2}} \tag{10-4}$$

is obtained where w_u is the vertical induced velocity on the upper side of the x, y plane and thus corresponds to the boundary condition for a nonlifting wing with a symmetrical section ($w_u = \frac{1}{2}\Delta w$). Eq. 10-4 is the first fundamental formula and was first applied in supersonic wing theory by Puckett [*17*]. As in subsonic flow, the source strength per unit area is proportional to the local slope of the wing.

It is also evident, from inspection of Eq. 3-12, that the velocity potential at the point x, y, z corresponding to a distribution of doublets over the x_1, y_1 plane is given by

$$\varphi = \frac{1}{2\pi}\frac{\partial}{\partial x}\iint_{\tau}\frac{z(x - x_1)\Delta\varphi(x_1, y_1)dx_1dy_1}{[(y - y_1)^2 + z^2]\sqrt{(x - x_1)^2 - \beta^2(y - y_1)^2 - \beta^2z^2}} \tag{10-5}$$

although, as in the case of the source, the appearance of the doublet is not immediately recognizable. Quite obviously the partial derivative should again be taken through the double integral sign. This time, however, such a manipulation must be performed with some caution and, depending on the order of integration,[4] an extra term may or may not appear.

[3] Such a manipulation is possible and introduces no extra terms since the inverse hyperbolic term vanishes everywhere on the Mach forecone.

[4] Since the order of integration is important in the subsequent analysis, the notation $\int dy \int dx f(x, y) \equiv \int[\int f(x, y)dx]dy$ is adopted: that is, the left member is used when the integration is made first with respect to x. When the notation $\iint f(x, y)dxdy$ is used, the order of integration is immaterial.

Consider the case when the integration is made first with respect to y_1. Introducing the notation

$$Y_1 = y + \frac{1}{\beta}\sqrt{(x - x_1)^2 - \beta^2 z^2}, \qquad Y_2 = y - \frac{1}{\beta}\sqrt{(x - x_1)^2 - \beta^2 z^2}$$

it is clear, from Fig. D,10a, that we have

$$\varphi = \frac{1}{2\pi}\frac{\partial}{\partial x}\int_{-\infty}^{x-\beta z} dx_1 \int_{Y_2}^{Y_1} dy_1 \frac{z(x - x_1)\Delta\varphi(x_1, y_1)}{[(y - y_1)^2 + z^2]\sqrt{(x - x_1)^2 - \beta^2(y - y_1)^2 - \beta^2 z^2}}$$

and that, furthermore, if in the y_1 integration x_1 is replaced by $x - \beta z$, the result is indeterminate. Such an indeterminate form can be evaluated,

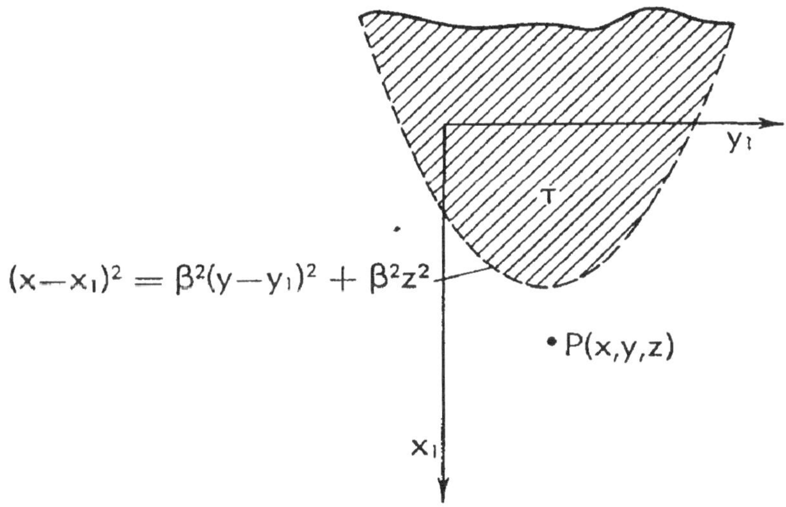

Fig. D,10a. The area τ, the portion of the $z_1 = 0$ plane in which sources can affect the flow at $P(x, y, z)$.

however, by excluding the limit $x - \beta z$ from the area of integration and then performing the appropriate limiting process. Hence, consider the integral

$$\varphi = \lim_{\epsilon\to 0}\frac{1}{2\pi}\frac{\partial}{\partial x}\int_{-\infty}^{x-\beta\sqrt{z^2+\epsilon^2}} dx_1 \int_{Y_2}^{Y_1} dy_1 \frac{z(x - x_1)\Delta\varphi(x_1, y_1)}{[(y - y_1)^2 + z^2]\sqrt{(x - x_1)^2 - \beta^2(y - y_1)^2 - \beta^2 z^2}}$$

which, by formal manipulation, becomes

$$\varphi = \lim_{\epsilon\to 0}\frac{z}{2\pi}\int_{y-\epsilon}^{y+\epsilon}\frac{\sqrt{z^2 + \epsilon^2}\,\Delta\varphi(x - \beta\sqrt{z^2 + \epsilon^2}, y_1)}{[(y - y_1)^2 + z^2]\sqrt{\epsilon^2 - (y - y_1)^2}}\,dy_1$$
$$+ \frac{1}{2\pi}\int_{-\infty}^{x-\beta z} dx_1 \frac{\partial}{\partial x}\int_{Y_2}^{Y_1} dy_1 \frac{z(x - x_1)\Delta\varphi(x_1, y_1)}{[(y - y_1)^2 + z^2]r_c}$$

where we have set $r_c = \sqrt{(x - x_1)^2 - \beta^2(y - y_1)^2 - \beta^2 z^2}$. Application of the mean value theorem to the first of these terms yields, for $-1 < \theta < 1$,

$$\lim_{\epsilon \to 0} \frac{z}{2\pi} \sqrt{z^2 + \epsilon^2}\, \Delta\varphi(x - \beta \sqrt{z^2 + \epsilon^2}, y + \theta\epsilon)$$

$$\int_{y-\epsilon}^{y+\epsilon} \frac{dy_1}{[(y - y_1)^2 + z^2] \sqrt{\epsilon^2 - (y - y_1)^2}}$$

$$= \lim_{\epsilon \to 0} \frac{1}{2} \Delta\varphi(x - \beta \sqrt{z^2 + \epsilon^2}, y + \theta\epsilon) = \frac{1}{2} \Delta\varphi(x - \beta z, y)$$

The second term can best be simplified through the introduction of a special notation for singular integrals. As developed in the appendix at the end of this chapter, the notation is defined by the relation

$$\frac{da}{dx} \fint_0^{a(x)} \frac{\partial}{\partial a} \frac{dy}{\sqrt{a - y}} = \frac{\partial}{\partial x} \int_0^{a(x)} \frac{dy}{\sqrt{a - y}}$$

and by means of it Eq. 10-5 can be written in the form

$$\varphi = \frac{1}{2} \Delta\varphi(x - \beta z, y) - \frac{z\beta^2}{2\pi} \int_\tau dx_1 \fint dy_1 \frac{\Delta\varphi(x_1, y_1)}{r_c^3} \tag{10-6}$$

If the first integration is made with respect to x_1, the result is

$$\varphi = \frac{1}{2\pi} \frac{\partial}{\partial x} \int_{-\infty}^{\infty} dy_1 \int_{-\infty}^{x - \beta\sqrt{(y-y_1)^2 + z^2}} dx_1 \frac{z(x - x_1)\Delta\varphi(x_1, y_1)}{[(y - y_1)^2 + z^2] r_c}$$

which reduces immediately to

$$\varphi = -\frac{z\beta^2}{2\pi} \int_\tau dy_1 \fint dx_1 \frac{\Delta\varphi(x_1, y_1)}{r_c^3} \tag{10-7}$$

Eq. 10-6 and 10-7 represent the two forms of our second fundamental formula and give the perturbation velocity potential in space due to a distribution of doublets in the x, y, plane.

An instructive generalization of these latter results can be made by transforming the x, y coordinates in Eq. 10-5 to an oblique set of coordinates ξ and η, such as those shown in Fig. D,10b. It can be shown by a process identical to the one just presented that if $\mu_1 = \sqrt{1 + m_1^2}$ and $\mu_2 = \sqrt{1 + m_2^2}$, where m_1, m_2 are constants, and

$$x = \eta \frac{m_1}{\mu_1} + \xi \frac{m_2}{\mu_2}$$

$$y = \eta \frac{1}{\mu_1} - \xi \frac{1}{\mu_2}$$

$$z = z$$

then for $0 \leqq m_1 < \beta,\ 0 \leqq m_2 < \infty$

$$\varphi = \frac{1}{2}\Delta\varphi(\xi_a, \eta_a) - \frac{z\beta^2(m_1 + m_2)}{2\pi\mu_1\mu_2}\int_\tau d\xi_1 ⨍ d\eta_1 \frac{\Delta\varphi(\xi_1, \eta_1)}{r_c^3} \qquad (10\text{-}8)$$

and for $0 \leqq m_2 < \beta,\ 0 \leqq m_1 < \infty$

$$\varphi = \frac{1}{2}\Delta\varphi(\xi_b, \eta_b) - \frac{z\beta^2(m_1 + m_2)}{2\pi\mu_1\mu_2}\int_\tau d\eta_1 ⨍ d\xi_1 \frac{\Delta\varphi(\xi_1, \eta_1)}{r_c^3} \qquad (10\text{-}9)$$

where r_c is now, of course, expressed in terms of ξ, ξ_1, η, and η_1 in the form

$$r_c^2 = \frac{(\eta - \eta_1)^2(m_1^2 - \beta^2)}{\mu_1^2} + \frac{2(\eta - \eta_1)(\xi - \xi_1)(m_1 m_2 + \beta^2)}{\mu_1\mu_2} + \frac{(\xi - \xi_1)^2(m_2^2 - \beta^2)}{\mu_2^2} - \beta^2 z^2$$

The values of ξ_a, η_a and ξ_b, η_b are best defined[5] geometrically: they are the points at which the trace of the forecone in the $z_1 = 0$ plane

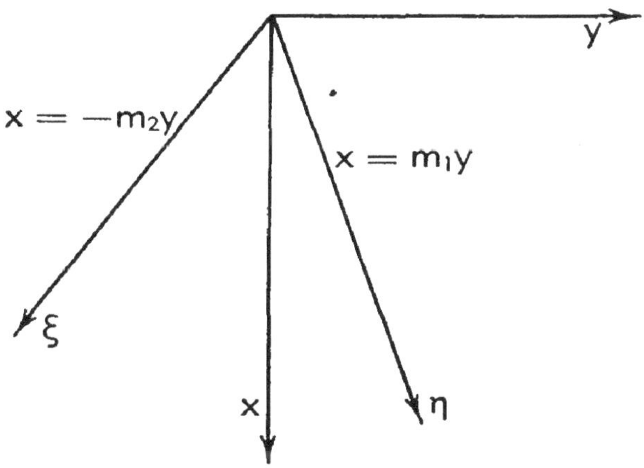

Fig. D,10b. Relation between x, y and ξ, η coordinates.

(given by one branch of the curve $r_c^2 = 0$, $z_1 = 0$) attains maximum values for ξ_1 and η_1 respectively; see Fig. D,10c.

As the ξ and η axes approach the Mach lines in the x, y plane, that is, as m_2 and m_1 approach β, the asymptotes of the hyperbola bounding τ in the ξ_1, η_1 plane become parallel to the ξ_1 and η_1 axes and the residue terms

[5] The mathematical expressions for these terms are

$$\xi_a = \xi - \frac{z\mu_2\sqrt{\beta^2 - m_1^2}}{m_1 + m_2}, \qquad \eta_a = \eta - \frac{\mu_1 z(m_1 m_2 + \beta^2)}{(m_1 + m_2)\sqrt{\beta^2 - m_1^2}}$$

$$\xi_b = \xi - \frac{z\mu_2(m_1 m_2 + \beta^2)}{(m_1 + m_2)\sqrt{\beta^2 - m_2^2}}, \qquad \eta_b = \eta - \frac{\mu_1 z\sqrt{\beta^2 - m_2^2}}{m_1 + m_2}$$

in Eq. 10-8 and 10-9 approach $\frac{1}{2}\Delta\varphi(\xi_a, -\infty)$ and $\frac{1}{2}\Delta\varphi(-\infty, \eta_b)$ respectively. These are the values of the discontinuity in the perturbation potential infinitely far ahead of and to the side of $P(\xi, \eta, z)$ and hence may be taken as zero. Consequently, when the ξ and η axes are placed along the traces of the characteristic or Mach cone the residue terms disappear[6] and the double integrals are independent of the order of integration. These characteristic coordinates play an important role

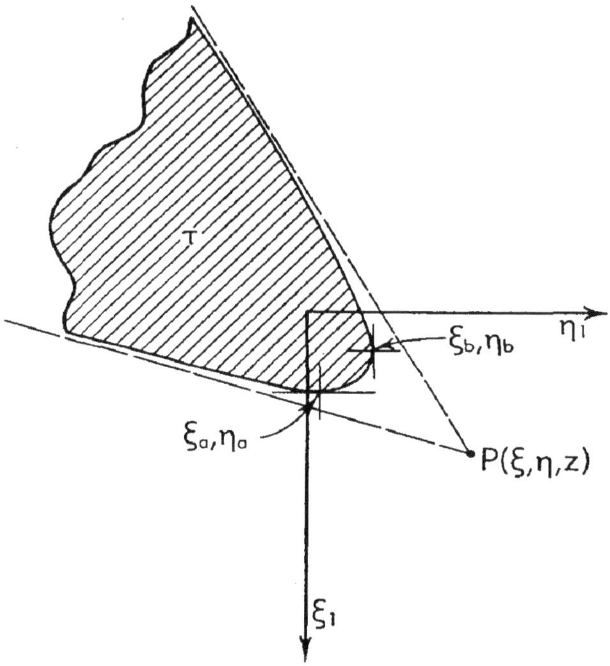

Fig. D,10c. The area τ in the ξ, η, z coordinate system.

in supersonic wing theory and are closely associated with the methods advanced by Evvard [*18*].

An interesting by-product of the theory is obtained by subtracting Eq. 10-9 and 10-8; thus, the difference between an integration of supersonic doublets made first in one order and then in the reverse order is equal to the difference in magnitude of the doublet distribution at two points in the plane.

The application of Eq. 3-12 in obtaining the velocity potential for a distribution of elementary vortices proceeds in a different manner than was the case for sources and doublets. One of the simplest ways of obtaining the desired results is to set Ω in Eq. 3-12 equal to the streamwise perturbation velocity u. Since thickness effects are to be deleted, we have $\Delta w = 0$ and from the irrotationality condition $\partial w/\partial x = \partial u/\partial z$ it follows that $\Delta(\partial u/\partial z) = 0$. Eq. 3-12 becomes

[6] The same is true if m_1 and m_2 are both greater than β. It is only when m_1 and m_2 are less than β that the residues appear.

$$u = \frac{1}{2\pi}\frac{\partial}{\partial x}\iint\limits_{\tau} \frac{z(x - x_1)\Delta u(x_1, y_1)}{[(y - y_1)^2 + z^2]r_c}\,dx_1dy_1 \tag{10-10}$$

Through use of the relation $\Delta p/\frac{1}{2}\rho_\infty U^2 = 2\Delta u/U$ and the definition $\varphi = \int_{-\infty}^{x} udx$ the perturbation potential for a distribution of elementary vortices can be written

$$\varphi = \frac{U}{4\pi}\iint\limits_{\tau} \frac{z(x - x_1)}{[(y - y_1)^2 + z^2]r_c}\,\frac{\Delta p(x_1, y_1)}{\frac{1}{2}\rho_\infty U^2}\,dx_1dy_1 \tag{10-11}$$

Eq. 10-11 is the third fundamental formula.

Reduction of the fundamental formulas to the plane of the wing. Most problems in wing theory are those in which certain data about a wing are specified and from these all remaining wing characteristics are to be determined. Usually, therefore, it is sufficient to restrict the equations developed in the above discussion to the special case when $z = 0$. Our present purpose is to find the effect of a source distribution on the pressure coefficient and the effect of a doublet or vortex distribution on the surface slope for the special case when z vanishes.

Such a reduction is especially easy in the case of sources and Eq. 10-4 becomes simply

$$\varphi(x, y, 0) = -\frac{1}{\pi}\iint\limits_{\tau} \frac{w_u(x_1, y_1)}{r_0}\,dx_1dy_1 \tag{10-12}$$

where $r_0 \equiv \sqrt{(x - x_1)^2 - \beta^2(y - y_1)^2}$. Reexpressed in terms of pressure coefficient, Eq. 10-12 becomes

$$C_p(x, y) = \frac{2}{\pi U}\frac{\partial}{\partial x}\iint\limits_{\tau} \frac{w_u(x_1, y_1)}{r_0}\,dx_1dy_1 \tag{10-13}$$

This latter relation is valid for a nonlifting wing with a symmetrical section. If λ_u is the slope of the wing's upper surface, we know $\lambda_u = w_u/U$, and a correspondence between wing slope and wing pressure coefficient is provided. Carrying the differential operator through the integral signs involves the same detailed operation we have used previously. The final results are

$$C_p(x, y) = \frac{2}{\beta}\lambda_u(x, y) - \frac{2}{\pi}\int\limits_{\tau} dx_1 \,⨍\, dy_1\,\frac{(x - x_1)\lambda_u(x_1, y_1)}{r_0^3} \tag{10-14}$$

and

$$C_p = -\frac{2}{\pi}\int\limits_{\tau} dy_1 \,⨍\, dx_1\,\frac{(x - x_1)\lambda_u(x_1, y_1)}{r_0^3} \tag{10-15}$$

The doublet formulas in Eq. 10-6 and 10-7 are of particular value when they are used to relate the vertical induced velocity in the plane of the wing to the jump in potential (i.e. the circulation) there. Such an expression can be derived by taking the partial derivative with respect to z of both sides of Eq. 10-6; there results

$$w(x, y, 0) = -\frac{\beta}{2}\Delta u(x, y) - \frac{\beta^2}{2\pi}\int_\tau dx_1 \,⨍\, dy_1 \frac{\Delta\varphi(x_1, y_1)}{r_0^3}$$

$$- \lim_{z\to 0}\frac{z\beta^2}{2\pi}\frac{\partial}{\partial z}\int_\tau dx_1 \,⨍\, dy_1 \frac{\Delta\varphi(x_1, y_1)}{r_0^3}$$

It can be shown that, since

$$⨍_{k_1}^{k_2} \frac{d\eta_1}{[(k_2 - \eta_1)(\eta_1 - k_1)]^{\frac{3}{2}}} = 0 \tag{10-16}$$

the last term in the expression vanishes. Finally, denoting $w(x, y, 0)$ by w_u (since there are no sources, $w_u = w_l$; in such cases we arbitrarily use the symbol w_u), we find

$$w_u = -\frac{\beta}{2}\Delta u(x, y) - \frac{\beta^2}{2\pi}\int_\tau dx_1 \,⨍\, dy_1 \frac{\Delta\varphi(x_1, y_1)}{r_0^3} \tag{10-17}$$

and a similar development for Eq. 10-7 gives

$$w_u = -\frac{\beta^2}{2\pi}\int_\tau dy_1 \,⨍\, dx_1 \frac{\Delta\varphi(x_1, y_1)}{r_0^3} \tag{10-18}$$

Equations similar to Eq. 10-17 and 10-18 can also be developed in terms of the ξ, η coordinates introduced in Fig. D,10b. In the special case when both the ξ and η axes lie on or inside the traces of the Mach cone; that is, for $\beta \leqq m_2 < \infty$, $\beta \leqq m_1 < \infty$

$$w_u = -\frac{\beta^2(m_1 + m_2)}{2\pi\mu_1\mu_2} ⨍⨍_\tau \frac{\Delta\varphi(\xi_1, \eta_1)}{r_0^3} d\xi_1 d\eta_1 \tag{10-19}$$

Eq. 10-19 is noteworthy because it is independent of the order of integration.

Finally it is necessary to reduce the formula involving vortex distributions, Eq. 10-11, to the x, y plane. This is of principal interest when it relates $w_u(x, y)$ to the wing loading $\Delta p/\frac{1}{2}\rho_\infty U^2$. Eq. 10-11 becomes

$$w_u(x, y) = \lim_{z\to 0}\frac{U}{4\pi}\frac{\partial}{\partial z}\iint_\tau \frac{z(x - x_1)}{[(y - y_1)^2 + z^2]r_c} \frac{\Delta p(x_1, y_1)}{\frac{1}{2}\rho_\infty U^2} dx_1 dy_1$$

Just as in the development of previous results, the order of integration is important and, if the first integration is made with respect to y_1, a residue term appears when the partial derivative is carried through the first integral. The type of analysis involved in deriving this term is now well established and leads to the expression

$$\frac{1}{U} w_u(x, y) = -\frac{\beta}{4} \frac{\Delta p(x, y)}{\frac{1}{2}\rho_\infty U^2} + \lim_{z\to 0} \frac{1}{4\pi} \int_\tau dx_1 \frac{\partial}{\partial z} \int dy_1 \frac{z(x - x_1)}{[(y - y_1)^2 + z^2]r_c} \frac{\Delta p(x_1, y_1)}{\frac{1}{2}\rho_\infty U^2}$$

The essential difficulty in reducing the second term in this equation is duplicated in the integral

$$I_0 = \lim_{z\to 0+} \int_{-a}^{b} \frac{\partial}{\partial z} \frac{zf(y, z)}{y^2 + z^2} dy, \qquad a, b > 0$$

It is shown in the appendix, see Eq. 14.1-16, that this can be written

$$I_0 = \,⨎_{-a}^{b} \frac{f(y, 0)}{y^2} dy + \pi f_z(0, 0)$$

(The notation ⨎ is also discussed in the appendix.) The method of evaluating this definite integral can be expressed quite simply; thus, if $G(y)$ is the indefinite integral of $f(y, 0)/y^2$ then for $a, b \neq 0$

$$⨎_{-a}^{b} \frac{f(y, 0)}{y^2} dy = G(b) - G(-a) \tag{10-20}$$

With the aid of this notation and the fact that the term corresponding to $\pi f_z(0, 0)$ is 0, the equation for w_u reduces to the form

$$\frac{1}{U} w_u(x, y) = -\frac{\beta}{4} \frac{\Delta p(x, y)}{\frac{1}{2}\rho_\infty U^2} + \frac{1}{4\pi} \int_\tau dx_1 ⨎ dy_1 \frac{(x - x_1)}{(y - y_1)^2 r_0} \frac{\Delta p(x_1, y_1)}{\frac{1}{2}\rho_\infty U^2} \tag{10-21}$$

and, when the order of integration is reversed, to the form

$$\frac{1}{U} w_u(x, y) = \frac{1}{4\pi} ⨎_\tau dy_1 \int dx_1 \frac{(x - x_1)}{(y - y_1)^2 r_0} \frac{\Delta p(x_1, y_1)}{\frac{1}{2}\rho_\infty U^2} \tag{10-22}$$

D,11. Direct Problems. The formulas developed in the preceding article enable us to solve a large class of problems encountered in wing theory. These problems can be divided into two principal types according to the manner in which the known data are related to the unknown function in the source, doublet, and vortex formulas. If the unknown function is to be determined by integrating known values, the problem is termed direct. On the other hand, if an unknown function is to be found such that the value of its integral is equal at every point to a known set of data, the problem is termed inverse. It is apparent that direct problems are much simpler than inverse ones, the former being essentially exercises in integration while the latter are problems requiring the inversion of an integral equation.

The two types of boundary-value problems encountered in the study of linearized wing theory that fall under the direct classification can be stated as follows:

Problem 1: The thickness case. The thickness of a wing that is symmetrical above and below a horizontal plane is given and the pressure distribution over the wing is to be determined. From a mathematical viewpoint, Eq. 10-14 or 10-15 apply, λ_u is given over the wing planform and C_p is to be determined.

Problem 2: The lifting case. The load distribution on a lifting plane, a surface without thickness, is given, and the slope of the surface that will support such a loading is to be determined. Again mathematically, Eq. 10-17 through 10-22 apply, $\Delta p/\frac{1}{2}\rho_\infty U^2$ or $\Delta\varphi$ is given over the wing planform and w_u is to be determined.

The following two applications involve solutions to these two types of problems.

Rectangular wing with biconvex section. As an example of a boundary-value problem listed as problem 1 above, consider a rectangular nonlifting wing with a chordwise section given by the equation

$$\lambda_u(x, y) = \frac{2t}{c_0^2}(c_0 - 2x) \tag{11-1}$$

where λ_u is the slope of the upper surface, c_0 is the chord, and t is the maximum thickness. The equation for the pressure on a wing such as shown in Fig. D,11a will change form in each of the four regions indicated. In region 1 the pressure will be the same as for a two-dimensional wing (i.e. a wing of infinite span), since the disturbances caused by the waves emanating from the side edges cannot enter this region. The remaining regions contain the three-dimensional effects, regions 2 and 3 experiencing disturbances from one edge or the other, and region 4 from both.

The equation for pressure coefficient on the wing follows immediately from Eq. 10-13, 10-14, or 10-15. Using Eq. 10-14, for example, there results for region 1,

$$C_p = \frac{4t}{\beta c_0^2}(c_0 - 2x) - \frac{4t}{\pi c_0^2}\int_0^x (c_0 - 2x_1)(x - x_1)dx_1 \, \mathop{{\int\!\!\!\!\!\!-}}\nolimits_{y-\frac{x-x_1}{\beta}}^{y+\frac{x-x_1}{\beta}} \frac{dy_1}{[(x - x_1)^2 - \beta^2(y - y_1)^2]^{\frac{3}{2}}}$$

From the result given in Eq. 10-16 it is readily apparent that the last term of the above expression is zero and the pressure coefficient in region 1 is

$$C_p = \frac{4t}{\beta c_0^2}(c_0 - 2x) \tag{11-2}$$

One can easily show that Eq. 10-15 yields the same result.

The evaluation of pressure coefficient in regions 2, 3, and 4 can be carried out in a similar fashion. A slightly different and sometimes very useful way of obtaining results in these and similar cases follows after

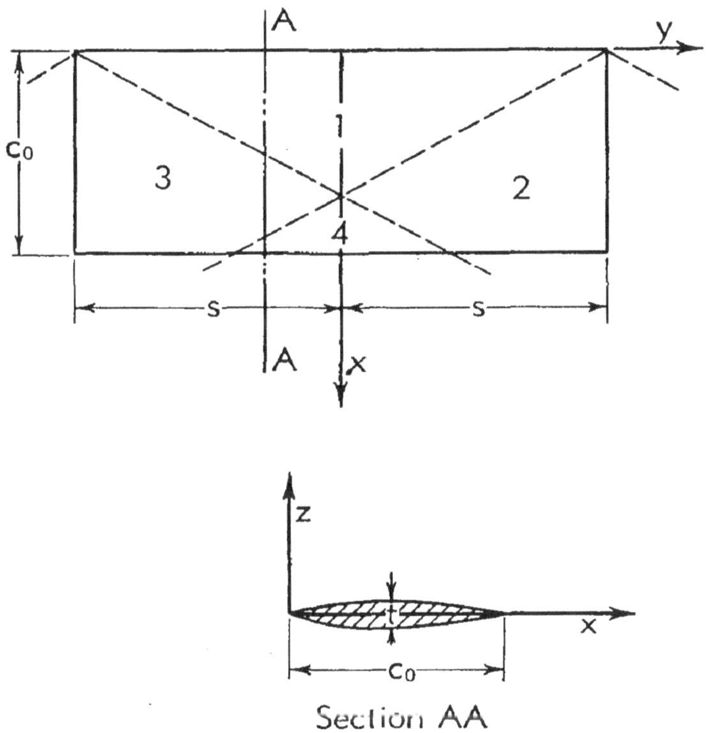

Fig. D,11a. Rectangular wing with biconvex section and regions used in describing its pressure distribution.

the introduction of complex variables. Consider the wing planform and regions shown in Fig. D,11b. The point P with coordinates x, y, 0, at which the pressure coefficient is to be found, is indicated and the region τ, over which the integrations in all of the previous equations have been made, is the area within the forecone from P and is labeled region b. The remaining area of the wing ahead of the line through P and parallel to the y_1 axis is labeled regions a and c. It is obvious that r_0 is a real number for all x_1, y_1 inside region b and a pure imaginary number for all x_1, y_1 lying in regions a and c. Further, it is clear from a physical viewpoint that terms in the integrands other than r_0 (or r_0^3) are always real on the wing planform. Hence, if the symbol τ_i is introduced as representing the area given by regions a, b, and c in Fig. D,11b, the area τ can be replaced by the area τ_i and the real part of the result will be the correct answer for pressure coefficient.[7]

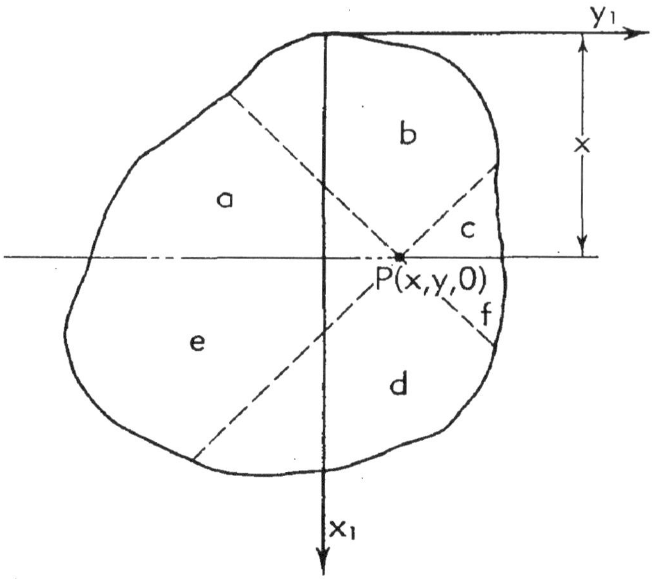

Fig. D,11b. Regions of wing bounded by fore and after Mach cone from point $P(x, y, 0)$.

We return now to the problem of the rectangular wing with a biconvex section. For all points on the wing, Eq. 10-14 becomes

$$C_p(x, y) = \frac{4t}{\beta c_0^2}(c_0 - 2x) - \frac{4t}{\pi c_0^2}\,\text{R.P.}\int_0^x dx_1 \,⨍_{-s}^{s} dy_1 \frac{(c_0 - 2x_1)(x - x_1)}{r_0^3}$$

and similarly Eq. 10-15 becomes

$$C_p(x, y) = -\frac{4t}{\pi c_0^2}\,\text{R.P.}\int_{-s}^{s} dy_1 \,⨍_0^{x} dx_1 \frac{(c_0 - 2x_1)(x - x_1)}{r_0^3}$$

where the letters R.P. indicate that the real part of the integral (a discussion of the integration technique is given in the appendix) is to be taken. Evaluation of the former equation gives

$$\begin{aligned} C_p = \frac{4t}{\beta c_0^2}(c_0 - 2x) + \frac{4t}{\pi c_0^2}\,\text{R.P.}\Bigg[&-\frac{c_0 - 2x}{\beta}\cos^{-1}\frac{\beta(s - y)}{x} \\ &+ 2(y - s)\cosh^{-1}\frac{x}{\beta(s - y)} - \frac{c_0 - 2x}{\beta}\cos^{-1}\frac{\beta(s + y)}{x} \\ &- 2(y + s)\cosh^{-1}\frac{x}{\beta(y + s)}\Bigg] \qquad (11\text{-}3) \end{aligned}$$

[7] The areas e and f could also be included in the definition of τ_i since r_0 is a pure imaginary within these regions also. However, region d must not be included since here r_0 is real. The definition of τ_i adopted here is the most convenient.

The real and imaginary parts of the $\cos^{-1}$ and $\cosh^{-1}$ terms in this equation are given in Table D,11 for all real values of the argument.

Eq. 11-3 contains, at once, the entire solution to the wing shown in Fig. D,11a. Thus, in region 1 none of the terms in the brackets has a real part so that the solution in this region is

$$C_p = \frac{4t}{\beta c_0^2}(c_0 - 2x)$$

which coincides with Eq. 11-2. In region 3 the solution can be written

$$C_p = \frac{4t}{\pi\beta c_0^2}\left\{(c_0 - 2x)\left[\pi - \cos^{-1}\frac{\beta(y+s)}{x}\right] - 2\beta(y+s)\cosh^{-1}\frac{x}{\beta(y+s)}\right\}$$

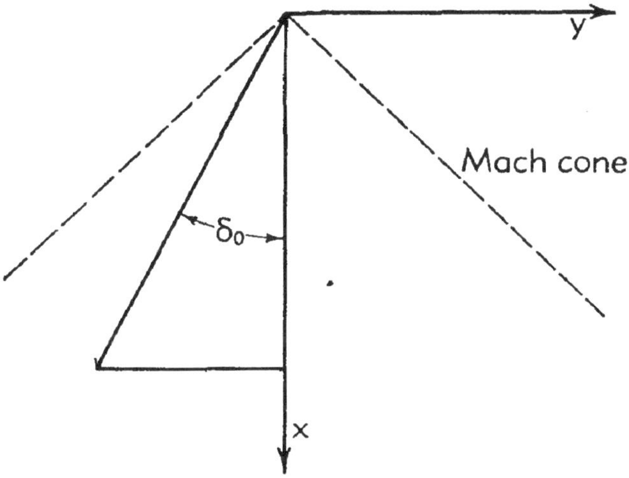

Fig. D,11c. Planform of right triangular wing.

The solution in region 2 follows from that given for region 3 by symmetry; and, finally, the expression for the pressure coefficient in region 4 is given by Eq. 11-3 wherein every term is real.

Table D,11. Real and imaginary parts of inverse trigonometric and hyperbolic terms.

Range of x	$-1 \geqq x$	$-1 \leqq x \leqq 1$	$1 \leqq x$
$\cosh^{-1} x = \int_1^x \frac{dt}{\sqrt{t^2 - 1}}$	$i\pi + \cosh^{-1}(-x)$	$i \cos^{-1} x$	$\cosh^{-1} x$
$\cos^{-1} x = \int_x^1 \frac{dt}{\sqrt{1 - t^2}}$	$\pi - i \cosh^{-1}(-x)$	$\cos^{-1} x$	$i \cosh^{-1} x$
$\sin^{-1} x = \int_0^x \frac{dt}{\sqrt{1 - t^2}}$	$-\frac{\pi}{2} + i \cosh^{-1}(-x)$	$\sin^{-1} x$	$\frac{\pi}{2} - i \cosh^{-1} x$

Triangular plate with uniform loading. The second type of boundary-value problem arising in the study of problems of the first kind can be illustrated by finding the camber and twist necessary to support a uniform loading on a wing of triangular planform. The wing is a right triangle with leading edge swept back of the foremost Mach cone, that is, with a subsonic leading edge (see Fig. D,11c). The solution can be obtained from, for example, Eq. 10-21. Hence, if $(\Delta p/\frac{1}{2}\rho_\infty U^2)_0$ is the magnitude of the uniform load and if δ_0 is the apex angle of the planform, we have

$$\frac{w_u}{U} = -\frac{\beta}{4}\left(\frac{\Delta p}{\frac{1}{2}\rho_\infty U^2}\right)_0 + \frac{1}{4\pi}\left(\frac{\Delta p}{\frac{1}{2}\rho_\infty U^2}\right)_0 \left[\int_0^{x+\beta y} (x - x_1)dx_1 \ {\int\!\!\!\!\!\!=}_{-x\tan\delta_0}^{0} \frac{dy_1}{(y-y_1)^2 r_0} + \int_{x+\beta y}^{\frac{x-\beta y}{1+\beta\tan\delta_0}} (x-x_1)dx_1 \ {\int\!\!\!\!\!\!=}_{-x\tan\delta_0}^{y+\frac{x-x_1}{\beta}} \frac{dy_1}{(y-y_1)^2 r_0} + \int_{\frac{x-\beta y}{1+\beta\tan\delta_0}}^{x} (x-x_1)dx_1 \ {\int\!\!\!\!\!\!=}_{y-\frac{x-x_1}{\beta}}^{y+\frac{x-x_1}{\beta}} \frac{dy_1}{(y-y_1)^2 r_0}\right]$$

and since

$$\int \frac{dy_1}{(y-y_1)^2 r_0} = \frac{r_0}{(x-x_1)^2(y-y_1)} + C$$

this becomes

$$\frac{w_u}{U} = \frac{1}{4\pi}\left(\frac{\Delta p}{\frac{1}{2}\rho_\infty U^2}\right)_0 \left[-\pi\beta + \int_0^{x+\beta y} \frac{\sqrt{(x-x_1)^2 - \beta^2 y^2}}{(x-x_1)y} dx_1 - \int_0^{\frac{x-\beta y}{1+\beta\tan\delta_0}} \frac{\sqrt{(x-x_1)^2 - \beta^2(y + x_1\tan\delta_0)^2}}{(x-x_1)(y + x_1\tan\delta_0)} dx_1\right]$$

Finally, the expression for w_u/U (which represents the local angle of attack, since $w_u/U = -\alpha(x, y)$) can be written

$$\frac{w_u}{U} = \frac{1}{4\pi}\left(\frac{\Delta p}{\frac{1}{2}\rho_\infty U^2}\right)_0 \left[\frac{\sqrt{x^2 - \beta^2 y^2}}{y} - \frac{1}{\tan\delta_0}\cosh^{-1}\left|\frac{x}{\beta y}\right| + \frac{\sqrt{\beta^2\tan^2\delta_0 - 1}}{\tan\delta_0}\cosh^{-1}\frac{x + y\beta^2\tan\delta_0}{\beta|x\tan\delta_0 + y|}\right] \quad (11\text{-}4)$$

Similar methods and complex variable techniques have been used to study prescribed loadings on other planforms [*7*,*19*].

D,12. Inverse Problems. As in the direct case, two types of physical problems arising in the study of linearized wing theory exemplify the set of boundary conditions for inverse problems. They can be stated as follows:

Problem 3: The thickness case. The pressure distribution over a wing that is symmetrical above and below a horizontal plane is given and the shape of the wing is to be determined. From a mathematical standpoint, Eq. 10-14 or 10-15 apply, C_p is given over the wing planform and λ_u is to be determined. Solutions to problems of this type are by no means unique, and many values of λ_u, when substituted in Eq. 10-14 or 10-15, will give the same expression for C_p over the wing surface. In fact, it is known that the subsonic-airfoil solution in the two-dimensional case is unique only if the prescribed pressure distribution is one leading to a real and closed wing section. In the general theory of three-dimensional wings, however, even these conditions are not sufficient to guarantee a unique shape from a given pressure distribution (although the reverse is always true, i.e. a given shape produces a unique pressure). This fact will be illustrated in the subsequent article dealing with quasi-conical supersonic flow problems.

Problem 4: The lifting case. The slope of a lifting plate is given and the resulting load distribution is to be determined. Mathematically, Eq. 10-17 through 10-22 apply, w_u is given over the wing planform and $\Delta p/\frac{1}{2}\rho_\infty U^2$ is to be determined. Again, solutions to these problems are not necessarily unique. It is necessary, for example, to assume that the Kutta condition applies to all trailing edges for which the normal component of free stream velocity is subsonic.

The classification of problems in aerodynamic wing theory as direct or inverse has been made with respect to the fundamental formulas presented in Art. 10. Hence, if those formulas are to form the analytic basis for the problems specified as 3 and 4 above, we are, by definition, led to the consideration of integral equations.

Singular integral equations. The equation

$$\int_{L_1} g(x)K(x, y)dx = w(y) \tag{12-1}$$

is called an *integral equation* of the first kind if both $w(y)$ and $K(x, y)$ are given functions and $g(x)$ is unknown. The term $K(x, y)$ is known as the kernel of the equation and the path of integration L_1 lies along the x axis (we consider only real variables, although the methods and results can be generalized to include complex variables) and, in general, depends on y. Further, Eq. 12-1 is a *singular integral equation* if either the path of integration has infinite extent or the kernel is infinite at a point or at points somewhere along this path. We consider only the latter of these two forms.

If both sides of Eq. 12-1 are multiplied by the function $H(\lambda, y)$ and integrated with respect to y along the interval L_2, the equation is said to have been transformed and the operator

$$\int_{L_4} H(\lambda, y)dy \tag{12-2}$$

is referred to as an integral transform. We now define the integral operator

$$\int_{L_3} H(\lambda, y)dy$$

as a *null transform* of order n to the function $K(x, y)$ in the interval L_3 if

$$\left(\frac{\partial}{\partial x}\right)^n \int_{L_3} H(\lambda, y)K(x, y)dy = 0 \tag{12-3}$$

where x and λ are on L_3. Some specific examples of null transforms are given later in Eq. 12-19 and 12-22.

Before discussing the actual solution of singular integral equations, we need the definitions of two more concepts, an *inherent singularity* and a *residual.* Consider the function $f(x, y)$ and let the point a, b lie somewhere in the x, y plane. Then an *inherent singularity* will be said to exist at the point a, b if

$$\lim_{\epsilon \to 0} \epsilon^2 f(\epsilon\xi + a, \epsilon\eta + b) \neq 0$$

where $x - a = \epsilon\xi$, $y - b = \epsilon\eta$. In other words, a square of width 2ϵ is first placed on the x, y plane with a, b at its center. Then, if the product of $f(x, y)$, evaluated at any point on the boundary of the square, and the area of the square is not zero in the limit $\epsilon \to 0$, the function $f(x, y)$ contains an inherent singularity at the point a, b. This concept can be generalized to include functions of more than two variables.

Finally, consider a double integration of the function $f(x, y)$ over the area S in the x, y plane. Form the integral of the same function over the same area but with the order of integrations reversed. The difference between the results of these two operations is defined as the residual R, thus

$$R \equiv \int_S dy \int dx\, f(x, y) - \int_S dx \int dy\, f(x, y) \tag{12-4}$$

Ordinarily R is zero, since the order of integration for a double integral is usually immaterial. In the manipulation of singular integral transforms, however, a nonvanishing residual often exists. The necessary condition for the existence of a residual is the existence of an inherent singularity in the area of integration over which the double integration is performed.

The following method can be used to evaluate the residual. Let the point a_1, b_1 be an inherent singularity in the area S. Then the residual R_1 at such a point is by definition

$$R_1 = \lim_{\epsilon \to 0} \left[\int_{b_1-\epsilon}^{b_1+\epsilon} dy \int_{a_1-\epsilon}^{a_1+\epsilon} dx\, f(x, y) - \int_{a_1-\epsilon}^{a_1+\epsilon} dx \int_{b_1-\epsilon}^{b_1+\epsilon} dy\, f(x, y) \right]$$

Setting $\epsilon\xi + a_1 = x$ and $\epsilon\eta + b_1 = y$, one can write

$$R_1 = \lim_{\epsilon \to 0} \left[\int_{-1}^{1} d\eta \int_{-1}^{1} d\xi\, \epsilon^2 f(\epsilon\xi + a_1, \epsilon\eta + b_1) - \int_{-1}^{1} d\xi \int_{-1}^{1} d\eta\, \epsilon^2 f(\epsilon\xi + a_1, \epsilon\eta + b_1) \right] \quad (12\text{-}5)$$

The total residual is the sum of the residuals from each inherent singularity in the area involved.

First let us inspect a method of inverting singular integral equations by means of null transforms. Consider Eq. 12-1; suppose that the kernel $K(x, y)$ is a function that tends to infinity as x approaches y, and let the point $x = y$ lie in L_1. Apply to both sides of this equation the integral transform $\int_{L_2} H(\lambda, y) dy$ so that

$$\int_{L_2} w(y) H(\lambda, y) dy = \int_{L_2} dy \int_{L_1} dx\, g(x) H(\lambda, y) K(x, y) \qquad (12\text{-}6)$$

Suppose that the area defined by x in L_1 and y in L_2 is bounded by a simple closed curve having the property that any line parallel to the x or y axis crosses its boundary at most twice. For such an area, it is always possible to write the reversed form of the double integral in Eq. 12-6 as

$$\int_{L_4} dx \int_{L_3} dy\, g(x) H(\lambda, y) K(x, y) \qquad (12\text{-}7)$$

where L_3 can be a function of x and λ, and L_4 can be a function only of λ.

We can now write for the residual the expression

$$R(\lambda) = \int_{L_2} dy \int_{L_1} dx\, g(x) H(\lambda, y) K(x, y) - \int_{L_4} dx \int_{L_3} dy\, g(x) H(\lambda, y) K(x, y) \quad (12\text{-}8)$$

By means of the definition given as Eq. 12-5, this can be rewritten in the form

$$\begin{aligned} R(\lambda) &= R_0(\lambda) \\ &+ \lim_{\epsilon \to 0} \left[\epsilon^2 \int_{-1}^{1} d\eta \int_{-1}^{1} d\xi\, g(\epsilon\xi + \lambda) K(\epsilon\xi + \lambda, \epsilon\eta + \lambda) H(\lambda, \epsilon\eta + \lambda) \right. \\ &\left. - \epsilon^2 \int_{-1}^{1} d\xi \int_{-1}^{1} d\eta\, g(\epsilon\xi + \lambda) K(\epsilon\xi + \lambda, \epsilon\eta + \lambda) H(\lambda, \epsilon\eta + \lambda) \right] \quad (12\text{-}9) \end{aligned}$$

where the two double integral terms on the right-hand side represent the residual from the point $x = y = \lambda$ and the term $R_0(\lambda)$ is the sum of the remaining residuals (if there are any) from other inherent singularities that may exist in the area of integration. Finally, Eq. 12-9 reduces to

$$R(\lambda) = R_0(\lambda) + g(\lambda)R^*(\lambda) \tag{12-10}$$

where

$$R^*(\lambda) = \lim_{\epsilon\to 0}\Big[\epsilon^2 \int_{-1}^{1} d\eta \int_{-1}^{1} d\xi\, K(\epsilon\xi + \lambda, \epsilon\eta + \lambda)H(\lambda, \epsilon\eta + \lambda) - \epsilon^2 \int_{-1}^{1} d\xi \int_{-1}^{1} d\eta\, K(\epsilon\xi + \lambda, \epsilon\eta + \lambda)H(\lambda, \epsilon\eta + \lambda)\Big] \tag{12-11}$$

If inherent singularities other than the one at $x = y = \lambda$ do exist in the area of integration, they must be at a point on the line $x = y$. The residual at such a point, say $x = y = a$, would be the product of $g(a)$ and $R^*(\lambda, a)$, the difference between the two appropriate double integrals. Hence, $R_0(\lambda)$ cannot contain $g(\lambda)$.

Placing Eq. 12-8 and 12-10 in Eq. 12-6, one obtains

$$\int_{L_2} w(y)H(\lambda, y)dy = \int_{L_4} g(x)dx \int_{L_3} dy\, H(\lambda, y)K(x, y) + R_0(\lambda) + g(\lambda)R^*(\lambda) \tag{12-12}$$

Now, if

$$\int_{L_3} H(\lambda, y)dy$$

is a null transform of zero order to the function $K(x, y)$, Eq. 12-12 becomes simply

$$g(\lambda) = \frac{1}{R^*(\lambda)}\Big[-R_0(\lambda) + \int_{L_2} w(y)H(\lambda, y)dy\Big] \tag{12-13}$$

which is an inversion of Eq. 12-1. Further, if

$$\int_{L_3} H(\lambda, y)dy$$

is a null transform of the first order to the function $K(x, y)$, so that

$$\int_{L_3} H(\lambda, y)K(x, y)dy = C(\lambda)$$

Eq. 12-12 becomes

$$\int_{L_2} w(y)H(\lambda, y)\, dy = C(\lambda) \int_{L_4} g(x)dx + R_0(\lambda) + g(\lambda)R^*(\lambda) \tag{12-14}$$

If L_4 is independent of λ this is again an inversion of Eq. 12-1. However, if L_4 contains λ then after dividing through by $C(\lambda)$ and taking the derivative with respect to λ, a first order differential equation in $g(\lambda)$ results. From this an inversion can be obtained.

In general, of course, the kernel $K(x, y)$ is so complicated that a null transform of any degree is not available. In such cases the transformed version of the integral equation 12-12 can still be useful. Let us set

$$F(\lambda) = \frac{1}{R^*(\lambda)}\left[-R_0(\lambda) + \int_{L_1} w(y)H(\lambda, y)dy\right] \tag{12-15}$$

and

$$K_1(\lambda, x) = \frac{1}{R^*(\lambda)}\int_{L_3} H(\lambda, y)K(x, y)dy \tag{12-16}$$

Then Eq. 12-12 can be written in the form

$$g(\lambda) = F(\lambda) - \int_{L_4} K_1(\lambda, x)g(x)dx \tag{12-17}$$

which is an integral equation of the second kind and can be resolved by iteration. Moreover it is important to notice that the kernel of Eq. 12-17 is *synthetic* and can be built into an arbitrary number of forms depending on the choice of $H(\lambda, y)$. If Eq. 12-17 is solved by an iterative process, the convergence of the resulting series depends on the magnitude of the greatest bound to the synthetic kernel $K_1(\lambda, x)$. Hence, a choice of $H(\lambda, y)$ would be that which minimizes the absolute value of the upper bound to K_1.

Obviously the optimum choice of $H(\lambda, y)$ is that which gives the null transform of zero order since then the greatest bound of the synthetic kernel is just zero, which is certainly a minimum. In the more difficult cases when the null transform is not apparent, $H(\lambda, y)$ can be chosen to contain, for example, a polynomial in y with arbitrary coefficients. These coefficients are then fixed by minimizing $|K_1(\lambda, x)|_{max}$. Such a method is a practical way of obtaining either numerically or analytically a rapidly converging series solution for $g(\lambda)$.

Most of the concepts outlined above are illustrated by application to the following examples. Consider first the equation

$$w(y) = \int_a^y \frac{g(x)dx}{\sqrt{y - x}}, \qquad a < y \tag{12-18}$$

This is a special form of Abel's integral equation and appears in the study of supersonic wing theory (see, for example, [*18,20*]). The solution to Abel's equation is familiar enough (it is to be found in such standard texts as *Modern Analysis* by Whittaker and Watson) but it is our purpose to rederive it by using null transforms. In the appendix (see Eq. 14.1-11) it is shown that

$$⨍_x^\lambda \frac{dy}{(\lambda - y)^{\frac{3}{2}}\sqrt{y - x}} = 0 \tag{12-19}$$

Or, in the terminology just introduced

$$\fint_x^\lambda \frac{dy}{(\lambda - y)^{\frac{3}{2}}}$$

is a null transform of order zero to the kernel $1/\sqrt{y - x}$ in the interval $x < y < \lambda$. Apply this transform to both sides of Eq. 12-18 and reverse the order of integration of the right-hand member. There results

$$\fint_x^\lambda \frac{w(y)dy}{(\lambda - y)^{\frac{3}{2}}} = \int_a^\lambda g(x)dx \fint_x^\lambda \frac{dy}{(\lambda - y)^{\frac{3}{2}} \sqrt{y - x}} + R(\lambda)$$

which reduces immediately to

$$\fint_x^\lambda \frac{w(y)dy}{(\lambda - y)^{\frac{3}{2}}} = R(\lambda) \tag{12-20a}$$

The only inherent singularity that exists in the area of integration (see Fig. D,12a) occurs at the point $x = y = \lambda$. The total residual is,

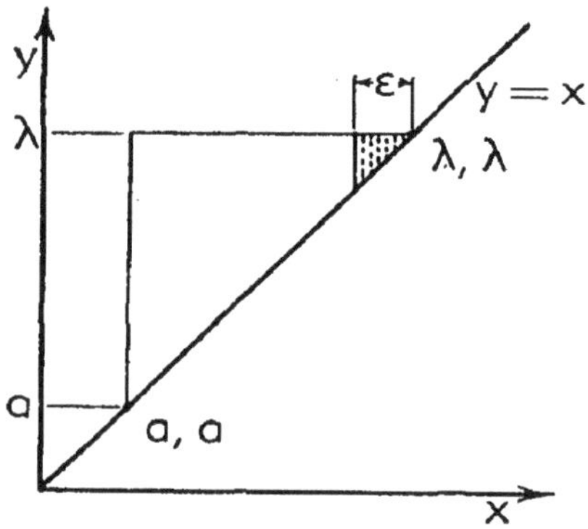

Fig. D,12a. Area of integration used for a certain singular integral.

therefore, obtained by integrating over the shaded area in the figure and finding the limit as ϵ goes to zero. Thus

$$R(\lambda) = \lim_{\epsilon \to 0} \left\{ \fint_{\lambda-\epsilon}^\lambda dy \int_{\lambda-\epsilon}^y \frac{g(x)dx}{\sqrt{y - x}\,(\lambda - y)^{\frac{3}{2}}} - \int_{\lambda-\epsilon}^\lambda g(x)dx \fint_x^\lambda \frac{dy}{\sqrt{y - x}\,(\lambda - y)^{\frac{3}{2}}} \right\}$$

and, since the right-hand term is zero for all ϵ, the transformations $\epsilon\xi = \lambda - x$ and $\epsilon\eta = \lambda - y$ reduce this to

$$R(\lambda) = \lim_{\epsilon \to 0} \fint_1^0 d\eta \int_1^\eta d\xi \frac{g(-\epsilon\xi + \lambda)}{\eta^{\frac{3}{2}} \sqrt{\xi - \eta}}$$

Finally, in the limit as ϵ goes to zero,

$$R(\lambda) = g(\lambda) \fint_1^0 \frac{d\eta}{\eta^{\frac{3}{2}}} \int_1^\eta \frac{d\xi}{\sqrt{\xi - \eta}} = -2\pi g(\lambda)$$

and, according to Eq. 12-20a, the inversion of Eq. 12-18 becomes

$$g(\lambda) = -\frac{1}{2\pi} {\int\!\!\!\!\!\!-}_{a}^{\lambda} \frac{w(y)dy}{(\lambda - y)^{\frac{3}{2}}} \tag{12-20b}$$

By applying the definition of the finite part one can rewrite Eq. 12-20b in the alternative way

$$g(\lambda) = \frac{1}{\pi} \frac{d}{d\lambda} \int_a^{\lambda} \frac{w(y)dy}{\sqrt{\lambda - y}}$$

which is the form of the inversion usually presented.

Next, consider the equation

$$w(y) = \int_a^b \frac{g(x)dx}{y - x} \tag{12-21}$$

This equation is basic to the study of boundary-value problems associated with the study of Laplace's equation in two dimensions. Hence, for the particular field of wing theory, it arises in the study of two-dimensional subsonic wing sections and in the study of problems amenable to the assumptions of slender wing theory (see Art. 19-24 of this section), since the field equation for both of these theories is the two-dimensional form of Laplace's equation. Again, the solution to Eq. 12-21 has been obtained by methods other than those used here (see e.g. [*21*]), but we will show how its inversion can be derived by using a null transform of the first order.

First, transform Eq. 12-21 to the form represented by Eq. 12-12, thus

$$\int\limits_{L_2} w(y)H(\lambda, y)dy = \int\limits_{L_1} g(x)dx \int\limits_{L_2} \frac{H(\lambda; y)dy}{y - x} + R(\lambda)$$

It is apparent that we need a null transform of the function $1/(y - x)$. This is provided by the integral identity

$$\int_a^b \frac{dy}{(y - x)\sqrt{(b - y)(y - a)}}$$

$$= \begin{cases} -\dfrac{\pi}{\sqrt{(x - b)(x - a)}}, & x > b \\ 0, & b > x > a \\ \dfrac{\pi}{\sqrt{(b - x)(a - x)}}, & x < a \end{cases} \tag{12-22}$$

from which it is easy to show that

$$\int_a^b \frac{\sqrt{(b - y)(y - a)}}{(y - x)(\lambda - y)} dy = \pi, \qquad a < \lambda < b; a < x < b$$

Therefore

$$\int_a^b [\sqrt{(b-y)(y-a)}/(\lambda - y)]dy$$

is a null transform of order one to the kernel $1/(y - x)$ in the interval $a < x < b$. Using this result, we can reduce our equation to the form

$$\int_a^b \frac{w(y)\sqrt{(b-y)(y-a)}}{\lambda - y}\,dy = \pi \int_a^b g(x)dx + R(\lambda) \qquad (12\text{-}23)$$

which will represent the inversion as soon as we have evaluated the residual.

Again the only inherent singularity in the area of integration is at the point $x = y = \lambda$ (see Fig. D,12b). Evaluating the residual by the method

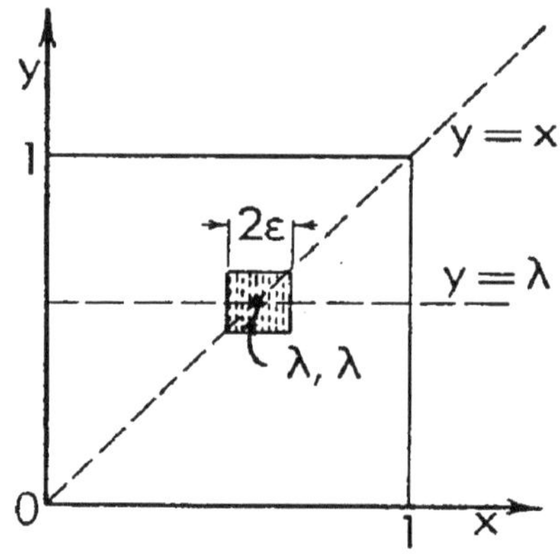

Fig. D,12b. Area of integration used for a certain singular integral.

previously outlined, we find

$$R(\lambda) = g(\lambda)\sqrt{(b-\lambda)(\lambda-a)}\left[\int_{-1}^{1} d\eta \int_{-1}^{1} \frac{d\xi}{\eta(\xi-\eta)} - \int_{-1}^{1} d\xi \int_{-1}^{1} \frac{d\eta}{\eta(\xi-\eta)}\right]$$

and since

$$\int_{-1}^{1} \frac{d\eta}{\eta(\xi-\eta)} = -\frac{1}{\xi}\ln\frac{1-\xi}{1+\xi}$$

this can be written

$$R(\lambda) = 2g(\lambda)\sqrt{(b-\lambda)(\lambda-a)}\int_{-1}^{1}\frac{d\eta}{\eta}\ln\frac{1-\eta}{1+\eta}$$
$$= -\pi^2 g(\lambda)\sqrt{(b-\lambda)(\lambda-a)}$$

Place this in Eq. 12-23 and the final inversion to Eq. 12-21 becomes

$$g(\lambda) = \frac{1}{\pi^2\sqrt{(b-\lambda)(\lambda-a)}}\left[\pi\int_a^b g(x)dx - \int_a^b \frac{w(y)\sqrt{(b-\lambda)(\lambda-a)}}{\lambda-y}\,dy\right] \qquad (12\text{-}24)$$

Since a and b are constants, the term

$$\int_a^b g(x)dx$$

is independent of x and can be regarded as an arbitrary constant. It is apparent, therefore, that the inversion given by Eq. 12-24 is not unique. Hence, in applying Eq. 12-24 to the solution of physical boundary-value problems it is not sufficient to specify the value of w along the y axis; some additional conditions must also be supplied. An example of such an

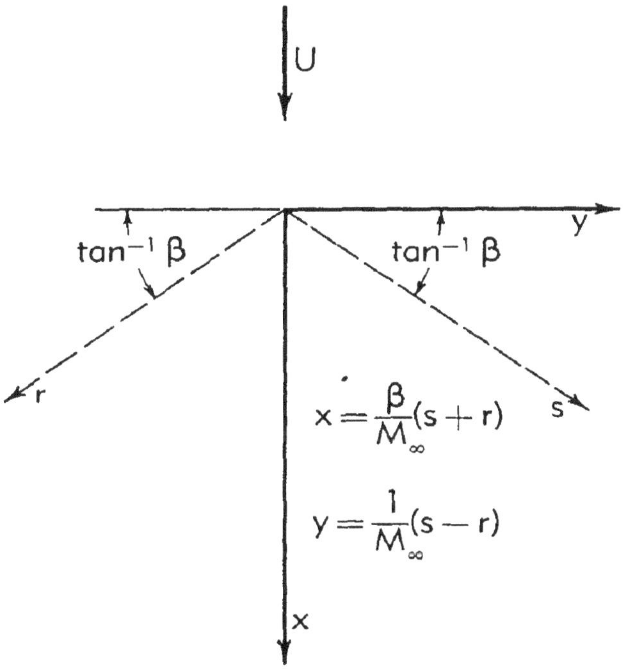

Fig. D,12c. Relation between x, y and r, s coordinate systems.

additional condition in the case of two-dimensional lifting surfaces is the assumption of the Kutta condition at a subsonic trailing edge.

The general concepts of the method just applied to the solution of singular integral equations with a single integral can also be used to invert double-integral equations with singular kernels. Success in solving these more complicated forms depends again on the discovery of an appropriate integral transform—in this case a double-integral transform—and the usefulness of these operators depends, in turn, on both the value of their integrand and, what is just as important, the space (now four-dimensional) of integration. Although the over-all problem of obtaining inversions is obviously much more difficult than in the case of single-integral equations, inversions can be obtained in many cases that are of importance in the study of the aerodynamic properties of wings flying at supersonic speeds.

Consider the equation that gives the vertical induced velocity in the plane of the wing in terms of the perturbation velocity potential on the upper side of a lifting surface in a supersonic free stream (Eq. 10-19). We will, henceforth, be concerned with the expression of this and the other fundamental equations presented in Art. 10 in terms of the characteristic (or r, s) coordinate system illustrated by Fig. D,12c. The physical significance of the r, s coordinates is made clear by visualizing

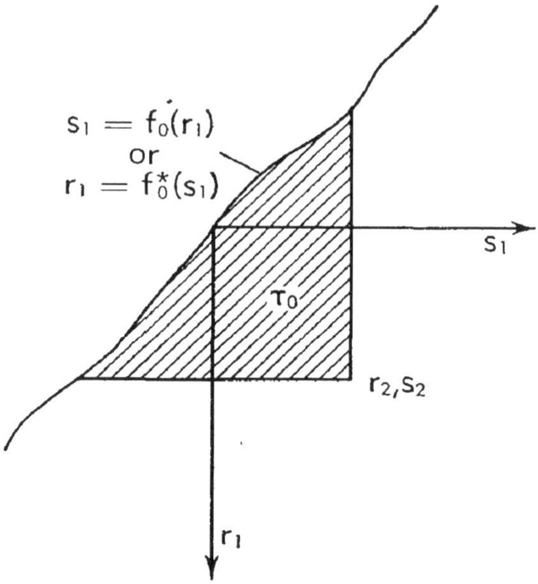

Fig. D,12d. Area τ for a completely supersonic edge in r_1, s_1 coordinate system.

the Mach cone shed by a small disturbance fixed in a supersonic free stream. If we locate the origin of our coordinate system at the disturbance and, as usual, let z represent vertical measurement, then one trace of this cone in the $z = 0$ plane forms the r axis and the other the s axis. These coordinates are a special form of the ξ, η system and follow from the latter when $\eta = s$, $\xi = r$, $\mu_1 = \mu_2 = M_\infty$, $m_1 = m_2 = \beta$, and $r_\theta = (2\beta/M_\infty)\sqrt{(r - r_1)(s - s_1)}$. Thus with reference to the r_2 and s_2 axes Eq. 10-19 becomes

$$w_u(r_2, s_2) = -\frac{M_\infty}{4\pi} \iint\limits_{\tau(r_2, s_2)} \frac{\varphi_u(r_1, s_1)dr_1ds_1}{[(r_2 - r_1)(s_2 - s_1)]^{3/2}} \tag{12-25}$$

the area $\tau(r_2, s_2)$ represents the area in the r_1s_1 plane within the forecone from the point r_2, s_2 at which the vertical induced velocity is being measured. The details of the solution to Eq. 12-25 will now be presented for two different types of boundaries to the area $\tau(r_2, s_2)$, that is, from an aerodynamic standpoint, for two different types of wing planforms.

First, consider the case when $\tau = \tau_0$, the area shown in Fig. D,12d. The traces of the Mach forecones from the point r_2, s_2, i.e. the lines

$r_1 = r_2$ and $s_1 = s_2$, form two bounds of the area while the third is given by the wing leading edge, the equation for which may be written as either $s_1 = f_0(r_1)$ or $r_1 = f_0^*(s_1)$. In the present case it is assumed, furthermore, that $s_1 = f_0(r_1)$ is a continuous monotonic function with a negative slope, or, physically, that the wing has a supersonic leading edge.[8]

The kernel of Eq. 12-25 is formed by the product of two functions, one independent of r_1 and r_2 and the other independent of s_1 and s_2. It follows from Eq. 12-19 that an integral operator containing the product of two linear functions of s_2 and r_2 to the $-\frac{1}{2}$ power could be a null transform of order zero provided the region of integration is chosen properly. One can show, in fact, that the operator

$$\iint\limits_{\tau_0(r,s)} \frac{dr_2\, ds_2}{\sqrt{(r - r_2)(s - s_2)}} = \frac{2\beta}{M_\infty} \iint\limits_{\tau_0(r,s)} \frac{dr_2\, ds_2}{r_0(r, s; r_2, s_2)} \tag{12-26}$$

is a null transform of order zero to the right-hand side of Eq. 12-25 where $\tau_0(r, s)$ is the same area that appears in the integral equation.

In order to show this, apply the operator given by Eq. 12-26 to Eq. 12-25. The resulting expression is

$$\begin{aligned}\iint\limits_{\tau_0(r,s)} &\frac{w_u(r_2, s_2)dr_2ds_2}{r_0(r, s; r_2, s_2)} \\ &= -\frac{2\beta^3}{\pi M_\infty^2} \iint\limits_{\tau_0(r,s)} dr_2\, ds_2 \,⨍⨍_{\tau_0(r_2,s_2)} \frac{\varphi_u(r_1, s_1)dr_1ds_1}{r_0(r, s; r_2, s_2)r_0^3(r_2, s_2; r_1, s_1)}\end{aligned} \tag{12-27}$$

Consider the right-hand side of this equation but with the r_2, s_2 integrals taken first. Then, by definition of the residual,

$$\begin{aligned}-\frac{2\beta^3}{\pi M_\infty^2} &\iint\limits_{\tau_0(r,s)} dr_2\, ds_2 \,⨍⨍_{\tau_0(r_2,s_2)} \frac{\varphi_u(r_1, s_1)dr_1ds_1}{r_0(r, s; r_2, s_2)r_0^3(r_2, s_2; r_1, s_1)} \\ &= -\frac{2\beta^3}{\pi M_\infty^2} \iint\limits_{[\tau_0(r,s)+\tau_0(r_2,s_2)]} \varphi_u(r_1, s_1)dr_1ds_1 \,⨍⨍ \frac{dr_2ds_2}{r_0(r, s; r_2, s_2)r_0^3(r_2, s_2; r_1, s_1)} \\ &\qquad + R(r, s)\end{aligned} \tag{12-28}$$

When the integration is made first with respect to r_2 and s_2, the area traversed by these two variables can be visualized with the aid of Fig. D,12e. In such a case the points r_1, s_1 and r, s are fixed and r_2, s_2 cover the area lying in the forecone from the point r, s and in the aftercone from the point r_1, s_1. This is represented by the shaded region in the figure. It is apparent from a study of Fig. D,12e that, when the edge $s_2 = f_0(r_2)$ is a monotonic function as previously defined, *the r_2 and s_2 integrals are*

[8] Supersonic and subsonic edges are defined and discussed in Art. 4.

always taken between the limits r_1 and r and s_1 and s respectively. Hence, according to Eq. 12-19, the operator given by Eq. 12-26 is a null transform, the integral term on the right-hand side of Eq. 12-28 vanishes, and Eq. 12-27 becomes

$$\iint_{\tau_0(r,s)} \frac{w_u(r_2, s_2)dr_2ds_2}{r_0(r, s; r_2, s_2)} = R(r, s) \tag{12-29}$$

The complete inversion of the integral equation can be realized when $R(r\ s)$ is expressed in terms of $\varphi_u(r, s)$. The evaluation of this residual

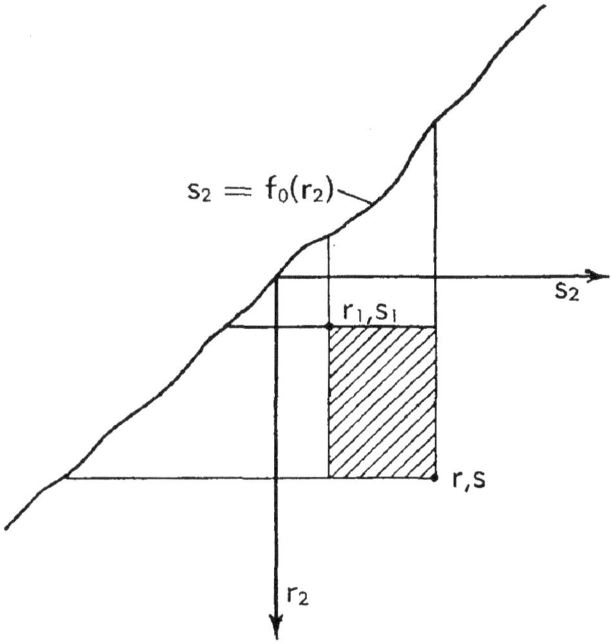

Fig. D,12e. Area of integration for fixed r_1, s_1; r, s.

follows along lines similar to those used in calculating the residual for the special form of Abel's integral equation studied previously. Now, however, there is no longer a single inherent singularity, rather the lines $r_1 = r_2 = r$ and $s_1 = s_2 = s$ are densely covered with them. First consider an integration made over the region close to the line $s_1 = s_2 = s$ (i.e. the sum of the areas a and b in Fig. D,12f). Make the substitutions $s_1 = s - \epsilon\sigma_1$ and $s_2 = s - \epsilon\sigma_2$ and take the limit as ϵ goes to zero. There results for the portion of the residual due to the inherent singularities $R_{a+b}(r, s)$ along the line $s_1 = s_2 = s$ the equation[9]

[9] The first term on the right-hand side of Eq. 12-28 contributes nothing to the residual since it always contains an integral equivalent to that in Eq. 12-19 and, therefore, is identically zero. This same phenomenon also appeared in the study of the residual appearing in the inversion of Abel's integral equation (see the development of Eq. 12-20b).

$$R_{a+b}(r, s) = -\frac{M_\infty^2}{8\pi\beta} \int_{f_0^*(s)}^{r} dr_2 \int_1^0 d\sigma_2 \, ⨍_{f_0^*(s)}^{r_2} dr_1$$

$$⨍_1^{\sigma_2} d\sigma_1 \frac{\varphi_u(r_1, s)}{(\sigma_1 - \sigma_2)^{\frac{3}{2}}(r_2 - r_1)^{\frac{3}{2}}\sqrt{\sigma_2(r - r_2)}} \qquad (12\text{-}30)$$

The integration with respect to σ_1 is immediate. Further, the r_1, σ_2 plane contains no inherent singularities in the region indicated so the σ_2 integra-

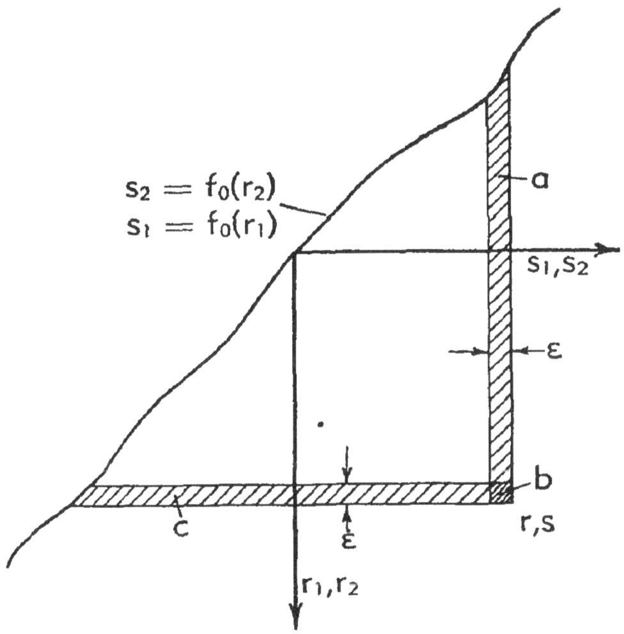

Fig. D,12f. Strips along which residuals are evaluated.

tion can be performed first. Thus

$$R_{a+b}(r, s) = \frac{M_\infty^2}{4\beta} \int_{f_0^*(s)}^{r} dr_2 \, ⨍_{f_0^*(s)}^{r_2} dr_1 \frac{\varphi_u(r_1, s)}{(r_2 - r_1)^{\frac{3}{2}}\sqrt{r - r_2}} \qquad (12\text{-}31)$$

It is now proposed to reverse the order of integration of the double integral in Eq. 12-31. But the r_1,r_2 plane contains an inherent singularity at the point $r_2 = r_1 = r$. Define the residual from this singularity as $R'(r, s)$ and it follows that

$$R_{a+b}(r, s) = \frac{M_\infty^2}{4\beta} \int_{f_0^*(s)}^{r} \varphi_u(r_1, s) dr_1 \, ⨍_{r_1}^{r} \frac{dr_2}{(r_2 - r_1)^{\frac{3}{2}}\sqrt{r - r_2}} + R'(r, s)$$

$$= R'(r, s)$$

since the integral term is zero by Eq. 12-19. The evaluation of $R'(r, s)$ follows the identical line of argument used in obtaining Eq. 12-20b.

Hence, setting $r_2 = r - \epsilon\rho_2$ and $r_1 = r - \epsilon\rho_1$ and letting ϵ go to zero, one finds

$$R_{a+b}(r, s) = R'(r, s) = \frac{M_\infty^2}{4\beta}\varphi_u(r, s)\int_1^0 d\rho_2 \int_1^{\rho_2} d\rho_1 \frac{1}{(\rho_1 - \rho_2)^{\frac{3}{2}}\sqrt{\rho_2}}$$

$$= -\frac{\pi M_\infty^2}{2\beta}\varphi_u(r,s) \tag{12-32}$$

The part of the residual in Eq. 12-29 contributed by the inherent singularities in the areas a and b in Fig. D,12f is, therefore, $-\pi M_\infty^2 \varphi_u/2\beta$.

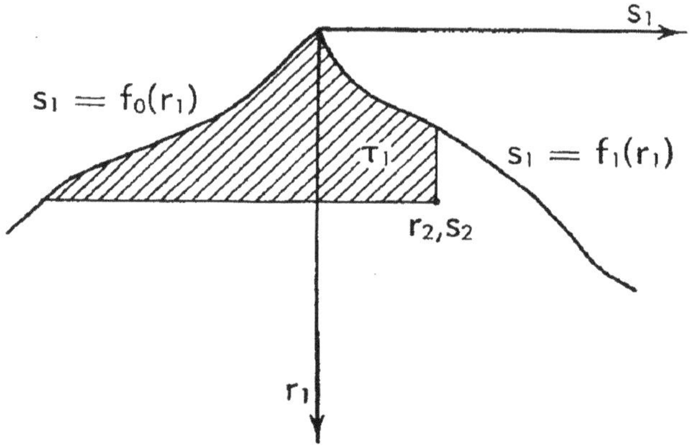

Fig. D,12g. Area τ for a part supersonic, part subsonic edge in the r_1, s_1 coordinate system.

A similar calculation shows that the singularities in the areas $c + b$ give the same result; and, finally, a calculation for the area b itself also yields $-\pi M_\infty^2 \varphi_u/2\beta$. The value of R in Eq. 12-29 is obtained by combining the results for the various areas. Hence

$$R = R_{a+b} + R_{b+c} - R_b = -\frac{\pi M_\infty^2}{2\beta}\varphi_u(r, s) \tag{12-33}$$

Combining the result expressed by the last equation with Eq. 12-29, we obtain the inversion of Eq. 12-25 when $\tau = \tau_0$, that is for a region influenced only by a supersonic edge. Thus,

$$\varphi_u(r, s) = -\frac{1}{\pi M_\infty}\iint_{\tau_0(r,s)} \frac{w_u(r_2, s_2)dr_2 ds_2}{\sqrt{(r - r_2)(s - s_2)}} \tag{12-34}$$

Consider next the equation

$$w_u(r_2, s_2) = -\frac{M_\infty}{4\pi}\iint_{\tau_1(r_2,s_2)} \frac{\varphi_u(r_1, s_1)dr_1 ds_1}{[(r_2 - r_1)(s_2 - s_1)]^{\frac{3}{2}}} \tag{12-35}$$

where $\tau_1(r_2, s_2)$ is the more complicated area shown in Fig. D,12g. Again the lines $r_1 = r_2$ and $s_1 = s_2$ form two bounds of the area. The remaining two boundaries are formed by the curves $s_1 = f_0(r_1)$ or in the inverse sense $r_1 = f_0^*(s_1)$ and $s_1 = f_1(r_1)$ or $r_1 = f_1^*(s_1)$ where f_0, as in the previous problem, is a monotonic curve with a negative slope. The curve $s_1 = f_1(r_1)$ is also a monotonic function, but with a positive slope. For convenience the origin is placed at the point of intersection of the f_0 and f_1 curves.

A physical interpretation of the area τ_1 with regard to problems in supersonic wing theory is simple. The lines $r_1 = r_2$ and $s_1 = s_2$ are the traces of the Mach forecone from the point r_2, s_2. The f_0 and f_1 curves are the edges of the wing planform, the line $s_1 = f_0(r_1)$ representing a supersonic leading edge, and the line $s_1 = f_1(r_1)$ representing a subsonic leading or trailing edge.

Proceeding exactly as before, we consider the operator

$$\iint\limits_{\tau_1(r,s)} \frac{dr_2 ds_2}{\sqrt{(r - r_2)(s - s_2)}}$$

and find whether or not it is a null transform of zero order to the kernel of Eq. 12-35, that is, whether or not the equality

$$\iint\limits_{[\tau_1(r,s) + \tau_1(r_2,s_2)]} \varphi(r_1, s_1) dr_1 ds_1 \iint \frac{dr_2 ds_2}{[(r_2 - r_1)(s_2 - s_1)]^{\frac{3}{2}} \sqrt{(r - r_2)(s - s_2)}} = 0$$

is satisfied. A study of Fig. D,12h (top) shows that it is not, since for a point r_1, s_1 located above the line $r_2 = f_1^*(s)$ a portion of the s_2 integral becomes

$$\int_{s_1}^{f_1(r_2)} \frac{ds_2}{(s_2 - s_1)^{\frac{3}{2}} \sqrt{s - s_2}}$$

which is not zero unless $f_1(r_2)$ is identically equal to s. For the same reason, portions of the r_2 integral will not vanish for the point r_1, s_1 so located.

However, the construction of a null transform of zero order can be accomplished by studying the above failure. Thus, if the s_2 integration were carried between the limits s_1 and s, for every location of the points r_1, s_1 and r, s, the operator $\int h(r, s, r_2) dr_2 \int (s - s_2)^{-\frac{1}{2}} ds_2$ would be a null transform of zero order regardless of the choice of h. The area of integration producing such a transform is simply the one shown in Fig. D,12h (middle); an area bounded by the lines $s_2 = s$, $r_2 = r$, $s_2 = f_0(r_2)$ and $r_2 = f_2^*(s)$, where $r_2 = f_2^*(s)$ is any line such that $f_2^*(s) \geqq f_1^*(s)$. This area is designated as $\tau_2(r, s)$. When such a transform is applied and the s_2 integration performed first, it is apparent from Fig. D,12h (bottom) that the limits of the s_2 integral are always s_1 and s even when the point r_1, s_1 is located above the line $r_2 = f_1^*(s)$. Hence, if the residual $R(r, s)$ is defined by the relation

$$- \frac{M_\infty}{4\pi} \iint\limits_{\tau_2(r,s)} \frac{h(r, s, r_2)dr_2ds_2}{\sqrt{s - s_2}} \iint\limits_{\tau_1(r_2,s_2)} \frac{\varphi_u(r_1, s_1)dr_1ds_1}{[(r_2 - r_1)(s_2 - s_1)]^{\frac{3}{2}}}$$

$$+ \frac{M_\infty}{4\pi} \int h(r, s, r_2)dr_2 \iint\limits_{[\tau_1(r_2,s_2)+\tau_2(r,s)]} \frac{\varphi_u(r_1, s_1)dr_1ds_1}{(r_2 - r_1)^{\frac{3}{2}}} \int_{s_1}^{s} \frac{ds_2}{\sqrt{s - s_2}\,(s_2 - s_1)^{\frac{1}{2}}}$$

$$= R(r, s) \quad (12\text{-}36)$$

(where $[\tau_1(r_2, s_2) + \tau_2(r, s)]$ represents the same integration volume as in

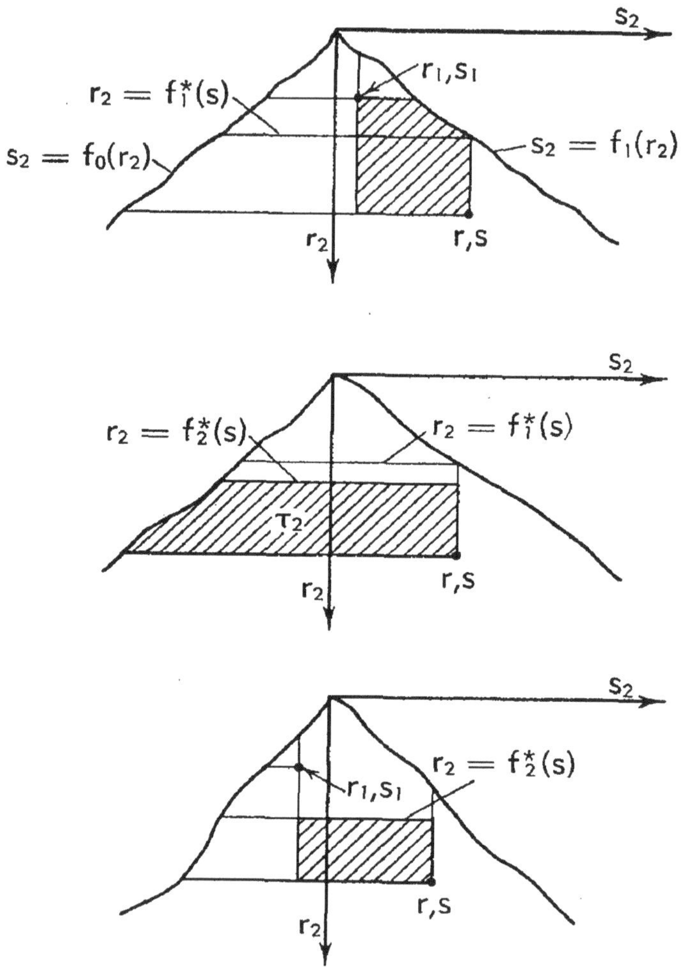

Fig. D,12h. Areas used in discussion of the doublet equation for the basic area τ_1 in the r_1, s_1 plane.

the first term) the second term on the left-hand side is zero, and Eq. 12-35 transforms into

$$\int_{f_2^*(s)}^{r} h(r, s, r_2)dr_2 \int_{f_0(r_2)}^{s} \frac{w_u(r_2, s_2)ds_2}{\sqrt{s - s_2}} = R(r, s) \qquad (12\text{-}37)$$

Next let us evaluate the residual in Eq. 12-36. Isolating the line of inherent singularities, we can write

$$R(r, s) = \lim_{\epsilon \to 0} -\frac{M_\infty}{4\pi} \int_{f(s)}^{r} h(r, s, r_2) dr_2 \int_{s-\epsilon}^{s} \frac{ds_2}{\sqrt{s - s_2}} \int_{f_1^*(s_2)}^{r_2} ds_1 \int_{s-\epsilon}^{s_2} dr_1 \frac{\varphi_u(r_1, s_1)}{[(r_2 - r_1)(s_2 - s_1)]^{\frac{1}{2}}}$$

which becomes

$$R(r, s) = \frac{M_\infty}{2} \int_{f_2^*(s)}^{r} h dr_2 \int_{f_1^*(s)}^{r_2} \frac{\varphi_u(r_1, s) dr_1}{(r_2 - r_1)^{\frac{1}{2}}} \tag{12-38}$$

It is now apparent that, if we set $f_2^*(s) = f_1^*(s)$ and $h = 1/\sqrt{r - r_2}$, Eq. 12-38 can be evaluated exactly as Eq. 12-31 was evaluated, so that

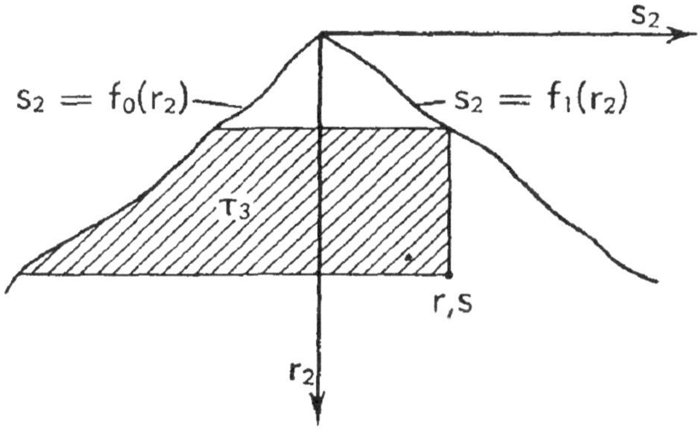

Fig. D,12i. Areas τ_3 in r_2, s_2 plane used to invert doublet equation for area τ_1 in r_1, s_1 plane.

for these values of f_2^* and h, $R(r, s) = -M_\infty \pi \varphi(r, s)$. Substituting into Eq. 12-37, one finds for an inversion of Eq. 12-35 the result

$$\varphi_u(r, s) = -\frac{1}{\pi M_\infty} \iint_{\tau_3(r,s)} \frac{w_u(r_2, s_2) dr_2 ds_2}{\sqrt{(r - r_2)(s - s_2)}} \tag{12-39}$$

where τ_3 is the shaded area in Fig. D,12i. This result was derived originally by Evvard [*18*] in a different manner.

Another very important kind of integral equation which can be solved directly by the proper choice of h in Eq. 12-38 is the "mixed" type problem, the boundary values of which are illustrated in Fig. D,12j. In this particular problem w_u is known over the portion of the r, s plane bounded by the curves $s = f_0(r)$ and $s = f_2(r)$ (the curve $s = f_2(r)$ is a monotonic function with a positive slope just like $s = f_1(r)$), while over another portion, bounded by the curves $s = f_2(r)$ and $s = f_1(r)$, the quantity

$$u_u(r, s) = \frac{M_\infty}{2\beta} \left(\frac{\partial}{\partial r} + \frac{\partial}{\partial s} \right) \varphi_u(r, s)$$

is given (this corresponds in aerodynamic applications to the specification of the vertical induced velocity over the former region and the loading over the latter). We also assume that $u_u(r, s)$ is continuous across the line $s = f_2(r)$ and that $\varphi_u(r, s)$ vanishes along the line $s = f_1(r)$; that is, $\varphi_u(f_1^*(s), s) = 0$.

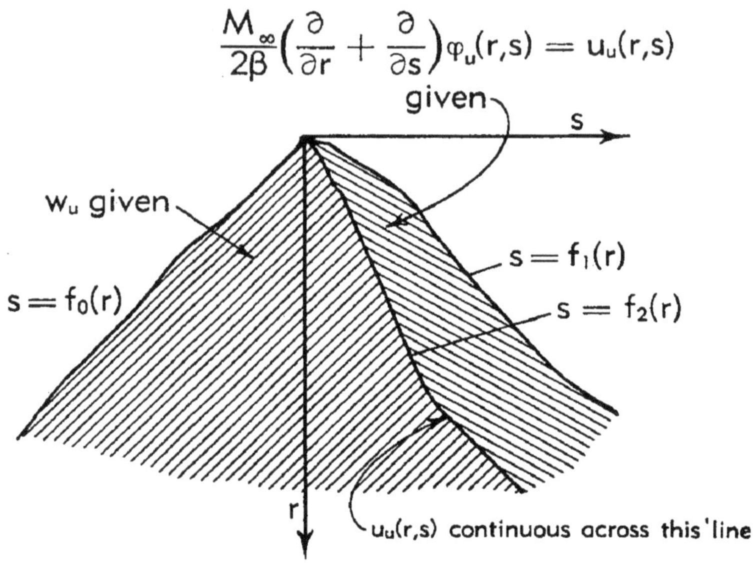

Fig. D,12j. Boundary conditions for mixed type problem.

Taking for the value of h the expression

$$h(r, s, r_2) = \frac{1}{\sqrt{r - r_2}}\left(\frac{\partial}{\partial r_2} + \frac{\partial}{\partial s}\right)$$

and for the area of integration the region $\tau_2(r, s)$ shown in Fig. D,12h (middle), we can write for the residual in Eq. 12-38

$$R(r, s) = \frac{M_\infty}{2}\int_{f_2^*(s)}^{r} \frac{dr_2}{\sqrt{r - r_2}}\left(\frac{\partial}{\partial r_2} + \frac{\partial}{\partial s}\right)\int_{f_1^*(s)}^{r_2} \frac{\varphi_u(r_1, s)dr_1}{(r_2 - r_1)^{\frac{3}{2}}}$$

The partial derivative can be taken through the second integral sign (first integrating by parts in the case of $\partial/\partial r_2$) since $\varphi_u(f_1^*(s), s) = 0$, so that

$$R(r, s) = \frac{M_\infty}{2}\int_{f_2^*(s)}^{r} dr_2 \int_{f_1^*(s)}^{r_2} dr_1 \frac{\left(\frac{\partial}{\partial r_1} + \frac{\partial}{\partial s}\right)\varphi_u(r_1, s)}{\sqrt{r - r_2}\,(r_2 - r_1)^{\frac{1}{2}}}$$

Part of the boundary condition is that $u_u(r, s)$ is given over the region indicated in Fig. D,12j, hence,

$$R(r, s) = \beta \int_{f_2^*(s)}^{r} dr_2 \int_{f_1^*(s)}^{f_2^*(s)} dr_1 \frac{u_u(r_1, s)}{\sqrt{r - r_2}\,(r_2 - r_1)^{\frac{3}{2}}}$$

$$+ \beta \int_{f_2^*(s)}^{r} dr_2 \int_{f_2^*(s)}^{r_2} dr_1 \frac{u_u(r_1, s)}{\sqrt{r - r_2}\,(r_2 - r_1)^{\frac{3}{2}}}$$

The latter of these two terms is similar to the one given in Eq. 12-31 so its evaluation is immediate. The first term may be simplified by reversing the order of integration (there is no residual since the point $r = r_1 = r_2$ is not included in the area of integration). Finally R becomes

$$R(r, s) = 2\beta \sqrt{r - f_2^*(s)} \int_{f_1^*(s)}^{f_2^*(s)} \frac{u_u(r_1, s)dr_1}{(r - r_1)\sqrt{f_2^*(s) - r_1}} - 2\pi\beta u_u(r, s)$$

and the inversion of this mixed type problem can be written

$$u_u(r, s) = \frac{\sqrt{r - f_2^*(s)}}{\pi} \int_{f_1^*(s)}^{f_2^*(s)} \frac{u_u(r_1, s)dr_1}{(r - r_1)\sqrt{f_2^*(s) - r_1}}$$

$$- \frac{1}{2\beta\pi} \int_{f_2^*(s)}^{r} \frac{dr_2}{\sqrt{r - r_2}} \left(\frac{\partial}{\partial r_2} + \frac{\partial}{\partial s}\right) \int_{f_0(r_2)}^{s} \frac{w_u(r_2, s_2)ds_2}{\sqrt{s - s_2}} \quad (12\text{-}40)$$

An equation which does not impose the condition that $u_u(r, s)$ be continuous across the line $s = f_2(r)$ can readily be developed by using the operator $h = 1/\sqrt{r - r_2}$. This leads directly to the result

$$\varphi_u(r, s) = \frac{\sqrt{r - f_2^*(s)}}{\pi} \int_{f_1^*(s)}^{f_2^*(s)} \frac{\varphi_u(r_1, s)dr_1}{(r - r_1)\sqrt{f_2^*(s) - r_1}}$$

$$- \frac{1}{\pi M_\infty} \int_{f_2^*(s)}^{r} dr_2 \int_{f_0(r_2)}^{s} ds_2 \frac{w_u(r_2, s_2)}{\sqrt{(s - s_2)(r - r_2)}} \quad (12\text{-}41)$$

which, of course, reduces to Eq. 12-39 when $f_1^*(s) = f_2^*(s)$.

The null transforms to the supersonic doublet Eq. 12-25 were constructed by applying an integral operator having an integrand which, when combined with the kernel, would produce the integral

$$\int \frac{dx}{(x - b)^{\frac{3}{2}}\sqrt{a - x}}$$

and having an area which, when combined with that of the integral equation and traversed in reverse order, would produce the limits a and b on the integral. If we are now given the supersonic source Eq. 10-12 expressed in r, s coordinates

$$\varphi_u(r_2, s_2) = -\frac{1}{\pi M_\infty} \iint_{\tau(r_2, s_2)} \frac{w_u(r_1, s_1)dr_1 ds_1}{\sqrt{(r_2 - r_1)(s_2 - s_1)}} \quad (12\text{-}42)$$

it is apparent that the null transform will differ from the transform used for the supersonic doublet equation only in the exponent of the integrand. Thus, for the area $\tau_0(r_2, s_2)$ the operator

$$\iint\limits_{\tau_0(r,s)} \frac{dr_2 ds_2}{[(r - r_2)(s - s_2)]^{\frac{3}{2}}}$$

is a null transform of order zero to Eq. 12-42, and its application yields the inversion

$$w_u(r, s) = -\frac{M_\infty}{4\pi} \iint\limits_{\tau_0(r,s)} \frac{\varphi_u(r_2, s_2) dr_2 ds_2}{[(r - r_2)(s - s_2)]^{\frac{3}{2}}} \tag{12-43}$$

For the area $\tau_1(r_2, s_2)$, the operator

$$\iint\limits_{\tau_2(r,s)} \frac{dr_2 ds_2}{[(r - r_2)(s - s_2)]^{\frac{3}{2}}}$$

is a null transform of order zero. Its application, under conditions like those specified in the development of Eq. 12-39, yields the inversion

$$w_u(r, s) = -\frac{M_\infty}{4\pi} \iint\limits_{\tau_3(r,s)} \frac{\varphi_u(r_2, s_2) dr_2 ds_2}{[(r - r_2)(s - s_2)]^{\frac{3}{2}}}$$

The latter transform can also be used to obtain inversions to Eq. 12-42 under conditions such as those imposed in the development of Eq. 12-40 and 12-41.

Wings with all edges supersonic. We now return to the study of the aerodynamic boundary-value problems stated at the beginning of this article and consider first conditions of the type specified in Problem 4. It is apparent that the solution to such problems can be obtained by inverting the double-integral equation relating the vertical induced velocity in the $z = 0$ plane to the value of φ_u there. This equation, which for brevity we refer to as the doublet equation, has been presented in terms of various coordinate systems, see Eq. 10-17, 10-18, 10-19, and 12-25. We will consider it in the latter form.

When all of the edges of the wing are supersonic, the forecone from any point on the wing never intersects the trailing edge. From another point of view this simply means that the waves spreading out from the trailing edge and signaling their existence to the flow field are all swept downstream and never pass over a part of the wing itself. Hence the load distribution over the wing planform is completely independent of the shape of the trailing edge so long as that edge remains everywhere supersonic. Since the leading edge is also supersonic, this means, mathematically, that the area over which the doublet distribution is to be integrated has, for every point on the wing, the special form τ_0 studied in the first part of this article and shown in Fig. D,12d. The solution to the double-integral equation under these conditions has therefore been obtained and presented as Eq. 12-34. Comparing the latter equation with the source integral equation (12-42), we see that for wings completely bounded by supersonic edges the source and doublet equations have an exact reciprocal relationship. In Cartesian coordinates this can be written

$$\begin{aligned} w_u(x, y) &= -\frac{\beta^2}{\pi}\int_{\tau_0} dy_1 \,⨍\, dx_1 \frac{\varphi_u(x_1, y_1)}{[(x-x_1)^2 - \beta^2(y-y_1)^2]^{\frac{3}{2}}} \\ \varphi_u(x, y) &= -\frac{1}{\pi}\iint_{\tau_0} \frac{w_u(x_1, y_1)dx_1dy_1}{\sqrt{(x-x_1)^2 - \beta^2(y-y_1)^2}} \end{aligned} \tag{12-44}$$

The relationship expressed by Eq. 12-44 was originally obtained in a far simpler manner by reasoning based on the physical structure of the flow field being considered. From a physical standpoint it is clear that the induced velocities on the upper surface of a wing cannot be influenced by the shape of any part of the lower surface if the edges of the wing are everywhere supersonic. In other words, the upper and lower surfaces are noninteracting. Hence the upper surface of the wing does not "know" whether it is the upper surface of a lifting plate that is supporting loading and has no thickness, or the upper surface of a wing section that is symmetrical above and below the $z = 0$ plane. Thus the source equation must be the inversion to the doublet equation and vice versa; and from a physical point of view the reciprocity between the two equations is obvious.

Eq. 12-44 represent the general solution to all of the direct and inverse boundary-value problems in thin-airfoil wing theory when the given wing planform is bounded by supersonic edges. Thus, in order to obtain the load distribution over a given plate, the known value of w_u is inserted and the double integral is evaluated. There are many cases, however, for which it is not necessary to carry out such integrations. For example, the total lift on a wing with all supersonic edges and a *straight* trailing edge can be found without ever finding the detailed loading. One method for solving this problem is presented below. Other methods can also be used (see Art. 25-28 on unsteady motion, where similar problems arise except that the time variable is included; also Art. 29-31 on reciprocity relations) to obtain the same results.

The total lift L developed by a surface having arbitrary twist and camber is given by the equation

$$\frac{L}{\frac{1}{2}\rho_\infty U^2} = \iint_{\text{wing planform}} \frac{\Delta p}{\frac{1}{2}\rho_\infty U^2}\, dydx$$

Since

$$\frac{\Delta p}{\frac{1}{2}\rho_\infty U^2} = \frac{4}{U}\frac{\partial \varphi_u}{\partial x}$$

and φ_u at the leading edge is zero, this equation can be written

$$\frac{L}{\frac{1}{2}\rho_\infty U^2} = \frac{4}{U}\int_{-s}^{s} \varphi_{u_{\text{t.e.}}} dy \tag{12-45}$$

where $\varphi_{u_{\text{t.e.}}}$ is the value of φ on the upper side of the trailing edge and $2s$ is the wing span measured parallel to the y axis (see Fig. D,12k).

Let the equation of the trailing edge be given by $x = a + y \tan \Lambda$ where Λ is its angle of sweep and a is some constant. Substitution of this

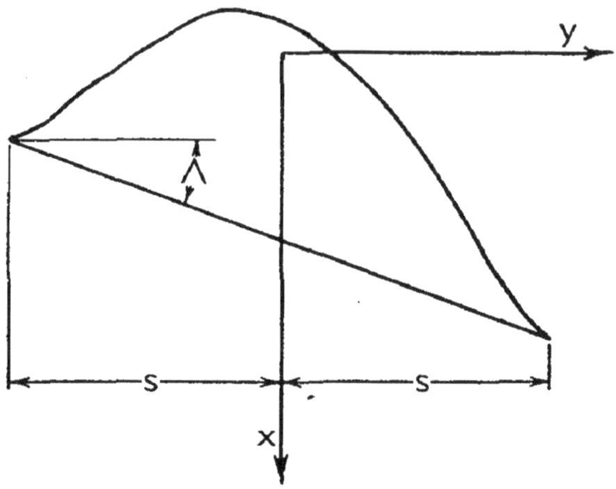

Fig. D,12k. Wing planform with all edges supersonic, trailing edge straight and swept at angle Λ.

expression into Eq. 10-12 yields

$$\varphi_{u_{\text{t.e.}}} = -\frac{1}{\pi}\iint_{\tau} \frac{w_u dy_1 dx_1}{\sqrt{(a + y \tan \Lambda - x_1)^2 - \beta^2(y - y_1)^2}}$$

Hence, according to Eq. 12-45

$$\frac{L}{\frac{1}{2}\rho_\infty U^2} = -\frac{4}{U\pi}\int_{-s}^{s} dy \iint_{\tau} \frac{w_u dy_1 dx_1}{\sqrt{(a + y \tan \Lambda - x_1)^2 - \beta^2(y - y_1)^2}}$$

Now reverse the order of integration so that the y integration is made first. To visualize the resulting region of integration, fix a point x_1, y_1 on the planform and note the range of y along the trailing edge which includes this point in its forecone. It is apparent that if the edges are everywhere supersonic the y integration when considered first will always be taken between the roots of the radical $\sqrt{(a + y \tan \Lambda - x_1)^2 - \beta^2(y - y_1)^2}$ and since

$$\int_{\lambda_1}^{\lambda_2} \frac{dy}{\sqrt{(\lambda_2 - y)(y - \lambda_1)}} = \pi$$

there results for the lift

$$\frac{L}{\frac{1}{2}\rho_\infty U^2} = \frac{-4}{U\sqrt{\beta^2 - \tan^2\Lambda}} \iint\limits_{\text{wing planform}} w_u dy_1 dx_1 \tag{12-46}$$

This equation can be put in a nicer form by introducing the average

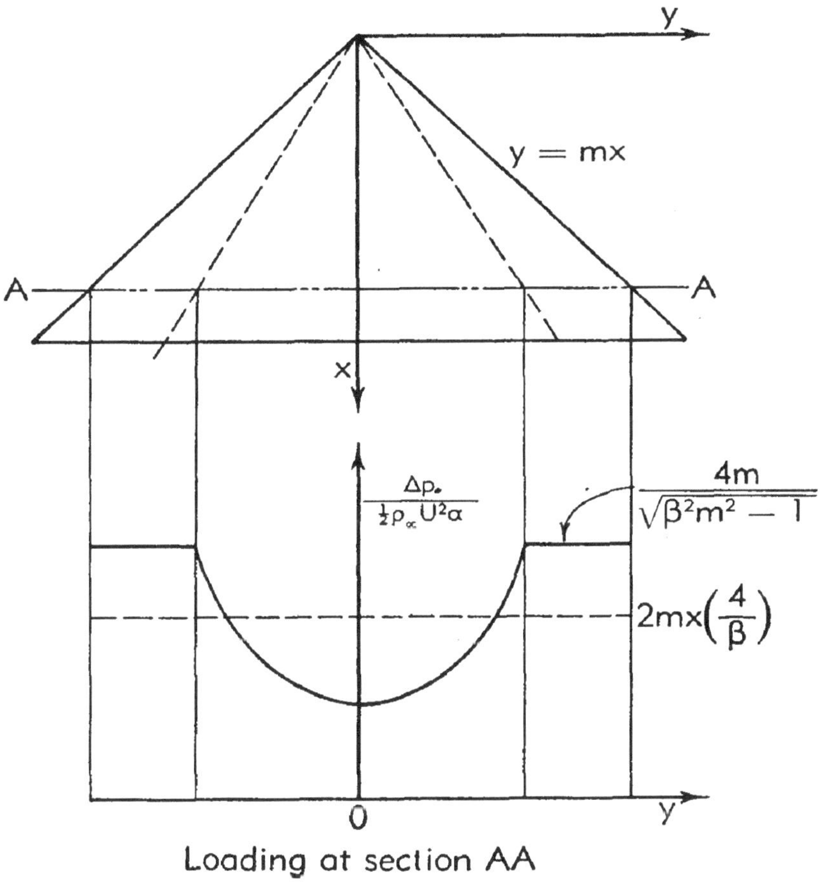

Fig. D,12l. Triangular flat plate and its load distribution for supersonic edges.

angle of attack of the wing $\bar{\alpha}$ such that

$$\bar{\alpha} = -\frac{1}{S}\iint\limits_{S}\left(\frac{w_u}{U}\right) dy_1 dx_1$$

S being the area of the wing. With the aid of this definition, the lift coefficient can be written[10]

$$C_L = \frac{4\bar{\alpha}}{\sqrt{\beta^2 - \tan^2\Lambda}} \tag{12-47}$$

[10] This equation was originally derived by Lagerstrom and Van Dyke [22].

which is the two-dimensional value for a wing having an angle of sweep Λ and an angle of attack equal to $\bar{\alpha}$. It is noteworthy that the total lift for these supersonic-edged wings is independent of the shape of the leading edge, depending only upon the trailing edge slope.

An example of an application of these results for wings with supersonic edges can be made using the triangular flat plate wing at an angle of attack α (Fig. D,12l). The loading can be evaluated by means of Eq. 10-12 in which w_0 is a constant equal to $-U\alpha$, and the distribution of this loading at a given chord is shown in the figure. Although the loading varies between $4m/\sqrt{\beta^2m^2 - 1}$ and $(4m/\sqrt{\beta^2m^2 - 1}) \sin^{-1} \sqrt{1 - (1/\beta m)^2}$, its integrated value, according to Eq. 12-46, yields for the lift coefficient the familiar two-dimensional value, $C_L = 4\alpha/\beta$.

D,13. Inverse Problems: Special Methods.

Generalized conical flow. An important type of problem arising in supersonic wing theory can be developed from the concept of a conical flow field. Such a field is one in which the perturbation velocity components and the induced pressures are constant in magnitude along any ray from the apex of the field. In this case, the perturbation potential may be written in the form

$$\varphi(x, y, z) = xf_1\left(\frac{y}{x}, \frac{z}{x}\right) = yf_2\left(\frac{x}{y}, \frac{z}{y}\right) = zf_3\left(\frac{x}{z}, \frac{y}{z}\right) \tag{13-1}$$

where φ is a homogeneous function of degree one in the three variables. An obvious generalization of this concept leads to the consideration of homogeneous potential fields of higher degree or, as they are sometimes called, quasi-conical fields. In this case we have

$$\varphi(x, y, z) = x^{k+1}F_1\left(\frac{y}{x}, \frac{z}{x}\right) = y^{k+1}F_2\left(\frac{x}{y}, \frac{z}{y}\right) = z^{k+1}F_3\left(\frac{x}{z}, \frac{y}{z}\right) \tag{13-2}$$

and φ is homogeneous of degree $k + 1$. Eq. 13-1 yields conical velocity fields, the degree of homogeneity being zero, while for Eq. 13-2 the velocity fields are homogeneous and of degree k.

The existence as well as the importance of such problems in supersonic theory can be illustrated by two examples. First consider the flat-plate triangular wing shown in Fig. D,13a (top). Since the angle of attack of the wing is everywhere constant, the boundary conditions affecting the solution at any point ahead of the trailing edge are homogeneous. It is possible to show that the loading will also be homogeneous by the following simple device. Consider a point P_1 located on some ray which passes through the apex of the wing. Now double the scale of the wing. The point P_1 transforms into P_2, Fig. D,13a (bottom), but since the basic differential equation is invariant under the transformation, the loadings at P_1 and P_2 are the same. Locate the point P_3 on the wing of the doubled scale in such a manner that it is on the same ray as P_2 but situated the same distance from the origin as P_1. The pressures at P_1 and P_3 are the same since nothing behind these points can affect the flow there and the conditions ahead of them are identical. Hence, the loading at P_2 is the same as at P_3 and by similar reasoning the loading all along the same ray is a constant. The flow is therefore conical. Further examples of flat plate planforms which lead to conical flows are shown in Fig. D,13b.

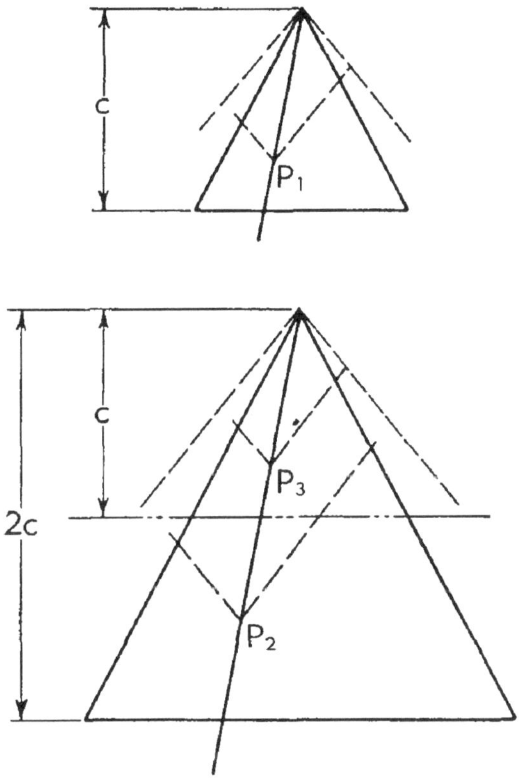

Fig. D,13a. Triangular flat plates referred to in discussion of conical flow.

Next consider a flat, unyawed, triangular wing rolling about its axis of symmetry. The boundary value for this type of problem is also homogeneous but of one degree higher than in the simple lifting case. It can be demonstrated that the loading on this wing is homogeneous and of degree one so that the field is quasi-conical.

At least two methods exist for solving quasi-conical boundary-value problems. In each case the success of the method depends on the fact that such problems are in effect problems involving only two independent variables. One of these methods commences with a modification of the original partial differential equation

$$\beta^2\varphi_{xx} - \varphi_{yy} - \varphi_{zz} = 0$$

If new variables are introduced such that

$$\frac{\beta y}{x} = \eta, \qquad \frac{\beta z}{x} = \zeta, \qquad \varphi(x, y, z) = cx^{k+1}\Omega(\eta, \zeta) \tag{13-3}$$

where c is an arbitrary constant, the transformed differential equation is

$$(\eta^2 - 1)\Omega_{\eta\eta} + 2\eta\zeta\Omega_{\eta\zeta} + (\zeta^2 - 1)\Omega_{\zeta\zeta} - k[2\eta\Omega_\eta + 2\zeta\Omega_\zeta - (k+1)\Omega] = 0 \tag{13-4}$$

It is possible to inspect the coefficients in Eq. 13-4 and determine the regions in which the equation is basically elliptic (i.e. essentially Laplace's

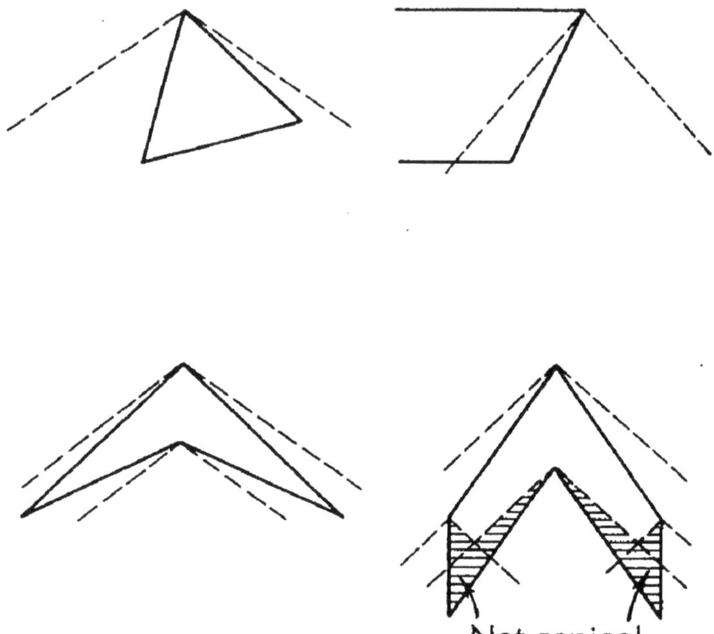

Fig. D,13b. Types of flat plate wing planforms that yield conical flow.

equation) or hyperbolic (i.e. the wave equation). These properties are governed by the coefficients of the second order derivatives and are therefore independent of the degree of homogeneity. Thus, for a quasi-conical flow field with apex at the origin of the coordinate system, the resultant equation is elliptic for all values of η, ζ satisfying the inequality

$$\eta^2\zeta^2 - (1 - \eta^2)(1 - \zeta^2) \leq 0$$

that is, for all points within the unit circle $\eta^2 + \zeta^2 = 1$, or, in other words, inside the Mach cone $x^2 - \beta^2 y^2 - \beta^2 z^2 = 0$. Outside the circle, or Mach cone, the equation is hyperbolic.

A convenient way of visualizing physically these properties of Eq. 13-4 can be obtained by considering the wing shown in Fig. D,13c. Let the boundary values of this wing be homogeneous and let the free stream be directed along the x axis at a Mach number equal to $\sqrt{2}$ so that the flow field is governed by the normalized form of the basic equation. Consider the plane $x = 1$. In this plane the variable $\eta = y/x$ becomes just $\eta = y$ and similarity $\zeta = z$. It is apparent, therefore, that the circle

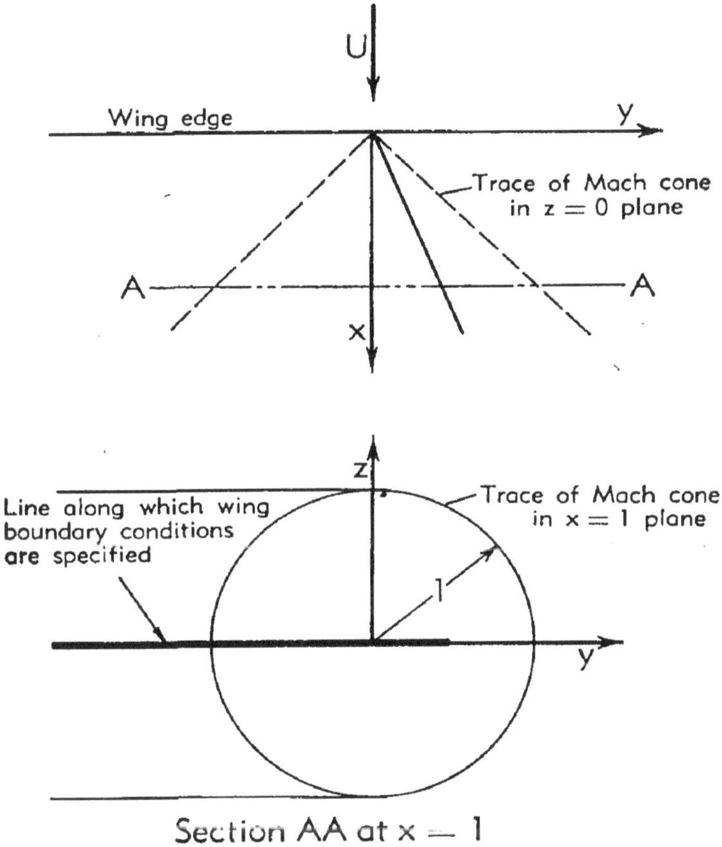

Fig. D,13c. Trapezoidal wing tip in supersonic flow.

$\eta^2 + \zeta^2 = 1$ becomes the circle $z^2 + y^2 = 1$ and that this in turn is the trace of the Mach cone from the origin. Hence Eq. 13-4 is elliptic within the Mach cone and hyperbolic without.

Eq. 13-4 cannot be readily analyzed in its present form. Hence, it is usually transformed[11] so that inside the circle $\eta^2 + \zeta^2 = 1$ it becomes Laplace's equation in two dimensions. In view of the information available on the solution of boundary-value problems for Laplace's equation, especially with regard to the use of the complex variable, such a transformation needs no justification. It is not our purpose, however, to

[11] In the case of conical flow, i.e. when $k = 0$, the appropriate transformation is the Chaplygin transformation applied to the conical flow problem by Busemann [23].

continue along these lines.[12] Rather, we shall reexamine the problem using an entirely different method, one that is well adapted to the type of analysis used in this chapter.

We are concerned with inverse problems for which the given boundary conditions are quasi-conical. The conditions may be stated as follows:

Lifting case: over all the x, y plane $\Delta w = w_u - w_l = 0$ and, except for the region occupied by the planform, $\Delta u = u_u - u_l = 0$. On the

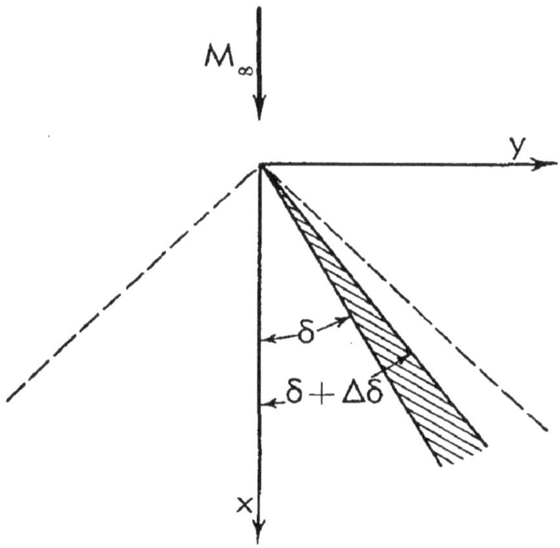

Fig. D,13d. Orientation of conical element.

planform, vertical induced velocity is specified in the form

$$w_u = w_l = Uy^k G_1\left(\frac{x}{y}\right) = Ux^k G_2\left(\frac{x}{y}\right) \tag{13-5}$$

where G_1 and G_2 are known polynomials in x/y.

Nonlifting case: over all the x, y plane $\Delta u = u_u - u_l = 0$ and, except for the region occupied by the planform, $\Delta w = w_u - w_l = 0$. On the planform, streamwise induced velocity is specified in the form

$$u_u = u_l = Uy^k N_1\left(\frac{x}{y}\right) = Ux^k N_2\left(\frac{x}{y}\right) \tag{13-6}$$

where N_1 and N_2 are known polynomials in x/y.

The method of approach for both types of problems is the same in that a basic element is found which has either a quasi-conical loading or thickness. This element extends radially from the apex of the field, as in Fig. D,13d, and is inclined at an angle δ to the x axis. In order to obtain

[12] The methods outlined above were originally applied to the study of conical flow problems by Busemann [*23*], and further developed by Lagerstrom [*24*], Germain [*25*], Hayes [*34*], and others.

such an element it is sufficient to subtract two planforms of prescribed loading or thickness, each planform having one side directed along the x axis while the other sides are inclined at angles δ and $\delta + \Delta\delta$ respectively. In application, the distribution of these elements over wings may be determined in such a manner as to satisfy the imposed conditions.

Consider first a radial lifting element that carries a load given by the expression

$$\frac{\Delta p}{\frac{1}{2}\rho_\infty U^2} = Cy^k \tag{13-7}$$

where C is a constant. Assume that δ is positive. The upwash field of a triangular planform can be found from Eq. 10-22,

$$\frac{w_u}{U} = \frac{1}{4\pi} \mathop{\not\!\!\int}_{\tau} dy_1 \int \frac{(x - x_1)}{(y - y_1)^2 \sqrt{(x - x_1)^2 - \beta^2(y - y_1)^2}} \frac{\Delta p(x_1, y_1)}{\frac{1}{2}\rho_\infty U^2} dx_1$$

where the integration with respect to x_1 is performed between the limits $x_1 = y_1 \cot \delta = y_1/m$ and $(x - x_1) = \pm\beta(y - y_1)$. This integration leads to the results

$$\left(\frac{\beta}{x}\right)^k \frac{w_u}{U} = \frac{\beta c}{4\pi} \mathop{\not\!\!\int}_0^{\theta\frac{1+\eta}{1+\theta}} \frac{\eta_1^k}{(\eta - \eta_1)^2} \left[\left(1 - \frac{\eta_1}{\theta}\right)^2 - (\eta - \eta_1)^2\right]^{\frac{1}{2}} d\eta_1, \tag{13-8a}$$

$$0 < \theta, \eta < \theta$$

$$= \frac{\beta c}{4\pi} \mathop{\not\!\!\int}_0^{\theta\frac{1-\eta}{1-\theta}} \frac{\eta_1^k}{(\eta - \eta_1)^2} \left[\left(1 - \frac{\eta_1}{\theta}\right)^2 - (\eta - \eta_1)^2\right]^{\frac{1}{2}} d\eta_1, \tag{138-b}$$

$$0 < \theta, \theta < \eta$$

where the substitutions

$$m\beta = \theta, \qquad \eta = \frac{\beta y}{x}, \qquad \eta_1 = \frac{\beta y_1}{x} \tag{13-9}$$

have been made. If δ is negative so that θ and m are also, the limits in Eq. 13-8 are reversed in order; and as the limits are now written, changes in sign are required in the equations.

In order to obtain the upwash field of the required lifting element, it is sufficient to perform a direct differentiation; for, if w_u of a planform with vertex angle δ is of the form

$$w_u = f(\theta, \eta)$$

it follows that when $0 < \theta$ the value of w_u for the element is

$$dw_u = f(\theta + \Delta\theta, \eta) - f(\theta, \eta) = \frac{\partial f}{\partial \theta} d\theta$$

and when $\theta < 0$ the value is

$$dw_u = f(\theta, \eta) - f(\theta + \Delta\theta, \eta) = -\frac{\partial f}{\partial \theta} d\theta$$

If this process is carried out in Eq. 13-8 and if the transformation

$$\eta_1 = \theta \frac{\eta - t}{\theta - t}$$

is introduced, the expressions for w_u become

$$\left(\frac{\beta}{x}\right)^k \frac{dw_u}{U} = \frac{\beta C d\theta}{4\pi} \theta^k \int_{-1}^{\eta} \frac{(\eta - t)^{k+1} dt}{(\theta - \eta)(\theta - t)^{k+1} t^2 \sqrt{1 - t^2}}, \qquad (13\text{-}10a)$$

$$-1 < \eta < \theta < 1$$

$$\left(\frac{\beta}{x}\right)^k \frac{dw_u}{U} = \frac{\beta C d\theta}{4\pi} \theta^k \int_{1}^{\eta} \frac{(\eta - t)^{k+1} dt}{(\theta - \eta)(\theta - t)^{k+1} t^2 \sqrt{1 - t^2}}, \qquad (13\text{-}10b)$$

$$-1 < \theta < \eta < 1$$

Eq. 13-10 provide the upwash fields for any radial element, regardless of the sign of θ. Integration by parts leads to the alternative forms

$$\left(\frac{\beta}{x}\right)^k \frac{dw_u}{U} = -\frac{\beta C d\theta (k + 1)\theta^k}{4\pi} \int_{-1}^{\eta} \frac{(\eta - t)^k \sqrt{1 - t^2}\, dt}{(\theta - t)^{k+2} t}, \qquad (13\text{-}11a)$$

$$-1 < \eta < \theta < 1$$

$$\left(\frac{\beta}{x}\right)^k \frac{dw_u}{U} = -\frac{\beta C d\theta (k + 1)\theta^k}{4\pi} \int_{1}^{\eta} \frac{(\eta - t)^k \sqrt{1 - t^2}\, dt}{(\theta - t)^{k+2} t}, \qquad (13\text{-}11b)$$

$$-1 < \theta < \eta < 1$$

If now the lifting elements cover the region between θ_1 and θ_2 and C is a function of θ determining the lift carried along the radial element at that position, upwash produced by the resultant planform is given by the relation

$$\left(\frac{\beta}{x}\right)^k \frac{w_u}{U} = -\frac{\beta(k + 1)}{4\pi} \int_{\theta_1}^{\theta_0} \theta^k C(\theta) H(\theta, \eta) d\theta \qquad (13\text{-}12)$$

where

$$H(\theta, \eta) = \int_{1}^{\eta} \frac{(\eta - t)^k \sqrt{1 - t^2}}{(\theta - t)^{k+2} t} dt, \qquad \theta_1 \leqq \theta < \eta$$

$$= \int_{-1}^{\eta} \frac{(\eta - t)^k \sqrt{1 - t^2}}{(\theta - t)^{k+2} t} dt, \qquad \eta < \theta \leqq \theta_0$$

The function $H(\theta, \eta)$ has a simple pole at $\theta = \eta$, and the integral expression for w_u in Eq. 13-12 is therefore evaluated as a Cauchy principal part.

The boundary condition to be satisfied by Eq. 13-12 is that $\beta^k w_u / x^k U$ is a polynomial of degree k in the variable η. It follows that the $(k + 1)$th derivative of the right-hand member of Eq. 13-12 must vanish. Thus

$$0 = \frac{\beta(k + 1)}{4\pi} \int_{\theta_1}^{\theta_0} \theta^k C(\theta) \left(\frac{\partial}{\partial \eta}\right)^{k+1} H(\theta, \eta) d\theta \qquad (13\text{-}13)$$

where the generalized finite part of an integral must be used. Continuing the calculation, one has, after taking the derivatives with respect to η,

$$0 = \frac{\sqrt{1-\eta^2}}{\eta} \,\text{⨎}\!\int_{\theta_1}^{\theta_0} \frac{\theta^k C(\theta)\,d\theta}{(\theta-\eta)^{k+2}}$$

which may be written in the form

$$\left(\frac{\partial}{\partial\eta}\right)^{k+1} \int_{\theta_1}^{\theta_0} \frac{\theta^k C(\theta)\,d\theta}{\theta-\eta} = 0$$

The function $\theta^k C(\theta)$ is thus to be found through inversion of the integral equation

$$\int_{\theta_1}^{\theta_0} \frac{\theta^k C(\theta)\,d\theta}{\theta-\eta} = \sum_{i=0}^{k} a_i \eta^i \tag{13-14}$$

Eq. 13-14 can be inverted by Eq. 12-24 and, provided $\theta^k C(\theta)$ is continuous in the interval $\theta_1 < \theta < \theta_0$, its expression is

$$\theta^k C(\theta) = \frac{1}{\pi^2\sqrt{(\theta_0-\theta)(\theta-\theta_1)}}\left[A - \int_{\theta_1}^{\theta_0} \frac{\sum_{i=0}^{k} a_i\eta^i \sqrt{(\theta_0-\eta)(\eta-\theta_1)}}{\theta-\eta}\,d\eta\right]$$

and this obviously leads to the form

$$\theta^k C(\theta) = \sum_{i=0}^{k+1} \frac{b_i\theta^i}{\sqrt{(\theta_0-\theta)(\theta-\theta_1)}} \tag{13-15}$$

where the coefficients b_i are functions of θ_0 and θ_1 but not functions of θ.

From Eq. 13-7, since $\theta = \beta y/x$, the loading on the planform is,

$$\frac{\Delta p}{\frac{1}{2}\rho_\infty U^2} = \left(\frac{x}{\beta}\right)^k \sum_{i=0}^{k+1} \frac{b_i\theta^i}{\sqrt{(\theta_0-\theta)(\theta-\theta_1)}} \tag{13-16}$$

where the coefficients b_i must be determined from known information about the surface geometry.

Consider next the identity

$$\text{⨎}\!\int_a^b \frac{dt}{(\theta-t)^{i+1}\sqrt{(b-t)(t-a)}} = \frac{(-1)^i}{i!}\left(\frac{\partial}{\partial\theta}\right)^i \int_a^b \frac{dt}{(\theta-t)\sqrt{(b-t)(t-a)}}$$

$$= 0, \qquad a < \theta < b \tag{13-17}$$

where i is zero or a positive integer. This expression implies the equation

$$\text{⨎}\!\int_{-1}^{1} \frac{(\eta-t)^k\sqrt{1-t^2}}{(\theta-t)^{k+2}t}\,dt = 0$$

since the latter form can be broken by the method of rational fractions into integrals like the left member of Eq. 13-17. Hence, we have

$$\neq\!\!\!\!\!\int_{-1}^{\eta} \frac{(\eta - t)^k \sqrt{1 - t^2}}{(\theta - t)^{k+2}t}\, dt = \neq\!\!\!\!\!\int_{1}^{\eta} \frac{(\eta - t)^k \sqrt{1 - t^2}}{(\theta - t)^{k+2}t}\, dt$$

where the finite part sign is needed on but one side of the equation, depending on the value of θ relative to η. From this equivalence and

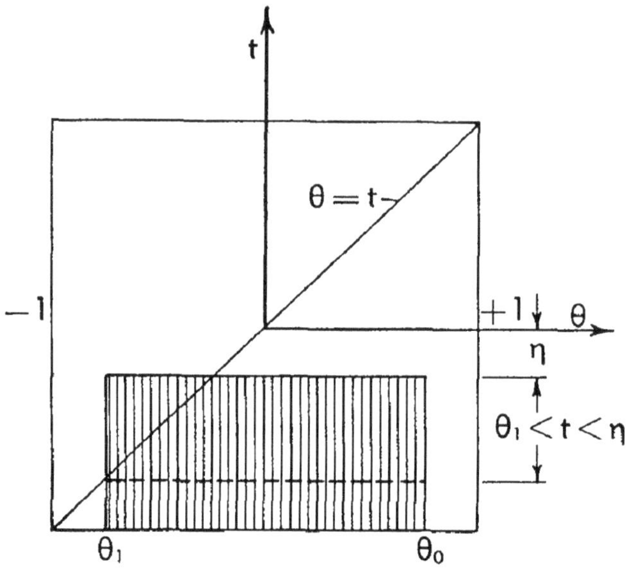

Fig. D,13e. Area of integration in t, θ plane.

Eq. 13-12, the following relations are obtained

$$\left(\frac{\beta}{x}\right)^k \frac{w_u}{U} = -\frac{\beta(k+1)}{4\pi} \int_{\theta_1}^{\theta_0} \theta^k C(\theta)d\theta\ \neq\!\!\!\!\!\int_{1}^{\eta} \frac{(\eta - t)^k \sqrt{1 - t^2}}{t(\theta - t)^{k+2}}\, dt \tag{13-18a}$$

$$\left(\frac{\beta}{x}\right)^k \frac{w_u}{U} = -\frac{\beta(k+1)}{4\pi} \int_{\theta_1}^{\theta_0} \theta^k C(\theta)d\theta\ \neq\!\!\!\!\!\int_{-1}^{\eta} \frac{(\eta - t)^k \sqrt{1 - t^2}}{t(\theta - t)^{k+2}}\, dt \tag{13-18b}$$

Consider now planforms with two subsonic leading edges so that the inequalities $-1 \leqq \theta_1 \leqq 0$ and $0 \leqq \theta_0 \leqq 1$ hold. Fig. D,13e shows the region of integration of Eq. 13-18b and, if the order of integration is reversed, one gets, after substituting from Eq. 13-15 and using the results of Eq. 13-17,

$$\left(\frac{\beta}{x}\right)^k \frac{w_u}{U} = -\frac{\beta(k+1)}{4\pi} \int_{-1}^{\theta_1} \frac{(\eta - t)^k \sqrt{1 - t^2}}{t}\, dt$$

$$\neq\!\!\!\!\!\int_{\theta_1}^{\theta_0} \frac{\sum_{i=0}^{k+1} b_i\theta^i}{(\theta - t)^{k+2}\sqrt{(\theta_0 - \theta)(\theta - \theta_1)}}\, d\theta, \qquad \theta_1 \leqq \eta \leqq \theta_0$$

and similarly,

$$\left(\frac{\beta}{x}\right)^k \frac{w_u}{U} = -\frac{\beta(k+1)}{4\pi} \int_1^{\theta_0} \frac{(\eta - t)^k \sqrt{1-t^2}}{t}\, dt$$

$$\int_{\theta_1}^{\theta_0} \frac{\sum_{i=0}^{k+1} b_i\theta^i}{(\theta - t)^{k+2}\sqrt{(\theta_0 - \theta)(\theta - \theta_1)}}\, d\theta, \qquad \theta_1 \leqq \eta \leqq \theta_0$$

Both of the last two expressions can be further simplified through consideration of the term

$$I = \int_{\theta_1}^{\theta_0} \frac{\sum_{i=0}^{k+1} b_i\theta^i}{(\theta - t)^{k+2}\sqrt{(\theta_0 - \theta)(\theta - \theta_1)}}\, d\theta = \frac{1}{(k+1)!}\left(\frac{\partial}{\partial t}\right)^{k+1} \int_{\theta_1}^{\theta_0} \frac{\sum_{i=0}^{k+1} b_i\theta^i}{(\theta - t)\sqrt{(\theta_0 - \theta)(\theta - \theta_1)}}\, d\theta$$

Since we have, in general,

$$\frac{\theta^n}{\theta - t} = \sum_{j=1}^{n} \theta^{n-j}\, t^{j-1} + \frac{t^n}{\theta - t}$$

the expression for I becomes

$$I = \frac{1}{(k+1)!}\left(\frac{\partial}{\partial t}\right)^{k+1} \sum_{i=0}^{k+1} b_i t^i \int_{\theta_1}^{\theta_0} \frac{d\theta}{(\theta - t)\sqrt{(\theta_0 - \theta)(\theta - \theta_1)}}$$

Finally, use of the identities,

$$\int_{\theta_1}^{\theta_0} \frac{d\theta}{(\theta - t)\sqrt{(\theta_0 - \theta)(\theta - \theta_1)}} = \begin{cases} -\dfrac{\pi}{\sqrt{(t - \theta_0)(t - \theta_1)}}, & \theta_0 < t \\ 0, & \theta_1 < t < \theta_0 \\ \dfrac{\pi}{\sqrt{(\theta_0 - t)(\theta_1 - t)}}, & t < \theta_1 \end{cases} \tag{13-19}$$

permits the reexpression of I and the relations for w_u become

$$\left(\frac{\beta}{x}\right)^k \frac{w_u}{U} = \frac{-\beta}{4k!} \int_{-1}^{\theta_1} \frac{(\eta - t)^k \sqrt{1-t^2}}{t} \left(\frac{\partial}{\partial t}\right)^{k+1} \frac{\sum_{i=0}^{k+1} b_i t^i}{\sqrt{(\theta_0 - t)(\theta_1 - t)}}\, dt \tag{13-20a}$$

$$\left(\frac{\beta}{x}\right)^k \frac{w_u}{U} = \frac{\beta}{4k!} \int_1^{\theta_0} \frac{(\eta - t)^k \sqrt{1 - t^2}}{t} \left(\frac{\partial}{\partial t}\right)^{k+1} \frac{\sum_{i=0}^{k+1} b_i t^i}{\sqrt{(t - \theta_0)(t - \theta_1)}} dt \tag{13-20b}$$

Eq. 13-20 are the fundamental relations for a lifting surface and it remains merely to determine the unknown coefficients b_i by equating coefficients of $\eta = \beta y/x$ on both sides of the expressions. In the present form of the equations, it appears that for $k > 0$ the number of equations obtainable exceeds the number of undetermined coefficients. No general theorems of determinacy will be presented here as to the uniqueness of the solutions, but applications to be made later will indicate the general technique of solution. For the symmetrical case, where $-\theta_1 = \theta_0$, the fundamental equations become

$$\left(\frac{\beta}{x}\right)^k \frac{w_u}{U} = \frac{-\beta}{4k!} \int_{-1}^{-\theta_0} \frac{(\eta - t)^k \sqrt{1 - t^2}}{t} \left(\frac{\partial}{\partial t}\right)^{k+1} \frac{\sum_{i=0}^{k+1} b_i t^i}{\sqrt{t^2 - \theta_0^2}} dt \tag{13-21a}$$

$$\left(\frac{\beta}{x}\right)^k \frac{w_u}{U} = \frac{\beta}{4k!} \int_1^{\theta_0} \frac{(\eta - t)^k \sqrt{1 - t^2}}{t} \left(\frac{\partial}{\partial t}\right)^{k+1} \frac{\sum_{i=0}^{k+1} b_i t^i}{\sqrt{t^2 - \theta_0^2}} dt \tag{13-21b}$$

We consider next a radial element emanating from the origin and possessing a quasi-conical thickness distribution

$$\left(\frac{dz}{dx}\right)_u = \lambda_u = Cy^k \tag{13-22}$$

The derivation of the induced pressure field associated with the element follows closely the analysis in the lifting case and the solution can therefore be written, as in Eq. 13-16, in the form

$$\lambda_u = \left(\frac{x}{\beta}\right)^k \sum_{i=0}^{k+1} \frac{a_i \theta^i}{\sqrt{(\theta_0 - \theta)(\theta - \theta_1)}} \tag{13-23}$$

where the unknown coefficients a_i are determined by equating coefficients of η on both sides of the expressions

$$\left(\frac{\beta}{x}\right)^k C_p = -\frac{2}{\beta k!} \int_{-1}^{\theta_1} \frac{t(\eta - t)^k}{\sqrt{1 - t^2}} \left(\frac{\partial}{\partial t}\right)^{k+1} \frac{\sum_{i=0}^{k+1} a_i t^i}{\sqrt{(\theta_0 - t)(\theta_1 - t)}} dt \tag{13-24a}$$

$$\left(\frac{\beta}{x}\right)^k C_p = \frac{2}{\beta k!} \int_1^{\theta_0} \frac{t(\eta - t)^k}{\sqrt{1 - t^2}} \left(\frac{\partial}{\partial t}\right)^{k+1} \frac{\sum_{i=0}^{k+1} a_i t^i}{\sqrt{(t - \theta_0)(t - \theta_1)}} dt \tag{13-24b}$$

Eq. 13-20 and 13-24 suffice for the solution of lifting and nonlifting problems in quasi-conical flow fields. If both the planform and boundary conditions are symmetrical about the x axis, these equations can be simplified. Suppose first that the left members of these equations are odd functions of η and, of course, that $\theta_1 = -\theta_0$. It follows from physical considerations that the loading or surface slope will be an odd function of η and that consequently the unknown coefficients a_i or b_i must vanish for even values of the subscript i. If the transformation $t = -\tau$ is made in either pair of equations, they reduce to single equalities which can be written

$$\left(\frac{\beta}{x}\right)^k \frac{w_u}{U} = \frac{\beta}{4k!} \int_1^{\theta_0} \frac{(\eta - t)^k \sqrt{1 - t^2}}{t} \left(\frac{\partial}{\partial t}\right)^{k+1} \frac{\sum_{j=0}^{[k/2]} b_{2j+1} t^{2j+1}}{\sqrt{t^2 - \theta_0^2}} dt \tag{13-25}$$

and

$$\left(\frac{\beta}{x}\right)^k C_p = \frac{2}{\beta k!} \int_1^{\theta_0} \frac{t(\eta - t)^k}{\sqrt{1 - t^2}} \left(\frac{\partial}{\partial t}\right)^{k+1} \frac{\sum_{j=0}^{[k/2]} a_{2j+1} t^{2j+1}}{\sqrt{t^2 - \theta_0^2}} dt \tag{13-26}$$

where the notation $[k/2]$ in the summation denotes the largest integer contained in $k/2$.

In the same fashion, a simplification can be achieved for even symmetry about the x axis. Again using the bracket notation to indicate largest integer, the resulting equations become

$$\left(\frac{\beta}{x}\right)^k \frac{w_u}{U} = \frac{\beta}{4k!} \int_1^{\theta_0} \frac{(\eta - t)^k \sqrt{1 - t^2}}{t} \left(\frac{\partial}{\partial t}\right)^{k+1} \frac{\sum_{j=0}^{[(k+1)/2]} b_{2j} t^{2j}}{\sqrt{t^2 - \theta_0^2}} dt \tag{13-27}$$

and

$$\left(\frac{\beta}{x}\right)^k C_p = \frac{2}{\beta k!} \int_1^{\theta_0} \frac{t(\eta - t)^k}{\sqrt{1 - t^2}} \left(\frac{\partial}{\partial t}\right)^{k+1} \frac{\sum_{j=0}^{[(k+1)/2]} a_{2j} t^{2j}}{\sqrt{t^2 - \theta_0^2}} dt \tag{13-28}$$

The yawed, lifting, triangular wing. Eq. 13-16 and 13-20 lead directly to the determination of angle-of-attack loading on a yawed triangular wing, Fig. D,13f. From the boundary conditions we have $w_u = -U\alpha$ on the planform, hence $k = 0$, and from Eq. 13-16 loading is as follows:

$$\frac{\Delta p}{\frac{1}{2}\rho_\infty U^2} = \frac{b_0 + b_1\theta}{\sqrt{(\theta_0 - \theta)(\theta - \theta_1)}}.$$

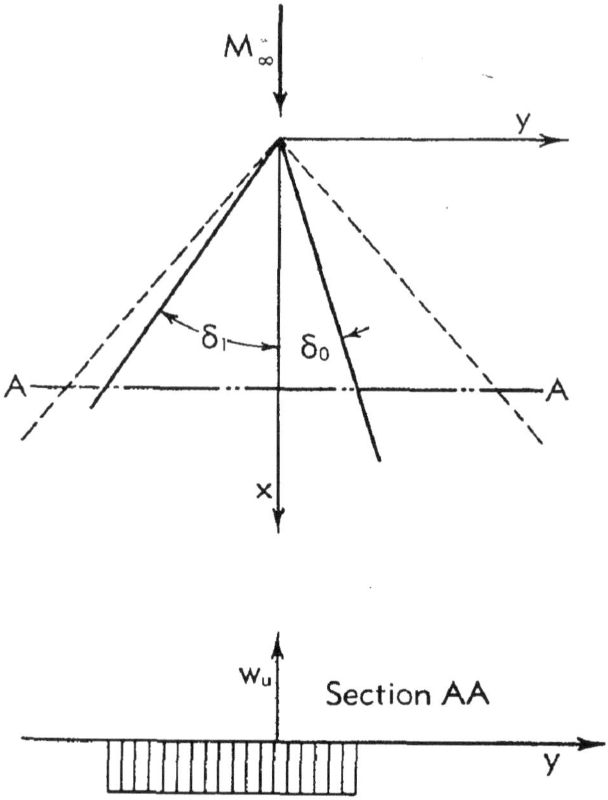

Fig. D,13f. Triangular wing planform and boundary conditions for w.

Eq. 13-20 lead to the equalities

$$-U\alpha = \frac{\beta}{4} \int_{1}^{\theta_0} \frac{\sqrt{1-t^2}}{t} \frac{\partial}{\partial t} \frac{b_0 + b_1 t}{\sqrt{(t-\theta_0)(t-\theta_1)}} dt \tag{13-29a}$$

$$-U\alpha = -\frac{\beta}{4} \int_{-1}^{\theta_1} \frac{\sqrt{1-t^2}}{t} \frac{\partial}{\partial t} \frac{b_0 + b_1 t}{\sqrt{(\theta_0-t)(\theta_1-t)}} dt \tag{13-29b}$$

After integration by parts, these relations become

$$U\alpha = -\frac{\beta}{4} \int_{1}^{\theta_0} \frac{(b_0 + b_1 t)dt}{t^2 \sqrt{(1-t^2)(t-\theta_0)(t-\theta_1)}}$$

$$U\alpha = \frac{\beta}{4} \int_{-1}^{\theta_1} \frac{(b_0 + b_1 t)dt}{t^2 \sqrt{(1-t^2)(\theta_0-t)(\theta_1-t)}}$$

Hence

$$U\alpha = \frac{\beta}{4} [b_0 L_0(\theta_0, \theta_1) + b_1 L_1(\theta_0, \theta_1)]$$

$$U\alpha = \frac{\beta}{4} [b_0 L_0(-\theta_1, -\theta_0) - b_1 L_1(-\theta_1, -\theta_0)] \tag{13-30}$$

where

$$L_0(\theta_0, \theta_1) = -\int_1^{\theta_0} \frac{dt}{t^2\sqrt{(1-t^2)(t-\theta_0)(t-\theta_1)}}$$

and

$$L_1(\theta_0, \theta_1) = -\int_1^{\theta_0} \frac{dt}{t\sqrt{(1-t^2)(t-\theta_0)(t-\theta_1)}}$$

The evaluation of these last two elliptic integrals is tedious but it is apparent that Eq. 13-30 can be solved for b_0 and b_1 in terms of the func-

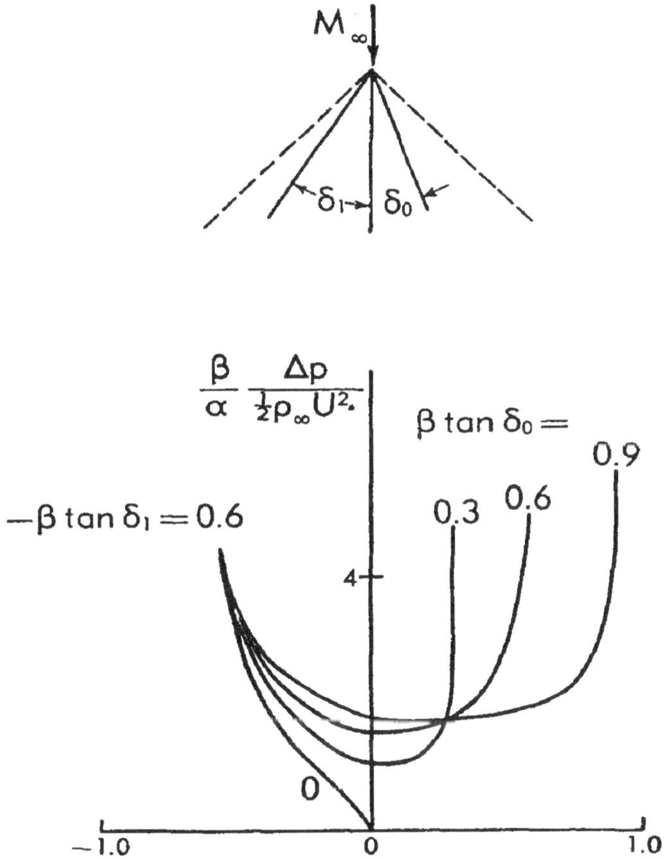

Fig. D,13g. Load distributions for various unsymmetric triangular wings.

tions $L_0(\theta_0, \theta_1)$ and $L_1(\theta_0, \theta_1)$. Substitution of the values into the equation for $\Delta p/\frac{1}{2}\rho_\infty U^2$ leads to the expression for load coefficient

$$\frac{\Delta p}{\frac{1}{2}\rho_\infty U^2} = \frac{2\alpha}{\beta E}\left(\frac{2G}{\theta_0-\theta_1}\right)^{\frac{1}{2}}\left[\frac{(\theta_0+\theta_1)\theta - 2\theta_0\theta_1}{\sqrt{(\theta_0-\theta)(\theta-\theta_1)}}\right] \tag{13-31}$$

where E is the complete elliptic integral of the second kind with modulus $\sqrt{1-G^2}$ and

$$G = \frac{1-\theta_0\theta_1-\sqrt{(1-\theta_0^2)(1-\theta_1^2)}}{\theta_0-\theta_1}$$

Typical load distributions over the wing are shown in Fig. D,13g for $\beta \tan \delta_1 = -0.6$ and for $\beta \tan \delta_0$ equal to 0, 0.3, 0.6, and 0.9.

In the particular case when $\theta_1 = -\theta_0$, the value of G becomes θ_0 and the resultant loading on the unyawed wing becomes

$$\frac{\Delta p}{\frac{1}{2}\rho_\infty U^2} = \frac{4\alpha\theta}{\beta E \sqrt{\theta_0^2 - \theta^2}} \tag{13-32}$$

where the modulus of E is $\sqrt{1 - \theta_0^2}$. This latter problem could, of course, have been solved directly from Eq. 13-27 in a much simpler fashion. Its solution was the first exact solution obtained for a lifting surface with subsonic leading edges and was given at about the same time by a number of authors [*26,27,28,29*].

The rolling triangular wing. Consider next the case of any unyawed triangular wing rolling about its line of symmetry [*30*]. If the angular rate of roll is P radians per second, the boundary conditions on the wing require that upwash be given by $w_u = -Py$. In this case $k = 1$ and, from Eq. 13-16, the loading on the wing is given by the expression

$$\frac{\Delta p}{\frac{1}{2}\rho_\infty U^2} = \frac{x}{\beta} \frac{b_1\theta}{\sqrt{\theta_0^2 - \theta^2}} \tag{13-33}$$

since the loading must obviously be antisymmetrical. Hence, in Eq. 13-25 the value of $[k/2]$ is zero and b_1 must satisfy the relations

$$\frac{P}{U} = -\frac{\beta b_1}{4} \int_1^{\theta_0} \frac{\sqrt{1 - t^2}}{t} \left(\frac{\partial}{\partial t}\right)^2 \frac{t}{\sqrt{t^2 - \theta_0^2}}\, dt \tag{13-34a}$$

$$0 = b_1 \int_1^{\theta_0} \sqrt{1 - t^2} \left(\frac{\partial}{\partial t}\right)^2 \frac{t}{\sqrt{t^2 - \theta_0^2}}\, dt \tag{13-34b}$$

Performing one differentiation with respect to t in each of the two integrals, and then integrating by parts, leads one to the expressions

$$\frac{P}{U} = \frac{\beta b_1 \theta_0^2}{4} \int_1^{\theta_0} \frac{dt}{t^2 \sqrt{1 - t^2}\,(t^2 - \theta_0^2)^{\frac{3}{2}}} = \frac{\beta b_1 \theta_0}{4} \frac{\partial}{\partial \theta_0} \int_1^{\theta_0} \frac{dt}{t^2 \sqrt{(1 - t^2)(t^2 - \theta_0^2)}}$$

$$0 = b_1 \theta_0^2 \int_1^{\theta_0} \frac{t\,dt}{\sqrt{1 - t^2}\,(t^2 - \theta_0^2)^{\frac{3}{2}}} = b_1 \theta_0 \frac{\partial}{\partial \theta_0} \int_1^{\theta_0} \frac{t\,dt}{\sqrt{(1 - t^2)(t^2 - \theta_0^2)}}$$

The last equation reduces to an identity while the former one becomes

$$\frac{P}{U} = -\frac{\beta b_1 \theta_0}{4} \frac{\partial}{\partial \theta_0} \left(\frac{E}{\theta_0^2}\right) = \frac{\beta b_1}{4\theta_0^2} \left[\frac{2E - \theta_0^2(E + K)}{1 - \theta_0^2}\right] \tag{13-35}$$

where the modulus of the complete elliptic integrals E and K is $\sqrt{1 - \theta_0^2}$. From Eq. 13-33 and 13-35 the resultant loading is

$$\frac{\Delta p}{\frac{1}{2}\rho_\infty U^2} = \frac{4P\theta_0^2 x\theta}{\beta^2 U \left[\frac{2 - \theta_0^2}{1 - \theta_0^2} E - \frac{\theta_0^2}{1 - \theta_0^2} K\right]} \tag{13-36}$$

Fig. D,13h shows the proper distribution of this loading and the nature of the boundary conditions.

The pitching triangular wing. If an unyawed triangular wing is pitching about its vertex, the boundary conditions become $w_u = -Qx$

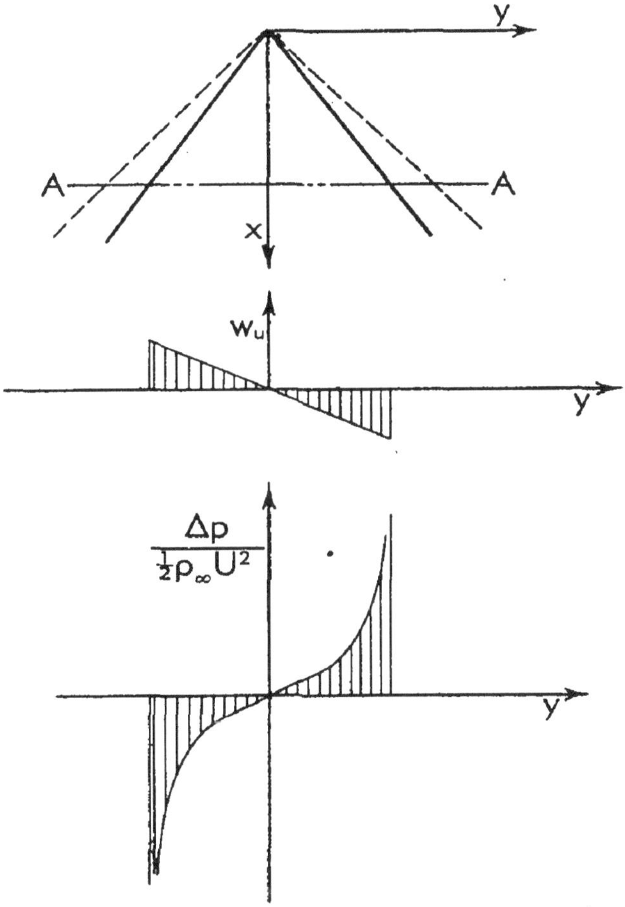

Fig. D,13h. Planform, boundary condition, and load distribution for triangular rolling wing.

where Q is the rate of pitch in radians per second. From Eq. 13-16 the expression for the loading is

$$\frac{\Delta p}{\frac{1}{2}\rho_\infty U^2} = \frac{x}{\beta} \frac{b_0 + b_2\theta^2}{\sqrt{\theta_0^2 - \theta^2}} \tag{13-37}$$

and the relations involving the undetermined constants are

$$0 = \int_1^{\theta_0} \frac{\sqrt{1 - t^2}}{t} \left(\frac{\partial}{\partial t}\right)^2 \frac{b_0 + b_2 t^2}{\sqrt{t^2 - \theta_0^2}}\, dt \tag{13-38a}$$

$$\frac{Q}{U} = \frac{1}{4} \int_1^{\theta_0} \sqrt{1 - t^2} \left(\frac{\partial}{\partial t}\right)^2 \frac{b_0 + b_2 t^2}{\sqrt{t^2 - \theta_0^2}}\, dt \tag{13-38b}$$

The latter equations reduce to the form

$$0 = b_0 + 2\theta_0^2 b_2$$

$$-\frac{4(1 - \theta_0^2)Q}{U} = b_0(E - K) + b_2(E - \theta_0^2 K)$$

where the modulus of the complete elliptic integral is $\sqrt{1 - \theta_0^2}$. Hence,

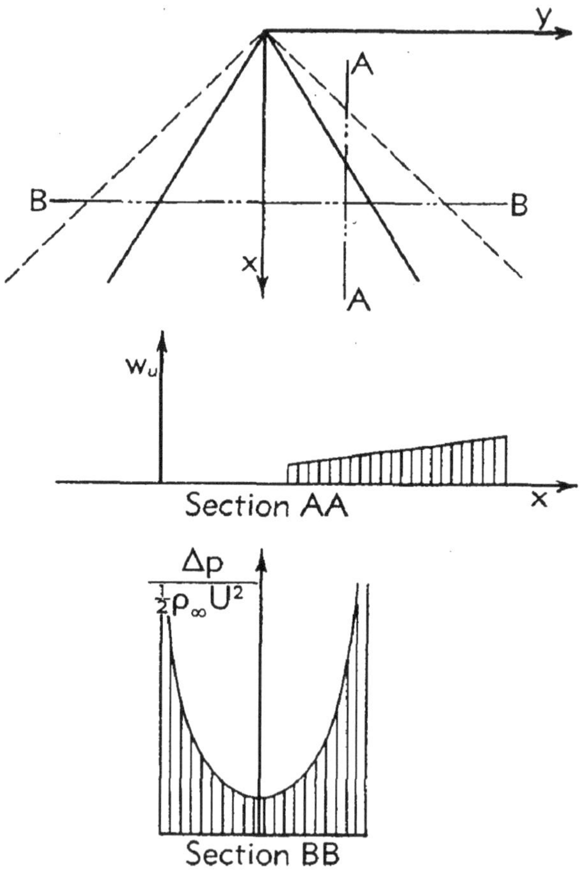

Fig. D,13i. Planform, boundary condition, and load distribution for triangular wing pitching about its apex.

the resultant loading can be expressed as

$$\frac{\Delta p}{\frac{1}{2}\rho_\infty U^2} = \frac{4xQ}{U\beta\sqrt{\theta_0^2 - \theta^2}} \frac{2\theta_0^2 - \theta^2}{\left[\frac{\theta_0^2 K}{1 - \theta_0^2} + \frac{1 - 2\theta_0^2}{1 - \theta_0^2} E\right]} \tag{13-39}$$

Fig. D,13i shows the wing, the distribution of w_u over the wing, and the loading.

Triangular wing with uniform pressure. In the calculation of thickness distributions corresponding to prescribed pressure distributions, it is instructive to consider first the simple case in which the pressure is a constant over an unyawed triangular planform. In this case, C_p is a constant, k is zero, and from Eq. 13-23 and 13-28 the following relations hold

$$\left(\frac{dz}{dx}\right)_u = \lambda_u = \frac{a_0}{\sqrt{\theta_0^2 - \theta^2}} \tag{13-40}$$

$$C_p = \frac{2}{\beta} \int_1^{\theta_0} \frac{t}{\sqrt{1 - t^2}} \frac{\partial}{\partial t} \frac{a_0}{\sqrt{t^2 - \theta_0^2}}\, dt \tag{13-41}$$

The value of a_0 may be found in a manner quite similar to that used in the previous examples. Thus, after differentiation, Eq. 13-41 becomes

$$\frac{\beta}{2} C_p = -a_0 \int_1^{\theta_0} \frac{t^2 dt}{\sqrt{1 - t^2}\,(t^2 - \theta_0^2)^{\frac{3}{2}}} = -\frac{a_0}{\theta_0} \frac{\partial}{\partial \theta_0} \int_1^{\theta_0} \frac{t^2 dt}{\sqrt{(1 - t^2)(t^2 - \theta_0^2)}}$$

and after integration, this yields

$$\frac{\beta C_p}{2} = \frac{a_0}{\theta_0} \frac{\partial}{\partial \theta_0} E = \frac{a_0(K - E)}{1 - \theta_0^2} \tag{13-42}$$

where the modulus of the elliptic integral is $\sqrt{1 - \theta_0^2}$. If a_0 is eliminated from Eq. 13-40 and 13-42, the slope of the wing on the upper surface is

$$\left(\frac{dz}{dx}\right)_u = \frac{\beta C_p(1 - \theta_0^2)}{2(K - E)\sqrt{\theta_0^2 - \theta^2}} = \frac{C_p(1 - \theta_0^2)}{2(K - E)} \frac{x}{\sqrt{m^2x^2 - y^2}}$$

The ordinate of the upper surface results from the integration

$$z_u = \int_{y/m}^{x} \left(\frac{dz}{dx}\right)_u dx$$

and is

$$z_u = \frac{(1 - \theta_0^2)C_p}{2m^2(K - E)} \sqrt{m^2x^2 - y^2} = \frac{(1 - \theta_0^2)C_p x}{2\beta m^2(K - E)} \sqrt{\theta_0^2 - \theta^2} \tag{13-43}$$

These results are equivalent to those of Squire [*31*], who considered the surface shape in Eq. 13-43 and found the value of C_p.

Triangular wing with linear pressure gradient. We now propose to determine the thickness distribution for an unyawed triangular wing for which pressure varies linearly in the streamwise direction. Setting C_p equal to bx, it follows that slope of the upper surface is expressible as

$$\left(\frac{dz}{dx}\right)_u = \frac{x}{\beta} \frac{a_0 + a_2\theta^2}{\sqrt{\theta_0^2 - \theta^2}} \tag{13-44}$$

since the solution is obviously symmetrical about the stream axis. Since $k = 1$, Eq. 13-28 becomes

$$\beta b = \frac{2}{\beta} \int_1^{\theta_0} \frac{t(\eta - t)}{\sqrt{1 - t^2}} \left(\frac{\partial}{\partial t}\right)^2 \frac{a_0 + a_2t^2]}{\sqrt{t^2 - \theta_0^2}}\, dt$$

Equating the coefficients of η, we are led to the two relations

$$0 = \int_1^{\theta_0} \frac{t}{\sqrt{1-t^2}} \left(\frac{\partial}{\partial t}\right)^2 \frac{a_0 + a_2 t^2}{\sqrt{t^2 - \theta_0^2}}\, dt \tag{13-45a}$$

$$\frac{\beta^2 b}{2} = -\int_1^{\theta_0} \frac{t^2}{\sqrt{1-t^2}} \left(\frac{\partial}{\partial t}\right)^2 \frac{a_0 + a_2 t^2}{\sqrt{t^2 - \theta_0^2}}\, dt \tag{13-45b}$$

It is of interest to compare the series of equations just developed with the corresponding equations in the problem of the pitching triangular wing. Formally the algebraic steps are the same, and it is to be expected, just as in the case of Eq. 13-38, that two simultaneous equations will be obtained and that their solutions will provide the constants a_0 and a_2. In the present case, however, Eq. 13-45a can be shown to vanish identically and as a result only one equation in two unknowns remains. This means that an infinite number of possible solutions exists. The following calculations will confirm these remarks.

Consider Eq. 13-45a and introduce the transformation

$$t^2 = \tau, \qquad \theta_0^2 = \tau_0 \tag{13-46}$$

Then, by means of the relations

$$\frac{\partial}{\partial t} = 2\sqrt{\tau}\,\frac{\partial}{\partial \tau}, \qquad \frac{\partial^2}{\partial t^2} = 2\left(2\tau\,\frac{\partial^2}{\partial \tau^2} + \frac{\partial}{\partial \tau}\right) \tag{13-47}$$

the equation becomes

$$\begin{aligned}
0 &= a_0 \int_1^{\tau_0} \frac{d\tau}{\sqrt{1-\tau}} \left(2\tau\,\frac{\partial^2}{\partial\tau^2} + \frac{\partial}{\partial\tau}\right) \frac{1}{\sqrt{\tau - \tau_0}} \\
&\qquad + a_2 \int_1^{\tau_0} \frac{d\tau}{\sqrt{1-\tau}} \left(2\tau\,\frac{\partial^2}{\partial\tau^2} + \frac{\partial}{\partial\tau}\right) \frac{\tau}{\sqrt{\tau - \tau_0}} \\
&= a_0 \left[\frac{\partial^2}{\partial\tau_0^2} \int_1^{\tau_0} \frac{2\tau\, d\tau}{\sqrt{(1-\tau)(\tau-\tau_0)}} - \frac{\partial}{\partial\tau_0} \int_1^{\tau_0} \frac{d\tau}{\sqrt{(1-\tau)(\tau-\tau_0)}}\right] \\
&\qquad + a_2 \left[\frac{\partial^2}{\partial\tau_0^2} \int_1^{\tau_0} 2\tau \sqrt{\frac{\tau-\tau_0}{1-\tau}}\, d\tau - \frac{\partial}{\partial\tau_0} \int_1^{\tau_0} \sqrt{\frac{\tau-\tau_0}{1-\tau}}\, d\tau \right. \\
&\qquad \left. + 2\tau_0 \frac{\partial^2}{\partial\tau_0^2} \int_1^{\tau_0} \frac{\tau\, d\tau}{\sqrt{(1-\tau)(\tau-\tau_0)}} - \tau_0 \frac{\partial}{\partial\tau_0} \int_1^{\tau_0} \frac{d\tau}{\sqrt{(1-\tau)(\tau-\tau_0)}}\right]
\end{aligned}$$

Each of the above integrals can be evaluated directly, and the bracketed terms are in both cases identically zero.

It remains now to calculate the terms in Eq. 13-45b. This can be carried out in terms of the elliptic integrals E and K both having, as before, the modulus $\sqrt{1-\theta_0^2}$. The final expression in which bx has been replaced by C_p becomes

$$C_p = \frac{2x}{\beta^2(1-\theta_0^2)^2} \{a_0[2K - E(3-\theta_0^2)] + a_2[(\theta_0^4 + \theta_0^2)K - 2\theta_0^2 E]\} \tag{13-48}$$

From Eq. 13-44, the ordinate of the upper surface of the wing is

$$z_u = \frac{a_0 x}{2\beta^2 m^2}\sqrt{m^2x^2 - y^2} + \frac{y^2}{2\beta^2 m^3}(a_0 + 2\beta^2 m^2 a_2)\cosh^{-1}\frac{mx}{y} \quad (13\text{-}49)$$

Squire [31] considered the thickness distribution that is obtained by neglecting the arc hyperbolic function in Eq. 13-49. His results correspond

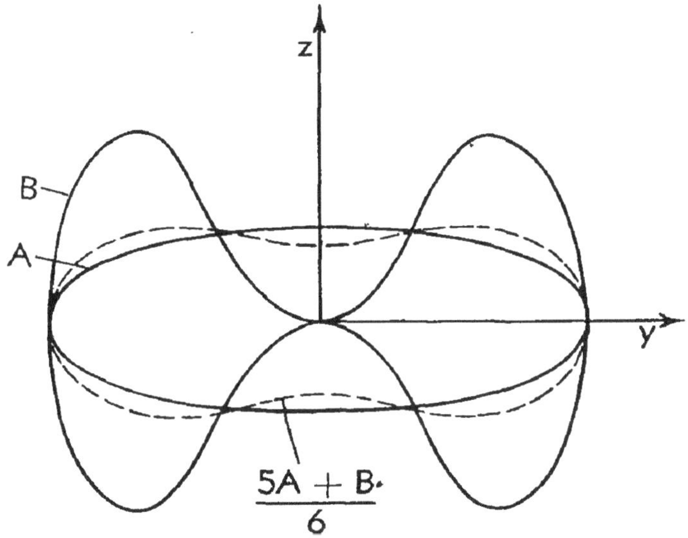

Fig. D,13j. Various spanwise section shapes that give the same pressure distribution on a triangular wing.

to the case when a_0 is $-2\beta^2 m^2 a_2$ and are specifically

$$z_u = \frac{a_0 x}{2\beta^2 m^2}\sqrt{m^2x^2 - y^2} \quad (13\text{-}50a)$$

$$C_p = \frac{a_0 x}{\beta^2(1 - \theta_0^2)^2}[(3 - \theta_0^2)K - E(4 - 2\theta_0^2)] \quad (13\text{-}50b)$$

where a_0/β^2 can be related to the thickness-chord ratio of the wing and the apex angle. It is apparent, however, that a multiplicity of thickness distributions exists possessing the same pressure distribution. For example, if a_0 is zero the surface ordinate and pressure coefficient are, respectively,

$$z_u = \frac{a_2}{m} y^2 \cosh^{-1}\frac{mx}{y} \quad (13\text{-}51a)$$

$$C_p = \frac{2a_2 x\theta_0^2}{\beta^2(1 - \theta_0^2)^2}[(1 + \theta_0^2)K - 2E] \quad (13\text{-}51b)$$

The values of a_0 and a_2 in Eq. 13-50b and 13-51b can be adjusted so that βC_p for both wings has a given streamwise gradient. In Fig. D,13j, cross sections of the two resultant surfaces are drawn, section A corresponding to Eq. 13-50a and section B corresponding to Eq. 13-51a.

Also shown is a third section having the same pressure distribution and formed by combining A and B in the ratio 5 to 1. It is interesting to note that it is not possible, by subtracting the two sections, to get a real wing with zero pressure coefficient since the resulting negative ordinates would require the surface to cross itself.

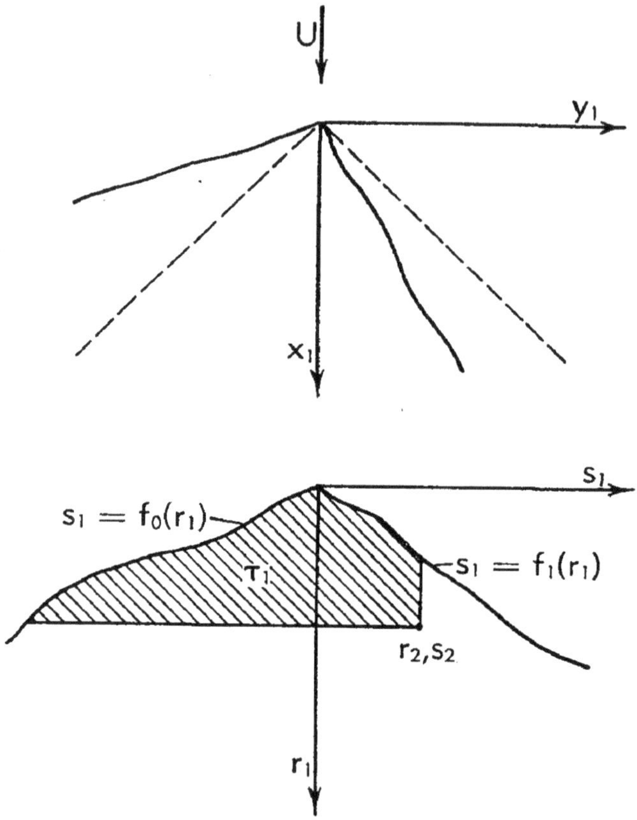

Fig. D,13k. Wing tip with one subsonic and one supersonic edge in both the x_1, y_1 and r_1, s_1 planes.

The general case. The solution to a wing traveling at a supersonic speed and having a completely general planform has not, as yet, been discovered. However, the solutions to certain quite broad classes of wings, other than those just discussed, can be written, these classes being determined by the inversions which have been obtained for the basic integral equation. Three approaches to these more general problems will be considered: first, the inversion of an integral equation which has been developed for the wing as a whole; second, the inversion of an integral equation which has been developed to cancel (by superposition) the lift over certain portions of a previously developed "basic" wing;[13] and

[13] For other approaches to this method of lift cancellation see [*20,32*]. A similar technique has also been used successfully to find solutions for nonconical problems by superimposing conical fields [*24,33*].

third, the inversion of an integral equation by iteration techniques when the actual null transform cannot be found. In this article we are concerned only with lifting surface problems, although the extension to nonlifting cases is apparent.

First let us consider a lifting surface with specified twist and camber having a leading edge which is entirely supersonic on one side of a point

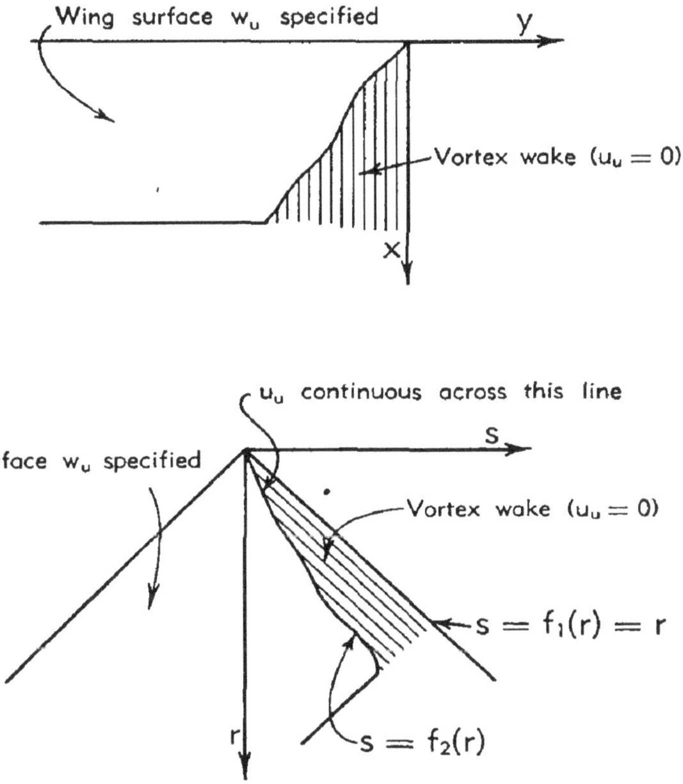

Fig. D,13l. Boundary conditions for constant chord wing with raked-in tip.

and entirely subsonic on the other. Such a wing is illustrated by Fig. D,13k (top). Introduce the r, s coordinate system presented in Fig. D,12c and designate the supersonic edge by $s = f_0(r)$ and the subsonic edge by $s = f_1(r)$ as in Fig. D,13k (bottom). The integral equation for such a wing can be developed just as Eq. 12-25 was derived so that

$$w_u = -\frac{M_\infty}{4\pi}\iint_{\tau_1}\frac{\varphi_u(r_1, s_1)dr_1ds_1}{[(r_2 - r_1)(s_2 - s_1)]^{\frac{3}{2}}} \tag{13-52}$$

where τ_1 is the portion of the wing in the forecone from r_2, s_2. The inversion to Eq. 13-52 for the integration area τ_1 has already been presented in the section on integral equations and is given by Eq. 12-39.

If the above problem is modified so that the subsonic edge becomes a trailing edge, the boundary conditions are changed. In such a case the vortex wake which streams back of the trailing edge must be considered along with the application of the Kutta condition to the flow at the trailing edge. The boundary conditions which satisfy these conditions are illustrated in Fig. D,13l. They yield a mixed type problem in which w_u is given over a portion of the $z = 0$ plane and u_u is given over another portion, along with the condition that u is continuous along the line $s = f_2(r)$. These boundary values are a special case of those considered in the development of Eq. 12-40. Therefore, since $u_u = 0$ for $f_2^*(s) > r > f_1^*(s)$,

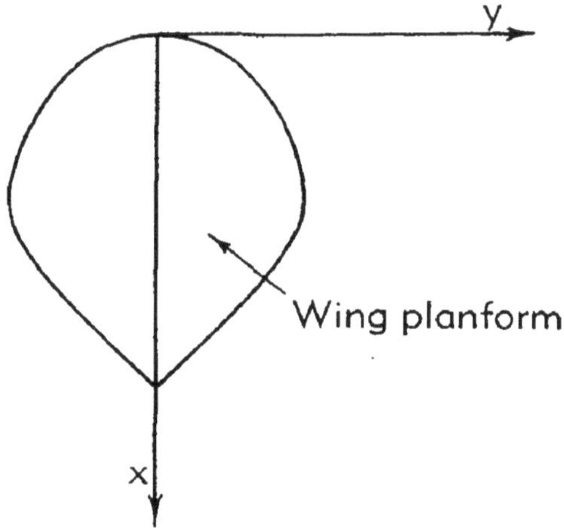

Fig. D,13m. Planform of particular wing receiving detailed load distribution study in [18].

the solution can immediately be written

$$u_u(r, s) = -\frac{1}{2\beta\pi}\int_{f_1^*(s)}^{r}\frac{dr_2}{\sqrt{r - r_2}}\left(\frac{\partial}{\partial r_2} + \frac{\partial}{\partial s}\right)\int_{f_0(r_2)}^{s}\frac{w_u(r_2, s_2)ds_2}{\sqrt{s - s_2}} \qquad (13\text{-}53)$$

for the wing with a supersonic leading edge and a subsonic trailing edge.

The solutions just discussed in connection with the problems sketched in Fig. D,13k and D,13l have been applied to several appropriate wing planforms. One example of a flat plate planform which has received attention is shown in Fig. D,13m. Evvard [18] has obtained the complete solution for the load distribution over such a flat plate flying at an angle of attack, rolling about the y axis or pitching about the two-thirds chord line. The example is included in order to illustrate the generality of the planform shapes which can be analyzed by the method of approach just presented.

The second method for solving these more general problems has been termed the "lift cancellation" method. The basic idea on which it is founded is simply that two solutions to a linear partial differential equation when added together form a third solution to the same equation. (In fact, such a concept is essential to the definition of a linear system.) In application to lifting surface problems this means that the solutions for two different wing problems can be added to form the solution to a third, the boundary values, to which this third solution applies, being obtained by superimposing the boundary values of the first two.

A simple illustration of the technique involved is provided by the following example. Consider a two-dimensional flat plate wing such as is shown in Fig. D,13n(1). The value of w_u over such a wing traveling at

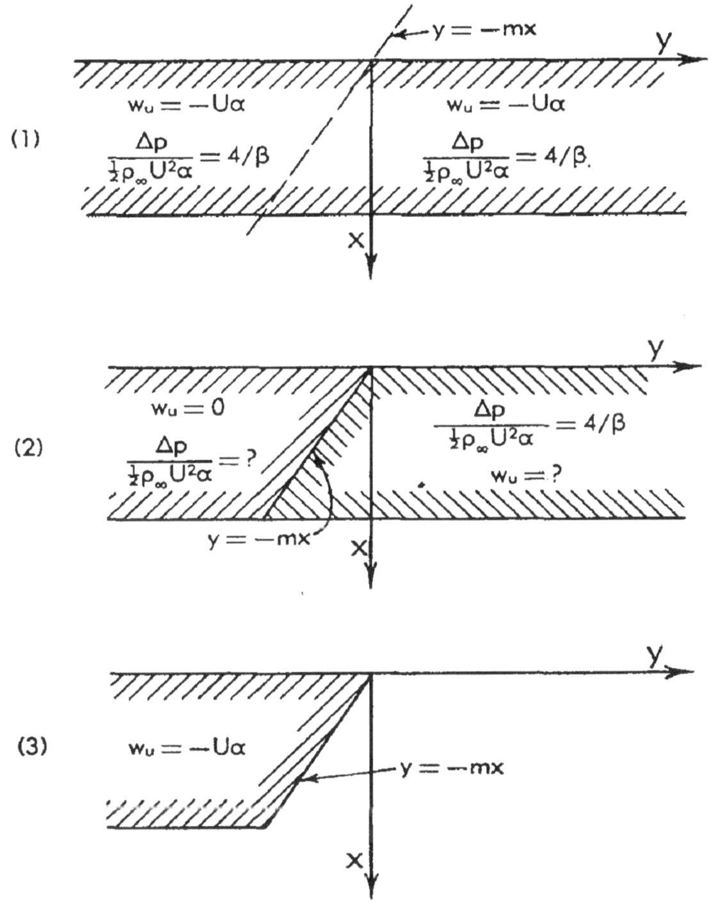

Fig. D,13n. Boundary conditions for cancellation element that give solution for trapezoidal flat plate when combined with solution for two-dimensional flat plate.

a supersonic speed is equal to $-U\alpha$ and the corresponding loading is well known to be that given by the equation $\Delta p/\frac{1}{2}\rho_\infty U^2 = 4\alpha/\beta$. Now consider a second wing such as is shown in Fig. D,13n(2). Wing 2 has an infinite span and has the same chord and is traveling at the same speed as wing 1. Further, wing 2 is constructed in such a way that to the left of a certain line ($y = -mx$), w_u equals zero and to the right of the same line the loading is the same as that on wing 1, i.e. $\Delta p/\frac{1}{2}\rho_\infty U^2 = 4\alpha/\beta$. The values of the loading over the right-hand region and the vertical induced velocity over the left-hand region of such a wing are as yet unknown.

However, if the boundary values for wing 2, are subtracted from those for wing 1, there result the boundary values for wing 3, namely, $w_u = -U\alpha$ over the shaded region and the loading is zero everywhere off the wing. Similarly, subtracting the solution representing the flow field for wing 2 from that representing wing 1 gives the solution to the flow field for wing 3.

Let us consider the problem of finding the loading over the left-hand region of wing 2. In the r, s coordinate system (Fig. D,13o) the boundary-

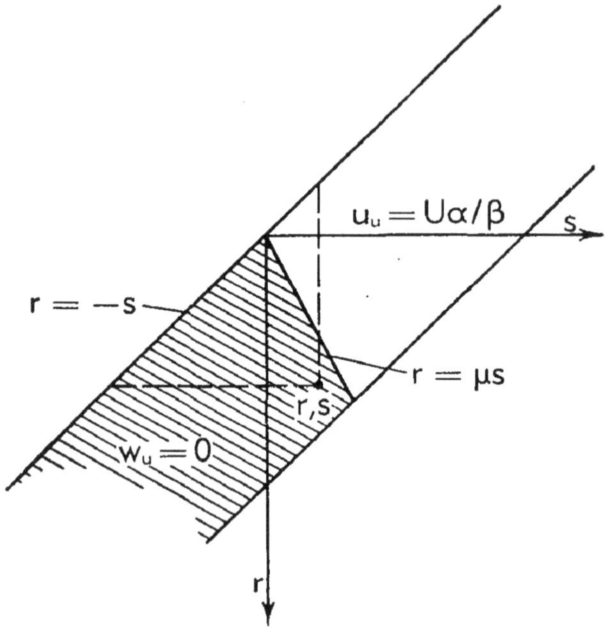

Fig. D,13o. Boundary conditions for cancellation element in r, s plane.

value problem becomes a special case of that shown in Fig. D,12j. Hence, if the loading (i.e. the value of u_u) is to be continuous across the line $y = mx$, the solution in terms of u_u is provided by Eq. 12-40 and can be written

$$u_u = \frac{\alpha U \sqrt{r - \mu s}}{\beta\pi} \int_{-s}^{\mu s} \frac{dr_1}{(r - r_1)\sqrt{\mu s - r_1}}$$

In the x, y coordinate system this can be expressed in terms of the loading coefficient as

$$\frac{\Delta p}{\frac{1}{2}\rho_\infty U^2} = \frac{8\alpha}{\beta\pi} \tan^{-1} \sqrt{\frac{x + \beta y}{\beta(mx - y)}}$$

Subtracting the loading given by the latter equation from that over the original two-dimensional wing (wing 1 in Fig. D,13n) one finds for the loading in region A of Fig. D,13p (top) the value

$$\frac{\Delta p}{\frac{1}{2}\rho_\infty U^2} = \frac{8\alpha}{\beta\pi}\tan^{-1}\sqrt{\frac{\beta(mx - y)}{x + \beta y}} \tag{13-54}$$

Eq. 13-54 gives a loading coefficient which is continuous across the line $y = mx$. Since this line is a trailing edge in Fig. D,13p (top), continuity of pressure across it (and hence zero pressure along it) is necessary if the Kutta condition is to be fulfilled. However, if the line $y = mx$ is a leading edge as in Fig. D,13p (bottom), then this continuity is no longer required

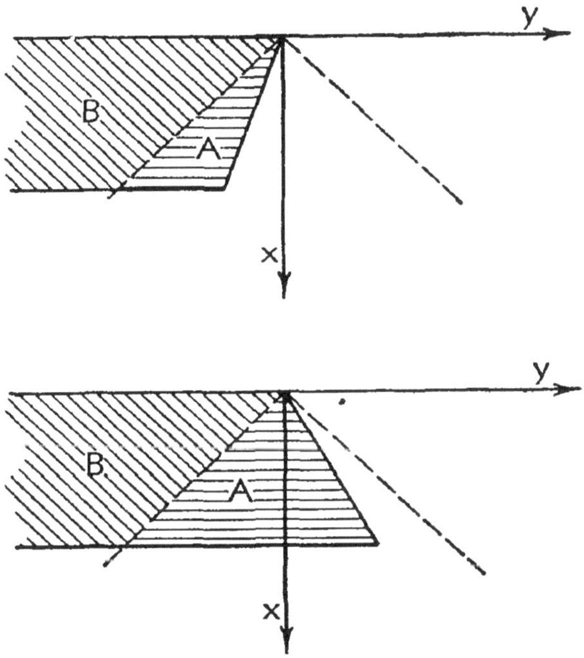

Fig. D,13p. **Trapezoidal wings with raked-in and raked-out tips.**

and, in fact, it is apparent that, if the edge is subsonic, the loading will become increasingly large as the edge is approached. For such a case the solution given by Eq. 12-41 must be considered. Using the same superposition procedure as was used in obtaining Eq. 13-54, one finds for region A of Fig. D,13p (bottom), the result

$$\frac{\Delta p}{\frac{1}{2}\rho_\infty U^2} = \frac{8\alpha}{\pi\beta}\left[\frac{\beta m}{1 + \beta m}\sqrt{\frac{x + \beta y}{\beta(mx - y)}} + \tan^{-1}\sqrt{\frac{\beta(mx - y)}{x + \beta y}}\right] \tag{13-55}$$

The example just discussed, although quite simple, serves further to illustrate the dual role played by Eq. 12-40 and 12-41. These expressions can be used on the one hand to solve directly boundary-value problems which represent wings such as those illustrated in Fig. D,13p, by specifying that w_u is a certain function over the wing and u_u is zero off the wing; or indirectly, by finding a cancellation field for which w is zero over the region to be occupied by the wing and the loading is a given function elsewhere.

An example of more complicated lifting surface problems which cannot be solved directly by Eq. 12-40 and 12-41 but can be solved indirectly using cancellation techniques is provided in the following: Let

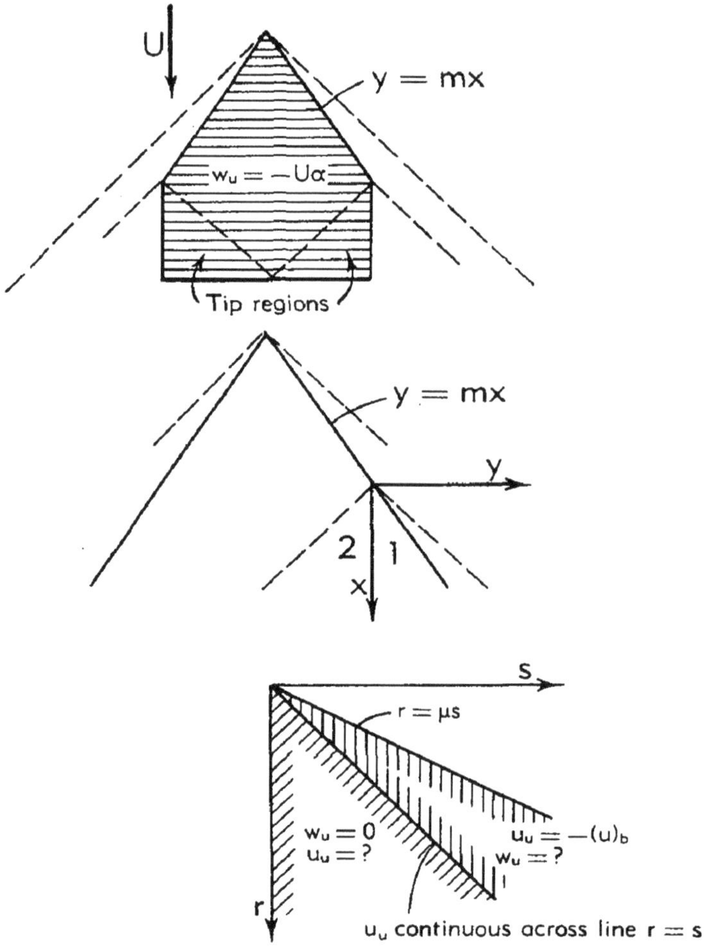

Fig. D,13q. Boundary conditions for cancellation element used to straighten out triangular wing tip.

us find the loading over the tip region of the cutback, flat plate, triangular wing flying at a supersonic speed shown in Fig. D,13q (top). First, consider the basic wing shown in Fig. D,13,q (middle). Over this wing w_u is everywhere constant and by the methods presented in the analysis of conical flow fields the value of the loading coefficient everywhere over the wing has been determined. Consequently the value of u_u is known over the basic planform. Designate the value of this streamwise induced velocity by $(u)_b$. Now, if we superpose on this field a solution for a wing flying at the same speed and having for its boundary value that

$u_u = -(u)_b$ over the region marked 1 in Fig. D,13q, (middle) and $w_u = 0$ over the region marked 2, the result will be a solution to the tip region of the wing shown in Fig. D,13q (top). The boundary conditions for this cancellation wing[14] are shown in the r, s coordinates in Fig. D,13q (bottom).

Again the problem is a special case of the mixed type presented in Fig. D,12j and the solution can be written immediately from Eq. 12-40 as

$$u_u(r, s) = \frac{\sqrt{r - \mu s}}{\pi} \int_s^{\mu s} \frac{-(u)_b(r_1, s)dr_1}{(r - r_1)\sqrt{\mu s - r_1}} \tag{13-56}$$

Adding this solution to that already known for the basic wing, we can find the correct value for u_u on the tip region of the wing illustrated in Fig. D,13q (top). The analysis for this example has been carried out by Mirels [*20*].

Finally, the third method for solving general inverse wing theory problems is to apply to the integral equation a *partially null transform*, where "partially null" is used in the sense that, when the operator is applied and the order of integration reversed, the transformed kernel does not vanish everywhere, as is the case for a completely null transform, but only over a certain portion of the integration space. The result of this transformation is still an integral equation but it has two advantages over the original one. First, it is an integral equation of the second kind to which iteration techniques can be applied. Second, and more important, the transformed kernel (by virtue of the fact that the transform is partially null) *can be made to vanish in the neighborhood* of the point where the unknown function is being evaluated. The latter property can be used not only to increase the rate of convergence of the series resulting from the iteration, but also, in certain cases, to express the solution in a series having a *finite number of terms*. These remarks will be clarified by studying the following two examples.

Let us consider the problem of finding the loading at the point x, y on the flat wing tip in Fig. D,13r. If the r, s coordinate system is used and the operator

$$\int_{f_2^*(s)}^{r} \frac{dr_2}{\sqrt{r - r_2}} \left(\frac{\partial}{\partial r_2} + \frac{\partial}{\partial s}\right) \int_{f_2(r)}^{s} \frac{ds_2}{\sqrt{s - s_2}}$$

where the area of integration is region 4 in Fig. D,13r (bottom), is applied to the doublet equation

$$w_u(r, s) = -\frac{M_\infty}{4\pi} \iint_{\tau_4} \frac{\varphi_u(r_1, s_1)dr_1ds_1}{[(r - r_1)(s - s_1)]^{\frac{3}{2}}}$$

where τ_4 is the sum of the areas labeled 1 through 4 in Fig. D,13r (bottom),

[14] The loading is assumed to drop to zero along the side edge of the wing. Hence the value of u_u is continuous there.

two results can be anticipated from a study of the analysis that lead to Eq. 12-40. One, since the loading off the wing is zero, the residual will be $-2\pi\varphi_u(r, s)$; the other, the transformed integral with the s_2 integration performed first is zero for all points r_1, s_1 lying in regions 2 and 4 in Fig. D,13r. These results follow directly from the discussion of the similar integral transform applied to the doublet equation with the τ_1 area.

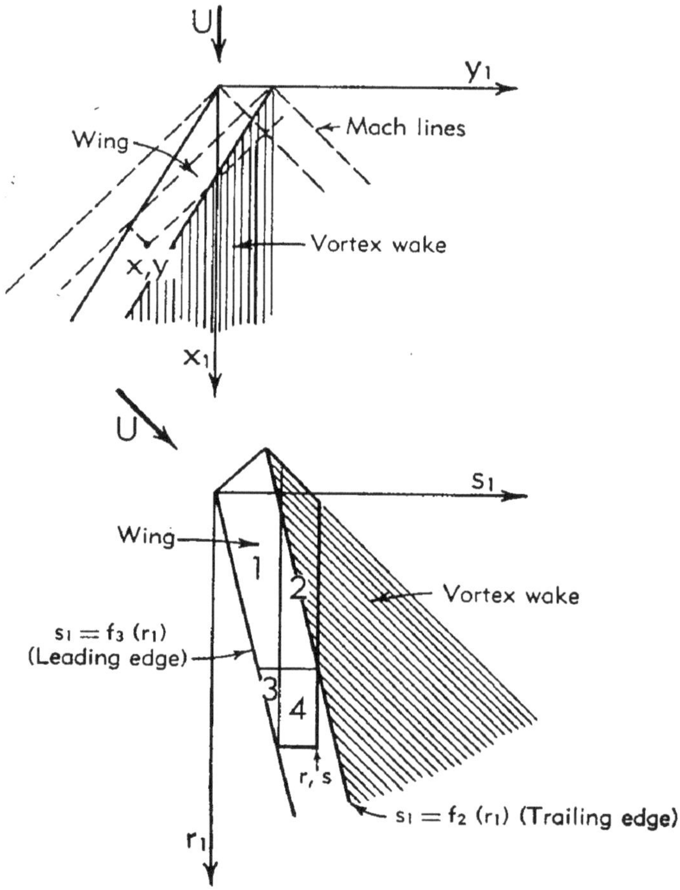

Fig. D,13r. Sweptback wing tip and its vortex wake in both the x_1, y_1 and r_1, s_1 planes.

Without proceeding further, therefore, it is apparent that the original integral equation has been reduced to, first, an integral of the known function w_u over region 4 in Fig. D,13r and, second, an integral of the unknown function φ_u over the regions 1 and 3 in the figure (and it can be shown that the integration over region 3 will also vanish). But region 1 is ahead of the Mach forecone from r, s and cannot be affected by the solution at r, s. Hence, by repeating the above process for regions farther and farther up (toward the origin) the wing, the problem must eventually be reduced to one of finding the solution for an area such as τ_1; in other words, to a problem involving only two edges, one subsonic and the other supersonic. Since the latter problem is solved, the former is also (theoretically at least) solved. Further, the solution has been obtained in a finite number of steps.

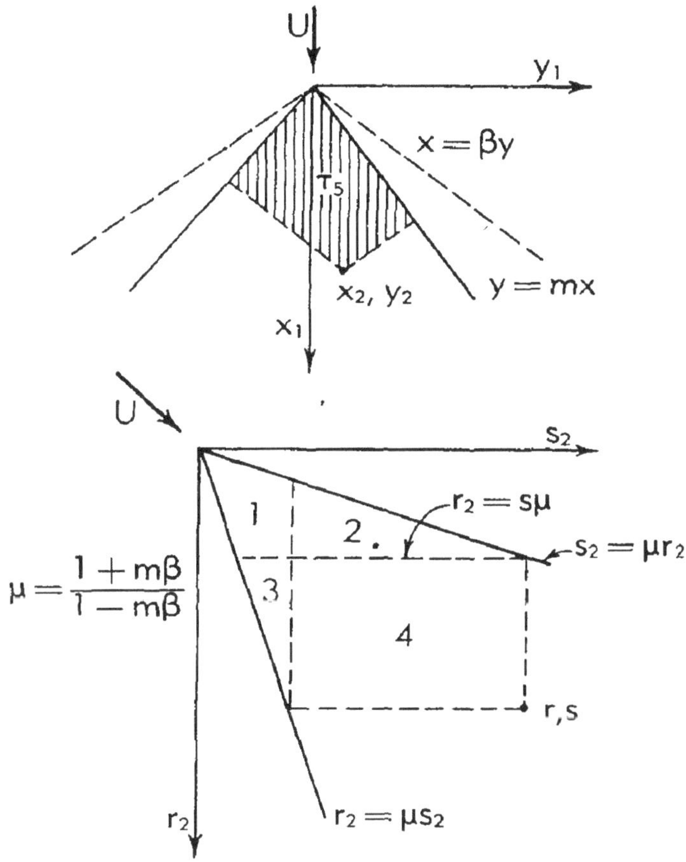

Fig. D,13s. Triangular wing in both x_1, y_1 and r_2, s_2 planes.

For the final example, consider the problem of finding the loading on a flat triangular wing flying at a supersonic speed but with both leading edges subsonic (see Fig. D,13s). To the supersonic doublet equation apply the transform

$$-\frac{1}{\pi M_\infty}\int_{s/\mu}^{r} dr_2 \int_{r/\mu}^{s} ds_2 \frac{1}{\sqrt{(r - r_2)(s - s_2)}}$$

where the region of integration is region 4 in Fig. D,13s (bottom). There results

$$\begin{aligned} -\frac{1}{\pi M_\infty}\int_{s/\mu}^{r} dr_2 \int_{r/\mu}^{s} ds_2 \frac{w_u(r_2, s_2)}{\sqrt{(r - r_2)(s - s_2)}} \\ = \frac{1}{4\pi^2}\int_{s/\mu}^{r} dr_2 \int_{r/\mu}^{s} ds_2 \iint_{\tau_5(r_2, s_2)} \frac{\varphi_u(r_1, s_1)\, dr_1\, ds_1}{\sqrt{(r - r_2)(s - s_2)}\,[(r_2 - r_1)(s_2 - s_1)]^{3/2}} \end{aligned} \tag{13-57}$$

where $\tau_5(r_2, s_2)$ is the original area of integration also shown in the figure. Since w_u is a constant equal to $-U\alpha$, the left-hand side of Eq. 13-57 becomes

$$\frac{4U\alpha}{\pi M_\infty}\sqrt{\left(r-\frac{s}{\mu}\right)\left(s-\frac{r}{\mu}\right)}$$

where $\mu = (1 + m\beta)/(1 - m\beta)$, m being the slope of the leading edge in the x, y plane. Inverting the order of integration and evaluating the residual, one finds that Eq. 13-57 can be written

$$\varphi_u(r, s) = \frac{4U\alpha\sqrt{\left(r-\frac{s}{\mu}\right)\left(s-\frac{r}{\mu}\right)}}{\pi M_\infty} - \frac{\sqrt{\left(r-\frac{s}{\mu}\right)\left(s-\frac{r}{\mu}\right)}}{\pi^2} \iint\limits_{\tau_6(r,s)} \frac{\varphi_u(r_1, s_1)dr_1ds_1}{(r-r_1)(s-s_1)\sqrt{\left(\frac{r}{\mu}-s_1\right)\left(\frac{s}{\mu}-r_1\right)}} \tag{13-58}$$

where $\tau_6(r, s)$ is the area shown as region 1 in Fig. D,13s (bottom).

Eq. 13-58 is not, of course, the inversion of the supersonic doublet equation for the triangular flat plate. In fact, it simply represents the transformation of the doublet equation, which is a singular integral equation of the first kind, to a singular integral equation of the second kind. However, it provides a good illustration of the advantages that can be gained by using the latter form, namely, Eq. 13-58 can be solved by iteration, and the area of integration $\tau_6(r, s)$ does not include the point or the neighborhood of the point r, s.

Starting the iteration, we take as a first approximation to φ_u the value

$$\varphi_1(r, s) = \frac{4U\alpha}{\pi M_\infty}\sqrt{\left(r-\frac{s}{\mu}\right)\left(s-\frac{r}{\mu}\right)} \tag{13-59}$$

and as the second, the value

$$\varphi_2(r, s) = \frac{4U\alpha}{\pi M_\infty}\sqrt{\left(r-\frac{s}{\mu}\right)\left(s-\frac{r}{\mu}\right)} \left[1 - \frac{1}{\pi^2}\iint\limits_{\tau_6} \frac{\sqrt{\left(r_1-\frac{s_1}{\mu}\right)\left(s_1-\frac{r_1}{\mu}\right)}\,dr_1ds_1}{(r-r_1)(s-s_1)\sqrt{\left(\frac{r}{\mu}-s_1\right)\left(\frac{s}{\mu}-r_1\right)}}\right] \tag{13-60}$$

and so on. By means of the substitutions

$$\mu r_1 = \frac{s + r\mu\rho_1 + s\sigma_1}{1 + \rho_1 + \sigma_1}$$
$$\mu s_1 = \frac{r + r\rho_1 + s\mu\sigma_1}{1 + \rho_1 + \sigma_1} \tag{13-61}$$

the double integral in the equation for φ_2 can be evaluated. Thus, after some manipulation, one finds[15]

$$\iint_{\tau_6} \frac{\sqrt{\left(s_1 - \frac{r_1}{\mu}\right)\left(r_1 - \frac{s_1}{\mu}\right)}\, dr_1 ds_1}{(r - r_1)(s - s_1)\sqrt{\left(\frac{r}{\mu} - s_1\right)\left(\frac{s}{\mu} - r_1\right)}} = 2\pi[2E_1 - (1 - k_1^2)K_1] - \pi^2$$

where K_1 and E_1 are the complete elliptic integrals with modulus $1/\mu$. Using the identity

$$2E_1 - (1 - k_1^2)K_1 = \frac{2E}{(1 + m\beta)}$$

where E has the modulus $\sqrt{1 - m^2\beta^2}$, and returning to the Cartesian coordinate system, one finds for the values of φ_1 and φ_2

$$\varphi_1 = \frac{4U\alpha}{\pi(1 + m\beta)} \sqrt{m^2x^2 - y^2} \tag{13-62}$$

$$\varphi_2 = \frac{4U\alpha}{\pi(1 + m\beta)} \sqrt{m^2x^2 - y^2} \left[2 - \frac{4E}{\pi(1 + m\beta)}\right] \tag{13-63}$$

This process of iteration could be continued, and φ_u would be expressed as a converging series of terms containing the parameter $m\beta$. However, since the terms in the series expansion are all independent of x and y, or, in the characteristic system, of r and s, it is more efficient to write φ_u in the form

$$\varphi_u(r, s) = \frac{M_\infty A}{1 + m\beta} \sqrt{m^2x^2 - y^2} = A\sqrt{\left(s - \frac{r}{\mu}\right)\left(r - \frac{s}{\mu}\right)}$$

and determine the magnitude of A by substituting this expression into Eq. 13-58. There results the equality

$$A = \frac{4U\alpha}{\pi M_\infty} - \frac{A}{\pi^2}\left[\frac{4\pi E}{1 + m\beta} - \pi^2\right]$$

from which it follows that

$$A = \frac{U\alpha}{M_\infty E}(1 + m\beta)$$

Finally, therefore, the velocity potential on the upper surface of a triangular wing with subsonic leading edges can be written

$$\varphi_u = \frac{U\alpha}{E} \sqrt{m^2x^2 - y^2} \tag{13-64}$$

[15] Notice that the area of integration, i.e. the area τ_6, is bounded by the lines which represent the four roots to the radicals in the integrand. Notice also that the value of this double integral is independent of r and s and depends only on the parameter μ, the slope of the leading edge in the r, s plane.

and the familiar expression for the loading coefficient (see, for example, Eq. 13-32) follows immediately

$$\frac{\Delta p}{\frac{1}{2}\rho_\infty U^2} = \frac{4\alpha m^2 x}{E\sqrt{m^2x^2 - y^2}}$$

The purpose of examining this particular problem was not so much to obtain the solution presented as Eq. 13-64 as to show how the supersonic doublet equation could be transformed to a singular integral equation of the second kind and how this equation could, in turn, be solved by applying an iteration process. Such a method has far more general applications than are given here and is by no means limited to problems in which the flow is conical or quasi-conical.

D,14. Evaluation of Aerodynamic Characteristics. In the preceding articles in this chapter we have been concerned with the evaluation of the velocity potential and the resulting induced velocities at a given point in the flow field. By means of these quantities we were able to derive expressions for the local pressure and loading coefficients over the surface of the wing. We have now to interpret these results in terms of the total forces and moments acting on the entire wing. Just as in the two-dimensional case discussed in Art. 5-9, we find that such an interpretation can be accomplished in two ways: one, by integrating the forces and moments over the airfoil itself, and the other, by taking a momentum balance across a control surface. In the present case this surface is located at a large distance from the position of the wing.

Conversion of load and pressure distributions to aerodynamic characteristics. If we assume that the local values of loading and pressure have been obtained by one of the methods presented in the above articles, the lift, drag, and pitching moment on the given wing can be evaluated by an appropriate integration of such results over the area occupied by the wing planform. Thus it is apparent that by extending the definition given in Art. 7 in the derivation of Eq. 7-1, the lift coefficient ($C_L = L/\frac{1}{2}\rho_\infty U^2 S$, where L is the total lift, $\frac{1}{2}\rho_\infty U^2$, the free stream dynamic pressure, and S, the wing area) is given by the equation

$$C_L = \frac{1}{S}\iint\limits_S \frac{\Delta p}{\frac{1}{2}\rho_\infty U^2}\,dydx \tag{14-1}$$

Similarly, the pitching moment coefficient ($C'_m = M'/\frac{1}{2}\rho_\infty U^2 Sc$, where M' is the total moment measured about the foremost wing position, positive when the trailing edge is forced down, and c is a characteristic chord) is given by the equation

$$C'_m = -\frac{1}{Sc}\iint\limits_S \frac{x\Delta p}{\frac{1}{2}\rho_\infty U^2}\,dydx \tag{14-2}$$

The equation for the drag coefficient, however, is not so apparent and, as we discovered in the study of the two-dimensional results, the force obtained by integrating the load and pressure distributions must be modified by the force at the leading edge whenever singularities in the induced velocities occur there. Hence, dividing the drag into two parts, one due to the camber-line shape and the other due to the symmetrical thickness shape, we can write for the former

$$C_D = -\frac{1}{S}\iint\limits_S \lambda \frac{\Delta p}{\frac{1}{2}\rho_\infty U^2}\,dydx + (C_D)_t \tag{14-3}$$

and for the latter

$$C_D = \frac{2}{S}\iint\limits_S \lambda_u (C_p)_u dydx + (C_D)_d \tag{14-4}$$

where $(C_D)_t$ and $(C_D)_d$ are the forces generated by the infinite velocities at the leading edges of the wing. The former of these, $(C_D)_t$, is referred to as "leading edge thrust," and the latter, $(C_D)_d$, as "leading edge drag."

The evaluation of the lift and the pitching moment by these methods is discussed in no further detail since they have been reduced to the problem of integrating known results over a given area. Similarly, the discussion of the drag is limited to a brief study of the thrust and drag terms just defined.

Leading edge thrust. In the calculation of the suction force along the leading edge of a two-dimensional wing in Art. 9, it is shown that such a force depended only on the nature of the induced velocity u in the vicinity of the leading edge. Extending this concept to the present case, we can surmise that the suction force at a point along the leading edge of a three-dimensional wing depends only on the nature of the induced velocity normal to the edge at the point in question. Hence, Eq. 9-15 becomes

$$\frac{v_n}{U} = \frac{G_n}{\beta_n \sqrt{x_n}} \tag{14-5}$$

where the subscript n refers to the normal component as shown in Fig. D,14a. Correspondingly, Eq. 9-16, which gives the drag coefficient per unit span, becomes

$$\frac{d}{dy_n}(C_D)_n = -\frac{2\pi}{\beta_n}\frac{G_n^2}{S} \tag{14-6}$$

Finally, since $(C_D)_t = (C_D)_n dy/dy_n$, the total thrust force can be expressed in the form

$$(C_D)_t = -\frac{2\pi}{S}\int\limits_{\text{span}} \frac{G_n^2}{\beta_n}\,dy \tag{14-7}$$

An instructive example of the use of Eq. 14-7 is given by its application to the problem of finding the drag of a flat triangular wing with subsonic leading edges. The loading over such a wing is given by Eq. 13-32 and can be written in the form

$$\frac{\Delta p}{\frac{1}{2}\rho_\infty U^2} = \frac{4u_u}{U} = \frac{4\alpha x \tan^2 \delta_0}{E\sqrt{x^2 \tan^2 \delta_0 - y^2}} \tag{14-8}$$

where δ_0 is the semiapex angle of the wing. Consider the force developed along the right edge of the wing. The velocity potential can be obtained

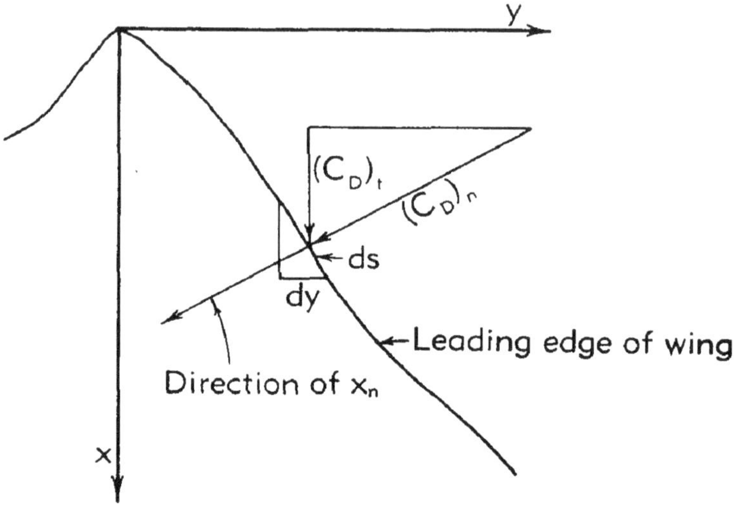

Fig. D,14a. Vector orientation referred to in discussion of **leading edge thrust.**

from the equality

$$\varphi_u = \int_{\text{l.e.}}^{x} u_u dx$$

and expressed in terms of a coordinate system the axes of which are normal and parallel to the right edge of the wing (see Fig. D,14b). There results

$$\varphi_u = \frac{U\alpha}{E}\frac{1}{\cos\delta_0}\sqrt{x_n[x_n(\sin^2\delta_0 - \cos^2\delta_0) + 2y_n \sin\delta_0 \cos\delta_0]} \tag{14-9}$$

Since $v_n = \partial\varphi_u/\partial x_n$, the value of G_n is simply

$$G_n = \frac{\alpha\beta_n}{2E}\sqrt{\frac{2y}{\cos\delta_0}} \tag{14-10}$$

where β_n is a constant equal to $\cos\delta_0\sqrt{1-\beta^2\tan^2\delta_0}$. The final result for $(C_D)_t$, then, can be written

$$(C_D)_t = \frac{\pi\alpha^2\tan^2\delta_0}{E^2}\sqrt{1-\beta^2\tan^2\delta_0} \tag{14-11}$$

where E is the elliptic integral of the second kind with modulus

$$\sqrt{1 - \beta^2 \tan^2 \delta_0}$$

Since the wing is a flat plate, the first term on the right-hand side of Eq. 14-3 is simply αC_L. Finally, the lift coefficient for such a wing can be shown (by placing Eq. 14-8 into Eq. 14-1) to be

$$C_L = \frac{\pi \alpha A}{2E} \tag{14-12}$$

where A is the wing aspect ratio (equal to 4 tan δ_0 for a triangular wing).

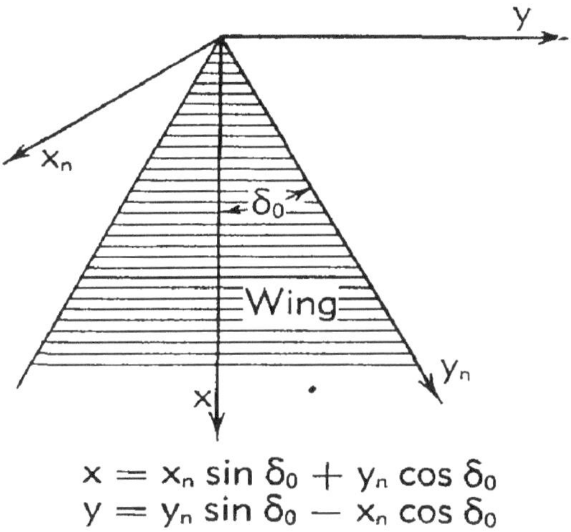

Fig. D,14b. Relation between x, y and x_n, y_n coordinates.

Therefore, the total drag on a triangular wing with subsonic leading edges can be written

$$C_D = \frac{\alpha C_L}{2}\left[2 - \frac{\sqrt{1 - \left(\frac{\beta A}{4}\right)^2}}{E}\right] \tag{14-13}$$

Leading edge drag. The problem of evaluating leading edge drag for nonlifting wings having a symmetrical thickness distribution and a blunt leading edge follows along lines identical to those given above for the lifting case. Thus, by using the two-dimensional results presented as Eq. 9-17 and 9-18, one can readily show that

$$(C_D)_d = \frac{2\pi}{S} \int_{\text{span}} \frac{G_n^2}{\beta_n} dy \tag{14-14}$$

Momentum methods. Consider a surface S which encloses a volume containing an aerodynamic body.[16] The vectorial force[17] **F** on the body

[16] Hereafter in this article the term body will be used to refer to a wing, a fuselage, or any combination of the two.

[17] We will not consider the evaluation of the moments by this method.

can be determined by balancing the momentum across S. In general there results

$$\mathbf{F} = -\iint\limits_S (p - p_\infty)d\mathbf{S} - \iint\limits_S \rho\mathbf{V}[(\mathbf{U} + \mathbf{V}) \cdot d\mathbf{S}] \qquad (14\text{-}15)$$

where vector notation is used and the subscript ∞ indicates free stream conditions, p and ρ are the local static pressure and density, and $\mathbf{V}$ is the local perturbation velocity vector.

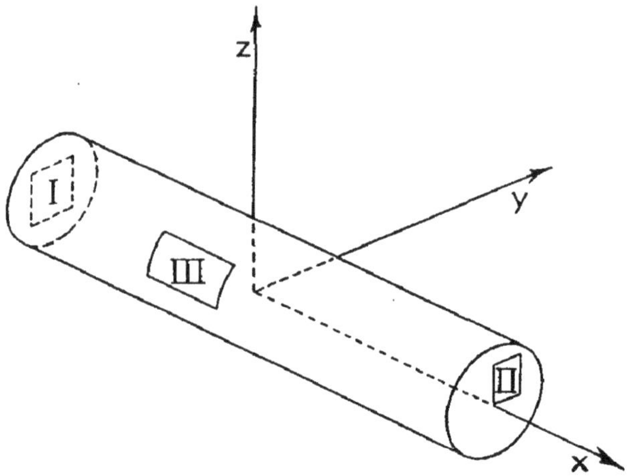

Fig. D,14c. Areas I, II, and III of cylindrical control surface used in momentum balance.

For our present purposes the surface S will be taken as the cylinder shown in Fig. D,14c. One face of the cylinder, denoted as I in the figure, in the plane x equals some negative constant; the other, denoted as II, in the plane x equals some positive constant. The radius of the cylinder, r, is large enough so that the body is entirely enclosed within the cylindrical surface which, in turn, is denoted as III.

Restricting our interest to linearized theory, we can set

$$\frac{\rho}{\rho_\infty} = 1 - M_\infty^2 \frac{u}{U}$$

and

$$p - p_\infty = -\rho_\infty[Uu + \tfrac{1}{2}(u^2 + v^2 + w^2)] + \tfrac{1}{2}\rho_\infty M_\infty^2 U^2$$

Further, limiting ourselves only to the component of force which yields drag,[18] we find for the total drag of the enclosed body

$$D = \tfrac{1}{2}\rho_\infty \iint\limits_{\text{I+II}} [(M_\infty^2 - 1)u^2 + v^2 + w^2]dydz - \rho_\infty \iint\limits_{\text{III}} uv_r r d\theta dx \qquad (14\text{-}16)$$

where v_r is the radial component of velocity and $r = \sqrt{y^2 + z^2}$.

[18] The lift follows immediately by taking the vertical component of force.

Eq. 14-16 is valid for any location of the three surfaces as long as they completely enclose the body. However, it is convenient to place the surfaces I and II infinitely far ahead of and behind the body respectively and to allow the radius of the cylinder to become very large. With such an arrangement it is clear that the integral over surface I is zero. Further, we can define the part of the drag contributed by the integration taken

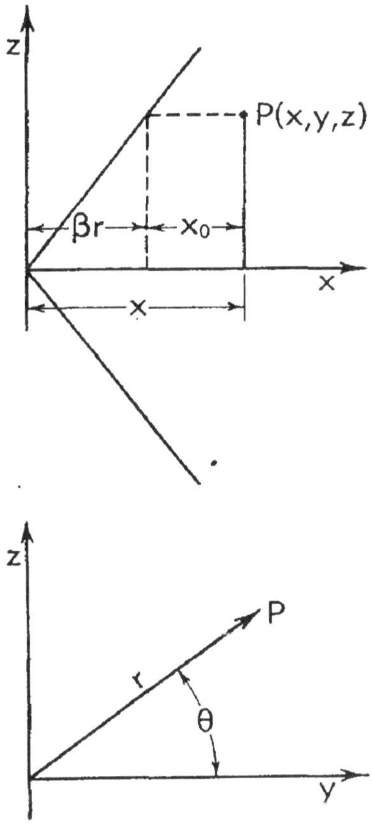

Fig. D,14d. Definition of x_0 and the r, θ coordinates.

over surface II as the *vortex drag* and the part contributed by that taken over surface III as the *wave drag*. The remaining portion of this article is devoted to the discussion of these two concepts.

Wave drag. The wave drag D_w has been defined as that portion of the total drag given by the equation

$$D_w = \lim_{r \to \infty} \left(-\rho_\infty r \int_0^{2\pi} d\theta \int_{-\infty}^{\infty} uv_r \, dx \right) \tag{14-17}$$

In subsonic flow a unit source placed at the origin produces the potential $\varphi = -1/4\pi \sqrt{x^2 + \beta^2 r^2}$. It is apparent that the product of the induced velocities u and v_r is of the order $1/r^5$ for large values of r. This fact, used in Eq. 14-17, establishes the well-known result that the wave drag for a body traveling at subsonic speeds is zero. In supersonic flow a unit source placed at the origin produces the potential $\varphi = -1/2\pi\sqrt{x^2 - \beta^2 r^2}$. It is again apparent that the product of the induced velocities u and v_r is of the order $1/r^5$ for large values of r *except in the neighborhood of the line* $x = \beta r$ where the order is indeterminate. Physically this line represents the intersection of the Mach wave emanating from the source point at the origin with the cylindrical control surface over which the integration of Eq. 14-17 is to be taken. We will find that the velocities induced along this Mach wave are of such an order that, when placed in Eq. 14-17, a finite value of the drag results.

In order to study the wave drag of bodies flying at supersonic speeds, we first require the concept of equivalent source position [34].

Consider, as in Fig. D,14d, a point P with coordinates x, y, z lying in the induced field of a supersonic body. The Mach forecone from P is given by the relation

$$x - x_1 = \beta\sqrt{(y - y_1)^2 + (z - z_1)^2} \tag{14-18}$$

where x_1, y_1, z_1 are running coordinates of a point on the surface of the cone. Introducing polar coordinates

$$y = r\cos\theta, \qquad z = r\sin\theta$$

and rewriting the abscissa of P in the form

$$x = x_0 + \beta r$$

it follows that the trace of the forecone in the $z_1 = 0$ plane is, in the limit as r approaches infinity,

$$x_1 = \beta y_1 \cos\theta + x_0 \tag{14-19}$$

It is, moreover, possible to show that the effect of the velocity potential at the point P as r approaches infinity is the same for all points x_1, y_1, 0 for which $x_1 - \beta y_1 \cos\theta = \text{const}$. The value of this effect is

$$\varphi = -\frac{1}{2\pi\sqrt{2\beta r(x_0 - x_1 + \beta y_1 \cos\theta)}} \tag{14-20}$$

and follows from the asymptotic evaluation of the supersonic source potential for large values of r. The potential at P for the source at the point x_1, y_1, 0 is thus the same as for the source shifted along the forecone trace to $x_1 - \beta y_1 \cos\theta$, 0, 0, the intercept of the trace on the x axis.

When lifting surfaces are to be analyzed, a similar treatment for the doublet potential can be constructed. In this article, however, we shall study only the wave drag due to thickness. For an analysis of the lifting case see [35].

Let us proceed now to the calculation of drag of a symmetrical body at zero angle of attack. Consider the general expression for the velocity potential given in Eq. 10-4. Introducing first the transformations (see Fig. D,14e), where $\tan \mu = \beta \cos \theta$,

$$\xi = -y_1 \tan \mu + x_1$$
$$\eta = y_1 \sec \mu \tag{14-21}$$

and

$$\bar{w}_u(\xi, \eta) = w_u(\xi + \eta \sin \mu, \eta \cos \mu)$$

Eq. 10-4 becomes

$$\varphi = -\frac{1}{\pi} \iint_\tau \frac{\cos \mu \bar{w}_u(\xi, \eta) d\xi d\eta}{\sqrt{(x - x_1)^2 - \beta^2(y - y_1)^2 - \beta^2 z^2}}$$

Since, however, it has been shown that φ evaluated infinitely far away from the wing does not change if a source is moved along the line

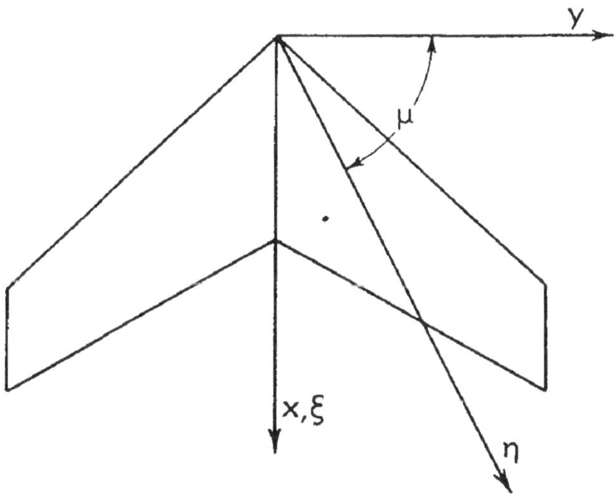

Fig. D,14e. Relation between the x, y and ξ, η coordinate systems.

$\xi = \text{const}$, it follows that the source strengths can be integrated along these lines. The second integration is then along $\eta = 0$ where, from Eq. 14-21, $\xi = x_1$ and the value of the potential at an infinite distance is

$$\varphi = -\frac{1}{\pi} \int \frac{\cos \mu dx_1}{\sqrt{(x - x_1)^2 - \beta^2 y^2 - \beta^2 z^2}} \int \bar{w}_u(x_1, \eta) d\eta$$

Setting

$$f(x_1, \mu) = -2 \cos \mu \int \bar{w}_u(x_1, \eta) d\eta$$

it follows that

$$\varphi(x, r, \theta) = \frac{1}{2\pi} \int_0^{x - \beta r} \frac{f(x_1, \mu) dx_1}{\sqrt{(x - x_1)^2 - \beta^2 r^2}} \tag{14-22}$$

and this is the same as the potential for a body of revolution with source strength per unit length given by $f(x_1)$. The induced velocities corresponding to the potential in Eq. 14-22 are found to be, after first integrating by parts and using the notation

$$\frac{\partial}{\partial x_1} f(x_1, \mu) = f'(x_1, \mu)$$

together with the relation $f(0, \mu) = 0$,

$$u = \frac{\partial \varphi}{\partial x} = \frac{1}{2\pi} \int_0^{x-\beta r} \frac{f'(x_1, \mu) dx_1}{\sqrt{(x - x_1)^2 - \beta^2 r^2}} \tag{14-23}$$

and

$$v_r = \frac{\partial \varphi}{\partial r} = -\frac{1}{2\pi r} \int_0^{x-\beta r} \frac{(x - x_1) f'(x_1, \mu)}{\sqrt{(x - x_1)^2 - \beta^2 r^2}} dx_1 \tag{14-24}$$

Asymptotic values of the velocity components for large values of r are readily seen to be, after first setting $x = x_0 + \beta r$,

$$u = \frac{1}{2\pi \sqrt{2\beta r}} \int_0^{x_0} \frac{f'(x_1, \mu)}{\sqrt{x_0 - x_1}} dx_1 \tag{14-25}$$

and

$$v_r = -\frac{1}{2\pi} \sqrt{\frac{\beta}{2r}} \int_0^{x_0} \frac{f'(x_1, \mu)}{\sqrt{x_0 - x_1}} dx_1 \tag{14-26}$$

Eq. 14-25 and 14-26 may be used together with Eq. 14-17 to give for the value of the wave drag the expression

$$D = \frac{\rho_\infty}{8\pi^2} \int_0^{2\pi} d\theta \int_0^\infty dx_0 \int_0^{x_0} \frac{f'(x_1, \mu)}{\sqrt{x_0 - x_1}} dx_1 \int_0^{x_0} \frac{f'(x_2, \mu)}{\sqrt{x_0 - x_2}} dx_2 \tag{14-27}$$

Assume that the body is of finite length so that $f'(x) = 0$ for $x > l$, and reverse the order of integration. There results the relation

$$D = -\frac{\rho_\infty}{8\pi^2} \int_0^{2\pi} d\theta \int_0^l \int_0^l f'(x_1, \mu) f'(x_2, \mu) \ln |x_1 - x_2| dx_1 dx_2 \tag{14-28}$$

If Eq. 14-28 had been derived for a body of revolution, then $f(x)$ would have been independent of the angle μ and in that case the expression for drag would reduce to the form

$$D = -\frac{\rho_\infty}{4\pi} \int_0^l \int_0^l f'(x_1) f'(x_2) \ln |x_1 - x_2| dx_1 dx_2 \tag{14-29}$$

This expression was first given by von Kármán [*11*]. It will appear later in the discussion of bodies of revolution in Art. 18.

Vortex drag. The vortex drag has been defined as that part of the total drag given by the equation

$$D_V = \tfrac{1}{2}\rho_\infty \iint_{\text{II}} [(M_\infty^2 - 1)u^2 + v^2 + w^2] dy dz \tag{14-30}$$

where the integration is to be carried out over a plane located at a very large distance behind the body, the so-called Trefftz plane. If x be this distance, in the limit as x goes to infinity the streamwise variations of velocity and potential in the plane of integration disappear. However, as a result of Helmholtz' fundamental investigations, we know that a vortex cannot end in a frictionless fluid; see, e.g., [*36*, pp. 165–166]. Hence, if the body develops any net lift, the resulting vorticity must still be streaming through the reference plane[19] and in the vicinity of these vortices, the induced velocities v and w are certainly not zero. It follows, therefore, that the vortex drag can be calculated by the equation

$$D_V = \tfrac{1}{2}\rho_\infty \int_{-\infty}^{\infty} \int_{-\infty}^{\infty} (v^2 + w^2)dydz \tag{14-31}$$

where the integration is taken over the Trefftz plane. Similarly, the partial differential equation which governs the flow field in the Trefftz plane is derived from the basic linearized equation by setting the φ_{xx} term equal to zero. There results

$$\varphi_{yy} + \varphi_{zz} = 0 \tag{14-32}$$

It should be noted that Eq. 14-31 and 14-32 are both independent of Mach number. Hence, all bodies which have the identical spanwise vorticity distribution have the same value for their vortex drag *regardless of the Mach number at which they are flying.* It is also apparent that a body which develops no net section lift has no vortex wake and has a zero value for its vortex drag.

Let us now apply the two-dimensional form of Green's theorem to the Trefftz plane. If S represents the area in this plane over which two functions σ and φ together with their first and second derivatives are bounded and continuous, and if C represents the contour which encloses this area, then we can write

$$\iint\limits_S \sigma\left(\frac{\partial^2\varphi}{\partial y^2} + \frac{\partial^2\varphi}{\partial z^2}\right) dS + \iint\limits_S \left(\frac{\partial\sigma}{\partial y}\frac{\partial\varphi}{\partial y} + \frac{\partial\sigma}{\partial z}\frac{\partial\varphi}{\partial z}\right) dS = -\int\limits_C \sigma\frac{\partial\varphi}{\partial n}\, ds \tag{14-33}$$

where n is the inward normal to the contour. Consider the case when φ equals the velocity potential. Then, according to Eq. 14-32 the first term on the left-hand side of Eq. 14-33 vanishes. Further, consider the case when σ equals φ. We are then provided with the identity

$$\iint\limits_S (v^2 + w^2)dS = -\int\limits_C \varphi\frac{\partial\varphi}{\partial n}\, ds \tag{14-34}$$

[19] We assume, of course, that any transient phenomenon produced by the manner in which the body started its motion does not affect the plane in question.

It follows immediately from Eq. 14-31 that the vortex drag can be written

$$D_V = -\frac{1}{2}\rho_\infty \int_C \varphi \frac{\partial \varphi}{\partial n} ds \tag{14-35}$$

If the vorticity streams through the Trefftz plane in a flat sheet in the $z = 0$ plane, the contour C can be chosen as that pictured in Fig. D,14f. Since the integrand around the circle AD vanishes like R^{-3} as R becomes

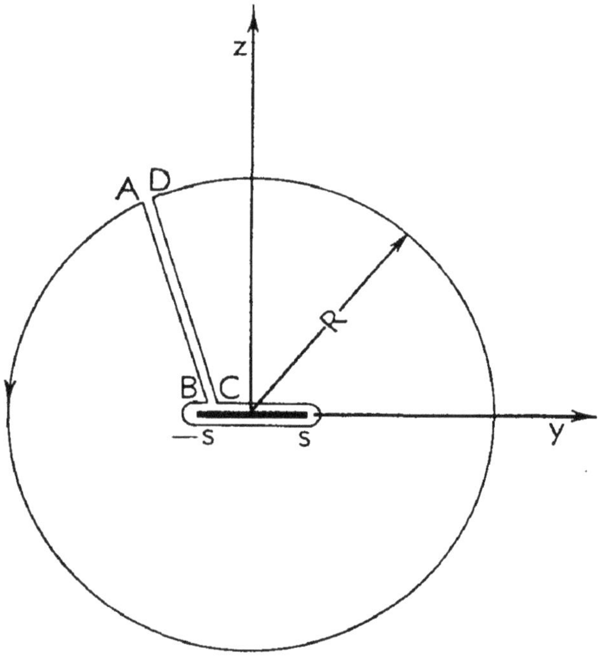

Fig. D,14f. Integration contour in Trefftz plane.

infinite, and since the integrals taken along the lines AB and CD cancel, the only contribution to the vortex drag comes from the integration taken around the slit BC. But the value of the potential φ is equal and opposite above and below a vortex sheet, hence

$$D_V = -\tfrac{1}{2}\rho_\infty \int_{-s}^{s} (\Delta\varphi)_{x=\infty} w_{x=\infty} dy \tag{14-36}$$

where $w_{x=\infty}$ is the vertical induced velocity along the y axis from $-s$ to s (the projection of the wing span) in the Trefftz plane and $(\Delta\varphi)_{x=\infty}$ is the jump in velocity potential along the same interval. We have already used the fact that the strength of a trailing vortex does not vary, however, so $(\Delta\varphi)_{x=\infty} = (\Delta\varphi)_{\text{t.e.}}$ and this, in turn, is simply the total circulation Γ developed by a given chordwise section of the wing. Further, it can be shown that the term $w_{x=\infty}$ is given by the expression

$$w_{x=\infty} = -\frac{1}{2\pi} \int_{-s}^{s} \frac{(d\Gamma/dy_1)}{y - y_1} dy_1 \tag{14-37}$$

Placing Eq. 14-37 in Eq. 14-36 and integrating by parts, using the fact that $\Gamma(s) = \Gamma(-s) = 0$, one finally obtains the classical result

$$D_V = -\frac{\rho_\infty}{4\pi}\int_{-s}^{s}\int_{-s}^{s}\Gamma'(y)\Gamma'(y_1)\ln|y - y_1|dydy_1 \tag{14-38}$$

for the vortex drag of a lifting surface.

For an example of the vortex drag produced by an actual wing, we return to the triangular wing traveling at a supersonic speed and having subsonic leading edges. Such a wing was studied in Art. 13 in the analysis of generalized conical flow and again in the first part of this article in connection with leading edge thrust. The spanwise variation of circulation, Γ, follows from a streamwise integration of Eq. 14-8 and can be written

$$\Gamma = \frac{2\alpha U}{E}\sqrt{s^2 - y^2} \tag{14-39}$$

Using Eq. 14-37, we find that $w_{x=\infty} = -\alpha U/E$. Hence, by Eq. 14-36 the total vortex drag for the triangular wing can be written

$$D_V = \frac{2}{\rho_\infty U^2}\frac{1}{S}\frac{\pi m\alpha^2}{E^2} \tag{14-40}$$

Using Eq. 14-12 for the lift coefficient, we can express the drag in coefficient form by the equation

$$(C_D)_V = \frac{C_L\alpha}{2}\left(\frac{1}{E}\right) \tag{14-41}$$

The difference between the total drag, given by Eq. 14-13, and the vortex drag, given by Eq. 14-41 is equal to the wave drag. It will be noted that, as the Mach number approaches 1, β approaches 0 and E approaches 1 and the wave drag, therefore, approaches zero.

D,14.1. Appendix.

THE FINITE PART OF AN INTEGRAL.

Definition. Consider the simple equality

$$\frac{\partial}{\partial x}\int_a^x \frac{dy}{\sqrt{x - y}} = \frac{1}{\sqrt{x - a}}$$

The finite part sign[20] can be introduced by the definition

$$⨍_a^x \frac{\partial}{\partial x}\frac{dy}{\sqrt{x - y}} \equiv \frac{\partial}{\partial x}\int_a^x \frac{dy}{\sqrt{x - y}} \tag{14.1-1}$$

from which it follows

$$⨍_a^x \frac{dy}{[x - y]^{\frac{3}{2}}} = \frac{-2}{\sqrt{x - a}}$$

[20] The concept of the finite part was originally introduced and generalized by Hadamard [37].

The natural extension of this idea is to consider

$$J = \fint_a^x \frac{\partial}{\partial x} \frac{A(y)dy}{\sqrt{x-y}} \equiv \frac{\partial}{\partial x} \int_a^x \frac{A(y)dy}{\sqrt{x-y}} \tag{14.1-2}$$

where $A(y)$ is continuous at $x = y$ and is integrable elsewhere in the range of integration. The evaluation of J can be reduced to a form that requires only the definition introduced by Eq. 14.1-1. Thus by adding and subtracting the same term, J can be written

$$J = \frac{\partial}{\partial x}\left[\int_a^x \frac{A(y) - A(x)}{\sqrt{x-y}}\, dy + A(x) \int_a^x \frac{dy}{\sqrt{x-y}}\right]$$

It follows that

$$\fint_a^x \frac{A(y)dy}{[x-y]^{3/2}} = \int_a^x \frac{A(y) - A(x)}{[x-y]^{3/2}}\, dy + A(x) \fint_a^x \frac{dy}{[x-y]^{3/2}} \tag{14.1-3}$$

The final generalization of the definitions given by Eq. 14.1-1 and 14.1-2 is accomplished by considering the nth derivative of the integrals and furthermore by allowing a functional dependence on x of the integrand $A(y)$. First consider the definition

$$\fint_a^x \left(\frac{\partial}{\partial x}\right)^n \frac{A(y)dy}{\sqrt{x-y}} = (-1)^n \frac{1 \cdot 3 \cdots (2n-1)}{2^n} \fint_a^x \frac{A(y)dy}{[x-y]^{n+\frac{1}{2}}}$$

$$\equiv \left(\frac{\partial}{\partial x}\right)^n \int_a^x \frac{A(y)dy}{\sqrt{x-y}} \tag{14.1-4}$$

from which it follows that

$$\fint_a^x \frac{A(y)dy}{[x-y]^{n+\frac{1}{2}}} = \int_a^x \frac{A(y) - B(x, y)}{[x-y]^{n+\frac{1}{2}}}\, dy + \fint_a^x \frac{B(x, y)dy}{[x-y]^{n+\frac{1}{2}}} \tag{14.1-5}$$

where

$$B(x, y) = A(x) - A'(x)(x-y) + \cdots + \frac{(-1)^{n-1}A^{(n-1)}(x)}{(n-1)!}(x-y)^{n-1}$$

and

$$\fint_a^x \frac{dy}{[x-y]^{i+\frac{1}{2}}} = \frac{(-1)^i 2^i}{1 \cdot 3 \cdots (2i-1)} \left(\frac{\partial}{\partial x}\right)^i \int_a^x \frac{dy}{\sqrt{x-y}}$$

$$= \frac{-2}{(2i-1)(x-a)^{i-\frac{1}{2}}}$$

Finally, replacing $A(y)$ by $A(x, y)$, Eq. 14.1-4 again defines uniquely a finite part integral provided that

$$\lim_{y \to x} \left[\sqrt{x-y} \left(\frac{\partial}{\partial x}\right)^n A(x, y)\right] = 0$$

Methods of evaluation. In application, one is confronted with the problem of evaluating the finite part integrals. This can always be done, of course, by means of the definitions already given. Often, however, such evaluations can be simplified by using one of the two following processes.

The first process is readily outlined. Consider the expression

$$\int_a^x \frac{f(x, y)dy}{[x-y]^{3/2}} = \lim_{\xi \to x} \int_a^\xi \frac{f(x, y) - f(x, x)}{[x-y]^{3/2}}\, dy - \frac{2f(x, x)}{\sqrt{x-a}}$$

and set the indefinite integral

$$\int \frac{f(x, y)dy}{[x-y]^{3/2}}$$

equal to $F(x, y) + C$. It follows that

$$\int_a^x \frac{f(x, y)dy}{[x-y]^{3/2}} = -[F(x, a) + C] \tag{14.1-6}$$

where

$$C = \lim_{y \to x}\left[\frac{2f(x, y)}{\sqrt{x-y}} - F(x, y)\right] \tag{14.1-7}$$

The second technique avoids the necessity of evaluating the constant C. It depends on the use of the complex variable and is valid only when $f(x, y)$ is real in the interval $a \leqq y \leqq b$ where b is some number greater than x. Again set the indefinite integral

$$\int \frac{f(x, y)dy}{[x-y]^{3/2}}$$

equal to $F(x, y) + C$. Now if R.P. stands for real part of a function

$$\int_a^x \frac{f(x, y)dy}{[x-y]^{3/2}} = \text{R.P.} \int_a^b \frac{f(x, y)dy}{[x-y]^{3/2}} \equiv \text{R.P.}[F(x, b) - F(x, a)],$$

$$a < x < b \tag{14.1-8}$$

An application. An application of these methods gives the following simple but useful results. Let

$$Y = a + by + cy^2 = (\lambda_2 - y)(y - \lambda_1)(-c)$$

$$q = 4ac - b^2$$

then

$$\int_t^{\lambda_2} \frac{(C_0 + C_1 y)dy}{Y^{3/2}} = -\int_{\lambda_1}^{t} \frac{(C_0 + C_1 y)dy}{Y^{3/2}}$$

$$= \frac{-2}{q\sqrt{a + bt + ct^2}}[C_0(2ct + b) - C_1(bt + 2a)] \tag{14.1-9}$$

Notice that

$$\int_{\lambda_1}^{\lambda_2} \frac{(C_0 + C_1 y)dy}{Y^{3/2}} = 0 \tag{14.1-10}$$

which contains the very important identity

$$\fint_{\lambda_1}^{\lambda_2} \frac{dy}{(\lambda_2 - y)^{3/2}\sqrt{y - \lambda_1}} = \fint_{\lambda_1}^{\lambda_2} \frac{dy}{(y - \lambda_1)^{3/2}\sqrt{\lambda_2 - y}} = 0 \qquad (14.1\text{-}11)$$

The case of multiple integrals. When applied to the analysis of single integrals the above definitions of the finite part coincide with (or are a reexpression of) those given originally by Hadamard. Hence,

$$\overline{\left| \int_a^x \frac{f(x, y)dy}{[x - y]^{3/2}} \right.} = \fint_a^x \frac{f(x, y)dy}{[x - y]^{3/2}}$$

where $\overline{|\quad}$ was the symbol used by Hadamard to express the finite part concept. However, when applied to double integrals the two signs $\overline{|\quad}$ and $\fint$ are no longer equivalent.

Hadamard and others, such as Robinson [38], use the convention that the order of integration in the operation $\overline{|\quad}$ is reversible; that is

$$\overline{\left| \int dy \int dx\, f(x, y) \right.} = \overline{\left| \int dx \int dy\, f(x, y) \right.}$$

Such a convention excludes from the area of integration all singularities over which the order of integration is not reversible. These singular regions are then treated separately. This convention has the disadvantage that, in evaluating multiple integrals, the value of a given integral is not independent of succeeding integrals. Thus [37]

$$\overline{\left| \int_0^\xi \frac{d\eta}{(\xi - \eta)^{3/2}\sqrt{\eta}} \right.} = 0$$

but

$$\overline{\left| \int_0^x d\xi \int_0^\xi \frac{d\eta}{(\xi - \eta)^{3/2}\sqrt{\eta}} \right.} = -2\pi$$

The operator $\fint$ avoids the above difficulty. The value of an integrand defined by $\fint_0^a f(y)dy$ is independent of succeeding operations. For example

$$\fint_0^\xi \frac{d\eta}{(\xi - \eta)^{3/2}\sqrt{\eta}} = \int_0^x d\xi \fint_0^\xi \frac{d\eta}{(\xi - \eta)^{3/2}\sqrt{\eta}} = 0$$

On the other hand, although the use of the symbol $\fint$ makes each integration independent of the succeeding integrations, the order of integrating operations involving the sign $\fint$ cannot be reversed. Hence,

$$\int dy \fint dx\, f(x, y) \neq \int dx \fint dy\, f(x, y)$$

For example, it has been shown

$$\int_0^x d\xi \fint_0^\xi \frac{d\eta}{(\xi - \eta)^{3/2}\sqrt{\eta}} = 0$$

but the same integral taken over the same area in reversed order is

$$\int_0^x d\eta \, \fint_\eta^x \frac{d\xi}{(\xi - \eta)^{\frac{3}{2}} \sqrt{\eta}} = -2\pi$$

Throughout this section, we will use the symbol $\fint$ as defined above to express operations involving the finite part.

THE GENERALIZED PRINCIPAL PART OF AN INTEGRAL.

Definition. In general, if $a < y < b$, the integral

$$\int_a^b \frac{A(y_1)dy_1}{y - y_1}$$

is not defined. However, when the integral is so written without further qualification it is generally accepted that the singularity occurring in the the integrand is to be treated using Cauchy's principal part. That is, the equalities

$$\int_a^b \frac{A(y_1)dy_1}{y - y_1} = \mathcal{P} \int^b \frac{A(y_1)dy_1}{y - y_1}$$
$$= \lim_{\epsilon \to 0} \left[\int_a^{y-\epsilon} \frac{A(y_1)dy_1}{y - y_1} + \int_{y+\epsilon}^b \frac{A(y_1)dy_1}{y - y_1} \right] \quad (14.1\text{-}12)$$

or, alternatively,

$$\int_a^b \frac{A(y_1)dy_1}{y - y_1} = \mathcal{P} \int_a^b \frac{A(y_1)dy_1}{y - y_1} = \frac{\partial}{\partial y} \int_a^b A(y_1) \ln |y - y_1| dy_1 \quad (14.1\text{-}13)$$

are understood. To assure the convergence of the right-hand side of Eq. 14.1-12 and 14.1-13, it is sufficient (but not necessary) to assume that $A(y_1)$ is continuous at the point $y_1 = y$ and that elsewhere within the region of integration $A(y_1)$ is either continuous or possesses integrable singularities. The concept of the Cauchy principal part is so well known that the symbol on the integral is often omitted, as is done throughout this section.

To obtain a generalization of the principal part,[21] we simply extend the concept presented in Eq. 14.1-13. Thus

$$n! \fint_a^b \frac{A(y_1)dy_1}{(y_1 - y)^{n+1}} = -\left(\frac{\partial}{\partial y}\right)^{n+1} \int_a^b A(y_1) \ln |y - y_1| dy_1$$
$$= \left(\frac{\partial}{\partial y}\right)^n \int_a^b \frac{A(y_1)dy_1}{y_1 - y} \quad (14.1\text{-}14)$$

The generalized principal part can also be expressed in the form

$$\fint_a^b \frac{A(y_1)dy_1}{(y_1 - y)^{n+1}} = \int_a^b \frac{A(y_1) - B(y, y_1)}{(y_1 - y)^{n+1}} dy_1 + \fint_a^b \frac{B(y, y_1)}{(y_1 - y)^{n+1}} dy_1$$

[21] In this connection, see also [39].

where

$$B(y, y_1) = A(y) + \frac{A'(y)}{1}(y_1 - y) + \cdots + \frac{A^{(n-1)}(y)}{(n-1)!}(y_1 - y)^{n-1}$$

and

$$\fint_a^b \frac{dy_1}{(y_1 - y)^{i+1}} = \frac{1}{i!}\left(\frac{\partial}{\partial y}\right)^i \int_a^b \frac{dy_1}{y_1 - y} = \frac{-1}{i}\left[\frac{1}{(b-y)^i} - \frac{1}{(a-y)^i}\right], \quad i \geqq 1$$

The first n derivatives of $A(y_1)$ are assumed to exist and be single-valued at $y_1 = y$ while elsewhere in the range of integration $A(y_1)$ may possess integrable singularities. This definition is in a form that involves no extension beyond the concept of Cauchy's principal part.

It is also possible to include a functional dependency on y in the numerator of the integrands. Thus, replacing $A(y_1)$ by $A(y, y_1)$, Eq. 14.1-14 again defines uniquely a principal part integral provided the first n derivatives of $A(y, y_1)$ with respect to both y and y_1 exist at $y = y_1$.

Method of evaluation. Operations involving the symbol $\fint$ can always be evaluated by means of the definitions just given. However, another method can be used, which in practice is often the simplest to apply. If the indefinite integral of $A(y_1)/(y_1 - y)^{n+1}$ exists such that

$$\int \frac{A(y_1)dy_1}{(y_1 - y)^{n+1}} = G(y, y_1) + C$$

then the value of the generalized principal part can be found by following the conventional rules for substitution of limits. Thus

$$\fint_a^b \frac{A(y_1)dy_1}{(y_1 - y)^{n+1}} = G(y, b) - G(y, a), \qquad y \neq a, b \qquad (14.1\text{-}15)$$

The proof of this result can be obtained by mathematical induction.

An application. Consider the expression

$$I_0 = \lim_{z \to 0+} \int_a^b \frac{\partial}{\partial z}\left[\frac{zf(y_1, z)}{z^2 + (y - y_1)^2}\right] dy_1$$

which is basic to the development of equations for the induced effects of planar vortex distributions (see the development of Eq. 10-21 and 10-22). The term $f(y, z)$ and its derivative are bounded and continuous in the interval $a \leqq y \leqq b$. To simplify I_0, first integrate by parts

$$I_0 = \lim_{z \to 0+} \frac{\partial}{\partial z}\left[f(a, z) \tan^{-1}\frac{y - a}{z} - f(b, z)\tan^{-1}\frac{y - b}{z} + \int_a^b \frac{\partial f(y_1, z)}{\partial y_1}\tan^{-1}\frac{y - y_1}{z}\,dy_1\right]$$

Then, since

$$\lim_{z \to 0+} \int_a^b \frac{\partial f_z(y_1, z)}{\partial y_1}\tan^{-1}\frac{y - y_1}{z}\,dy_1 = \pi f_z(y, 0) - \frac{\pi}{2}[f_z(a, 0) + f_z(b, 0)]$$

where f_z denotes $\partial f/\partial z$, this becomes

$$I_0 = \frac{f(b, 0)}{y - b} - \frac{f(a, 0)}{y - a} - \int_a^b \frac{\partial f(y_1, 0)}{\partial y_1} \frac{dy_1}{y - y_1} + \pi f_z(y, 0)$$

But it can be readily shown that

$$\oint_a^b \frac{f(y_1,0)}{(y - y_1)^2} dy_1 = \frac{f(b, 0)}{y - b} - \frac{f(a, 0)}{y - a} - \int_a^b \frac{\partial f(y_1, 0)}{\partial y_1} \frac{dy_1}{y - y_1}$$

Hence, finally

$$\lim_{z\to 0+} \int_a^b \frac{\partial}{\partial z}\left[\frac{zf(y_1, z)}{z^2 + (y - y_1)^2}\right] dy_1 = \oint_a^b \frac{f(y_1, 0)}{(y - y_1)^2} dy_1 + \pi f_z(y, 0) \quad (14.1\text{-}16)$$

CHAPTER 4. AXIALLY SYMMETRIC SUPERSONIC STEADY STATE FLOW

D,15. Basic Potential Equation and Solutions. In previous applications to supersonic theory we have dealt with so-called planar problems, that is, problems in which boundary conditions have been expressible to the required degree of accuracy in an x, y plane. The potential field was then determined from a planar distribution of elementary solutions. For sufficiently slender bodies of revolution it is possible to derive an analogous method by specifying boundary conditions in the *vicinity* of the body axis and relating the potential field to a single line of sources and doublets. The present chapter is confined to the development of this phase of axial symmetry theory. It will be seen in the following chapter that the results are, in fact, part of a more general theory applicable to slender wings, bodies, and wing-body combinations.

Consider first the linearized form of the basic partial differential equation governing the flow field. As we have seen, in Cartesian coordinates this is

$$\beta^2\varphi_{xx} - \varphi_{yy} - \varphi_{zz} = 0 \quad (15\text{-}1)$$

provided that almost everywhere in the flow field the ratios of the perturbation velocities to the free stream velocity are small enough to be neglected compared to unity, i.e. provided

$$\left[\left(\frac{u}{U}\right)^2 + \left(\frac{v}{U}\right)^2 + \left(\frac{w}{U}\right)^2\right]^{\frac{1}{2}} \ll 1 \quad (15\text{-}2)$$

The equation for the pressure coefficient in a field having the property expressed by Eq. 15-2 can be written

$$C_p = -\frac{2u}{U} - \frac{v^2 + w^2}{U^2} \quad (15\text{-}3)$$

We have already mentioned in Art. 2 that, when the perturbation velocities are all of the same order in magnitude, it is logical and customary to neglect the v^2 and w^2 terms in expressing C_p; this has been done, for example, in Art. 1-14. However, one can readily visualize airplane shapes that would induce flow fields in which the perturbation velocities would have widely dissimilar magnitudes and, in particular, shapes that would induce values of $(v^2 + w^2)/U^2$ similar in magnitude to the value of $2u/U$. Further, these fields can still have the property given in Eq. 15-2. Thus, it can be anticipated (and solutions presented later will verify the fact) that flow disturbances created by slender bodies of revolution are governed[22] by Eq. 15-1 and should be interpreted as to pressure coefficient[23] by means of Eq. 15-3.

It is advantageous to make use of the property of axial symmetry inherent in the geometry of bodies of revolution. In order to do this, let us consider the coordinate system in which the basic partial differential equation is expressed. Eq. 15-1 is written in terms of a coordinate system wherein the free stream is directed parallel to the x axis. We refer to a set of coordinates so orientated as the wind axes. Obviously, if the body is not inclined to the free stream, the boundary conditions and hence the physical flow field are axially symmetric, and Eq. 15-1 and 15-3 can be expressed completely in terms of the two coordinates, $r = \sqrt{y^2 + z^2}$ and x. However, if the body is inclined to the free stream direction, it is equally obvious that the flow varies with the angle $\theta = \tan^{-1} y/z$ and is no longer axially symmetric. If we continue to use wind axes, the boundary conditions at infinity are independent of θ but the boundary conditions at the body are not. The difficulties resulting from this asymmetry can be reduced by introducing body axes, that is, a coordinate system in which the x axis is fixed along the centerline of the body and is, therefore, inclined with respect to the free stream at infinity.

One can easily show that Eq. 15-1 is not invariant to a rotation about the y or z axis. Hence, in general, we cannot change from wind to body axes without changing the form of the governing equation. However, it must be remembered that Eq. 15-1 is valid only to a certain order. In this sense. Eq. 15-1 is invariant to a rotation about the y or z axis if the new terms introduced by the rotation are similar in magnitude to those already neglected in its derivation. Thus, Eq. 15-1 is to the lowest order the governing partial differential equation for both wind and body axes provided the body is slender and the angle of attack is small.

Although Eq. 15-1 is invariant to a rotation in the sense just discussed, we must realize at once that the expression for pressure coefficient,

[22] A considerable amount of effort has been expended in the determination of the order of magnitude involved in body of revolution analysis [*40,41,42*]. This is discussed in detail in Sec. E.

[23] The fact that the expression for C_p induced by a body of revolution differs from that used in planar theory has already been mentioned in C,5.

Eq. 15-3, is not. If the rotation is α radians about the y axis, the equation for pressure coefficient with reference to the body axes becomes

$$C_p = -\frac{2}{U}(\varphi_x + \alpha\varphi_z) - \frac{1}{U^2}(\varphi_y^2 + \varphi_z^2) \qquad (15\text{-}4)$$

With reference to the wind axes it remains, of course, as Eq. 15-3. Henceforth, in this and the following chapter we will refer our solutions to the body axes unless otherwise specifically stated.

Introduce now the cylindrical coordinates x, r, and θ and let $\phi(x, r, \theta)$ be the velocity potential of the flow field; the perturbation potential

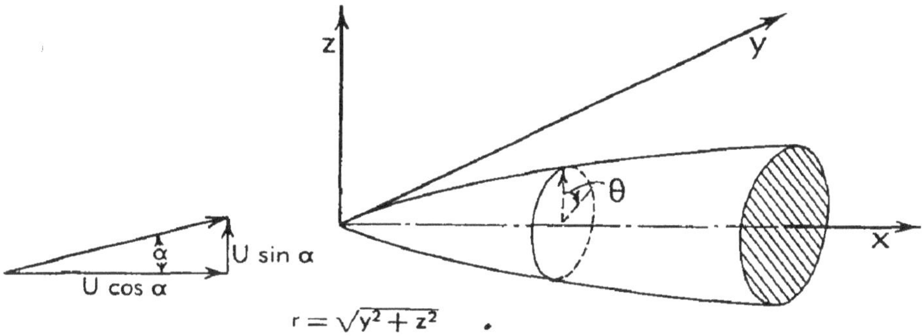

Fig. D,15. Body of revolution with body axes.

$\varphi(x, r, \theta)$ can then be defined (see Fig. D,15) by means of the equation

$$\phi(x, r, \theta) = Ux\cos\alpha + Ur\sin\alpha\cos\theta + \varphi(x, r, \theta) \qquad (15\text{-}5)$$

where α is the angle of attack. In terms of $\varphi(x, r, \theta)$, the partial differential equation becomes

$$\beta^2\varphi_{xx} - \frac{1}{r}\varphi_r - \varphi_{rr} - \frac{1}{r^2}\varphi_{\theta\theta} = 0 \qquad (15\text{-}6)$$

A solution to Eq. 15-6 representing a source distribution of strength f and a doublet distribution of strength σ follows immediately from Eq. 10-2. For distributions commencing at the origin and continuing along the x_1 axis there results

$$\varphi(x, r, \theta) = -\frac{1}{2\pi}\int_0^{x-\beta r}\frac{f(x_1)dx_1}{\sqrt{(x-x_1)^2 - \beta^2r^2}}$$
$$+ \frac{\beta^2 r\cos\theta}{2\pi}\int_0^{x-\beta r}\frac{\sigma(x_1)dx_1}{[(x-x_1)^2 - \beta^2r^2]^{\frac{3}{2}}}, \qquad \beta r \leqq x \qquad (15\text{-}7)$$

As is obvious from physical considerations, the potential induced by the sources is axially symmetric and that induced by the doublets is not.

In general the boundary conditions are that $\varphi(x, r, \theta)$ evaluated at infinity and the gradient of $\phi(x, r, \theta)$ normal to and evaluated on the surface of the body are zero. If $r = R(x)$ is the equation of the meridian section of the body, the latter conditions can be expressed analytically by the equation

$$-[U \cos \alpha + \varphi_x(x, R, \theta)] \frac{dR}{dx} + [U \sin \alpha \cos \theta + \varphi_r(x, R, \theta)] = 0$$

Neglecting terms of order higher than that to which the partial differential equation controls the flow, we can reduce this to

$$\varphi_r(x, R, \theta) = U \frac{dR}{dx} - U\alpha \cos \theta \tag{15-8}$$

Eq. 15-8 shows immediately that, to the order of our analysis, the boundary values due to thickness are satisfied by a source distribution [*43,44*] and those due to angle of attack by a doublet distribution [*45*].

D,16. Slender Pointed Bodies of Revolution. For slender pointed bodies of revolution it is possible to relate $f(x)$ and $\sigma(x)$ to the body geometry. In order to accomplish this, we limit attention to the vicinity of the body and approximate the flow field for small values of βr. Assuming a priori that $f(0) = \sigma(0) = 0$ for a pointed body, it follows that Eq. 15-7 can be written in the form

$$\varphi(x, r, \theta) = -\frac{1}{2\pi} \frac{\partial}{\partial x} \int_0^{x-\beta r} f(x_1) \cosh^{-1} \frac{x - x_1}{\beta r} \, dx_1 - \frac{\cos \theta}{2\pi r} \frac{\partial}{\partial x} \int_0^{x-\beta r} \frac{(x - x_1)\sigma(x_1) dx_1}{\sqrt{(x - x_1)^2 - \beta^2 r^2}}$$

For small values of βr the approximate expression for the perturbation potential is

$$\varphi(x, r, \theta) \cong -\frac{1}{2\pi} \frac{\partial}{\partial x} \int_0^x f(x_1) \ln \frac{2(x - x_1)}{\beta r} \, dx_1 - \frac{\cos \theta}{2\pi r} \sigma(x) \tag{16-1}$$

and radial component of velocity becomes

$$\varphi_r(x, r, \theta) \cong \frac{f(x)}{2\pi r} + \frac{\cos \theta \, \sigma(x)}{2\pi r^2} \tag{16-2}$$

Comparing Eq. 16-2 and 15-8, we see that

$$f(x) \cong 2\pi U R \frac{dR}{dx} = US'(x) \tag{16-3}$$

and that

$$\sigma(x) \cong -2\pi U\alpha R^2 = -2U\alpha S(x) \tag{16-4}$$

where S is the cross-sectional area of the body and is expressed as a function of x. It is apparent that the source and doublet strengths per unit length are proportional to the longitudinal rate of change of body area and to the cross-sectional area respectively. Hence, the perturbation velocity potential is expressible as

$$\frac{1}{U}\varphi(x, r, \theta) = \frac{S'(x)}{2\pi}\ln\frac{\beta r}{2} - \frac{1}{2\pi}\frac{\partial}{\partial x}\int_0^x S'(x_1)\ln(x - x_1)dx_1 + \frac{\alpha}{r}R^2(x)\cos\theta \quad (16\text{-}5)$$

The pressure coefficient is given in Cartesian coordinates by Eq. 15-4 and can be reexpressed in cylindrical coordinates in the form

$$C_p = -\frac{2\varphi_x}{U} - \frac{1}{U^2}\left[(U\alpha\cos\theta + \varphi_r)^2 + \left(U\alpha\sin\theta - \frac{1}{r}\varphi_\theta\right)^2 - U^2\alpha^2\right] \quad (16\text{-}6)$$

If C_p is evaluated on the surface of the body and Eq. 16-5 is used to determine the velocity component φ_θ, Eq. 16-6 becomes

$$C_p = -\frac{2\varphi_x}{U} - \left(\frac{dR}{dx}\right)^2 + \alpha^2(2\cos 2\theta - 1)$$

The latter expression divides naturally into two parts,

$$(C_p)_t = -\frac{S''(x)}{\pi}\ln\frac{\beta R}{2} - \left(\frac{dR}{dx}\right)^2 + \frac{1}{\pi}\frac{\partial^2}{\partial x^2}\int_0^x S'(x_1)\ln(x - x_1)dx_1 \quad (16\text{-}7)$$

and

$$(C_p)_L = \left[-4\frac{dR}{dx}\cos\theta + \alpha(2\cos 2\theta - 1)\right]\alpha \quad (16\text{-}8)$$

giving the effects of thickness and angle of attack respectively. It should be noted that the effect of thickness is given entirely by the source distribution and the effect of angle of attack is given entirely by the doublet distribution. The part due to angle of attack, Eq. 16-8, is the same as found in C,9 for subsonic stream speed.

D,17. Slender Pointed Bodies of Revolution: Angle-of-Attack Effects. It is possible, by means of Eq. 16-8, to find the effect of angle of attack on the body lift, drag, and pitching moment. The lift and pitching moment can be calculated from the loading coefficient which follows immediately from Eq. 16-8

$$\frac{\Delta p}{\frac{1}{2}\rho_\infty U^2} = (C_p)_{\text{lower}} - (C_p)_{\text{upper}} = 8\alpha\frac{dR}{dx}\cos\theta$$

The lift is

$$L = \int_0^l \int_{-\pi/2}^{\pi/2} \Delta p\cos\theta R d\theta dx$$

$$= \frac{2\alpha\rho_\infty U^2}{\pi}\int_0^l \frac{dS}{dx}dx\int_{-\pi/2}^{\pi/2}\cos^2\theta d\theta = \alpha\rho_\infty U^2 S_l \quad (17\text{-}1)$$

where S_l denotes base area, and the pitching moment about the vertex is

$$M' = -\int_0^l \int_{-\pi/2}^{\pi/2} \Delta p \; x \cos \theta R d\theta dx$$

$$= -\alpha\rho_\infty U^2 \int_0^l x \frac{dS}{dx} dx = -\alpha\rho_\infty U^2 \int_0^l (S_l - S_x) dx \qquad (17\text{-}2)$$

The component of force in the direction of the body axis may be denoted D_x and is easily calculated in the form

$$D_x = \frac{1}{2}\rho_\infty U^2 \int_0^l \int_0^{2\pi} (C_p)_L R \frac{dR}{dx} dx d\theta$$

$$= -\frac{\rho_\infty U^2 \alpha^2 \pi R^2(l)}{2} = -\frac{\rho_\infty U^2 \alpha^2}{2} S_l \qquad (17\text{-}3)$$

If we wish, however, to find the magnitude of the drag in the free stream direction (see Fig. D,15) it is necessary to include the contribution of the lift vector. It follows that

$$D = L \sin \alpha + D_x \cos \alpha \cong L\alpha + D_x = \frac{\rho_\infty U^2 \alpha^2}{2} S_l \qquad (17\text{-}4)$$

The load distribution, lift, drag and moment of the body are thus independent of the free stream Mach number. If the body is pointed at both ends, Eq. 17-1 indicates that the lift vanishes; for all other cases, the lift and drag coefficients may be expressed as

$$C_L = \frac{L}{\frac{1}{2}\rho_\infty U^2 S_l} = 2\alpha, \qquad C_D = \frac{D}{\frac{1}{2}\rho_\infty U^2 S_l} = \alpha^2 \qquad (17\text{-}5)$$

and are independent of Mach number and body shape. If the mean area, S_m, of the body is defined as the volume divided by the body length l, we can write Eq. 17-2 in the form [*45*]

$$M' = -\alpha\rho_\infty U^2 l S_l \left(1 - \frac{S_m}{S_l}\right) \qquad (17\text{-}6)$$

The distance d between the point of application of the resultant lift force and the vertex, is given by the equation

$$d = \left(1 - \frac{S_m}{S_l}\right) l \qquad (17\text{-}7)$$

This moment arm determines the center of pressure position and is independent of Mach number.

D,18. Slender Pointed Bodies of Revolution: Thickness Effects. When the angle of attack is zero, the wind axes and the body axes coincide and the pressure coefficient can be written in general as

$$C_p = -\frac{2\varphi_x}{U} - \left(\frac{\varphi_r}{U}\right)^2 \tag{18-1}$$

Since for such cases the flow is independent of the angle θ, the velocity potential is a function only of x and r and follows from Eq. 16-5 as[24]

$$\frac{1}{U}\varphi(x, r) = \frac{S'(x)}{2\pi}\ln\frac{\beta r}{2} - \frac{1}{2\pi}\int_0^x S''(x_1)\ln(x - x_1)dx_1 \tag{18-2}$$

As an application of these results, consider the case of a slender cone with vertex angle equal to $2\omega_0$. Since dR/dx is equal to $\tan\omega_0 \cong \omega_0$ and $S(x) = \pi\omega_0^2x^2$, we find from Eq. 18-2

$$\frac{\varphi(x, r)}{U} = \omega_0^2x\left(1 - \ln\frac{2x}{\beta r}\right) \tag{18-3}$$

It is apparent from the perturbation velocity components

$$\varphi_x = -U\omega_0^2\ln\frac{2x}{\beta r}, \qquad \varphi_r = U\omega_0^2\frac{x}{r}$$

that the flow field is conical in nature since the velocities are functions only of x/r.

Along the surface of the cone we have

$$\varphi_x = -U\omega_0^2\ln\left(\frac{2}{\beta\omega_0}\right), \qquad \varphi_r = U\omega_0$$

This particular example also serves to indicate the difference in the magnitudes of the velocity components in axial symmetry theory. Pressure coefficient on the surface of the cone is

$$(C_p)_t = 2\omega_0^2\ln\left(\frac{2}{\beta\omega_0}\right) - \omega_0^2 \tag{18-4}$$

and in Fig. D,18a a comparison is given between the present results and the exact solution obtained from shock wave theory [*46*].

The drag resulting from the constant pressure on the body surface is easily calculated as follows

$$D = \frac{1}{2}\rho_\infty U^2\int_0^l\int_0^{2\pi} C_pR\frac{dR}{dx}\,d\theta dx = \pi\rho_\infty U^2l^2\omega_0^2\left(\ln\frac{2}{\beta\omega_0} - \frac{1}{2}\right)$$

As a further example, consider the body obtained by revolving about

[24] In the calculation of φ_x from this equation it is usually convenient to use the relation

$$\frac{\partial}{\partial x}\int_0^x S''(x_1)\ln(x - x_1)dx_1 = S''(0)\ln x + \int_0^x S'''(x_1)\ln(x - x_1)dx_1$$

the axis the parabolic arc

$$R = \frac{2t}{l^2}(lx - x^2) \tag{18-5}$$

where l is the length and t is the maximum thickness of the body [47].

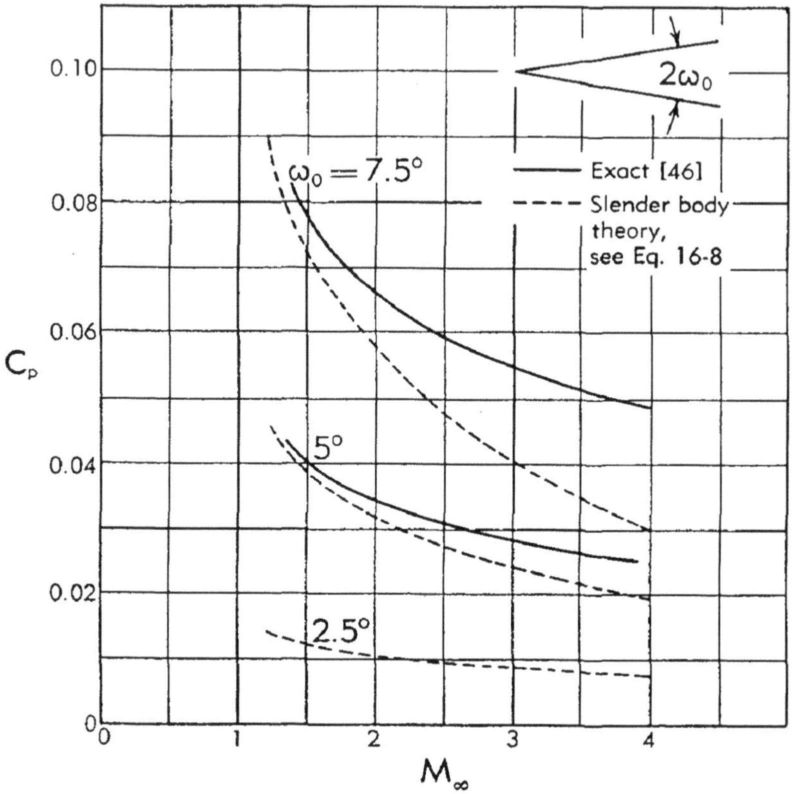

Fig. D,18a. Comparison of pressure coefficients given by exact and slender body theories.

Direct calculation gives for the velocity components on the surface

$$\varphi_x = -\frac{4Ut^2}{l^4}\left[3x(2l - 3x) + (l^2 - 6lx + 6x^2)\ln\frac{2x}{\beta R}\right]$$

$$\varphi_r = \frac{2Ut}{l^2}(l - 2x) \tag{18-6}$$

Fig. D,18b shows the calculated pressure distribution at a free stream Mach number 1.4 for a body thickness ratio t/l equal to 0.10. In this case the theory indicates a recompression, presumably without shock, over the rear portion of the body.

The solutions presented above also permit one to obtain a general and useful expression for the wave drag of nonlifting slender pointed bodies. Historically, the methods and results stem from the initial work of von Kármán [11] although the form to be given here appears to have been given first by Ward [40] and Frankl and Karpovich [48].

As we have seen, the body experiences a force in the axial direction given by the relation

$$D = -2\pi\rho_\infty \int_0^l \left(U\varphi_x + \frac{1}{2}\varphi_r^2\right)_{r=R} R\,\frac{dR}{dx}\,dx \tag{18-7}$$

and the drag can be evaluated by direct integration after substitution

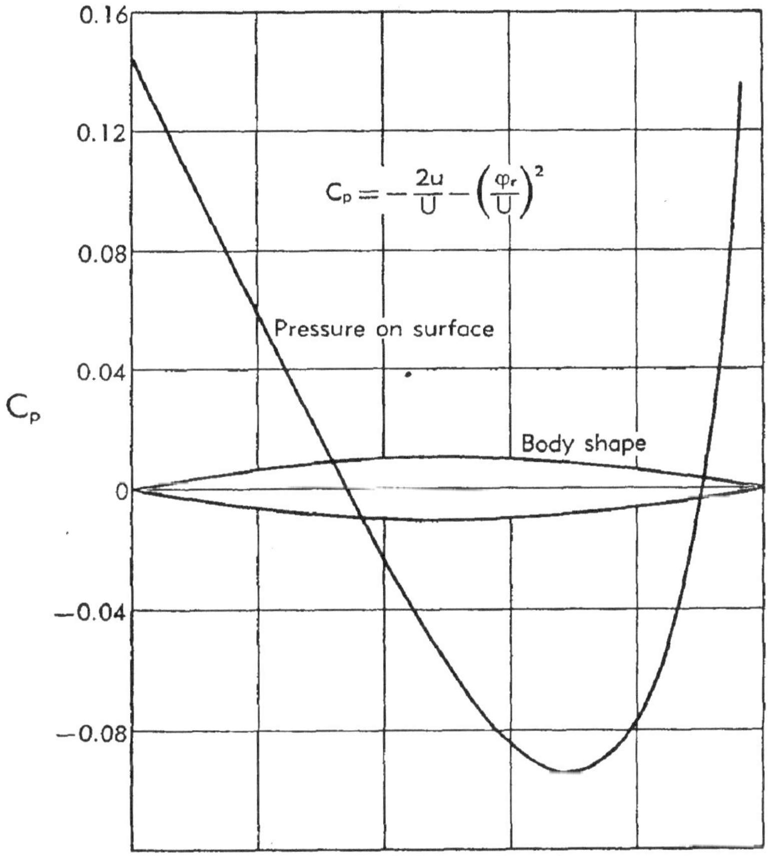

Fig. D,18b. Body of revolution and corresponding pressure distribution.

from Eq. 18-2. Eq. 18-7 then becomes

$$-\frac{2\pi D}{\rho_\infty U^2} = \frac{1}{2}\int_0^l \frac{d[S'(x)]^2}{dx}\ln\frac{\beta R}{2}\,dx + \pi\int_0^l \left(\frac{dR}{dx}\right)_1^2 S'(x)dx$$
$$-\int_0^l S'(x)dx\,\frac{\partial}{\partial x}\int_0^x S''(x_1)\ln(x - x_1)dx_1$$

or, after integrating the first and third integrals by parts

$$-\frac{2\pi D}{\rho_\infty U^2} = \frac{[S'(l)]^2}{2} \ln \frac{\beta R(l)}{2} - S'(l) \cdot \int_0^l S''(x_1) \ln (l - x_1)\, dx_1$$
$$- \int_0^l S''(x)dx \int_0^x S''(x_1) \ln (x - x_1)dx_1$$

The double integral in the latter equation can be put into a more symmetric form, as is obvious after interchanging the order of integration and averaging the sum of the original and derived expressions. The wave drag is, therefore,

$$D = \rho_\infty U^2 \left\{ \frac{[S'(l)]^2}{4\pi} \ln \frac{2}{\beta R(l)} + \frac{S'(l)}{2\pi} \int_0^l S''(x_1) \ln (l - x_1)dx_1 \right.$$
$$\left. - \frac{1}{4\pi} \int_0^l \int_0^l S''(x_1)S''(x) \ln |x - x_1| dxdx_1 \right\} \quad (18\text{-}8)$$

In the particular case when $dR/dx = \omega_0$, Eq. 18-8 gives for the wave drag of a slender cone a result in agreement with the value already calculated.

If $f(l) = 0$, so that either $R(l) = 0$ and the body closes, or the slope $dR(l)/dx = 0$ and the body is the bow of a semi-infinite cylinder, we get the von Kármán relation

$$D = -\frac{\rho_\infty U^2}{4\pi} \int_0^l \int_0^l S''(x_1)S''(x) \ln |x - x_1| dxdx_1 \quad (18\text{-}9)$$

It is interesting to observe that although Eq. 18-8 displays a variation with Mach number, this dependence does not appear in Eq. 18-9.

The von Kármán relation provides a remarkable correspondence between the wave drag of a body of revolution and the vortex drag of a finite span wing. If the span of the wing is b and the spanwise distribution of circulation, which is equal to the discontinuity in the potential at the trailing edge, is denoted by $\Gamma(y)$, the vortex drag is given by the formula (see Eq. 14-38)

$$D_V = -\frac{\rho_\infty}{4\pi} \int_{-b/2}^{b/2} \int_{-b/2}^{b/2} \Gamma'(y_1)\Gamma'(y) \ln |y - y_1| dydy_1 \quad (18\text{-}10)$$

The duality in terms resulting from Eq. 18-9 and 18-10 can be expressed as follows:

Body of revolution	*Finite span wing*
Wave drag	Vortex drag
l, length	b, span
Source strength or $U \times$ (gradient of area)	Spanwise circulation or loading$/\rho_\infty U$
$U \times$ (base area) $= US_l$	(Total lift)$/\rho_\infty U = L/\rho_\infty U$
Closure of body	No net lift on wing
$U \times$ (volume)	$[bL/2 -$ Rolling moment about mid-span$]/\rho_\infty U$
$U \times$ (volume of closed body)	$[-$Rolling moment with zero net lift$]/\rho_\infty U$

Aerodynamicists are familiar with the determination, based on the original investigations of Munk [49,50], of wing loadings that provide minimum vortex drag under prescribed conditions. It is obvious that by means of the existing duality each solution in wing theory is a possible optimal solution in body of revolution theory. However, since the details of the derivation are not difficult, we proceed here to a direct analysis of two particular body problems.

Consider a body (Fig. D,18c) of fixed length $2l$, the central portion of which is to be a circular cylinder of length $2L$. We seek the proper

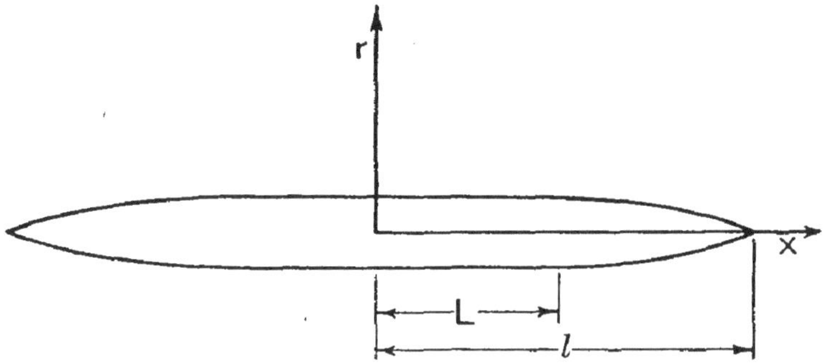

Fig. D,18c. Body of revolution of total length $2l$ having central cylindrical portion of length $2L$.

fairing of the bow and stern in order to attain a minimum wave drag and at the same time have the body enclose a specified volume G. It is important to note that the origin of the axial system in the figure and the length of the body are not the same as those used in the derivation of Eq. 18-9. However, the general nature of the drag equation and the given conditions are such that the final shape will be symmetrical fore and aft and the advantage of the new notation will soon be apparent.

In analytical terms, we are given

$$\frac{dS(0)}{dx} = \frac{dS(l)}{dx} = 0; \qquad \frac{d^2S}{dx^2} = \frac{dS}{dx} = 0, \qquad -L < x < L$$

and

$$\int_{-l}^{l} S(x)dx = G = \text{const}$$

and, since we are concerned with a conventional isoperimetric problem, it follows that

$$\delta\left[\int_{-l}^{l} S'(x)dx \int_{-l}^{l} \frac{S''(x_1)dx_1}{x - x_1} - \lambda \int_{-l}^{l} S(x)dx\right] = \delta[I_1 - \lambda I_2] = 0 \quad (18\text{-}11)$$

where I_1, I_2 denote the two integral expressions and λ is a constant to be determined later from the imposed conditions. Direct calculation yields

$$\delta I_1 = \int_{-l}^{l} \delta S'(x)dx \int_{-l}^{l} \frac{S''(x_1)dx_1}{x - x_1} + \int_{-l}^{l} \delta S''(x_1)dx_1 \int_{-l}^{l} \frac{S'(x)dx}{x - x_1}$$

and, after interchanging the variables in the second term and integrating by parts, we get the fundamental relation

$$\delta I_1 = \int_{-l}^{l} \delta S'(x)dx \int_{-l}^{l} \frac{S''(x_1)dx_1}{x - x_1} + \int_{-l}^{l} \delta S'(x)dx \frac{d}{dx} \int_{-l}^{l} \frac{S'(x_1)dx_1}{x - x_1}$$

$$= 2 \int_{-l}^{l} \delta S'(x)dx \int_{-l}^{l} \frac{S''(x_1)dx_1}{x - x_1} \tag{18-12}$$

For the purposes of the present problem, Eq. 18-12 may be integrated by parts once more and Eq. 18-11 becomes

$$\int_{-l}^{l} \delta S(x) \left[\frac{\partial}{\partial x} \int_{-l}^{l} \frac{S''(x_1)dx_1}{x - x_1} - \lambda \right] dx = 0$$

Thus the problem becomes one of inverting the integral equation

$$\frac{\partial}{\partial x} \int_{-l}^{l} \frac{S''(x_1)dx_1}{x - x_1} = \lambda \tag{18-13}$$

or

$$\int_{-l}^{l} \frac{S''(x_1)dx_1}{x - x_1} = \lambda x \tag{18-14}$$

since the symmetry in $S''(x_1)$ about the mid-point of the body necessitates odd symmetry in the right member.

Eq. 18-14 can be rewritten as

$$\int_{-l}^{-L} \frac{S''(x_1)dx_1}{x - x_1} + \int_{L}^{l} \frac{S''(x_1)dx_1}{x - x_1} = \lambda x$$

and, using the symmetry of $S''(x_1)$, this relation becomes

$$\int_{L}^{l} \frac{2xS''(x_1)dx_1}{x^2 - x_1^2} = \lambda x$$

Setting $x^2 = \xi$, $x_1^2 = \xi_1$, and $dx = d\xi/2\sqrt{\xi}$, we have

$$\lambda = \int_{L^2}^{l^2} \frac{\frac{d^2}{dx_1^2} S(\sqrt{\xi_1})d\xi_1}{\sqrt{\xi_1}\,(\xi - \xi_1)}$$

Since

$$\int_{L}^{l} \frac{d^2S}{dx^2} dx = \frac{dS(l)}{dx} - \frac{dS(L)}{dx} = 0$$

the inversion of this equation is, from Eq. 12-24,

$$\frac{1}{\sqrt{\xi}} \frac{d^2S(\sqrt{\xi})}{dx^2} = \frac{-4\lambda}{\sqrt{(l^2 - \xi)(\xi - L^2)}} \int_{L^2}^{l^2} \frac{\sqrt{(l^2 - \xi_1)(\xi_1 - L^2)}}{\xi - \xi_1} d\xi_1$$

and, after direct integration, we have

$$\frac{d^2S(x)}{dx^2} = \frac{-\lambda|x|(2x^2 - L^2 - l^2)}{2\pi\sqrt{(l^2 - x^2)(L^2 - x^2)}}$$

Moreover

$$\frac{dS(x)}{dx} = \frac{\lambda|x|}{2\pi x}\sqrt{(l^2 - x^2)(x^2 - L^2)}$$

and

$$S(x) = -\frac{\lambda l^3}{6\pi}\left\{-\frac{2L^2}{l^2}F(\sigma, k) + \left(1 + \frac{L^2}{l^2}\right)E(\sigma, k) - \frac{|x|[(l^2 - x^2)(x^2 - L^2)]^{\frac{1}{2}}}{l^3}\right\} \tag{18-15}$$

where $F(\sigma, k)$, $E(\sigma, k)$ are incomplete elliptic integrals of the first and second kind respectively and $k^2 = (l^2 - L^2)/l^2$, $\sigma^2 = (l^2 - x^2)/(l^2 - L^2)$.

Since

$$D = \frac{\rho_\infty U^2}{4\pi}\int_{-l}^{l} S'(x)dx \int_{-l}^{l} \frac{S''(x_1)dx_1}{x - x_1}$$

it follows, after using Eq. 18-14, that

$$D = -\frac{\lambda \rho_\infty U^2 G}{4\pi} \tag{18-16}$$

Setting $x = L$ in Eq. 18-15, we get

$$S_{\max} = -\frac{\lambda l^3}{6\pi}\left[E + \frac{L^2}{l^2}(E - 2K)\right]$$

so

$$\lambda = \frac{-6\pi S_{\max}}{l[(l^2 + L^2)E - 2L^2K]}$$

where K and E are complete elliptic integrals. The volume of the bow or stern is

$$G_1 = \int_L^l S(x)dx = \frac{\lambda}{2\pi}\left\{\frac{lL}{3}[(l^2 + L^2)E - 2L^2K] - \frac{\pi}{16}(l^2 - L^2)\right\}$$

$$= -LS_{\max} + \frac{3\pi}{16}\frac{(l^2 - L^2)^2 S_{\max}}{l[(l^2 + L^2)E - 2L^2K]}$$

and the total volume of the body is therefore

$$G = 2G_1 + 2LS_{\max} = \frac{3\pi}{8}\frac{(l^2 - L^2)^2 S_{\max}}{l[(l^2 + L^2)E - 2L^2K]}$$

From Eq. 18-16 and the subsequent results, drag can be expressed in the various following forms

$$D = \frac{9\rho_\infty U^2}{16}\frac{\pi S_{\max}^2 (l^2 - L^2)^2}{l^2[(l^2 + L^2)E - 2L^2K]^2}$$

$$= \frac{4\rho_\infty U^2}{\pi}\frac{G^2}{(l^2 - L^2)^2} \tag{18-17}$$

When $L = 0$, we get

$$G = \frac{3\pi}{8} lS_{\max}$$

$$D = \frac{9\rho_\infty U^2}{16} \pi \frac{S_{\max}^2}{l^2} \tag{18-18}$$

$$= \frac{4\rho_\infty U^2}{\pi} \frac{G^2}{l^4}$$

These latter results are in agreement with the published solutions of Sears [51] and Haack [52].

If the body has length $2l$ and the cylindrical center portion has a length $2L$, minimum wave drag, for a specified caliber can also be found. Only the results are given here:

$$2\frac{d^2S}{dx^2} = (L^2K^2 - E^2l^2)^{-1}S_{\max}\left\{\frac{4|x|K}{l}\sqrt{\frac{x^2 - L^2}{l^2 - x^2}} + \frac{4|x|l(E - K)}{[(l^2 - x^2)(x^2 - L^2)]^{\frac{1}{2}}} - 4[EF(\sigma, k) - KE(\sigma, k)]\right\}$$

where the modulus of the elliptic functions is $k = \sqrt{1 - (L/l)^2}$ and the argument is $\sigma = \sqrt{(l^2 - x^2)/(l^2 - L^2)}$;

$$2\frac{dS}{dx} = (L^2K^2 - E^2l^2)^{-1}S_{\max}\left\{4x[KE(\sigma, k) - EF(\sigma, k)] - \frac{4K}{l}\frac{x}{|x|}\sqrt{(l^2 - x^2)(x^2 - L^2)}\right\}$$

and

$$S = (L^2K^2 - E^2l^2)^{-1}S_{\max}\left\{(x^2E + L^2K)F(\sigma, k) - (El^2 + Kx^2)E(\sigma, k) + |x|\frac{K}{l}\sqrt{(l^2 - x^2)(x^2 - L^2)}\right\} \tag{18-19}$$

Drag is given by the relation

$$D = -\frac{\rho_\infty U^2 \pi S_{\max}^{2l}}{2(L^2K^2 - l^2E^2)} \tag{18-20}$$

When $L = 0$, this becomes

$$D = \frac{\rho_\infty U^2 \pi S_{\max}^2}{2l^2}$$

in agreement with results obtained by Haack. Keeping $(l - L)$ fixed, so that the nose length is constant, and allowing l and L to become very large, drag becomes

$$D = \frac{4\rho_\infty U^2 S_{\max}^2}{\pi(l - L)^2}$$

This is twice the drag of a given caliber ogive on an infinitely long cylinder. The latter result was given first by von Kármán [*11*]. The determination of minimum wave drag for bodies of revolution with cylindrical center sections has been studied extensively by Fuller and Briggs [*53*].

CHAPTER 5. SLENDER AIRPLANE THEORY

D,19. Basic Potential Equation for Slender Airplanes. The results of Art. 10-14 show that a striking contrast exists between the number of exact results obtainable in supersonic wing theory and those available from subsonic analysis. This is particularly true in the study of lifting surfaces. It is within the realm of practicability, for example, to calculate for Mach numbers greater than 1 the load distribution for flat lifting surfaces of almost arbitrary rectilinear planforms, that is, planforms bounded by straight lines. However, in many cases such calculations still involve recourse to numerical methods. For this reason they are tedious, and the results, by their very nature, are somewhat lacking in generality. In an attempt both to treat more complicated wing planforms and to achieve some simplification in the existing methods, two rather obvious possibilities can be considered. In the first case, when nearly rectangular wings of high aspect ratio are to be studied, the spanwise velocity gradients can be neglected so that the two-dimensional analysis of Art. 5-9 applies. Such an approach is a strip theory where longitudinal sections on the wing are assumed to possess their two-dimensional characteristics. The second case applies to wings that are either very slender in planform or are flying at speeds close to that of sound. In this latter case, the chordwise velocity gradients are assumed to be small and again a kind of two-dimensional analysis applies. Moreover, it turns out that not only can more complicated wing planforms be studied but even the much more complicated problems of wing-body interference can be treated. In fact, the ramifications of slender airplane theory are such that an entire idealized airplane configuration can be analyzed, including the wing, fuselage, and tail plane. This method which stems from the basic works of Munk [*54*] and R. T. Jones [*55*] and has been applied and extended by several authors (see, e.g. [*35,40,56,57*]) is to be the subject of Art. 19-24. It is also discussed, chiefly from the standpoint of subsonic flow, in C,9.

Let us now discover how the basic partial differential equation governing linearized supersonic flow fields can be simplified when the boundary conditions represent slender airplanes. Consider this equation in its normalized form:

$$\varphi_{xx} - \varphi_{yy} - \varphi_{zz} = 0 \tag{19-1}$$

Then our analysis is equivalent to specifying that $M_\infty = \sqrt{2}$ but, of course, the results can be extended throughout the supersonic Mach number range by applying the Prandtl-Glauert rule. An application of Green's theorem yields the solution, known as Volterra's solution,

$$\varphi = -\frac{1}{2\pi}\frac{\partial}{\partial x}\iint_{\tau}\left(\frac{\partial \varphi}{\partial \nu} - \varphi\frac{\partial}{\partial \nu}\right)\ln\frac{|x - x_1| + \sqrt{(x - x_1)^2 - r^2}}{r}\,dS \quad (19\text{-}2)$$

where dS is an element of surface area on the airplane or its trailing vortex sheet and $r^2 = (y - y_1)^2 + (z - z_1)^2$. Further, the area of integration τ is the portion of the airplane and its vortex wake within the forecone from the point x, y, z; and the gradient $\partial/\partial\nu$ is the conormal[25] to the airplane surface.

If n_1, n_2, and n_3 are the direction cosines of normals to the surface S with respect to the x, y, and z axes, respectively, then $\partial/\partial\nu$ can be written $-(n_1\partial/\partial x) + (n_2\partial/\partial y) + (n_3\partial/\partial z)$ and dS can be written $dx_1 ds_1/\sqrt{1 - n_1^2}$ where ds_1 is a differential length along the surface in a y, z plane. If the airplane is slender then n_1 is small almost everywhere on its surface. Hence, the first simplification is to neglect n_1 compared to either $\sqrt{n_2^2 + n_3^2}$ or unity and write Eq. 19-2 in the form

$$\varphi = -\frac{1}{2\pi}\frac{\partial}{\partial x}\iint_{\tau}\left(\frac{\partial \varphi}{\partial n} - \varphi\frac{\partial}{\partial n}\right)\ln\frac{|x - x_1| + \sqrt{(x - x_1)^2 - r^2}}{r}\,dx_1 ds_1$$

where $\partial/\partial n = (n_2\partial/\partial y) + (n_3\partial/\partial z)$ and is the surface normal in a y, z plane.

The second simplification follows from the observation that

$$\left[\frac{r}{(x - x_1)}\right]^2 \ll 1$$

over almost all of the area of integration *provided* the point x, y, z is on or in the vicinity of the airplane surface. This implies that the logarithmic term in Eq. 19-2 can be written $\ln 2(x - x_1)/r$ and that the area of integration is simply that ahead of the plane $x = x_1$. With these further simplifications, Eq. 19-2 becomes

$$\varphi = \frac{1}{2\pi}\frac{\partial}{\partial x}\int_0^x dx_1 \int_s ds_1\left(\frac{\partial \varphi}{\partial n} - \varphi\frac{\partial}{\partial n}\right)\ln r$$

$$- \frac{1}{2\pi}\frac{\partial}{\partial x}\int_0^x dx_1 \ln 2(x - x_1)\int_s \frac{\partial \varphi}{\partial n}\,ds_1$$

where s is the line bounding the cross-sectional area in a y, z plane. Since

$$\frac{1}{U}\int_s \frac{\partial \varphi}{\partial n}\,ds_1 = \frac{dS}{dx}$$

$S(x)$ being the cross-sectional area in a y, z plane, the equation for the

[25] See the discussion of Eq. 5-6.

velocity potential reduces to

$$\varphi = \frac{1}{2\pi}\int_s \left(\frac{\partial\varphi}{\partial n} - \varphi\frac{\partial}{\partial n}\right)\ln r ds_1 - \frac{1}{2\pi}\frac{\partial}{\partial x}\int_0^x S'(x_1)\ln 2(x - x_1)dx_1 \quad (19\text{-}3)$$

It is apparent that the first integral on the right-hand side of Eq. 19-3 is the general solution to Laplace's equation in two (the y and z) dimensions. We can represent this term, therefore, by the symbol $\varphi_2(x; y, z)$. The x dimension, which does not appear explicitly in the integral, enters as a parameter when the solution is adapted to particular boundary values. Hence, Eq. 19-3 can be written in the following alternative form wherein the Prandtl-Glauert rule has been applied to account for the Mach number effect:

$$\varphi(x, y, z) = \varphi_2(x; y, z) - \frac{U}{2\pi}\frac{\partial}{\partial x}\int_0^x S'(x_1)\ln\frac{2(x - x_1)}{\beta}dx_1 \quad (19\text{-}4)$$

This result was obtained, by somewhat different methods, by Ward [*40*]. It is interesting to notice that a similar analysis applied to the subsonic case yields the equation [*60*]

$$\varphi(x, y, z) = \varphi_2(x; y, z) - \frac{U}{2\pi}\frac{\partial}{\partial x}\int_0^\infty S'(x_1)\frac{x - x_1}{|x - x_1|}\ln\frac{2|x - x_1|}{\beta}dx_1 \quad (19\text{-}5)$$

A special case of Eq. 19-5, namely, that appropriate for bodies of revolution at incidence, is obtained in C,9.

The significance of these expressions can be interpreted readily. The three-dimensional velocity field induced by airplane shapes that are slender or flying at near sonic speeds is approximated *in the vicinity* of the airplane by a velocity field that satisfies the two-dimensional Laplace equation and the boundary conditions in transverse planes plus a longitudinal velocity field that depends on the streamwise rate of change of cross-sectional area and is independent of y and z.

D,20. Discussion of Boundary Conditions. The general description of the boundary conditions is simple enough: the gradient of the total velocity potential is consistent with the required free stream at infinity and is zero when it is evaluated normal to and on the surface of the airplane itself. Thus using ϕ for total velocity potential, φ for perturbation velocity potential, and referring our analysis to *body* axes in a free stream that makes an angle α with the x axis and γ with the y axis, we write

$$\phi(x, y, z) = U(x\cos\alpha\cos\gamma + y\sin\gamma + z\cos\gamma\sin\alpha) + \varphi(x, y, z)$$

such that on the airplane surface

$$U(n_1\cos\alpha\cos\gamma + n_2\sin\gamma + n_3\sin\alpha\cos\gamma) + n_1\varphi_x + n_2\varphi_y + n_3\varphi_z = 0$$

n_1, n_2, and n_3 again being the direction cosines of a normal to the airplane surface with respect to the x, y, and z axes, respectively. By the assumptions basic to slender airplane theory, the latter equation reduces to

$$U(n_1 + n_2\gamma + n_3\alpha) + \frac{\partial}{\partial n}\varphi_2(x; y, z) = 0 \tag{20-1}$$

where, as in the previous article, n is the normal to the line bounding a cross section in the y, z plane.

Eq. 20-1, which applies to general slender airplane boundary shapes, can be simplified for many aerodynamic problems. We consider three types of configurations that lead to such simplifications: first, a surface, such as a wing, which deviates only slightly from a plane; second, a surface which forms a body of revolution; and third, a surface which is a combination of the above two.

Consider first planar type problems. Let $h(x, y)$ be the distance a surface deviates from the $z = 0$ plane,[26] and let b represent the wing span. If the ratios $\alpha/(db/dx)$ and $(\partial h/\partial x)/(db/dx) \ll 1$, almost everywhere, it is consistent with our previous approximations to neglect the y component of the normal derivative along the wing surface and to project the resulting vertical derivative to the upper or lower surface of the $z = 0$ plane. In this way Eq. 20-1 becomes

$$U(n_1 + n_3\alpha) + n_3\left(\frac{\partial\varphi_2}{\partial z}\right)_{z=0} = 0$$

and since $n_1/n_3 = -\partial h/\partial x$, the boundary conditions for planar problems are expressed by as follows:

$$\left(\frac{\partial\varphi_2}{\partial z}\right)_{z=0} = -U\alpha + U\frac{\partial h}{\partial x} \tag{20-2}$$

Consider next a body of revolution (see Fig. D,20, middle). Then since

$$n_1 = \frac{-dR/dx}{\sqrt{1 + (dR/dx)^2}} \cong -\frac{dR}{dx}, \qquad n_3 = \frac{\cos\theta}{\sqrt{1 + (dR/dx)^2}} \cong \cos\theta$$

Eq. 20-1 becomes for $\gamma = 0$

$$\left(\frac{\partial\varphi_2}{\partial r}\right)_{r=R} = U\frac{dR}{dx} - U\alpha\cos\theta \tag{20-3}$$

which is the simplified expression of the boundary conditions pertaining to bodies of revolution.

Consider, finally, surfaces which are a combination of the above two as, for example, the one shown in Fig. D,20 (bottom). The rather obvious

[26] The $z = 0$ plane is chosen only for convenience. Also it should be noted that certain problems, such as the cruciform wing, require more than one plane but the concepts are the same.

extension of the above concepts is to apply Eq. 20-2 over the winged portion of the configuration and Eq. 20-3 over the body. It is now necessary, however, to discuss the relative magnitudes of the terms $\partial h/\partial x$, db/dx, and dR/dx since they are likely to appear in the solutions in

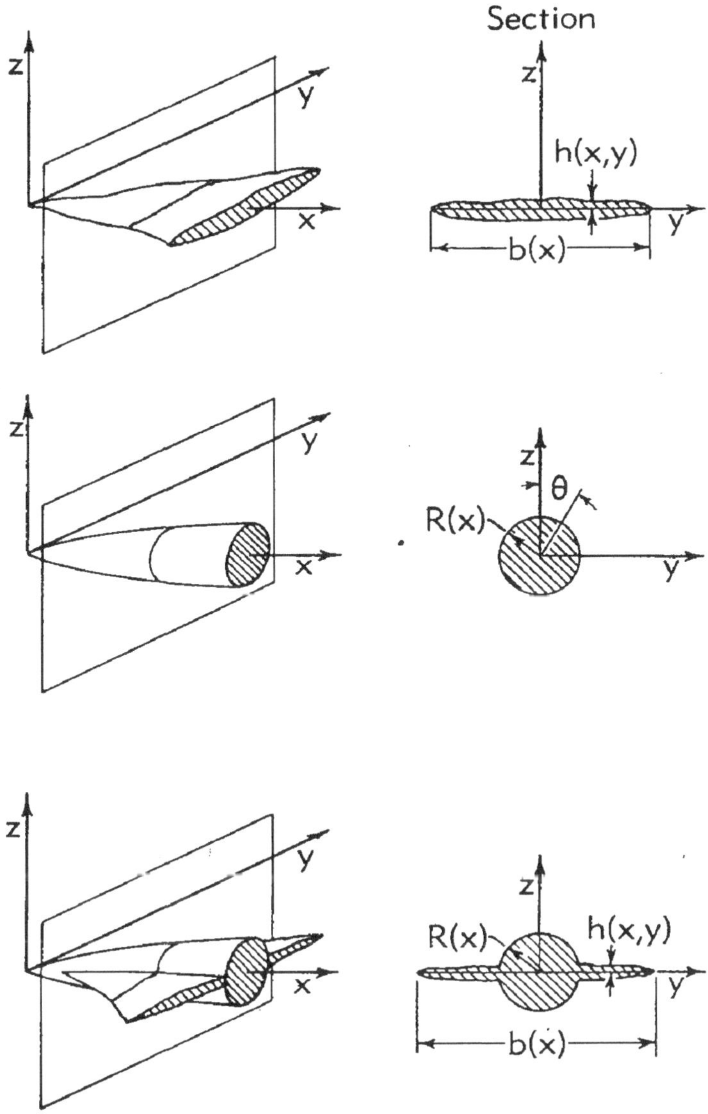

Fig. D,20. Slender wing, body of revolution, and wing-body combination.

various combinations. If the winged portion is to be treated as a planar problem, the magnitude of $\partial h/\partial x$ must be small enough to be neglected in comparison to db/dx. But this does not imply that $\partial h/\partial x$ can be neglected in comparison to dR/dx or that dR/dx can be neglected in comparison to db/dx. Therefore, the latter two approximations will not, in general, be made.

D,21. Evaluation of Lifting Forces. We have seen that the pressure coefficient with reference to body axes is

$$C_p = -\frac{2}{U}(\varphi_x + \alpha\varphi_z + \gamma\varphi_y) - \frac{1}{U^2}(\varphi_y^2 + \varphi_z^2) \tag{21-1}$$

and that the perturbation velocity potential can be written

$$\varphi(x, y, z) = \varphi_2(x; y, z) + A(x) \tag{21-2}$$

where $A(x)$ is given by the integral term in either Eq. 19-4 or 19-5. Further, by definition the loading coefficient is

$$\frac{\Delta p}{\frac{1}{2}\rho_\infty U^2} = (C_p)_l - (C_p)_u \tag{21-3}$$

where the subscripts u and l refer to the upper and lower surfaces of the airplane, respectively. It is immediately apparent from an inspection

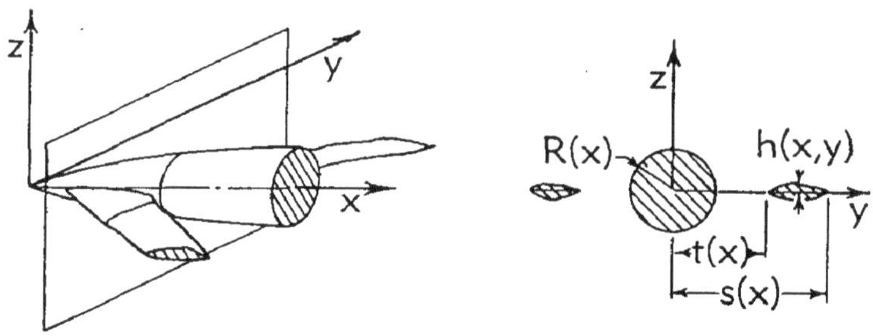

Fig. D,21. Sweptback wing on body of revolution.

of these equations that the loading is not affected by $A(x)$. Hence, the lift, pitching moment, and rolling moment can all be expressed entirely in terms of $\varphi_2(x; y, z)$.

Obviously the loading coefficient can always be calculated by means of the above equations. There are, however, many cases when such calculations can be simplified. One of these cases we propose to study in detail.

Consider the airplane configuration shown in Fig. D,21, which is composed of a sweptback wing mounted symmetrically on a body of revolution. The variations of the body radius and the sweep of the wing leading and trailing edges with x are arbitrary. If we make use of the boundary conditions expressed by Eq. 20-2 and 20-3, the pressure coefficients on the surface of the wing and body are respectively

$$C_p = -\frac{2\varphi_x}{U} + \alpha^2 - \left(\frac{\partial h}{\partial x}\right)^2 - \left(\frac{\varphi_y}{U}\right)^2 \tag{21-4a}$$

and

$$C_p = -\frac{2\varphi_x}{U} + \alpha^2 - \left(\frac{dR}{dx}\right)^2 - \frac{1}{U^2}\left(U\alpha\sin\theta - \frac{1}{R}\varphi_\theta\right)^2 \tag{21-4b}$$

We have seen that the loading coefficient is a function only of φ_2. For the configuration shown in Fig. D,21, we can write φ_2 in the form

$$\varphi_2(x;\, y,\, z) = \left[\alpha\varphi_a(t,\, s,\, R;\, y,\, z) + \frac{\partial h}{\partial x}\,\varphi_b(t,\, s,\, R;\, y,\, z) + \omega\varphi_c(t,\, s,\, R;\, y,\, z)\right] \tag{21-5}$$

where $R(x)$ is the body radius, $\omega = dR/dx$ and $s(x)$ and $t(x)$ are the distances from the x axis to the leading and trailing edges, respectively. We can think of $\alpha\varphi_a$ as the potential due to angle of attack and the quantity $(\partial h/\partial x)\,\varphi_b + \omega\varphi_c$ as that due to thickness. Hence, φ_a has odd symmetry with reference to the $z = 0$ plane and both φ_b and φ_c have even symmetry. Placing Eq. 21-5 into Eq. 21-4a and using these symmetry properties, we find on the wing

$$\frac{\Delta p}{\frac{1}{2}\rho_\infty U^2} = \frac{2\alpha}{U}\left[\frac{ds}{dx}\,\Delta\,\frac{\partial\varphi_a}{\partial s} + \frac{dt}{dx}\,\Delta\,\frac{\partial\varphi_a}{\partial t} + \omega\Delta\,\frac{\partial\varphi_a}{\partial R} + \frac{\omega}{U}\,\Delta\left(\frac{\partial\varphi_c}{\partial y}\,\frac{\partial\varphi_a}{\partial y}\right) + \frac{1}{U}\,\frac{\partial h}{\partial x}\,\Delta\left(\frac{\partial\varphi_b}{\partial y}\,\frac{\partial\varphi_a}{\partial y}\right)\right] \tag{21-6}$$

The orders of the terms on the right-hand side of Eq. 21-5 are controlled by the coefficients of the potential terms so the gradients of φ_a, φ_b, and φ_c with respect to s, t, R, y, and z can all be considered as unity. Thus, neglecting α and $\partial h/\partial x$ but not ω in comparison to the leading and trailing edge slopes, ds/dx and dt/dx, Eq. 21-6 becomes

$$\left(\frac{\Delta p}{\frac{1}{2}\rho_\infty U^2}\right)_{\text{wing}} = \frac{2\alpha}{U}\left[\Delta\,\frac{\partial\varphi_a}{\partial x} + \frac{\omega}{U}\,\Delta\left(\frac{\partial\varphi_c}{\partial y}\,\frac{\partial\varphi_a}{\partial y}\right)\right] \tag{21-7}$$

A similar derivation for the loading coefficient on the body yields the expression

$$\left(\frac{\Delta p}{\frac{1}{2}\rho_\infty U^2}\right)_{\text{body}} = \frac{2\alpha}{U}\left[\Delta\,\frac{\partial\varphi_a}{\partial x} + \frac{\omega}{UR^2}\,\Delta\left(\frac{\partial\varphi_c}{\partial\theta}\,\frac{\partial\varphi_a}{\partial\theta}\right)\right] \tag{21-8}$$

where Δ indicates the difference between a quantity on vertically opposed points of the upper and lower surface of the airplane.

We notice from the latter two equations that in general the angle of attack and thickness solutions have a coupling effect on the loading coefficient and therefore their contribution to the load distribution cannot be treated separately. It is equally as important to notice, however, two special cases in which the coupling effects vanish, in fact, cases in which the thickness terms disappear altogether. These special configurations are a body of revolution without wings and an airplane with a cylindrical body between the foremost and rearmost extent of the wing. In the former case the term $\partial\varphi_c/\partial\theta$ is zero (see Eq. 16-5) and in the latter $\omega = dR/dx$ is zero. Hence, in both cases the equation for the loading coefficient becomes

$$\frac{\Delta p}{\frac{1}{2}\rho_\infty U^2} = \frac{2\alpha}{U} \Delta \frac{\partial \varphi_a}{\partial x} = \frac{2}{U} \Delta \frac{\partial \varphi_2}{\partial x} = \frac{4}{U} \left(\frac{\partial \varphi_2}{\partial x}\right)_{\text{upper}} \tag{21-9}$$

If one is interested in obtaining the total lift on the airplane, the loading coefficient can be integrated over the surface and the lift follows according to Eq. 14-1. A much simpler way of finding the lift, however, comes from applying the momentum balance given by Eq. 14-15. Let the surfaces over which Eq. 14-15 is applied be a cylinder of infinite radius and two y, z planes closing the cylinder and located infinitely far ahead of and behind the airplane. Then, the lift force is given by

$$L = F_z = \int_{-\infty}^{\infty} \int_{-\infty}^{\infty} [\rho w(U + u)]_{x=\infty}\, dydz$$

which reduces to

$$L = \rho_\infty U \int_{-\infty}^{\infty} \int_{-\infty}^{\infty} (w)_{x=\infty}\, dydz$$

This can be simplified since $w = \partial\varphi/\partial z$ and $(\Delta\varphi)_{x=\infty}$ is the same as the jump in potential evaluated at the airplane trailing edge. Thus

$$L = \rho_\infty U \int_{\text{span}} \Delta\varphi_{\text{t.e.}} dy \tag{21-10}$$

Eq. 21-10 applies to all airplane shapes and, since thickness effects (e.g. the terms φ_b and φ_c in Eq. 21-5) always have even symmetry with respect to the $z = 0$ plane, it shows that the total lift is affected only by the part of the potential that multiplies the angle of attack (e.g. the term φ_a in Eq. 21-5.) Hence, even though the detailed load distribution depends upon both thickness and angle-of-attack solutions, its integrated value, the total lift, depends only on the latter. Further, it should be noted that in general the spanwise vortex distribution cannot be calculated from the spanwise variation of loading.

D,22. Examples of Planar Problems. The triangular planforms studied in Art. 13 offer an opportunity to check the accuracy of the relations just derived and also to show the relative simplicity of the new methods. Consider first a triangular wing traveling at a supersonic speed and having a thickness distribution of the type given in Eq. 13-43:

$$h(x, y) = \frac{\tau_0}{2mc_0} \sqrt{m^2x^2 - y^2} \tag{22-1}$$

In this form, the ordinate of the upper surface is given in terms of m the tangent of the semiapex angle; c_0, the root chord of the wing; and τ_0, the maximum thickness of the wing at the root chord.

It follows from Eq. 22-1 that the elliptic section in the plane $x = x_1$ has major and minor semiaxes equal to mx and $\tau_0 x/2c_0$ respectively.

The cross-sectional area and the surface slope are, respectively

$$S(x_1) = \frac{\pi\tau_0 m x_1^2}{2c_0}, \qquad \frac{\partial h}{\partial x} = \frac{m\tau_0 x}{2c_0\sqrt{m^2x^2 - y^2}} \tag{22-2}$$

From Eq. 19-3, the perturbation potential can be written

$$\varphi(x, y, z) = \frac{1}{2\pi}\int_{-mx}^{mx} \Delta w(x, y_1) \ln \sqrt{(y - y_1)^2 + z^2}\, dy_1 - \frac{1}{2\pi}\frac{\partial}{\partial x}\int_0^x S'(x_1) \ln \frac{2(x - x_1)}{\beta}\, dx_1$$

since $\Delta\varphi$ over a nonlifting surface is zero. Further, since

$$\Delta w = 2w_u = 2U(\partial h/\partial x)$$

$$\frac{\varphi(x, y, z)}{U} = \frac{m\tau_0 x}{2\pi c_0}\int_{-mx}^{mx} \frac{\ln\sqrt{(y - y_1)^2 + z^2}}{\sqrt{m^2x^2 - y_1^2}}\, dy_1 - \frac{m\tau_0 x}{2c_0}\ln\frac{2}{\beta} - \frac{m\tau_0}{2c_0}\int_{-mx}^{mx} \ln(x - x_1)dx_1 \tag{22-3}$$

It is possible to carry out the above integrations in complete detail and thus calculate the entire flow field in the vicinity of the wing. We limit ourselves, however, to the two planes $z = 0$ and $y = 0$. For the $z = 0$ case, one has

$$\frac{\varphi(x, y, 0)}{U} = \frac{\tau_0 m x}{2c_0}\left(1 - \ln\frac{4}{m\beta}\right), \qquad |y| < mx$$

$$= \frac{\tau_0 m x}{2c_0}\left(1 - \ln\frac{4x}{\beta(|y| + \sqrt{y^2 - m^2x^2})}\right), \qquad |y| > mx \tag{22-4}$$

Since we are assuming that the wing surface deviates only slightly from a plane, τ_0/c_0 is neglected relative to m and the expression for pressure coefficient can be written

$$C_p = -\frac{2\varphi_x}{U} \tag{22-5}$$

It follows that pressure coefficient in the plane of the wing is

$$C_p(x, y, 0) = \frac{\tau_0 m}{c_0}\left(\ln\frac{4}{m\beta} - 1\right), \qquad |y| < mx$$

$$= \frac{\tau_0 m}{c_0}\left(\frac{|y|}{\sqrt{y^2 - m^2x^2}} + \ln\frac{4x}{\beta(|y| + \sqrt{y^2 - m^2x^2})} - 1\right), \qquad |y| > mx \tag{22-6}$$

The pressure distribution on the wing is uniform and off the wing has a square root singularity at the leading edge. The more stringent analysis of

Art. 13 yields the expression

$$C_p(x, y, 0) = \frac{\tau_0 m}{c_0} \frac{K - E}{\sqrt{1 - \beta^2 m^2}}, \qquad |y| < mx \tag{22-7}$$

for pressure coefficient where the complete elliptic integrals K and E have the modulus $\sqrt{1 - m^2\beta^2}$. Since, for values of the modulus near 1, the asymptotic expressions

$$K \cong \ln \frac{4}{\beta m}, \qquad E \cong 1 \tag{22-8}$$

apply, the pressure coefficient in slender wing theory is seen to be a first order approximation.

For the $y = 0$ case, one has

$$\frac{\varphi(x, 0, z)}{U} = \frac{\tau_0 m x}{2c_0}\left[1 + \ln \frac{\beta m^2 x}{4(\sqrt{z^2 + m^2x^2} - z)}\right], \qquad z > 0 \tag{22-9}$$

and

$$C_p(x, 0, z) = \frac{\tau_0 m}{c_0}\left[\frac{z}{\sqrt{z^2 + m^2x^2}} - 1 - \ln \frac{\beta m^2 x}{4(\sqrt{z^2 + m^2x^2} - z)}\right], \qquad z > 0 \tag{22-10}$$

These latter results are of interest since, together with the previous expressions, they can provide the velocity potential or pressure coefficient associated with a particular cruciform wing formed by a combination of the original thickness distribution in the $z = 0$ plane and the same thickness distribution in the $y = 0$ plane. The boundary conditions for the resulting cruciform wing are satisfied by a summation of the separate velocity potentials. It follows that, if z is replaced by y in Eq. 22-10 and this expression is combined with Eq. 22-6, the pressure coefficient on one panel of the cruciform wing is

$$C_p(x, y, 0) = \frac{\tau_0 m}{c_0}\left[\frac{y}{\sqrt{y^2 + m^2x^2}} - 2 - \ln \frac{\beta^2 m^3 x}{16(\sqrt{y^2 + m^2x^2} - y)}\right], \qquad 0 < y < mx \tag{22-11}$$

Other wings considered in Art. 13 are also amenable to slender airplane analysis. If the surface ordinate is

$$h(x, y) = \frac{\tau_0 x}{2c_0^2 m}\sqrt{m^2x^2 - y^2} \tag{22-12}$$

the lateral section is again elliptic so that cross-sectional area and surface slope are, respectively,

$$S(x) = \frac{\pi m \tau_0 x^3}{2c_0^2}, \qquad \frac{\partial h}{\partial x} = \frac{\tau_0}{2mc_0^2} \frac{2m^2x^2 - y^2}{\sqrt{m^2x^2 - y^2}}$$

Pressure coefficient on the wing is

$$C_p = \frac{m\tau_0 x}{c_0^2}\left(3 \ln \frac{4}{\beta m} - 4\right) \tag{22-13}$$

as opposed to the expression

$$C_p = \frac{m\tau_0 x}{c_0^2(1 - \beta^2 m^2)}\left[(3 - \beta^2 m^2)K - (4 - 2\beta^2 m^2)E\right]$$

which is not subject to the restrictions of slender airplane theory. Again the results are in agreement if higher order terms in βm are neglected.

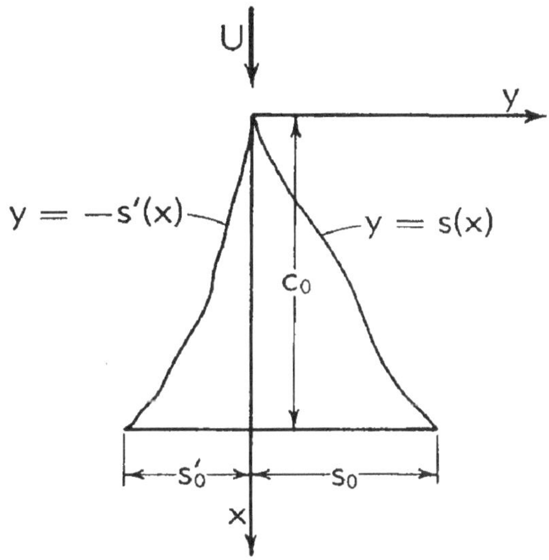

Fig. D,22a. Quasi-triangular wing.

It is possible to show that the pressure coefficient on a wing with thickness specified by the relation

$$h(x, y) = \frac{k}{m^2} y^2 \cosh^{-1} \frac{mx}{y}$$

is

$$C_p = 2kmx\left(\ln \frac{4}{\beta m} - 2\right).$$

Comparing this result with Eq. 22-13, we see that slender airplane theory retains the property of the more general linear theory (Art. 13) in that a given pressure distribution does not necessarily yield a unique thickness distribution.

Consider next a plate without thickness having a specified twist and angle of attack but with the restriction that its surface never deviates far from the $z = 0$ plane. Further, let the planform of this lifting surface be quasi-triangular in the sense that the right and left edges are not necessarily straight or symmetric about the x axis (see Fig. D,22a). In this case, since the configuration is without thickness, Eq. 19-3 reduces to[27]

$$\varphi(x, y, z) = \frac{z}{2\pi} \int_{-s'}^{s} \frac{\Delta\varphi(x, y_1)dy_1}{(y - y_1)^2 + z^2} = \frac{z}{\pi} \int_{-s'}^{s} \frac{\varphi_u(x, y_1)dy_1}{(y - y_1)^2 + z^2} \quad (22\text{-}14)$$

Since φ must be a continuous function everywhere except on the lifting surface and its vortex wake, $\varphi_u(x, s)$ and $\varphi_u(x, -s')$ are zero and an integration by parts gives

$$\varphi(x, y, z) = \frac{1}{\pi} \int_{-s'}^{s} v_u(x, y_1) \tan^{-1} \frac{y - y_1}{z} dy_1$$

where v_u is the perturbation velocity in the y direction on the upper surface of the wing. Since, moreover, s', s, and v_u are not functions of z, a differentiation with respect to z leads to the result

$$w(x, y, z) = -\frac{1}{\pi} \int_{-s'}^{s} \frac{(y - y_1)v_u(x, y_1)dy_1}{(y - y_1)^2 + z^2}$$

which, as z goes to zero, reduces to the final form

$$w(x, y, 0) = w_u(x, y) = -\frac{1}{\pi} \int_{-s'}^{s} \frac{v_u(x, y_1)dy_1}{y - y_1} \quad (22\text{-}15)$$

The particular type of configuration we are studying is a plate with a given twist and angle of attack, that is, a given value of $w_u(x, y)$ in the interval $-s' \leqq y \leqq s$. Applied to such problems, Eq. 22-15 is an integral equation that must be solved for v_u. We notice immediately, however, that this integral equation is identical to the one studied in Art. 12 as Eq. 12-21. Therefore, its inversion is given by Eq. 12-24, and since

$$\int_{-s'}^{s} v_u dy = \varphi_u(x, s) - \varphi_u(x, -s') = 0$$

its solution is unique and can be written

$$v_u(x, y) = \frac{1}{\pi\sqrt{(y + s')(s - y)}} \int_{-s'}^{s} \frac{w_u(x, y_1)}{y - y_1} \sqrt{(y_1 + s')(s - y_1)}\, dy_1 \quad (22\text{-}16)$$

Using the relation $\varphi_u = \int_s^y v_u dy$ and integrating by parts, this becomes

$$\varphi_u(x, y) = -\frac{2}{\pi} \int_{-s'}^{s} w_u(x, y_1) \ln \frac{\sqrt{(s - y_1)(s' + y)} + \sqrt{(s' + y_1)(s - y)}}{\sqrt{|y - y_1|(s' + s)}} dy_1 \quad (22\text{-}17)$$

[27] Eq. 22-14 is valid provided no point exists within or on the boundaries of the flow field such that

$$\lim_{\epsilon \to 0} \epsilon \frac{\partial \varphi}{\partial \epsilon} \neq 0$$

where ϵ is the radial distance from the point.

Again integrating by parts and adopting the notation

$$\nu(x, y_1) + \text{const} = \int w_u(x, y_1)dy_1 \tag{22-18}$$

one can reduce Eq. 22-17 to the form

$$\varphi_u(x, y) = -\frac{\sqrt{(s-y)(s'+y)}}{\pi}\int_{-s'}^{s}\frac{\nu(x, y_1)dy_1}{(y-y_1)\sqrt{(s-y_1)(s'+y_1)}} \tag{22-19}$$

where the constant in Eq. 22-18 contributes zero to Eq. 22-19. Since the total chordwise circulation, $\Gamma(y)$, is equal to $2\varphi_u$ evaluated at the trailing edge, Eq. 22-19 gives the span loading for a wing having a planform such as that shown in Fig. D,22a and an arbitrary twist and camber. The total lift on the wing can be computed from Eq. 21-10, hence

$$\frac{L}{\frac{1}{2}\rho_\infty U^2} = -\frac{2}{U}\int_{-s_0'}^{s_0}\frac{(s_0 - s_0' - 2y_1)\nu(c_0, y_1)}{\sqrt{(s_0-y_1)(s_0'+y_1)}}\,dy_1$$

Finally, this reduces to

$$\frac{L}{\frac{1}{2}\rho_\infty U^2} = -4\int_{-s_0'}^{s_0}\frac{w_u(c_0, y_1)}{U}\sqrt{(s_0-y_1)(s_0'+y_1)}\,dy_1 \tag{22-20}$$

If $w_u(c_0, y)$ is independent of y, we have

$$\frac{L}{\frac{1}{2}\rho_\infty U^2} = \frac{\pi\alpha_1(s_0+s_0')^2}{2} \tag{22-21}$$

where $\alpha_1 = -w_u(c_0)/U$.

The conclusion can be drawn from Eq. 22-20 that the total lift on any quasi-triangular wing is, by slender wing theory, a function only of the span and the streamwise slope of the trailing edge and is independent[28] of the shape of the leading edges and of the twist and camber of the rest of the wing. It is also apparent that Eq. 22-20 can be used to evaluate flap effectiveness provided the flaps are such that they do not violate the assumptions of the theory. For example, the total lift generated by the flaps shown in Fig. D,22b would be

$$-\frac{L}{\frac{1}{2}\rho_\infty U^2} = 4\int_{-s_0'}^{-e_0'}\frac{w_u(c_0, y_1)}{U}\sqrt{(s_0-y_1)(s_0'+y_1)}\,dy_1 + 4\int_{e_0}^{s_0}\frac{w_u(c_0, y_1)}{U}\sqrt{(s_0-y_1)(s_0'+y_1)}\,dy_1$$

The above equations simplify considerably when the condition is

[28] In visualizing this property one must not lose sight of the basic assumptions upon which the theory is founded. For example, it would not be consistent with these assumptions to consider a wing with a narrow deflected full span flap along the trailing edge since such a flap would induce forces similar to those on a high aspect ratio rectangular wing for which slender wing theory is not suited.

imposed that the wing is a flat plate. For such a case where the quantity $-w_u(c_0, y_1)/U$ is equal to the angle of attack, α, which is a constant, Eq. 22-19 becomes simply

$$\varphi_u(x, y) = \alpha U \sqrt{(s - y)(s' + y)} \tag{22-22}$$

and the loading coefficient is

$$\frac{\Delta p}{\frac{1}{2}\rho_\infty U^2} = 2\alpha \frac{(s' + y)\frac{ds}{dx} + (s - y)\frac{ds'}{dx}}{\sqrt{(s - y)(s' + y)}} \tag{22-23}$$

Eq. 22-23 reveals the important roles played by the shape of the side edges on the magnitude of the loading along a given lateral strip. Thus,

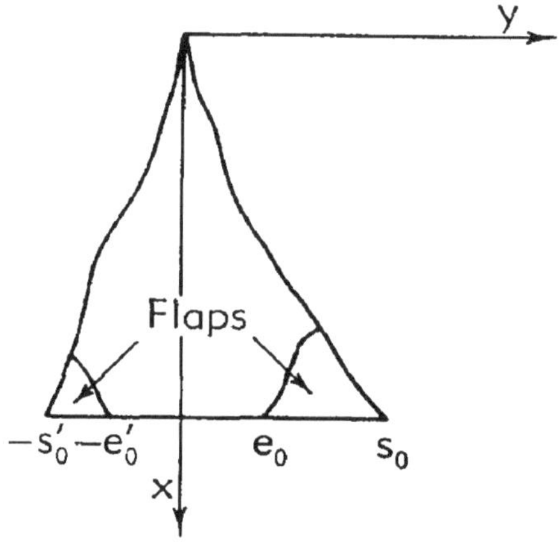

Fig. D,22b. Quasi-triangular wing-flap combination.

for portions of the wing bounded laterally by sides parallel to the free stream, ds/dx and ds'/dx vanish and the loading on such a region (see g. D,22c (left)) is everywhere zero. This result can be extended to ngs such as those shown in Fig. D,22c (middle) which have trailing edges of arbitrary shape located entirely behind the line of maximum span. It is apparent from a study of Eq. 22-15 that the vortex sheet (v_u is a measure of the vorticity) which trails behind the wing shown in Fig. D,22c (right) will induce the same value of w_u everywhere behind the wing that it did right at the trailing edge. Hence, for a flat plate the stream angle of the vortex wake is just α, the angle of attack of the wing.[29] To insert any flat plate also at an angle of attack α (e.g. to extend the wing behind the line of maximum span as in Fig. D,22c (middle)) is to insert a surface along a stream plane. Such a surface in no way disturbs the flow (the fluid is assumed, of course, to be frictionless) and experiences no loading. It is a result of slender wing theory, therefore, that the loading

[29] This assumes, of course, that the vortex wake has not "rolled up" enough to affect the results. A detailed treatment of the vortex behavior behind low aspect ratio wings is to be found in [58].

on a flat plate is zero at all points behind the line of maximum span, provided the trailing edge is entirely behind this line.

The loading on a triangular wing with semivertex angle equal to $\tan^{-1} m$ is given by Eq. 22-23 in the form

$$\frac{\Delta p}{\frac{1}{2}\rho_\infty U^2} = \frac{4\alpha m^2 x}{\sqrt{m^2x^2 - y^2}} \tag{22-24}$$

This equation differs from the one derived by means of exact linearized theory for the same wing at a supersonic Mach number only by the

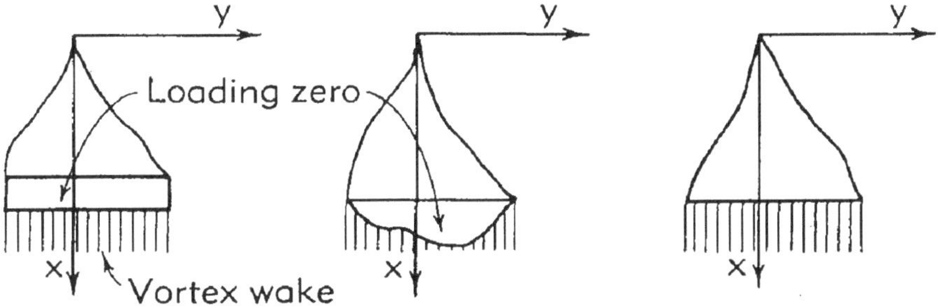

Fig. D,22c. Quasi-triangular wing with various side and trailing edge combinations.

factor $1/E$ (see Eq. 13-32). For small values of βm, $1/E$ is equal to

$$1 - \frac{1}{2}\beta^2 m^2 \left(\ln \frac{4}{\beta m} - \frac{1}{2}\right)$$

plus higher order terms. Hence, it can be seen that as m approaches zero (i.e. the body becomes slender) or as β approaches zero (i.e. the free stream Mach number nears one) the two theories coincide.

The span loading for the flat-plate quasi-triangular wing follows from Eq. 22-22, and written in terms of Γ is

$$\Gamma = 2\alpha U \sqrt{(s_0 - y)(s_0' + y)} \tag{22-25}$$

The total lift is

$$\frac{L}{\frac{1}{2}\rho_\infty U^2} = \frac{\alpha b^2 \pi}{2}$$

where b is the wing span equal to $s_0 + s_0'$.

It is clear from this equation that the total lift on a flat plate is a function only of the angle of attack, free stream dynamic pressure, and wing span. Expressed as a lift coefficient and in terms of the aspect ratio Æ this becomes

$$C_L = \frac{\pi}{2}\alpha Æ$$

The vortex drag can readily be determined for the flat quasi-triangular wing by means of Eq. 14-36, thus

$$D_V = -\tfrac{1}{2}\rho_\infty \int_{-s_0'}^{s_0} \Gamma w_{x=\infty}\, dy$$

As has already been pointed out, $w_{x=\infty}$, the upwash component in the Trefftz plane, is equal to $-U\alpha$ and Γ is given by Eq. 22-25. Combining and integrating we find

$$C_{D_V} = \frac{1}{2}\,\alpha C_L = \frac{C_L^2}{\pi \mathcal{R}}$$

The form of the expression on the right is known to apply to all span loadings of elliptic form.

D,23. Examples of Interference Problems. Although it is obvious from the previous discussion that slender airplane theory can be used

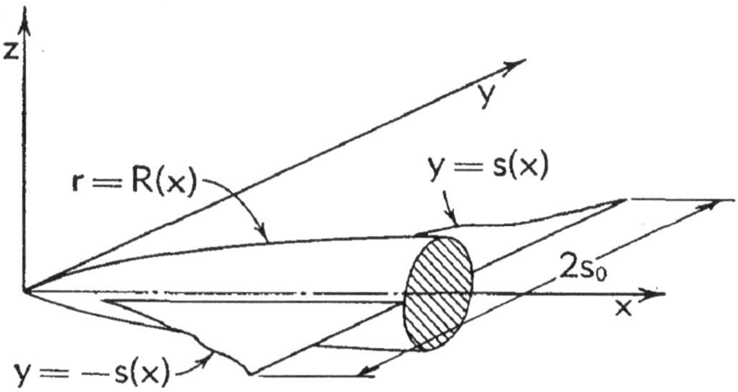

Fig. D,23a. Flat plate wing mounted on a body of revolution.

to calculate the effects of quite general airplane configurations on the pressure as well as the loading coefficient, we will, for simplicity, limit our further analysis to a development of interference effects on loading only. With such a limitation, we can neglect the term $A(x)$ in Eq. 21-2 and concern ourselves only with the evaluation of $\varphi_2(x; y, z)$. In terms of the total velocity potential[30]

$$\phi_2(x; y, z) = U\alpha z + \varphi_2(x; y, z) \tag{23-1}$$

this means we must satisfy the equation[31]

$$\phi_{yy} + \phi_{zz} = 0 \tag{23-2}$$

subject to the appropriate boundary conditions.

As a specific example, let us consider the combination of a flat plate mounted on a body of revolution as shown in Fig. D,23a. If the free

[30] In the following we will make use of the total velocity potential and the total velocity components

$$V = \frac{\partial \phi}{\partial y} = \frac{\partial \varphi}{\partial y}$$

$$W = \frac{\partial \phi}{\partial z} = U\alpha + \frac{\partial \varphi}{\partial z}$$

[31] For convenience we discard the subscript 2 for the remainder of this article.

stream makes an angle α with the x axis, the boundary conditions for such a configuration can be written

$$\begin{aligned} &\text{(i)} && V_r = 0; && r = R(x), \quad -\pi < \theta < \pi \\ &\text{(ii)} && W = 0; && z = 0, \quad R^2 \leqq y^2 \leqq s^2 \\ &\text{(iii)} && V = 0,\ W = U\alpha; && r = \infty, \quad -\pi < \theta < \pi \end{aligned} \tag{23-3}$$

The analysis of these and similar boundary conditions is greatly simplified by using the techniques associated with the study of the com-

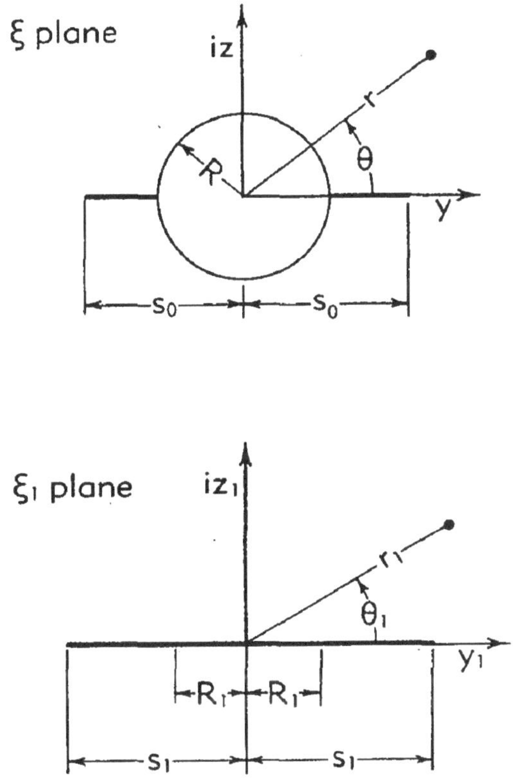

Fig. D,23b. Quasi-triangular wing and body boundaries in complex ξ and ξ_1 planes.

plex variable. For example, it is well known that any analytic function of a complex variable is a solution to Eq. 23-2. Further, if we set parameter $\xi = y + iz = re^{i\theta}$ and $\xi_1 = y_1 + iz_1 = r_1e^{i\theta_1}$ then it is also well known that the Joukowski transformation

$$\xi_1 = \xi + \frac{R^2(x)}{\xi} \tag{23-4}$$

maps the circle $r = R(x)$ in the ξ plane (see Fig. D,23b) onto a portion of the y_1 axis in the ξ_1 plane and maps the portion of the y axis lying outside of this circle in the ξ plane[32] onto the rest of the y_1 axis. The method of attack will be to transform the boundary conditions from the ξ to the ξ_1 plane and obtain the solution in the latter plane; then we will transform this solution back to the ξ, or physical, plane for the completion of the problem.

Before proceeding with our analysis, it is convenient at this point to consider a few of the more useful transformations the application of which is necessary in the following: thus, as a consequence of Eq. 23-4 we have

$$y = \frac{y_1 + \sqrt{y_1^2 - R_1^2}}{2}, \qquad y_1 \geqq R_1, \qquad z_1 = 0$$

$$y = \frac{y_1 - \sqrt{y_1^2 - R_1^2}}{2}, \qquad y_1 \leqq -R_1, \qquad z_1 = 0 \tag{23-5}$$

and for $z = 0$

$$y_1 = y + \frac{R^2}{y}, \qquad s_1 = s + \frac{R^2}{s}, \qquad R_1 = 2R \tag{23-6}$$

From the basic theory underlying the use of complex variables, induced velocities in the two planes are related by the expression

$$V - iW = (V_1 - iW_1)\frac{d\xi_1}{d\xi} \tag{23-7}$$

from which, since in polar coordinates

$$\frac{d\xi_1}{d\xi} = \left[1 - \left(\frac{R}{r}\right)^2 \cos 2\theta\right] + i\left(\frac{R}{r}\right)^2 \sin 2\theta \tag{23-8}$$

it follows that

$$V = V_1\left[1 - \left(\frac{R}{r}\right)^2 \cos 2\theta\right] + W_1\left(\frac{R}{r}\right)^2 \sin 2\theta$$

$$W = W_1\left[1 - \left(\frac{R}{r}\right)^2 \cos 2\theta\right] - V_1\left(\frac{R}{r}\right)^2 \sin 2\theta \tag{23-9}$$

$$V_r = [V_1 \cos\theta + W_1 \sin\theta]\left[1 - \left(\frac{R}{r}\right)^2\right] + 2W_1\left(\frac{R}{r}\right)^2 \sin\theta$$

Finally, if the values of V_1 and W_1 are known only along certain portions of the y_1 axis, their values throughout the ξ_1 plane can be determined by applying the principles of analytic continuation. Thus, if

$$f(\xi) = V(y, z) - iW(y, z) \tag{23-10a}$$

then

$$f(\xi) = V(\xi, 0) - iW(\xi, 0) \tag{23-10b}$$

[32] The transformation is double-valued, the regions outside and inside the circle $r = R(x)$ both mapping onto the entire plane. We are concerned here only with the outside region.

Returning now to the boundary-value problem specified for the ξ plane by Eq. 23-3, we see that in the ξ_1 plane this transforms to

$$\begin{aligned} &\text{(i)} && W_1 = 0, && 0 \leqq y_1^2 \leqq R_1^2 \\ &\text{(ii)} && W_1 = 0, && R_1^2 \leqq y_1^2 \leqq s_1^2 \\ &\text{(iii)} && V_1 = 0,\ W_1 = U\alpha, && r_1 = \infty,\ -\pi < \theta_1 < \pi \end{aligned} \tag{23-11}$$

Finally these boundary conditions can be expressed in terms of the perturbation velocity components in the ξ_1 plane. Defining the perturbation velocity potential φ_1 by the equation

$$\phi_1(x;\, y_1,\, z_1) = U\alpha z_1 + \varphi_1(x;\, y_1,\, z_1) \tag{23-12}$$

we can reexpress Eq. 23-11 as

$$\begin{aligned} &\text{(i)} && w_1 = -U\alpha, && 0 \leqq y_1^2 \leqq R_1^2 \\ &\text{(ii)} && w_1 = -U\alpha, && R_1^2 \leqq y_1^2 \leqq s_1^2 \\ &\text{(iii)} && w_1 = v_1 = 0, && r_1 = \infty,\ -\pi < \theta_1 < \pi \end{aligned} \tag{23-13}$$

Eq. 23-13, however, represent precisely the same boundary conditions which have already been treated in the preceding article for the case of

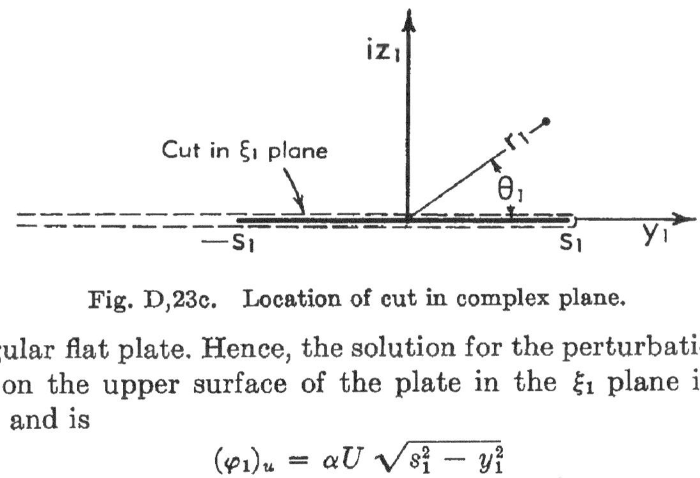

Fig. D,23c. Location of cut in complex plane.

the triangular flat plate. Hence, the solution for the perturbation velocity potential on the upper surface of the plate in the ξ_1 plane is given by Eq. 22-22 and is

$$(\varphi_1)_u = \alpha U \sqrt{s_1^2 - y_1^2} \tag{23-14}$$

Since $(\varphi_1)_u$ in ξ_1 space lies entirely in the $z_1 = 0$ plane, $(\varphi_1)_u = (\phi_1)_{z_1=0}$ and the total velocity components in the $z_1 = 0$ plane between $\pm s_1$ are

$$V_1(y_1,\, 0) = \frac{-\alpha U y_1}{\sqrt{s_1^2 - y_1^2}}, \qquad W_1(y_1,\, 0) = 0$$

By means of Eq. 23-10, these can be extended throughout space, thus

$$V_1(y_1,\, z_1) - iW_1(y_1,\, z_1) = -\alpha U \frac{\xi_1}{\sqrt{s_1^2 - \xi_1^2}} \tag{23-15}$$

If the ξ_1 space is cut as shown in Fig. D,23c so that θ_1 varies between $\pm\pi$, one can see that Eq. 23-15 satisfies the boundary conditions given by Eq. 23-11.

By application of Eq. 23-7, the solution given as Eq. 23-15 can be transformed to ξ space and solutions for any configuration represented by Fig. D,23a can be obtained. It is not difficult to show, for example, that on the wing

$$\Delta\phi = 2\alpha U \sqrt{\left(s + \frac{R^2}{s}\right)^2 - \left(y + \frac{R^2}{y}\right)^2}, \qquad R^2 \leqq y \leqq s^2 \quad (23\text{-}16a)$$

and on the body[33]

$$\Delta\phi = 2\alpha U \sqrt{\left(s + \frac{R^2}{s}\right)^2 - 4R^2 \cos\theta}, \qquad -\pi < \theta < 0 \quad (23\text{-}16b)$$

According to Eq. 21-10, therefore, the total lift for a wing-body combination such as that shown in Fig. D,23a is

$$\frac{L}{\frac{1}{2}\rho_\infty U^2 \alpha} = 2\pi \frac{s_0^4 - s_0^2 R_0^2 + R_0^4}{s_0^2} \quad (23\text{-}17)$$

If we limit ourselves to the study of wings mounted on circular cylinders as shown in Fig. D,23d, then the loading can be calculated directly from Eq. 23-16. For the wing there results

$$\frac{\Delta p}{\frac{1}{2}\rho_\infty U^2} = 4\alpha \frac{ds}{dx} \frac{\left[1 - \left(\frac{R_0}{s}\right)^2\right]\left[s + \frac{R_0^2}{s}\right]}{\left[\left(s + \frac{R_0^2}{s}\right)^2 - \left(y + \frac{R_0^2}{y}\right)^2\right]^{\frac{1}{2}}}, \qquad R_0^2 \leqq y^2 \leqq s^2 \quad (23\text{-}18a)$$

and for the body

$$\frac{\Delta p}{\frac{1}{2}\rho_\infty U^2} = 4\alpha \frac{ds}{dx} \frac{\left[1 - \left(\frac{R_0}{s}\right)^2\right]\left[s + \frac{R_0^2}{s}\right]}{\left[\left(s + \frac{R_0^2}{s}\right)^2 - 4y^2\right]^{\frac{1}{2}}}, \qquad 0 \leqq y^2 \leqq R_0^2 \quad (23\text{-}18b)$$

The variation of the loading coefficient along a section of a quasi-triangular wing mounted on a cylinder for which $(R_0/s)^2 = 0.1$ is shown in Fig. D,23d (bottom). The loading over the same wing without a body is also shown for the purpose of comparison.

The lift on the wing and body can be determined separately. The lift on the wing is

$$\frac{L_w}{\frac{1}{2}\rho_\infty U^2 \alpha} = \frac{2\pi}{s_0^2}(s_0^4 + R_0^4) - 2\left(\frac{s_0^2 + R_0^2}{s_0}\right)^2 \sin^{-1}\frac{2R_0 s_0}{s_0^2 + R_0^2} - 4\frac{R_0}{s_0}(s_0^2 - R_0^2) \quad (23\text{-}19)$$

[33] It should be noted that Eq. 23-16b cannot be used to calculate the loading on the body if the body radius is a function of x since for such cases $\partial\Delta\varphi/\partial x \neq \Delta\partial\varphi/\partial x$

The lift on the body is separated into two parts, that developed on the nose and that developed behind the juncture of the body and the wing leading edge. (It should be remembered that the portion of the body on which the wing is mounted is everywhere a circular cylinder.) The lift

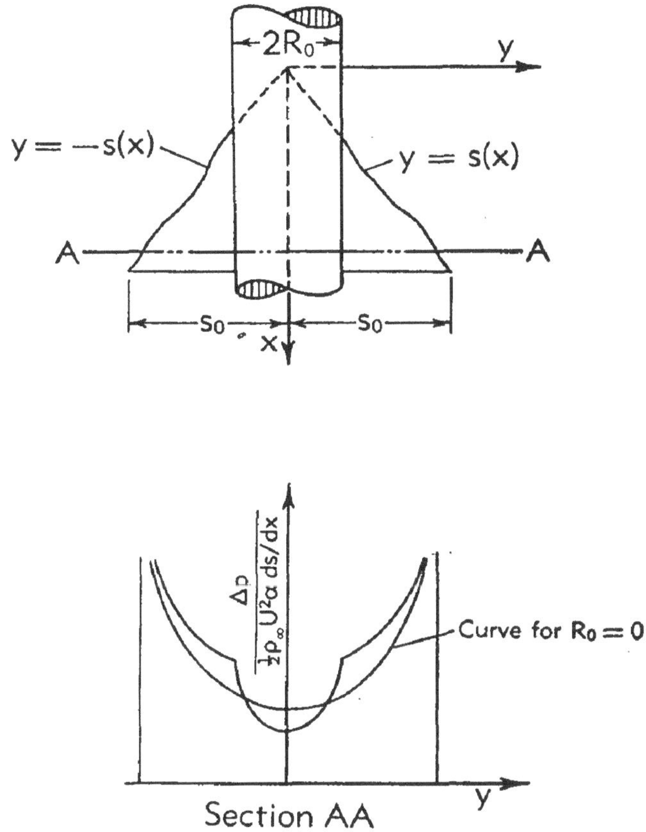

Fig. D,23d. Quasi-triangular wing mounted on cylinder and its load distribution due to angle of attack.

on the nose of the body has already been evaluated in Art. 17; hence, according to Eq. 17-1,

$$\frac{(L_b)_n}{\frac{1}{2}\rho_\infty U^2\alpha} = 2\pi R_0^2 \tag{23-20}$$

The lift on the remaining cylindrical portion of the body is

$$\frac{(L_b)_c}{\frac{1}{2}\rho_\infty U^2\alpha} = 4\frac{R_0}{s_0}(s_0^2 - R_0^2) + 2\left(\frac{s_0^2 + R_0^2}{s_0}\right)^2 \sin^{-1}\frac{2R_0s_0}{R_0^2 + s_0^2} - 4\pi R_0^2 \tag{23-21}$$

Obviously, the sum of these three terms equals the total lift as given by Eq. 23-17.

Designate the lift on a wing with no body by L_0 and the lift on a wing-body combination (the wing planform being in both cases the same) by

L. The ratio L/L_0 gives an indication of the interference effects. Such a ratio is shown in Fig. D,23e for various values of R_0/s_0.

Finally, let us consider briefly the airplane configuration shown in Fig. D,23f, consisting of a sweptback flat plate wing mounted on a circular cylinder. The analysis of this problem proceeds in a manner similar to that used in the study of the triangular wing and body. In fact, the solution over section AA, any section ahead of the trailing edge-fuselage juncture (see Fig. D,23f), is identical to that derived in the

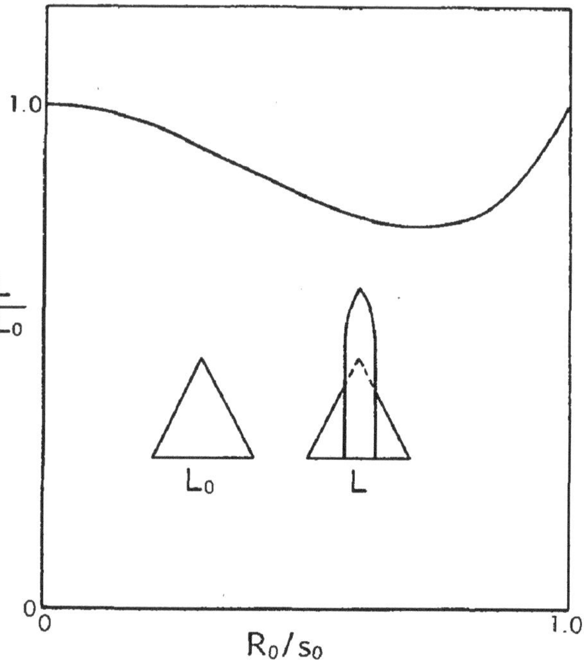

Fig. D,23e. The ratio of the lift on a wing-body combination to that on a wing alone as a function of body radius to wing semispan ratio.

quasi-triangular case. Solutions behind this point, such as those along section BB, are altered by the presence of the gap between the wing and the body. In the transverse y, z space, section BB presents a configuration like the one shown in Fig. D,23g. The boundary conditions in this plane are similar to those given in Eq. 23-3 except, first, (ii) is replaced by the condition that W shall be zero along the portion of the y axis for which $t^2 \leqq y^2 \leqq s^2$, and second, the vorticity in the region $R_0^2 < y^2 < t^2$ shall be consistent with the circulation on the section of the wing ahead of it.

The Joukowski transformation can be used again to map the wing panels and body in the ξ plane onto the y_1 axis in the ξ_1 plane. As shown in Fig. D,23g, the result in this case forms along the y_1 axis three distinct slits on which the boundary conditions must be satisfied. These conditions are readily obtained from those given for the ξ plane; and just as Eq. 23-13 followed from Eq. 23-3, so does it follow for the combination

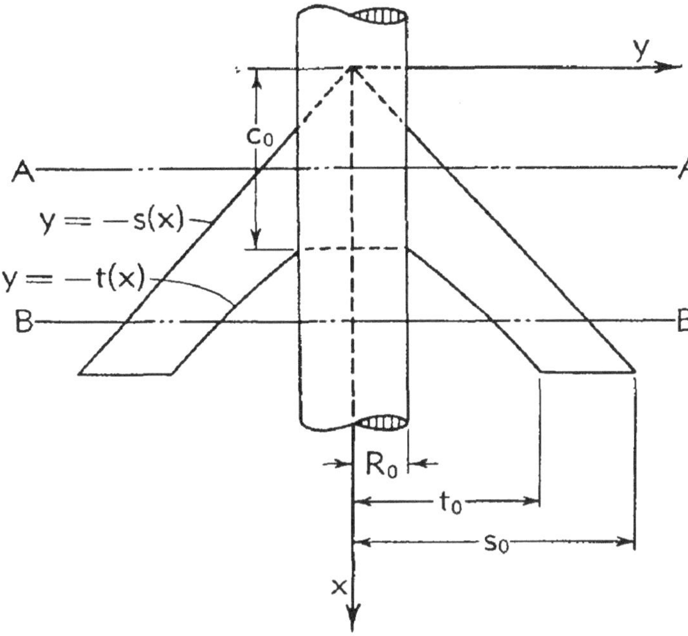

FIG. D,23f. Sweptback wing mounted on a cylinder.

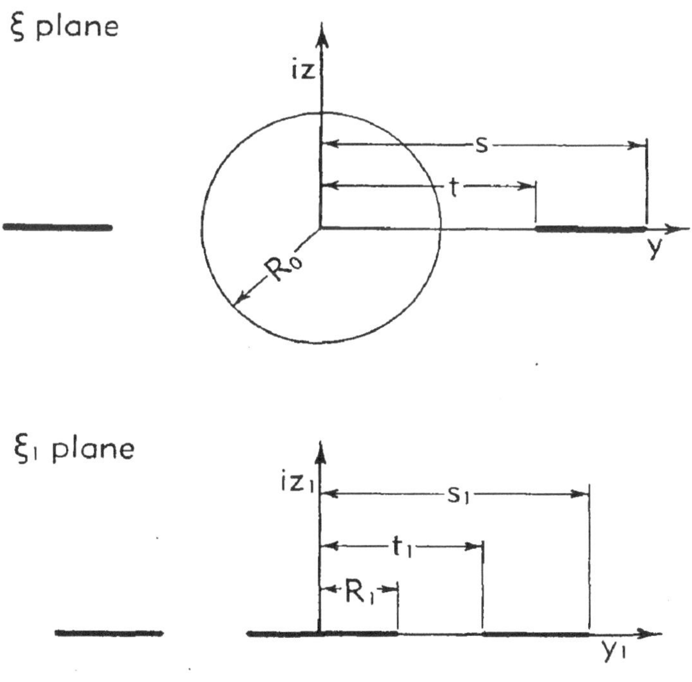

Fig. D,23g. Sweptback wing and body boundaries in the complex ξ and ξ_1 planes.

shown in Fig. D,23g that

$$
\begin{array}{llll}
\text{(i)} & w_1 = -U\alpha, & 0 \leqq y_1^2 \leqq R_1^2 & \\
\text{(ii)} & w_1 = -U\alpha, & t_1^2 \leqq y_1^2 \leqq s_1^2 & \quad (23\text{-}22) \\
\text{(iii)} & w_1 = v_1 = 0, & r_1 = \infty,\ -\pi < \theta_1 < \pi &
\end{array}
$$

The question of the vorticity in the region between the body and the fuselage can be treated from two different viewpoints. The more direct

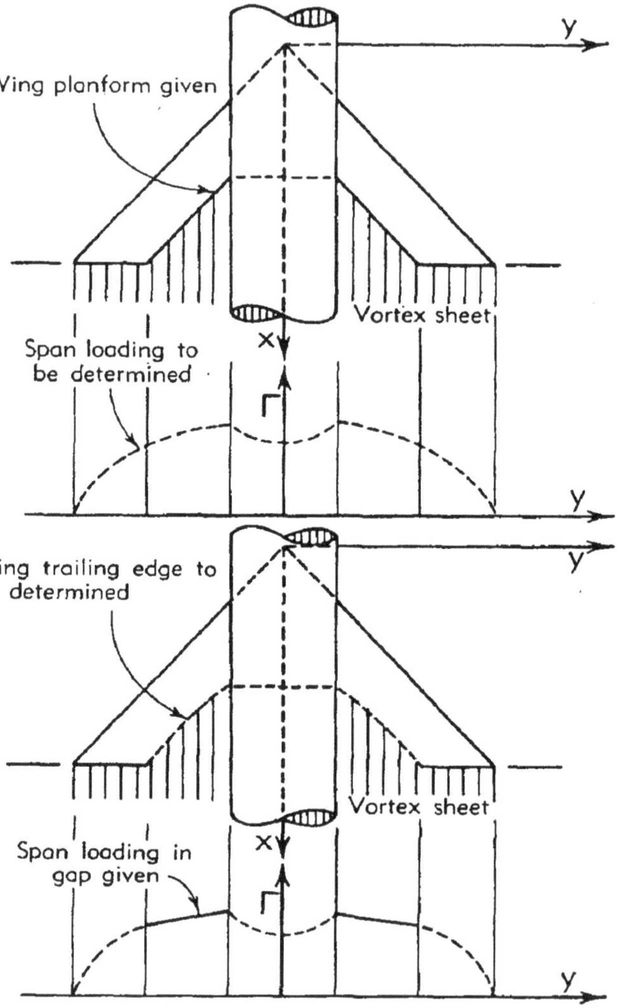

Fig. D,23h. Two different methods by means of which sweptback wings can be analyzed.

of these is that in which the wing planform is specified, as, for example, the one shown in Fig. D,23h (top). In this case the span load variation, $\Gamma(y)$, is to be determined and the gradient of this span-loading curve $d\Gamma/dy$ provides a measure of the vorticity streaming back from the trailing edge and filling the gap between the body and the wing with a vortex sheet. Such a sheet passes through every transverse plane such as the one shown in Fig. D,23g and must be regarded in the analysis. A second and less direct viewpoint stems from the specification of the span load distribution in the gap (see Fig. D,23h, bottom). Corresponding to this distribution there is a certain planform, or, assuming a straight leading edge, a certain trailing edge shape. For example, one can assume that Γ

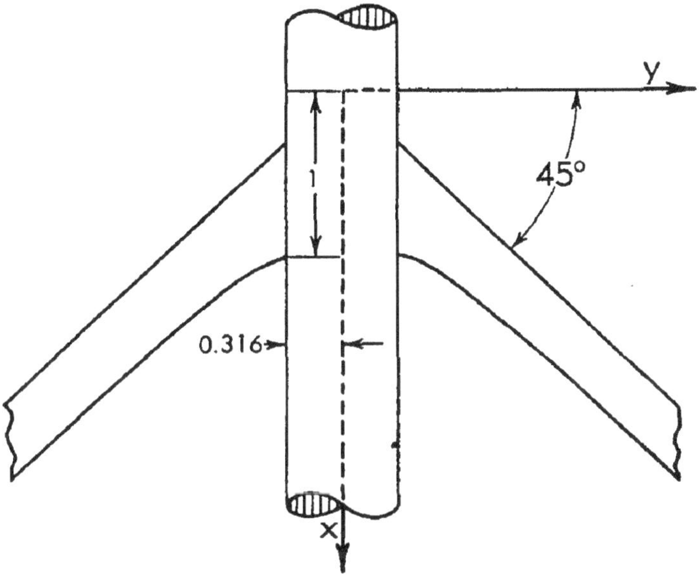

Fig. D,23i. Sweptback nearly-constant-chord wing mounted on a cylinder.

in the gap is given by a power series in y_1; thus

$$\Gamma(y_1) = \sum_{n=0}^{m} b_n y_1^n; \qquad R_1^2 < y_1^2 < t_1^2$$

The resulting trailing edge shape will then be determined in terms of the coefficients b_n and, by adjusting these coefficients, a variety of trailing edge shapes can be obtained.

Reference [*59*] contains a study of the above problem for the case in which the span loading for $R_0^2 < y^2 < t^2$ is a constant, or, in other words, when there is no vorticity between the body and the wing trailing edge for $y^2 < t_0^2$. The trailing edge shape which produces such a flow for $R_0/mc_0 = 0.316$ is shown in Fig. D,23i, in which it can be seen that the chord is nearly constant except in the vicinity of the root. The solution for the total transverse velocity components for sections such as BB in Fig. D,23f is

$$V_1(y_1, z_1) - iW_1(y_1, z_1) = U\alpha \sqrt{\frac{\xi_1^2(\xi_1^2 - t_1^2)}{(s_1^2 - \xi_1^2)(\xi_1^2 - R_1^2)}} \qquad (23\text{-}23)$$

where the branch points are avoided by cutting the ξ_1 plane as shown in Fig. D,23c. The equation for the total lift on the entire configuration is

$$\frac{L}{\frac{1}{2}\rho_\infty U^2 \alpha} = 2\pi \left(\frac{s_0^4 + R_0^4}{s_0^2} - \frac{t_0^4 + R_0^4}{t_0^2} + R_0^2 \right) \tag{23-24}$$

Various interference effects determined from Eq. 23-24 are shown in Fig. D,23j.

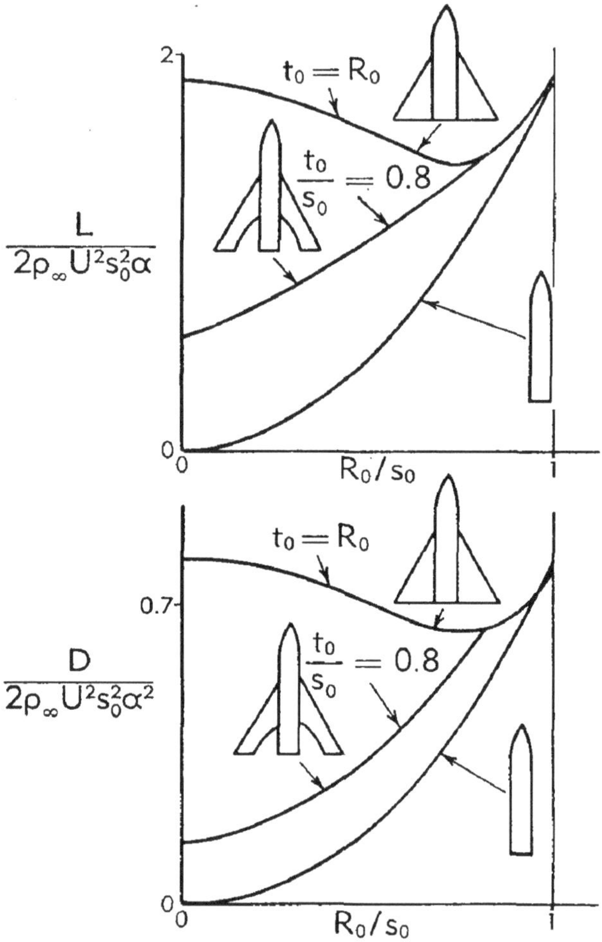

Fig. D,23j. Values of lift and drag on various types of wing-body combinations.

D,24. Concluding Remarks. The examples presented in this chapter have been chosen for the purpose of showing the versatility of slender wing analysis in application to the wing and wing-body shapes of current interest in missile and high speed aircraft design. The theory appears to offer attractive possibilities when used semiempirically. For example, the predicted lift of a wing-body combination, as determined from Fig. D, 23e, should be improved when used in conjunction with experimental or more accurately calculated values of wing loading. Furthermore, the original analysis of sweptback wings can be refined, from an engineering standpoint, by introducing in a judicious manner Mach number corrections determined for simpler configurations by more exact theory. The development of such ideas is, of course, outside the range of our presentation.

Space limitations have prevented discussion of cruciform-body and multi-wing-body combinations as well as lifting wing-tail-body combinations. In [*59*] the latter problem has been treated for the extreme cases in which the vortex wake from the wing remains a flat sheet, as originally predicted by the theory, and when the vortex sheet has rolled up into its asymptotic form [*58*] by the time it crosses the tail plane. A final field of application, as yet incompletely developed, lies in the study of lateral edge conditions and the possible modification of the Kutta condition at the side edges of wings.

CHAPTER 6. UNSTEADY LIFT

D,25. Basic Equations and Wing Boundary-Value Problems. Art. 25-29 are concerned with the study of the forces and moments on thin wings undergoing prescribed motions in time and space. It will be possible to refer only to certain general methods and give indications of the more important results; for further details and applications see VII,F and VIII,C.

As we have seen in Art. 2, the governing partial differential equation with reference to a system of axes fixed relative to the still air at infinity is the three-dimensional wave equation. Thus for the perturbation velocity potential we can write

$$\varphi_{xx} + \varphi_{yy} + \varphi_{zz} = \varphi_{tt} \tag{25-1}$$

where the time variable t' has been replaced by the variable t/a_∞, where a_∞ is the speed of sound in the undisturbed air. From the nature of Eq. 25-1 it is apparent that the wing is, in effect, a source of acoustic disturbances and that the flow field can be simulated by an appropriate superposition of spherical sound waves.

In the problems to be studied, the wings are considered to deviate only slightly from the $z = 0$ plane. Further, these wings are divided into two classes: plates having a certain twist and angle of attack, and wings with vertically symmetrical thickness. These idealizations are usual in linearized theory and they imply that the boundary conditions for the wing can be specified in terms of the induced velocities (or their jump) in the $z = 0$ plane. Hence, for a wing of given planform the boundary values are one of the four sets outlined at the beginnings of Art. 11 and 12, with the added stipulation that the appropriate set of conditions concerning the wing apply only over the part of the $z = 0$ plane occupied by the wing and its vortex wake at any given time.

Given one of the above boundary conditions, the immediate problem is to find the upwash in the plane of the wing or the value of the pressure coefficient there. Since we are limiting our discussion to planar problems, the pressure coefficient can be written (see Eq. 2-6)

$$C_p = \frac{-2}{U_0 M_0} \frac{\partial}{\partial t} \varphi(x, y, z, t) \tag{25-2a}$$

where U_0 is the reference speed for C_p, say the speed of flight of the wing, and M_0 denotes U_0/a_∞. Similarly, the load coefficient is

$$\frac{\Delta p}{\frac{1}{2}\rho_\infty U_0^2} = \frac{4}{U_0 M_0} \frac{\partial}{\partial t} \varphi(x, y, 0^+, t) \tag{25-2b}$$

A general solution to Eq. 25-1 that relates the potential at a point x, y, z, t to the jump in the potential and its gradients across the $z = 0$ plane can be derived by means of Green's theorem [*61*] just as Eq. 3-4 and 3-12 were derived for Laplace's equation and the two-dimensional wave equation. This expression, known as Kirchhoff's solution, is $\varphi(x, y, z, t)$

$$= \frac{-1}{4\pi} \iint_{S_0} \left\{ \frac{1}{r_0} \left[\Delta \frac{\partial \varphi}{\partial z_1} \right] + [\Delta\varphi] \frac{\partial}{\partial z} \left(\frac{1}{r_0} \right) - \frac{1}{r_0} \frac{\partial r_0}{\partial z} \left[\Delta \frac{\partial \varphi}{\partial t} \right] \right\} dS \tag{25-3}$$

where $r_0 = \sqrt{(x - x_1)^2 + (y - y_1)^2 + z^2}$. The brackets [ ] about the functions have a special meaning which is defined as follows: if the value of a function f at a fixed point P depends on the coordinates x_1, y_1, z_1, and t of a moving point Q so that

$$f = f(x_1, y_1, z_1, t)$$

then

$$[f] \equiv f(x_1, y_1, z_1, t - r)$$

where r is the distance from P to Q. The quantity $[f]$ is referred to as the retarded value of f.

The terms in Eq. 25-3 may be interpreted physically as source and doublet distributions. The perturbation velocity potential of a source of strength $\bar{f}(t')$, where t' is time, is

$$\varphi = \frac{-1}{4\pi r} \bar{f}\left(t' - \frac{r}{a_\infty}\right) = \frac{-1}{4\pi r} f(t - r) = \frac{-1}{4\pi r} [f(t)]$$

so the first term of the integral in Eq. 25-3 represents a distribution of sources with strengths fixed by the retarded value of the jump in vertical induced velocity at the plane of the wing. On the other hand, the perturbation velocity potential of a doublet of strength $\bar{g}(t')$ with its axis normal to the x, y plane is

$$\varphi = \frac{-1}{4\pi}\frac{\partial}{\partial z}\left\{\frac{1}{r}\bar{g}\left(t' - \frac{r}{a_\infty}\right)\right\} = -\frac{1}{4\pi}\frac{\partial}{\partial z}\left\{\frac{1}{r}g(\mathrm{t} - r)\right\}$$

$$= \frac{-1}{4\pi}\left\{[g(\mathrm{t})]\left(\frac{\partial}{\partial z}\frac{1}{r}\right) - \frac{1}{r}\frac{\partial r}{\partial z}\left[\frac{\partial g}{\partial \mathrm{t}}\right]\right\}$$

The second and third terms of the integral thus represent a distribution of doublets, and in the absence of thickness effects these doublets simulate a lifting surface.

The area S_a in Eq. 25-3 is the portion of the x, y plane that includes all of the sources and doublets which can influence the potential at x, y, z, t. Let us discuss this region of influence in some detail.

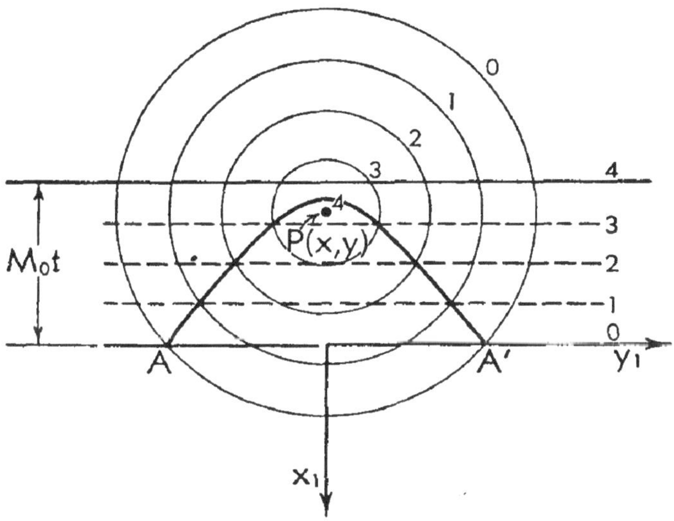

Fig. D,25a. Trace of intersection of inverse sound waves with moving source line.

Suppose that a line of sources is placed along the y axis. The strength of these sources is zero for $\mathrm{t} < 0$, but at $\mathrm{t} = 0$ the sources are "turned on" and, at the same time, the line starts moving along the negative x_1 axis with the velocity U_0. After t' seconds have passed, the source line has traveled a distance $M_0\mathrm{t}$ as shown in Fig. D,25a. Suppose next that there are two recorders placed at the point $P(x, y, 0)$ located somewhere ahead of the y_1 axis; one of these recorders is responsive to light and the other to sound. The light recorder at any given time will record the sources to be lying in a straight line just as they would appear visually at each particular instant, but the situation is entirely different for the sound recorder. The latter "hears" only[34] those sources which are located such that their spherical sound waves are just reaching the recorder at the given instant of time. The locus of all the points which,

[34] It is necessary to understand the nature of a spherical sound wave. Such a wave moves outward from its origin at a velocity a_∞ inducing disturbances in the flow field only on points that are momentarily on the spherical surface itself.

at a time t' ago, emitted sound waves that are now reaching the point $P(x, y, 0)$ is itself a sphere and for convenience this sphere will be referred to as an *inverse sound wave*. The traces of these inverse sound waves in the $z = 0$ plane are drawn in Fig. D,25a as concentric circles about the point $P(x, y)$. Finally, the intersection of an inverse sound wave[35] of radius $\mathrm{t} - \tau$ with the line representing the position of the sources at a time $t' - \tau'$ seconds ago gives the position of the sources just now signaling their presence to the sound recorder at P. For example, when the source line started, it was lying along the y_1 axis. With reference to the present time, t', this was a distance $a_\infty t'$ removed. Hence, the intersection of the circle about P of radius t with the y_1 axis fixes the two points A and A', the spherical sound waves of which are just now reaching P. A continuation of this process yields, for the locus of all points from which waves emanated that are just now touching the point P, the part of the ellipse shown in the figure.

In the sense that the light recorder, because of the very high velocity of light, has "seen" a straight line of sources, so also the sound recorder, because of the relatively slow velocity of sound, has "heard" an elliptic line of sources. Extending this concept to include a sheet of sources distributed over the surface of the wing, one can refer to the outline of that part of the wing which generates disturbances measured by the light recorder as the *planform* (i.e. the visual planform), and to the outline of that part of the wing which affects the sound recorder as the *acoustic planform*. In a mathematical sense, the acoustic planform is the area S_a over which the integration of Eq. 25-3 is to be made.

The equation for the acoustic planform can be formulated by means of the two expressions

$$(x - x_1)^2 + (y - y_1)^2 + z^2 = (\mathrm{t} - \tau)^2 \tag{25-4}$$

$$f(x_1, y_1, \tau) = 0 \tag{25-5}$$

where x, y, and z are the coordinates of the point at which the induced effects are to be measured, x_1, y_1 are the variable points of the sources, t' is time now, and $t' - \tau'$ is time ago. Eq. 25-4 represents the inverse sound waves and Eq. 25-5 represents the position of the visual planform $t' - \tau'$ seconds ago. If $\mathrm{t}^2 - z^2 < 0$, the acoustic planform does not exist since for such a case no source on the x_1, y_1 plane has had time to transmit its effect to the point x, y, z. On the other hand, if the circle (in the x_1, y_1 plane) given by the equation $(x - x_1)^2 + (y - y_1)^2 = \mathrm{t}^2 - z^2$ lies entirely within the area occupied by the visual planform of the wing at the beginning of the motion, the acoustic planform is just this circle itself.

[35] The inverse sound wave has for its analogue in steady lifting surface theory the Mach forecone. In that theory a disturbance outside the Mach forecone cannot affect the values of the induced velocities at the point where they are being measured. Similarly, in the present study a source located outside the inverse sound wave of radius t cannot affect the values of any measurement made at the point P.

For any other situation the acoustic planform is formed in part, or in whole, by the curve found by eliminating τ from Eq. 25-4 and 25-5 and in part, or not at all, by an arc of the circle.

Kirchhoff's solution of the wave equation can be applied to arbitrary wing planforms undergoing arbitrary maneuvers involving small deviations from the $z = 0$ plane. Solutions to direct problems are obtained by integrating the retarded values of the source and doublet distributions over the acoustic planforms. Inverse problems, as usual, present a more formidable task, and general methods of inverting the resulting double integral equations have not yet been developed. There are, however, many types of planforms and maneuvers which lead to boundary-value problems that are simpler to analyze. In the following we will consider several important classes of such problems.

Gardner's method. One method which can be used to solve certain inverse problems is that proposed by Gardner [*62*]. We will outline this method below as it applies to a rectangular wing tip, but its application, as pointed out in Gardner's original paper, is more general.

Consider the set of equations

$$\psi_{tt} - \psi_{xx} - \psi_{\xi\xi} = 0 \tag{25-6}$$

$$\psi_{\xi\xi} - \psi_{yy} - \psi_{zz} = 0 \tag{25-7}$$

One can see that a function $\psi(x, y, z, t, \xi)$ which satisfies both Eq. 25-6 and 25-7 must also satisfy Eq. 25-1 in any $\xi =$ constant plane. Hence, choosing for convenience the $\xi = 0$ plane, we can set

$$\left(\frac{\partial\psi}{\partial\xi}\right)_{\xi=0} = \varphi(x, y, z, t) \tag{25-8}$$

where φ is the velocity potential used in Eq. 25-1.

The advantage in exchanging the single three-dimensional wave equation for two simultaneous, two-dimensional wave equations becomes apparent in the study of the inverse boundary-value problems in wing theory. When the governing equations are put in the form of Eq. 25-6 and 25-7, the powerful methods for solving the inverse boundary-value problems introduced by Evvard [*18*] and given in Art. 12 can be used directly. The advantage is illustrated by the following study of the loading developed on the tip of a rectangular plate traveling with a uniform supersonic velocity and making small but arbitrary vertical movements.

Let

$$\left(\frac{\partial\psi}{\partial z}\right)_{z=0} = W(x, y, t, \xi) \tag{25-9}$$

Then Eq. 25-6 can be written, after differentiation,

$$W_{tt} - W_{xx} - W_{\xi\xi} = 0 \tag{25-10}$$

Since the wing is traveling at a supersonic speed, the trailing edge cannot affect the flow on the wing, and the boundary conditions for any section $y > 0$ (see Fig. D,25b (top)) are that

$$\left(\frac{\partial W}{\partial \xi}\right)_{\xi=0} = \left(\frac{\partial \varphi}{\partial z}\right)_{z=0} = w_u = w_l \tag{25-11}$$

where w_u is a given function of x, y, and t. For $y < 0$, the value of w_u is unknown and, for $y \leqq 0$, $(\varphi_t)_{z=0}$ is zero.

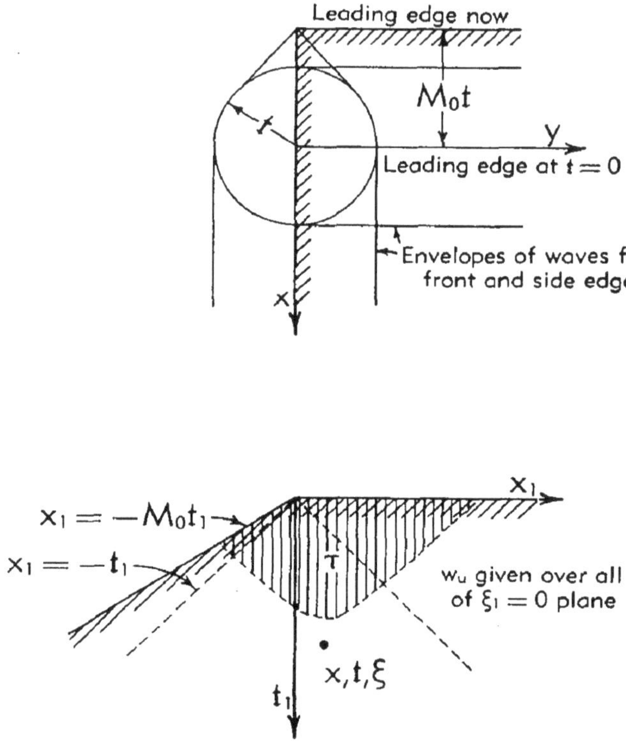

Fig. D,25b. Trace of various waves and wave envelopes in x, y and x_1, t_1 planes.

The solution to Eq. 25-10 for the boundary conditions (25-11) follows immediately from an analogy with steady state linearized wing problems. Thus Eq. 25-10 is the same as that which governs a wing in steady supersonic flow and the boundary conditions are analogous to those applying to a wing with all supersonic edges (see Fig. D,25b, bottom). Hence, we can write for $y > 0$

$$W(x, y, \mathrm{t}, \xi) = \frac{-1}{\pi} \iint_{\tau} \frac{w_u(x_1, y, \mathrm{t}_1) dx_1 d\mathrm{t}_1}{\sqrt{(\mathrm{t} - \mathrm{t}_1)^2 - (x - x_1)^2 - \xi^2}} \tag{25-12}$$

where τ is the portion of the x_1, t_1 plane shown in Fig. D,25b (bottom).

The solution for W given by Eq. 25-12 now forms the boundary value for Eq. 25-7. The value of W is not given for $y < 0$ but this is analogous to the boundary-value specification for a rectangular plate flying at a steady supersonic speed in which the value of the upwash is not specified off the wing. Hence, referring to Eq. 12-39, we can write the solution to Eq. 25-7 in the $z = 0$ plane for the boundary conditions given by Eq. 25-12 as

$$\psi(x, y, 0, \xi, t) = \frac{-1}{\pi} \iint_{\tau_1} \frac{W(x, y_1, t, \xi_1)dy_1 d\xi_1}{\sqrt{(\xi - \xi_1)^2 - (y - y_1)^2}}$$

where τ_1 is the area shown in Fig. D,25c. Finally, the solution for φ in

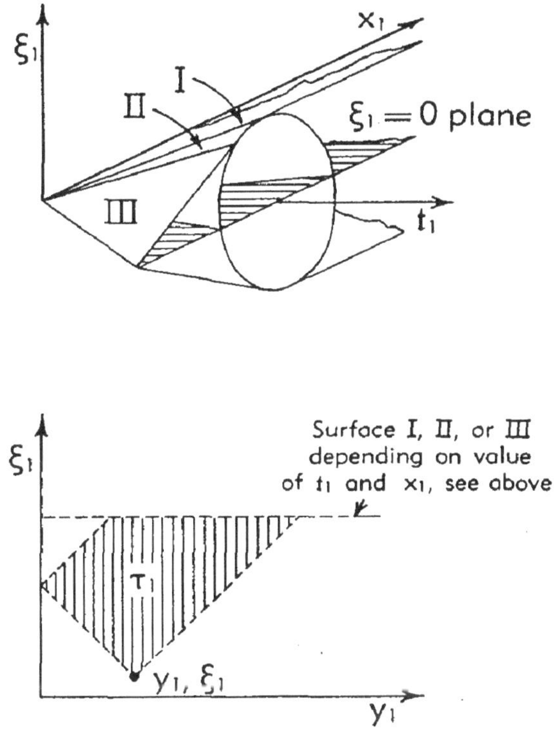

Fig. D,25c. Integration regions involving ξ_1.

the $z = 0$ plane is found in the form

$$\varphi(x, y, 0, t) = \lim_{\xi \to 0} \frac{-1}{\pi} \frac{\partial}{\partial \xi} \iint_{\tau_1} \frac{W(x, y_1, t, \xi_1)dy_1 d\xi_1}{\sqrt{(\xi - \xi_1)^2 - (y - y_1)^2}} \quad (25\text{-}13)$$

Eq. 25-13 and 25-12 represent the solution to Eq. 25-1 under the boundary conditions specified. An application of this method is given in Art. 28. For other applications of essentially the same method see Goodman [*63,64*].

Homogeneous boundary-value problems. In Art. 13 a homogeneous perturbation potential was introduced and the concept applied to three-dimensional steady state problems. The obvious extension to the three-dimensional unsteady case is to assume that the velocity potential $\varphi(x, y, z, t)$ governing the flow field is also homogeneous and of degree n, that is, can be written in the form

$$t^n\Phi\left(\frac{x}{t}, \frac{y}{t}, \frac{z}{t}\right)$$

The simplification brought about in the analysis of such problems is the same as for the steady state case, namely, that the number of mathematical "dimensions" has been reduced.

Examples of homogeneous boundary-value problems occurring in three-dimensional unsteady flow are easy to construct. For example, it can be shown that the pressure in any region affected by only two intersecting edges of a straight-sided flat plate traveling at a uniform subsonic or supersonic speed is homogeneous and of degree zero, i.e. satisfies the equation

$$\frac{\Delta p}{\frac{1}{2}\rho_\infty U_0^2} = \frac{\Delta p}{\frac{1}{2}\rho_\infty U_0^2}\left(\frac{x}{t}, \frac{y}{t}\right)$$

Similarly, it can be shown that the pressure in any region affected by only two intersecting edges of a straight-edged flat plate traveling at a uniform subsonic or supersonic speed and rotating at a constant rate of pitch or roll is homogeneous and of degree one, i.e. satisfies the equation

$$\frac{\Delta p}{\frac{1}{2}\rho_\infty U_0^2} = y\,\frac{\Delta p}{\frac{1}{2}\rho_\infty U_0^2}\left(\frac{x}{t}, \frac{y}{t}\right)$$

It should be noted that both statements are equally true for the steady state case when all transient effects have disappeared. In supersonic wing theory they lead to conical and quasi-conical flows, respectively. In general, homogeneous flow occurs when the boundary conditions after a change in scale are proportional to themselves.

Consider next the modification of the basic partial differential equation (25-1) under the assumption that the flow is homogeneous. If the pressure is homogeneous and of degree zero, then, by Eq. 25-2, the velocity potential will be homogeneous and of degree one. If the notation

$$\frac{x}{t} = x_3, \qquad \frac{y}{t} = y_3, \qquad \frac{z}{t} = z_3$$

$$\varphi(x, y, z, t) = t\phi(x_3, y_3, z_3)$$

is used, then Eq. 25-1 becomes

$$(1 - x_3^2)\phi_{x_3x_3} + (1 - y_3^2)\phi_{y_3y_3} + (1 - z_3^2)\phi_{z_3z_3} - 2x_3y_3\phi_{x_3y_3} - 2x_3z_3\phi_{x_3z_3} - 2y_3z_3\phi_{y_3z_3} = 0 \quad (25\text{-}14)$$

and a linear partial differential equation with three independent variables is therefore obtained.

In the general theory of partial differential equations of second order, the character of an equation is determined from the geometric nature of a related quadric surface [*65*]. The character of Eq. 25-14 can be shown from such considerations to depend on the sign of $1 - x_3^2 - y_3^2 - z_3^2$. It is immediately apparent, however, that within the unit sphere in the x_3, y_3, z_3 space the sign of $1 - x_3^2 - y_3^2 - z_3^2$ is everywhere positive, and outside the sign is everywhere negative. It follows from such considerations that outside the unit sphere Eq. 25-14 is hyperbolic and inside the unit sphere it is elliptic.

Boundary-value problems involving noninteracting surfaces. The simplification to the four-dimensional theory brought out by the introduction of homogeneous flow was more apparent than real, since the resulting partial differential equation, although containing one less dimension, was unwieldy. Another class of wing problem which is simplified both in theory and in practice by reasoning from a physical insight of the flow behavior is that in which the wing has a supersonic edge (i.e. an edge which is traveling with a supersonic normal component of velocity).

When the acoustic planform is affected only by a supersonic edge, it is obvious that the flow on the upper surface of the wing is independent of that on the lower surface. Hence, the solution to such problems can always be written in terms only of sources as follows:

$$\varphi = -\frac{1}{2\pi}\iint\limits_{S_a} \frac{[w_u]}{r_0}\, dS \tag{25-15}$$

where $w_u(x_1, y_1) = U_0\lambda(x_1, y_1)$, λ being the local slope of the surface in the direction of U_0. Since Eq. 25-15 is equally valid for symmetrical nonlifting surfaces and lifting plates, its value and simplicity is evident. Such an analysis has been applied toward the calculation of the load distribution due to a sudden change in angle of attack over flat plate triangular wings with supersonic edges [*1,66*].

If the wing planform is further specialized in that it moves along the negative x axis, has all portions of the leading edge supersonic, and has a straight trailing edge cut normal to the direction of motion, certain methods have been developed by means of which the total lift and pitching moment can be evaluated with relative simplicity. Thus, if a detailed knowledge of the load distribution over wings with planforms of the type just described is not required, and only the spanwise integral of the loading is needed, the following analysis can be employed.

Consider again Eq. 25-1 and integrate each term with respect to y between the limits minus and plus infinity.[36] There results the equation

$$\int_{-\infty}^{\infty} \frac{\partial^2\varphi}{\partial x^2}\, dy + \int_{-\infty}^{\infty} \frac{\partial^2\varphi}{\partial y^2}\, dy + \int_{-\infty}^{\infty} \frac{\partial^2\varphi}{\partial z^2}\, dy - \int_{-\infty}^{\infty} \frac{\partial^2\varphi}{\partial t^2}\, dy = 0$$

[36] The basic idea for this solution was given by P. A. Lagerstrom in his lectures at the California Institute of Technology. See also Miles [*83*] and Froehlich [*84*].

If $y = y_l(x, z, t)$ and $y = y_r(x, z, t)$ are the equations of the Mach wedges streaming back from the leading edges on the left and right sides of the wing, respectively, then, since φ is continuous across these wedges but φ_x, φ_y, φ_z, and φ_t are not,

$$\frac{\partial^2}{\partial x^2}\int_{y_l}^{y_r}\varphi dy = \frac{\partial y_r}{\partial x}u_r - \frac{\partial y_l}{\partial x}u_l + \int_{y_l}^{y_r}\frac{\partial^2\varphi}{\partial x^2}dy$$

where u_r and u_l are the values of the discontinuity in u on the right and left wedges, respectively. Values of the terms involving w and φ_t are similar to those involving u and

$$\int_{y_l}^{y_r}\frac{\partial^2\varphi}{\partial y^2}dy = v_r - v_l$$

Finally, therefore, if

$$\psi = \int_{y_l}^{y_r}\varphi dy$$

then

$$\frac{\partial^2\psi}{\partial x^2} + \frac{\partial^2\psi}{\partial z^2} - \frac{\partial^2\psi}{\partial t^2} = \left(\frac{\partial y_r}{\partial x}u_r + \frac{\partial y_r}{\partial z}w_r - \frac{\partial y_r}{\partial t}\varphi_{t_r} - v_r\right) - \left(\frac{\partial y_l}{\partial x}u_l + \frac{\partial y_l}{\partial z}w_l - \frac{\partial y_l}{\partial t}\varphi_{t_l} - v_l\right)$$

The terms enclosed in the parenthesis in the last equation represent the directional derivatives of the velocity potential taken along the conormal of the foremost disturbance surfaces. Since φ is constant on the surface, these terms are zero. Finally, therefore, Eq. 25-1 has been reduced in terms of ψ to the expression

$$\psi_{tt} - \psi_{xx} - \psi_{zz} = 0 \tag{25-16}$$

Consider now the boundary conditions. These conditions specify that w, the vertical induced velocity in the $z = 0$ plane, is given over the region of the plane occupied by the wing. But $(\partial\psi/\partial z)_{z=0}$ becomes

$$\int_{-\infty}^{\infty}wdy$$

Further, since the edges of the wing are supersonic, w must be zero off the wing planform (for values of x not back of the trailing edge) and the integration need extend only over the planform itself. Hence, for wings with trailing edges normal to the direction of motion, the boundary conditions affecting all regions ahead of the trailing edge can be expressed in terms of a known variation of $(\partial\psi/\partial z)_{z=0}$.

These boundary conditions are equivalent to those studied in steady state supersonic wing theory. In fact, the lifting surface analogue is shown in Fig. D,25d, being a wing tip of specified twist and camber in a supersonic free stream having a Mach number equal to $\sqrt{2}$. The solution to this problem can, therefore, be written immediately as

$$\psi = -\frac{1}{\pi}\iint_{\sigma} \frac{(\partial\psi/\partial z)_{z=0}\, dt_1 dx_1}{\sqrt{(t - t_1)^2 - (x - x_1)^2}} \tag{25-17}$$

where σ is the area on the shaded surface in the figure lying ahead of the forecone traces given by the equation $(t - t_1)^2 = (x - x_1)^2$. In application the evaluation of Eq. 25-17 is considerably simpler than that of the more general expression, Eq. 25-15.

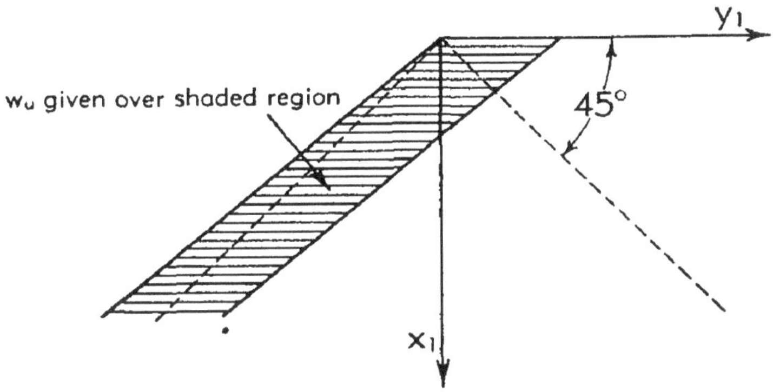

Fig. D,25d. Boundary conditions for wing tip with supersonic edges.

An instructive alternative method for solving problems involving the unsteady forces on three-dimensional wings with supersonic leading edges and a straight trailing edge normal to the direction of motion can also be given. Consider the two-dimensional wing in Fig. D,25e(1). Let this wing have an angle of attack $\alpha(t)$ which varies with time in an arbitrary manner. To such an angle-of-attack variation there corresponds a certain lift which also varies with time. Hence, if L' represents the total lift on an airfoil of very high aspect ratio and $c'_{l\alpha}$ represents the section lift-curve slope, then[37]

$$\frac{L'}{\frac{1}{2}\rho_\infty U_0^2 \alpha} = c c'_{l\alpha}(c, t)(\text{span})$$

Next, it is clear by reason of symmetry that the total lift on wings (2) and (3) in the figure are equal. Then, since the analysis is based on a linear partial differential equation, by superposition principles the total lift on wing (2) or (3) is just half of that on wing (1). In another sense, the lift coefficient for the whole wing based on the deflected area is the same for all three cases. A suitable superposition of wings (1), (2), and (3) will give wing (4), which then has the same lift coefficient based on the deflected area. Finally, because of the supersonic stream, wing (5) can

[37] The primes on the quantities indicate that two-dimensional values are taken, or that a high aspect ratio wing is considered and tip effects are neglected.

be obtained from (4); hence it also has the lift coefficient common to the other wings. It is of course necessary that the variation of α with time be the same in each case.

The preceding process can be extended one step further to the development of the lift due to a single element. From Fig. D,25f it can be seen

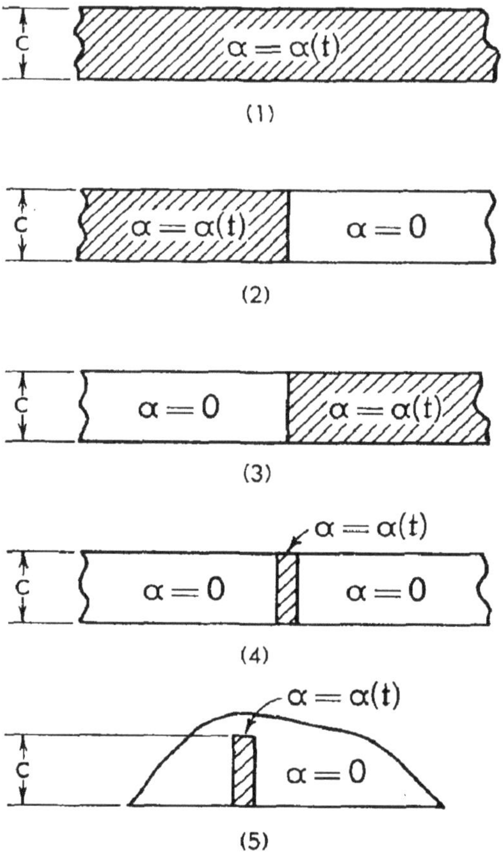

Fig. D,25e. Various wing boundary conditions used in superposition to obtain single deflected element.

that

$$\frac{L'}{\frac{1}{2}\rho_\infty U_0^2\alpha} = \frac{cc'_{l\alpha}(c,\ t) - (c - \Delta c)c'_{l\alpha}(c - \Delta c,\ t)}{\Delta c}\,\Delta S$$

where ΔS is the area of the deflected element and c is the distance from ΔS to the trailing edge. By the usual limiting process the latter equation becomes

$$\frac{L'}{\frac{1}{2}\rho_\infty U_0^2\alpha} = \frac{\partial}{\partial c}\,[cc'_{l\alpha}(c,\ t)]dS$$

Finally, if a wing is composed of a distribution of these elements, then, for an axial system centered at the apex of the wing leading edge, there results

$$C_L = -\frac{1}{S}\iint_S \alpha(x, y)\frac{\partial}{\partial x}[(c_0 - x)c'_{l\alpha}(c_0 - x, t)]dydx \qquad (25\text{-}18)$$

where c_0 is the maximum chord as indicated in Fig. D,25g. In the development of Eq. 25-18 each element is assumed to have the same variation of motion with time.

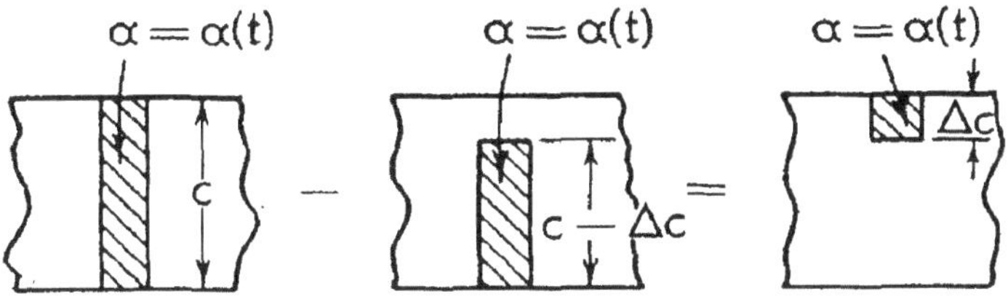

Fig. D,25f. Elemental strip used in superposition to obtain single deflected element.

Notice that, when the wing is a flat plate flying at a steady speed so that all transient effects have disappeared, $c'_{l\alpha}$ is independent of t and of the chord length, being, in fact, equal to $4/\beta$. Then Eq. 25-18 becomes

$$C_L = \frac{4}{\beta S}\iint_S \alpha(x, y)dxdy = \frac{4\bar{\alpha}}{\beta}$$

where $\bar{\alpha}$ is the average angle of attack of the wing. Such results for the steady state case were presented originally by Lagerstrom and Van Dyke

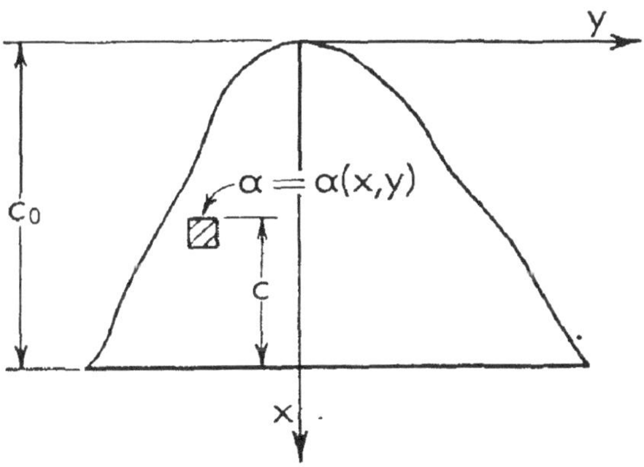

Fig. D,25g. Quasi-triangular wing with supersonic edges and a single deflected element.

[22]. When $\alpha(x, y)$ is independent of x (as for a flat wing sinking or rolling), Eq. 25-18 becomes

$$C_L = \frac{1}{S}\int_{-b/2}^{b/2} \alpha(y)cc'_{l\alpha}(c, t)dy \qquad (25\text{-}19)$$

where c is the local chord which is, in general, a function of y. Eq. 25-19 simply indicates that longitudinal strip theory is exact for such wings, a fact shown originally by Miles [67].

It is apparent that the calculation of the unsteady lift on three-dimensional wings with supersonic leading edges and straight trailing edges perpendicular to the free-stream direction has been reduced to an integration involving the relatively simple results for a two-dimensional wing undergoing the same unsteady motion. The lift on a flat triangular wing rising and sinking with a harmonic motion has been computed by such means by Miles [68].

Two-dimensional unsteady flow. The simplification brought about when the flow field is independent of one dimension is obvious. In such

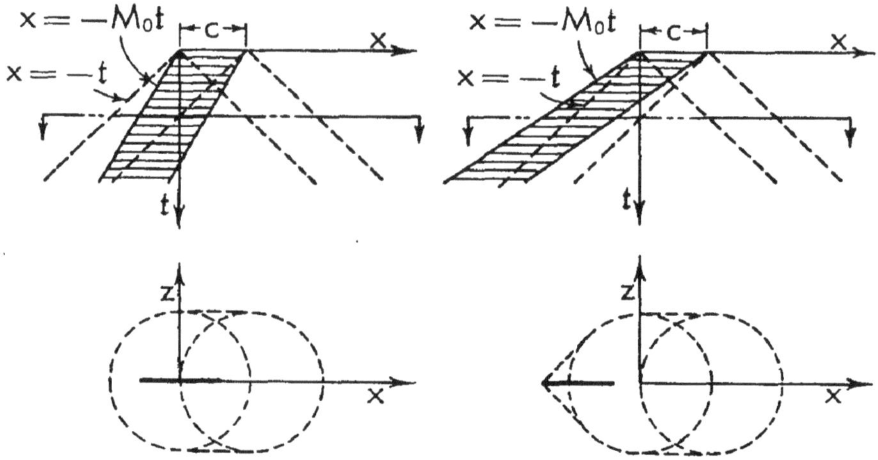

Fig. D,25h. Traces of two-dimensional wing sections in the x, t plane moving at subsonic and supersonic speeds. Left, subsonic. Right, supersonic.

cases Eq. 25-1 reduces to the two-dimensional wave equation. Further, the boundary conditions for wings moving in the $z = 0$ plane are directly analogous to those applying to three-dimensional steady state wing problems. Hence, the solutions to both the direct and inverse problems presented in Art. 10-14 transform into solutions of pertinent two-dimensional unsteady flow problems.

For example, consider the flow in the center portion of a high aspect ratio rectangular wing that starts from rest at $t = 0$ and moves steadily at a constant forward speed U_0. Neglecting the spanwise gradient of velocity, we find the governing partial differential equation to be

$$\varphi_{tt} - \varphi_{xx} - \varphi_{zz} = 0 \tag{25-20}$$

The boundary conditions are illustrated in Fig. D,25h. The figure shows the trace in the $z = 0$ plane of a typical section moving in the negative x direction at either a subsonic speed, Fig. D,25h (top left) or a supersonic speed, Fig. D,25h (top right). Since $t = a_\infty t'$, the slope dx/dt of the leading edge trace is $-M_0$. If the wing is a plate sinking steadily (without pitching), $w_u(x, t) = w_l(x, t) = -U_0\alpha$ where α is the angle of attack. If, on the other hand, the wing is a plate pitching about its leading edge at a constant rate $\dot{\theta} = d\theta/dt'$, then $w_u(x, t) = w_l(x, t) = -(x + M_0 t)\dot{\theta}$ everywhere on the wing.

The dotted lines in Fig. D,25h represent the traces of the pressure (sound) waves which start from the wing leading and trailing edges at $t = 0$ and move outward, as time increases, at a velocity equal to the speed of sound. At a given time these waves form circles in an x, z plane (cylinders in x, y, z space) as shown in Fig. D,25h (bottom left) and D,25h (bottom right). The envelopes of all such waves which started along the wing chord at $t = 0$ are also shown in Fig. D,25h (bottom left) and D,25h (bottom right), and at a given time it is only the air within these envelopes that is disturbed by the motion of the wings. The traces in the x, t plane of these initial waves all have slopes in the chosen coordinate system equal to ± 1.

The analogy between two-dimensional unsteady lift problems, such as those just described, and three-dimensional steady state lifting surface problems[38] is now readily seen. The partial differential equations which govern the flow fields, Eq. 25-20 and 10-1, are identical if in the three-dimensional case x is replaced by t, y by x, and β by 1. In both cases the boundary conditions are given for w_u and, after the appropriate notation change, the equations for the pressure coefficient (see Eq. 2-6) are proportional. Thus, the solution to the problem outlined in connection with Fig. D,25h is directly analogous to the solution for the loading on a three-dimensional wing tip flying at a Mach number equal to $\sqrt{2}$ and having either subsonic side edges, as in Fig. D,25h (top left) or supersonic side edges, as in Fig. D,25h (top right).

Other types of two-dimensional boundary-value problems of interest in unsteady flow analysis are illustrated in Fig. D,25i. Fig. D,25i (top) represents a wing section, restrained so that it cannot move vertically, traveling at a supersonic speed and entering a vertical gust. The gust is situated to the left of the line $x = 0$ and its structure, which is arbitrary, determines the value of w_u over the region occupied by the wing. Fig. D,25i (middle) represents a wing which has been traveling with a constant angle of attack and at a steady supersonic speed for $t < 0$ and at $t = 0$ starts to decelerate. The rate at which it decelerates determines the shape of the leading and trailing edge traces in the x, t plane. Finally, Fig. D,25i (bottom) represents a wing with a deflected flap, the flap being deflected to the angle δ at $t = 0$ and held constant at that angle thereafter.

Slender wings in unsteady flow. When the wing planform is slender in the sense that its length in the streamwise direction is large compared

[38] This analogy can be extended and it can be shown that a two-dimensional unsteady incompressible flow field is analogous to a three-dimensional steady flow field having a free stream Mach number equal to 1. Hence, the concepts of slender airplane theory see (Art. 19–23) can be interpreted in terms of unsteady motion in an incompressible field and vice versa. A detailed discussion of this analogue is given in [*1*].

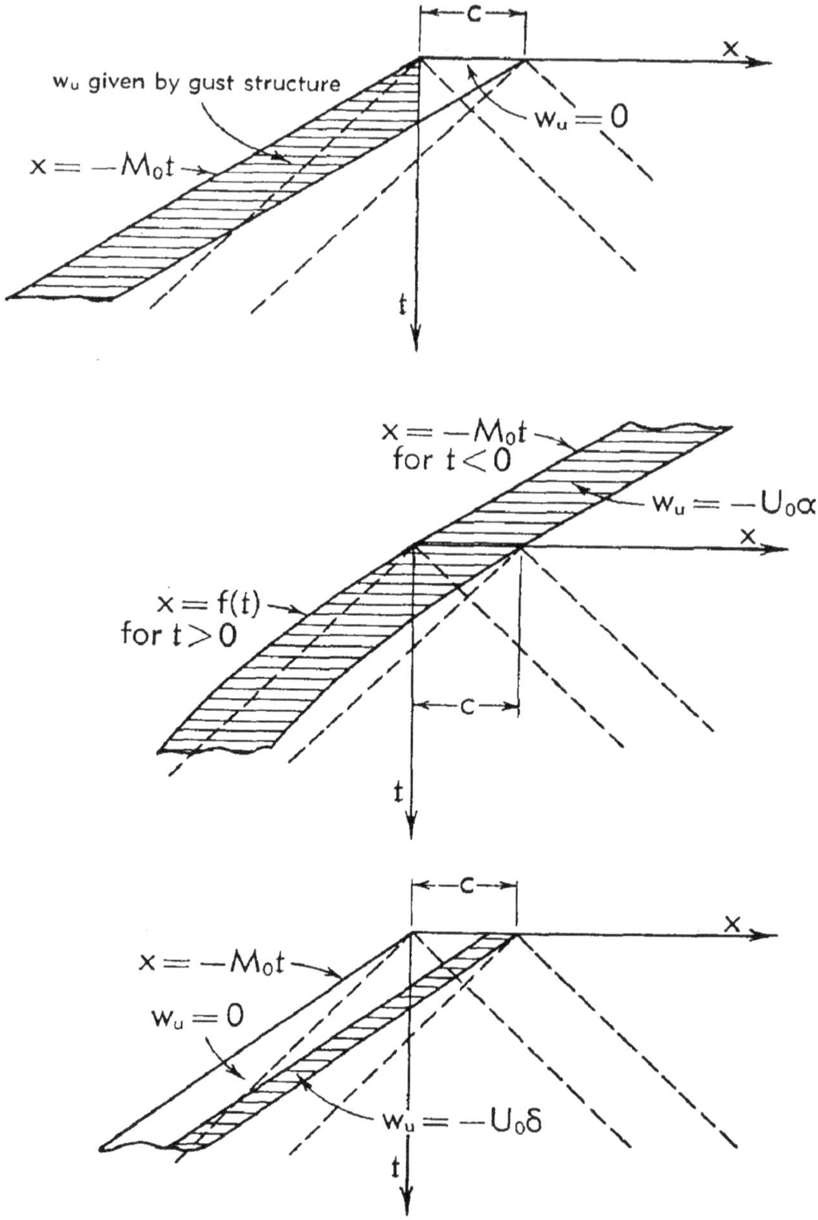

Fig. D,25i. Various wing boundary-value problems in two-dimensional (space) unsteady flow. Top, airfoil entering gust. Middle, decelerating airfoil. Bottom, sudden flap deflection.

to its span, an estimation of the loading on it can be obtained by neglecting in the partial differential equation the gradient of the induced velocity component in the stream direction. Thus, if the wing is moving in the negative x direction, Eq. 25-1 reduces to

$$\varphi_{yy} + \varphi_{zz} = \varphi_{tt} \tag{25-21}$$

which is again the wave equation but in two space dimensions. Since Eq. 25-21 is independent of x, study can be made independently in each plane x = const. This is an extension of steady state slender wing theory. Fig. D,25j (left) shows a typical section in the y,t plane. If the wing is a flat plate at a constant angle of attack, the value of φ_z over the part of the y,t plane occupied by the wing is a constant and the jump in φ across the vortex wake must be consistent with its value at the wing trailing edge.[39]

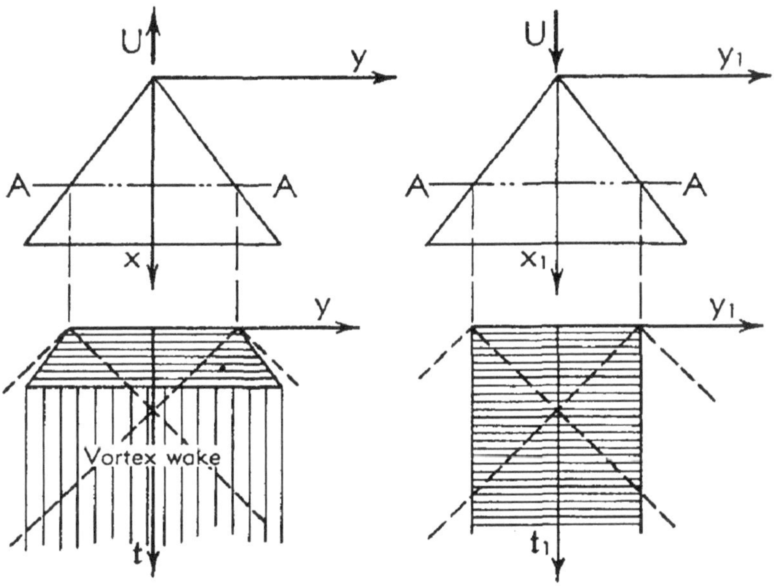

Fig. D,25j. Trace of slender wing in x, y and y, t. planes. Left, x fixed at station AA, wing moves away. Right, axes fixed on wing, air moves by.

There is another way by means of which the effects of unsteady motion on slender wing pressures can be estimated. If, instead of using the stationary x,y,z,t coordinate system, the reference axes are fixed on the wing by the simple set of transformations

$$\begin{aligned} x_1 &= x + M_0 t \\ t_1 &= t \\ y_1 &= y \\ z_1 &= z \end{aligned}$$

Eq. 25-1 becomes

$$(1 - M_0^2)\varphi_{x_1x_1} - 2M_0\varphi_{x_1t_1} + \varphi_{y_1y_1} + \varphi_{z_1z_1} = \varphi_{t_1t_1}$$

Again, if the induced velocity components in the stream direction are neglected, the simplified equation

$$\varphi_{y_1y_1} + \varphi_{z_1z_1} = \varphi_{t_1t_1}$$

[39] The analogy with three-dimensional, steady-state, supersonic, lifting surface theory is apparent.

results. The latter equation is identical in form to Eq.25-21. However, now the axes are fixed on the wing and a typical section in the y_1,t_1 plane is similar to that shown in Fig. D,25j (right). In this case, a flat plate wing is represented by a constant value of φ_{z_1} over the entire shaded area in the figure, and elsewhere φ must be continuous.

D,26. The Indicial Functions. By definition, an indicial function is the response to a disturbance that has the form of a step function.

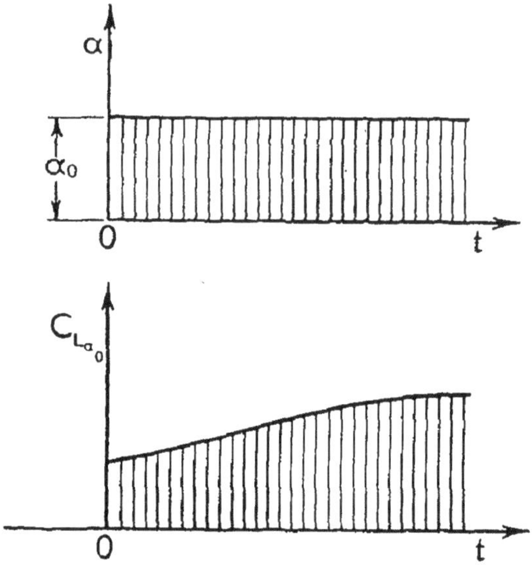

Fig. D,26a. Step variation of angle of attack α and the corresponding indicial response of the wing lift coefficient.

For example, if the angle of attack of a wing varies with time as shown in Fig. D,26a, the corresponding lift response, also shown in Fig. D,26a, is designated as the indicial lift coefficient due to angle of attack. In the following we consider four basic indicial functions, namely, the indicial lift and pitching moment responses due to step functions in angle of attack and pitch. In coefficient form these are designated[40] $C_{L\alpha}$, $C'_{m\alpha}$, C'_{Lq}, and C'_{mq}. (Indicial functions are also treated in VII,F and VIII,C.)

The primes on the coefficients indicate that the axes about which the pitching motion occurs and about which the pitching moments are measured both coincide either with the leading edge of the wing or with a line through the apex normal to the root chord of the wing. The equations which transform these functions to those for a wing pitching about an axis a distance ac back from the leading edge and having its moment center a distance bc back from the leading edge are simply

[40] The variable q to which the lift and pitching moment coefficients of the pitching wing are referred is equal to $\dot{\theta}c/U_0$, a dimensionless form of the expression for rate of pitch.

$$
\begin{aligned}
C_{L_\alpha} &= C_{L_\alpha} \\
C_{m_\alpha} &= C'_{m_\alpha} + bC_{L_\alpha} \\
C_{L_q} &= C'_{L_q} - aC_{L_\alpha} \\
C_{m_q} &= C'_{m_q} + bC'_{L_q} - aC'_{m_\alpha} - abC_{L_\alpha}
\end{aligned} \tag{26-1}
$$

It is apparent that the boundary conditions by which the indicial loading on sinking or pitching wings can be determined have already

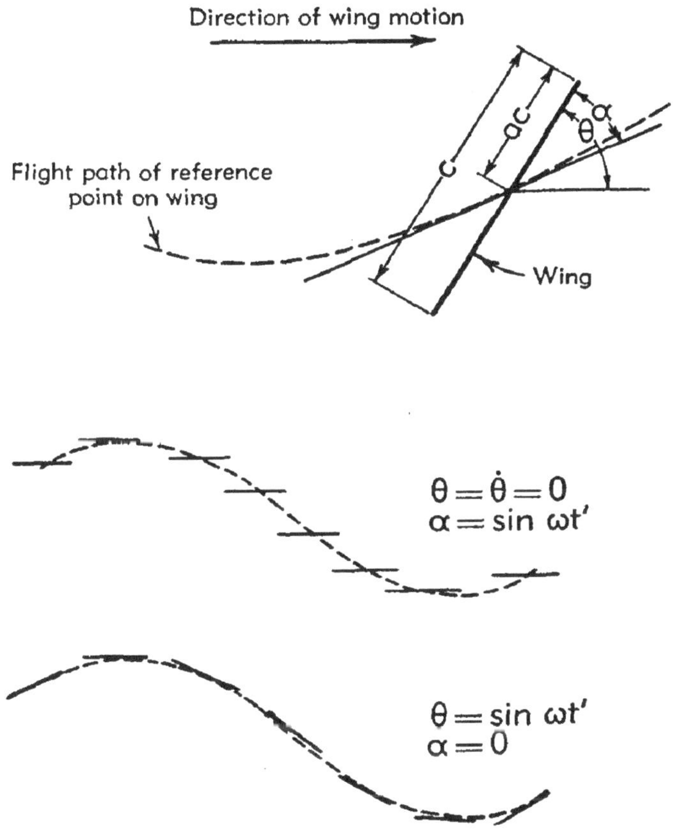

Fig. D,26b. **Wing oscillating in angle of attack at a constant angle of pitch (referred to the horizontal) and vice versa.**

been discussed for two-dimensional wings in connection with Fig. D,25h. Thus, regardless of whether the wing is flying at subsonic or supersonic speed, the variation of $(w)_{z=0}$ over the portion of the $z = 0$ plane occupied by the wing will be

$$w_u = w_l = -U_0\alpha \tag{26-2a}$$

for a sinking wing and

$$w_u = w_l = -(x + M_0 t)\dot{\theta} \tag{26-2b}$$

for a pitching wing. Further, the loading will be zero for all parts of the $z = 0$ plane not occupied by the wing.

The difference between θ and α is illustrated in Fig. D,26b. The angle of attack α is the angle between the flat wing surface and the tangent to the flight path of some point fixed at a distance ac back from the wing leading edge. For example, in applications involving the dynamic behavior of an entire airplane the distance ac would usually be taken as the distance back to the center of gravity of the airplane. The angle θ is the angle between the flat wing surface and the horizontal. Fig. D,26b (middle) shows a wing undergoing a sinusoidal angle-of-attack variation with a zero angle of pitch throughout. For convenience and in order to distinguish from the pitching wing, such a wing will be referred to as a sinking wing. Figure D,26b (bottom) shows a wing undergoing a sinusoidal angle-of-pitch variation taken about the axis ac back from the leading edge. In the case shown the wing has a constant (zero) angle of attack.

Adaptation of the indicial functions to maneuvers. Since the theory is linear, the lift and moment on a wing undergoing an arbitrary maneuver (in which both α and $\dot{\theta}$ remain small) can be calculated from the indicial functions by the principle of superposition. Thus, it can be shown (see e.g. [*69*]) that the lift and moment induced on the wing by an arbitrary maneuver are given by Duhamel's integral

$$\begin{aligned} C_L &= \frac{d}{dt'} \int_0^{t'} [C_{L\alpha}(t' - t_1')\alpha(t_1') + C_{L_q}(t' - t_1')q(t_1')]dt_1' \\ C_m &= \frac{d}{dt'} \int_0^{t'} [C_{m\alpha}(t' - t_1')\alpha(t_1') + C_{m_q}(t' - t_1')q(t_1')]dt_1' \end{aligned} \tag{26-3}$$

One simple maneuver which has been studied by applying the above concepts is that of a wing entering a gust. The motion of a wing subjected to arbitrary external lifting forces is governed by Newton's second law of motion. Let w be the vertical velocity of the wing and m its mass; then

$$m \frac{dw}{dt'} = \sum \text{forces} \tag{26-4}$$

The forces to be summed are those resulting from the lift on the wing and the impressed force due to the gust; the effect of pitching is neglected.

If the wing is restrained so that it can move neither upward nor downward, the time variation of the lift induced on it when it enters a sharp-edged gust having a uniform upward velocity w_0 will be designated $C_{L_g}(t')$. The boundary conditions involved in the problem of finding $C_{L_g}(t')$ are discussed in connection with Fig. D,25i (top). Applying Eq. 26-4, we find the equation of the vertical motion of an unrestrained wing flying into this gust is

$$m \frac{dw}{dt'} - \frac{1}{2} \rho_\infty U_0^2 S \frac{d}{dt'} \int_0^{t'} C_{L\alpha}(t' - t_1')\alpha(t_1')dt_1' = \frac{w_0}{U_0} \frac{1}{2} \rho_\infty U_0^2 S C_{L_g}(t') \tag{26-5}$$

Eq. 26-5 is an integral equation of the second kind with a variable upper limit. Its solution gives directly the gust load induced on the wing [70,81].

Flutter derivatives. By definition, a flutter derivative is the response to a harmonic oscillation in vertical displacement (angle of attack) or pitch, such oscillation having continued so long that all transient effects of its origin are damped out and the induced forces and moments are periodic. In the notation used in [71] the two-dimensional harmonic responses are expressed in terms of the derivatives L_1, L_2, L_3', L_4' and M_1', M_2', M_3', M_4'. The relation between the two-dimensional or section indicial responses and these flutter derivatives follows from Eq. 26-3 and can be written

$$
\begin{aligned}
L_1 + iL_2 &= \lim_{t' \to \infty} \frac{ie^{-i\omega t'}}{4k} \frac{d}{dt'} \int_0^{t} c_{l\alpha}(t' - t_1') e^{i\omega t_1'} dt_1' \\
M_1' + iM_2' &= \lim_{t' \to \infty} \frac{-ie^{-i\omega t'}}{2k} \frac{d}{dt'} \int_0^{t'} c_{m\alpha}'(t' - t_1') e^{i\omega t_1'} dt_1' \\
L_3' + iL_4' &= \lim_{t' \to \infty} \frac{e^{-i\omega t'}}{4k^2} \cdot \frac{d}{dt'} \int_0^{t'} [c_{l\alpha}(t' - t_1') + 2ikc_{l_q}'(t' - t_1')] e^{i\omega t_1'} dt_1' \\
M_3' + iM_4' &= \lim_{t' \to \infty} \frac{-e^{-i\omega t'}}{2k^2} \frac{d}{dt'} \int_0^{t'} [c_{m\alpha}'(t' - t_1') + 2ikc_{m_q}'(t' - t_1')] e^{i\omega t_1'} dt_1'
\end{aligned}
\tag{26-6}
$$

where ω is the frequency and k is the reduced frequency, $\omega c/2U_0$. The primes on the quantities again indicate that the wing is pitching about and the moments are measured about the leading edge of the wing. By means of Eq. 26-6 the indicial lift and pitching moment curves can be used to study the flutter stability boundaries as expressed in terms of L_1, L_2, etc. in [71].

The response in lift and moment for harmonically oscillating wings can be expressed in terms of their absolute values and phase shift with reference to the period of the forcing disturbance. Thus, if an angle-of-attack variation given by the equation $\alpha = e^{i\omega t'}$ is impressed on the wing, the response in section lift would be $|c_{l\alpha}| e^{i(\omega t' + \psi)}$. In terms of the indicial functions this can be written

$$
|c_{l\alpha}| e^{i\psi} = \lim_{t' \to \infty} e^{-i\omega t'} \frac{d}{dt'} \int_0^{t'} c_{l\alpha}(t' - t_1') e^{i\omega t_1'} dt_1'
$$

which can be reexpressed in terms of the flutter derivatives as

$$
|c_{l\alpha}| e^{i\psi} = 4k(L_2 - iL_1) = 4k \sqrt{L_1^2 + L_2^2}\, e^{-i \tan^{-1}(L_1/L_2)}
$$

Similarly, each of the stability derivatives can be expressed in terms of the indicial functions and the flutter derivatives as follows:

Term	*Maximum value*	*Phase angle*	
c_{l_α}	$4k\sqrt{L_1^2 + L_2^2}$	$-\tan^{-1}\left(\frac{L_1}{L_2}\right)$	
c'_{m_α}	$2k\sqrt{M_1'^2 + M_2'^2}$	$-\tan^{-1}\left(\frac{M'_1}{M'_2}\right)$	(26-7)
$ik(c_{l_{\dot\alpha}} + c'_{l_q})$	$2k\sqrt{L_3'^2 + L_4'^2}$	$\tan^{-1}\left(\frac{L'_4}{L'_3}\right)$	
$ik(c'_{m_{\dot\alpha}} + c'_{m_q})$	$k^2\sqrt{M_3'^2 + M_4'^2}$	$\tan^{-1}\left(\frac{M'_4}{M'_3}\right)$	

where it is assumed that unit absolute values of angle of attack and pitch are impressed and the term $\dot\alpha = (c/U_0)(d\alpha/dt')$.

A wing undergoing slow harmonic oscillations. The boundary conditions for these problems are exactly the same as those used for evaluating the flutter derivatives, being, in fact, a special case of the latter when the frequency of oscillation is very low. This added simplification is important, however, because the frequency of angle of attack and pitch oscillations for the entire airplane is, relative to the flutter frequencies, very low. Further, special methods based on the assumption that the wing is oscillating slowly have been devised. If the lift and moment are expanded in powers of the frequency, solutions for the coefficients of the lowest order terms have been found for both triangular and rectangular wings flying at supersonic speeds (see, e.g. [*68,72,73,74*]).

Consider the case of L_1 and L_2 as given by Eq. 26-6. This expression can be written

$$L_1 + iL_2 = \lim_{t'\to\infty} \frac{ie^{-i\omega t'}}{4k}\frac{d}{dt'}\int_0^{t'} [c_{l_\alpha}(\infty) - \Delta c_{l_\alpha}(t')]e^{i\omega(t'-t'_1)}dt_1$$

where $\Delta c_{l_\alpha}(t')$ is the difference between the indicial lift and its asymptotic value, thus

$$\Delta c_{l_\alpha}(t') = c_{l_\alpha}(\infty) - c_{l_\alpha}(t')$$

Since $\Delta c_{l_\alpha}(\infty)$ is zero, it follows

$$L_1 + iL_2 = \frac{i}{4k}\left[c_{l_\alpha}(\infty) - i\omega\int_0^\infty \Delta c_{l_\alpha}(t'_1)e^{-i\omega t'_1}dt'_1\right]$$

or, in terms of the reduced frequency parameter $k = \omega c/2U_0$ and the number of chord lengths traveled $\tau_0 = U_0 t'/c$,

$$L_1 + iL_2 = \frac{i}{4k}\left[c_{l_\alpha}(\infty) - 2ik\int_0^\infty \Delta c_{l_\alpha}e^{-2ik\tau_0}d\tau_0\right]$$

The indicial functions for supersonic edged wings reach their steady state value in a finite time. Hence, L_1 and L_2 for such wings can be expanded immediately in powers of k. We find

$$4L_1 = 2\int_0^{\tau_s} \Delta c_{l_\alpha}(\tau_0)d\tau_0 - 4k^2\int_0^{\tau_s} \tau_0^2\Delta c_{l_\alpha}(\tau_0)d\tau_0 + \cdots \quad (26\text{-}8a)$$

$$4kL_2 = c_{l_\alpha}(\infty) - 4k^2\int_0^{\tau_s} \tau_0\Delta c_{l_\alpha}(\tau_0)d\tau_0 + \cdots \quad (26\text{-}8b)$$

where τ_s is the number of chord lengths the wing must travel for the indicial response to reach its final value. Similar expressions for the other

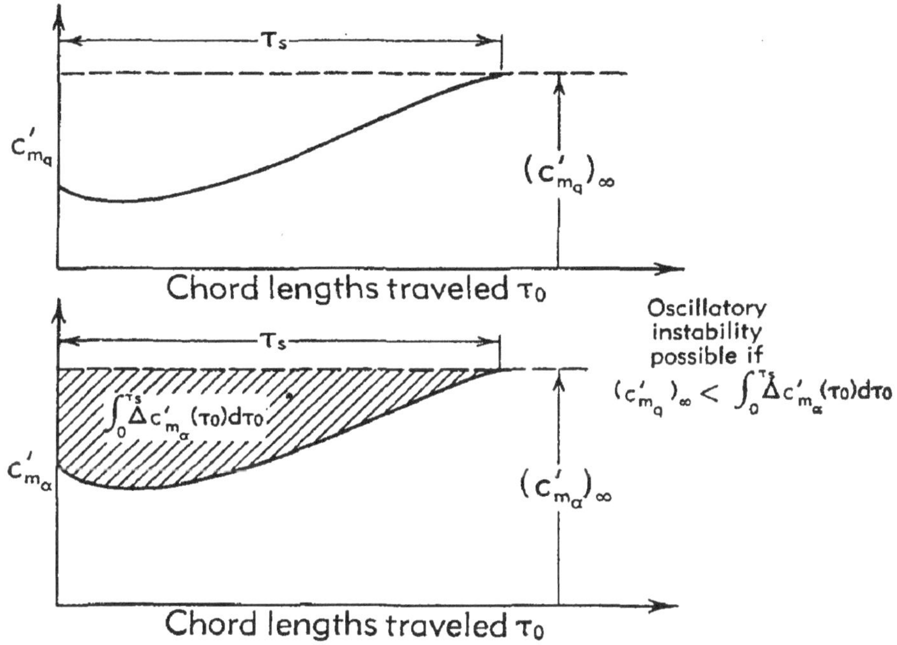

Fig. D,26c. Condition for oscillatory instability.

coefficients can be evaluated readily. For example, the expression for M_4' becomes

$$-kM_4' = (c'_{m_\alpha})_\infty - \int_0^{\tau_s} \Delta c'_{m_\alpha}(\tau_0)d\tau_0 - 2k^2\int_0^{\tau_s} [2\tau_0\Delta c'_{m_\alpha}(\tau_0) - \tau_0^2\Delta c'_{m_\alpha}(\tau_0)]d\tau_0 + \cdots \quad (26\text{-}9)$$

It is interesting to examine these equations briefly with regard to the problem of one-degree-of-freedom oscillatory instability. As was shown in [71], an oscillatory stability boundary is given by the equation $M_4 = 0$. Hence, a wing pivoted about its leading edge can be neutrally stable if $M_4' = 0$. Such a condition arises only if the frequency is very low[41] for which case the expansion for kM_4' applies. Hence, to a first order the stability boundary is given by the condition

$$(c'_{m_\alpha})_\infty = \int_0^{\tau_s} \Delta c'_{m_\alpha}(\tau_0)d\tau_0 \quad (26\text{-}10)$$

[41] That is, the moment of inertia is large. Experimental verification of the existence of this type of instability can be found in [75].

This leads to the interpretation of the indicial curves shown in Fig. D,26c, namely, an airfoil pivoted at its leading edge can have an oscillatory instability if the steady state value of c'_{m_θ} is less than the area between the steady state value of $c'_{m\alpha}$ and its indicial value when expressed in chord lengths traveled. The above criterion is valid for indicial functions applying to wings flying at subsonic speeds, although the expansions in Eq. 26-8 and 26-9 are not.

D,27. Discussion of Some Two-dimensional Unsteady Problems. It was pointed out in Art. 25 that the analysis of two-dimensional

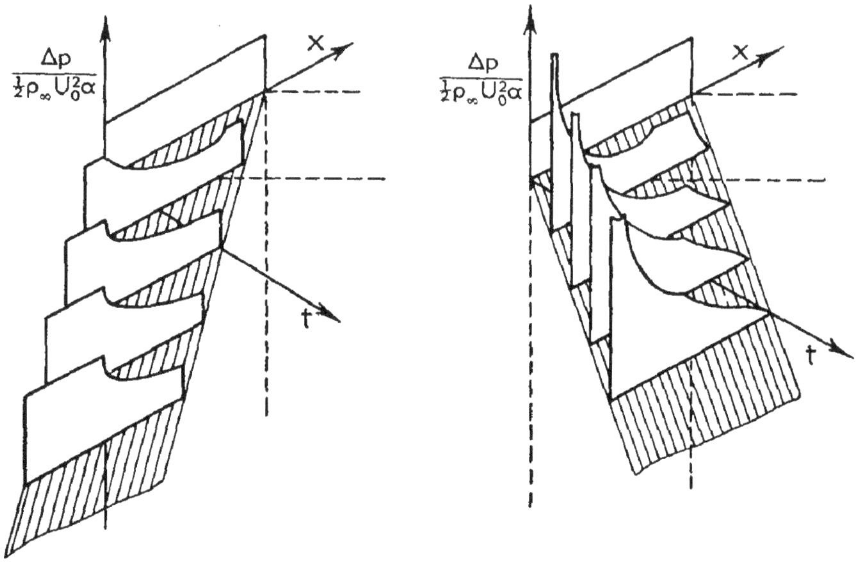

Fig. D,27a. Variation of indicial load coefficient on airfoils flying at supersonic or subsonic speeds. Left, supersonic. Right, subsonic.

unsteady lift problems is analogous to the study of three-dimensional steady state lifting surface problems. The evaluation on this basis of the indicial lift and pitching moment on sinking and pitching two-dimensional wings has been carried out in [*1*] for both subsonic and supersonic Mach numbers.

The variation of the loading over a section of a sinking wing traveling at a supersonic speed is shown in Fig. D,27a (left). The initial value of the loading coefficient is $4\alpha/M_0$, and the way in which it changes with time to become the familiar steady state value of $4\alpha/\sqrt{M_0^2 - 1}$ is indicated. For the purpose of comparison, the variation of the load over a sinking wing section traveling at a subsonic speed is also shown in Fig. D,27a (right). Again the initial value of the loading coefficient is $4\alpha/M_0$, but in this case as time passes the loading approaches (asymptotically) the familiar subsonic steady state value given by

$$\frac{\Delta p}{\frac{1}{2}\rho_\infty U_0^2} = \frac{4\alpha}{\sqrt{1 - M_0^2}} \sqrt{\frac{c - x}{x}}$$

The variation of the total section lift coefficient with both time and Mach number is shown in Fig. D,27b. At the two extremes, $\tau_0 = 0$ and $\tau_0 = \infty$, it is apparent that the total lift obeys the simple rules noted above for the loading. Thus, at $\tau_0 = \infty$ the lift coefficient is $4\alpha/\sqrt{M_0^2 - 1}$ and $2\pi\alpha/\sqrt{1 - M_0^2}$ for supersonic and subsonic Mach numbers, respectively. At $M_0 = 1$ the theory gives an infinite value for the lift.

At $\tau_0 = 0$ the lift coefficient is $4\alpha/M_0$ for all Mach numbers. From this we make two important observations. First, at $M_0 = 0$ the initial

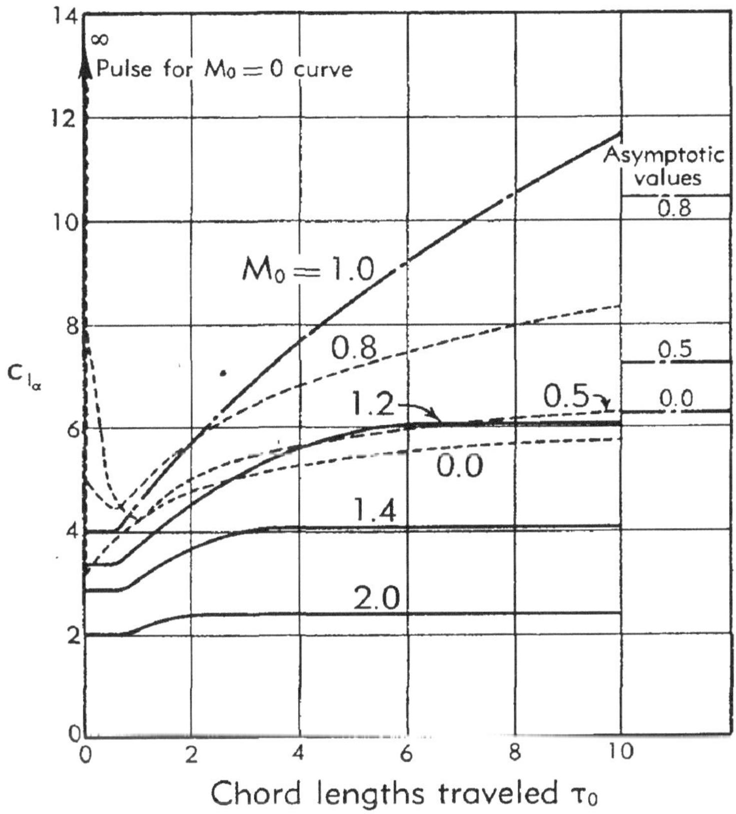

Fig. D,27b. Variation of indicial lift coefficient on a sinking wing with chord lengths traveled, for various Mach numbers.

value of the indicial lift is infinite, as indicated in Fig. D,27b by the arrow along the c_{l_α} axis. (The initial *variation* of the lift in this incompressible case is given by a pulse function. That is, although $c_{l_\alpha}(0)$ is infinite, $c_{l_\alpha}(0+) = \pi$ and

$$\lim_{\epsilon \to 0} \int_0^\epsilon c_{l_\alpha}(\tau_0) d\tau_0 = \frac{\pi}{2}$$

The study of the incompressible case is classical [*76*], also VII,F.) Second, at $M_0 = 1$ the initial value of the lift is similar in magnitude to the initial value at neighboring Mach numbers. In fact, at $M_0 = 1$ the variation of lift in the range from zero to around 10 chord lengths is not remarkably different in magnitude or shape from the variation of the other indicial curves in the interval $0 < M_0 < 2$. The analytical expression for the section lift coefficient in this range is simply {*1,35*}

$$c_l = 4\alpha, \qquad 0 \leqq \tau_0 \leqq \tfrac{1}{2}$$

$$c_l = \frac{4\alpha}{\pi}\left(2\sqrt{2\tau_0 - 1} + \cos^{-1}\frac{\tau_0 - 1}{\tau_0}\right), \qquad \tau_0 \geqq \frac{1}{2} \tag{27-1}$$

As time increases the magnitude of the lift grows indefinitely and eventually violates the basic assumptions upon which the theory is founded.

The indicial responses shown in Fig. D,27b were used in Eq. 26-6 (except for the case $M_0 = 0.5$) to calculate the flutter derivatives L_1 and L_2 and these, in turn, were used according to Eq. 26-7 to compute the maximum value of c_{l_α} attained by an oscillating wing. The results are shown in Fig. D,27c (top). At $k = 0$ the angle-of-attack input is simply a step function and $|c_{l_\alpha}|$ is given by the final value on the indicial response curves. As the frequency is increased, however, it is apparent that the shape of the $|c_{l_\alpha}|$ curves undergoes a complete change. Fig. D,27c (bottom) indicates this more clearly. The curve for $k = 0$ shows the familiar steady state behavior, the Prandtl-Glauert correction applies, and the value of the lift near $M_0 = 1$ is extremely high. For $k = 0.5$, however, the Prandtl-Glauert rule obviously no longer applies and the maximum lift does not even occur at $M_0 = 1$ but rather near $M_0 = 0.75$. As the frequency of the angle-of-attack oscillation increases, $|c_{l_\alpha}|$ approaches the initial value of the indicial response $4/M_0$, the latter being represented by the hyperbola labeled $k = \infty$ in the figure.

At $M_0 = 1$ the value of $k(L_1 + iL_2)$ has a relatively simple expression.[42] Thus, placing Eq. 27-1 into Eq. 26-6, one can show

$$k(L_1 + iL_2) = \left[\frac{\cos k + \sin k}{\sqrt{2\pi k}} + S(k) - C(k)\right] + i\left[\frac{\cos k - \sin k}{\sqrt{2\pi k}} + S(k) + C(k)\right] \tag{27-2}$$

where $S(k)$ and $C(k)$ are Fresnel's integrals [*78*] defined by the integral

$$C(k) - iS(k) = \int_0^k \frac{e^{-it}dt}{\sqrt{2\pi t}}$$

The magnitudes of $|c_{l_\alpha}|$ shown in Fig. D,27c agree with the numerical results presented in [*71*] for supersonic Mach numbers (as they should, since the methods used are operationally equivalent). The values for subsonic Mach numbers are in good agreement (where a comparison

[42] Extensive calculations for the case $M_0 = 1$ of all the flutter derivatives mentioned in the previous article as well as those due to flap deflection are given in [*77*].

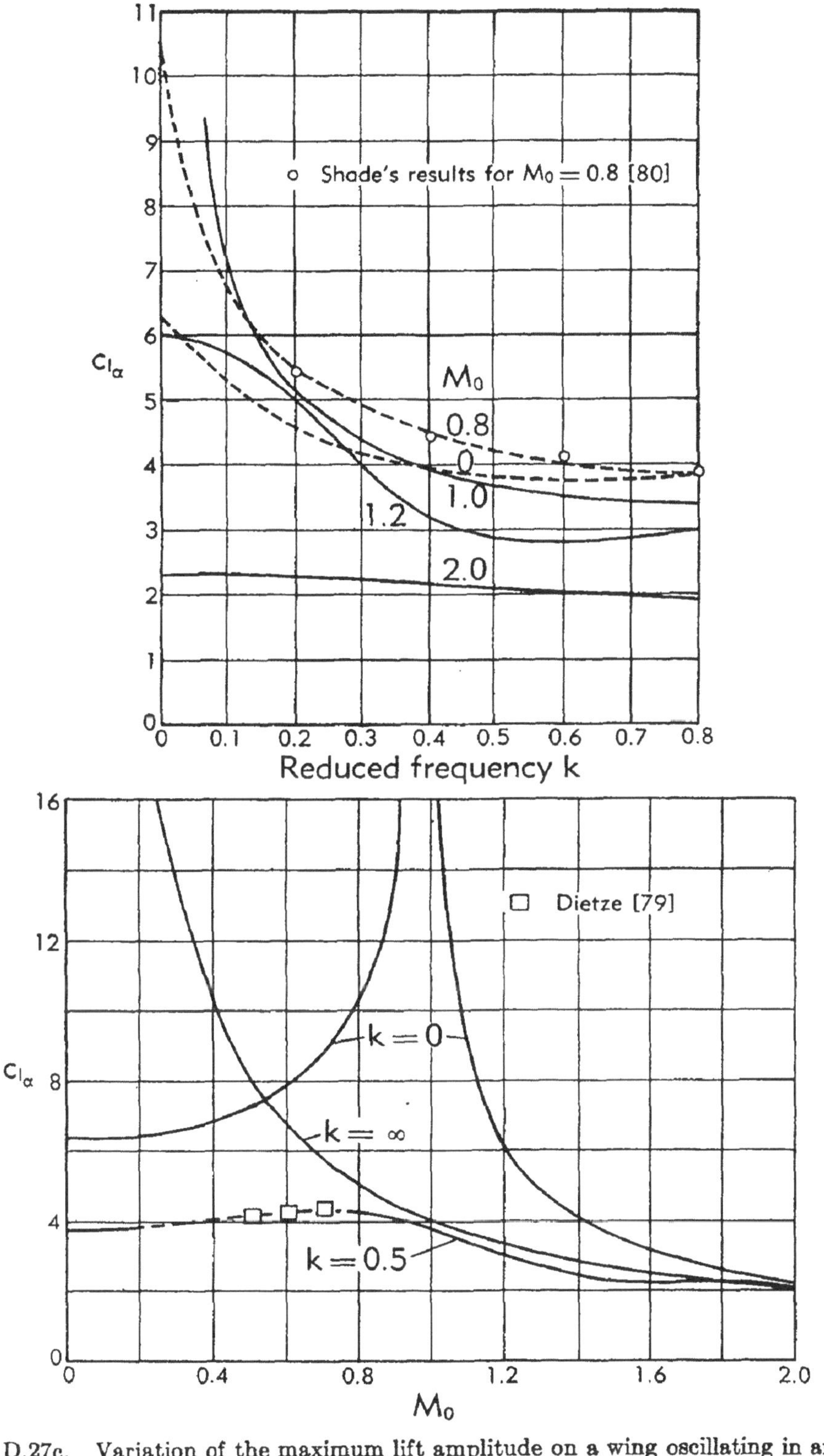

Fig. D,27c. Variation of the maximum lift amplitude on a wing oscillating in angle of attack (without pitching) with reduced frequency and Mach number.

could be made) with those obtained by the methods of Dietze and Schade [79,80]. The comparison in the subsonic case is more significant since the calculations contain approximations that are not equivalent.

The time variation of lift on a two-dimensional wing penetrating a sharp-edged gust with a uniform upward velocity w_0 has also been obtained [70,81]. The variation of the lift on a wing that is restrained so that it

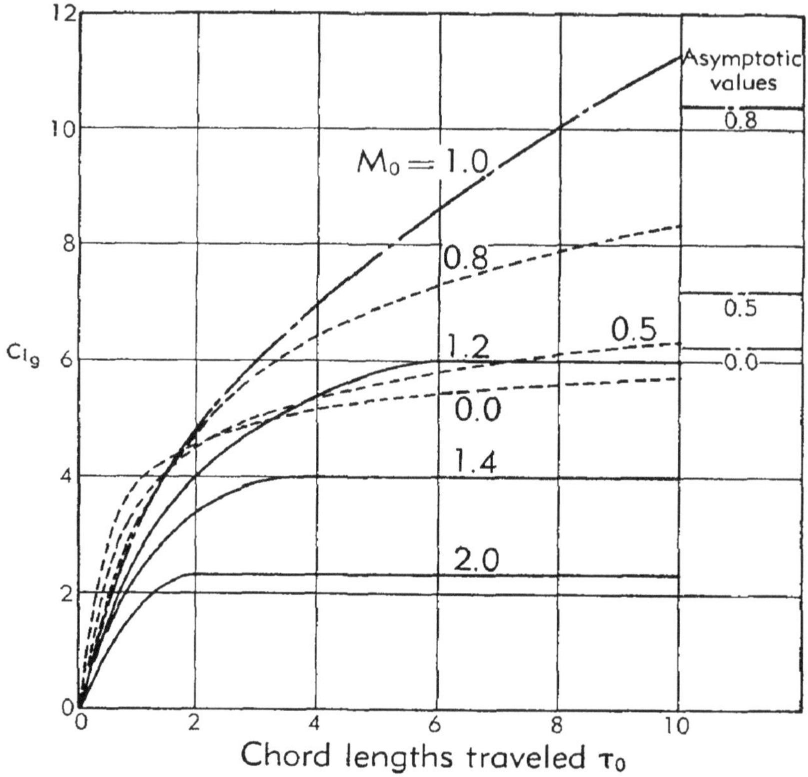

Fig. D,27d. Response of a restrained wing entering a sharp-edged gust, for several Mach numbers.

cannot move vertically (has infinite mass) is shown in Fig. D,27d for various Mach numbers. These results can be extended by means of Eq. 26-5 to give the variation of lift c_l^* on an unrestrained wing (with pitching neglected) entering a sharp-edged gust. Thus [81]

$$c_l^* = c_{l_g}(\tau_0) - \frac{1}{\mu}\int_0^{\tau_0} c_{l_\alpha}(\tau_0 - \tau_1)c_{l_g}(\tau_1)d\tau_1 + \frac{1}{\mu^2}\int^{\tau_0} c_{l_\alpha}(\tau_0 - \tau_1)d\tau_1 \int_0^{\tau_1} c_{l_\alpha}(\tau_1 - \tau_2)c_{l_g}(\tau_2)d\tau_2 - \cdots \quad (27\text{-}3)$$

where $\mu = 2m/\rho_\infty c^2$. The variation of c_l^* with chord lengths traveled τ_0 is shown in Fig. D,27e (top) for $\mu = \infty$, 300, 100, and 60 and a Mach number equal to 1.2. The maximum values of c_l^* obtained from similar

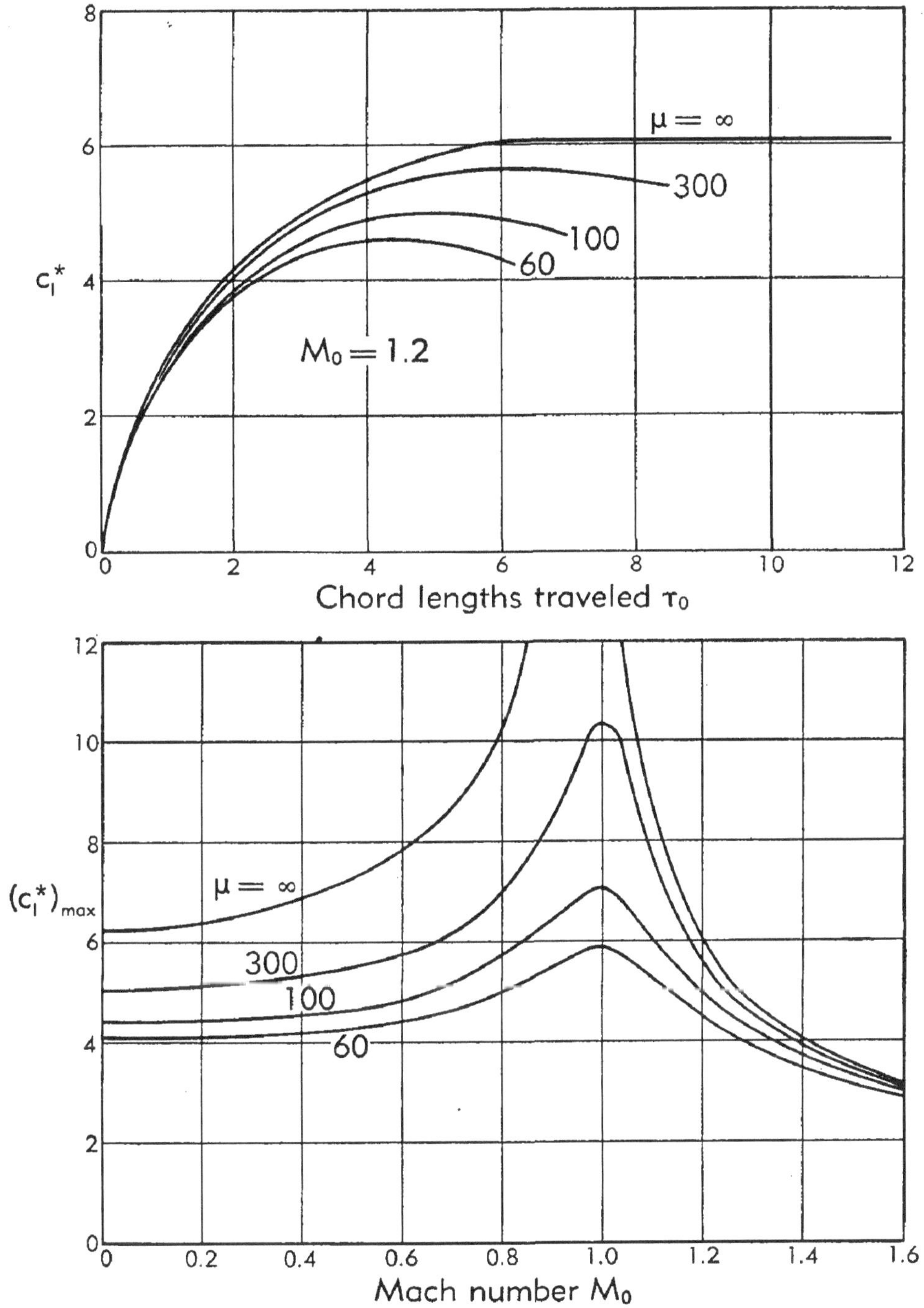

Fig. D,27e. Response of unrestrained wing entering a sharp-edged gust. Top, variation of response with chord lengths traveled, for various wing-air density ratios. Bottom, variation of maximum response with Mach number, for various wing-air density ratios.

calculations made for a range of flight Mach numbers and wing inertia factors are shown in Fig. D,27e (bottom). Notice that a wing for which $\mu = 100$ develops only about 57 per cent more lift in flying into a gust at $M_0 = 1$ than it does flying into the same gust at $M_0 = 0$.

D,28. Examples of Some Three-dimensional Unsteady Problems. The problems involving all three space dimensions as well as time are generally quite complicated. We can consider three simple examples, however, illustrating the use of sources and doublets in such problems and also indicating the applicability of Gardner's method.

The starting lift of a wing. One of the simplest and yet most general results derivable on the basis of the theory outlined in Art. 25 is the initial value of pressure on a wing surface starting impulsively from rest with a velocity U_0. The discussion will be made for a wing without thickness although the results will be seen to apply to the thickness cases as well.

Consider a surface with a planform such as that shown in Fig. D,28a. The acoustic planform with reference to a point $P(x, y)$ on the surface is a small circle of radius t. Since no point on the wing outside this circle can influence the pressure at P, the upper surface, except for a band of width t around the edge of the wing, is independent of the lower surface. It is, therefore, evident that the boundary value to be satisfied has been treated in Art. 25 under the subheading "Boundary-value problems involving noninteracting surfaces." The solution follows directly from Eq. 25-15.

Using polar coordinates with origin at the point $P(x, y)$, one has[43]

$$\varphi \cong -\frac{w(x, y)}{2\pi} \int_0^{2\pi} d\theta \int_0^{t} dr = -tw(x, y)$$

so that

$$\left(\frac{\Delta p}{\frac{1}{2}\rho_\infty U_0^2}\right)_{t=0} = \frac{4}{U_0 M_0}\left(\frac{\partial \varphi}{\partial t}\right)_{t=0} = -\frac{4w(x, y)}{U_0 M_0}$$

If $\alpha = -w(x, y)/U_0$ is the local slope of the wing, the expression for the load coefficient becomes

$$\left(\frac{\Delta p}{\frac{1}{2}\rho_\infty U_0^2}\right)_{t=0} = \frac{4\alpha(x, y)}{M_0}$$

The starting lift coefficient can, therefore, be written

$$C_L = \frac{4\bar{\alpha}}{M_0} \tag{28-1}$$

[43] The mean value theorem gives

$$\varphi = -\frac{w(\xi, \eta)}{2\pi} \iint_{S_a} \frac{dS}{r_0}$$

where ξ and η lie somewhere in S_a. Hence, as t approaches 0, ξ and η approach x and y respectively.

where $\bar{\alpha}$ is the average angle of attack of the surface and is defined by the equation

$$\bar{\alpha} = \frac{1}{S} \iint_S \alpha dy dx$$

S being the area of the wing planform.

Moving vortex. A further interesting application of Kirchhoff's formula is the development of the moving rectilinear vortex, where a vortex is represented by a line of constant circulation. The moving vortex is associated with a lifting surface[44] which starts impulsively at t = 0 and travels at a uniform velocity away from a fixed axial system. Since a

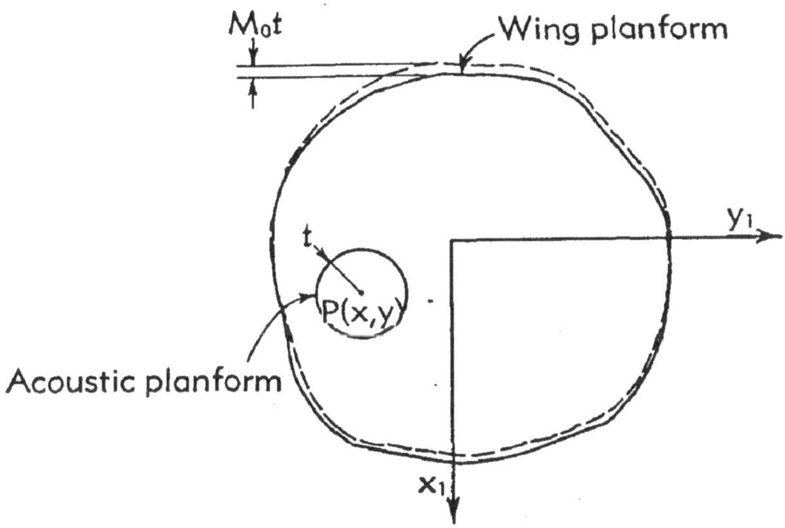

Fig. D,28a. Wing planform at $t = 0$ and wing and acoustic planforms a short time later.

vortex cannot end in a frictionless fluid (Helmholtz's result), the moving vortex must be adjoined by two trailing vortices and they, in turn, by a final starting vortex (see Fig. D,28b) which remains in the vicinity of the position at which the motion commenced. For the present, the assumptions are made, first, that the trailing vortices are far enough away so that they do not affect[45] the flow around the center of the moving and starting vortices (i.e. the flow is two-dimensional), and second, that the starting vortex remains fixed on the line along which the moving vortex first appeared. With these assumptions, the boundary conditions can be determined from the definition that the vortex system described is equivalent to a distribution of doublets of constant strength over the

[44] Associated in the sense that a distribution of these vortices will satisfy the boundary conditions for the lifting surface. The moving vortex is often referred to as the bound vortex.

[45] For a solution giving the effect of the trailing vortices see [*82*].

region of the $z = 0$ plane bounded by the moving vortex, the stationary vortex, and the vortices joining them at their extremities. Such a sheet of doublets gives rise to a constant potential difference, $\Delta\varphi$, across the sheet, while the normal derivative of the potential is continuous so that $\Delta(\partial\varphi/\partial z) = 0$. The solution is then given by Eq. 25-3. Thus

$$\varphi = \varphi_1 + \varphi_2 \tag{28-2}$$

where

$$\varphi_1 = -\frac{\Delta\varphi}{4\pi}\iint\limits_{S_a}\left(\frac{\partial}{\partial z}\frac{1}{r_0}\right)dS \tag{28-3}$$

and

$$\varphi_2 = \frac{1}{4\pi}\iint\limits_{S_a}\frac{1}{r_0}\left[\frac{\partial\Delta\varphi}{\partial t}\right]\frac{\partial r_0}{\partial z}\,dS \tag{28-4}$$

The evaluation of φ_1 imposes no difficulties except in integration but the calculation of φ_2 requires some special consideration. Since $\Delta\varphi$ is a

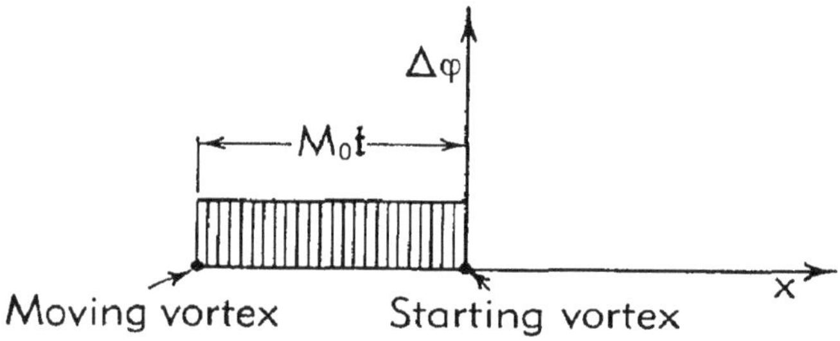

Fig. D,28b. Relation between $\Delta\varphi$ and the moving and starting vortices.

constant, $\partial\Delta\varphi/\partial t$ is zero everywhere except along the line of the moving vortex, and there it is infinite. In order to evaluate such an indeterminate form, assume that the potential difference rises linearily to its value $\Delta\varphi$ in the time interval ϵ (see Fig. D,28c). Then the value of $\partial\Delta\varphi/\partial t$ is $\Delta\varphi/\epsilon$ in the narrow strip between the two ellipses formed by the intersection of the inverse sound wave

$$(x - x_1)^2 + (y - y_1)^2 + z^2 = (t - \tau)^2$$

with the lines

$$x_1 = -M_0\tau$$

$$x_1 = -M_0(\tau - \epsilon)$$

When the equations for these two ellipses are formulated and placed in Eq. 28-4 there results an equation which can be represented by the expression

$$\varphi_2 = \frac{\Delta\varphi}{4\pi\epsilon}\left[F(t) - F(t - \epsilon)\right]$$

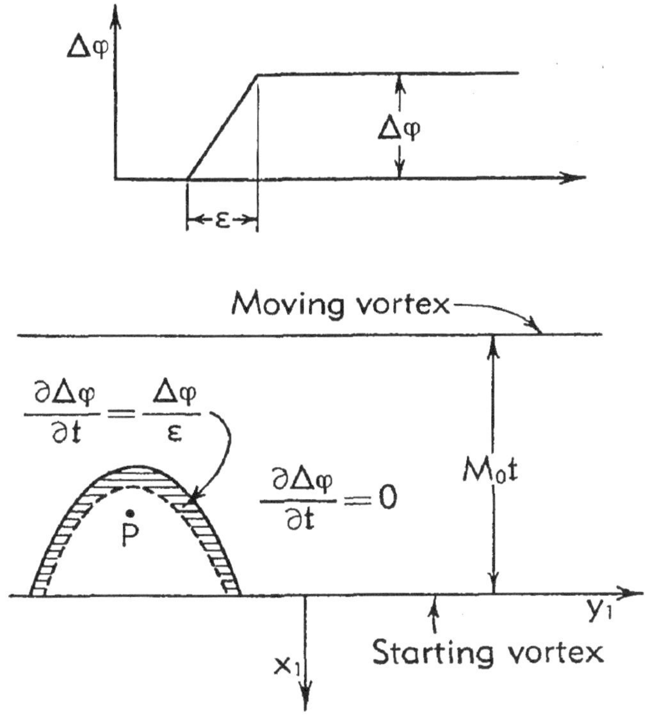

Fig. D,28c. Acoustic planform at $P(x, y)$ of $\Delta\varphi$ distribution with initial constant (and then zero) gradient.

where

$$F(\mathrm{t}) = \iint\limits_{S_a} \frac{1}{r_0} \frac{\partial r_0}{\partial z} dS$$

and S_a is the same acoustic planform used in Eq. 28-3. But since

$$\lim_{\epsilon \to 0} \frac{F(\mathrm{t}) - F(\mathrm{t} - \epsilon)}{\epsilon} = \frac{\partial F}{\partial \mathrm{t}}$$

the equation for φ_2 becomes

$$\varphi_2 = \frac{\Delta\varphi}{4\pi} \frac{\partial}{\partial \mathrm{t}} \iint\limits_{S_a} \frac{1}{r_0} \frac{\partial r_0}{\partial z} dS$$

The calculation of φ_1 and φ_2 over the acoustic planform given by the area within the curve

$$(x - x_1)^2 + (y - y_1)^2 + z^2 = \left(\mathrm{t} + \frac{x_1}{M_0}\right)^2$$

or between this curve and the y_1 axis gives for the regions defined by Fig. D,28d

Region 1

$$\varphi = \frac{\Delta\varphi}{2\pi}\left[\tan^{-1}\frac{(x + M_0 t)\sqrt{t^2 - x^2 - z^2}}{z(M_0 x + t)} - \tan^{-1}\frac{x}{zt}\sqrt{t^2 - x^2 - z^2}\right] \quad (28\text{-}5)$$

Region 2

$$\varphi = \frac{\Delta\varphi}{2\pi}\left[\pi\frac{z}{|z|} - \tan^{-1}\frac{x}{zt}\sqrt{t^2 - x^2 - z^2} + \tan^{-1}\frac{(x + M_0 t)\sqrt{t^2 - x^2 - z^2}}{z(M_0 x + t)}\right] \quad (28\text{-}6)$$

Region 3

$$\varphi = \frac{\Delta\varphi}{2}\frac{z}{|z|} \quad (28\text{-}7)$$

It is to be noted that Eq. 28-5 gives the potential everywhere within the cylinder in which disturbances can be affected in subsonic flow as well as over a large portion of the disturbed region in supersonic flow.

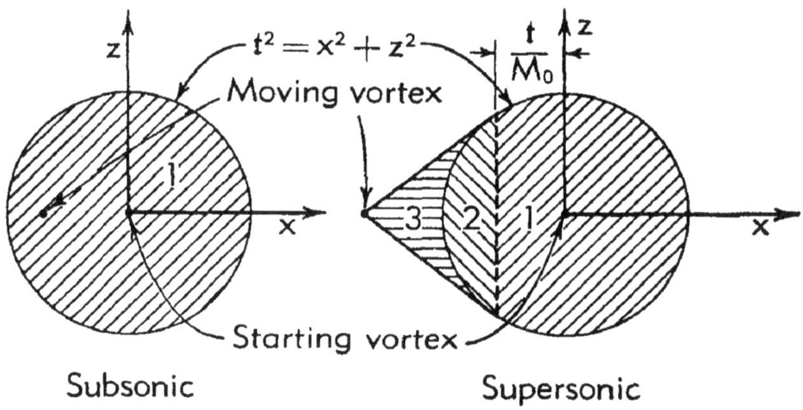

Fig. D,28d. Regions of influence for moving vortices traveling at subsonic and supersonic speeds.

The foregoing results reduce to more familiar expressions when their asymptotic expressions for large values of time are determined. A study of the two terms in Eq. 28-5 shows that as t becomes large the expression becomes

$$\varphi = \frac{\Delta\varphi}{2\pi}\left(\frac{\pi}{2}\frac{z}{|z|} - \tan^{-1}\frac{x}{z}\right)$$

This result provides the velocity potential for a stationary vortex in an incompressible medium, the jump in the value of the potential occurring along the plane of the vortex motion.

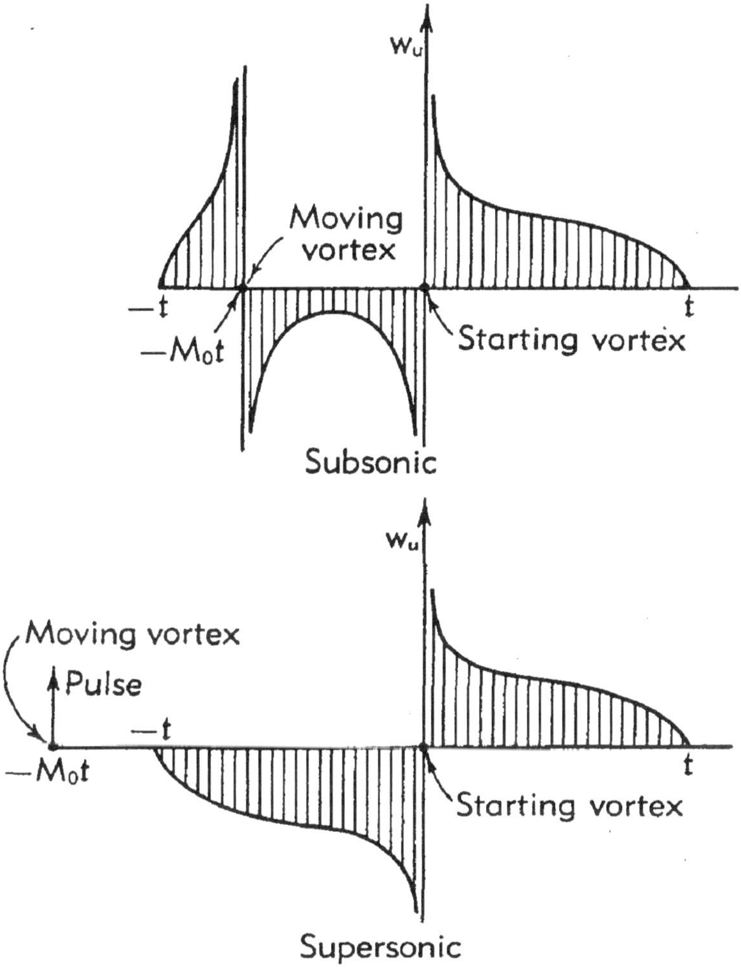

Fig. D,28e. Vertical induced velocity along $z = 0$ plane for starting vortex and subsonic and supersonic moving vortices.

If the axes are fixed in the moving vortex, new coordinates ξ, ζ, τ need to be introduced as defined by the equations

$$\xi = x + M_0 t$$

$$\zeta = z$$

$$\tau = t$$

As τ becomes large, Eq. 28-5 reduces to the form

$$\varphi = \frac{\Delta\varphi}{2\pi}\left(-\frac{\pi}{2}\frac{z}{|z|} + \tan^{-1}\frac{\xi}{\zeta\sqrt{1 - M_0^2}}\right)$$

when the free-stream Mach number is less than one. This result agrees with that of Glauert [3] for a vortex obtained from the steady state linearized theory of compressible flow, so that in the vicinity of the moving vortex the induced velocity field is identical to the perturbation field of a fixed vortex in a subsonic free stream.

It is interesting to inspect the vertical velocity induced by the vortex pair in the $z = 0$ plane. From Eq. 28-5 and 28-6 it follows that for $x^2 < \mathrm{t}^2$ the vertical velocity w for both subsonic and supersonic speeds is given by the relation

$$w = -\frac{\Delta\varphi}{2\pi}\left[\left(\frac{M_0 x + \mathrm{t}}{x + M_0 \mathrm{t}}\right) - \frac{\mathrm{t}}{x} - \frac{1}{\sqrt{\mathrm{t}^2 - x^2}}\right]$$

$$= \frac{\Delta\varphi}{2\pi}\frac{M_0\sqrt{\mathrm{t}^2 - x^2}}{x(x + M_0\mathrm{t})} \tag{28-8}$$

Fig. D,28e shows the variation of w along the x axis for a moving vortex traveling at both subsonic and supersonic speeds. For both cases in the vicinity of the starting vortex w behaves as it does near a two-dimensional steady state fixed vortex, having large values of downwash on one side and large values of upwash on the other. The velocity distribution around the moving vortex is quite different, however, depending on whether its Mach number is greater or less than one. In the subsonic case the distribution is similar to that for the starting vortex, but in the supersonic case the velocity is a pulse existing at the position of the vortex, in agreement with the results of Art. 5. Again in both cases the velocity falls to zero at the primary wave front (i.e. at $x = \pm\mathrm{t}$ in the figure).

A final remark can be made concerning Eq. 28-5 in connection with incompressible theory. In this case, the value of a_∞ becomes infinitely large and, if the equation is written in the form

$$\varphi = \frac{\Delta\varphi}{2\pi}\left[\tan^{-1}\frac{(x + U_0 t')\sqrt{a_\infty^2 t'^2 - x^2 - z^2}}{z\left(\dfrac{U_0 x}{a_\infty} + a_\infty t'\right)} - \tan^{-1}\frac{x\sqrt{a_\infty^2 t'^2 - x^2 - z^2}}{z a_\infty t'}\right]$$

it is apparent that, as a_∞ goes to infinity the velocity potential becomes

$$\varphi = \frac{\Delta\varphi}{2\pi}\left(\tan^{-1}\frac{x + U_0 t'}{z} - \tan^{-1}\frac{x}{z}\right) \tag{28-9}$$

This latter equation gives the velocity potential for a moving and a fixed vortex pair in an incompressible fluid.

A rectangular wing tip traveling at a supersonic speed. Consider the corner of a plate (the trailing edge induces no disturbance on the wing since the speed is supersonic) such as that shown in Fig. D,28f. The wing starts suddenly to move with a velocity U_0 at $\mathrm{t} = 0$. After a certain time has passed the envelopes of the starting waves from the two edges and the corner have formed the boundaries of the eight regions shown in the figure. For example, the trace in the $z = 0$ plane of the starting spherical wave from the corner is shown as a circle bounding region 5.

Let us consider the wing to be a plate of given twist and apply Gardner's method, outlined in Art. 25, to find the solution in region 8.

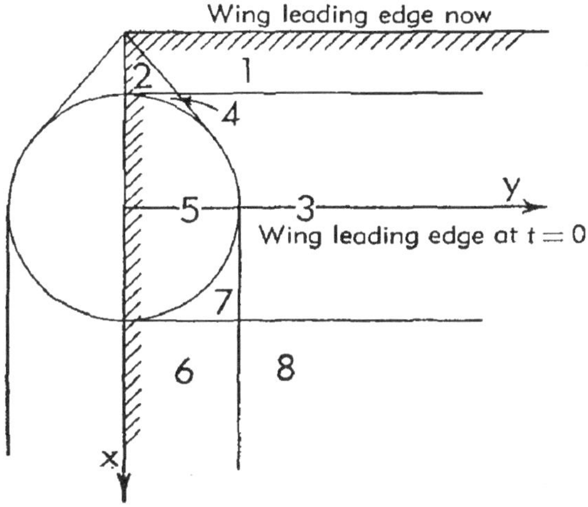

Fig. D,28f. Various regions in which analytic expressions for loading on starting rectangular wing tip change form.

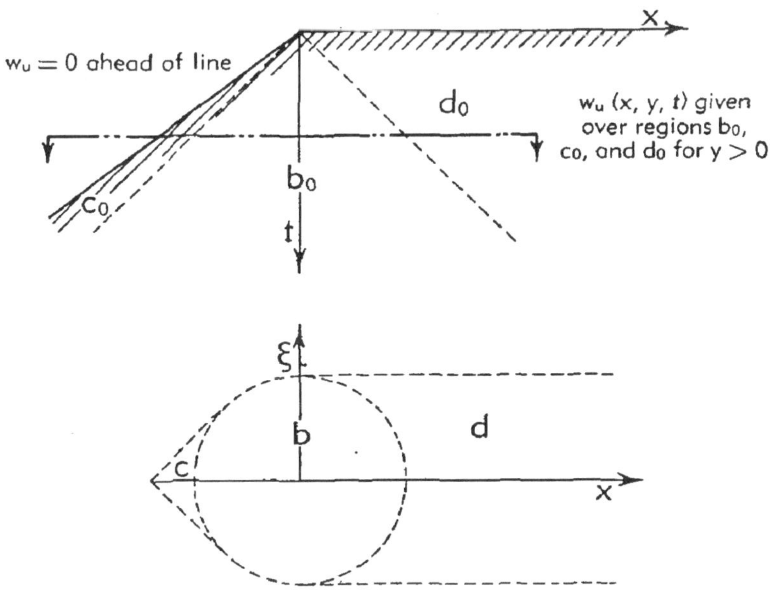

Fig. D,28g. Boundary values in $\xi = 0$ plane and wave envelope in an x, ξ plane.

We study first the t, x plane for $z = 0$ and for y equal to a given positive constant. Eq. 25-10 applies; the solution is given by Eq. 25-12; and the boundary conditions are specified over regions b_0, c_0, and d_0 of the $\xi = 0$ plane as shown in Fig. D,28g (top). A study reveals that the solution in region 8 of Fig. D,28f depends on the evaluation of $W(x, y, t, \xi)$ throughout region d of Fig. D,28g (bottom). For this case the area of integration in the x_1, t_1 plane is bounded by the hyperbola $t_1 = t - \sqrt{(x - x_1)^2 + \xi^2}$ and the x_1 axis. Hence, for this region Eq. 25-12 becomes

$$W_a(x, y, t, \xi) = -\frac{1}{\pi} \int_{x-\sqrt{t^2-\xi^2}}^{x+\sqrt{t^2-\xi^2}} dx_1$$

$$\int_0^{t-\sqrt{(x-x_1)^2+\xi^2}} \frac{w_u(x_1, y, t_1)}{\sqrt{(t - t_1)^2 - (x - x_1)^2 - \xi^2}} dt_1 \quad (28\text{-}10)$$

Consider next the ξ, y, z plane, controlled by Eq. 25-7. The solution to this equation in terms of $\varphi(x, y, 0, t)$ is given by Eq. 25-13 where, for

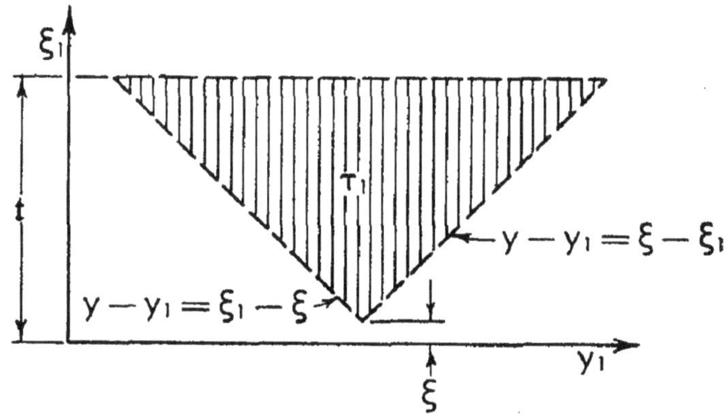

Fig. D,28h. Region of integration τ_1 when the point $(x, y, 0, t)$ is on the wing and outside the wave envelopes from the side edge.

region 8, the area of integration is that shown in Fig. D,28h. If the sequence of integration is taken to be $d\xi_1$ first followed by dy_1 the operation $\lim_{\xi\to 0} \partial/\partial\xi$ gives (see Art. 10)

$$\varphi(x, y, 0, t) = -\frac{1}{\pi} \int_{y-t}^{y} dy_1 \int_{y-y_1}^{t} d\xi_1 \frac{\xi_1 W_a(x, y_1, t, \xi_1)}{[\xi_1^2 - (y - y_1)^2]^{\frac{3}{2}}}$$

$$-\frac{1}{\pi} \int_{y}^{y+t} dy_1 \int_{y_1-y}^{t} d\xi_1 \frac{\xi_1 W_a(x, y_1, t, \xi_1)}{[\xi_1^2 - (y - y_1)^2]^{\frac{3}{2}}} \quad (28\text{-}11)$$

Let us consider the first set of integrals in Eq. 28-11. Further, let us consider, for the present, only the $d\xi_1$ integral in this set. Thus, writing for W_a its expression given by Eq. 28-10, we study the group

$$I = \frac{-1}{\pi} \int_{y-y_1}^{t} d\xi_1 \int_{x-\sqrt{t^2-\xi_1^2}}^{x+\sqrt{t^2-\xi_1^2}} dx_1 \int_0^{t-\sqrt{(x-x_1)^2+\xi_1^2}} dt_1 Q w_u(x_1, y_1, t_1) \quad (28\text{-}12)$$

where

$$Q = \frac{\xi_1}{[\xi_1^2 - (y - y_1)^2]^{\frac{3}{2}} \sqrt{(t - t_1)^2 - (x - x_1)^2 - \xi_1^2}} \quad (28\text{-}13)$$

The dx_1 and $d\xi_1$ integrals can be interchanged because no residual exists in the x_1, ξ_1 plane. There results

$$I = \frac{-1}{\pi} \int_L dx_1 \oint_A d\xi_1 \int dt_1 Q w_u(x_1, y_1, t_1) \tag{28-14}$$

where A is the area shown in Fig. D,28i and L is the interval bounded by $x - \sqrt{t^2 - (y - y_1)^2}$ and $x + \sqrt{t^2 - (y - y_1)^2}$. Now set

$$I_0 = \frac{-1}{\pi} \int_L dx_1 \int_A dt_1 \oint d\xi_1 Q w_u(x_1, y_1, t_1) \tag{28-15}$$

Then

$$I - I_0 = R \tag{28-16}$$

where R is the residual (see Art. 12) and is not zero since an inherent singularity (see Art. 12) exists at the point P.

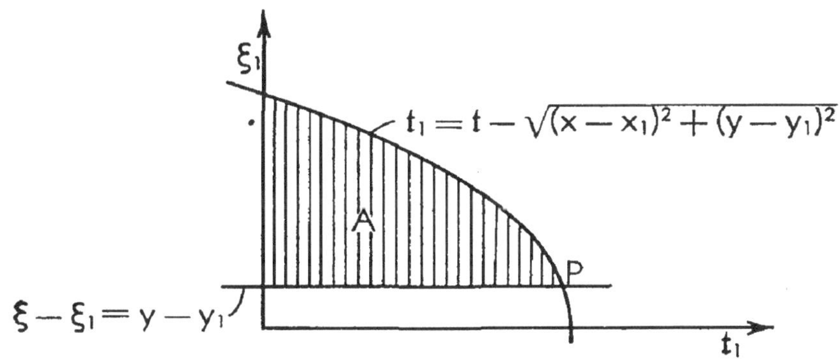

Fig. D,28i. Area of integration in the ξ_1, t_1 plane.

Making the substitution $\xi_1^2 = \zeta$ and recalling the identity given by Eq. 14.1-11, we notice that for the area shown in Fig. D,28i the quantity $\oint Q d\xi_1$ is zero. Hence, $I_0 = 0$ and, by Eq. 28-16, $I = R$. The residual can be evaluated according to the methods outlined in Art. 12. One finds, finally,

$$I = R = \frac{1}{2} \int_L dx_1 \frac{w_u(x_1, y_1, t - r)}{r} \tag{28-17}$$

where

$$r = \sqrt{(x - x_1)^2 + (y - y_1)^2}$$

Substituting Eq. 28-17 into Eq. 28-11 and making use of the fact that the second set of integrals in the latter equation leads to the same result as was found for the former, we can write

$$\varphi(x, y, 0, t) = \frac{-1}{2\pi} \int_{y-t}^{y+t} dy_1 \int_{x-\sqrt{t^2-(y-y_1)^2}}^{x+\sqrt{t^2-(y-y_1)^2}} dx_1 \frac{w_u(x_1, y_1, t - r)}{r} \tag{28-18}$$

It is apparent that the area of integration in Eq. 28-18 is a circle. The acoustic planform with reference to any point x, y in region 8 of Fig. D,28f is this same circle. Hence, Eq. 28-18 can be rewritten

$$\varphi(x, y, 0, t) = \frac{-1}{2\pi} \iint\limits_{S_0} \frac{[w_u]}{r} \, dy_1 dx_1$$

which is identical to Eq. 25-15 in the $z = 0$ plane. Thus, we note that for the special region under consideration the residual in the ξ_1, t_1 plane gives the integrand of the part of Kirchhoff's solution representing a source distribution. This result is obviously correct since region 8 is not affected by either the leading or the side edge.

CHAPTER 7. RECIPROCITY RELATIONS AND REVERSE FLOW THEOREMS IN AERODYNAMICS

D,29. Introduction. One of the most significant achievements in the development of supersonic analysis has been a strengthening of methods that provide useful techniques for wing theory throughout all flight speeds. An outstanding example of this trend is found in the evolution of reverse flow theorems. Attention to such relations was initiated by von Kármán [*85*] who first noted the invariance of drag with forward and reversed directions of flight for a nonlifting symmetrical wing at supersonic speeds. Hayes' [*34,86*] early investigations of reverse flow theorems were concerned with supersonic flow as were also results given by Munk [*87*], Harmon [*88*], and Flax [*89*]. Brown [*90*], however, showed that lift curve slope remains the same for wings in forward and reverse flight in both the subsonic and supersonic regime and the subsequent investigations of Ursell and Ward [*91*], R. T. Jones [*92*], Hayes, and Flax [*96,97*] were concerned consistently with compressible flow fields in general. In these later papers, equivalences were derived which were essentially reciprocity relations between two modes of motion of a dynamic system and, as will be shown subsequently, the results can be applied to steady as well as unsteady flight. Moreover, when approached in this manner, the theory takes its place alongside other reciprocity relations in electrostatics, elasticity, and dynamics. Various references to such theorems are given by Lord Rayleigh [*14*] in his two volumes on the theory of sound and by Clerk Maxwell [*93*] in his work on electricity and magnetism. A paper by Lamb [*94*] is of particular interest since it contains the essential idea underlying reverse flow theorems in wing theory. Lamb remarks, as had Lord Rayleigh previously, that reciprocity relations between sound sources do not apply directly in a moving atmosphere.

He points out, however, that the reciprocity can be restored if the direction of the wind is also reversed. The derivations to be given here are taken from a paper by Heaslet and Spreiter [*117*].

As an introduction to the theory, we can restate Green's theorem for linear partial differential equations of second order with the independent variables $x_1, x_2, \ldots, x_m$ that may be thought of as rectangular coordinates in a space of m dimensions. Denote differentiation of the function $\psi(x_1, x_2, \ldots, x_m)$ with respect to the variables x_i and x_j by the subscript notation

$$(\psi)_{ij} \equiv \frac{\partial^2\psi}{\partial x_i \partial x_j}$$

and consider the differential equation

$$L(\psi) \equiv \sum_{i=1}^{m} \sum_{j=1}^{m} A_{ij}(\psi)_{ij} + B\psi = 0 \tag{29-1}$$

where, for the purposes at hand, $A_{ij} = A_{ji}$ are made independent of x_i and x_j and B is a constant. Such equations fall within the class of self-adjoint equations, and by Green's theorem it is possible to relate two arbitrary functions ψ and Ω by the integral expression

$$\iiint [\Omega L(\psi) - \psi L(\Omega)]dV = -\iint [\Omega D_n\psi - \psi D_n\Omega]dS \tag{29-2}$$

where the left member is a volume integral over a prescribed region in m-dimensional space and the right member is a surface integral over the hypersurface S enclosing the given region. Eq. 29-2 certainly holds for any region in which ψ and Ω and their first and second derivatives are continuous. The directional derivatives D_n are defined in terms of the direction cosines $n_1, n_2, \ldots, n_m$ of the inner normal to the surface S. Thus, setting

$$\sum_{j=1}^{m} n_j A_{ij} = N\nu_i$$

where $\nu_1, \nu_2, \ldots, \nu_m$ are the direction cosines of the so-called conormal, the directional derivative is defined by the expression

$$D_n\psi = N \sum_{i=1}^{m} \nu_i(\psi)_i = N(\psi)_\nu = N\frac{\partial\psi}{\partial\nu} \tag{29-3}$$

If, finally, ψ and Ω are assumed to satisfy Eq. 29-1, the left side of Eq. 29-2 vanishes and the resulting expression

$$\iint \Omega D_n\psi dS = \iint \psi D_n\Omega dS \tag{29-4}$$

is a general reciprocity relation which establishes the functional dependence between two arbitrary solutions of Eq. 29-1.

An interesting interpretation of Eq. 29-4 has been given by several writers (see, e.g. [*15*]) and applies to the particular case when ψ and Ω are identified with the perturbation velocity potential φ in the theory of incompressible fluid flow. The governing equation is Laplace's equation in three dimensions

$$\varphi_{xx} + \varphi_{yy} + \varphi_{zz} = 0$$

and the reciprocity relation takes the form, for two possible solutions φ and φ',

$$\iint \varphi \frac{\partial \varphi'}{\partial n} dS = \iint \varphi' \frac{\partial \varphi}{\partial n} dS \tag{29-5}$$

where the directional derivatives are now along the true normals to the surface S enclosing a three-dimensional volume. It is known that any actual state of motion of a liquid for which a single-valued velocity potential exists can be produced instantaneously by the application of a properly chosen system of impulsive pressures. These impulsive pressures are directly proportional to the velocity potential plus an arbitrary constant which may in the present case be associated with the pressure of the uniform stream. Eq. 29-5 is thus seen to represent summations of the cross products of impulse and normal velocity in two possible motions of a conservative system and is a special case of the dynamical theorem

$$\sum_{r=1}^{\infty} p_r \dot{q}'_r = \sum_{r=1}^{\infty} p'_r \dot{q}_r \tag{29-6}$$

where p_r, $\dot{q}_r$, and p'_r, $\dot{q}'_r$ are generalized components of impulse and velocity in any two possible motions of the system [*14*]. The identification, at any instant of time, of the perturbation velocity potential $\varphi(x, y, z, t')$ with an impulsive pressure can also be extended to include the entire regime of linearized compressible flow as defined by the wave equation.

If Eq. 29-4 is applied to the differential equation

$$(M_\infty^2 - 1)\psi_{xx} - \psi_{yy} - \psi_{zz} = 0$$

the reciprocity relation becomes

$$\iint \Omega \frac{\partial \psi}{\partial \nu} dS = \iint \psi \frac{\partial \Omega}{\partial \nu} dS \tag{29-7}$$

where the conormal has direction cosines proportional to $-\sqrt{M_\infty^2 - 1}\,\eta_1$, η_2, η_3.

D,30. Reverse Flow Theorems in Wing Theory. Consider a thin wing at possibly a small angle of attack and situated in the immediate vicinity of a flat surface which is designated the x, y plane. For sufficiently small thickness and angle of attack the perturbation velocity potential or any of the perturbation velocity components of the wing satisfy the equation

$$\beta^2\psi_{xx} + \psi_{yy} + \psi_{zz} = 0 \tag{30-1}$$

where $\beta^2 = 1 - M_\infty^2 = 1 - (U/a_\infty)^2$. Eq. 30-1 applies in forward or reverse flow provided the free stream Mach numbers M_∞ and M'_∞ of the two flows are equal in magnitude. In Fig. D,30a the wing with planform P is indicated along with the possible vortex wakes as the configuration would appear if the free streams were superposed. It will be assumed that the wing chord is finite and that the profiles are closed.

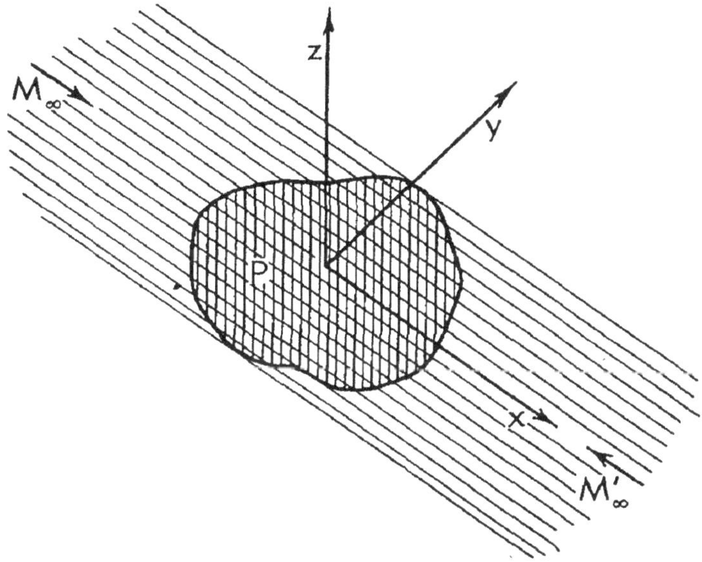

Fig. D,30a. Wing planform and vortex wake in forward and reversed flow.

We wish to apply the reciprocity relation (29-4) to these two flow fields. In the case of subsonic flow we therefore consider two hemispherical regions of large radius lying above and below the plane of the wing. The surface integrals of Eq. 29-4 will thus extend over a hemisphere with center at the wing and a flat surface that lies immediately adjacent to the $z = 0$ plane. This latter surface is subsequently to be brought into coincidence with the plane of the wing but must be considered first in its displaced position since only then can the flow be assumed free of possible singularities in perturbation velocities and their gradients. As in wing theory, in general, the attenuation of the perturbation potential and its gradients may be assumed of such a nature that the contributions of the integrals over the hemispherical surfaces vanish in the limit as the radius becomes infinitely large. It remains to consider the integrals over the surfaces at $z = 0 + \epsilon$ and $z = 0 - \epsilon$. Denote these surfaces as σ_u and σ_l respectively where the subscripts u and l specify values above and below the $z = 0$ plane. Eq. 29-4 takes the two forms

$$\iint_{\sigma_u} \psi \frac{\partial \psi'}{\partial n} dS = \iint_{\sigma_u} \psi' \frac{\partial \psi}{\partial n} dS, \qquad n \text{ directed upward} \qquad (30\text{-}2a)$$

and

$$\iint_{\sigma_l} \psi \frac{\partial \psi'}{\partial n} dS = \iint_{\sigma_l} \psi' \frac{\partial \psi}{\partial n} dS, \qquad n \text{ directed downward} \qquad (30\text{-}2b)$$

where the primes denote conditions for the reversed flow.

In Eq. 30-2, ψ is now replaced by the perturbation velocity potential $\varphi(x, y, z)$ and ψ' by the x or streamwise component of perturbation velocity $u(x, y, z)$. By virtue of the irrotationality of the flow, the gradients of u and w are related by the expression $\partial u/\partial n = \pm \partial w/\partial x$, and Eq. 30-2a and 30-2b can both be written in the form

$$\iint \varphi \frac{\partial w'}{\partial x} dS = \iint u'w dS$$

Integration by parts over either of the surfaces σ_u or σ_l gives

$$\int_{-\infty}^{\infty} \varphi w' \Big]_{-\infty}^{\infty} dy - \iint uw'dS = \iint u'wdS \qquad (30\text{-}3)$$

At $x = \infty$ the potential φ for the forward flow vanishes and at $x = -\infty$ the upwash w' in the reverse flow vanishes so that the first term on the left is zero. The remaining double integrals have for their surfaces of integration the displaced planes σ_u and σ_l. In order to obtain a concise form of the reverse flow theorem it suffices to subtract the integrals extended over σ_l from the integrals over σ_u and let the planes approach coincidence with the plane of the wing. Since w, w' and u, u' are continuous everywhere except possibly in the immediate vicinity of the wing, the integration region can be restricted to planes slightly above and below the wing but extending beyond the wing edges. Provided the singularities at the edges can be disregarded, the analytic expression of the reverse flow theorem becomes

$$-\iint_P (u_u w_u' - u_l w_l')dS = \iint_P (u_u' w_u - u_l' w_l)dS \qquad (30\text{-}4)$$

for either lifting surfaces or symmetrical wings where P is the planform of the wing in the $z = 0$ plane.

It remains to discuss the effect of edge singularities. In the case of a lifting surface, square-root singularities in both u and w can occur at the leading and trailing edges just on and off the wing, respectively. In the combined flow fields, the limiting process would then yield residue terms analogous to the leading edge thrust of a lifting plate (Art. 9). If, however, the Kutta condition is imposed at the trailing edge for both the flow in forward and reverse direction, a combination of singularities does not occur and Eq. 30-4 is valid. If the edges of a symmetrical wing are rounded so as to produce square-root singularities in w on the wing, square-root singularities in u occur just off the wing and terms corresponding to the leading edge drag (Art. 9) appear. If the geometry of the wing is fixed in the forward and reverse flow, however, the effect of these terms is canceled and Eq. 30-4 still applies. It is important to point out that it is not necessary to impose the Kutta condition in order to get a reverse flow relation provided the nature of the singularities at the wing edges is known.

The two sides of Eq. 30-4 are expressed in terms of the same coordinate system but it is often preferable to associate with each of the two streams an x axis extending in the stream direction. To this end introduce now the subscripts 1 and 2 to denote the forward and reverse flow and the two coordinate systems. Thus, in general $x_1 = -x_2 + \xi$, $y_1 = -y_2 + \eta$, $z_1 = z_2$ where ξ and η are arbitrary constants. Eq. 30-4 then becomes

$$\iint_{P_1} [u_{u1}(x_1, y_1)w_{u2}(x_1, y_1) - u_{l1}(x_1, y_1)w_{l2}(x_1, y_1)]dx_1dy_1$$
$$= \iint_{P_2} [u_{u2}(x_2, y_2)w_{u1}(x_2, y_2) - u_{l2}(x_2, y_2)w_{l1}(x_2, y_2)]dx_2dy_2 \quad (30\text{-}5)$$

In the case of a symmetrical nonlifting wing, theory yields the relations $u_u = u_l$, $w_u = -w_l$ and for lifting surfaces $u_u = -u_l$, $w_u = w_l$. It follows that in either case Eq. 30-5 reduces to the form

$$\iint_{P_1} u_1(x_1, y_1)w_2(x_1, y_1)dx_1dy_1 = \iint_{P_2} u_2(x_2, y_2)w_1(x_2, y_2)dx_2dy_2 \quad (30\text{-}6)$$

where the velocity components can be evaluated on either the upper or lower surface of the planform. Furthermore, since both integrals are evaluated over the same area and the wing profiles have been assumed closed, the linearized pressure relation $p - p_\infty = -\rho_\infty Uu$ permits one to rewrite Eq. 30-6 as

$$\iint_{P_1} p_1(x_1, y_1)w_2(x_1, y_1)dx_1dy_1 = \iint_{P_2} p_2(x_2, y_2)w_1(x_2, y_2)dx_2dy_2 \quad (30\text{-}7)$$

If, instead of specifying boundary conditions in a single plane, it is necessary to treat a system of planes, the expression of the reversed flow theorem is of the same general form as Eq. 30-4. Provided the Kutta condition is imposed at the trailing edges of all lifting surfaces, the relation becomes

$$\iint_{\Sigma_1} (V_n)_1u_2dS_1 = \iint_{\Sigma_2} (V_n)_2u_1dS_2 \quad (30\text{-}8)$$

where the area of integration Σ extends over both sides of all wing surfaces, V_n is the component of perturbation velocity normal to and directed away from the wing, and the subscripts 1 and 2 refer to forward and reverse flow and the two systems of axes.

In the case of supersonic flow, the development of the reciprocal relation for reverse flows parallels closely the subsonic case. For either planar or multiplanar systems the conormal in Eq. 29-3 is, in fact, the normal as long as the surface of integration is a plane parallel to the x axis. In the case of the single wing, for example, Eq. 30-2 apply where the surfaces σ are slightly removed from the plane of the wing. In the limit

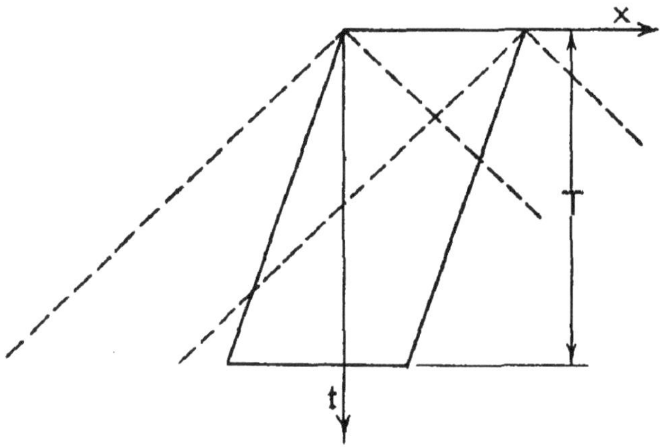

Fig. D,30b. Trace of airfoil path in x, t plane up to time T.

as σ approaches the $z = 0$ plane, the reverse flow theorem takes the form of Eq. 30-4 and 30-7 provided the singularities at the leading edges are evaluated by a proper limiting process. It is necessary to include such considerations for wings with subsonic leading and trailing edges when the Kutta condition is not imposed. For supersonic type edges, or when the Kutta condition is applied at subsonic trailing edges, the integration is straightforward. Eq. 30-8 relates the two possible flows in the case of multiplanar systems.

In the case of unsteady motion, the basic equation is

$$\psi_{tt} - \psi_{xx} - \psi_{yy} - \psi_{zz} = 0 \tag{30-9}$$

where $t = a_\infty t'$, a_∞ is the speed of sound in the undisturbed region of the field, t' is time, and ψ is the perturbation velocity potential or any of the perturbation velocity components. Thin wings at small angles of attack are to be considered and the motion is assumed to take place in the x, y plane. The visualization of the time and geometry relations is relatively easy for two-dimensional wings moving at a uniform speed and has been developed in Art. 26. In Fig. D,30b the swept-out region in the x, t plane applies to the motion of two wings, the first wing having started at time zero and moved to the left at a constant velocity U_0 until the time

$T' = T/a_\infty$; the second wing is assumed to start at the final position of the first wing and to move backward in time until it has reached the initial position of the first wing. With these concepts in mind, it follows from Eq. 29-4 that the relation

$$\iint \varphi \frac{\partial^2 \varphi'}{\partial z \partial t} dxdt = \iint \frac{\partial \varphi'}{\partial t} wdxdt \tag{30-10}$$

holds where φ and φ' are the perturbation potentials for the forward and reverse motions, the region of integration is determined by the area occupied in the x, t plane by the wing, and the Kutta condition is assumed to apply to the trailing edge when in subsonic flight.

If the left side of Eq. 30-10 is integrated by parts, the general relation becomes

$$-\iint \frac{\partial \varphi}{\partial t} w'dxdt = \iint \frac{\partial \varphi'}{\partial t} wdxdt \tag{30-11}$$

If the motion of a three-dimensional wing is to be studied, Eq. 30-10 and 30-11 must be modified to include an integration with respect to y.

Two further changes in Eq. 30-11 serve to simplify applications. In the first place, a symmetry is restored by resorting to two distinct systems of axes; in the second place, the pressure relation

$$p - p_\infty = -\rho_\infty \frac{\partial \varphi}{\partial t'} = -\rho_\infty a_\infty \frac{\partial \varphi}{\partial t} \tag{30-12}$$

where p_∞ denotes undisturbed pressure, permits the introduction of pressure p in the integrands. The final expression for the three-dimensional case is, in the form given originally in [*117*],

$$\begin{aligned}&\iiint p_1(x_1, y_1, t_1)w_2(x_1, y_1, t_1)dx_1dy_1dt_1\\ &\qquad = \iiint p_2(x_2, y_2, t_2)w_1(x_2, y_2, t_2)dx_2dy_2dt_2\end{aligned} \tag{30-13}$$

where the two motions now follow the same path in reverse directions but are referred to the two sets of oppositely oriented axes satisfying the relations

$$x_1 = -x_2 + \xi, \qquad y_1 = -y_2 + \eta, \qquad z_1 = z_2, \qquad t_1 = -t_2 + \tau \tag{30-14}$$

and where ξ, η, and τ are arbitrarily fixed quantities. It is obvious that ξ, η, and τ appear implicitly in Eq. 30-13. For a large class of motions, however, the dependence on τ is such that a further simplification can be achieved in the statement of a reciprocal theorem. Consider the case in which the wings have traveled a time $T'(T' = T/a_\infty)$ and a distance $U_0T' = M_0T$. Let the two systems of coordinate axes be placed such that $t_1 = 0$ sets the starting time of the forward motion and $t_2 = 0$ sets the starting time of the reverse motion. Eq. 30-13 then becomes

$$\int_0^T dt_1 \iint\limits_{P_1(t_1)} p_1(x_1, y_1, t_1)w_2(x_1, y_1, t_1)dx_1dy_1$$
$$= \int_0^T dt_2 \iint\limits_{P_2(t_2)} p_2(x_2, y_2, t_2)w_1(x_2, y_2, t_2)dx_2dy_2$$

where the functions w_1 and w_2 have an implicit dependence upon T. If w_1 and w_2 remain constant for $x_1 + M_0t_1 = \text{const}$ or $x_2 + M_0t_2 = \text{const}$, differentiation with respect to T yields

$$\iint\limits_{P_1(T)} p_1(x_1, y_1, T)w_2(x_1, y_1, T)dx_1dy_1$$
$$= \iint\limits_{P_2(T)} p_2(x_2, y_2, T)w_1(x_2, y_2, T)dx_2dy_2 \qquad (30\text{-}15)$$

The value of T appearing in this result now takes the role of an auxiliary parameter and Eq. 30-15 and 30-7 are equivalent in form. In this way steady and unsteady motions can be treated simultaneously.

By means of Duhamel's integral and the linearity of the wave equation, the response of a wing to arbitrary variations in time of angle of attack, rate of roll, and rate of pitch can be expressed in terms of responses to unit or step variations at time zero of angle of attack, rate of pitch, and rate of roll. These latter indicial motions and their indicial responses satisfy the requirements underlying the derivation of Eq. 30-15. Therefore, relations derived from this equation apply not only to wings in steady or indicial motions but to more general unsteady motions derivable from indicial motions provided the time histories of the motions in forward and reverse flight are identical. Theorems to be given in the following article will speak specifically of steady and indicial motions but it is to be understood that the indicial results apply equally to the derivable class of unsteady motions.

D,31. Applications. In this article, representative examples of theorems for wings in forward and reverse flow will be given for steady and indicial motions. These applications are exact within the framework of linear theory and involve no further restrictions on the wing planform or Mach number. It may be helpful to think of the results as falling into two groups. At the outset, reversal theorems are given, that is, theorems relating the aerodynamic characteristics of identical wings executing identical motions in forward and reverse flight. Subsequently, relations are established between the aerodynamic properties of wings in forward and reverse flight that have similar planforms but, in general, dissimilar angles of attack and thickness distributions. In distinction to the reversal theorems, the latter results are referred to as reciprocal theorems.

Drag of symmetrical nonlifting wings. The drag of a symmetrical sharp-edged wing in linear theory may be determined by integrating over the planform the product of the pressure and the slope in the x direction of the wing surface; when the wing has blunt edges with slopes having square-root singularities, these singularities yield an added contribution. In general, therefore, the drag D of a symmetrical section is given by

$$D = D_e + 2 \iint\limits_P p \left(\frac{dz}{dx}\right)_u dS \tag{31-1}$$

where D_e is the drag attributable to the edges.

If the subscripts 1 and 2 refer to the same wing in forward and reverse flow, respectively, as in Eq. 30-14, local slopes are related as follows:

$$\left[\frac{dz_1(x_1, y_1, T)}{dx_1}\right]_u = -\left[\frac{dz_2(x_2, y_2, T)}{dx_2}\right]_u \tag{31-2}$$

Eq. 31-1 and 31-2 together with Eq. 30-15 yield

$$D_1 - (D_e)_1 = 2 \iint\limits_{P_1} p_1 \left(\frac{dz_1}{dx_1}\right)_u dS = -2 \iint\limits_{P_1} p_1 \left(\frac{dz_2}{dx_2}\right)_u dS$$

$$= -2 \iint\limits_{P_2} p_2 \left(\frac{dz_1}{dx_1}\right)_u dS = 2 \iint\limits_{P_2} p_2 \left(\frac{dz_2}{dx_2}\right)_u dS = D_2 - (D_e)_2$$

Since the geometry of the wing is fixed, the edge contributions are the same, $(D_e)_1 = (D_e)_2$, and consequently [92]

$$D_1 = D_2 \tag{31-3}$$

Lift on flat plate wings. The lift L of a wing is determined by integrating the differential pressure $\Delta p = p_l - p_u$ over the wing planform, thus

$$L = \iint\limits_P \Delta p dS \tag{31-4}$$

For flat plate wings, the local angle of attack of the wing surface is a constant

$$\alpha_1(x_1, y_1, T) = \alpha_1 = \text{const}, \qquad \alpha_2(x_2, y_2, T) = \alpha_2 = \text{const} \tag{31-5}$$

From Eq. 30-15 we therefore have

$$L_1\alpha_2 = \iint\limits_{P_1} \Delta p_1 \alpha_2 dS_1 = \iint\limits_{P_2} \Delta p_2 \alpha_1 dS_2 = L_2\alpha_1$$

or

$$\frac{L_1}{\alpha_1} = \frac{L_2}{\alpha_2} \tag{31-6}$$

This result, which generalizes a theorem given originally by Brown [90], demonstrates the invariance of lift per unit angle of attack in steady or indicial motion.

Damping in roll of flat plate wings. The rolling moment L' exerted on a wing, following the usual sign convention, is

$$L' = -\iint_P y\Delta p dS \tag{31-7}$$

The local angle of attack due to rotation about the x axis is

$$\alpha_1(x_1, y_1, T) = \frac{p_1'y_1}{U}, \qquad \alpha_2(x_2, y_2, T) = \frac{p_2'y_2}{U} = \frac{-p_2'y_1}{U} \tag{31-8}$$

where p' is the angular velocity of roll, assumed constant. We therefore have

$$\begin{aligned}\frac{p_2'L_1'}{U} &= \frac{-p_2'}{U}\iint_{P_1} y_1\Delta p_1 dS_1 = \frac{p_2'}{U}\iint_{P_1} y_2\Delta p_1 dS_1 = \frac{p_1'}{U}\iint_{P_2} y_1\Delta p_2 dS_2\\ &= \frac{-p_1'}{U}\iint_{P_2} y_2\Delta y_2 dS_2 = \frac{p_1'L_2'}{U}\end{aligned}$$

or

$$\frac{L_1'}{p_1'} = \frac{L_2'}{p_2'} \tag{31-9}$$

It follows that the rolling moment per unit angular rolling velocity of flat plate wings in steady or indicial motion is the same in forward and reverse flight.

Damping in pitch of flat plate wings. Consider a wing, first in forward flight and pitching with a uniform angular velocity q_1 about a lateral axis and second, in reverse flight and pitching with angular velocity q_2 about another lateral axis. Place each wing in a coordinate system such that the y axis coincides with the axis of rotation and designate the distance to the moment axis by x_0 with proper subscript. In such a coordinate system, the pitching moment M'_{x_0} exerted on a wing, following the usual sign convention, is

$$M'_{x_0} = -\iint_P (x - x_0)\Delta p dS \tag{31-10}$$

The local angle of attack distributions due to rotations about the y axis are

$$\alpha_1 = \frac{q_1}{U}x, \qquad \alpha_2 = \frac{q_2}{U}x_2 = \frac{q_2}{U}(\xi - x_1)$$

We therefore have

$$\begin{aligned}\frac{q_2}{U}[(M_1')_{x_{01}} + (\xi - x_{01})L_1] &= \frac{q_2}{U}\iint_{P_1} [(x_{01} - x_1) + (\xi - x_{01})]\Delta p_1 dS\\ &= \frac{q_2}{U}\iint_{P_1} (\xi - x_1)\Delta p_1 dS_1 = \frac{q_2}{U}\iint_{P_1} x_2\Delta p_1 dS_1\\ &= \frac{q_1}{U}\iint_{P_2} x_1\Delta p_2 dS_2 = \frac{q_1}{U}[(M_2')_{x_{02}} + (\xi - x_{02})L_2]\end{aligned}$$

or

$$\frac{(M_1')_{x_{01}} + (\xi - x_{01})L_1}{q_1} = \frac{(M_2')_{x_{02}} + (\xi - x_{02})L_2}{q_2} \tag{31-11}$$

This equation indicates that the pitching moment due to pitching velocity is, in general, not the same for wings in forward and reverse flight. However, if $x_{01} = x_{02} = \xi$, the pitching moment per unit angular pitching velocity of flat plate wings in steady or indicial motion is invariant.

The above reversal theorems are general in character. The following reciprocal theorems are designed to show the advantages that can be realized in specific calculations related to wings with nonuniform angle-of-attack distributions or in nonuniform flow fields.

Lift of arbitrarily cambered wings. Consider two wings of identical planform in flight in opposite directions. If the first wing is arbitrarily cambered and twisted and the second wing is flat we have

$$\alpha_1 = \alpha_1(x_1, y_1, T), \qquad \alpha_2 = \text{const}$$

Application of Eq. 30-15 yields the result

$$\alpha_2 L_1 = \iint\limits_{P_1} \alpha_2 \Delta p_1 dS_1 = \iint\limits_{P_2} \alpha_1 \Delta p_2 dS_2$$

or

$$L_1 = \iint\limits_{P_2} \alpha_1 \left(\frac{\Delta p_2}{\alpha_2}\right) dS_2 \tag{31-12}$$

The lift in steady or indicial motion of a wing having arbitrary twist and camber is therefore equal to the integral over the planform of the product of the local angle of attack and the loading per unit angle of attack at the corresponding point of a flat plate wing of identical planform in flight in the reverse direction.

Consider, for example, the loading per unit angle of attack of an airfoil in steady subsonic flight

$$\frac{\Delta p_2}{\alpha_2} = \frac{4}{\beta}\left(\frac{1}{2}\rho_\infty U^2\right)\sqrt{\frac{c - 2x_2}{c + 2x_2}} = \frac{4}{\beta}\left(\frac{1}{2}\rho_\infty U^2\right)\sqrt{\frac{c + 2x_1}{c - 2x_1}}$$

Substitution of this result into Eq. 31-11 provides Munk's well-known integral formula for lift of an arbitrarily cambered airfoil:

$$L_1 = \frac{-4}{\beta}\left(\frac{1}{2}\rho_\infty U^2\right)\int_{-c/2}^{c/2} \frac{dz_1}{dx_1}\sqrt{\frac{c + 2x_1}{c - 2x_1}}\, dx_1 \tag{31-13}$$

The corresponding formula for the lift of a tier of curved airfoils may also be derived similarly from the expression for the loading on an equivalent tier of flat airfoils.

The expression for the lift on a supersonic-edged wing with a straight trailing edge (see Fig. D,12k) can be derived simply from Eq. 31-12. The load distribution per unit angle of attack of the reverse wing is, from sweep theory,

$$\frac{\Delta p_2}{\alpha_2} = \left(\frac{1}{2}\rho_\infty U^2\right)\frac{4\cos\Lambda}{\sqrt{M_\infty^2\cos^2\Lambda - 1}}$$

and the lift is

$$L_1 = \frac{-4}{\sqrt{M_\infty^2 - \sec^2\Lambda}}\left(\frac{1}{2}\rho_\infty U^2\right)\iint\limits_{P_1}\frac{dz_1}{dx_1}\,dx_1dy_1 \qquad (31\text{-}14)$$

as previously given in Eq. 12-46.

The above two examples were for steady state flow but it is possible to calculate, for example, the build-up of lift with time of an arbitrarily cambered two-dimensional supersonic airfoil starting impulsively from rest at a constant speed since the flat plate loading is as given in Art. 27.

Lift on a wing in a nonuniform downwash field. The reciprocal theorem of Eq. 31-12 also provides a particularly good method of determining the lift on a wing in certain nonuniform downwash fields of known structure. Such problems arise whenever a wing acts in the presence of other wings, bodies, or propellers but is always of prime concern in the determination of the lift on a tail acting in the downwash field of a wing. In most such problems, the velocities at the position of the tail may be considered constant in the longitudinal direction and to vary in the spanwise direction; thus

$$\alpha_1 = \alpha_1(y, T), \qquad \alpha_2 = \text{const}$$

and

$$L_1 = \iint\limits_{P_2}\alpha_1\left(\frac{\Delta p_2}{\alpha_2}\right)dS_2 = \int_{-s}^{s}\alpha_1\left(\frac{l_2}{\alpha_2}\right)dy_1 \qquad (31\text{-}15)$$

where l_2 is the span load distribution associated with the load distribution.

The lift in a downwash field which varies across the span is thus equal to the integral over the span of the product of the local angle of attack and the span loading at the corresponding spanwise station of a flat plate wing of identical planform in flight in the reverse direction.

As an example, consider the problem of determining the lift on a wing at a geometrical angle of attack of zero resulting from the presence of an infinite line vortex of strength Γ extending in the flight direction. The wing will be considered here to have such a planform that its span loading when in flight in the reverse direction is elliptic. Let the vortex line be a lateral distance b from the center line of the wing and a vertical distance c. The span loading of the wing in reverse flight is given by

$$\frac{l_2}{\alpha_2} = \frac{2}{\pi}\frac{L_2}{s\alpha_2}\sqrt{1 - \frac{y_2^2}{s^2}}$$

and the local angle of attack of the original wing due to the presence of the vortex is

$$\alpha_1 = \frac{\Gamma}{2\pi U} \frac{y - b}{[(y - b)^2 + c^2]}$$

Application of Eq. 31-15 yields for the lift

$$L_1 = \frac{\Gamma}{\pi U s} \frac{L_2}{\alpha_2} \left\{ \frac{-b}{s} + \sqrt{\frac{1}{2}\left[\sqrt{\left(1 + \frac{c^2}{s^2} - \frac{b^2}{s^2}\right)^2 + \frac{4b^2c^2}{s^4}} - \left(1 + \frac{c^2}{s^2} - \frac{b^2}{s^2}\right)\right]} \right\}$$

By superposition, the lift induced on the wing by a number of such vortices may be calculated.

It is obvious that an almost limitless number of results of practical significance can be derived by means of reciprocity relations. The methods also furnish a powerful tool for the determination of wings with optimal drag characteristics. A discussion of such problems will be found in VII,A.

As a concluding remark, it should be mentioned that, although the above results were limited to applications of Eq. 30-15, it is also possible to apply the relation (30-13) advantageously. For example, one can relate the build-up of lift on a wing entering a gust and the build-up of circulation on the same wing moving impulsively at a constant angle of attack in the reverse direction. In two-dimensional incompressible flow this establishes a direct connection between the circulation function of Wagner and the gust lift calculated by von Kármán and Sears. A proof of the agreement between these latter results has been given by Sears [*95*] and is now seen to apply for arbitrary stream Mach numbers.

CHAPTER 8. SMALL PERTURBATION THEORIES OF TRANSONIC AND HYPERSONIC FLOWS

D,32. Introduction. The entire development of the preceding articles has been based on the linearized potential equation for fluid flow and, in application to steady flight, the Prandtl-Glauert equation has governed the resulting flow fields. The nature of the solutions has also verified, a posteriori, that for thin wings and bodies at small angles of attack such an analysis is valid for flight Mach numbers sufficiently removed from 1 and not of excessive magnitude. It remains to consider these two latter regimes, that is, transonic and hypersonic Mach numbers, and, as soon will become evident, the treatment of such cases forces one to abandon the simplification inherent in linearized theory. We propose to state here the basic equations necessary to a full treatment of such problems and to indicate the manner in which similarity rules for these flow fields can be derived.[46]

[46] The authors are indebted to W. A. Mersman, John R. Spreiter, and Milton D. Van Dyke of the Theoretical Aerodynamics Section, Ames Laboratory, NACA, for the use of unpublished investigations that have been incorporated into this chapter.

In conventional vector notation we have (cf. III,A):

Euler's momentum equation

$$-\frac{1}{\rho}\operatorname{grad} p = \frac{\partial \mathbf{q}}{\partial t'} - \mathbf{q} \times \operatorname{curl} \mathbf{q} + \operatorname{grad} \frac{q^2}{2} \tag{32-1}$$

Continuity equation

$$\frac{\partial \rho}{\partial t'} + \operatorname{div}(\rho \mathbf{q}) = \frac{\partial \rho}{\partial t'} + \rho \operatorname{div} \mathbf{q} + \mathbf{q} \cdot \operatorname{grad} \rho = 0 \tag{32-2}$$

To these we may add the following thermodynamic relations and the definition of the speed of sound:

Equation of state

$$p = \rho \mathcal{R} T, \qquad \left[\mathcal{R} = c_p - c_v = (\gamma - 1)c_v = \frac{\gamma - 1}{\gamma} c_p\right] \tag{32-3}$$

Internal energy

$$e = c_v T = \frac{1}{\gamma - 1}\frac{p}{\rho} \tag{32-4}$$

Enthalpy

$$h = c_p T = \gamma e = \frac{\gamma}{\gamma - 1}\frac{p}{\rho} \tag{32-5}$$

Entropy

$$s = c_v \ln \frac{p}{\rho^\gamma} \tag{32-6}$$

Speed of sound

$$a^2 = \frac{dp}{d\rho} \tag{32-7}$$

In Eq. 32-3 to 32-6 the terms c_p and c_v are respectively the specific heats of the fluid at constant pressure and constant volume and both are assumed constant. Their difference, $c_p - c_v$, is equal to the gas constant $\mathcal{R}$.

Eliminating p and ρ from Eq. 32-1 by means of Eq. 32-5 and 32-6, we get the relation

$$T \operatorname{grad} s + \mathbf{q} \times \operatorname{curl} \mathbf{q} = \frac{\partial \mathbf{q}}{\partial t'} + \operatorname{grad}\left(h + \frac{q^2}{2}\right) \tag{32-8a}$$

which, for steady states, takes the form

$$T \operatorname{grad} s + \mathbf{q} \times \operatorname{curl} \mathbf{q} = \operatorname{grad}\left(h + \frac{q^2}{2}\right) \tag{32-8b}$$

and is known as Crocco's equation (III,A).

The assumptions up to the present are:

1. Viscosity is neglected.
2. Heat conduction is neglected.
3. External force fields are excluded.
4. The equation of state for a perfect gas applies.

The term in the right-hand member of Eq. 32-8b is the sum of the internal and kinetic energy per unit mass, and for aerodynamic applications we assume uniform conditions infinitely far ahead of the induced field of motion together with the conservation of energy so that the quantity $h + q^2/2 = \text{const}$. For such a flow field with constant stagnation enthalpy we have

$$-T \text{ grad } s = \mathbf{q} \times \text{curl } \mathbf{q} \tag{32-9}$$

and Euler's equation for steady state flow can be written in the alternative form

$$-\frac{1}{\rho} \text{ grad } p = T \text{ grad } s + \text{grad } \frac{q^2}{2} \tag{32-10}$$

From Eq. 32-9 a number of conclusions can be drawn. In the first place, if s is not a constant so that the flow field is nonisentropic, then curl $\mathbf{q}$ cannot vanish and the field cannot be irrotational. Second, if curl $\mathbf{q}$ is zero and potential flow results then the flow is necessarily isentropic. Third, if in an isentropic flow field curl $\mathbf{q}$ is not zero then by virtue of the vector product it follows that free vortex lines must follow the streamlines. Finally, scalar multiplication of Eq. 32-9 by $\mathbf{q}$ leads to the relation

$$\mathbf{q} \cdot \text{grad } s = 0 \tag{32-11}$$

Thus, along each streamline the flow is isentropic and along each streamline we have p/ρ^γ equal to a constant although the constant need not be the same for different streamlines. Since, however, we have limited ourselves to uniform upstream conditions, it follows that in this case the compressible flow field is isentropic everywhere and possesses a velocity potential *provided*, of course, that the differential relations exist and consequently that the field is continuous along streamlines.

It remains now to investigate the possibility of such discontinuities occurring. It can be shown (A,5; III,E,2) that discontinuities (shock waves) appear only when the local velocity of the medium is greater than the local velocity of sound a where $a^2 = \gamma p/\rho$. In general, these shock waves are curved and, since the gain in entropy occurs downstream of the wave and is dependent on the angle of the shock relative to the stream, it follows from Eq. 32-9 that the flow field is not irrotational. Thus, along each streamline behind a curved shock wave the flow is isentropic but the field behind the wave no longer has constant entropy.

It is possible, however, to estimate the magnitude of the change in the entropy. From the shock-wave relations (see III,E,2) it follows directly that, if Δs is the increment in s and Δq_n is the increment in the velocity q_n normal to the shock wave, then

$$\Delta s = c_p \ln \left(\frac{1 + \frac{\gamma - 1}{2} \frac{\Delta q_n}{q_n}}{1 - \frac{\gamma - 1}{2} \frac{\Delta q_n}{q_n}} \right)^{1/\gamma} \left(1 - \frac{\Delta q_n}{q_n}\right)$$

If Δq_n is small relative to q_n, this becomes

$$\Delta s \cong c_p \left(\frac{\gamma^2 - 1}{2}\right) \left(\frac{\Delta q_n}{q_n}\right)^3 \tag{32-12}$$

Neglecting third and higher order powers of perturbation velocities we can therefore say that the flow field has constant entropy in either subsonic or supersonic fields for which the velocity perturbations remain small,[47] that a velocity potential exists, and that p/ρ^γ has a fixed value determined by the relation

$$\frac{p}{\rho^\gamma} = \frac{p^0}{(\rho^0)^\gamma}$$

where the superscript 0 refers to stagnation conditions. Eq. 32-10 appears finally in the form

$$-\frac{1}{\rho} \operatorname{grad} p = \operatorname{grad} \frac{q^2}{2} \tag{32-13}$$

In order to derive the potential equation for the flow field, we use Eq. 32-13 and the steady state form of Eq. 32-2 together with the isentropic pressure-density relation. Eliminating p and ρ from these equations, we have

$$\operatorname{div} \mathbf{q} - \frac{1}{a^2} \mathbf{q} \cdot \operatorname{grad} \frac{1}{2} q^2 = 0$$

where

$$a^2 = (a^0)^2 - \frac{\gamma - 1}{2} q^2$$

Since $\mathbf{q} = \operatorname{grad} \phi$, where ϕ is the velocity potential of the field, we may write the differential equation in the form

$$\operatorname{div} \operatorname{grad} \phi - \frac{1}{a^2} (\operatorname{grad} \phi) \cdot \operatorname{grad} \left[\frac{1}{2} (\operatorname{grad} \phi)^2\right] = 0 \tag{32-14a}$$

or, in terms of Cartesian coordinates,

$$\phi_{xx}(a^2 - \phi_x^2) + \phi_{yy}(a^2 - \phi_y^2) + \phi_{zz}(a^2 - \phi_z^2) - 2\phi_{yz}\phi_y\phi_z - 2\phi_{xz}\phi_x\phi_z - 2\phi_{xy}\phi_x\phi_y = 0 \tag{32-14b}$$

[47] The orders of magnitude involved in the transition from Eq. 32-10 to Eq. 32-13 have been treated in more detail by Liepmann [*98*].

Assume now that we are to deal with bodies of small thickness and angle of attack situated in a free stream of velocity U directed along the positive x axis. We may, furthermore, introduce a perturbation velocity potential φ such that $\varphi_x = u$, $\varphi_y = v$, $\varphi_z = w$ and defined by the relation

$$\phi = Ux + \varphi \tag{32-15}$$

The free stream Mach number M_∞ is equal to the ratio of U and a_∞ the speed of sound in the free stream. The velocity components and a_∞ are related to the local speed of sound a by the equation

$$a^2 = a_\infty^2 - \frac{\gamma - 1}{2}(2Uu + u^2 + v^2 + w^2) \tag{32-16}$$

Introducing these variables into Eq. 32-14b, we express the differential equation in the form

$$\begin{aligned}
&(1 - M_\infty^2)\varphi_{xx} + \varphi_{yy} + \varphi_{zz} \\
&= \frac{\varphi_{xx}}{a_\infty^2}\left[(\gamma + 1)U\varphi_x + \frac{\gamma + 1}{2}\varphi_x^2 + \frac{\gamma - 1}{2}(\varphi_y^2 + \varphi_z^2)\right] + \frac{2}{a_\infty^2}\varphi_{xy}\varphi_y(U + \varphi_x) \\
&+ \frac{\varphi_{yy}}{a_\infty^2}\left[(\gamma - 1)U\varphi_x + \frac{\gamma - 1}{2}(\varphi_x^2 + \varphi_z^2) + \frac{\gamma + 1}{2}\varphi_y^2\right] + \frac{2}{a_\infty^2}\varphi_{xz}\varphi_z(U + \varphi_x) \\
&+ \frac{\varphi_{zz}}{a_\infty^2}\left[(\gamma - 1)U\varphi_x + \frac{\gamma - 1}{2}(\varphi_x^2 + \varphi_y^2) + \frac{\gamma + 1}{2}\varphi_z^2\right] + \frac{2}{a_\infty^2}\varphi_{yz}\varphi_y\varphi_z
\end{aligned} \tag{32-17}$$

In order to linearize this equation, it is necessary to assume that u/U, v/U, $w/U \ll 1$ and that the gradients are sufficiently small relative to the free stream velocity divided by some reference length of the body that powers and cross products of the velocity terms and their gradients can be neglected. As a result one has the familiar Prandtl-Glauert equation

$$(1 - M_\infty^2)\varphi_{xx} + \varphi_{yy} + \varphi_{zz} = 0 \tag{32-18}$$

In the expression of a general equation for transonic flow fields, where wing and body shapes are not necessarily slender, the decrease in the term $(1 - M_\infty^2)$ relative to $(\gamma + 1)U\varphi_x/a_\infty^2$ must be taken into consideration. The nature of the linearized solutions in the vicinity of $M_\infty = 1$ indicates, however, that we may justifiably assume the streamwise gradients of potential φ and u are large in comparison with the lateral gradients of potential or velocity components and, as a consequence, only the first term on the right side of Eq. 32-17 is retained. The result is

$$(1 - M_\infty^2)\varphi_{xx} + \varphi_{yy} + \varphi_{zz} = \frac{(\gamma + 1)M_\infty^2}{U}\varphi_x\varphi_{xx} \tag{32-19}$$

The boundary conditions associated with Eq. 32-19 in wing theory are specified as in linear analysis and require uniform flow far ahead of the body as well as tangential flow at the wing surface. Imposing the latter condition in the plane $z = 0$, we have

$$\text{at } x = -\infty, \qquad \varphi_x = \varphi_y = \varphi_z = 0 \tag{32-20}$$

$$\text{at } z = 0, \qquad \left(\frac{\varphi_z}{U}\right)_{z=0} = \frac{dz}{dx} = \tau \frac{\partial}{\partial(x/c)} f\left(\frac{x}{c}, \frac{y}{b}\right) \tag{32-21}$$

where the shape of the wing is fixed by the relation

$$z = c\tau f\left(\frac{x}{c}, \frac{y}{b}\right) \tag{32-22}$$

In general, a variation of τ represents a simultaneous change of thickness ratio, camber, and angle of attack. In the special case of a nonlifting wing with symmetrical sections, τ is proportional to the thickness ratio; for inclined flat plate wings of vanishing thickness, τ is proportional to the angle of attack. The Kutta condition of smooth flow at all subsonic trailing edges also applies. Pressure coefficient is evaluated in general by the formula

$$C_p = -\frac{2}{U} \varphi_x \tag{32-23}$$

The above statement of the transonic problem has the advantage that it is valid throughout the Mach number range, exclusive of hypersonic flow. At subsonic and supersonic Mach numbers, Eq. 32-19 is of the same order of accuracy as the Prandtl-Glauert equation although more difficult to solve. At $M_\infty = 1$ it is identical to a form widely used in transonic analysis. The present approach is closely related to investigations by Oswatitsch and Berndt [*99,100*]. Other workers [*101–106*] have based their theories on the assumption that all velocities are small perturbations around the critical velocity of sound a^*. In our case, perturbation potential φ is defined by

$$\varphi = -Ux + \phi \tag{32-24}$$

In the latter scheme one has a perturbation potential φ' defined by

$$\varphi' = -a^*x + \phi \tag{32-25}$$

and the resulting differential equation is

$$\varphi'_{yy} + \varphi'_{zz} = \frac{\gamma + 1}{a^*} \varphi'_x \varphi'_{xx} \tag{32-26}$$

Spreiter [*108*] has distinguished between the two developments by referring to them respectively as the U statement and the a^* statement of the problem. In the a^* statement, boundary conditions are

$$\text{at } x = -\infty, \qquad \varphi'_x = \frac{(M_\infty^2 - 1)a^*}{\gamma + 1}, \qquad \varphi'_y = \varphi'_z = 0$$

$$\text{at } z = 0, \qquad (\varphi'_z)_{z=0} = a^*\left(\frac{dz}{dx}\right)$$

and pressure coefficient is

$$C_p = \frac{-2}{a^*} [\varphi'_x - (\varphi'_x)_{x=-\infty}]$$

D,33. Similarity Rules for Transonic Flow. At the present time, direct analysis of the transonic equation has been devoted almost exclusively to the two-dimensional form (i.e. where $\varphi_{yy} = 0$). An important simplification of the two-dimensional theory is brought about through the use of the hodograph transformation (Sec. F) since the resulting equation is linear although of mixed type. In the present article we confine ourselves to the derivation of the similarity rules in wing theory, the treatment being based on the work of Spreiter [*107,108*]. A discussion of axially symmetric flow is given in [*99*].

It is assumed that Eq. 32-19 is transformed into a second set of primed variables. Introducing proportionality factors denoted by s with appropriate subscripts, we have

$$x' = s_x x, \qquad y' = s_y y, \qquad z' = s_z z, \qquad \varphi' = s_\varphi \varphi, \qquad U' = s_U U$$

$$(M'_\infty)^2(\gamma + 1)' = \Gamma' = s_\Gamma \Gamma = s_\Gamma(\gamma + 1)M_\infty^2, \qquad \beta' = s_\beta \beta \quad (33\text{-}1)$$

where $\beta'^2 = |1 - M'^2_\infty|$ and $\beta^2 = |1 - M_\infty^2|$. (The variable φ' is, of course, not to be confused with the φ' of the previous article.) From the differential equation in the primed system:

$$\pm \frac{s_x^2}{s_\beta^2 s_\varphi} \beta'^2 \frac{\partial^2 \varphi'}{\partial x'^2} + \frac{s_y^2}{s_\varphi} \frac{\partial^2 \varphi'}{\partial y'^2} + \frac{s_z^2}{s_\varphi} \frac{\partial^2 \varphi'}{\partial z'^2} = \begin{cases} 0 & \text{Linear theory} \\ \dfrac{s_U s_x^3}{s_\Gamma s_\varphi^2} \dfrac{\Gamma'}{U'} \dfrac{\partial \varphi'}{\partial x'} \dfrac{\partial^2 \varphi'}{\partial x'^2} & \text{Nonlinear theory} \end{cases}$$

we get the following relations between the stretching factors

$$\frac{s_y s_\beta}{s_x} = 1, \qquad \frac{s_y}{s_z} = 1$$

$$\frac{s_\varphi}{s_U s_x} = \begin{cases} \dfrac{\lambda}{\lambda'} & \text{Linear theory} \\ \dfrac{s_\beta^2}{s_\Gamma} & \text{Nonlinear theory} \end{cases}$$

where, for the linear theory, λ/λ' is an arbitrary constant which can be equated to s_β^2/s_Γ if desired.

From the nature of the transformation, it follows that aspect ratios AR and AR' in the two systems are related by the equations

$$AR' = \frac{s_y}{s_x} AR = \frac{AR}{s_\beta} = \frac{\beta}{\beta'} AR$$

and, consequently, we have

$$\beta' AR' = \beta AR \qquad (33\text{-}2)$$

It remains to see that the boundary conditions are satisfied. Eq. 32-20 impose no added restraints; Eq. 32-21, however, yields the relation

$$(\varphi'_{z'})_{z'=0} = U'\tau' \frac{\partial}{\partial(x'/c')} f'\left(\frac{x'}{c'}, \frac{y'}{b'}\right) = s_U U \tau' \frac{\partial}{\partial(x'/c')} f'\left(\frac{x'}{c'}, \frac{y'}{b'}\right)$$

Since, however, we have

$$(\varphi'_{z'})_{z'=0} = \frac{s_\varphi}{s_z} (\varphi_z)_{z=0} = \frac{s_\varphi}{s_z} U\tau \frac{\partial}{\partial(x/c)} f\left(\frac{x}{c}, \frac{y}{b}\right)$$

it follows that, if the distribution functions $f'(x'/c', y'/b')$ and $f(x/c, y/b)$ are the same, the following relations hold

$$\tau' = \frac{s_\varphi}{s_U s_z}\tau = \frac{s_\varphi s_\beta}{s_U s_x}\tau = \begin{cases} \dfrac{\lambda}{\lambda'}\dfrac{\beta'}{\beta}\tau & \text{Linear theory} \\[2ex] \dfrac{s_\beta^3}{s_\Gamma}\tau = \left(\dfrac{\beta'}{\beta}\right)^3 \dfrac{\Gamma}{\Gamma'}\tau & \text{Nonlinear theory} \end{cases}$$

The similarity conditions thus become

$$\begin{aligned} \frac{\beta'}{\lambda'\tau'} &= \frac{\beta}{\lambda\tau} \qquad \text{Linear theory} \\ \frac{\beta'}{(\Gamma'\tau')^{\frac{1}{3}}} &= \frac{\beta}{(\Gamma\tau)^{\frac{1}{3}}} \qquad \text{Nonlinear theory} \end{aligned} \tag{33-3}$$

Eq. 33-2 and 33-3 establish the correspondence between the related flow fields.

Pressure coefficients in the two systems satisfy the equation

$$C'_p = \frac{-2}{U'}\varphi'_{x'} = \frac{s_\varphi}{s_U s_x}\left(\frac{-2}{U}\varphi_x\right) = \frac{s_\varphi}{s_U s_x} C_p$$

from which we conclude

$$C'_p = \begin{cases} \dfrac{\lambda}{\lambda'} C_p & \text{Linear theory} \\[2ex] \dfrac{s_\beta^2}{s_\Gamma} C_p = \left(\dfrac{\tau'}{\tau}\right)^{\frac{2}{3}} \left(\dfrac{\Gamma}{\Gamma'}\right)^{\frac{1}{3}} C_p & \text{Nonlinear theory} \end{cases}$$

The functional relation for pressure coefficient in linear theory is therefore

$$C_p = \frac{1}{\lambda}\bar{P}\left[\frac{\beta}{\lambda\tau}, \beta Æ; \frac{x}{c}, \frac{y}{b}\right] \tag{33-4}$$

and in transonic theory is

$$C_p = \frac{\tau^{\frac{2}{3}}}{\Gamma^{\frac{1}{3}}} P\left[\frac{\beta}{(\Gamma\tau)^{\frac{1}{3}}}, \beta Æ; \frac{x}{c}, \frac{y}{b}\right] \tag{33-5}$$

In linear theory, pressure coefficient must be proportional to τ and in that case Eq. 33-4 can be reduced to the form

$$C_p = \frac{\tau}{\beta}\bar{P}_1\left[\beta Æ; \frac{x}{c}, \frac{y}{b}\right]$$

An important consequence of the above derivation is obtained from the observation that Eq. 33-4 and 33-5 are in complete agreement provided we set $\lambda = (\Gamma/\tau^2)^{\frac{1}{3}}$. The latter result may therefore be thought of as applying throughout the entire Mach number range, a fact that is useful in the correlation of experimental data.

From Eq. 33-5 the following functional relations can be obtained for the lift, drag, and pitching moment coefficients of wings

$$C_L = \frac{\tau^{\frac{2}{3}}}{\Gamma^{\frac{1}{3}}} L\left[\frac{\beta}{(\Gamma\tau)^{\frac{1}{3}}}, \beta Æ\right]$$

$$C_D = \frac{\tau^{\frac{5}{3}}}{\Gamma^{\frac{1}{3}}} D\left[\frac{\beta}{(\Gamma\tau)^{\frac{1}{3}}}, \beta Æ\right] \tag{33-6}$$

$$C_m = \frac{\tau^{\frac{2}{3}}}{\Gamma^{\frac{1}{3}}} M\left[\frac{\beta}{(\Gamma\tau)^{\frac{1}{3}}}, \beta Æ\right]$$

It should be pointed out that the parameters $\beta/(\Gamma\tau)^{\frac{1}{3}}$ and $\beta Æ$ can be combined in an arbitrary manner. In the actual plotting of data it is sometimes convenient to use the parameters $\beta/(\Gamma\tau)^{\frac{1}{3}}$ and $(\Gamma\tau)^{\frac{1}{3}}Æ$.

Since the transonic equation involves the existence of a velocity potential, the derived similarity rules might be assumed limited to regions of flow lying between discontinuities or shock waves. It can be shown [*108*], however, that for weak shocks, the same basic parameters govern the flow so that any possible question as to the validity of the similarity rules throughout the field is resolved.

D,34. Hypersonic Flow. In linearized theory it was assumed at the outset that the thickness ratio, camber, and angle of attack of the wing were sufficiently small to justify the assumption that perturbation velocity components were small relative to the free stream velocity. This implies that the parameter τ introduced in Eq. 32-22 is small relative to 1 ($\tau \ll 1$). A further restriction on the Mach number of the free stream and the wing geometry is implicit in the assumption of potential flow and as a consequence we may assume $M_\infty\tau \ll 1$. This inequality assures us that the vertical extent of the wing is well within the foremost Mach cone and permits the use of the linearized potential equation in which all shock waves are weak and are replaced by Mach cones and lines. It remains, finally, to consider the case of high Mach number flow where the shock waves remain close to the wing and the entire disturbance field is sharply limited in lateral extent. Characterizing this regime by the condition that τ and $1/M_\infty$ are of the same order of magnitude, that is

$$\tau \ll 1, \qquad M_\infty \tau = O(1) \tag{34-1}$$

we propose to study, first, flow through the initial shock wave and, second, flow in the remainder of the field. Under the assumed conditions, it will be possible to derive the fundamental equations of this high Mach number rotational flow. Although inconsistencies in the definition of the term hypersonic appear in the literature, we shall use it here with the meaning implied by Eq. 34-1. In the following discussion we confine our attention principally to the two-dimensional and axially symmetric

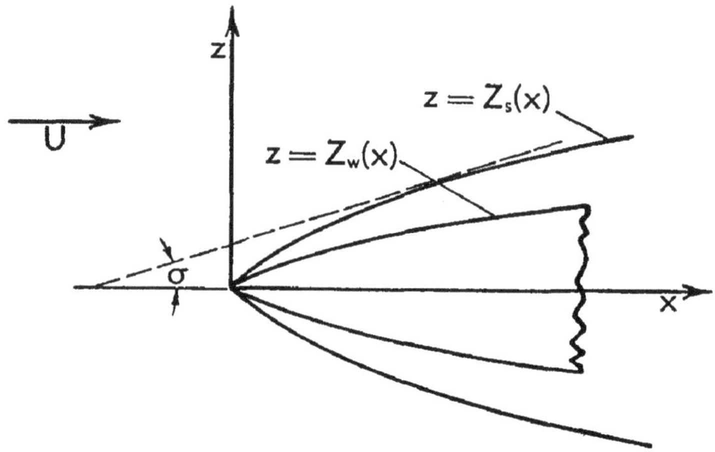

Fig. D,34. Airfoil and enveloping shock wave in hypersonic flow.

aspects of such flow fields and derive results analogous to those given by Goldsworthy [*109*].

If z is the lateral coordinate in either the two-dimensional or axially symmetric flow and x the streamwise coordinate, the body shape is assumed given in the form

$$z = Z_w(x) \tag{34-2}$$

and the boundary conditions are,

at $x = -\infty$, $$u = w = 0 \tag{34-3}$$

on the body, $$\frac{dZ_w}{dx} = \frac{w}{U + u} \tag{34-4}$$

where u and w are the streamwise and lateral perturbation velocity components. Since u is small, Eq. 34-4 and 34-2 yield the result $w = O(U\tau)$ and the order of magnitude for w is assumed to apply throughout the disturbed flow region.

In Fig. D,34 the body is indicated along with the bow shock wave $z = Z_s(x)$. If $\tan \sigma = dZ_s/dx$ is the slope of the bow wave and θ is the deflection angle of the stream, where $w_1/(U + u_1) = \tan \theta$, the Rankine-

Hugoniot relations (III,D) for shock waves may be written

$$U \cos \sigma = (U + u_1) \cos \sigma + w_1 \sin \sigma$$

$$\rho_\infty U \sin \sigma = \rho_1[(U + u_1) \sin \sigma - w_1 \cos \sigma]$$

$$\rho_\infty U^2 \sin^2 \sigma + p_\infty = \rho_1[(U + u_1) \sin \sigma - w_1 \cos \sigma]^2 + p_1$$

$$\frac{1}{2} U^2 + \frac{\gamma}{\gamma - 1} \frac{p_\infty}{\rho_\infty} = \frac{1}{2} [(U + u_1)^2 + w_1^2] + \frac{\gamma}{\gamma - 1} \frac{p_1}{\rho_1}$$

where the subscripts ∞ and 1 denote conditions ahead of and behind the shock wave. Algebraic manipulation leads to the results

$$\frac{\gamma + 1}{2} \frac{\sin \theta \sin \sigma}{\cos (\sigma - \theta)} = \sin^2 \sigma - \frac{1}{M_\infty^2}$$

$$p_1 - p_\infty = \rho_\infty U^2 \left[\sin^2 \sigma - \frac{1}{M_\infty^2} \right] \frac{2}{\gamma + 1}$$

and under the imposed conditions the latter expressions reduce to

$$\frac{w_1}{U} = \frac{\frac{2}{\gamma + 1} \left[\left(\frac{dZ_s}{dx} \right)^2 - \frac{1}{M_\infty^2} \right]}{\frac{dZ_s}{dx}} \tag{34-5}$$

$$\frac{p_1}{p_\infty} = \frac{2\gamma}{\gamma + 1} M_\infty^2 \left(\frac{dZ_s}{dx} \right)^2 - \frac{\gamma - 1}{\gamma + 1} \tag{34-6}$$

where the following orders of magnitude hold

$$w_1 = O(U\tau), \quad Z_s = O\left(\frac{1}{M_\infty}\right), \quad u = O\left(\frac{U\tau}{M_\infty}\right), \quad \frac{p_1}{p_\infty} = O(M_\infty \tau) \tag{34-7}$$

and neglected terms correspond to a relative error of $O(\tau^2)$. Eq. 34-5 and 34-6 are in the form given by Goldsworthy; for some purposes, however, it may be preferable to express the reduced equations as

$$\rho_\infty U \frac{dZ_s}{dx} = \rho_1 \left(U \frac{dZ_s}{dx} - w_1 \right) \tag{34-8}$$

$$\rho_\infty \left(U \frac{dZ_s}{dx} \right)^2 + p_\infty = \rho_1 \left(U \frac{dZ_s}{dx} - w_1 \right)^2 + p_1 \tag{34-9}$$

$$\frac{1}{2} \left(U \frac{dZ_s}{dx} \right)^2 + \frac{\gamma}{\gamma - 1} \frac{p_\infty}{\rho_\infty} = \frac{1}{2} \left(U \frac{dZ_s}{dx} - w_1 \right)^2 + \frac{\gamma}{\gamma - 1} \frac{p_1}{\rho_1} \tag{34-10}$$

It remains to consider the equations of motion for steady state flow. If $\lambda = 0$ for two-dimensional flow and $\lambda = 1$ for axially symmetric flow, we have from Eq. 32-2 and 32-1

$$(U + u)\frac{\partial \rho}{\partial x} + w\frac{\partial \rho}{\partial z} + \rho\left(\frac{\partial u}{\partial x} + \frac{\partial w}{\partial z}\right) + \lambda\frac{\rho w}{z} = 0$$

$$(U + u)\frac{\partial u}{\partial x} + w\frac{\partial u}{\partial z} + \frac{1}{\rho}\frac{\partial p}{\partial x} = 0$$

$$(U + u)\frac{\partial w}{\partial x} + w\frac{\partial w}{\partial z} + \frac{1}{\rho}\frac{\partial p}{\partial z} = 0$$

and to these we add the energy relation

$$(U + u)\frac{\partial}{\partial x}\left(\frac{p}{\rho^\gamma}\right) + w\frac{\partial}{\partial z}\left(\frac{p}{\rho^\gamma}\right) = 0$$

obtained from Eq. 32-11 and 32-6.

Assuming now that the orders of magnitude in Eq. 34-7 hold, we get, to the same order of accuracy as before,

$$U\frac{\partial \rho}{\partial x} + w\frac{\partial \rho}{\partial z} + \rho\frac{\partial w}{\partial z} + \lambda\frac{\rho w}{z} = 0 \tag{34-11}$$

$$U\frac{\partial w}{\partial x} + w\frac{\partial w}{\partial z} + \frac{1}{\rho}\frac{\partial p}{\partial z} = 0 \tag{34-12}$$

$$U\frac{\partial p}{\partial x} + w\frac{\partial p}{\partial z} - \gamma\frac{p}{\rho}\left(U\frac{\partial \rho}{\partial x} + w\frac{\partial \rho}{\partial z}\right) = 0 \tag{34-13}$$

where the boundary conditions are

$$\frac{dZ_w}{dx} = \frac{w}{U} \tag{34-14}$$

and u can be determined from the modified energy equation

$$Uu + \frac{1}{2}w^2 = \frac{\gamma}{\gamma - 1}\left(\frac{p_\infty}{\rho_\infty} - \frac{p}{\rho}\right) \tag{34-15}$$

Eq. 34-8 to 34-10 together with Eq. 34-11 to 34-15 suffice for the determination of a two-dimensional or axially symmetric hypersonic flow field including effects of entropy changes through the shock wave. Furthermore, it is possible to find, by the methods of the last article, a similarity rule provided γ remains the same for the two flow fields. The rule states that, if two bodies have the same thickness distributions with deflections characterized respectively by τ and τ', the flow patterns at Mach numbers M_∞ and M'_∞ will be similar if

$$M_\infty\tau = M'_\infty\tau' \tag{34-16}$$

This result was first obtained by Tsien [*110*] through consideration of a potential equation which, for two-dimensional flow, takes the form

$$\varphi_{zz}\left[1 - (\gamma - 1)M_\infty\frac{\varphi_z}{a_\infty} - \frac{(\gamma + 1)}{2}\frac{\varphi_z^2}{a_\infty^2}\right] = M_\infty^2\varphi_{xx} + 2M_\infty\frac{\varphi_z}{a_\infty}\varphi_{xz}$$

with the boundary conditions

at $x = -\infty$, $\qquad \varphi_x = \varphi_z = 0$

on the body, $\qquad \varphi_z = U_\infty \dfrac{dZ_w}{dx}$

The fact that Tsien's results apply to flow with shock waves and rotation was established first by Hayes [*111*].

An interesting analogy between the hypersonic equations and the nonlinear equations of unsteady flow in one less dimension was also pointed out by Hayes. This can be seen after introducing the variable $t' = x/U$ in the fundamental equations since one obtains precisely the governing equations of unsteady unidimensional flow ($\lambda = 0$) and plane flow with cylindrical symmetry ($\lambda = 1$). These ideas can also be extended to include three-dimensional hypersonic flow.

It was shown in the last article that similarity rules for the subsonic, transonic, and supersonic regimes can be unified. It appears impossible to include hypersonic flow, but as pointed out by Van Dyke [*112*] supersonic and hypersonic rules can be unified for slender bodies that differ from one another only by a uniform expansion or contraction of all dimensions normal to the free stream direction. Thus, since for large Mach numbers the difference between M_∞ and $\sqrt{M_\infty^2 - 1}$ is negligible, the similarity parameter may be taken to be $\tau \sqrt{M_\infty^2 - 1}$ and the combined rule for pressure coefficient, for example, becomes

$$C_p = \tau^2 \bar{P}\left[\tau \sqrt{M_\infty^2 - 1}; \frac{x}{c}, \frac{y}{b}\right] \tag{34-17}$$

The corresponding rules for drag and lift coefficient are

$$C_D = \tau^{2+k} D(\tau \sqrt{M_\infty^2 - 1}) \tag{34-18}$$

$$C_L = \tau^{1+k} L(\tau \sqrt{M_\infty^2 - 1}) \tag{34-19}$$

where $k = 1$ if some planform area is taken for reference and $k = 0$ if some cross-sectional area is used. Since hypersonic similarity requires that γ be fixed, this restriction applies to the combined form of the rules. Further discussions of similarity rules are given in [*113,114,115*].

The prediction of pressure distributions on airfoils at high supersonic speeds by means of shock expansion methods has been studied by Eggers and Syvertson [*116*] and further approximations to this method obtained. Since temperatures encountered in flight at very high Mach numbers cause the behavior of air to depart from that of an ideal gas, attention is given to the resultant change in the properties of the air and the effect on the calculated pressure distributions.

D,35. Cited References.

1. Lomax, H., Heaslet, M. A., Fuller, F. B., and Sluder, L. Two- and three-dimensional unsteady lift problems in high-speed flight. *NACA Report 1077*. 1952.
2. Prandtl, L. Allgemeine Überlegungen über die Strömung zusammendruckbaren Flussigkeiten. *Atti del Convegno della Fondazione Alessandro Volta 1935*, 169–198 (1936). Also *NACA Technical Memorandum 805*. 1936.
3. Glauert, H. The effect of compressibility on the lift of an aerofoil. *ARC Reports and Memoranda 1135*. 1927.
4. Munk, M. Elements of the wing section theory and of the wing theory. *NACA Report 191*. 1924.
5. Jeffreys, H. On aerofoils of small thickness. *Proc. Roy. Soc. A121*, 22–28 (1928).
6. Volterra, V. *Leçons sur l'intégration des équations différentielles aux dérivées partielles*. Paris, 1912.
7. Heaslet, M. A., Lomax, H., and Jones, A. L. Volterra's solution of the wave equation as applied to three-dimensional supersonic airfoil problems. *NACA Report 889*. 1947.
8. Heaslet, M. A., and Lomax, H. The use of source-sink and doublet distributions extended to the solution of boundary-value problems in supersonic flow. *NACA Report 900*, 1948.
9. Ackeret, J. Air forces on airfoils moving faster than sound. *Z. Flugtech u. Motorluftschiffahrt 16*, 72–74 (1925). Also *NACA Technical Memorandum 317*. 1925.
10. Busemann, A. Aerodynamischer Auftrieb bei Überschallgeschwindigkeit. *Atti del Convegno della Fondazione Alessandro Volta 1935*, 328–360 (1936). Also *Luftfahrtforschung 12*, 210–220 (1935).
11. de Kármán, Th. The problem of resistance in compressible fluids. *Atti del Convegno della Fondazione Alessandro Volta 1935*, 223–326 (1936).
12. Jones, R. T. Wing planforms for high speed flight. *NACA Report 863*. 1945.
13. Taylor, G. I. Applications to aeronautics of Ackeret's theory of airfoils moving at speeds greater than that of sound. *ARC Reports and Memoranda 1467*. 1932.
14. Rayleigh, Lord. *Theory of Sound*, Dover, 1945.
15. Lamb, H. *Hydrodynamics*, 6th ed., Dover, 1945.
16. Jones, R. T. Leading edge singularities in thin airfoil theory. *J. Aero. Sci. 17*, 307–310 (1950).
17. Puckett, A. E. Supersonic wave drag of thin airfoils. *J. Aero. Sci. 13*, 475–484 (1946).
18. Evvard, J. C. Use of source distributions for evaluating theoretical aerodynamics of thin finite wings at supersonic speeds. *NACA Report 951*. 1950.
19. Baldwin, B. S. Triangular wings cambered and twisted to support specified distributions of lift at supersonic speeds. *NACA Technical Note 1816*. 1949.
20. Mirels, H. Lift-cancellation techniques in linearized supersonic-wing theory. *NACA Technical Note 2145*. 1950.
21. Söhngen, H. Die Lösungen der Integralgleichung $g(x) = \frac{1}{2\pi} \oint_{-a}^{a} \frac{f(\xi)d\xi}{x - \xi}$. *Math. Zeits. 45*, 245–264 (1939).
22. Lagerstrom, P. A., and Van Dyke, M. D. General considerations about planar and non-planar lifting systems. *Douglas Aircraft Co. Report SM 13432*. 1949.
23. Busemann, A. Infinitesimal conical supersonic flow. *NACA Technical Memorandum 1100*. 1947.
24. Lagerstrom, P. A. Linearized supersonic theory of conical wings. *NACA Technical Note 1658*. 1948.
25. Germain, P. The general theory of conical motions and its application to supersonic aerodynamics. *ONERA Publication 34*, 1949.
26. Stewart, H. J. The lift of a delta wing at supersonic speeds. *Quart. Appl. Math. 4*, 246–254 (1946).
27. Robinson, A. Lift and drag of a flat delta wing at supersonic speeds. *Roy-Air Est. Rept. 1791*. 1946.
28. Brown, C. E. Theoretical lift and drag of thin triangular wings at supersonic speeds. *NACA Report 839*. 1946.

29. Gurevich, M. I. On the problem of the thin triangular wing moving with supersonic speed. *Prikladnaya Matematika i Mekbanika 11*, 395–396 (1947).
30. Brown, C. E., and Adams, Mac C. Damping in pitch and roll of triangular wings at supersonic speeds. *NACA Report 892*. 1947.
31. Squire, H. B. An example in wing theory at supersonic speed. *ARC Reports and Memoranda 2549*. 1947 (corrected, 1949).
32. Goodman, T. R. The lift distribution on conical and non-conical flow regions of thin finite wings in a supersonic stream. *J. Aero. Sci. 16*, 365–374 (1949).
33. Cohen, D. Formulas for the supersonic loading, lift, and drag of flat sweptback wings with leading edges behind the Mach lines. *NACA Report 1050*. 1951.
34. Hayes, W. D. Linearized supersonic flow. *North American Aviation Report AL-222*. 1947.
35. Heaslet, M. A., Lomax, H., and Spreiter, J. R. Linearized compressible-flow theory for sonic flight speeds. *NACA Report 956*. 1950.
36. Milne-Thomson, L. M. *Theoretical Aerodynamics*. Van Nostrand, 1947.
37. Hadamard, J. *Lectures on Cauchy's Problem in Linear Partial Differential Equations*. Yale University Press, 1928.
38. Robinson, A. On source and vortex distributions in the linearised theory of steady supersonic flow. *The College of Aeronautics, Cranfleld, Report 9*. 1947.
39. Ribner, H. S. Some conical and quasi-conical flows in linearized supersonic wing theory. *NACA Technical Note 2157*. 1950.
40. Ward, G. N. Supersonic flow past slender pointed bodies. *Quart. J. Mech. and Appl. Math. 2*, 75–97 (1949).
41. Lighthill, M. J. Supersonic flow past bodies of revolution. *ARC Reports and Memoranda 2003*. 1945.
42. Laitone, E. V. The linearized subsonic and supersonic flow about inclined slender bodies of revolution. *J. Aero. Sci. 14*, 631–642 (1947).
43. von Kármán, Th., and Moore, N. B. The resistance of slender-bodies moving with supersonic velocities with special reference to projectiles. *Trans. ASME 54*, 303–310 (1932).
44. Ferrari, C. Campi di corrente ipersonora attorno a solidi di rivoluzione. *L'Aerotecnica 17*, 507–518 (1937).
45. Tsien, H. S. Supersonic flow over an inclined body of revolution. *J. Aero. Sci. 5*, 480–483 (1938).
46. Kopal, Z. Supersonic flow of air around cones. *Mass. Inst. of Technol. Technical Report 1*. 1947.
47. Jones, R. T., and Margolis, K. Flow over a slender body of revolution at supersonic velocities. *NACA Technical Note 1081*. 1946.
48. Frankl, F. I., and Karpovich, E. A. *Gas Dynamics of Slender Bodies* (in Russian). Gos. Izd-Vo Tekhn.-Teoret. Lit-Ry, Moscow, 1948.
49. Munk, M. M. Isoperimetrische Aufgaben aus der Theorie des Fluges. *Inaug.-Dissertation*, Göttingen, 1919.
50. Munk, M. M. The minimum induced drag of aerofoils. *NACA Report 121*, 1921.
51. Sears, W. R. On projectiles of minimum wave drag. *Quart. Appl. Math. 14*, 361–366 (1947).
52. Haack, W. Geschossformen kleinsten wellenwiderstandes, *Lilienthal-Gesellschaft, Bericht 139*, 14–28 (1941).
53. Fuller, F. B., and Briggs, B. R. Minimum wave drag of bodies of revolution with a cylindrical center section. *NACA Technical Note 2535*. 1951.
54. Munk, M. M. The aerodynamic forces on airship hulls. *NACA Report 184*. 1924.
55. Jones, R. T. Properties of low-aspect-ratio pointed wings at speeds below and above the speed of sound. *NACA Report 835*. 1946.
56. Spreiter, J. R. The aerodynamic forces on slender plane- and cruciform-wing and body combinations. *NACA Report 962*. 1950.
57. Ribner, H. S. The stability derivatives of low-aspect ratio wings at subsonic and supersonic speeds. *NACA Tecynical Note 1432*. 1947.
58. Spreiter, J. R., and Sacks, A. H. The rolling up of the trailing vortex sheet and its effect on the downwash behind wings. *J. Aero. Sci. 18*, 21–32 (1951).

59. Lomax, H., and Byrd, P. Theoretical aerodynamic characteristics of a family of slender wing-tail-body combinations. *NACA Technical Note 2554.* 1951.
60. Heaslet, M. A., and Lomax, H. The calculation of pressure on slender airplanes in subsonic and supersonic flow. *NACA Technical Note 2900.* 1953.
61. Baker, B. B., and Copson, E. T. *The Mathematical Theory of Huygens' Principle.* Clarendon Press, Oxford, 1939.
62. Gardner, C. Time dependent linearized supersonic flow past planar wings. *Comm. on Pure and Appl. Math. 3,* 33–38 (1950).
63. Goodman, T. R. Aerodynamics of a supersonic rectangular wing striking a sharp-edged gust. *J. Aero. Sci. 18,* 519–526 (1951).
64. Goodman, T. R. The quarter-infinite wing oscillating at supersonic speeds. *Cornell Aero. Lab. Report 36.* 1951. Also *Quart. Appl. Math. 10,* 189–192 (1952).
65. Bateman, H. *Partial Differential Equations of Mathematical Physics,* pp. 133–134. Dover, 1944.
66. Strang, W. J. Transient lift of three-dimensional purely supersonic wings. *Proc. Roy. Soc. A202,* 54–80 (1950).
67. Miles, J. W. On simple planforms in supersonic flow. *J. Aero Sci. 17,* 127 (1950).
68. Miles, J. W. On harmonic motion of wide delta airfoils at supersonic speeds. *Naval Ordnance Test Station, Inyokern, Technical Memorandum RRB*-36. 1950.
69. Churchill, R. V. *Modern Operational Mathematics in Engineering.* McGraw-Hill, 1944.
70. Heaslet, M. A., and Lomax, H. Two-dimensional unsteady lift problems in supersonic flight. *NACA Report 945.* 1949.
71. Garrick, I. E., and Rubinow, S. I. Flutter and oscillating air force calculations for an airfoil in a two-dimensional supersonic flow. *NACA Report 846.* 1946.
72. Watkins, C. E. Effect of aspect ratio on the air forces and moments of harmonically oscillating thin rectangular wings in supersonic potential flow. *NACA Technical Note 2064.* 1950.
73. Watkins, C. E. Air forces and moments on triangular and related wings with subsonic leading edges oscillating in supersonic potential flow. *NACA Technical Note 2457.* 1951.
74. Miles, J. W. The oscillating rectangular airfoil at supersonic speeds. *Quart. Appl. Math. 9,* 47–64 (1951).
75. Runyan, H. L. Single-degree-of-freedom-flutter calculations for a wing in subsonic potential flow and comparison with an experiment. *NACA Technical Note 2396.* 1951.
76. Wagner, H. Über die Entstehung des dynamischen Auftriebes von Tragflugeln. *ZAMM 5,* 17-35 (1925).
77. Nelson, H. C., and Berman, J. H. Calculations on the forces and moments for an oscillating wing-aileron combination in two-dimensional potential flow at sonic speed. *NACA Technical Note 2590.* 1952.
78. Jahnke, E., and Emde, F. *Tables of Functions.* Dover, 1943.
79. Dietze, F. The air force the harmonically oscillating aerofoil in a compressible medium at subsonic speeds (two-dimensional problem). *ARC Report 10,219, 0.633.* 1946. Translated from the German: *ZWB Reports 1733 and 1733/2.* 1943.
80. Schade, T. The numerical solution of Possio's integral equation for an oscillating aerofoil in a two-dimensional subsonic stream. *ARC Reports 9506, 0.560; 10108, 0.560b and Addendum.* 1946.
81. Lomax, H. Lift developed on unrestrained rectangular wings entering gusts at subsonic and supersonic speeds. *NACA Technical Note 2925.* 1953.
82. Lomax, H., Heaslet, M. A., and Fuller, F. B. Three-dimensional, unsteady-lift problems in high-speed flight—Basic concepts. *NACA Technical Note 2256.* 1951.
83. Miles, J. W. On nonsteady motion of delta wings. *J. Aero. Sci. 16,* 568–569 (1949).
84. Froehlich, J. E. Nonstationary motion of purely supersonic wings. *J. Aero. Sci. 18,* 298–310 (1951).
85. von Kármán, T. Supersonic aerodynamics—principles and applications. *J. Aero. Sci. 14,* 373–409 (1947).
86 Hayes, W. D. Reversed flow theorems in supersonic aerodynamics. *North American Aviation Report AL-755.* 1948.

87. Munk, M. M. The reversal theorem of linearized supersonic airfoil theory. *J. Appl. Phys. 21,* 159–161 (1950).
88. Harmon, S. M. Theoretical relations between the stability derivatives of a wing in direct and in reverse supersonic flow. *NACA Technical Note 1943.* 1949.
89. Flax, A. H. Relations between the characteristics of a wing and its reverse in supersonic flow. *J. Aero. Sci. 16,* 496–504 (1949).
90. Brown, C. E. The reversibility theorem for thin airfoils in subsonic and supersonic flow. *NACA Technical Note 1944.* 1949.
91. Ursell, F., and Ward, G. N. On some general theorems in the linearized theory of compressible flow. *Quart. J. Mech. and Appl. Math. 3,* 326–348 (1950).
92. Jones, R. T. The minimum drag of thin wings in frictionless flow. *J. Aero. Sci. 18,* 307–310 (1950).
93. Maxwell, J. C. *A Treatise on Electricity and Magnetism.* Clarendon Press, Oxford, 1881.
94. Lamb, H. On reciprocal theorems in dynamics. *Proc. London Math. Soc. 19,* 144–151 (1888).
95. Sears, W. R. Operational methods in the theory of airfoils in nonuniform motion. *J. Franklin Inst. 230,* 95–111 (1940).
96. Flax, A. H. General reverse flow and variational theorems in lifting-surface theory. *Cornell Aero. Lab. Report 34.* 1951. Also *J. Aero. Sci. 19,* 361–374 (1952).
97. Flax, A. H. Reverse flow and variational theorems for lifting surfaces in nonstationary compressible flow. *Cornell Aero. Lab. Report 42,* 1952. Also *J. Aero. Sci. 19,* 352–353 (1952).
98. Liepmann, H. W. On the relation between wave drag and entropy increase. *Douglas Aircraft Co. Report SM-13726.* 1950.
99. Oswatitsch, K., and Berndt, S. B. Aerodynamic similarity at axisymmetric transonic flow around slender bodies. *Kgl. Tek. Högskol. Handl. Aero. Technical Note 15,* Roy. Inst. Tech., Stockholm, Sweden. 1950.
100. Berndt, S. B. Similarity laws for transonic flow around wings of finite aspect ratio. *Kgl. Tek. Högskol. Handl. Aero. Technical Note 14,* Roy. Inst. Tech., Stockholm, Sweden. 1950.
101. Guderley, K. G. Considerations of the structure of mixed subsonic-supersonic flow patterns. *Wright Field Report F-TR-2168-ND.* 1947.
102. von Kármán, Th. The similarity law of transonic flow. *J. Math. and Phys. 26,* 182–190 (1947).
103. Kaplan, C. On similarity rules for transonic flows. *NACA Technical Note 1527.* 1948.
104. Cole, J. D. Drag of finite wedge at high subsonic speeds. *J. Math. and Phys. 30,* 79–93 (1951).
105. Vincenti, W. G., and Wagoner, C. B. Transonic flow past a wedge profile with detached bow wave—general analytical method and final calculated results. *NACA Technical Note 2339.* 1951.
106. Vincenti, W. G., and Wagoner, C. B. Transonic flow past a wedge profile with detached bow wave—details of analysis. *NACA Technical Note 2588.* 1951.
107. Spreiter, J. R. Similarity laws for transonic flow about wings of finite span. *NACA Technical Note 2273.* 1951.
108. Spreiter, J. R. On the application of transonic similarity rules. *NACA Technical Note 2726.* 1952.
109. Goldsworthy, F. A. Two-dimensional rotational flow at high Mach number past thin airfoils. *Quart. J. Mech. and Appl. Math. 5,* 54–63. 1952.
110. Tsien, H. S. Similarity laws of hypersonic flows. *J. Math. and Phys. 25,* 247–251 (1946).
111. Hayes, W. D. On hypersonic similitude. *Quart. Appl. Math. 5,* 105–106 (1947).
112. Van Dyke, M. D. The combined supersonic-hypersonic similarity rule. *J. Aero. Sci. 18,* 499–500 (1951).
113. Neice, S. E., and Ehret, D. M. Similarity laws for slender bodies of revolution in hypersonic flows. *J. Aero. Sci. 18,* 527–531 (1951).
114. Hamaker, F. M., Neice, S. E., and Eggers, Jr., A. J. The similarity law for hyper-

sonic flow about slender three-dimensional shapes. *NACA Technical Note 2443*. 1951.
115. Ehret, D. M., Rossow, V. J., and Stevens, V. I. An analysis of the applicability of the hypersonic similarity law to the study of flow about bodies of revolution at zero angle of attack. *NACA Technical Note 2250*. 1950.
116. Eggers, Jr., A. J. and Syvertson, C. A. Inviscid flow about airfoils at high supersonic speeds. *NACA Technical Note 2656*. 1952.
117. Heaslet, M. A., and Spreiter, J. R. Reciprocity relations in aerodynamics. *NACA Technical Note 2700*. 1952.

SECTION E

HIGHER APPROXIMATIONS

M. J. LIGHTHILL

E,1. Introduction. The theories of high speed aerodynamics described in this volume, on the basis of which a quantitative study of the characteristically "nonlinear" effects is made in the present section, are those which neglect the diffusion of temperature by conduction and the diffusion of shear by viscosity. Thus they apply to phenomena in which the regions of boundary layer type, in which the diffusion mechanisms are important, are so thin that the said adiabatic theories give a good approximation to the flow. The influence of the diffusion phenomena may often be sketched in, to improve the theory, after it has been completed. For example, in calculating the drag on a body which moves rapidly through the atmosphere, it is usually necessary to include an estimate of the skin friction due to the laminar or turbulent boundary layer. Frequently, too, at supersonic speeds, it is necessary to take into account the modification of the adiabatic pressure distribution, which results from interaction between a rear shock and the boundary layer.

But the theories which should guide one, in making these final modifications to the adiabatic theory, are described in Vol. IV and V. The present volume is confined to a study of the application of the adiabatic theory to those phenomena in which the diffusion mechanisms are not fundamental, and to the comparison of the resulting predictions with experimental results.

In this restricted class of high speed flow phenomena, the mathematical apparatus necessary to the theory is simplified. The only basic physical concepts which need be considered, besides those already appearing in the theory of the thermodynamics of a gas in equilibrium, are the velocity field at any instant (a concept first used by Euler) and the associated concepts of momentum and kinetic energy. The basic relations, from which all equations used in the theory are deduced, are as follows:

1. The mass of a portion of fluid does not change as its position and shape alter.

2. Its momentum changes at a rate equal to the resultant of the pressure forces acting on it. (Gravity may be neglected in high speed aerodynamics.)

3. Its energy, i.e. the sum of its kinetic and internal energies, changes at a rate equal to that at which pressure forces do work upon it.

4. At any point of the gas, at any time, the pressure, temperature, and the specific thermodynamic variables (specific volume, entropy, internal energy, and enthalpy) are related by the equation of state of the gas. (Thermodynamic relaxation times are unimportant for the flow of *air*, except at very high temperatures, though not necessarily for the high speed flow of other gases such as CO_2 or Freon.)

The theory further postulates that at points not on a shock or a vortex sheet the velocity and the thermodynamic variables are continuous functions. At these points the relations (1) and (2) imply Euler's equations of continuity and momentum, while the relations (3) and (4) imply that the rate of change of specific entropy for a particle of fluid is zero.

A shock, on the adiabatic theory, may be regarded as a (possibly moving) surface of zero thickness. As a particle of fluid crosses a shock, its component of velocity tangential to the shock remains continuous, but the component normal to the shock, with the pressure, temperature and specific thermodynamic variables, all change discontinuously, though subject to relations (1) to (4) above. The changes are fully discussed in III,E; here it is only necessary to remark that the specific entropy increases discontinuously, although it has been seen that except at a shock it cannot change at all. One is still justified, however, in calling the theory adiabatic, since the microscopically thin region, in which the heat exchanges producing this increase in entropy occur, is ignored.

A vortex sheet (otherwise called "contact discontinuity"), on the other hand, in this theory, is a surface of discontinuity through which no fluid passes. Thus the normal component of velocity relative to the sheet is zero on both sides. It follows from (2) above that the pressure is continuous at the vortex sheet; but the tangential component of the velocity, as well as the temperature and the specific thermodynamic variables, are, in general, all discontinuous there.

The relations described above hold throughout the gas, subject to appropriate conditions assumed to hold initially, or at large distances from the phenomena of interest, and subject also to a boundary condition at the surfaces of any solid bodies in the field of flow. This permits a relative velocity between the gas and the portion of body surface in contact with it, but requires that it be tangential to the surface. It is of interest to notice that the discontinuity, between the surface conditions given by the theory and those obtaining in practice, has all the properties of a vortex sheet which were listed above. In fact the theory simply replaces the diffuse vorticity in the boundary layer by a sheet of concentrated vorticity enveloping the body.

The question of existence and uniqueness of solutions to high speed flow problems, if tackled by the adiabatic theory, is discussed in Sec. B. The rather restricted class of exact analytic solutions is fully described in Sec. F. In the present section the study of *approximate* solutions to the equations and boundary conditions of the adiabatic theory, begun in Sec. C and D, is continued. In Sec. C and D every problem was treated by the roughest approximate method which could reasonably be applied. In most cases this consisted mathematically of a complete linearization of the problem. It is these cases which are reconsidered in the present section, to take into account "nonlinear" effects. The two problems studied in Sec. D in which the only applicable approximation was nonlinear, namely, transonic and hypersonic flow, are not treated to any higher approximation in this section.

In a sense, then, this section constitutes a critique of the process of linearizing problems of adiabatic flow. For every approximate theory discussed herein is superior in at least one respect to a completely linearized theory; that is, each is closer to the theoretically exact adiabatic flow; thus they indicate shortcomings of the linearized theories and constitute methods of (at least partially) overcoming these shortcomings. It does not, of course, follow that the improved approximate theories will always be nearer in all respects to the real flow in which the diffusion of shear and heat plays a part. It may happen occasionally that the influence of adiabatic nonlinearity on some measurable quantity is partly canceled out by the influence of the boundary layer. But in the general flow picture the two sources of error distort in quite different ways the simplified version given by the linearized theory. The object of the present section is so to improve the theory that these errors which are due to the nonlinearity of the equations of adiabatic flow are reduced as far as possible.

Expressed physically, the linearization process is always equivalent to assuming that the flow, in some frame of reference (in which the fluid speeds are expected to be relatively small), is governed by the equations of sound. The physical significance of the nonlinearity, which the higher approximations will try to take account of, consists in three differences from acoustic theory, as follows:

5. The velocity of sound is not uniform, but varies with the other thermodynamic variables (for air, in proportion to the square root of the temperature) throughout the flow.

6. At any point disturbances are propagated at the local speed of sound *relative to the local fluid velocity*, not relative to the fixed frame of reference.

7. Because, as a result of (6), in continuous flow compression waves would overtake the waves ahead of them, a discontinuous wave, the shock, must be assumed to appear, always just in time to prevent this happening. The speed of the shock (unlike that of discontinuities in the

acoustic theory) is supersonic relative to the fluid velocity ahead of it, and subsonic relative to the fluid velocity behind it. It can reflect a part of the energy of sound waves which overtake it, and transmit a part of the energy of those it overtakes, and it adds to the entropy of the fluid through which it passes.

It is literally true that the properties (5) and (6) embody all the difference, in the equations of motion for continuous flow, between the exact adiabatic theory and the acoustic theory. For those of the adiabatic theory constitute five linear homogeneous equations in the first derivatives of the pressure, the density, and the three velocity components; the coefficients in the equations are functions of these latter variables. If in these coefficients the pressure and density are replaced by uniform values, with the fluid speed replaced by zero, the equations of sound result. (These, as was seen in C,10, do not imply irrotationality; but they state that, if any vorticity is present, it remains where it is independently of any pressure waves which may pass, which in turn are propagated just as if no vorticity were present.) But at any point, of course, the equations of the adiabatic theory are true, with the local values for the coefficients. Hence, locally, propagation is acoustic relative to the local velocity, and with the local value for the speed of sound. This will be the guiding principle throughout the following discussion.

In all the problems treated in this section, the flow is steady in a frame of reference fixed with respect to the solid boundaries. (But of course this is not normally the frame in which, on the linearized small perturbation theory, the equations of sound hold.) We restrict our subject matter thus because on the one hand one-dimensional unsteady motion has been fully treated in III,C, while on the other hand the determination of nonlinear effects in two-dimensional unsteady motion is on the verge of being too complicated a problem for really useful results to be expected in practice, except perhaps in relation to the calculation of flutter derivatives (VII,F). It should also be remembered that, in high speed flow, changes have to be very rapid indeed if there are to be "transient" effects significant compared with the basic transition through a succession of steady states.

In Art. 2 subsonic flow problems are considered. In these there can be no shock and therefore the nonlinearity consists only in items (5) and (6) above. (Actually, the limitation on all the methods used in Art. 2 is not so much the appearance of fluid moving with the local speed of sound, but rather the approach of conditions in which the appearance of a shock becomes necessary.) Now for flow about thin obstacles and similar problems, the small perturbation theory uses (essentially) the equations of sound in a frame which moves with the fluid at large distances from the obstacle. But for subsonic flow about thick obstacles one might expect nearly as good an approximation by using them in a frame fixed with respect to the obstacle. This yields, in steady flow, Laplace's equation. In other words, it gives the same result as is obtained by assuming from the start an infinite velocity of sound, as in the "incompressible fluid" theory. Thus there are two alternative linearizations, one simpler than the other, from which a search for a better approximate solution may begin, which fact leads to a multiplicity of possible approaches to the subject. However, the various improved methods of Art. 2 are all methods of successive approximation which take into account, with increasing accuracy (but in a different order of importance in the various methods), the finiteness of the speed of sound, its variation within the field of flow, and the convection of sound with the fluid.

In Art. 3, et seq., which deal with supersonic problems, the difficulties associated with items (5) and (6) similarly arise, although it will be found that they can be rather more easily surmounted, so that a wider variety of problems may be treated. However, item (7) presents difficulties of an altogether new type. These would make the analytic evaluation of supersonic flows, to really high degrees of approximation, formidable indeed. However, careful study will show, in these articles, that approximations at least better than the small perturbation theory can often be obtained by quite simple processes. Frequently these are based on neglecting those properties of shocks which were collected into the last sentence of item (7).

For the purpose of evaluating most of the quantities of greatest practical importance, the theories to be described merely improve approximations already given by the linearized small perturbation theory. But there are other aspects of the flow as to which information derived from linearized theory is absent, or else completely erroneous, but which can nevertheless be evaluated by the more exact theories. Many of these are apparent from the three points of difference listed above. Almost always, for example, the strength of any shock occurring in the flow falls into this category.

Again, the shape of the solid boundaries may be such as to produce small disturbances on the whole but large disturbances in some particular region (e.g. the rounded leading edge of an airfoil). Then the linearized theory may simply give disturbances tending to infinity in this region, while more exact theories give useful finite approximations to the disturbance distribution.

But there are yet further aspects of flow on which the linearized theory gives quite erroneous information. These may be described generically as "the behavior of any wave after it has traveled a distance of many wavelengths," where the word "many" is to be understood in a sense whose import varies inversely with the amplitude of the wave. For when the wave has traveled such a distance the nonlinear effects (1) and (2), namely, the convection of sound with the fluid and the nonuniformity of its speed—to say nothing of the influence of shocks—may clearly, by their accumulation, have altered the shape of the wave altogether from that given by linearized theory. This fact is relevant to the steady supersonic flow past an obstacle if the flow at some distance from the flight path is required, since this consists of a wave (headed by a shock, and containing at least one more shock farther back) which is propagated diagonally outward from the flight path. The change of wave form due to effects (1) and (2) is found to be of the nature of a spreading out; at the same time the shocks decay in strength by interaction with the wave, whence the problem is often referred to as that of determining the decay of shocks.

Mathematically, the failure of linearized theory in such a case is explained by the fact that an approximation to a partial differential equation, while yielding adequate results in a limited region, may yield a worse and worse approximation to the *solution* farther and farther from where the boundary conditions determining the solution were applied. There is an interesting parallel in Stokes' theory of flow at low Reynolds number past a solid sphere (IV,B). The inertial stresses, being small compared with the viscous stresses, are neglected. The solution which can then be obtained gives a correct first approximation to the surface pressure distribution, and hence the famous drag formula. But the flow far from the sphere is given quite incorrectly by the theory, on which it is symmetric fore and aft of the sphere, from which one could deduce (by simple momentum considerations) the absence of drag. (The general position of approximate theories of supersonic aerodynamics will be found to be similar, e.g. in Art. 4.) The deficiency may be removed by using Oseen's alternative approximation (IV,B) to the equations of motion at low Reynolds number, which is equally accurate near the sphere and much closer far away from it. The solution of Oseen's equations gives the same first approximation to the surface pressures as that of Stokes's but also gives a correct approximate behavior (involving a wake) at large distances from the sphere.

Often the true behavior of high speed flow, in cases such as those discussed above (for example near a shock, or in a small region of large disturbance on the body surface, or in a wave which has traveled a large distance), cannot be calculated by the kind of process, for obtaining further approximations beyond the small perturbation theory, which seems mathematically most natural. In fact, such a process often diverges in such regions, although converging elsewhere, thus making the local failure of the first approximation hardly surprising.

In the present section, on the other hand, some attempt will be made to show that methods based on sound physical thinking, and especially on the principles (5), (6), and (7) above, with the commentaries following them, will not usually lead one into such difficulties. In fact, such physical approaches are used throughout, except in certain cases where it is convenient, as leading the reader on from the simpler to the more complex, to develop a subject historically. This mode of presentation was adopted as giving a unifying theme which would be generally intelligible, and as holding out most hope for stimulating research into extensions of the theory.

But it should be remarked that there exists, as an alternative, a mathematical approach to these problems, which constitutes a semi-mechanical process of ferreting out the answer in any case of this kind in which physical intuition (though perhaps only temporarily) fails one. This is the author's "technique for rendering approximate solutions to physical problems uniformly valid" [*31*], which has been applied successfully to problems of all of the three types mentioned at the beginning of the second last paragraph. The principle of this technique may be described quite briefly as follows (though the fact that it works, and the practical following out of the principle, are both more complicated). At the same time as successive approximations are made to the dependent variables, they must also be made to one of the *independent* variables or to a suitably chosen combination of them, say x. If y is representative of the other independent variables, then each successive approximation, both to the dependent variables and to x, is a function of y and of a new variable z. The first approximation to x is taken as z itself. Now if x were taken identical with z (so that, effectively, there were no transformation of independent variables), then in the "difficult" region successive approximations would diverge. Hence the successive approximations to x as a function of y and z must be so chosen that those to the dependent variables no longer diverge, as far as those which are treated indicate, anywhere in the region of physical interest. For rendering the initial linearized approximation uniformly valid it is necessary only to proceed to the second approximation to x as a function of y and z. (This may be determined so that the second approximations to the dependent variables are not diverging, even in the difficult region, without actual calculation of these second approximations.) To obtain a uniformly valid *second* approximation to the dependent variables, x would have to be evaluated to the *third* approximation. The interpretation of the final solution is that the expression for x as a function of y and z *defines* the new variable z in terms of the physically significant variables x and y, whence the expressions for the dependent variables in terms of x and z become physically significant. The difference between x and z is only significant near the difficult region.

This method, which, for reasons explained two paragraphs back, will *not* be employed in the present section, is explained in full detail in the reference cited above, and further exemplified in papers by Whitham [*32,66*], Lighthill [*33,56*], Legras [*57,58*], and Tan [*59*]. This mathematical approach will occasionally be referred to below as confirming results based on other points of view.

We may close this article with remarks on the chronology of the material presented in Sec. E. The work of Art. 2 and 3 may fairly be called "classical," if account is taken of the relative novelty of all the subject matter of the series. On the other hand, the material presented in Art. 4 to 8 is taken partly from work published in 1948–1950, and even more from work published after February 1951. This shows how rapidly the subject is expanding, and holds out hopes that its ultimate extent and utility will be very great.

E,2. Subsonic Flows. This article is concerned with the steady adiabatic flow of a uniform subsonic stream past an obstacle not thin enough for the compressibility effect to be estimated satisfactorily from the Prandtl-Glauert small perturbation theory. For a given shape of obstacle (relative to the direction of the stream), any two such flows (with different body sizes, and different values at infinity of the velocity and thermodynamic variables) are of course similar if the Mach number $M_\infty = q_\infty/a_\infty$ of the uniform stream is the same in both. The methods of this article for studying the variation of the flow pattern with M_∞ are not applicable for all values of M_∞ less than 1, but evidence will be presented that they are applicable, in principle, for all values of M_∞ less than that value for which a continuous flow without a shock is no longer possible. It has been seen (A,7 and D,33) that this value of M_∞ is less than 1 by an amount which increases with the thickness ratio of the obstacle, at first like the two-thirds power thereof. However, it exceeds the Mach number at which local supersonic speeds first appear, and it will be shown below that the methods given can evaluate certain continuous transonic flows with reasonable accuracy.

The methods work by evaluating a sequence of successive approximations to the exact adiabatic flow, starting either from the "incompressible" flow with $M_\infty = 0$ or from the Prandtl-Glauert approximation to the flow. For the methods to be of value, the sequence must converge in some sense, and, since each approximate flow pattern is continuous, it is fairly evident that there will be no such convergence when the exact flow contains a shock. But when there is convergence it may often be quite adequate to use only the second approximation, and this is usually done; indeed, the calculation of higher approximations than this is in most cases prohibitively lengthy. At the end of the article the results of the methods are compared with those of the Kármán-Tsien method to be described in F,5.

Expansion in powers of the Mach number. A type of approach by successive approximations, which is natural mathematically, though (as suggested in Art. 1) not necessarily the best for many of the physical problems with which we have to deal, consists in evaluating the successive partial sums in a series expansion of, say, the velocity potential ϕ in powers of some small parameter. The earliest approach by successive approximations, made independently by Janzen [1] and Rayleigh [2], used just such an expansion of ϕ, namely, one in powers of M_∞. The first of their approximations, consisting of only the first term in the series, is the potential in the limit of zero Mach number, when the problem is linear and governed by Laplace's equation. The second approximation embodies the earliest appearance of compressibility effects, as M_∞ increases, and so on. In this approach there is clearly no restriction on the thickness of the obstacle.

It follows from Kelvin's circulation theorem that, in the steady continuous adiabatic flow of a uniform stream past an obstacle, the said velocity potential ϕ exists. Also, from the equation of continuity, and from Bernoulli's equation, with a constant value for the adiabatic index γ, the potential ϕ satisfies a partial differential equation

$$[a_\infty^2 + \tfrac{1}{2}(\gamma - 1)(q_\infty^2 - q^2)]\nabla^2\phi = \nabla\phi \cdot \nabla(\tfrac{1}{2}q^2) \tag{2-1}$$

where $q^2 = (\nabla\phi)^2$ is the square of the fluid speed. Here, the coefficient of $\nabla^2\phi$ is the square of the local speed of sound, and so embodies the difference from acoustic theory (5) of Art. 1. The right-hand side is what would become $\partial^2\phi/\partial t^2$ in axes moving with the local fluid velocity (remembering that the flow is steady in the fixed frame of reference). Thus it embodies the difference (6), i.e. the convection of sound.

Eq. 2-1 suggested the form of expansion used by Janzen and Rayleigh, namely

$$\phi = q_\infty[\phi_0 + M_\infty^2\phi_1 + M_\infty^4\phi_2 + \cdots] \tag{2-2}$$

Here ϕ_0, ϕ_1, ϕ_2, . . . are taken to be independent of the Mach number $M_\infty = q_\infty/a_\infty$ as well as of the main stream speed q_∞ itself.

Substitution of the series (2-2) in (2-1) gives

$$\begin{aligned}\{1 + \tfrac{1}{2}(\gamma - 1)M_\infty^2[1 - (\nabla\phi_0 + M_\infty^2\nabla\phi_1 + \cdots)^2]\}\\ (\nabla^2\phi_0 + M_\infty^2\nabla^2\phi_1 + M_\infty^4\nabla^2\phi_2 + \cdots)\\ = M_\infty^2(\nabla\phi_0 + M_\infty^2\nabla\phi_1 + \cdots) \cdot \nabla[\tfrac{1}{2}(\nabla\phi_0 + M_\infty^2\nabla\phi_1 + \cdots)^2]\end{aligned} \tag{2-3}$$

By expanding each side of Eq. 2-3 into a regular power series in M_∞^2, and equating successively the coefficients of 1, M_∞^2, M_∞^4, . . . in these two power series, the equations

$$\nabla^2\phi_0 = 0 \tag{2-4}$$

$$\nabla^2\phi_1 = \nabla\phi_0 \cdot \nabla[\tfrac{1}{2}(\nabla\phi_0)^2] \tag{2-5}$$

$$\begin{aligned}\nabla^2\phi_2 = \tfrac{1}{2}(\gamma - 1)[(\nabla\phi_0)^2 - 1]\nabla^2\phi_1\\ + \nabla\phi_1 \cdot \nabla[\tfrac{1}{2}(\nabla\phi_0)^2] + \nabla\phi_0 \cdot \nabla(\nabla\phi_0 \cdot \nabla\phi_1)\end{aligned} \tag{2-6}$$

. . . .

result. By Eq. 2-4 it is verified mathematically that the "incompressible" flow is the limit of the compressible flow for infinite velocity of sound.

Since the boundary condition of zero normal velocity at the surface of the obstacle is independent of M_∞, all the coefficients $\phi_0, \phi_1, \phi_2, \ldots$ in the expansion (Eq. 2-2) for ϕ must have zero normal derivative at this surface. Since also, at large distances from the obstacle, $\phi - q_\infty x$ must have a vanishingly small gradient (where the x axis is chosen in the direction of the undisturbed stream), the conditions at infinity on $\phi_0, \phi_1, \phi_2, \ldots$ must be that the gradients of $\phi_0 - x, \phi_1, \phi_2, \ldots$ all tend to zero.

The solution of Eq. 2-4, 2-5, 2-6, . . . , which are of Poisson's type, under these boundary conditions, presents no difficulty of principle if the body shape is one for which a Green's function is known. For two-dimensional shapes this means that the conformal mapping onto a circle is known. But the method is very laborious except for the two simplest shapes, the circular cylinder and sphere, and no attempt has been made for other shapes than these to calculate the potential more accurately than to the second approximation $\phi = q_\infty(\phi_0 + M_\infty^2 \phi_1)$. Therefore, although of course the flow past circular cylinders and spheres bears no resemblance to the adiabatic flow unless special methods of boundary layer control are applied, there is valuable information to be derived from the explicit formulas available in these two cases. Firstly, light may be shed on the question of convergence of the Janzen-Rayleigh process and similar processes. Secondly, the order of magnitude and general behavior of the compressibility correction may be expected to be similar for the real flow around bluff bodies with substantial wakes.

For the circular cylinder the formulas

$$\phi_0 = (r + r^{-1}) \cos \theta$$

$$\phi_1 = (\tfrac{13}{12} r^{-1} - \tfrac{1}{2} r^{-3} + \tfrac{1}{12} r^{-5}) \cos \theta + (-\tfrac{1}{4} r^{-1} + \tfrac{1}{12} r^{-3}) \cos 3\theta \qquad (2\text{-}7)$$

where (r, θ) are polar coordinates with the center of the circle as origin and the direction of the undisturbed stream as axis, and the unit of length is the radius of the cylinder, were given by Janzen [*1*]. The values of ϕ_2 and ϕ_3 are given in a survey paper by Eser [*3*], ϕ_2 having been first calculated correctly by Imai [*4*] (and incorrectly by other writers who subsequently agreed with Imai) and ϕ_3 independently by Eser and Imai. It is sufficient here to state that the maximum compressibility correction to the velocity field is at the two symmetrically placed positions of maximum velocity, where the speed is

$$q_\infty\{2 + 1.17M_\infty^2 + [2.34 + 0.59(\gamma - 1)]M_\infty^4 + [6.06 + 3.44(\gamma - 1) + 0.50(\gamma - 1)^2]M_\infty^6 + \cdots\} \qquad (2\text{-}8)$$

(In Eq. 2-8 the terms in $(\gamma - 1)$ are those due to the variability of the speed of sound; the others are those due to the convection of sound waves with the fluid (as well as to the finiteness of the speed of sound, which Laplace's equation neglects), corresponding to the first and second points of difference between the exact adiabatic theory and the theory of sound which were noted in Art. 1.) On the other hand, near the stagnation points, the compressibility correction reduces the velocity by a factor $1 - 0.42M_\infty^2 - \cdots$.

For $M_\infty = 0.4$, with $\gamma = 1.40$ (its value for air), the four displayed terms of the series within braces in Eq. 2-8 are $2 + 0.187 + 0.066 + 0.031 = 2.284$; the rate of fall of these terms indicates convergence. The sum to infinity is unlikely to be short of 2.318, the value of q/q_∞ which, under these conditions, indicates a local Mach number of 1. Roughly, we may say then that for flow at $M_\infty = 0.4$ past a circular cylinder the Janzen-Rayleigh process converges although the critical velocity is just attained at the ends of the diameter perpendicular to the flow.

For a value of M_∞ rather greater than this the process presumably starts to diverge. A closer estimate of this value emerges from the method of Taylor and Sharman discussed below.

For the flow past a sphere the formulas

$$\phi_0 = (r + \tfrac{1}{2}r^{-2}) \cos \theta$$

$$\phi_1 = (\tfrac{1}{3}r^{-2} - \tfrac{1}{5}r^{-5} + \tfrac{1}{24}r^{-8})P_1(\cos \theta) + (-\tfrac{3}{10}r^{-2} + \tfrac{27}{55}r^{-4} - \tfrac{3}{10}r^{-5} + \tfrac{3}{176}r^{-8})P_3(\cos \theta) \tag{2-9}$$

where (r, θ) are spherical polar coordinates with the center of the sphere as origin and the direction of the undisturbed stream as axis, and the unit of length is the radius of the sphere, were given by Rayleigh [2]. The value of ϕ_2 was found by Tamada [5] and is quoted by Eser [3]. The maximum compressibility correction occurs (as for the cylinder) at the position of maximum velocity, where the speed is

$$q_\infty\{1.5 + 0.313M_\infty^2 + [0.330 + 0.062(\gamma - 1)]M_\infty^4 + \cdots\} \tag{2-10}$$

On the other hand, near the stagnation points, the compressibility correction reduces the velocity, by a factor $1 - 0.25M_\infty^2 - \cdots$.

For $M_\infty = 0.55$, with $\gamma = 1.40$, the three displayed terms of the series within braces in Eq. 2-10 are $1.5 + 0.095 + 0.033$. The series probably converges to a sum not far short of 1.71, the value of q/q_∞ which, under these conditions, indicates a local Mach number of 1. As for the circular cylinder, there is probably a range of values of M_∞ exceeding 0.55 in which the Janzen-Rayleigh process converges to a continuous transonic flow pattern. It is reasonable to expect (see below) that this is bounded above by the Mach number at which shock-free flow becomes impossible.

A difference, in the order of magnitude of the compressibility correction to the flow past bluff bodies, between two-dimensional and axially symmetric flow patterns, has emerged. For the circular cylinder the correction factor to be applied to the surface velocity distribution varied between limits roughly $1 \pm \frac{1}{2}M_\infty^2$ for small M_∞. Since this is in agreement with the value $(1 - M_\infty^2)^{-\frac{1}{2}} = 1 + \frac{1}{2}M_\infty^2 + \cdots$ given by the small perturbation theory for thin airfoils, a compressibility correction of this order may be regarded as the rule in two-dimensional problems. On the other hand, for a sphere the correction factor varies between limits roughly $1 \pm \frac{1}{4}M_\infty^2$. Thus compressibility effects are considerably smaller for bluff bodies of revolution than for corresponding two-dimensional bodies, and in particular the value of M_∞ for which local supersonic speeds first appear is higher (e.g. 0.55 for the sphere as against 0.40 for the circular cylinder). But for bodies of revolution the small perturbation theory (C,5) is more complicated than it is for airfoils and a direct comparison with the correction for a sphere is in consequence not possible.

Both Janzen [*1*] and Rayleigh [*2*] attached a further significance to their calculations, which has not so far been mentioned. They observed that the flow pattern in both these problems was symmetrical about a plane perpendicular to the stream, not only in the first approximation ($M_\infty = 0$) but also in the second, which they had calculated, and indeed in all higher approximations, as they were able to show by induction. Hence, for values of M_∞ for which the process converges, the fluid pressures on the front of the obstacle are canceled out by those behind and there is no drag. Thus d'Alembert's so-called paradox, that, in a steady continuous flow of an inviscid fluid past a body, the body can experience no resultant force in the direction of the stream, is extended to compressible fluids.

Here, the essential discovery of Janzen and Rayleigh was that a steady continuous flow existed for some range of values of M_∞. For, this being so, then by considering the rate of change of momentum of the fluid lying (at one instant) between the body and a surface Σ at a large distance, the drag is seen to be

$$-\int_\Sigma p\,dydz - \int_\Sigma \rho(u - q_\infty)(u\,dydz + v\,dzdx + w\,dxdy) \qquad \text{(2-11)}$$

and it follows from Bernoulli's equation in the form

$$\frac{dp}{\rho} + \frac{1}{2}d(u^2 + v^2 + w^2) = \text{const}$$

that the only terms inside the integrals in Eq. 2-11 which are not at least quadratic in the disturbance velocities $u - q_\infty$, v, and w cancel one another out. But the disturbance velocities must fall off sufficiently rapidly at large distances from the obstacle for the integral of their squares over a large surface to tend to zero, by the properties of harmonic functions (for since, at large distances, the disturbances are small, they satisfy the equation of small perturbations, which is Laplace's equation trivially transformed). Hence the drag is zero.

But the existence of such a flow was not obvious. It might have been conjectured that, when a body moves at an appreciable subsonic Mach number, a steady flow is never set up, but the body continues to send sound pulses to infinity. The radiation of this energy to infinity would only be possible if work were done by the body against a drag. The flow pattern would not be symmetrical about any plane perpendicular to the stream, since the source of sound is moving. But, although at supersonic speeds a drag due to this cause does appear (Art. 4), the methods of the present article show that for values of M_∞ below a certain critical value the disturbance set up by the obstacle constitutes a standing wave of constant energy. Of course, as the body is accelerated from rest to its final uniform motion, work must be done to set up this standing wave; this is the work associated with the concept of "virtual mass." Also, if the acceleration is large (i.e. of the order of the square of the velocity of sound divided by the diameter of the body) a comparable amount of additional work is done in sending out a sound pulse [*60,61*]; but this pulse ceases to be emitted when the body speed becomes uniform.

It remains, as far as the Janzen-Rayleigh method is concerned, to mention that for thinner shapes, for which the adiabatic flow is realized more closely in experiment, only two-dimensional flow patterns have been evaluated, and those only to the second approximation $\phi = q_\infty(\phi_0 + M_\infty^2\phi_1)$. The shapes treated were the ellipse and the symmetrical Joukowsky airfoil (Kaplan [*6*] and others quoted by Eser [*3*], who however gives the formulas for the ellipse incorrectly). But since with these restrictions, at least in the absence of circulation, the method is inferior to the simpler Kármán-Tsien method (see below) it is not necessary to discuss these results, or the general formulation [*7*] of the solution for ϕ_1, in two-dimensional problems without circulation. However, passing to problems with circulation, some light is thrown on the influence of compressibility on lift curve slope, for airfoils of varying thickness, by calculations of Tomotika and Umemoto [*8*] on the flow at incidence past a Joukowsky airfoil with circulation fixed by the Kutta-Joukowsky condition. Expanding lift/$(\text{lift})_{M_\infty=0}$ in a series $1 + \frac{1}{2}M_\infty^2A + \cdots$, the coefficient A is graphed against the thickness ratio t of the airfoil, for the three angles of attack 5°, 10°, and 15°, in Fig. E,2a. On the small perturbation theory A is 1. Thus the figure illustrates the addition to the compressibility effect on lift due to large disturbances, a subject further discussed below.

Any future work on the Janzen-Rayleigh method would advantageously be confined to three-dimensional problems since, while several other methods for tackling two-dimensional problems exist (as will be seen) the effect of compressibility in other classes of problems, including even axisymmetrical flow at Mach numbers below that at which a shock appears, is known only very sketchily indeed. (Note added in proof: A first step in this direction has been taken by Longhorn [62].)

Successive approximation to the density field. The first attempt to improve on the cumbersome Janzen-Rayleigh method, in the case of two-dimensional flow, was made by Taylor and Sharman [9]. Their procedure for deriving successive approximations to the flow pattern has a

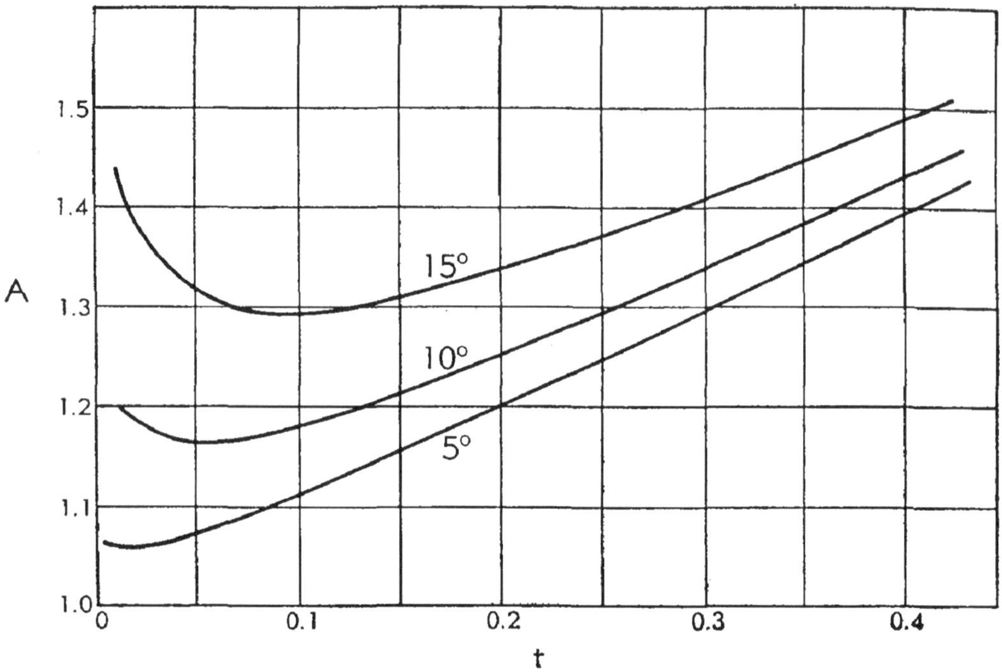

Fig. E,2a. The coefficient A of $\frac{1}{2}M_\infty^2$, in the ratio of lift at Mach number M_∞ to lift at zero Mach number, for a symmetrical Joukowsky airfoil of thickness ratio t at three angles of attack.

more direct physical significance. They work not with Eq. 2-1 for the potential but with the fundamental equations of continuity, irrotationality, and energy conservation

$$\frac{\partial(\rho u)}{\partial x} + \frac{\partial(\rho v)}{\partial y} = 0, \qquad \frac{\partial v}{\partial x} - \frac{\partial u}{\partial y} = 0 \tag{2-12}$$

$$\rho = \rho_\infty \left[1 + \frac{1}{2}(\gamma - 1)\frac{q_\infty^2 - q^2}{a_\infty^2}\right]^{1/(\gamma-1)} \tag{2-13}$$

The idea is to guess a rough approximation to the density field, and, using this value for ρ, to solve Eq. 2-12 for u and v under the given boundary conditions. Eq. 2-13 then (with q derived from these approximate values of u and v) yields a better approximation for the density field. Eq. 2-12 can be solved again using this new value for ρ, and so on. If the process converges, the limit must be the steady adiabatic flow.

The solution of Eq. 2-12 may be accomplished by means of a tank of variable depth, small compared with its length and breadth, and filled with an electrolyte of uniform conductivity. Then Eq. 2-12 holds if (u, v) is proportional to the electric intensity field and ρ is proportional to the depth (analogy A). The obstacle should then be represented by a solid insulator immersed in the electrolyte, and the main stream by a uniform current flow (parallel to it) through the tank, which the obstacle disturbs exactly as it does the fluid stream in the aerodynamic problem.

Alternatively, Eq. 2-12 holds if $(\rho v, \ -\rho u)$ is proportional to the electric intensity field and ρ^{-1} is proportional to the depth (analogy B). The obstacle should then be represented by a perfect conductor, and the main stream by a uniform current flow perpendicular to it. This second analogy method of calculation, though less natural than analogy A, has the advantage that it is possible to obtain a flow with circulation. For it is easily seen that the circulation

$$\oint (udx + vdy)$$

about the body is directly proportional to the current (if any) supplied by the conductor representing the body, and a nonzero value of this quantity can therefore be obtained by connecting the said conductor, through an appropriate resistance, to one terminal of the battery which generates the current flow.

It should be noticed that in both analogies it is sufficient at each stage to measure only the magnitude of the intensity field at various points, for the purpose of deducing the next approximation to the density distribution. A detailed treatment of the use of the electrolytic tank as a tool in aeronautical research is given in IX, H,2.

Taylor and Sharman always started with uniform depth of electrolyte, corresponding to the approximation $\rho = \rho_\infty$. Their first approximation to the velocity field is therefore the same as that of Janzen and Rayleigh, namely, the incompressible pattern. Subsequent approximations are not identical in the two methods; however, the order of magnitude of the error should be the same at parallel stages. Thus the error in taking $\rho/\rho_\infty = 1$ is of order M_∞^2, and so it may be inferred that (u, v) is given by the first approximation with an error of order M_∞^2. The use of these values in Eq. 2-13 now gives ρ/ρ_∞ with an error of order M_∞^4, giving a second approximation for (u, v) after Eq. 2-12 has been solved, with an error of order M_∞^4, and so on.

However the process just described makes it physically much clearer why there is an exactly steady continuous adiabatic flow (at any rate for values of M_∞ small enough for the process to converge), a point whose significance was explained above.

It also has the practical advantage over the Janzen-Rayleigh method that it can be applied with equal ease to any cylindrical body, including airfoils. Further, although the complete step of measuring the electric intensity field, calculating the density field from it by Eq. 2-13, and reshaping the bottom of the tank accordingly, is laborious, the subsequent repetitions of this process are only equally laborious, while in the Janzen-Rayleigh method they are progressively more so. One result of the latter advantage is that a sufficiently large number of terms can be calculated to make it possible to determine with much greater certainty between convergence and divergence. Thus, for the circular cylinder, the process converged for $M_\infty = 0.4$ and diverged slowly for $M_\infty = 0.5$, and by studying the behavior of the approximations Taylor was able to estimate that the radius of convergence was at about $M_\infty = 0.45$.

There is not perfect accord between the measurements of Taylor and Sharman and the subsequent calculations of the higher Janzen-Rayleigh approximations which were quoted above. The second approximations agree fairly closely, as Taylor and Sharman pointed out, but the convergence for $M_\infty = 0.4$ is rather surprisingly more rapid in their method, in which the third approximation is found to differ little from the second. Some improvement in rate of convergence would be expected since the successive approximations are chosen on sound physical, rather than arbitrary mathematical, grounds. But the velocity near the surface is in consequence given smaller than in the higher Janzen-Rayleigh approximations. The agreement is excellent at a distance of half a radius or more from the surface, taking into account the slight constraining effects of the walls of the tank, each of which measured about nine cylinder diameters. But it gradually falls off as the surface is approached, so that the maximum surface value for q/q_∞, obtained by extrapolation from values measured off the surface, is 2.20, whereas the calculations quoted above from Eser [*3*] indicate a value of at least 2.30 for this quantity. The discrepancy is probably partly due to the large gradients of velocity in this neighborhood which must have made the measurements difficult even though great care was taken. Also, the bottom of the tank had a 15 per cent gradient at this point, so that the accuracy with which the electric field is two-dimensional and satisfies Eq. 2-12 must have been considerably reduced. However, even on the measured values, there is extremely rapid convergence although a Mach number of 0.95 is attained in the field, and the existence of convergence even in transonic flows can, in consequence, hardly be doubted. It would be plausible, in any case, to claim this from the fact that the method is found to converge when $M_\infty = 0.4$, for which it is now believed, from the Janzen-Rayleigh method, that the critical velocity is just achieved.

Finally, if it is agreed that convergence ceases when M_∞ exceeds, by a certain small but nonzero amount, the "lower critical" Mach number (for which the critical speed is first attained in the flow), then, since experimental results and calculations by the hodograph method (Sec. F) suggest that the existence of continuous adiabatic flow ceases under similar conditions, and since in any case convergence could not be expected to continue after such continuous flow had broken down, it is reasonable, until evidence to the contrary is found, to suppose that the two events occur simultaneously. However, the methods could hardly be used for values of M_∞ just before that at which a shock becomes necessary, as convergence would doubtless be too slow.

A successive approximation method for small disturbances to a uniform stream. Another method for treating flows which are subsonic at least at large distances was described by Görtler [*10*] and by Hantzsche and Wendt [*11*]. The limitations on convergence appear to be similar to those just stated, but the method is nevertheless greatly superior to those described above in problems where it is suitable, namely, for evaluating the flow past obstacles whose breadth is considerably smaller than their length, but not so much smaller that the Prandtl-Glauert small perturbation theory can be applied. This latter theory, which neglects the squares of the perturbations, is treated as a first approximation to the flow, and higher approximations are deduced by neglecting only the cubes of the perturbations, and so on.

Putting $\phi = q_\infty(x + \varphi)$, so that $q_\infty\varphi$ represents the potential of disturbances to the main stream, Eq. 2-1 may be approximated, neglecting the cubes of disturbances, as

$$\nabla^2\varphi - M_\infty^2\varphi_{xx} = (\gamma - 1)M_\infty^2\varphi_x\nabla^2\varphi + 2M_\infty^2\nabla\varphi\cdot\nabla\varphi_x \qquad (2\text{-}14)$$

(Eq. 2-14 would be exact if φ_x, in its two occurrences on the right, were replaced by $\varphi_x + \frac{1}{2}(\nabla\varphi)^2$ and $\varphi_x + \frac{1}{4}(\nabla\varphi)^2$ respectively.)

If φ_1 is the first approximation to the disturbance potential φ, given by the small perturbation theory, in other words by approximating the right-hand side of Eq. 2-14 by zero, under the boundary condition of zero normal velocity at solid surfaces, then the second approximation φ_2 is obtained by solving Eq. 2-14 for φ, under the same boundary condition, with φ replaced throughout the right-hand side by φ_1. For the third approximation φ_3 the same process is gone through, but with the right-hand side of Eq. 2-14 rendered exact as indicated in the sentence following it, and with φ replaced therein by φ_2. This approximation is as far as any author has pursued the calculations, but it is clear how they would be continued. At each stage an equation of Poisson's type has to be solved, in the region outside the body stretched in the direction of motion by a factor $(1 - M_\infty^2)^{-\frac{1}{2}}$. The analysis is greatly simplified by Kaplan's discovery [*53*] of general expressions for particular integrals of the equations for φ_2 and φ_3, in terms of the φ's already found.

Hantzsche and Wendt [*11*], by calculating φ_2 for various plane flows, were able to give a general picture of the influence of thickness, camber, and angle of attack on the compressibility effect. Actually (unlike

Görtler [*10*]), they work with the stream function ψ rather than the potential ϕ, and this has definite advantages in the present method, in which it is much easier to apply a boundary condition in the form $\psi = 0$ than one in the form $\partial\phi/\partial n = 0$.

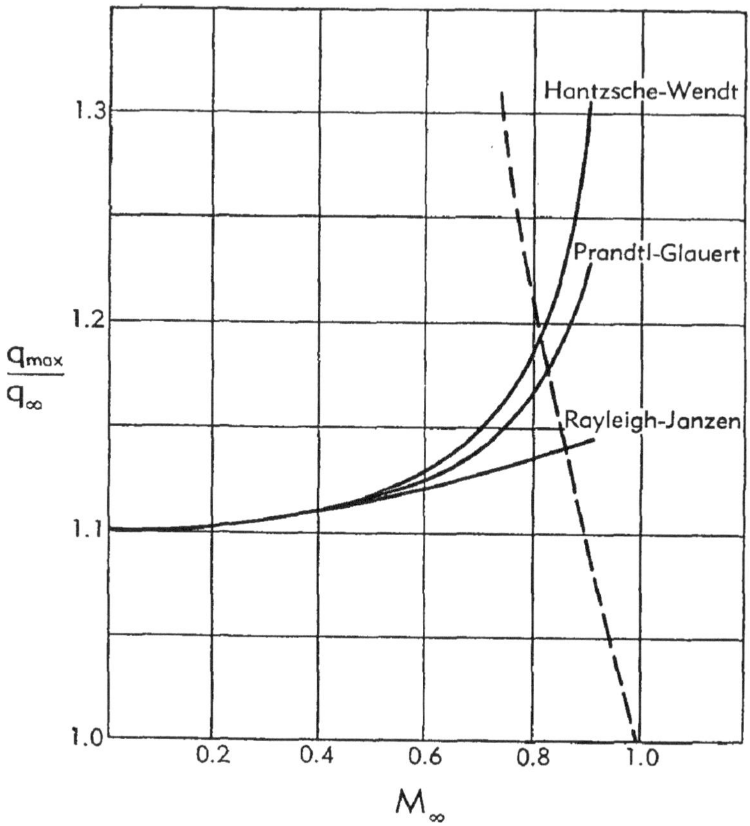

Fig. E,2b. Maximum surface velocity in symmetrical flow at Mach number M_∞ past an ellipse of thickness ratio 0.1.

For example, they calculate the maximum velocity $q_{\max}$ in the symmetrical flow past an ellipse of thickness ratio ϵ as

$$q_{\max} = q_\infty\left[1 + \frac{\epsilon}{\sqrt{1 - M_\infty^2}} + \frac{\epsilon^2 M_\infty^2(1 - \frac{2}{5}M_\infty^2)}{2(1 - M_\infty^2)^2} + O(\epsilon^3)\right] \quad (2\text{-}15)$$

It is likely that, for a given value of M_∞ less than 1, the series (2-15) converges when the thickness ratio ϵ is small enough for a continuous flow to be possible. For M_∞ near 1 the order of the second and third terms in the series indicates that the limiting value of ϵ is likely to be of order $(1 - M_\infty^2)^{\frac{3}{2}}$; and this is confirmed by the transonic similarity theory (D,33). The ratio of maximum surface velocity to main stream speed, $q_{\max}/q_\infty$, given by Eq. 2-15, is graphed against Mach number for $\epsilon = 0.1$ in Fig. E,2b. It is compared with the value $1 + \epsilon/\sqrt{1 - M_\infty^2}$ given by the Prandtl-Glauert small perturbation theory, and with the value given by

Kaplan's calculation [6] of the second Janzen-Rayleigh approximation. The broken line represents critical conditions, where the local Mach number is 1; thus it is the graph of q^*/q_∞ against M_∞ as given by Bernoulli's equation.

It is fairly clear that the second Hantzsche-Wendt approximation is superior to the second Janzen-Rayleigh approximation. For the first approximation is certainly better, since it yields Glauert's equation rather than Laplace's. And the procedure to the second approximation is more accurate in two respects: (1) in the equation for φ_2 in terms of φ_1 a term $-M_\infty^2\varphi_{2xx}$ is included where the Janzen-Rayleigh method would have $-M_\infty^2\varphi_{1xx}$; (2) the variation of the velocity of sound is taken into account by including the term $(\gamma - 1)M_\infty^2\varphi_{1x}\nabla^2\varphi_1$. These are likely to be far more significant than the third point of difference, which works the other way: (3) the term $2M^2\nabla\varphi_1 \cdot \nabla\varphi_{1x}$ is less accurate than the term $2M^2\nabla\varphi_1 \cdot \nabla[\varphi_{1x} + \frac{1}{4}(\nabla\varphi_1)^2]$ occurring in the equation for the Janzen-Rayleigh second approximation as rewritten in terms of φ.

Probably, for bodies whose thickness ratio is of the same order as 1/10, which is that of the ellipse of Fig. E,2b, the second Hantzsche-Wendt approximation is reasonably accurate in neglecting its cube for all subcritical Mach numbers; e.g. for $M_\infty = 0.81$ the terms inside the square brackets in Eq. 2-15 are $1 + 0.171 + 0.020$. Rapid convergence is indicated, to a value close to the critical value 1.198. This indicates not only that the lower critical Mach number for symmetrical flow past an ellipse of thickness ratio 1/10 is 0.81, but also that all calculations of φ_2 for shapes of practical importance should be useful.

Hantzsche and Wendt [11], by calculating φ_2 for a circular arc airfoil of zero thickness, were able to estimate the influence of camber on the compressibility effect. Its influence on the maximum velocity achieved is similar to that of thickness, increasing this quantity above the Prandtl-Glauert value. However, it has no influence on the lift at a given angle of attack, which is given by Glauert's law even to this second approximation if only the camber and not the thickness of an airfoil is taken into consideration.

To estimate the influence of thickness on lift, φ_2 was calculated for a symmetrical Joukowsky airfoil. The method of calculation makes it necessary to assume that the airfoil, when stretched in the direction of motion by a factor $(1 - M_\infty^2)^{-\frac{1}{2}}$, is still a Joukowsky airfoil, but this is true to a good approximation for fairly thin airfoils. For thickness ratio ϵ, the ratio of the lift coefficient to its value for zero Mach number is

$$\frac{C_L}{(C_L)_{M_\infty=0}} = \frac{1}{\sqrt{1 - M_\infty^2}} + \frac{4\epsilon}{3\sqrt{3}} \frac{(1 - M_\infty^2) - (1 - M_\infty^2)^{\frac{3}{2}} + \frac{1}{4}(\gamma + 1)M_\infty^4}{(1 - M_\infty^2)^2} \quad (2\text{-}16)$$

The actual value of $(C_L)_{M_\infty=0}$, with the theoretical value of the circulation sufficient to give a finite velocity at the trailing edge without any boundary layer, is $(1 + 4\epsilon/3\sqrt{3})2\pi \sin \alpha$. Thus Eq. 2-16 states that, while the basic flat plate lift coefficient $2\pi \sin \alpha$ is increased by the factor $(1 - M_\infty^2)^{-\frac{1}{2}}$ as a result of compressibility, the *addition to the lift coefficient due to thickness* is increased by the factor

$$\frac{1 - M_\infty^2 + \frac{1}{4}(\gamma + 1)M_\infty^4}{(1 - M_\infty^2)^2} \tag{2-17}$$

Kaplan [*12*] gives reasons for believing that this latter result is true, to within the Hantzsche-Wendt approximation, for all airfoils.

However, the concept of an "additional lift due to thickness" is one of the conclusions of potential theory bearing least relation to reality. The boundary layer thickness increases so much with airfoil thickness (owing to the larger positive pressure gradient over the rear portion of the surface) that the reduction due to this in the cast-off vorticity necessary to achieve smooth flow at the trailing edge easily balances the increase due to thickness derived from potential theory. For some sections the lift curve slope is less than 6; only for a few modern "low drag" sections does it exceed 2π. As for wind tunnel observations on the increase of lift curve slope with Mach number (which, it may be remarked, continues noticeably beyond the lower critical value), the Prandtl-Glauert curve is usually within the limits of experimental error, and there appears to be no conclusive evidence that the rate of increase which it gives is ever exceeded. It should be remembered, however, that, if ever effective methods of boundary layer control are used, then additional lifts as discussed above should be observed.

Görtler [*10*] applied the method under discussion to flow along a wavy wall, and by simplifying the boundary condition was able to proceed to the third approximation φ_3. The boundary condition adopted for the disturbance potential φ was

$$\varphi = K \cos \frac{2\pi x}{l} \qquad \text{on} \quad y = 0 \tag{2-18}$$

which indicates approximately a solid wavy wall with wavelength l on $y = 0$; the flow is to be determined for $y > 0$, and since the velocity is q_∞ at infinity, φ has zero gradient at large distances from the wall. The precise shape of the wall remains to be determined. For $M_\infty = 0.9$ and $K = 0.035l$ the first three approximations were calculated. They appear to be converging, and the flow indicated has a local supersonic region attached to each place where the wall is convex to the stream. In Fig. E,2c a section of the flow, comprising a single wavelength, is illustrated by a plot of isobars: the figures on these represent the ratio of the fluid speed q to the critical speed q^*, a ratio which exceeds 1 only if the flow is locally supersonic. The wall shape is not sinusoidal; the troughs are considerably deeper and very much "sharper" than the peaks.

Görtler points out that, since in the local supersonic regions the streamlines must diverge with increasing velocity, whereas the approximation φ_1 is essentially of subsonic type with convergence of streamlines in accelerating flow, it is clear that, when the present method is applied to transonic problems, the new terms in the second approximation φ_2 must be of the same order of magnitude as the old terms (i.e. those in φ_1). On

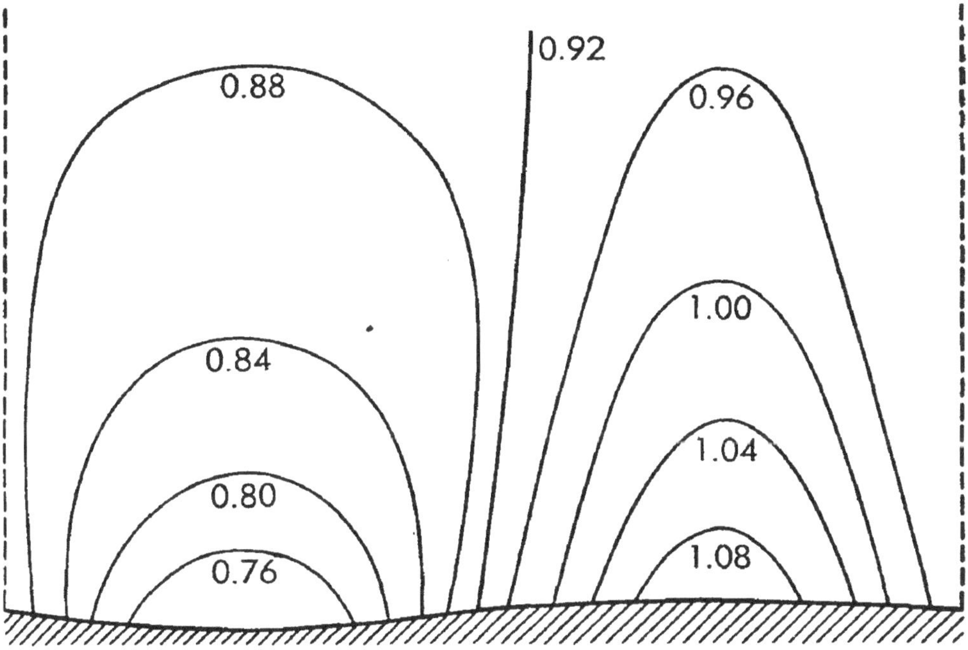

Fig. E,2c. Isobars in the flow at Mach number 0.9 past a wavy wall.

the other hand it might be hoped that φ_3 would represent only a small correction to φ_2, and this is observed in the example computed.

In a footnote, Görtler remarks that the solution was also attempted with an alternative boundary condition, analogous to Eq. 2-18 but prescribing $\partial\varphi/\partial y$ on $y = 0$ rather than φ. However, there was no indication of convergence for values of the parameters such that local supersonic regions were present. The physical significance of this may be that the corresponding wall shape would be much more nearly sinusoidal, and that for such shapes (as against that of Fig. E,2c) the breakdown of continuous flow occurs for a lower value of M_∞.

Expansion in powers of a thickness parameter. Before passing from the Hantzsche-Wendt method it should be remarked that a method identical in its aims, and indeed appearing at first sight to constitute the carrying of the method of Hantzsche and Wendt to its logical conclusion, had been tried by them and also by Kaplan [*12*] but subsequently discarded. It is directly analogous to the Janzen-Rayleigh method in that the disturbance potential φ is supposed expanded in powers of some parameter, say ϵ, defining the magnitude of the disturbance. For flow past an obstacle it would be convenient to take ϵ as the thickness ratio of the obstacle. The flow past a whole class of obstacles, obtained by shrinking the given obstacle by a uniform factor, but only in the direction perpendicular to the flow, is considered. Then the disturbance potential φ depends on ϵ, which varies from 0 to ∞ for this class of flows, and an expansion

$$\varphi = \epsilon\varphi^{(1)} + \epsilon^2\varphi^{(2)} + \cdots \tag{2-19}$$

can reasonably be attempted, where $\varphi^{(1)}, \varphi^{(2)}, \ldots$ are functions of x and y depending on M_∞. They are determined by substituting Eq. 2-19 in the equation for the disturbance potential *and in the boundary conditions*, and equating coefficients of powers of ϵ. It is this apparently logical conception, of approximating to the boundary condition as well as to the equation of motion, which causes the method to fail in important cases. If the equation of the obstacle, supposed two-dimensional, is $y = \epsilon Y(x)$, then the boundary condition is

$$\frac{v}{u} = \frac{\partial\varphi/\partial y}{1 + \partial\varphi/\partial x} = \epsilon Y'(x) \quad \text{on} \quad y = \epsilon Y(x) \tag{2-20}$$

To expand this completely in powers of ϵ, it is necessary to replace the values of any function, say $\partial\varphi/\partial y$, on $y = \epsilon Y(x)$, in terms of the values of it and its derivatives on $y = 0$, namely, by

$$\left[\frac{\partial\varphi}{\partial y} + \epsilon Y(x)\frac{\partial^2\varphi}{\partial y^2} + \frac{1}{2}\epsilon^2 Y^2(x)\frac{\partial^3\varphi}{\partial y^3} + \cdots\right]_{y=0} \tag{2-21}$$

As a result the boundary conditions on $\varphi^{(1)}, \varphi^{(2)}, \ldots,$ obtained by equating coefficients of $\epsilon, \epsilon^2, \ldots$ in condition (2-20), become

$$\frac{\partial\varphi^{(1)}}{\partial y} = Y'(x), \qquad \frac{\partial\varphi^{(2)}}{\partial y} = Y'(x)\frac{\partial\varphi^{(1)}}{\partial y} - Y(x)\frac{\partial^2\varphi^{(1)}}{\partial y^2}, \cdots \quad \text{on } y = 0 \tag{2-22}$$

The equations to be solved for $\varphi^{(1)}, \varphi^{(2)}, \ldots$ are almost identical with those in the Hantzsche-Wendt method, but the boundary conditions are applied on a straight strip in the direction of flow, coextensive with the obstacle.

Such boundary conditions are in general easier to apply. However, in the important case when the obstacle, though thin, has a blunt nose, the flow over which has a stagnation point, the method fails. It does this even for $M_\infty = 0$, i.e. when used as a method for obtaining the incompressible potential flow around airfoils. The first approximation $\epsilon\varphi^{(1)}$ then represents the first order airfoil theory, already given, for camber lines only, by Glauert [*13*, p. 87]. The flow is reasonably approximated (for small ϵ) on this theory except near the leading edge. Here there is a square root singularity of $\varphi^{(1)}$, and its gradient (representing the velocity) is infinite. However, for many purposes the errors in this neighborhood do not matter, and the over-all picture, which is known by a comparison with Theodorsen's more exact airfoil theory [*14*] to be a correct first approximation, is all that is required. But when $\varphi^{(2)}$ is calculated, using the second of conditions (2-22), it seems likely from the form of this condition that a more severe singularity will appear in $\varphi^{(2)}$ than in $\varphi^{(1)}$. Actually, however, the presence of this more severe singularity can be avoided in $\varphi^{(2)}$ by fixing the origin at the leading edge, so that $Y(x) = 0$ there; but it is unavoidable in $\varphi^{(3)}$. It may be presumed that the order of the singularity grows with successive terms. Hence, at least sufficiently near the leading edge, the process cannot converge. Therefore, since the process is not tending to a limit satisfying all the prescribed conditions, there is no reason why a later term should be more accurate than an earlier term.

Hence the method is not applicable, at any rate without modification, to bodies with blunt noses. For these it would therefore appear to be necessary to use the Hantzsche-Wendt method sketched above, in which the boundary condition is applied exactly on the surface of the body. The only successful investigation on the lines just described has been Kaplan's on the flow past a smooth hump in a wall [*52*], for which the difficulties mentioned do not arise.

Lighthill [*56*] has shown (using the method described at the end of Art. 1) that the logical process of expansion in powers of ϵ gives, when $M_\infty = 0$, and when terms in ϵ and ϵ^2 only are included, an approximation valid uniformly, even near the leading edge, provided the deduced flow field is shifted downstream through a distance equal to half the radius of curvature of the profile at the leading edge. It is possible that a similar result may be true for the case of general subsonic Mach number, which however is not investigated in the paper cited, and the point is still unsettled. If this were so, it would produce a method for calculating second order compressibility effects, due to the thickness, camber, and incidence of airfoils, simpler than the Hantzsche-Wendt process. But, for still higher order terms, the method of expansion (2-19) is certainly impossible, since Hantzsche showed [*15*] that for flow past an elliptic cylinder terms in $\epsilon^4 \ln \epsilon$ appear.

Comparison with the Kármán-Tsien method. Considering the results of all the methods so far described as a whole, it may be claimed that a fair picture of shock-free subsonic flow has been achieved. Enough particular solutions have been found to make possible an adequate guess in any given problem of how the compressibility effect predicted by potential theory will depart from that given by the small perturbation theory.

However, the material has in fact been little used in practical high speed flow work, because it is somewhat extensive. It is much easier to take a single formula and apply it under all conditions, and such a formula was provided by Tsien [*16*], who worked on an idea of von Kármán's.

The derivation of the Kármán-Tsien approximation is described in detail in F, 5 but its importance (due to its intrinsic interest as well as to the fact that it is by far the most widely used formula relating to two-dimensional subsonic flows) is so great that a comparison of its results with those of the other methods (described above) is here made.

The approximation made to the equations of motion by von Kármán was equivalent to taking full account of the finiteness of the velocity of sound and of the convection of sound by the local motion of the fluid, but adopting a quite different mode of variation of the speed of sound with the density (and hence with the local fluid speed) from the true one. Thus the velocity of sound is taken proportional to a negative power of the density, $a/a_\infty = (\rho/\rho_\infty)^{-1}$ instead of to a small positive power, $a/a_\infty = (\rho/\rho_\infty)^{\frac{1}{2}(\gamma-1)}$. The equation of motion (2-1) is therefore modified by the coefficient a^2 of $\nabla^2\phi$ being replaced by $a_\infty^2 - (q_\infty^2 - q^2)$; in short, γ is therein replaced by -1.

It is at once clear that exact solutions of the resulting equation for the velocity field (which can be obtained fairly easily, although, as will be seen, the obstacle shape cannot be specified in advance) must give the first approximation to the effect of small nonzero Mach number quite correctly. In other words, they are at least the equal of the second Janzen-Rayleigh approximation. For it was seen above that this was influenced only by the terms due to the convection of sound on the right of Eq. 2-1, and not at all by the variation of the speed of sound with q. In fact Eq. 2-5 for ϕ_1 is unaltered by replacing γ by -1.

But actually the exact solution with γ replaced by -1 is probably always better than the second Janzen-Rayleigh approximation, because more accuracy is gained by accurately admitting the convection of sound than is lost by allowing the speed of sound to vary incorrectly. That this is so for very thick obstacles is indicated by Eq. 2-8 for the maximum speed over a circular cylinder, which is closer to its exact value with γ replaced by -1 than with all the terms in M_∞^4, M_∞^6, . . . omitted. For very thin obstacles, on the other hand, the equation of von Kármán becomes identical with that of Prandtl and Glauert (which again is independent of γ), and the conclusions are again superior to those of the second Janzen-Rayleigh approximation (which replaces $(1 - M_\infty^2)^{-\frac{1}{2}}$ by $1 + \frac{1}{2}M_\infty^2$).

For obstacles which are only fairly thin the equation probably gives results intermediate in accuracy between the Prandtl-Glauert small perturbation theory and the second Hantzsche-Wendt approximation.

For on this latter approximation, the disturbance potential φ is replaced by φ_1 in the right-hand side of Eq. 2-14 and hence also $\nabla^2\varphi$ may be replaced by $M_\infty^2\varphi_{1xx}$. The relative importance of the two terms is therefore probably as $(\gamma - 1)M_\infty^4 : 2M_\infty^2$, and so the change of γ from 1.4 to -1 probably does not actually reverse the effect of φ_2 for any $M_\infty < 1$, and for smaller M_∞ may give a good approximation to φ_2. Roughly then, the replacement of γ by -1 deals accurately with the compressibility effect on very thin obstacles, but approximates *as for small Mach number* to the additional compressibility effect due to the finiteness of the disturbances.

The solution of Eq. 2-1 with γ replaced by -1, which is here discussed only for plane flows without circulation, is as follows (F, 5): If the velocity vector in incompressible potential flow about an obstacle is (u_i, v_i) at x_i, y_i—here the subscript i stands for "incompressible"—then the velocity vector in "compressible flow with γ replaced by -1" about a certain obstacle of modified shape, with the same speed in the undisturbed stream, is

$$(u, v) = \frac{(u_i, v_i)}{1 + \frac{1}{2}[1 - (1 - M_\infty^2)^{-\frac{1}{2}}]\left(\frac{u_i^2 + v_i^2}{q_\infty^2} - 1\right)} \tag{2-23}$$

at the point x, y given by

$$x + iy = x_i + iy_i - \frac{M_\infty^2}{[1 + (1 - M_\infty^2)^{\frac{1}{2}}]^2}\int\left(\frac{u_i + iv_i}{q_\infty}\right)^2 (dx_i - idy_i) \tag{2-24}$$

Eq. 2-24 represents a mapping of the x_i, y_i plane onto the x, y plane. Streamlines correspond under this mapping, and therefore the flow in the x, y plane is the flow about an obstacle obtained by carrying out the mapping on the area outside the original obstacle.

It is seen that the formulation of the solution is brief. Yet the solution is certainly just as accurate as the second Janzen-Rayleigh approximation (even for thick obstacles, and more so for thin ones), for which the analytic solution for a general shape of obstacle [7] takes a far more complicated form. This is essentially because a deformation of the body shape (and, with it, of the whole physical plane), as well as of the velocity variations, has been permitted.

However, the exact solution just stated is not the one advocated by Tsien and used widely in aeronautics. The Tsien approximation is to adopt the formula (2-23) defining the compressibility correction to the velocity vector, but to ignore the deformation of the obstacle and the flow plane by the mapping (2-24). To gauge the significance of this, observe that, although the obstacle is shrunk by the mapping (2-24), a uniform shrinkage (i.e. a mere change of scale) would not alter the velocity distribution on potential theory. It is the nonuniformity of shrinkage in the mapping that causes errors in neglecting it.

Now a small elementary arc of a streamline is shrunk by a factor

$$1 - \frac{M_\infty^2}{[1 + (1 - M_\infty^2)^{\frac{1}{2}}]^2} \frac{(u_i^2 + v_i^2)}{q_\infty^2} \tag{2-25}$$

(but unaltered in direction) in the mapping (2-24). Thus for Mach numbers as high as 0.8, for which the fraction involving the Mach number in Eq. 2-25 is $\frac{1}{4}$, the shrinkage near the point of maximum velocity is considerable, while there is none at the stagnation point. (On the other hand, small elementary arcs normal to the streamlines are *expanded* in the mapping by a factor identical with Eq. 2-25 but with the sign of the second term changed; this is less relevant, however, from the practical point of view, since one is mainly interested in conditions on the surface of the obstacle, which constitutes a streamline.) The obstacle as modified by the mapping therefore has a greater thickness ratio.

The correctness to order M_∞^2, for bodies of arbitrary thickness, of the complete solution given by the "velocity correction formula" (2-23) and the mapping (2-24), is lost by dropping the latter, since the mapping changes the flow field by amounts of order M_∞^2. However, the Tsien approximation was put forward only for dealing with the case of fairly thin obstacles, which constitutes the major field of practical application of potential flow theory.

But even here, and even if we temporarily ignore the difficulties associated with the perturbations not being small for practical obstacles with leading edge stagnation points, there is a serious loss of accuracy due to dropping the mapping (2-24). For if, as before, ϵ denotes the thickness ratio of the obstacle and hence the order of magnitude of the disturbances, the expression (2-25) indicates that differences in relative shrinkage can be of order ϵM_∞^2. Since the main part of the disturbance potential, derived from the small perturbation theory, is of order ϵ, it follows that changes in it due to the differential shrinkage are of order $\epsilon^2 M_\infty^2$. But the difference between the Prandtl-Glauert and second Hantzsche-Wendt approximations is itself of order $\epsilon^2 M_\infty^2$. Hence the Kármán-Tsien approximation retains no advantage over the Prandtl-Glauert approximations as far as strict mathematical orders of magnitude are concerned, the error in both being of order $\epsilon^2 M_\infty^2$ in the disturbance velocities (while the full solution of Eq. 2-23 with Eq. 2-24 contains the correct term in $\epsilon^2 M_\infty^2$, the error being of order $\epsilon^2 M_\infty^4$).

However, this discussion of mathematical orders of magnitude is misleading if it leads one to suppose that the velocity field is given just as accurately by the small perturbation theory as by the Kármán-Tsien formula. In practically all tests of airfoils in high speed wind tunnels the surface pressure distribution as deduced from the Kármán-Tsien formula is within the limits of experimental error for subcritical Mach numbers, while that deduced from the Prandtl-Glauert formula is not so for the higher Mach numbers (say $M_\infty \geqq 0.5$).

It is commonly suggested that this is true, in spite of the identity in mathematical order of magnitude of the two errors, for two reasons: (1) the fraction containing M_∞ in Eq. 2-25 remains rather smaller for subcritical Mach numbers than its order of magnitude "M_∞^2" would suggest, because it is approximately $\frac{1}{4}M_\infty^2$ for small M_∞ and is only 0.25 for $M_\infty = 0.8$, which is about the highest lower critical Mach number occurring in practice; (2) the errors due to allowing the velocity of sound a to vary in an incorrect manner, and due to neglecting the mapping (2-24), are of opposite sign and partly cancel out, although probably the latter is larger at lower Mach numbers and the former at higher Mach numbers. There is some evidence that the canceling suggested in the latter explanation does occur on the suction side of the airfoil, which is probably the critical region. For one may argue that the equations solved are the equations of continuity and irrotationality (2-12), with an incorrect relationship between density and fluid speed replacing Eq. 2-13. This incorrect relationship (obtained by replacing γ by -1) consistently overestimates the density for given q/q_∞ other than unity, by a factor increasing approximately like $1 + \frac{1}{8}(\gamma + 1)(q^2 - q_\infty^2)^2/a_\infty^4$, as q departs from q_∞ on either side. But on the suction side the error due to this might be partly canceled by the error due to ignoring the mapping (2-24), owing to the fact that the area of a small rectangle in the network of streamlines and equipotentials, on a scale such that it is unaltered for $q = q_\infty$, is changed in the mapping by a factor less than 1 by a certain multiple of $q_\infty^2(q^2 - q_\infty^2)/a_\infty^4$. These two factors, when multiplied to give the change in mass in such a small rectangle, tend to cancel in part, doing so completely for $q = 1.19q_\infty$ at small Mach numbers and for $q = 1.35q_\infty$ at $M_\infty = 0.8$. But the importance of this point can at present only be described as conjectural.

Actually, it is with measurements of pressure, not of velocity, that the theoretical results are compared. In using the expression (2-23) for the velocity field to deduce the density field and hence the pressure field, Tsien [*16*] recommends that the pressure-density relation should be taken as that obtained by integrating the assumed relation between a^2 and ρ, and taking $p = p_\infty$ in the undisturbed stream. This gives a linear relation $p/p_\infty = 1 + \gamma(1 - \rho_\infty/\rho)$ between pressure p and specific volume $1/\rho$. The relation between the pressure coefficient $C_p = (p - p_\infty)/\frac{1}{2}\rho_\infty q_\infty^2$ and the fluid speed is then not the exact relation for adiabatic flow, namely,

$$C_p = \frac{[1 + \frac{1}{2}(\gamma - 1)M_\infty^2(1 - q^2/q_\infty^2)]^{\gamma/(\gamma-1)} - 1}{\frac{1}{2}\gamma M_\infty^2} \tag{2-26}$$

but is what is obtained by replacing γ by -1 in Eq. 2-26. This latter relation, with Eq. 2-23, gives the following very simple relation between the pressure coefficient in two-dimensional compressible flow, C_p, and that in the incompressible flow past the same obstacle, C_{pi}:

$$C_p = \frac{C_{pi}}{(1 - M_\infty^2)^{\frac{1}{2}} + \frac{1}{2}[1 - (1 - M_\infty^2)^{\frac{1}{2}}]C_{pi}} \tag{2-27}$$

This is the formula mentioned above as so extensively used in aeronautics, and later as fitting wind tunnel measurements in subcritical conditions to within the experimental error.

The reader may perhaps imagine at this stage that the further error due to replacing γ by -1 in Eq. 2-26 is of importance in the discussion of the total deviation of the Kármán-Tsien pressure field from the exact field which would occur in adiabatic flow. For the Kármán-Tsien theory is essentially one relating to the velocity and density fields; it establishes by kinematic considerations the irrotational flow in which the density and fluid speed are connected in a certain way. Thus it would seem inadvisable, from a general point of view, to add to errors in Eq. 2-23 by using an inaccurate form of Eq. 2-26; but on the other hand a surprising accuracy in the final result might seem explicable by supposing that the previous errors are partly canceled by the error in using the said form of Eq. 2-26.

However, there is relatively little substance in these suspicions, for the simple reason that the error in replacing γ by -1 in Eq. 2-26 is always rather small. When Eq. 2-26 is expanded in powers of the disturbances, the coefficients of the first and second powers are independent of γ and the error in changing γ from 1.4 to -1 in the third term is $\frac{1}{10}M_\infty^4(1 - q^2/q_\infty^2)^3$. For small M_∞ or ϵ this term of order $M_\infty^4\epsilon^3$ is smaller than the errors of order $M_\infty^4\epsilon^2$ and $M_\infty^2\epsilon^2$ which were mentioned above. And indeed, even in a rather extreme case when $q^2/q_\infty^2 = 2.5$ and $M_\infty = 0.8$, expression (2-26) is -1.174 for $\gamma = 1.4$ but -1.111 for $\gamma = -1$, a difference of only 5.4 per cent.

But the formula relating the lower critical Mach number M_{cr} to the negative pressure coefficient $(C_p)_{min}$ with largest numerical magnitude on the surface, namely

$$(C_p)_{min} = \frac{2}{\gamma M_{cr}^2}\left[\left(\frac{2 + (\gamma - 1)M_{cr}^2}{\gamma + 1}\right)^{\gamma/(\gamma-1)} - 1\right] \tag{2-28}$$

cannot be simplified in such a way. Tsien recommends that Eq. 2-28 should be used in its exact form in conjunction with the experimentally verified formula (2-27). This gives a relation between the minimum pressure coefficient $(C_{pi})_{min}$ in the *low speed* potential flow past an obstacle and the lower critical Mach number M_{cr} for flow past the obstacle, which is shown graphically in Fig. E,2d. This curve is extensively used in airfoil design, since (as follows from statements already made above) it is almost always in agreement with wind tunnel experiments to within the orders of accuracy obtainable.

To sum up this article, considerable information on the development of flow patterns with Mach number until just before shock waves appear is obtainable; but the methods described are useless for investigating its further development. Indeed, as will be seen from other parts of this work, our theoretical understanding of this further development, at least up to where the Mach number of the main stream reaches unity, is very restricted indeed.

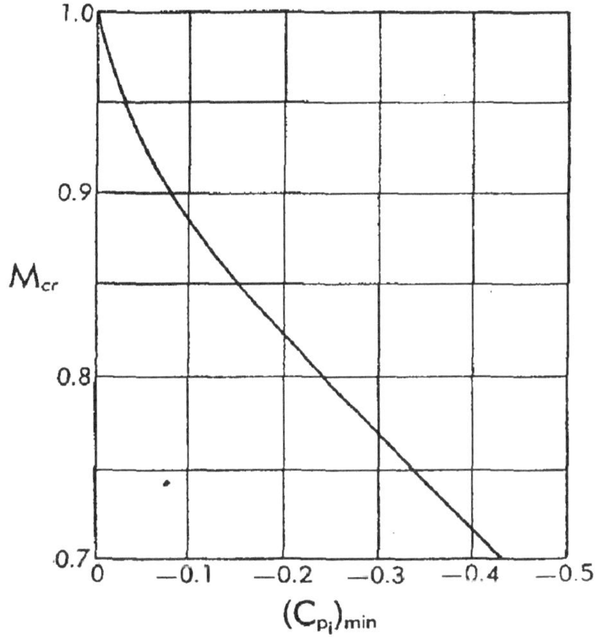

Fig. E,2d. Tsien's curve of lower critical Mach number against minimum pressure coefficient for incompressible flow.

E,3. Supersonic Two-dimensional Airfoil Theory: Surface Pressures. The theory of higher approximations than the acoustic "small perturbation" theory is more extensive in the supersonic than in the subsonic regime. For this reason it is here divided into the part relating to two-dimensional flow (Art. 3, 4, and 5), for which a remarkably detailed picture can be obtained, and the part relating to three-dimensional (and especially axisymmetrical) flows, of which our understanding is less complete. In Art. 3, 4, and 5, then, we describe those improvements which have been made to the picture given by Ackeret's small perturbation theory of the steady two-dimensional flow of a uniform supersonic stream past an airfoil, without taking account of the diffusion of shear and heat in the boundary layer.

The reason why the nonlinearity associated with the variable speed of sound, and with the convection of sound waves by the local motion of the fluid, is more easily dealt with in supersonic than in subsonic flow, is that every point in the flow has only a limited domain of influence (the downstream characteristic conoid). This already makes the small perturbation theory simpler. (For example, numerous calculations on exact lifting surface theory have been made in the supersonic regime (Sec. D), but these are almost prohibitively lengthy for the simplest planforms in the subsonic regime.) It is not surprising that this greater simplicity should penetrate also to the higher approximations.

Nevertheless, there is a third source of nonlinearity in supersonic flow, as mentioned in Art. 1, namely, the shock (the need for whose presence springs from a combination of the other two sources). As will be seen below, an attempt to estimate the effect of those properties of the shock which pressure discontinuities in the acoustic theory do not possess, especially the partial reflection of waves which overtake it and the gain in specific entropy of the fluid passing through it, adds considerably to the complexity of the theory, although here it has been carried fairly far for two-dimensional flow past airfoils. However, it will also be found that the properties of the shock just mentioned can be neglected while retaining a much better approximation to the flow field than is given by the small perturbation theory.

It should be remembered that in supersonic flow the use of a two-dimensional theory of flow past an isolated section may be of value not merely as an approximation, but also for an accurate study of the flow past wings, whether in a wind tunnel or in free flight. For in a two-dimensional wind tunnel test there will be no tunnel interference provided that the shocks from the leading edge have reflections at the walls which intersect behind the airfoil; for then the flow within the curvilinear quadrilateral formed by the shocks, and in particular the surface pressure distribution, will be given correctly by a two-dimensional theory in which the incident stream is supposed uniform and infinitely wide. Or again, in the uniform supersonic flight of a wing of uniform cross section and rectangular planform, a portion of the flow, namely, that which is outside the domains of influence of the endpoints of the leading edge, will be two-dimensional.

During the whole of this article and of Art. 4 we confine ourselves to the study of airfoil sections which are sharply pointed at both their leading and trailing edges. Further, we suppose that the angle of attack and the Mach number M_∞ of the incident stream are so limited that any deflections of the stream through angles concave to the flow, which have to be made at the leading and trailing edges, can be achieved with the aid of attached shocks with supersonic flow on both sides of them. The condition necessary for this is fully studied in III,E,2; for given Mach number, the deflection must not be too large; for given deflection, the Mach number must not be too small. If we make the restriction that the shocks are to be attached, we might just as well require also that the flow behind them be supersonic, since this greatly assists the theory, while (for all Mach numbers) it only limits further the range of possible deflections by less than half a degree (the maximum percentage change in the greatest permitted deflection being 8 per cent, attained in the limit $M_\infty \to 1$, and the maximum total change 0°25′, attained for $M_\infty = 1.4$). Only in Art. 5 is the effect of a blunt leading edge discussed.

It follows from the assumption just introduced that there is no mutual influence between the flow over the upper and lower surfaces of the airfoil. For the incident stream is uniform ahead of the sharp leading edge, where it divides into two streams, above and below the airfoil, which cannot interact until they reunite behind the trailing edge. Since the flow is assumed supersonic even behind the trailing edge, this interaction by union has a domain of influence entirely downstream of the trailing edge, and it cannot therefore affect the surface pressures. Hence for the purpose of calculating these, which forms the subject of the present article, the flow at and beyond the trailing edge need not be considered. Only in Art. 4 is the complete flow pattern discussed. (But it must be mentioned that the principle just established is the point at which the real flow is worst approximated by a theory neglecting the boundary layer. The conditions at the trailing edge do influence the pressures upstream of it, by propagation up the boundary layer, particularly when it is laminar. In making predictions from the adiabatic theory it is therefore important to make allowance for this, as discussed in IV,B.)

Prandtl-Meyer flow. It is seen, then, that on adiabatic theory the problem of determining the flow over, say, the upper surface is equivalent to that of determining how a uniform supersonic stream parallel to a plane wall is disturbed at and behind a corner (corresponding to the leading edge), where the wall bends abruptly and then curves in an arbitrary two-dimensional manner. This problem was already considered by Prandtl [*17*] in 1907, in connection with the flow of steam in Laval nozzles. He points out that the flow round a corner concave to the flow can be achieved by means of a plane oblique shock (provided that the required deflection is less than a quantity dependent on the Mach number to which reference was made above), but that some other mechanism is necessary to produce flow round a corner convex to the flow. He argues that, since the state of affairs behind the corner cannot influence the flow around it, the boundary conditions define no length scale proper to the problem. Therefore, it is reasonable to seek a solution in which every physical quantity is constant along radii through the corner, and which depends only on the angular polar coordinate with the corner as pole. He states that he has found solutions of this type to the equations of potential flow, with a polytropic equation of state (which includes, for air, the condition of constant entropy), in which streamlines are bent toward the wall while the gas expands. The complete deflection around the corner is achieved within a wedge-shaped region pointing downstream and away from the wall. Beyond this, if the wall remained straight, then the flow would remain uniform (Fig. E,3a).

Though diagrams of such flows were given in Prandtl's paper [17], the formulas for the velocity and thermodynamic variables as functions of the polar angle (see G,4) were first published by Th. Meyer in the subsequent year, and the predicted phenomenon (which corresponds very closely with the observed supersonic flow around corners) is therefore called the Prandtl-Meyer expansion.

Prandtl [17] observes that, inside the wedge of expansion, the velocity component perpendicular to a radius vector is always equal to the local speed of sound. This is readily interpreted in terms of the dictum that the equation of irrotational motion is at any one point identical with the equation of sound, in axes moving with the local velocity of the fluid, and with the local value for the speed of sound. Each radius vector then

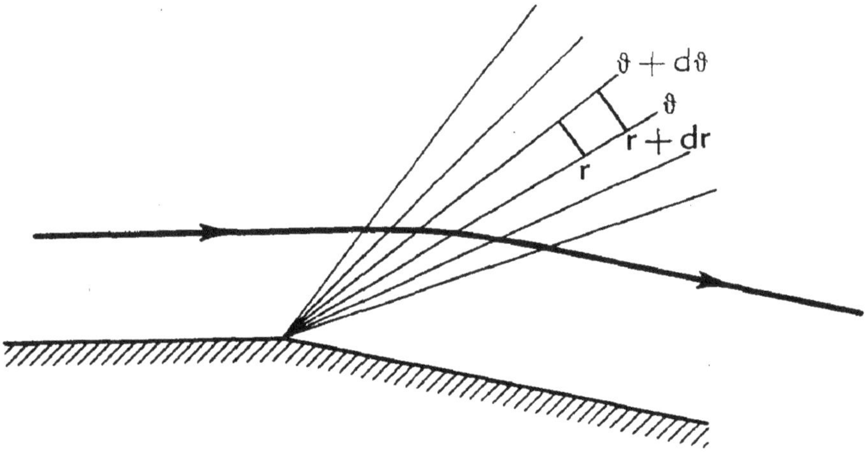

Fig. E,3a. The Prandtl-Meyer expansion.

appears as a plane acoustic wave, which is stationary because its propagation normal to itself at a speed equal to the local value of the speed of sound (which is uniform along it) is exactly balanced by its convection downstream at a velocity whose component normal to the wave is equal and opposite to the velocity of propagation.

This result can be expressed by saying that the angle β between the direction of motion at any point and the local radius vector satisfies $\sin \beta = a/q$ (in terms of the local fluid speed q and speed of sound a). The angle β satisfying this condition is called the local Mach angle; its cosecant is the local Mach number. If θ is the direction of motion (measured in the positive sense from the direction of the undisturbed stream, so that it is negative in the Prandtl-Meyer expansion) and ϑ the polar angle (i.e., in the same sense, the direction of the radius vector), we can write

$$\vartheta = \beta + \theta \tag{3-1}$$

The pressure p, density ρ, local speed of sound a, local fluid speed q, and hence also the Mach angle $\beta = \sin^{-1}(a/q)$, are all functions of one another, by the condition of uniform specific entropy (which applies because the undisturbed stream is uniform) and by Bernoulli's equation. We can find how one of them is related to ϑ (and hence how all of them are, and hence also, by Eq. 3-1, how they are related to the direction of motion θ) by expressing that the mass flowing into a small volume element of space in unit time is equal to that flowing out of it. If the element (Fig. E,3a) is bounded by two radius vectors ϑ and $\vartheta + d\vartheta$ and by two circular arcs with center at the corner and radii r and $r + dr$, and has unit depth perpendicular to the plane of flow, then the mass entering this element (on balance) across the radius vectors is $[d(\rho a)/d\vartheta]d\vartheta dr$, since the velocity normal to a radius vector is a. But, since the velocity and density are uniform along any radius vector, and the velocity component in this direction is $(q^2 - a^2)^{\frac{1}{2}}$ by Pythagoras theorem, the mass flowing out (on balance) across the circular arcs is entirely due to the fact that one is longer than the other by an amount $drd\vartheta$; the said mass is therefore $\rho(q^2 - a^2)^{\frac{1}{2}}drd\vartheta$, and hence

$$\frac{d(\rho a)}{d\vartheta} = \rho(q^2 - a^2)^{\frac{1}{2}} \tag{3-2}$$

With Bernoulli's equation, Eq. 3-2 gives

$$\vartheta = \lambda \tan^{-1}\left[\frac{\lambda a}{(q^2 - a^2)^{\frac{1}{2}}}\right] + \text{const} = \lambda \tan^{-1}(\lambda \tan \beta) + \text{const} \tag{3-3}$$

where $\lambda^2 = (\gamma + 1)/(\gamma - 1)$. Hence by Eq. 3-1 the relation between the direction θ and the magnitude q of the velocity takes the form

$$\theta + \beta - \lambda \tan^{-1}(\lambda \tan \beta) = \text{const} \tag{3-4}$$

$$= \beta_\infty - \lambda \tan^{-1}(\lambda \tan \beta_\infty) \tag{3-5}$$

Here the constant in Eq. 3-3 and 3-4 has been determined by the conditions on the leading radius vector of the expansion, where the flow is still undisturbed, and hence $\theta = 0$.

It was gradually realized, and probably first stated by Ackeret [*18*], that the above arguments, with only slight modifications, are sufficient to determine the deformation of the uniform stream by an arbitrary (two-dimensional) bending of the wall, provided that no shock occurs. For then there is no vorticity or variation of entropy, and a flow is again possible consisting of a single stack of plane acoustic waves, each with all the physical quantities uniform over it. To render each wave stationary, the velocity component normal to it must equal the local speed of sound. The generalization from the Prandtl-Meyer flow around a corner consists only in the fact that the stack of plane waves is not assumed concurrent in a line (the corner, when the wall turns). However the angle ϑ may still be given a meaning, as the angle which the plane wave makes (measured in the positive sense) with the direction of the undisturbed stream. The condition for stationary waves may then again be written in the form of Eq. 3-1. Further, the argument concerning mass flow balance which led to Eq. 3-2 is valid word for word if "plane wave" is read for "radius vector" and if the "corner" (which is taken as the center of the circular arcs bounding the volume element) is replaced by the point (i.e. line, in three dimensions) where the two adjacent plane waves meet.

Thus Eq. 3-3 and 3-5 hold for two-dimensional flow through an arbitrary stack of plane acoustic waves, kept stationary by a balance between their propagation upstream and their convection downstream. This flow

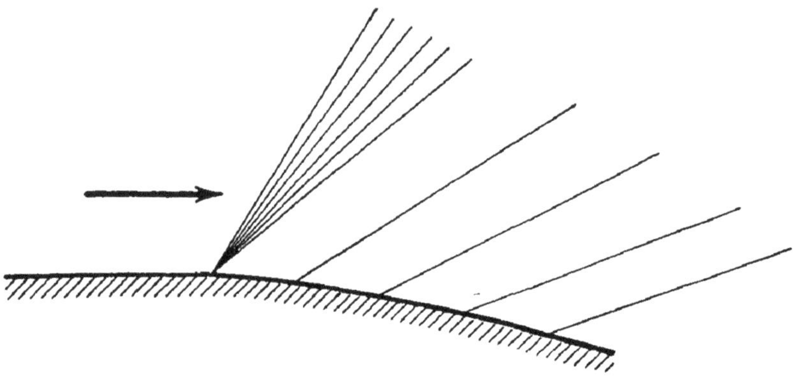

Fig. E,3b. Supersonic flow past a convex wall.

pattern can be made to fit the boundary conditions due to any type of bending of the wall. For these are equivalent to prescribing the direction of motion θ at the wall. Thus, provided that the plane wave through a given point of the wall has the right value of θ, the boundary condition is satisfied. This determines the wave uniquely by Eq. 3-5 and 3-1, which give the Mach angle β on it and the direction of the wave ϑ as a function of θ for a given value of β_∞. Only where the value θ on the wall changes discontinuously (at a corner) is it necessary to have a wedge of expansion where the waves run together.

Such a flow is illustrated in Fig. E,3b. Observed flows in Laval nozzles correspond closely with the predictions of the theory (provided, as is clear from the one-dimensional theory (III,B) of Laval nozzles, that the pressure at the downstream end is low enough to permit the expansion in question without the formation of a normal shock across the nozzle).

Eq. 3-4 (with the constant arbitrary) actually represents the general form of the equations of the characteristics of one of the two systems in any steady two-dimensional supersonic potential flow at constant entropy. (The equations of the characteristics of the other system are the same with $-\theta$ for θ.) Thus the plane waves (or at least the lines in which they cut the plane of the paper in Fig. E,3b) are characteristics. They are also often called Mach lines, since they make the Mach angle β

(whose significance was first realized by E. Mach) with the local direction of flow. (In the more general case of rotational flows this distinguishes them from the third system of characteristics, the streamlines.)

The general type of flow pattern which has here been revealed (and illustrated in Fig. E,3b) is called a "simple wave," because there is one system of characteristics along which the values of physical quantities are propagated unchanged. The properties of simple waves are derived in a different manner, from the general theory of characteristics, in G,6. There, too, a very rapid geometrical method of plotting them, based on Eq. 3-1 and 3-5, is given.

There is a very strong analogy between the simple wave defined above for steady two-dimensional flows, and the simple wave (or "progressive wave") occurring in the theory of nonsteady one-dimensional gas dynamics (III,C). Indeed Fig. E,3b would make an excellent illustration of the latter theory (qualitatively, of course, not quantitatively), provided that it be interpreted as a space-time diagram, with distance measured vertically and time horizontally. The wall would then be reinterpreted as a piston in a long tube which is being very rapidly withdrawn, in some prescribed manner, and emits expansion waves, each traveling with constant velocity along the lines indicated in the figure as Mach lines. The similarity is analytic as well as geometric; thus the equation of any Mach line is $y - x \tan \vartheta = \text{const}$, and the value of the constant is functionally related to, say, β, and hence

$$\beta = f[y - x \tan (\beta + \theta)] \tag{3-6}$$

Eq. 3-6 and 3-5 are directly analogous to the equations for a one-dimensional progressive wave which Earnshaw derived in 1860, namely,

$$q = f[x - (a + q)t], \qquad a - \tfrac{1}{2}(\gamma - 1)q = a_\infty \tag{3-7}$$

The same term "simple wave" came to be used for both phenomena when this analogy was realized. In both cases the solutions were inferred from physical considerations, but it may easily be verified analytically that they solve the appropriate equations of motion.

Notice that the assumption of uniform specific entropy (which follows from that of a uniform undisturbed stream) is essential for the existence of a simple wave. Without it, even if the velocities were constant along Mach lines, the speed of sound would not be constant; hence the Mach lines would not be plane waves, and therefore would not propagate unchanged unless some sort of reflection along the other system of characteristics was permitted.

Fig. E,3c shows what happens when the simple wave theory is applied in order to determine the effect of bends in the wall concave to the stream. There is no difficulty in applying the boundary conditions as before. A unique mathematical solution of the form of Eq. 3-6 (taken with Eq. 3-5) still emerges. But in the corresponding physical flow picture (Fig. E,3c) the Mach lines converge as we follow their course away from the wall. This is because ϑ (as given by Eq. 3-5) is an increasing function of θ; as the flow turns, the direction of the Mach lines turns in the same sense.

As a result, in the region of compression, any two adjacent Mach lines cross each other at some point. The locus of such intersections is the envelope of the Mach lines. This is found to be a cusped figure, shown as a heavy line in Fig. E,3c. Through each point *outside* it there is one Mach line which defines the value of each physical quantity at the point. Through each point *inside* it there are three Mach lines, giving three values for each physical quantity at the point, which is nonsense.

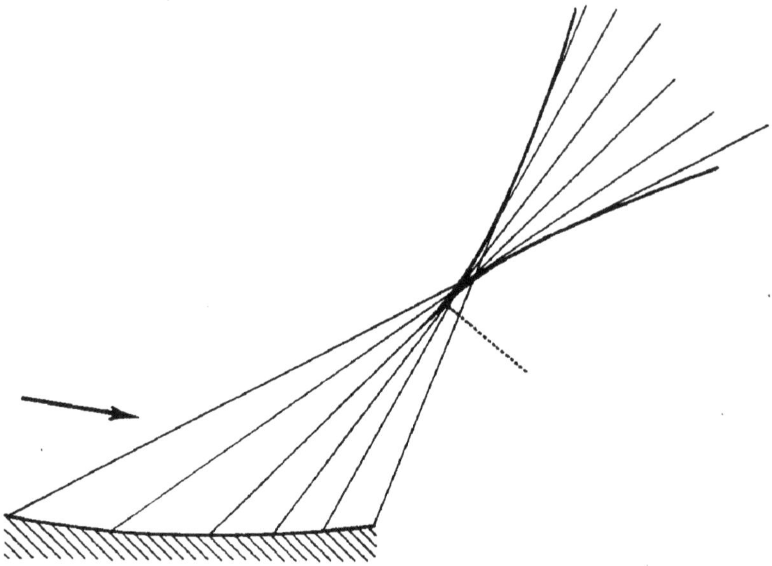

Fig. E,3c. Supersonic flow past a concave wall.

We may pause before discussing what really happens to remark that, whatever it is, provided that the flow remains supersonic, its influence must be confined to the influence domain of the cusp, and in particular the simple wave solution must remain accurate below the dotted line in Fig. E,3c, which represents the characteristic of the other family which passes through the cusp.

But the step to determining what really happens is easily taken in the light of the theory of the formation of shocks in one-dimensional unsteady flow (III,C), especially since the flow plane in the present problem has qualitatively identical properties to the space-time plane in the other problem. The steady two-dimensional flow pattern may be thought of as formed progressively, as the influence of the wall spreads out into the fluid at the speed of sound. The large gradients at the base of the cusped figure are resisted by viscous stresses normal to the cusp. These keep the characteristics from crossing, but they will still get closer and closer, and so need larger and larger viscous stresses to keep them apart. The result is a shock which originates at (or at least very close to) the cusp, with zero strength (i.e. at the Mach angle to the local flow, and thus parallel to the cusp), and whose strength increases as its path is followed away from the wall. The beginning of the path, at least, lies between the two arms of the cusped figure, and near it one may confidently predict that the flow is given approximately by assuming that the physical quantities are still constant along those parts of Mach lines which lie between the wall and their intersection with the shock. (That there is only one such part of a Mach line through a point within the cusped figure can be seen by examination of Fig. E,3c.) The reason for expecting this to be so is that the flow so postulated satisfies the equations of motion under the conditions of zero vorticity and uniform entropy; these conditions are not satisfied exactly behind a shock, but are satisfied to good approximation behind a weak shock; and therefore the flow is possible if there is some curve along which the shock can lie such that the Rankine-Hugoniot shock conditions will correctly link the two values of each physical quantity at the curve (as calculated from the two portions of simple wave, borne on Mach lines entering the cusped figure respectively across one branch or the other). This is not exactly true, unless some propagation from the shock along characteristics of the other family than that shown in Fig. E,3c is permitted; this propagation may be thought of as a result of reflection of the rear simple wave at the shock and transmission of the front one through it; but, again, it may be neglected to a good approximation. On these approximations the position of the shock within the simple wave may be calculated by a method due to Friedrichs [*19*] which is described in another connection in Art. 4. But the problem of the formation of a shock in the midst of an envelope of Mach lines is not further discussed in this section, since the study of the flow past a wall which bends in the direction concave to the flow is not of frequent application to supersonic airfoil theory except when the bending is *sudden*, as at the leading and trailing edges.

But some idea of the behavior of the flow at, say, the leading edge, in the case where the necessary deflection of the stream there is concave to the flow, can be obtained by making the curvature of the wall tend to infinity in the flow studied in the last paragraph (and illustrated in Fig. E,3c), while the length of wall with this curvature is made to tend to zero. In the limit the region covered thrice by Mach lines is bounded (see Fig. E,3d) by two straight lines, of which that on the right is parallel to the Mach lines in the undisturbed stream. Again the shock must occur somewhere in the midst of this region; and only parts of Mach lines between the wall and the shock are of physical significance (other parts of Mach lines are shown dashed in Fig. E,3d). But now the shock is of finite nonzero strength from where it starts at the wall; the portion where it grows from zero strength has vanished in the limit, and now the strength decreases as distance from the wall increases, in a manner to be investigated in Art. 4. However, provided that the shock strength in question is not too large, it may still be a good approximation to neglect the gain in specific entropy at the shock, and the "reflection" of the rear simple wave at the shock, and so take the flow behind the shock as given correctly by the same simple wave theory which is applicable when all the bends in the wall are convex to the stream.

The simple wave theory of airfoil flow. This approximation, when applied at all such corners which occur, affords a complete approximate

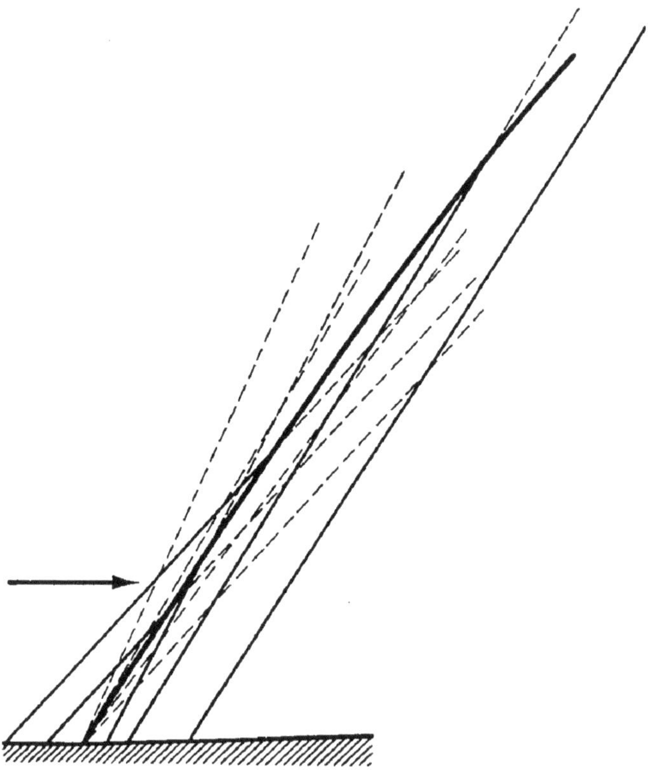

Fig. E,3d. Supersonic flow around a concave corner.

theory of the supersonic flow past an airfoil, which is illustrated in Fig. E,3e for a case where the angle of attack is such that shocks are needed at all four corners. (For larger angles of attack, the shocks in the top left and bottom right-hand corners would be replaced by Prandtl-Meyer expansions.) Between the leading and trailing edge shocks, both above and below the airfoil, there is a simple wave flow. If the direction of motion θ at any point is measured in the positive sense in the upper simple wave, and in the negative sense in the lower simple wave (i.e. in both cases in the sense away from the solid surface), then the local Mach angle β at any point is given in terms of θ by Eq. 3-5. Since θ is known on the airfoil surface, being simply the inclination of the surface to the stream, the values of β on the surface are deduced at once, and hence also the pressure distribution, by the relation

$$p = \frac{p_\infty \left(\dfrac{\sin^2 \beta}{\gamma - \cos 2\beta}\right)^{\gamma/(\gamma-1)}}{\left(\dfrac{\sin^2 \beta_\infty}{\gamma - \cos 2\beta_\infty}\right)^{\gamma/(\gamma-1)}} \tag{3-8}$$

which follows from Bernoulli's equation when the specific entropy is taken as uniform.

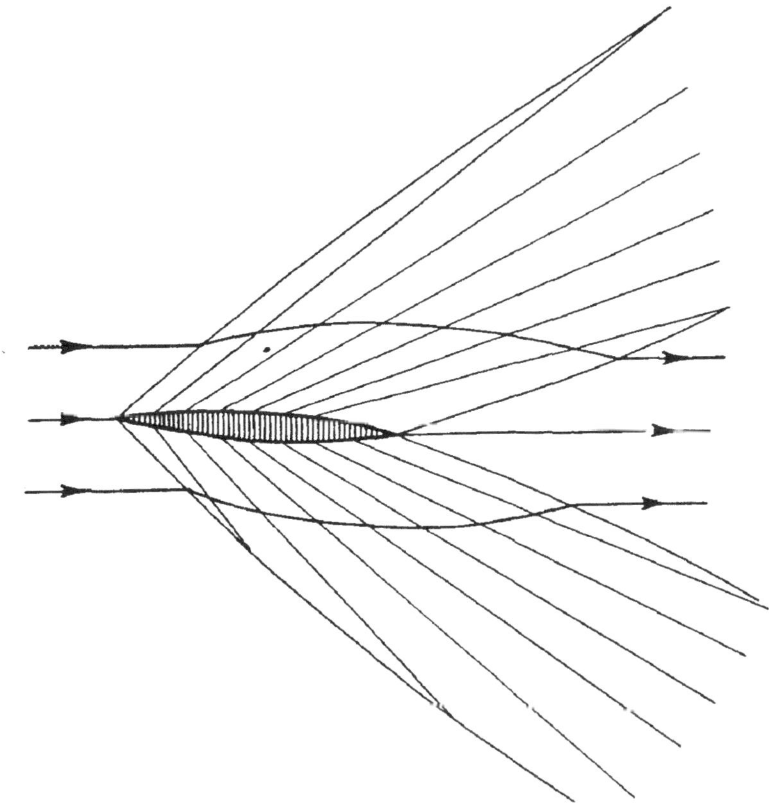

Fig. E,3e. Flow about an airfoil.

The shapes of the shocks remain to be found. No shapes will cause all the Rankine-Hugoniot conditions to be satisfied at the shocks exactly; this is why the simple wave solution is only approximate; but Friedrichs's determination [*19*] of the shock shapes, to within the order of accuracy to which the conditions can be made to apply, is given in Art. 4. This development in the supersonic airfoil theory came late because it is not necessary for the purpose of estimating the surface pressures.

The flow behind the trailing edge shocks must be uniform and parallel to the main stream on the approximate simple wave theory. For, where the streams from above and below the airfoil unite, the pressure must be continuous. If here the direction of motion, measured in the positive sense, is θ, then the pressure is given as a certain function of θ, by Eq. 3-5 and 3-8, for the upper stream, and as the same function of $-\theta$ for the lower stream. These are equal only if $\theta = 0$. The argument applies all along the boundary separating the two streams. On this θ has, and therefore all the physical quantities have, the same values as in the incident stream; hence they have them everywhere behind the trailing edge shock, since quantities are propagated (up to this shock) unchanged along Mach lines. Fig. E,3e, like the previous figures in this article, represents a flow actually calculated from the theory; it includes some representative streamlines. Notice how the stream tube area increases with the rise in velocity along the airfoil surface.

It is fairly easy to estimate the size of the error in the prediction of surface pressure distribution given by simple wave theory. This is because the pressure at the leading edge itself can be calculated exactly from the equations for an oblique shock when the deflection is known. It is true that for a given deflection two alternative shocks are theoretically possible, one stronger than the other (III,E,2.) But it is now fully established experimentally and theoretically that the stronger of the two never occurs except when there is sufficient blockage downstream to make the greater pressure increase necessary in the sense of the steady one-dimensional theory of nozzle flow. (Of course in such a case a region of subsonic flow downstream protests against the resulting low pressure by sending a shock upstream to cause the oblique shock pattern to alter.) In the flow past an isolated airfoil there is of course no blockage, and it is therefore practically certain that the stronger of the two theoretically possible shocks never appears.

In Fig. E,3f and E,3g, the conclusions of the simple wave theory are summarized respectively for $\theta > 0$ and $\theta < 0$. In Fig. E,3f they are critically examined. In this figure the ordinate is the pressure coefficient $C_p = (p - p_\infty)/\frac{1}{2}\rho_\infty q_\infty^2$. The abscissa is M_∞, the Mach number of the incident stream. The curves give the relation between C_p and M_∞ for various positive values of θ, the direction of the tangent to the surface. The plain lines denote the relation given by simple wave theory, from Eq. 3-5 and 3-8. As a comparison, the true value of C_p at the leading edge, as calculated from oblique shock theory, is indicated by a dashed line. The dotted lines will be explained presently. The dash-dotted lines denote the values given by the Ackeret small perturbation theory. (These are included only for the two smallest values of θ in order not to overcrowd the figure, and because the curves for the larger values of θ are simply obtained by scaling up those shown.) The simple wave curves terminate on the left where the magnitude of the deflection first requires the fluid speed to fall below the critical value q^*; no further deflection by a simple wave is possible. Similarly, the curves relating to the oblique shock terminate at the lowest Mach number for which the given deflection is possible.

It is seen that there is remarkably good agreement even for $\theta = 20°$, over the range of Mach numbers for which this deflection is possible by

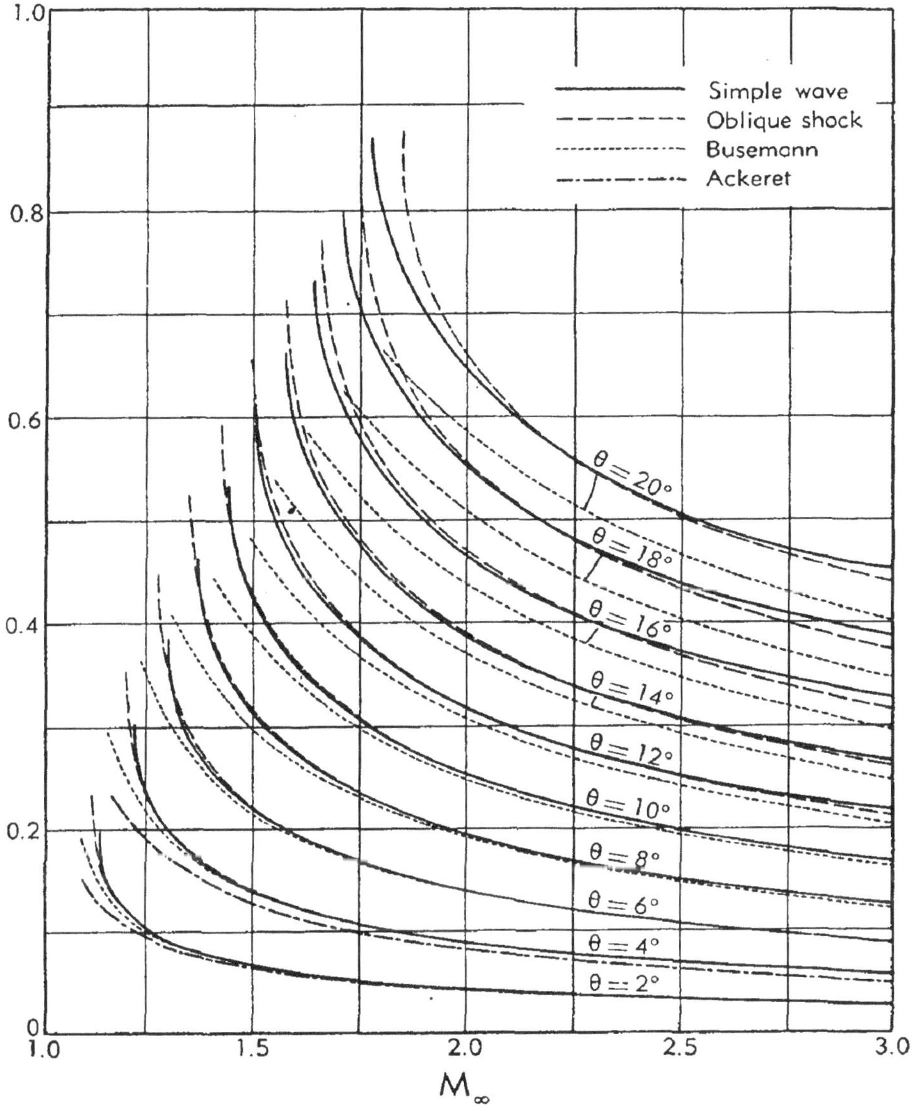

Fig. E,3f. Pressure coefficient produced by flow deflection through an angle θ, plotted against the initial Mach number M_∞.

means of an oblique shock or by a simple wave. For $\theta = 8°$ the curves are practically indistinguishable. This suggests that the expressions for C_p as function of θ, as given by the theories of the simple wave and the oblique shock, might with advantage be expanded in powers of θ, in order to see up to what power they are in agreement. This was first done

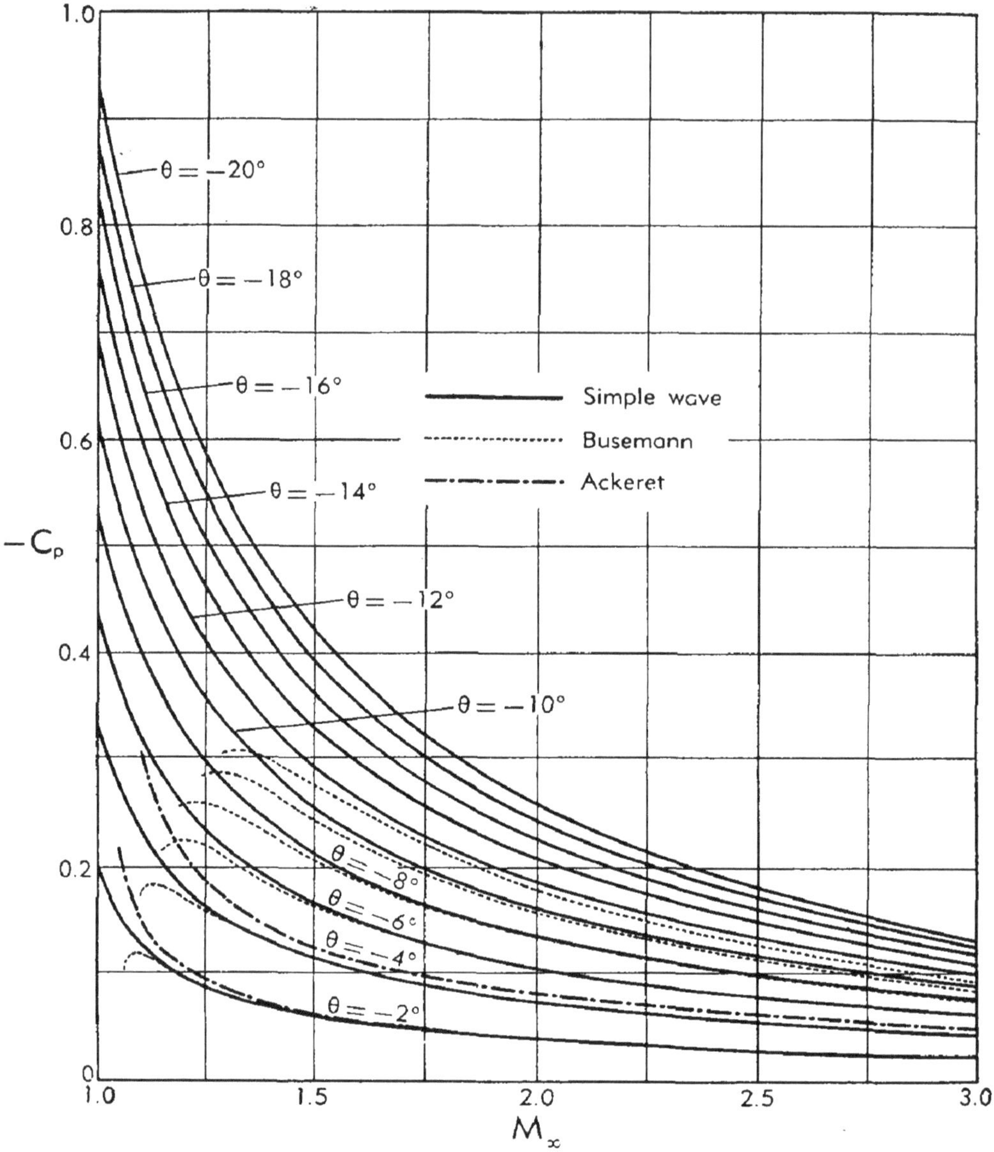

Fig. E,3g. Pressure coefficient produced by flow deflection through a negative angle θ, i.e. convex to the stream.

by Busemann [*20*]. He obtained that, on simple wave theory,

$$C_p = C_1\theta + C_2\theta^2 + C_3\theta^3 + \cdots \tag{3-9}$$

while on oblique shock theory

$$C_p = C_1\theta + C_2\theta^2 + (C_3 - D)\theta^3 + \cdots \tag{3-10}$$

where

$$C_1 = \frac{2}{(M_\infty^2 - 1)^{\frac{1}{2}}}, \qquad C_2 = \frac{(M_\infty^2 - 2)^2 + \gamma M_\infty^4}{2(M_\infty^2 - 1)^2} \tag{3-11}$$

and C_3 and D were given incorrectly, probably as a result of copying errors. Kahané and Lees [*29*] first gave them correctly as

$$C_3 = \frac{M_\infty^4\{[2(\gamma+1)M_\infty^2 + 2\gamma^2 - 7\gamma - 5]^2 - 4\gamma^4 + 28\gamma^3 + 11\gamma^2 - 8\gamma - 3\} + 2(\gamma+1)(3M_\infty^2 - 4)^2}{24(\gamma+1)(M_\infty^2 - 1)^{\frac{7}{2}}} \tag{3-12}$$

$$D = \frac{(\gamma+1)M_\infty^4[(5-3\gamma)M_\infty^4 - (12-4\gamma)M_\infty^2 + 8]}{48(M_\infty^2 - 1)^{\frac{7}{2}}} \tag{3-13}$$

The asymptotic values of the coefficients as $M_\infty \to 1$ are

$$C_1 \sim \frac{2}{(M_\infty^2 - 1)^{\frac{1}{2}}}, \qquad C_2 \sim \frac{\gamma+1}{2(M_\infty^2 - 1)^2}, \qquad C_3 \sim \frac{(\gamma+1)^2}{3(M_\infty^2 - 1)^{\frac{7}{2}}},$$

$$D \sim \frac{(\gamma+1)^2}{48(M_\infty^2 - 1)^{\frac{7}{2}}} \tag{3-14}$$

These are in agreement with the transonic similarity theory (A,7;D,33) under which Eq. 3-9 is replaced by

$$C_p = \frac{2}{\gamma+1}\{M_\infty^2 - 1 - [(M_\infty^2 - 1)^{\frac{3}{2}} - \tfrac{3}{2}(\gamma+1)\theta]^{\frac{2}{3}}\} \tag{3-15}$$

and Eq. 3-10 by

$$\theta = \tfrac{1}{2}C_p[M_\infty^2 - 1 - \tfrac{1}{4}(\gamma+1)C_p]^{\frac{1}{2}} \tag{3-16}$$

The asymptotic values of the coefficients as $M_\infty \to \infty$ are

$$C_1 \sim \frac{2}{M_\infty}, \qquad C_2 \sim \frac{\gamma+1}{2}, \qquad C_3 \sim \frac{\gamma+1}{6}M_\infty,$$

$$D \sim \frac{(\gamma+1)(5-3\gamma)}{48}M_\infty \tag{3-17}$$

These are in agreement with the hypersonic similarity theory (A,6;D,34) under which Eq. 3-8 is replaced by

$$C_p = \frac{2}{\gamma M_\infty^2}\left\{\left[1 + \frac{1}{2}(\gamma-1)M_\infty\theta\right]^{2\gamma/(\gamma-1)} - 1\right\} \tag{3-18}$$

and Eq. 3-10 by

$$C_p = \frac{1}{M_\infty^2}\left\{\left[4(M_\infty\theta)^2 + \left(\frac{\gamma+1}{2}\right)^2(M_\infty\theta)^4\right]^{\frac{1}{2}} + \frac{\gamma+1}{2}(M_\infty\theta)^2\right\} \tag{3-19}$$

The coefficients C_1, C_2, C_3 are positive for all M_∞ for any gas, but the coefficient D (which is such that $D\theta^3$ is the first approximation to the overestimate of leading edge pressure coefficient given by simple wave theory for $\theta > 0$) changes sign as M_∞ varies. For air ($\gamma = 1.40$), D is negative when $1.25 < M_\infty < 2.54$ and positive outside this range. But even where D is positive it is only a small fraction, say 10 per cent, of C_3.

This goes far toward explaining the remarkable agreement between the solid and broken lines in Fig. E,3f, even for $\theta = 20°$.

The error in the simple wave theory at the leading edge, which has just been investigated, is taken as an estimate of the error at all stations along the surface; there is no reason to suppose that the error would increase markedly from this value, since the shock is at its strongest at the leading edge. This supposition is supported by a more detailed investigation below.

It was not clear a priori that the error should have been of order θ^3 for small θ. It is of course true that the specific entropy change at the shock is known (III,E,2) to be proportional to the cube of its strength for small strengths, and the strength is proportional to θ; in fact the jump s in specific entropy from its value in the main stream (which may be taken as zero) is

$$s \sim c_p \frac{\gamma^2 - 1}{12} \frac{M_\infty^6}{(M_\infty^2 - 1)^{\frac{3}{2}}} \theta^3 \tag{3-20}$$

as $\theta \to 0$. But this jump is not the whole explanation of the difference (approximately $D\theta^3$) between the pressure coefficient obtained by continuous compression and that obtained by a shock, which is partly due to the very fact that the change at the shock is discontinuous, so that the shock is propagated faster than sound.

However, once it is established that both pressure and entropy are in error, on simple wave theory, only by terms of order θ^3, it follows that the same is true of all the thermodynamic variables, including enthalpy. Hence further, by Bernoulli's equation, the same is true of the fluid speed. Thus in all respects simple wave theory is correct to order θ^2, but incorrect to order θ^3.

Busemann's second order approximation. After observing this fact, Busemann [*20*] remarked that it was a reasonable simplification of the problem to neglect all terms of order θ^3, and adopt only the first two terms of Eq. 3-9 and 3-10,

$$C_p = C_1\theta + C_2\theta^2 \tag{3-21}$$

as what has become known as the Busemann approximation to the surface pressure distribution. It is compared with the values given by simple wave theory and oblique shock theory for $\theta > 0$ in Fig. E,3f. It does not remain as close to either as they do to each other, and it diverges from them badly both near the lowest Mach number for which the deflection is possible, and also when M_∞ becomes large. This is because, as Eq. 3-14 and 3-17 show, the radius of convergence of the series of Eq. 3-9 and 3-10 is proportional to $(M_\infty^2 - 1)^{\frac{3}{2}}$ for M_∞ near 1 and to M_∞^{-1} for large M_∞. In Fig. E,3g, the Busemann approximation (3-21) is compared, for $\theta < 0$, with the simple wave theory. The divergence for M_∞ near 1 (due to the small radius of convergence of the series) is of course just as bad as in Fig. E,3f, in spite of the fact that in the present case C_p behaves quite regularly in this region. For the larger expansion angles $-\theta$ the Busemann curves had to be omitted from Fig. E,3g to prevent overcrowding, but the divergence continues to increase. At the higher Mach numbers the absolute errors in the Busemann values for C_p are similar to those for an equal compression angle $+\theta$ (Fig. E,3f).

However, for an ordinary airfoil flying at a Mach number less than 3, where the maximum deflection concave to the stream is such as to permit, with a little to spare, attainment by an oblique shock, the maximum deflection convex to the stream is unlikely to be much greater, and Busemann's approximation can be applied with reasonable accuracy. It is a great improvement on the Ackeret small perturbation theory, in which $C_p = C_1\theta$, which is now seen to be an underestimate of the pressure everywhere except perhaps on portions of surface parallel to the stream. Actually, if we set out to improve on the small perturbation theory after the method of Hantzsche and Wendt (Art. 2), or by a direct expansion (2-19) of the disturbance potential in powers of a thickness parameter, the second approximation obtained is the Busemann approximation, and higher ones are the successive partial sums of the series (3-9) representing the simple wave theory. But, of course, there is no need to do this since the whole exact series can be obtained at once by the simple wave theory, and since in any case both approaches ignore the changes produced by the appearance of a shock.

In using the Busemann approximation (3-21), θ is interchanged at will with $\tan\theta$, the slope of the airfoil surface, since the order of the approximation is not thereby altered. If at "zero angle of attack," when the line joining the leading and trailing edges is parallel to the incident stream, the upper surface is $y = y_u(x)$ and the lower surface is $y = y_l(x)$, then at angle of attack α we therefore put $\theta = -\alpha + y'_u(x)$ on the upper surface and $\theta = \alpha - y'_l(x)$ on the lower surface. Hence Eq. 3-21 becomes

$$C_p = C_1[-\alpha + y'_u(x)] + C_2[\alpha^2 - 2\alpha y'_u(x) + y'^2_u(x)] \tag{3-22}$$

on the upper surface, and

$$C_p = C_1[\alpha - y'_l(x)] + C_2[\alpha^2 - 2\alpha y'_l(x) + y'^2_l(x)] \tag{3-23}$$

on the lower surface. Averaging the difference between Eq. 3-22 and 3-23 along the chord from $x = 0$ (leading edge) to $x = c$ (trailing edge), where $y_u(0) = y_l(0) = y_u(c) = y_l(c)$, to obtain the lift coefficient C_L, we obtain

$$C_L = 2C_1\alpha + \frac{C_2}{c}\int_0^c [y'^2_l(x) - y'^2_u(x)]dx \tag{3-24}$$

Multiplying Eq. 3-22 and 3-23 by the slopes $-\alpha + y'_u(x)$ and $\alpha - y'_l(x)$

respectively, adding, and averaging along the chord to obtain the drag coefficient C_D, we obtain

$$C_D = \frac{C_1}{c}\int_0^c [y_l'^2(x) + y_u'^2(x)]dx + 2C_1\alpha^2$$
$$+ \frac{3C_2\alpha}{c}\int_0^c [y_l'^2(x) - y_u'^2(x)]dx + \frac{C_2}{c}\int_0^c [y_u'^3(x) - y_l'^3(x)]dx \quad (3\text{-}25)$$

The second term in Eq. 3-24 represents the lift at zero angle of attack, which at subsonic speeds would be associated with camber. But we see at once that the influence of camber is different in three important respects from that which is familiar from subsonic flow theory: (1) for an airfoil with a flat lower surface ($y_l(x) \equiv 0$) and curved upper surface, i.e. with *positive* camber, the lift at zero angle of attack is *negative;* (2) if the airfoil which has just been described is modified by *doubling* its camber, so that the upper and lower surfaces coincide, and the airfoil is a curved plate, the lift at zero angle of attack is not doubled but reduced to zero; (3) any effect on lift due to camber is only proportional to its square.

It is clear that any attempt to obtain high maximum lift at supersonic speeds by means of camber would be a mistake. For (1) the camber would have to be negative, which would have a bad influence if the airfoil were ever to be used at subsonic speeds; (2) the deflection on the lower surface, at large angles of attack, would be increased, leading to shock detachment, which is a sort of stall, since it implies rapid flow around the leading edge, and hence probably boundary layer separation on the upper surface; (3) the effect in any case would only be rather small.

The correct method of designing for large maximum lift is to aim at making a large value of α, and hence of the first term in Eq. 3-24, possible. This requires that the distance perpendicular to the stream between the leading and trailing edges be maximized, subject to the deflection of the stream being nowhere too great, and to the wing having sufficient thickness for purposes of strength. This requires a symmetrical airfoil section. It is true that here a large negative value of θ has been avoided equally with a large positive value. But it is in practice true that a strong trailing edge shock can have just as deleterious an effect on lift as a strong leading edge shock, an effect that begins for a deflection considerably below the theoretical maximum (which exceeds that pertaining to the leading edge shock), owing to shock boundary layer interaction.

Thus considerations of maximum lift indicate a symmetrical section as desirable, just as do considerations of low drag according to the Ackeret theory (D,7). For a symmetrical section, the lift coefficient predicted by Eq. 3-24 is $2C_1\alpha$, identical to that predicted by the Ackeret theory. Further, the drag coefficient due to angle of attack is the same, namely, $2C_1\alpha^2$. Only the drag coefficient at zero lift, C_{D_0}, given by the first and last terms of Eq. 3-25, contains a new term due to Busemann's corrected theory of the pressure distribution. This is

$$\frac{2C_2}{c}\int_0^c y_u'^3(x)dx \tag{3-26}$$

This affects the conclusion of the Ackeret theory that the profile of given thickness ratio ϵ whose drag coefficient is least is the diamond shape, symmetrical fore and aft as well as about the stream direction. Taking into account the term (3-26), the profile of least drag is a diamond shape but with the position of maximum thickness a distance $(C_2/2C_1)\epsilon c$ behind the half-chord position. (See Fig. E, 3h, below, for a graph of C_2/C_1.) But this change is hardly of practical significance. In the important case of a section symmetrical about two perpendicular lines, the term (3-26) vanished and the lift and drag coefficients as given by the Ackeret theory are correct even on the more accurate Busemann theory.

As a consequence, the predictions of lift and drag given by the Ackeret theory are well borne out in wind tunnel tests on sharp-nosed sections (which have in practice usually been chosen with this double symmetry) at moderate angles of incidence. The measured drag is greater than the predicted drag by an amount which can reasonably be attributed to skin friction. The lift curve slope is given fairly accurately in cases where the boundary layer is turbulent at the trailing edge. But even in these cases the lift falls away from the theoretical value at higher angles of incidence, before full stalling sets in, as a result of the recompression at the trailing edge on the upper surface beginning farther upstream. This effect also reduces the increase of drag with incidence, and indeed makes it much nearer the theoretical value than would be expected, by cancellation with the increased skin friction drag.

But it must be remembered that it is somewhat fortuitous that the small perturbation theory has such success in these predictions. The actual pressure distribution is given much less accurately by it. But since the pressure coefficient is underestimated all around the body, it is not surprising that for a body with double symmetry the resultant lift and drag are given to greater accuracy. However, the position of the aerodynamic center is sensitive to these errors (for example equal forces on the upper surface near the leading edge and on the lower surface near the trailing edge, though with zero resultant, form a couple) and on the Busemann theory it is not at the half-chord position which the Ackeret theory predicts. Thus the nose-up pitching moment coefficient, taken about the half-chord position, is

$$C_M = \frac{C_1}{c^2}\int_0^c x[y_u'(x) + y_l'(x)]dx - \frac{2C_2\alpha}{c^2}\int_0^c x[y_u'(x) - y_l'(x)]dx + \frac{C_2}{c^2}\int_0^c \left(x - \frac{1}{2}c\right)[y_u'^2(x) - y_l'^2(x)]dx \tag{3-27}$$

and even for sections symmetrical only in the ordinary sense (i.e. about the stream direction) the first and third terms vanish, while the second is proportional to the area S of the section, giving $C_M = (2C_2S/c^2)\alpha$. Since $C_L = 2C_1\alpha$ for symmetrical sections, the position of the aerodynamic center predicted by the Busemann theory is therefore a distance

$$\frac{C_2}{C_1}\frac{S}{c} = \frac{(M_\infty^2 - 2)^2 + \gamma M_\infty^4}{4(M_\infty^2 - 1)^{3/2}}\frac{S}{c} \tag{3-28}$$

ahead of the half-chord position. Here S/c may be interpreted as the mean thickness of the airfoil. The coefficient C_2/C_1, which governs the

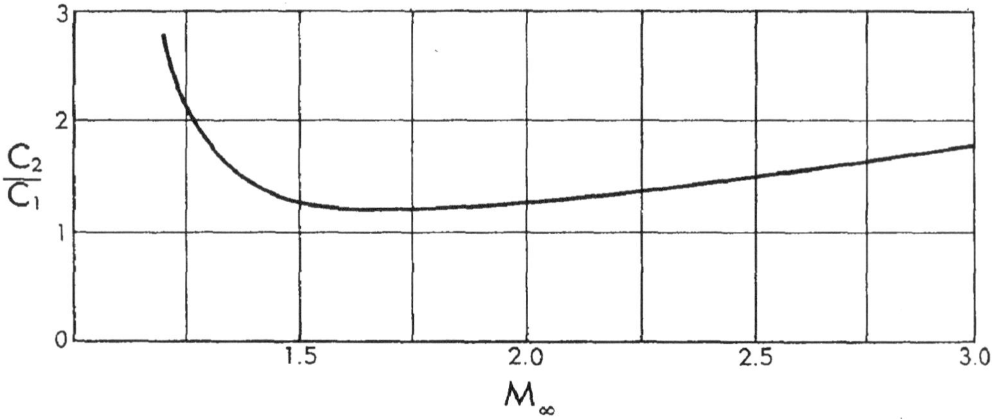

Fig. E,3h. The coefficient of the ratio of sectional area to chord, in the distance of the aerodynamic center ahead of the half-chord position.

variation of aerodynamic center with Mach number for supersonic airfoils, is plotted in Fig. E,3h.

The view taken above is that since the simple wave theory is not accurate (even when the boundary layer is ignored), as a comparison of Eq. 3-9 and 3-10 shows, the further approximation (3-21) might as well be made, since it retains the essential features of the departure from the small perturbation theory, and leads to conveniently simple expressions for the force and moment coefficients. However, for certain airfoil sections (as Epstein [*21*] first showed) one can calculate exact surface pressures (on the assumptions stated at the beginning of the article) which are worth comparing with the Busemann values.

The shock-expansion theory. These sections have a portion near the leading edge consisting of a wedge with both faces plane. When this is so the flow immediately behind the leading edge, say above the airfoil, is uniform, and the values of physical quantities in this flow are given exactly by the oblique shock theory. If the straight portion of the upper surface is AB, where A is the leading edge, then the region of uniform flow is a triangle whose sides are (1) AB, (2) the plane shock AC through A which produces the given deflection, say θ_l, and (3) the Mach line BC through B which is the boundary of the domain of influence of B. Since conditions in ABC are uniform, the Mach line BC is a straight line at the local Mach angle $\beta_l = \csc^{-1} M_l$ to the direction of flow, and so at an angle $\beta_l + \theta_l$ to the incident stream. The pressure coefficient is, say, C_{pl}, given as a function of θ_l and M_∞ by the broken lines in Fig. E,3f.

Now the turning of this uniform flow around the upper surface of the airfoil must be given by the simple wave theory (but, of course with the initial values M_l, θ_l, C_{pl}) provided that the surface lies entirely outside the domain of influence of the point C, where uniformity ceases and (for example) the shock starts to weaken. This may well be the case even if AB is only a small fraction of the chord because the shock AC and Mach line BC make only a small angle with each other, so that C is already a considerable distance downstream.

If the plain lines in Fig. E,3f and E,3g are denoted by $C_{pw}(M_\infty, \theta)$, where w stands for "simple wave," then the pressure on the upper surface behind B will be given, on the hypothesis of the last paragraph, as

$$p = p_l + \tfrac{1}{2}\rho_l q_l^2 C_{pw}(M_l, \theta - \theta_l) \tag{3-29}$$

Hence, subtracting p_∞ and dividing by $\frac{1}{2}\rho_\infty q_\infty^2$, it follows that

$$C_p = C_{pl} + M_l^2(M^{-2} + \tfrac{1}{2}\gamma C_{pl})C_{pw}(M_l, \theta - \theta_l) \tag{3-30}$$

The evaluation of Eq. 3-30 from Fig. E,3f (broken lines) for C_{pl} and Fig. E,3g (plain lines) for C_{pw} requires only a knowledge of M_l, which is deducible from M_∞ and θ_l according to the theory of the oblique shock.

As an example of the results of this method, the flow past an airfoil of "diamond" (i.e. "double wedge," or "rhombus") shape is considered, at variable angle of attack α, for a particular thickness ratio $\epsilon = 0.04$ and Mach number $M_\infty = \sqrt{2} = 1.414$. The results are illustrated in Fig. E,3i. The pressure coefficients on the four sides of the rhombus, with the lift coefficient C_L, drag coefficient C_D and moment coefficient about the half-chord position C_M are shown as plain lines. The approximation of Busemann to each quantity is shown as a dashed line, and that of Ackeret as a dotted line. In the case of C_L and C_D these two coincide (as discussed above), and a dotted line is used. On the other hand C_M, on the Ackeret theory, is identically zero.

It is of some interest to observe what the expansion of Eq. 3-30 in powers of θ and θ_l would be. Omitting all terms of the fourth or higher order in θ and θ_l combined, it is

$$C_p = -D\theta_l^3 + C_1\theta + C_2\theta^2 + C_3\theta^3 \tag{3-31}$$

(No details of this calculation are given, since the reason for Eq. 3-31 becomes clear after the consideration of the flow as a whole to be given in Art. 4.) This supports the statement made above that the error in the simple wave theory was well estimated by its approximate leading edge value $D\theta_l^3$.

When the airfoil surface immediately behind the leading edge is curved, the theory just described (which one may call the shock-expansion theory or Epstein theory) is of course inaccurate, since the shock then starts to weaken from its inception, by interaction with Mach lines, and hence

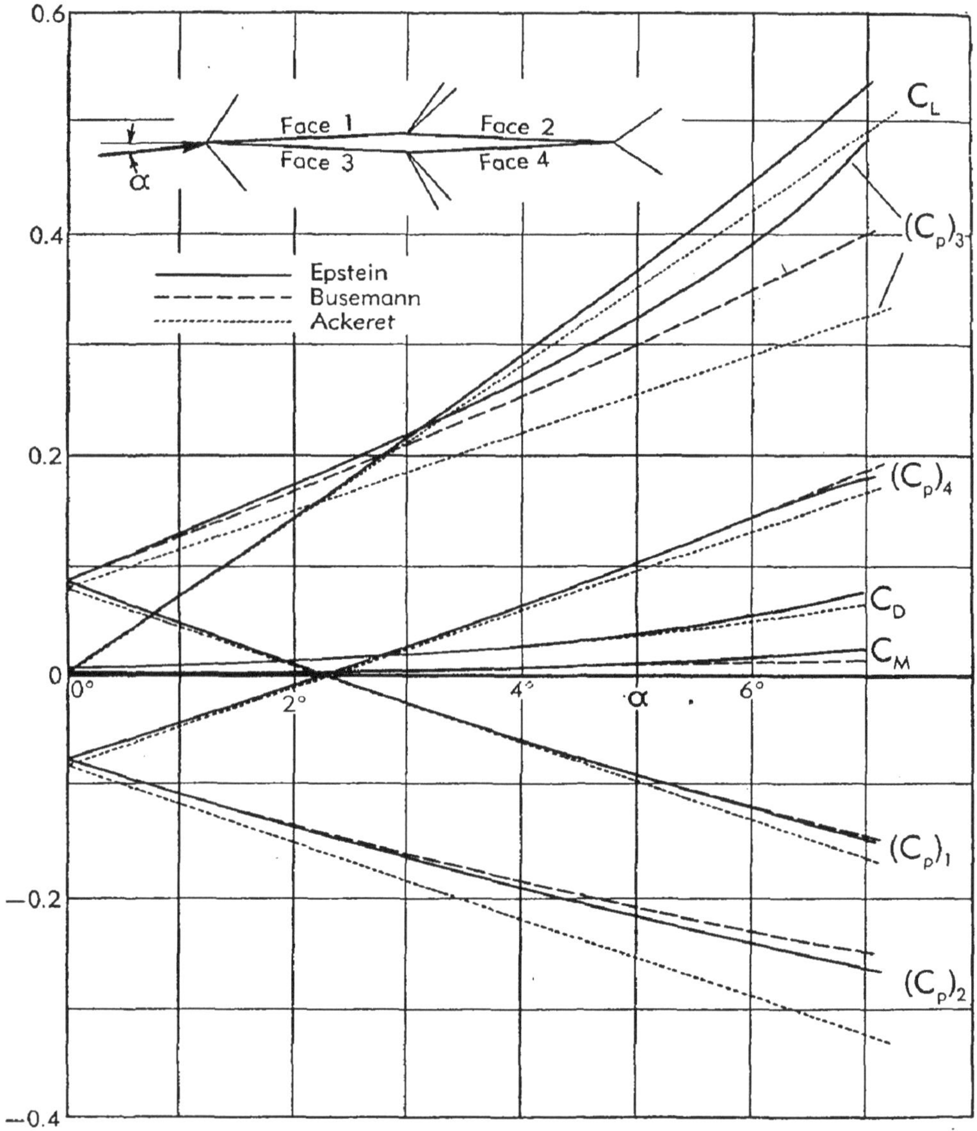

Fig. E,3i. Lift, drag, and moment coefficients, with pressure coefficients on the four faces, for a double wedge airfoil of thickness ratio 0.04 at Mach number 1.414, plotted against angle of attack.

there is no region of uniform specific entropy in which a simple wave theory can be exactly true. However the approximation (3-31) remains true if terms of the fourth order in θ and θ_l are neglected. This is because, though the shock strength is of order θ_l, its gradient (which produces the entropy gradient and hence, by a general theorem on steady flow (III,A), vorticity) is only of order θ_l^2. This in turn is because the shock makes a small angle, of order θ_l, with the Mach lines behind it, so that unit length of shock is only weakened by characteristics emanating from a length of surface of order θ_l, in which θ only decreases by an amount of order θ_l^2.

Thus the errors due to vorticity in Eq. 3-30, where θ_l represents the deflection at the leading edge when this is concave to the stream (and is zero when the deflection is convex so that no shock is formed), are of the order of the fourth power of the maximum deflection. A theory taking into account these errors to a first approximation was already produced by Donov [*22*] in 1939; however, only a review of Donov's paper is available to the present writer. A more systematic theory has recently been given by Thomas [*23*]. Translating his ideas into the notation of the present article, he supposes the pressure coefficient at any point expanded in a series

$$C_p = c_0(\theta) + (s - s_l)c_1(\theta) + (s - s_l)^2c_2(\theta) + \cdots \qquad (3\text{-}32)$$

to take into account that the pressure is not the same function of the deflection on every streamline (i.e. line on which the specific entropy s is uniform). Then the function to be determined is $c_0(\theta)$, the pressure coefficient on the surface (on which, since it is a streamline, the entropy takes its leading edge value s_l uniformly). Thomas shows that, if as an approximation the series (3-32) is terminated by the term in $(s - s_l)^n$, then the equations of motion and conditions at the shock lead to an ordinary differential equation of the nth order for the required function $c_0(\theta)$. Doubtless only the case $n = 1$ could ever be worked out. Then the equation involves explicitly the curvature at the leading edge, since it is this which determines the extent of the entropy gradient. As Thomas' equations stand there appear to be insufficient initial conditions to determine a solution uniquely, but this deficiency could probably be remedied.

However, such work is irrelevant to the practical conclusions of this article, which are as follows:

1. Under the conditions stated at the beginning of the article, the "simple wave" theory of the surface pressures (which does not need to take any account of the shocks) is, as Fig. E,3f shows, perfectly adequate for all practical purposes, up to a Mach number of 3.

2. For *most* purposes the further simplification due to Busemann is adequate. Fig. E,3f and E,3g show in detail when this is so, and Fig. E,3i illustrates the fact in a particular case. This Busemann approximation implies some interesting conclusions concerning lift, drag, and moment which are displayed in Eq. 3-22 to 3-28 and in the discussion accompanying them.

3. There is some evidence that, even though at very high Mach

numbers the simple wave theory becomes inaccurate, the shock-expansion theory, even for shapes for which it is not exact, may be considerably better.

E,4. Supersonic Two-dimensional Airfoil Theory: Complete Flow Pattern. The subject of this article, the determination of the complete flow pattern when a uniform supersonic stream flows steadily and two-dimensionally past an airfoil, is, on the whole, of general scientific, rather than immediate practical, significance. This is not because in realizations of the problem the surface pressures are the only quantities of importance; for clearly flow conditions, almost anywhere, resulting from the disturbance to the stream made by a wing may be of importance in determining the reaction on other aircraft components or on measuring instruments. Nor is it because the theory neglects the action of viscosity and conduction of heat (principally in the boundary layers and wake), since after the theory has completed its prediction of the flow pattern it is comparatively easy to sketch in the modifications due to these influences. It is rather because (1) in flight the flow is very different from a two-dimensional flow—except on and near the wing surface; (2) in wind tunnel experiments the flow is not at all as though the incident stream were infinitely wide—except on and near the wing surface; indeed most of the complete flow field is cut off by the walls, and the rest modified by reflections.

Nevertheless, the theory is here set out, and for the following reasons. (1) It represents a first step toward the determination of complete flow patterns for more complex problems of supersonic flow (e.g. the analogous problem with three-dimensional axisymmetry) which represent experimental conditions much more closely. (2) Even in this problem no exact solution is possible, but the various approximate approaches may be checked against one another. Since they agree, confidence in all is increased, and in more complex problems one may choose among them the most convenient approach. This increase in confidence is desirable because an infinite flow field is considered and therefore errors can pile up as a wave progresses; also certain theories of phenomena of this kind (involving decaying shocks) assume uniform specific entropy, while others take into account entropy changes and get the same answers, but get different answers when they neglect entropy changes. This state of affairs demands careful study. (3) The theory is closely connected with problems in unsteady flow involving decaying shocks. The connection is greatest with one-dimensional problems, as was noted in Art. 3, but some hope arises of extending the theory to decaying spherical shocks produced by explosions. (4) The contrast with subsonic flow patterns can be fully investigated in this case. It is of especial interest, since a supersonic problem is determined when the upstream state of the fluid is given, to find out to what extent the fluid returns (as it would in subsonic flow) to that state. It is also a question whether the work done by the airfoil against drag all goes (ultimately) into heating up the fluid behind the airfoil, or whether (as will be shown to be the case) some of this energy is transported to an indefinite distance from the line of flight. (5) Finally, as will be seen in Art. 5, the theory is useful in the problem of determining the influence of a blunted leading edge on the flow past an airfoil.

Analysis of the decay of shocks which are not so strong that the dissipation of energy within them is a significant fraction of the internal energy of the fluid was first given, in a special problem of unsteady one-dimensional motion, by Chandrasekhar [*24*], and in one of steady two-dimensional motion by the author [*25*]. A systematic theory for general problems of both these types was later produced by Friedrichs [*19*].

The Friedrichs theory. When applied to the supersonic two-dimensional flow past airfoils this Friedrichs theory is an approximation to the "simple wave theory" which was stated, but not completely worked out, in Art. 3. It remained to determine the front and rear shocks. When this has been done the "simple wave" theory states that the flow between them (either above or below the airfoil) is the simple wave, at the entropy of the undisturbed stream, which is compatible with the boundary condition at the surface. Behind the rear shocks, the fluid returns to the state of the incident stream.

It is possible, as will be shown forthwith, to determine uniquely, on this theory, the positions of the front and rear shocks which will cause *one* of the Rankine-Hugoniot relations for stationary oblique shocks, say that between slope of shock and deflection θ, to be satisfied *exactly* all along it. Then it was seen in Art. 3 that the conditions on the pressure change are not satisfied exactly, the error being of order θ^3 for small θ; and the same is true of the change in any of the physical quantities. Since this is so, Friedrichs simplifies the calculation by neglecting θ^3 throughout. Therefore, the Friedrichs theory is the Busemann theory extended to include the determination of the shocks and the whole flow between them. The Friedrichs theory will first be displayed, and then critically examined and amplified.

In the simple wave springing from, say, the upper surface, the equation of the Mach line on which the deflection takes the value θ is

$$y - x \tan(\beta + \theta) = Y(\theta) \qquad (4\text{-}1)$$

where β is given in terms of θ by Eq. 3-5, and $Y(\theta)$ is a known function, namely, the value of the left-hand side at the point on the airfoil surface where the deflection of the stream is θ. (It is also the value of y, always negative, where the characteristic produced meets the line $x = 0$ normal to the stream through the leading edge.) With this value for $Y(\theta)$, Eq. 4-1 holds for all points (x, y) in the simple wave. Now the angle η which the front shock makes with the direction of the incident stream at a point where it produces a deflection $\theta > 0$ (and therefore meets the characteristic (4-1)), and also (for $\theta < 0$) the angle η which the rear shock makes with the stream direction at a point where it produces a deflection $-\theta$ (and therefore meets the characteristic (4-1), converting the values of the physical quantities thereon into those for the undisturbed stream) are given by the same function $\eta(\theta)$, as the theory of the shock polar diagram (III,E,2) shows. Hence to determine the shocks (as suggested above) so that this relation between η and θ is satisfied exactly, one must solve

$$\frac{dy}{dx} = \tan\,[\eta(\theta)] \tag{4-2}$$

where y is related to x by Eq. 4-1, to obtain the relation between x and θ on each shock. Eq. 4-1 will then complete the specification of the curves in terms of the parameter θ.

Eq. 4-1 and 4-2 give the linear differential equation

$$\{\tan\,(\beta + \theta) - \tan\,[\eta(\theta)]\}\frac{dx}{d\theta} + x\frac{d}{d\theta}\tan\,(\beta + \theta) = -Y'(\theta) \tag{4-3}$$

for the value of x on the front shock where the deflection by the shock is θ, and also for the value of x on the rear shock where the deflection by the shock is $-\theta$. An explicit solution of Eq. 4-3 could be written down with little trouble. But, for reasons explained above, there is little gained by including terms of order θ^3 in the calculations; hence Friedrichs simplifies the matter by neglecting these. Now, neglecting terms in θ^3, we have, for the slope of the Mach lines $\tan\,(\beta + \theta)$ and the slope of the shock $\tan\,[\eta(\theta)]$, the equations

$$\begin{aligned} \tan\,(\beta + \theta) &= \frac{1}{(M_\infty^2 - 1)^{\frac{1}{2}}} + 2A\theta + B\theta^2 \\ \tan\,[\eta(\theta)] &= \frac{1}{(M_\infty^2 - 1)^{\frac{1}{2}}} + A\theta + C\theta^2 \end{aligned} \tag{4-4}$$

where

$$\begin{aligned} A &= \frac{\gamma + 1}{4}\,\frac{M_\infty^4}{(M_\infty^2 - 1)^2}, \qquad B = \frac{(\gamma + 1)M_\infty^4}{2(M_\infty^2 - 1)^{\frac{7}{2}}}\,(\gamma M_\infty^2 + 1) \\ C &= \frac{(\gamma + 1)M_\infty^4}{32(M_\infty^2 - 1)^{\frac{7}{2}}}\,[(\gamma + 1)M_\infty^4 + 4(\gamma - 1)M_\infty^2 + 8] \end{aligned} \tag{4-5}$$

Some remarks on the value of C are needed before proceeding further. It has been given wrongly in the leading papers on the present theory. Thus Lighthill [*25*] states that a shock bisects the directions of the Mach lines ahead of and behind it if terms of order θ^3 be neglected; this means that $\eta(\theta)$ is the mean of $\beta + \theta$ and β_∞, which is obviously true, by Eq. 4-4, if θ^2 be neglected, but is actually false if only θ^3 be neglected. Also, Friedrichs [*19*], in his expansion for $\eta(\theta)$, which he calls β, (Eq. 14-5 of the cited paper) gives a quite incorrect coefficient of θ^2; he appears to have erred by applying Prandtl's relation, between the normal components of velocity ahead of and behind the shock, in the form which is appropriate only when there is no tangential component. Actually this coefficient C of θ^2 in Eq. 4-4 for the shock slope tan $[\eta(\theta)]$ has one rather surprising property. This is that, when generalized from a perfect gas with constant specific heats to a fluid with arbitrary thermodynamic properties, it does not depend simply on M_∞ and on the shape of the isentrope (in the pressure-volume diagram) through the state of the incident stream, but also contains a term proportional to the rate of change of pressure with entropy (say at constant volume). Thus, though the expressions for all the flow quantities behind the shock, neglecting θ^3, are uninfluenced by the change in specific entropy at the shock, this is not true of the *slope* of the shock. Martin [26], in treating problems of one-dimensional unsteady flow (which are equivalent to our hypersonic case) to a second approximation, gives the equation corresponding to Eq. 4-5 incorrectly because to simplify the calculation he freely assumes no change of specific entropy at the shock; in consequence his theory of the reflection of expansion waves at shocks is erroneous; in reality, to his order of approximation and for his special value $\frac{5}{3}$ for γ, there is no such reflection. But Eq. 4-5 for C has now been checked independently from first principles by several workers, and also checked against the transonic and hypersonic similarity theories (as was done for C_1, C_2, C_3, and D in Art. 3), and therefore it may be accepted with full confidence.

Eq. 4-3, with the values of Eq. 4-4 substituted in it, has the exact solution

$$x = \frac{1}{A\theta_f^2}\left(1 + \frac{B - C}{A}\,\theta_f\right)^{-2C/(B-C)} \int_{\theta_f}^{\theta_l} \theta\left(1 + \frac{B - C}{A}\,\theta\right)^{(3C-B)/(B-C)} dY(\theta) \qquad (4\text{-}6)$$

for the front shock. From here onward θ_f, a function of x, is used to signify the value of θ at the front shock (on the simple wave side of it) at a distance from the leading edge whose component measured downstream is x; then $x = 0$ when θ_f takes its leading edge value θ_l, which has been used in obtaining Eq. 4-6. The same equation holds, with θ_r, the value of θ at the rear shock, written for θ_f, θ_t for θ_l, and with the integral increased by the constant

$$cA\theta_t^2\left(1 + \frac{(B - C)\theta_t}{A}\right)^{2C/(B-C)}$$

where c and θ_t are the values of x and θ at the trailing edge.

It follows at once from Eq. 4-6 that, as $\theta_f \to 0$, x tends to infinity like a multiple of θ_f^{-2} (since the integral tends to a nonzero limit). The same result is true concerning θ_r. Thus on both shocks the shock strength, as indicated by the deflection, varies as the *inverse square root* of the distance from the leading edge when this distance is large compared with the airfoil chord. It follows also, by Eq. 4-1 and Eq. 4-4, that, since, e.g., x is proportional to θ_f^{-2}, $y - x(M_\infty^2 - 1)^{-\frac{1}{2}}$ is proportional to θ_f^{-1}, and hence to $x^{\frac{1}{2}}$, for large x. Thus each shock is asymptotically a parabola with axis at the Mach angle $\beta_\infty = \tan^{-1}[(M_\infty^2 - 1)^{-\frac{1}{2}}]$ to the incident stream. The two shocks become farther and farther apart, as their course is followed outward, like the square root of the distance from the leading edge. The conclusion from this that their slope differs from that of the Mach lines in the undisturbed stream by an amount asymptotically proportional to $x^{-\frac{1}{2}}$, and so to the strength, is a check on the above arguments.

These conclusions had previously been derived by the present author [*25*] from the fact that the shock direction bisects (at least to a first approximation) the angle between the Mach lines in the undisturbed stream and those in the simple wave behind the shock. Since, at great distances, the Mach lines are "as if" concurrent (because, at least, they all intersect the airfoil surface, which is small compared with the distances involved), the shocks must be parabolas with axis parallel to the Mach lines in the undisturbed stream and focus at the point where the Mach lines in the simple wave approximately concur, by the property of a parabola which causes it to be used as a mirror section.

For giving quantitative conclusions Eq. 4-6 and its companion equation for the rear shock may be approximated (without loss of accuracy greater than has been incurred already) by neglecting factors differing from unity by terms of order the square of any deflection occurring. One obtains

$$
\begin{aligned}
x &= \frac{1}{A\theta_f^2}\int_{\theta_f}^{\theta_l}\left(\theta + \frac{3C - B}{A}\,\theta^2\right)dY(\theta) - \frac{2C}{A^2\theta_f}\int_{\theta_f}^{\theta_l}\theta\, dY(\theta) \\
x &= \frac{1}{A\theta_r^2}\int_{\theta_r}^{\theta_t}\left(\theta + \frac{3C - B}{A}\,\theta^2\right)dY(\theta) + \frac{c\theta_t^2}{\theta_r^2} - \frac{2C}{A^2\theta_r}\int_{\theta_r}^{\theta_t}\theta\, dY(\theta)
\end{aligned}
\tag{4-7}
$$

By Eq. 4-1, $Y(\theta)$ may be regarded (to this approximation) as the value of

$$
y_u(x) - x\left[\frac{1}{(M_\infty^2 - 1)^{\frac{1}{2}}} + 2Ay_u'(x)\right] \tag{4-8}
$$

at the point where $y_u'(x) = \theta$. (In this article, it is not desired, as it was in Art. 3, to vary the angle of attack. Hence the equations of the upper and lower surfaces will be taken as $y = y_u(x)$, $y = y_l(x)$, where y is measured normal to the stream, even when the angle of attack is not zero.) Eq. 4-7 and 4-8, with 4-1, complete Friedrichs's analytical approximation to the flow pattern, by specifying parametrically the position of the shocks and giving the value of the stream deflection at any point of each.

It is of particular interest to determine the coefficients in the asymptotic behavior of the shocks as $x \to \infty$. This is achieved by letting θ_f, $\theta_r \to 0$ in Eq. 4-7. One obtains for the front shock that

$$\theta_f \sim \left[\frac{y_p}{(M_\infty^2 - 1)^{\frac{1}{2}}} + K \int_0^{x_p} y_u'^2(x)dx\right]^{\frac{1}{2}} (Ax)^{-\frac{1}{2}} \tag{4-9}$$

where subscript p signifies conditions at what Friedrichs calls the "peak" of the upper surface of the airfoil, namely, the point where the flow is parallel to the undisturbed stream, and where

$$K = \frac{3C - B}{A(M_\infty^2 - 1)^{\frac{1}{2}}} + A - 1 = \frac{(5\gamma - 3)M_\infty^4 - 4(\gamma - 1)M_\infty^2}{8(M_\infty^2 - 1)^2} > 0$$

Similarly, for the rear shock,

$$-\theta_r \sim \left[\frac{y_p - y_t}{(M_\infty^2 - 1)^{\frac{1}{2}}} - K \int_{x_p}^{c} y_u'^2(x)dx\right]^{\frac{1}{2}} (Ax)^{-\frac{1}{2}} \tag{4-10}$$

The strengths $\Delta p/p$ of the shocks for $x \to \infty$ are obtained by multiplying Eq. 4-9 and 4-10 respectively by $\gamma M_\infty^2(M_\infty^2 - 1)^{-\frac{1}{2}}$. It is of some interest to note that the dependence of the strength of the front shock on y, the distance from the airfoil perpendicular to the stream, is to the first order as $2\gamma \sqrt{[y_p/(\gamma + 1)y]}$, which is independent of Mach number. The asymptotic shock shapes, by Eq. 4-1 and 4-4, are

$$y - \frac{x}{(M_\infty^2 - 1)^{\frac{1}{2}}} \sim 2(Ax)^{\frac{1}{2}} \left[\frac{y_p}{(M_\infty^2 - 1)^{\frac{1}{2}}} + K \int_0^{x_p} y_u'^2(x)dx\right]^{\frac{1}{2}} \tag{4-11}$$

for the front shock and

$$y - \frac{x}{(M_\infty^2 - 1)^{\frac{1}{2}}} \sim -2(Ax)^{\frac{1}{2}} \left[\frac{y_p - y_t}{(M_\infty^2 - 1)^{\frac{1}{2}}} - K \int_{x_p}^{c} y_u'^2(x)dx\right]^{\frac{1}{2}} \tag{4-12}$$

for the rear shock. Thus, to a first approximation, the shocks form part of the same parabola if the angle of attack is zero, since then $y_t = 0$. To a second approximation this is not true, according to Eq. 4-11 and 4-12, since the additional terms within brackets have opposite signs.

Friedrichs [*19*] was careful not to suggest that the coefficients in Eq. 4-9 to 4-12 are correct, neglecting cubes of disturbances, but described them merely as approximations. Actually (as shown in the critical discussion below) the coefficients in Eq. 4-9 and 4-11 are correct in this sense, but the second order terms in the coefficients in Eq. 4-10 and 4-12 are wrong for really large x, and will be emended. In other words the Friedrichs theory gives the asymptotic behavior of the *rear* shock correct only to a *first* approximation, although it gives that of the front shock correct to a second approximation.

The variation of pressure throughout the field is given in terms of the variation of the deflection θ by Eq. 3-21. As $x \to \infty$, the variation of C_p between the shocks, along a line $x =$ const perpendicular to the stream, is therefore given (using Eq. 4-1 and 4-4) by

$$\frac{\partial C_p}{\partial y} \sim C_1 \frac{\partial \theta}{\partial y} \sim C_1 \frac{1}{2Ax} = \frac{4(M_\infty^2 - 1)^{\frac{3}{2}}}{(\gamma + 1)M_\infty^4} \frac{1}{x} \tag{4-13}$$

Thus, for a given Mach number, the pressure gradient between the shocks becomes independent of the airfoil shape, and inversely proportional to the distance from the airfoil, when this distance is large enough. The pressures at the two shocks, and also their distance apart, are sensitive to the airfoil shape, but the uniform pressure gradient between the shocks is unaffected by it at large distances from the surface. In

Fig. E,4a. Pressure distribution in an N wave.

Fig. E,4a the pressure between the shocks is plotted for a large constant value of x with the front shock on the left. Owing to the shape of this curve the configuration is called an N wave. For given x and M_∞ the heights of the uprights of the N depend on airfoil shape but the slope of the diagonal portion is uniquely determined. Similar conclusions apply to a plot of pressure for a large constant value of y.

Effects of entropy variation. It might be supposed that to obtain a higher approximation than the Friedrichs theory, which would give some insight into the flow behind the rear shock, would be very laborious, since the variation in entropy in the simple wave would have to be taken into account. Actually it is not so, owing to the following useful fact. Variations of entropy may, to a first approximation, be neglected when variations of the pressure p or the direction of motion θ are being evaluated. Only in deducing other quantities, e.g. the density ρ or the fluid speed q, need the entropy variations be taken into account.

Put more precisely, the principle states that a difference in p or θ between adjacent points is only influenced by the entropy difference to an amount of the order of this entropy difference *multiplied by* the difference in p or θ. The most rigorous proof is from the equations of the characteristics of steady plane flow with nonuniform entropy (G,18); these state that

$$\begin{aligned} dp \pm (\rho q^2 \tan \beta) d\theta = 0 \qquad &\text{when} \qquad \frac{dy}{dx} = \tan(\theta \pm \beta) \\ ds = 0 \qquad &\text{when} \qquad \frac{dy}{dx} = \tan \theta \end{aligned} \tag{4-14}$$

a form probably first given by R. E. Meyer [27]. If Eq. 4-14 are solved with ρ, q, and β given as functions of p by Bernoulli's equation with a *uniform* value for the specific entropy s, then the error is of the order of the error in s multiplied by the change in θ (not simply of the order of the error in s itself as it would be if any quantity other than p or θ were considered). In particular, in the flow behind a shock the error is of the order of the fourth power of the shock strength.

Physically the result follows, as far as the pressure is concerned, from the fact that at any point the local equations are those of sound, and that pressure is propagated acoustically independently of small vorticity or variations in entropy. The result for the direction θ depends more explicitly on the fact that the flow is steady.

Now the result does not mean that the simple wave theory is true to a higher order than might first have been supposed; for the result applies only to the equations of motion, not to the conditions which must be satisfied at the shocks. These are incompatible with the conditions in the simple wave. It was found in Art. 3, to a first approximation, just what the discrepancy was, in terms of p and θ. At any point immediately behind the front shock, by Eq. 3-9 and 3-10, the pressure is less by approximately $\frac{1}{2}\rho_\infty q_\infty^2 D\theta^3$ than what it would be on the simple wave theory. This causes an additional wave, superposed on the simple wave, and propagated along the second system of Mach lines $dy/dx = \tan(\theta - \beta)$ (see Fig. E,4b); this additional wave may be regarded as the reflection of the simple wave at the front shock. The interaction between the simple wave and its reflection may here be neglected as being of the order of the product of their strengths, and so of the order of the fourth power of disturbances.

Using subscript s for values of quantities in a simple wave, the equation $dp_s - (\rho q^2 \tan\beta)d\theta_s = 0$ holds everywhere, not merely (4-14) on the Mach lines $dy/dx = \tan(\theta - \beta)$, since on the other Mach lines $dp_s = d\theta_s = 0$. Similarly, using subscript R for values of quantities in the additional pressure and direction fields due to the reflection of the simple wave at the shock, the equation $dp_R + (\rho q^2 \tan\beta)d\theta_R = 0$ holds everywhere. Since we are not interested in quantities of smaller order than θ_R (i.e. than the cube of the disturbances) this may be integrated approximately as

$$p_R + (\rho q^2 \tan\beta)\theta_R = 0 \tag{4-15}$$

Now at the front shock, $p_s + p_R$ is related to $\theta_s + \theta_R$ as in simple wave theory but with the additional term $-\frac{1}{2}\rho_\infty q_\infty^2 D\theta_f^3$, with subscript f (as above) signifying the value at the shock. But to our approximation the value of p corresponding to $\theta_s + \theta_R$ in simple wave theory would be $p_s + (\rho q^2 \tan\beta)\theta_R$. Hence

$$p_R - (\rho q^2 \tan\beta)\theta_R = -\tfrac{1}{2}\rho_\infty q_\infty^2 D\theta_f^3 \tag{4-16}$$

Eq. 4-15 and 4-16 show that, in the reflection of the simple wave at the front shock, the additional pressure p_R and deflection θ_R (constant along the second system of Mach lines) are given in terms of the value $\theta = \theta_f$

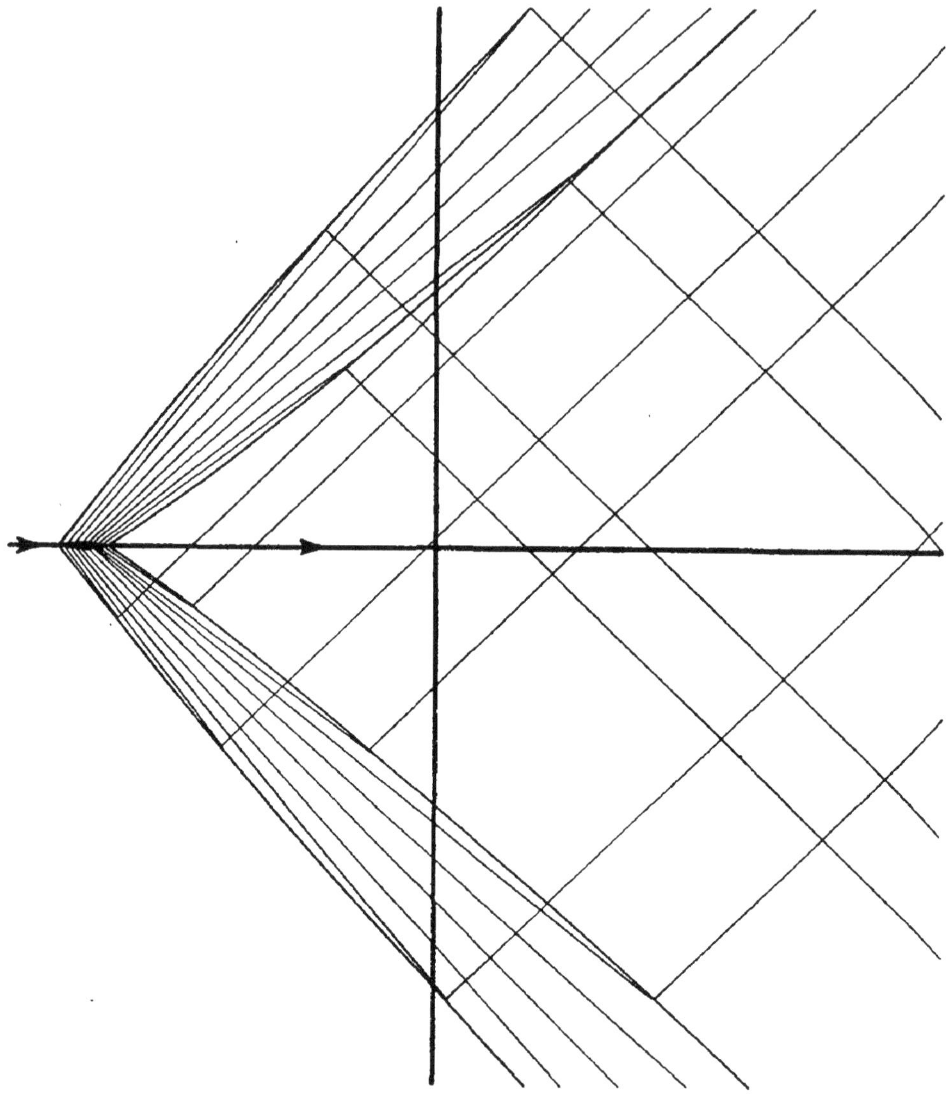

Fig. E,4b. Supersonic two-dimensional airfoil theory: complete flow pattern.

where the Mach line emerges from the front shock as

$$p_R = -\tfrac{1}{4}\rho_\infty q_\infty^2 D\theta_f^3, \qquad \theta_R = \tfrac{1}{4}(M_\infty^2 - 1)^{\frac{1}{2}} D\theta_f^3 \tag{4-17}$$

Part of this reflected wave is in turn reflected at the airfoil surface, after which it goes to modify the simple wave by adding additional third order terms.

In this doubly reflected wave, the pressure is $p_{RR} = p_R$ and the deflection $\theta_{RR} = -\theta_R$. Actually the Mach lines of the second system which hit the airfoil spring from a part of the shock where the deflection does not vary significantly from the leading edge value θ_l. Hence the modification to the simple wave consists approximately of the constant change in pressure $-\frac{1}{4}\rho_\infty q_\infty^2 D\theta_l^3$, which is negative except when $1.25 < M_\infty < 2.54$. At the airfoil surface the additional pressure to that given on simple wave theory is $p_R + p_{RR} = 2p_R = -\frac{1}{2}\rho_\infty q_\infty^2 D\theta_l^3$, agreeing with the conclusions of Art. 3.

Lighthill [28] studied conditions just behind a stationary shock and obtained an exact reflection coefficient for the reflection of the simple waves at the shock. It increases with θ_f initially like θ_f^3 but later somewhat more slowly; for $M_\infty = 2$, the reflection coefficient for the maximum shock deflection is two thirds of that given by the approximate theory above. Of course to higher terms than those retained above there would be an interaction between the simple wave and its reflection at the shock, so that although the theory mentioned gives the exact form of the latter at the shock it cannot predict how it will propagate. Hence the approximate theory, which appears here for the first time, is both adequate and the best that can be done with the amount of labor which the problem merits.

An error in the cited paper [28] should be noted here. Owing to an algebraic slip the limiting value of the reflection coefficient for infinite Mach number was obtained incorrectly as -1 instead of its true value of -0.14, which follows from the work of Eggers and Syvertson [67]. The actual value shows that shock waves are always good absorbers of incident sound waves, even in extreme cases of large Mach number and large deflection angle, and helps to explain the point noted at the end of Art. 3, that the shock-expansion theory is rather good even for hypersonic flow.

We now continue our further analysis of the flow field into the region behind the rear shock, and study the wave which this shock transmits back along the Mach lines of the second system. (It cannot transmit waves behind it along Mach lines of the first system, since these point upstream from it; see Fig. E,4b.) If subscript T signifies conditions in the transmitted wave, the pressure and deflection behind the shock are $p_\infty + p_T$, θ_T. Here (as in Eq. 4-15) $p_T + (\rho q^2 \tan \beta)\theta_T = 0$. Now the pressure and direction ahead of the shock are $p_s + p_R$, $\theta_s + \theta_R$ (where as in all the arguments above the simple wave value p_s includes the reinforcement to the simple wave provided by the doubly reflected wave p_{RR}). Further, the pressure ratio $(p_s + p_R)/(p_\infty + p_T)$ is related to the negative deflection $\theta_s + \theta_R - \theta_T$ exactly as in the ordinary shock equations. In particular this relation differs from the relation given by simple wave theory by a term $-\frac{1}{2}\rho_\infty q_\infty^2 D\theta_r^3$. Approximating to this fact as before, it is easily deduced that

$$p_s + p_R - p_T = p_s + (\rho q^2 \tan \beta)(\theta_R - \theta_T) - \tfrac{1}{2}\rho_\infty q_\infty^2 D\theta_r^3 \quad (4\text{-}18)$$

Hence $p_T = p_R + \frac{1}{4}\rho_\infty q_\infty^2 D\theta_r^3$, i.e.

$$p_T = -\tfrac{1}{4}\rho_\infty q_\infty^2 D(\theta_f^3 - \theta_r^3), \qquad \theta_T = \tfrac{1}{4}(M_\infty^2 - 1)^{\frac{1}{2}} D(\theta_f^3 - \theta_r^3) \tag{4-19}$$

The wave transmitted by the rear shock is thus not identical with that reflected from the front shock; the former is *reinforced* (since $\theta_f > 0$ and $\theta_r < 0$) by an additional wave which can be regarded as the transmission of the simple wave at the rear shock. The mechanism of the process is that a shock, in catching up with acoustic waves, can touch off a small part of their energy as a wave traveling in the opposite direction (the rest of the energy being absorbed in weakening the shock).

The flow behind the rear shocks therefore consists (Fig. E,4b) of a wave transmitted through the upper rear shock, which is propagated along Mach lines pointing downward, and a similar wave transmitted through the lower rear shock, which is propagated along Mach lines pointing upward. Both waves constitute third order effects and are therefore propagated closely according to the small perturbation theory, as far as p and θ are concerned. (Since the entropy variations are of the same order as the pressure waves, the density and fluid speed are *not* obtained from the ordinary small perturbation theory, but from p and the known behavior of s, given by its increment, (Eq. 3-20), at each shock and subsequent uniformity along streamlines.) The values of p and θ propagated along a given Mach line are given by Eq. 4-19, where θ_f and θ_r are the values of θ in the simple wave where the Mach line meets the shocks; these values are given by Eq. 4-7.

The most interesting thing about this tail pressure wave is its breadth. For by Eq. 4-7 and 4-8, the front shock (for example) decays to one half of its initial strength, so that $\theta_f = \frac{1}{2}\theta_l$, in a distance of order θ_l^{-1} chords. This may be taken as a measure of the breath of the tail pressure wave, since the amplitude of waves originating from the shock falls to one eighth of its maximum in this distance. In the limit of infinite distance downstream, both waves have disappeared from the flight path (diagonally) and the pressure and flow direction have returned to their values in the incident stream.

Some attention has been given in the literature to the state immediately behind the trailing edge. At this point the waves are those originating in parts of the shocks still at their initial strengths. Hence the pressure p and the direction of flow θ, measured in the positive sense from the direction of the undisturbed stream (as it has been above for the upper surface but not for the lower surface), are given by

$$\begin{aligned} p &= -\tfrac{1}{4}\rho_\infty q_\infty^2 D[y_u'^3(0) - y_u'^3(c) - y_l'^3(0) + y_l'^3(c)] \\ \theta &= \tfrac{1}{4}(M_\infty^2 - 1)^{\frac{1}{2}} D[y_u'^3(0) - y_u'^3(c) + y_l'^3(0) - y_l'^3(c)] \end{aligned} \tag{4-20}$$

For any airfoil symmetrical about two perpendicular planes, the value of θ given by Eq. 4-20 is zero, since even for nonzero angle of attack $y'_u(0) = y'_l(c)$ and $y'_l(0) = y'_u(c)$. Thus even to a third approximation the streams above and below such an airfoil reunite in a stream locally parallel to the incident stream. This is in marked contrast to subsonic flow past a lifting airfoil, in which the stream is depressed immediately behind the trailing edge. Lighthill [*25*], as a result of a mistake, obtained a nonzero downwash at the trailing edge to this third approximation, but Kahane and Lees [*29*] calculated it by Epstein's method (Art. 3) for the flat plate and double wedge sections, and obtained a fourth order value for the downwash, which they tabulated. It is essentially due to the difference in $(M_\infty^2 - 1)^{\frac{1}{2}}D$ (due to difference in Mach number) at the front and rear shocks. It is always a minute fraction of a degree.

Of course all quantities other than p and θ may be discontinuous at the trailing edge on the present theory, since the entropy increments at the shocks on the two surfaces may not be the same, and this influences all these other flow quantities. A weak vortex sheet (contact discontinuity) may therefore stretch downstream from the trailing edge.

Lift related to downwash. Now that the picture of the flow is complete it is possible to investigate how the airfoil is putting momentum and energy into the fluid by resisting the aerodynamic forces. Besides its independent interest, this provides a check on the work, in both its algebraic and conceptual aspects.

In the case of the lift force, we have to express that this is equal to the rate at which downward momentum ($-\rho v$ per unit mass perpendicular to the incident stream) is transported downstream. Thus the lift $\mathfrak{L}$ per unit span is given by

$$\mathfrak{L} = -\int_{-\infty}^{\infty} \rho uv dy \tag{4-21}$$

along any line $x = \text{const} > c$ perpendicular to the stream and downstream of the airfoil. The question is: For given x, what ranges of y contribute most to the integral (4-21)?

The answer is at once clear. For any x, the "simple wave" regions between the shocks, both above and below the airfoil, predominate in the momentum integral (4-21). For in those regions the disturbance is of the first order, and elsewhere it is of the third order. Also, when x is many multiples of c, so that the momentum crossing a surface far downstream is considered, the disturbance has been reduced like $x^{-\frac{1}{2}}$ but nevertheless the width of the simple wave region has increased like $x^{\frac{1}{2}}$.

The verification that Eq. 4-21 is correct to the first order for any x, even if the integral be taken only in the simple wave regions, is an easy matter. For consider the value of the integral in the upper simple wave region. To the first order it is

$$-\int_{\theta_r}^{\theta_f} \rho_\infty q_\infty^2 \theta[2Ax + Y'(\theta)]d\theta \tag{4-22}$$

where dy has been substituted in terms of $d\theta$ by Eq. 4-1 and 4-4. By Eq. 4-7 this is, still to a first approximation,

$$-\rho_\infty q_\infty^2 \left[(\tfrac{1}{2}\theta_f^2 - \tfrac{1}{2}\theta_r^2)2Ax + \int_{\theta_t}^{\theta_l} \theta Y'(\theta)d\theta - A\theta_f^2 x + A\theta_r^2 x \right] \quad (4\text{-}23)$$

of which all but the middle term cancels out. Substituting from Eq. 4-8 for $Y(\theta)$, Eq. 4-23 becomes

$$-\rho_\infty q_\infty^2 \int_0^c y_u'(x) \frac{dx}{(M_\infty^2 - 1)^{\frac{1}{2}}} = -\rho_\infty q_\infty^2 (M_\infty^2 - 1)^{-\frac{1}{2}} y_t \quad (4\text{-}24)$$

which is half the lift given by Ackeret's small perturbation theory. The other half comes from the simple wave region below the airfoil.

Thus to the first order it is true that a supersonic airfoil achieves lift by transmitting downwash downstream in two waves stretching out diagonally behind it. On the other hand, of course, a subsonic airfoil transmits downwash downstream in all directions (proportional to the cosine of the direction), and also transmits upwash upstream.

The corresponding investigation carried to the second order of approximation is more complicated but in some ways even more interesting. It will be sufficient here, in order to bring out the significant points, to display the work only for a value of x so large that in the simple wave the asymptotic formulas (4-9) etc. apply, and that also the squares of disturbances may be neglected there. (It is not of course inconsistent to retain the squares of the main disturbances and neglect those of the much smaller disturbances which obtain for very large x.)

Then the integral (4-21), taken over only those values of y within the upper simple wave region, is, to the second order of approximation, still given by Eq. 4-22. But the relations between θ_f, θ_r, and x are now Eq. 4-9 and 4-10. Also the integral of $\theta Y'(\theta)$ between the limits becomes negligible as they approach one another for large x. Thus Eq. 4-22 becomes

$$-\rho_\infty q_\infty^2 Ax(\theta_f^2 - \theta_r^2) = -\rho_\infty q_\infty^2 \left[\frac{y_t}{(M_\infty^2 - 1)^{\frac{1}{2}}} + K \int_0^c y_u'^2(x)dx \right] \quad (4\text{-}25)$$

Adding Eq. 4-25 to the similar term arising from the simple wave below the airfoil, we obtain a second order expression for the lift which ignores any possible contribution from downwash behind the rear shock and may therefore be called the Friedrichs-theory contribution to the lift, $\mathfrak{L}_{\text{Friedrichs}}$. This is

$$\mathfrak{L}_{\text{Friedrichs}} = \rho_\infty q_\infty^2 \left\{ -\frac{2y_t}{(M_\infty^2 - 1)^{\frac{1}{2}}} + K \int_0^c [y_l'^2(x) - y_u'^2(x)]dx \right\} \quad (4\text{-}26)$$

But the actual lift is as in Eq. 4-26 but with $\frac{1}{2}C_2$ replacing K (Eq. 3-24). Now K is not equal to $\frac{1}{2}C_2$. Hence, either the Friedrichs theory of the flow between the shocks is wrong to the second order, or a significant part of the integral (4-21) comes from behind the rear shock. Actually the latter is true. (Here the parallel with the Stokes-Oseen theory, noted in Art. 1, is very strong.)

This is because, though the disturbances behind the rear shock are of the order of the cube of the maximum disturbances, they form (as has been seen) a wave whose breadth increases as the maximum disturbance decreases, like its inverse first power. Thus the contribution to the integral (4-21), for large x, from the wave transmitted through the upper rear shock, is, by Eq. 4-19,

$$-\rho_\infty q_\infty^2 \tfrac{1}{4}(M_\infty^2 - 1)^{\frac{1}{2}} D \int (\theta_f^3 - \theta_r^3) dy \tag{4-27}$$

But the wave remains unchanged in shape as it is propagated downstream, and a vertical height dy of wave springs from a stretch of shock with horizontal component $dx = \frac{1}{2}(M_\infty^2 - 1)^{\frac{1}{2}} dy$. Hence $(M_\infty^2 - 1)^{\frac{1}{2}} dy$ may be replaced by $2dx$, if the integral in Eq. 4-27 is now taken along the two shocks, and dx in turn may then be expressed in terms of θ_f or θ_r by Eq. 4-7. This gives for Eq. 4-27 (rejecting third order terms), after integration by parts,

$$-\rho_\infty q_\infty^2 \frac{1}{2} D \left\{ \int_0^{\theta_t} 3\theta_f^2 \left[\frac{1}{A\theta_f^2} \int_{\theta_f}^{\theta_t} \theta dY(\theta) \right] d\theta_f \right.$$

$$\left. - \int_0^{\theta_t} 3\theta_r^2 \left[\frac{1}{A\theta_r^2} \int_r^{\theta_t} \theta dY(\theta) \right] d\theta_r \right\} = -\rho_\infty q_\infty^2 \frac{3D}{2A} \int_{\theta_t}^{\theta_t} \theta^2 dY(\theta)$$

$$= -\rho_\infty q_\infty^2 \frac{3D}{2A(M_\infty^2 - 1)^{\frac{1}{2}}} \int_0^c y_u'^2(x) dx \tag{4-28}$$

With this additional term, and with the corresponding one from the wave transmitted through the lower rear shock, the calculated value for the lift integral (4-21) is no longer the erroneous one of Eq. 4-26. The coefficient K is replaced therein by

$$K + \frac{3D}{2A(M_\infty^2 - 1)^{\frac{1}{2}}} \tag{4-29}$$

The final question (whether these considerations of rate of transmission of downward momentum downstream indicate consistency for our complete picture of the flow, and in particular for the Friedrichs theory of the position of the shocks) therefore resolves into calculating whether the function (4-29), of M_∞ and γ, is equal to $\frac{1}{2}C_2$. This is true, as is quickly verified from Eq. 3-10 and 3-12.

This indicates that the Friedrichs theory, which neglects third order terms, is correct to this order although the lift and the rate of transport of downwash downstream calculated from it are inconsistent. For the latter fact is only true because the broad third order pressure wave behind the rear shock makes a second order contribution to the downwash. Precisely the same conclusion is reached when the above calculation is gone through in the more complicated case of general x, rather than in the limiting case $x \to \infty$.

Drag related to downstream flow. When we turn to the problem of correlating drag with downstream flow pattern, we see that it may be put (unlike the problem of lift) in a multiplicity of forms. First we may work with a frame of reference fixed in the airfoil, and equate the drag $\mathfrak{D}$ per unit span to the difference between the flux of longitudinal momentum across planes normal to the stream respectively upstream and downstream of the airfoil, plus the difference between the normal force due to pressure across these planes. This gives

$$\mathfrak{D} = \int_{-\infty}^{\infty} (p_\infty + \rho_\infty q_\infty^2 - p - \rho u^2) dy \tag{4-30}$$

an integral to be evaluated, like Eq. 4-21, for constant $x > c$. On the other hand we may take a frame so that the undisturbed fluid is at rest, and equate the drag to the rate of gain of momentum of the fluid between two such planes (due to the flow pattern moving forward), again plus the difference in normal force. This gives

$$\mathfrak{D} = \int_{-\infty}^{\infty} [p_\infty - p - \rho u(u - q_\infty)] dy \tag{4-31}$$

Also, in this frame, we may equate the rate at which the airfoil does work on the fluid (per unit span), namely $\mathfrak{D}q_\infty$, to the rate of gain of total energy (kinetic and internal) of the fluid between two such planes, minus the sum of the rates of work done by the pressures on the planes, i.e.

$$\mathfrak{D} = \int_{-\infty}^{\infty} \frac{1}{q_\infty} \left\{ p(u - q_\infty) + \rho u \left[e - e_\infty + \frac{1}{2}(u - q_\infty)^2 + \frac{1}{2} v^2 \right] \right\} dy \tag{4-32}$$

Finally, we may work with "wave energy" (in the sense used in acoustics) rather than total energy, i.e. ignore that part of the internal energy associated with compression by the pressure p_∞ of the undisturbed stream. This gives

$$\mathfrak{D} = \int_{-\infty}^{\infty} \frac{1}{q_\infty} \left\{ (p - p_\infty)(u - q_\infty) + \rho u \left[e - e_\infty + \frac{p_\infty}{\rho} - \frac{p_\infty}{\rho_\infty} + \frac{1}{2}(u - q_\infty)^2 + \frac{1}{2} v^2 \right] \right\} dy \tag{4-33}$$

Of the four forms, Eq. 4-31 and 4-33 are to be preferred as concentrating as far as is possible only on departures of quantities from their values in the incident stream. It is therefore especially interesting that, while of course all the four integrals are equal, in the case of Eq. 4-31 and 4-33 the *integrands* are equal everywhere. This follows from Bernoulli's equation

$$e + \frac{p}{\rho} + \frac{1}{2}u^2 + \frac{1}{2}v^2 = e_\infty + \frac{p_\infty}{\rho_\infty} + \frac{1}{2}q_\infty^2 \tag{4-34}$$

Thus the problems of relating the drag to the distributions of flux of additional longitudinal momentum, and of flux of wave energy, are identical.

On the other hand the integrand in Eq. 4-30 differs from that in Eq. 4-31 and 4-33 by q_∞ times $(\rho u - \rho_\infty q_\infty)$. This latter term represents the mass flow across unit area (compared with its upstream value), and its integral over a plane normal to the stream is, of course, zero. Similarly, the integrand in Eq. 4-32 differs from that in Eq. 4-31 and 4-33 by $(p_\infty/\rho_\infty q_\infty)$ times the same mass flow.

Hence two problems arise. One is that of investigating the distribution of flux of additional longitudinal momentum, or (what is the same, when divided by q_∞) that of flux of wave energy, and verifying that its integral over a downstream plane is equal to the drag. The other is that of investigating the distribution of excess mass flow, and verifying that its integral over a downstream plane is zero. This is of significance from the point of view of energy because of the work done by the atmospheric pressure when the density changes.

The latter problem is here dealt with first, as being more directly analogous to the problem of downwash distribution which has just been treated. As in that problem, first consider the distribution for arbitrary $x > c$, but with squares of all disturbances neglected. Then the excess mass flow across the line $x = const$ where it intersects the upper simple wave region is determined (as Eq. 4-24 was determined) to be

$$\rho_\infty q_\infty y_t \tag{4-35}$$

while that across the part where it intersects the lower simple wave region is equal in magnitude but opposite in sign. There is no contribution, to the first order, from behind the rear shock. Thus, for a lifting airfoil (for which Eq. 4-35 is negative), the increased volume above the dividing streamline downstream of the airfoil is, to the first order, never compensated. The deficiency in mass flow which it produces is concentrated, to the first order, in the wave stretching out diagonally downstream between the shocks. This wave contains the same deficiency of mass flow at any station downstream because its width increases as its amplitude decays. The decreased volume below the dividing streamline similarly causes an equal excess of mass flow propagated diagonally downstream in the lower simple wave. There is a positive downstream flux of total energy $-p_\infty y_t$ in this latter wave, due to the work done by the pressure of the incident stream in compressing the air, and a negative flux equal in magnitude in the upper wave. The total flux of total energy has no first order term, but the continual further separation by the airfoil's motion of regions with excess and deficient total energy is doubtless associated with the lift on the airfoil.

Passing to the second order distribution of excess mass flow we confine ourselves (as we did with the downwash) to its determination for very large x, for which all disturbances have become much weaker than their initial values. Then the calculation of the contribution from the simple waves is straightforward, giving (cf. Eq. 4-26)

$$\rho_\infty q_\infty \left[y_t + K \int_0^c y_u'^2(x)dx \right] + \rho_\infty q_\infty \left[-y_t + K \int_0^c y_l'^2(x)dx \right]$$
$$= \rho_\infty q_\infty (M_\infty^2 - 1)^{\frac{1}{2}} K \int_0^c \left[y_u'^2(x) + y_l'^2(x) \right] dx \qquad (4\text{-}36)$$

where the left-hand terms refer respectively to the upper and lower simple wave. Thus, to the second order, the excess mass flow in the simple wave regions adds up to a positive nonzero quantity. Again the question arises whether this is compensated by reduced mass flow behind the rear shock, or whether the Friedrichs theory is wrong to the second order.

Now behind the rear shock, variations of entropy are of the same order as other disturbances, and must be considered in calculating ρu. Using that, by the equation of state and Bernoulli's equation,

$$\frac{d\rho}{\rho} = \frac{1}{\gamma}\left(\frac{dp}{p} - \frac{ds}{c_v}\right), \qquad qdq = -dh = -\, Tds - \frac{1}{\rho}\,dp \qquad (4\text{-}37)$$

we obtain that, neglecting the squares of disturbances,

$$\rho u - \rho_\infty q_\infty = \frac{M_\infty^2 - 1}{q_\infty}(p - p_\infty) - \rho_\infty q_\infty \frac{s}{c_p}\left[1 + \frac{1}{(\gamma - 1)M_\infty^2}\right] \qquad (4\text{-}38)$$

Eq. 4-38 shows how additional entropy reduces the mass flow. The excess mass flow behind the rear shocks consists of three parts, one from the pressure wave transmitted through each shock, and one from the increased entropy of the fluid. The former two are calculated as was Eq. 4-28, giving a value

$$-\rho_\infty q_\infty \frac{3D}{2A} \int_0^c [y_u'^2(x) + y_l'^2(x)]dx \qquad (4\text{-}39)$$

for their sum. The latter is

$$-\rho_\infty q_\infty \left[1 + \frac{1}{(\gamma - 1)M_\infty^2}\right] \int_{-\infty}^{\infty} \frac{s}{c_p}\,dy \qquad (4\text{-}40)$$

for some $x > c$. But the specific entropy s is conserved along streamlines, and therefore the integral (4-40) may be taken along each of the four shocks and added up, using Eq. 3-20 to relate the entropy gain to the deflection at a shock. Then the contributions from the two upper shocks are

$$-\rho_\infty q_\infty \left[1 + \frac{1}{(\gamma - 1)M_\infty^2}\right] \int \frac{\gamma^2 - 1}{12} \frac{M_\infty^6}{(M_\infty^2 - 1)^{\frac{3}{2}}} (\theta_f^3 - \theta_r^3) \frac{dx}{(M_\infty^2 - 1)^{\frac{1}{2}}} \tag{4-41}$$

This is of the same form as Eq. 4-27 and is similarly evaluated, giving finally for the whole excess mass flow behind the rear shocks

$$-\rho_\infty q_\infty \left\{\frac{(\gamma + 1)M_\infty^4}{4A(M_\infty^2 - 1)^{\frac{5}{2}}}[1 + (\gamma - 1)M_\infty^2] + \frac{3D}{2A}\right\} \int_0^c [y_u'^2(x) + y_l'^2(x)]dx \tag{4-42}$$

This negative term exactly balances the positive one (4-36) since a simple calculation shows that the coefficient in braces is precisely $(M_\infty^2 - 1)^{\frac{1}{2}}K$.

Thus the excess mass flow in the simple waves is principally compensated by deficient mass flow due to the gain in entropy of the fluid. The pressure wave in the region behind the rear shock may fall on either side of the balance, according to the Mach number of the incident stream. For moderate Mach numbers ($1.25 < M_\infty < 2.54$ for small disturbances, when $D < 0$), the pressure wave adds to the excess mass flow; for others it detracts from it.

We pass next to the two identical problems of determining the distributions of flux of additional longitudinal momentum and flux of wave energy and relating these to drag in accordance with Eq. 4-31 and 4-33. This is done to the second order, neglecting cubes of all disturbances, since no first order terms arise. Again, it is particularly easy to do it so far downstream that the squares of the local disturbances may be neglected. The form of the integrand in Eq. 4-33 then makes it especially clear that the only term involving the first power of the disturbances is the Ts term due to specific entropy increment in the specific internal energy e (since $de = Tds - pd\rho/\rho^2$). Hence

$$\mathfrak{D} = \int_{-\infty}^{\infty} \rho_\infty Ts\, dy \tag{4-43}$$

which is evaluated to the second order, as was Eq. 4-40, as an integral taken along all four shocks,

$$\rho_\infty q_\infty^2 \frac{(\gamma + 1)M_\infty^4}{12(M_\infty^2 - 1)^2} \int (\theta_f^3 - \theta_r^3)dx = \frac{\rho_\infty q_\infty^2}{(M_\infty^2 - 1)^{\frac{1}{2}}} \int_0^c [y_u'^2(x) + y_l'^2(x)]dx \tag{4-44}$$

agreeing with the Ackeret value.

For general values of $x > c$, the integral (4-31) or (4-33) consists of a portion (4-43), which however is less than its value for $x \to \infty$ because it only comprises the entropy increment at the part of the shock ahead of the plane normal to the stream which is considered, and also a portion due to isentropic disturbances in the simple waves. Actually an easy calculation shows that in any simple wave the integrand in Eq. 4-31 or 4-33 is $\rho_\infty q_\infty^2 \theta^2$, to the second order. Integrated with respect to y between the shocks, this yields precisely that part of the integral (4-43) due to the parts of the shocks downstream of the plane considered. Thus the total remains the same, and equal to the drag.

The expression (4-43) is not absolutely exact since the squares of entropy increments have been neglected. But an exact expression is obtainable. For in the limit as $x \to \infty$ all disturbances in p and θ vanish, and the integral of their square vanishes. Hence the drag is exactly equal to the integral (4-31) or (4-33), with the integrand regarded as a function of the specific entropy, and with the pressure and flow direction at their initial values. Since entropy is constant along a streamline, the expression $\rho u dy$ at constant x in either integral may be replaced by $\rho_\infty q_\infty dy$ along the front shock. For the rear shock it is similarly the element of mass flow across the shock, i.e. it is $d\psi$ if ψ is the stream function. Hence Eq. 4-31 may be written as

$$\mathfrak{D} = \int \{q_\infty - [q_\infty^2 + 2h(p_\infty, 0) - 2h(p_\infty, s)]^{\frac{1}{2}}\} d\psi \tag{4-45}$$

taken along the rear shocks, or along any curve stretching right across the stream behind them. Here the velocity u has been expressed in terms of the specific enthalpy $h(p, s)$ by Bernoulli's equation, with $p = p_\infty$ and $\theta = 0$, but with the specific entropy s given its value behind the rear shock. Eq. 4-43 above is merely the first term when Eq. 4-45 is expanded in powers of s.

The analysis of drag in terms of wave energy is convenient because wave energy is proportional to amplitude squared, and hence for example an N wave, for which ultimately wavelength times amplitude is constant, has wave energy tending to zero. However, the analysis might sometimes be misleading; it depends essentially on the fact that the total space available to the fluid remains unchanged as it flows past the airfoil, for only then will the total work done by the pressure of the incident stream be zero. For example, in the flow past a "half-body" shape, which stretches with finite thickness to infinity, Eq. 4-33 would be wrong for this reason. Also Eq. 4-30 and 4-31 would be wrong because the momentum flux and normal forces for the upstream and downstream planes, which are being subtracted, have been taken within the same integral, while actually the limits of integration are different for each. Only Eq. 4-32 remains true in this problem, since on the one hand the total energy flux is considered therein, and on the other hand there is no such flux across the upstream plane. These considerations show that there is some fundamental importance in Eq. 4-32, which relates drag to total energy flux. Therefore, we describe the distribution of the integrand, which is at once deducible by adding wave energy flux and $(p_\infty/\rho_\infty q_\infty)$ times the excess mass flow. The total energy flux in the simple waves, far downstream, is

$$p_\infty(M_\infty^2 - 1)^{\frac{1}{2}}K \int_0^c [y_u'^2(x) + y_l'^2(x)]dx \tag{4-46}$$

(At positive angle of attack the contribution to Eq. 4-46 from the upper simple wave is less than that for the lower one; the contributions may even be negative and positive respectively.) On the other hand, the total energy flux behind the rear shock is equal to the drag (4-44) minus the term (4-46). It is interesting to observe that a proportion

$$\frac{p_\infty(M_\infty^2 - 1)^{\frac{1}{2}}K}{\rho_\infty q_\infty^2(M_\infty^2 - 1)^{-\frac{1}{2}}} = 1 - \left(\frac{\gamma + 1}{8\gamma}\right)\left(\frac{3M_\infty^2 - 4}{M_\infty^2 - 1}\right) \tag{4-47}$$

of the drag goes to sending energy to infinity diagonally along the simple waves. For $M_\infty \rightarrow \infty$ this is only $\frac{5}{14}$ of the total, a fraction already calculated by the author [30] in his discussion of the analogous problems in one-dimensional unsteady flow, from which the above account of momentum and energy balance has been adapted. But as M_∞ decreases the fraction (4-47) increases, until for $M_\infty < \sqrt{\frac{4}{3}} = 1.154$ it exceeds unity, and the energy flux behind the rear shock is negative because that due to the rear pressure wave is negative and exceeds in magnitude the essentially positive contribution from the increased entropy. This latter contribution is a fraction

$$\frac{M_\infty^2 - 1}{\gamma M_\infty^2} \tag{4-48}$$

of the whole, which is $\frac{5}{7}$ for $M_\infty \rightarrow \infty$ but much less for smaller M_∞.

Of course, actually, the pressure wave is also propagated diagonally to infinity, so that the proportion of total energy flux so propagated is not Eq. 4-47 but is 1 minus Eq. 4-48. With regard to this pressure wave there is one additional complication which still has to be mentioned. This is that it overtakes the rear shocks. For Eq. 4-12 of the rear shocks makes it clear that every Mach line, originating however far behind it, must at last intersect it when $x^{\frac{1}{2}}$ has become sufficiently large. Actually the rear pressure waves are very broad, but as the rear shocks are followed backwards they must gradually be intersected by more and more of the pressure waves (see Fig. E,4b). Now these waves are already of the third order, but so is the reflection coefficient for reflection at the shock. Hence their reflection at the shock constitutes a negligible wave. But their energy cannot be transmitted through the shock since the Mach lines ahead of the shock both point upstream. Hence their energy goes simply into *changing the shock path from that given by the Friedrichs theory*, making the shock decay less or more rapidly according as the said energy is positive or negative (i.e. according as $1.25 < M_\infty < 2.54$ or not). Finally, then, all the energy, which was shed by the N wave as a reflected and transmitted pressure wave, is reabsorbed in the N wave by the overtaking of the rear shocks by acoustic waves traveling behind them. When this has happened the proportion of total energy flux in the N wave is simply 1 minus Eq. 4-48.

From this knowledge of the ultimate total energy flux in the N wave, after the whole broad tail pressure wave has caught up with the rear shock, the final asymptotic position and strength of the rear shock can be calculated. It is exactly as in Eq. 4-10 and 4-12 but with K replaced by

$$\frac{1 + (\gamma - 1)M_\infty^2}{M_\infty^2 - 1} \tag{4-49}$$

Thus the shape of the rear shock gradually alters slightly, so that in its asymptotic form the coefficient of

$$\int_0^c y_u'^2(x)dx$$

changes from K to the expression (4-49), as the rear pressure wave overtakes it. This is the one point in which our critical analysis of the Friedrichs theory has found it deficient. It gives the front shock correctly to a second approximation, but after a long time begins to give the rear shock correctly only to a first approximation, although by a modified theory we can determine the total departure from the theoretical form to the second approximation. Of course the check relating lift and downwash is not affected, since the downwash in the tail pressure wave is absorbed in the N wave as the process goes on.

E,5. Supersonic Two-dimensional Airfoil Theory: Influence of a Blunt Leading Edge. In this article some attempt is made to extend the theory of Art. 3 and 4 to the flow past airfoils with blunt leading edges. It is important to notice that this includes all real airfoil sections. It is not possible to make a leading edge of true wedge shape. With wedge angles such as the theory indicates are desirable, careful construction will produce a leading edge "thickness" (i.e. perpendicular distance between nearest parts of surface which are truly inclined) at best of the order of 5 microns. Further, the edge so produced is very fragile indeed (note that, of course, *cutting* edges are not designed of wedge section) and, after a few minutes in a supersonic wind tunnel, it may have deteriorated so that the effective leading edge thickness is roughly of the order of 20 microns. If this happens, the spanwise distribution of leading edge thickness becomes very irregular, and therefore it is probably wise to aim at a value not less than 20 microns in the original design. Of course in many low-precision experiments, with so-called sharp leading edges, the leading edge thickness is 100 microns or more. But it may still be a very small fraction of the airfoil thickness. Thus considerable interest is attached to predictions of how a *slight* blunting of the leading edge will affect the flow, when the conclusions of this article constitute a small correction to the theories of Art. 3 and 4.

In the steady flow of a uniform supersonic stream past any airfoil with a blunt leading edge, the front shock is detached, and there is a region of subsonic flow between it and the leading edge. The theoretical

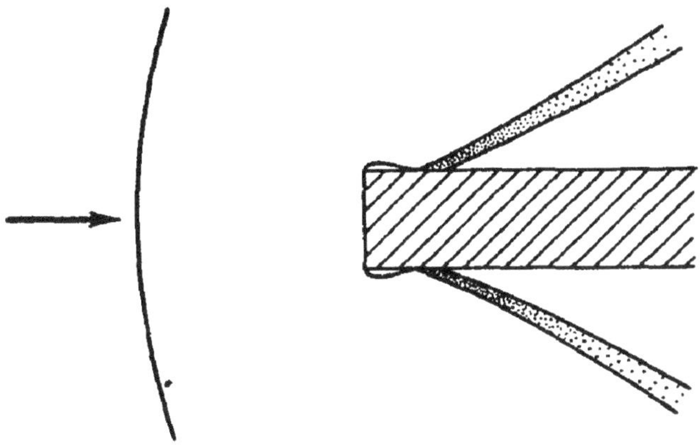

Fig. E,5a. Boundary layer separation in supersonic flow past a square-nosed plate.

prediction of the complete transonic flow field is a very complicated matter (Sec. H). Here, only the region of supersonic flow is discussed theoretically; the influence on it of the subsonic region is taken from experimental data.

The data used are from experiments at the NPL by Holder and Chinneck [*63*] on symmetrical supersonic flow past flat plates with blunt leading edges of elliptical shape. The axis ratio of the ellipses varied from zero (giving a square nose) through unity (semicircular nose) to 8. It will be explained shortly why, although in each case the body shape behind the blunt leading edge was flat, the data can nevertheless be used to predict flows around more general airfoil shapes.

It must be stated at once that, especially for the square-nosed plate, boundary layer effects are observed near the leading edge (where, for the normal development of the Blasius boundary layer, they would be negligible). Thus the flow pattern cannot be completely independent of Reynolds number. The precise effect observed is that the boundary layer appears to separate and become reattached. (This separation doubtless occurs because, if it did not, the flow would overexpand around the shoulder of the plate so much that a large positive pressure gradient must follow.) Where reattachment takes place, the main stream turns rapidly through an angle concave to the flow, producing an oblique shock. Fig. E,5a, which illustrates this behavior for a square-nosed plate, is a fairly close copy of a photograph taken with "micro-schlieren" apparatus by the authors cited above.[1] (The large apparent thickness of the second shock is of course due to the fact that the dead air region is not very accurately two-dimensional.)

However, for the plates whose leading edge had a more "streamline" shape, say semicircular, or semielliptical with axis ratio exceeding 1, the phenomenon just described was much less marked. Only the shock could be observed, not the separation region which produces it. Shadow photographs indicated that this subsidiary shock is very weak in comparison

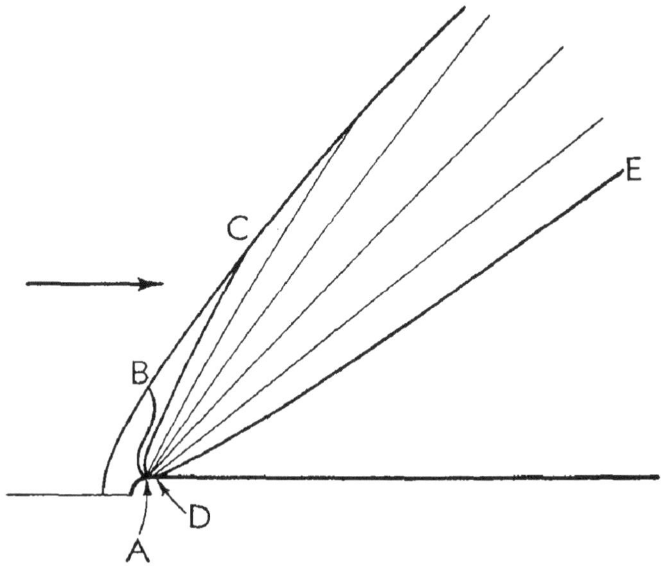

Fig. E,5b. Symmetrical supersonic flow past a round-nosed plate.

with the main front shock. The effective deformation of the surface, produced by the dead air region between it and the separated boundary layer, is apparently unimportant. That is to say, the separation and resulting shock, while of intrinsic interest, appear to have little influence on the rest of the flow. Hence the effect of the boundary layer will be neglected in what follows, except in occasional critical remarks. This will be justified provided that the shoulders at the nose are not too sharp (as in the square-nosed plate). Some attempt to make them more streamlined is clearly of practical importance, since the extra shock (if significantly strong) and its associated entropy change must be connected, as Art. 4 shows, with an increase in drag; and it will be seen below that the drag associated with the main front shock is also reduced by streamlining the leading edge, given its thickness.

Ignoring the boundary layer effects the flow pattern appears to be as

[1] However, Mair and Bardsley [*64*] present evidence that such separation may not occur when the Reynolds number based on leading edge thickness is less than about 1,000.

sketched in Fig. E,5b, which shows only the upper half of the symmetric flow pattern. The shock shape shown is that observed at $M_\infty = 1.8$ for a semicircular leading edge. The sonic line AB is drawn in conjecturally from study of the schlieren photographs. The point B is, however, known for certain from the shock inclination producing sonic conditions behind it. The first Mach line emanating from the surface is AC. Other Mach lines behind it form a fan ending with the final Mach line DE (in reality, as remarked above, a weak shock). Little disturbance is visible behind DE on schlieren photographs, but surface pressure measurements show that slight variations subsist for a considerable length beyond D. The precise nature of these varies considerably with nose shape.

Now it is clear that the flow to the left of AC cannot be influenced by changes in the shape of the boundary downstream of A, for the effect of these would be propagated along Mach lines to the right of AC, which will not cross AC if the boundary curvature is convex to the stream. Conversely, the supersonic flow to the right of AC is determined by the conditions on AC itself and by the shape of the boundary downstream of A. If conditions on AC were uniform, and AC were a straight line, then the flow to the right of AC would be a simple wave, determined by the boundary shape. As it is, the nonuniformity in AC, both in slope and in the distribution of flow quantities along it, causes a certain amount of propagation along Mach lines pointing in toward the body.

The study of the flow is made easier if we use the result of Art. 4 that, in spite of the higher specific entropy of the fluid near the surface, which has passed through a normal shock, the variations of pressure p and flow direction θ can be treated to good approximation as if the specific entropy were uniformly that of the incident stream. Then the pressure wave propagated in toward the airfoil by reason of the nonuniformity on AC may be regarded, together with its reflection at the surface, as simply superimposed on the simple wave which would be produced by the part of the surface downstream of A in continuous shock-free deformation of the incident stream. (The analogy is very strong between this superposed pressure wave and the wave which was shown in Art. 4 to be reflected from an attached shock.)

To a good enough approximation it may be assumed that the effect of this additional pressure wave on the surface pressures (which is double the pressure distribution in the wave, owing to the reflection at the surface) depends only on the nose shape (i.e. we ignore its interaction with the simple wave). Hence experiments, such as were done at the NPL, at three Mach numbers and for six nose shapes, with a perfectly *flat* surface to the right of D, can be used to determine the pressure wave in the general case, at least as far as its influence on surface pressures to the right of D is concerned. For in the simple wave there would, for the NPL shapes, be no departure from the undisturbed pressure downstream of D, and therefore the said *departures* constitute the whole of what must, in other problems, be added to the pressure distribution predicted by simple wave theory.

In Fig. E,5c this excess pressure distribution downstream of D is shown, as measured at the NPL for $M_\infty = 1.6$ and elliptic noses of axis ratios 0 (square nose), $\frac{1}{2}$, 1, 2, 4, and 8. The distributions, and their

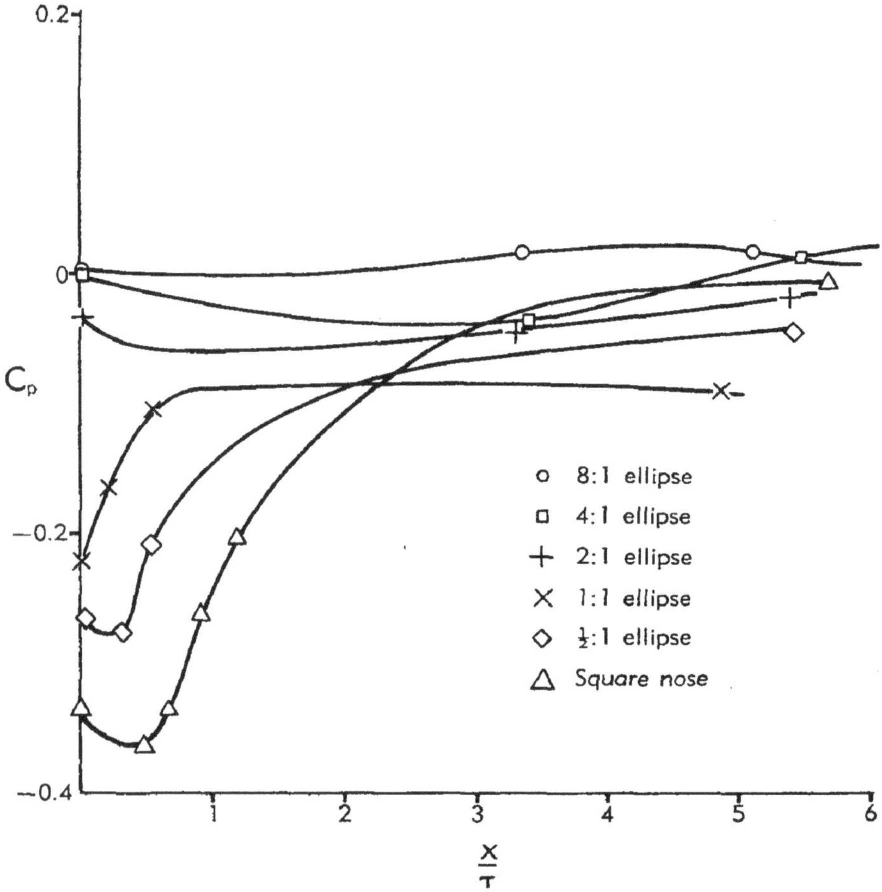

Fig. E,5c. Additional pressure distribution on a blunt-nosed airfoil due to departure of the flow from simple wave form, for different nose shapes.

dependence on nose axis ratio, were closely similar for $M_\infty = 1.42$ and $M_\infty = 1.79$. It is worth remarking that these Mach numbers are all in the range where the coefficient D of Art. 3 and 4 is negative, so that any part of the wave arising simply from reflection of a simple wave at the front shock would be *compressive* ($C_p > 0$).

It must next be inquired what effect the blunted nose is likely to have on the aerodynamic forces, and particularly on the drag. Again it is clear that the forces due to the nose will, in the main, be additive to those due to the simple wave flow behind it, the principal contribution being from the region upstream of D in Fig. E,5b. The importance of the additional pressure wave behind D, which is plotted in Fig. E,5c, will, however, depend on how much the airfoil continues to thicken in this region, and so will not be additive. The drag due to the pressures on the nose itself is deducible at once from the NPL experiments, and is plotted against M_∞ for the six different nose shapes in Fig. E,5d. The ordinate is

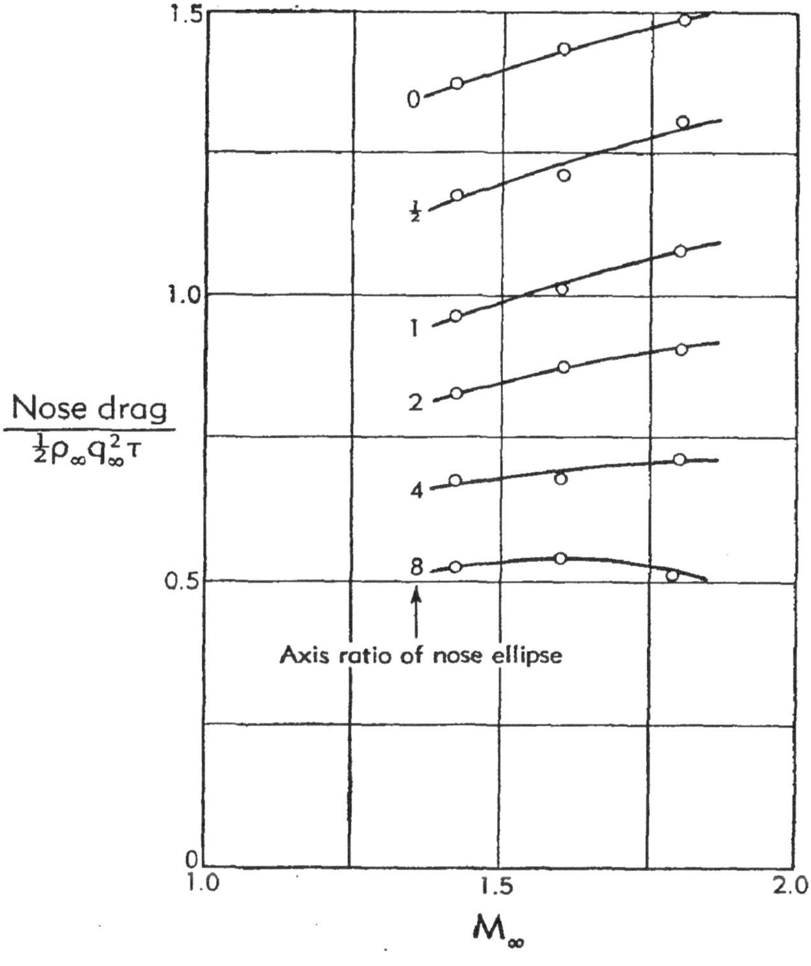

Fig. E,5d. Nose drag at Mach number M_∞, for different nose shapes with thickness τ.

drag/$(\frac{1}{2}\rho_\infty q_\infty^2 \tau)$, where τ is the thickness of the leading edge. For an airfoil the contribution of the nose drag to C_D will therefore be (τ/c) times the ordinate in Fig. E,5d.

It is seen that this quantity varies critically with the axis ratio of the nose ellipse, and that, for values of τ exceeding $0.002c$, significant drag reduction can be obtained by making the axis ratio large rather than using, say, a square nose. Some understanding of the rapid rise of drag as the nose axis ratio decreases can be obtained by plotting pressure coefficient against flow direction over the various noses, for one Mach number (actually 1.79), on a single figure, Fig. E,5e. It is seen that for axis ratios 2, 4, and 8 the points are scattered about Ackeret's straight line $C_p = C_1\theta$, even in the subsonic region, though for smaller axis ratios they begin to fall consistently below it. At least there is some indication that pressure is more closely related to flow direction than to various characteristic distances (plotted against which, the pressure graphs are widely separated in the different cases). Hence for given thickness τ, the drag would be expected to be least if the flow direction *averaged with*

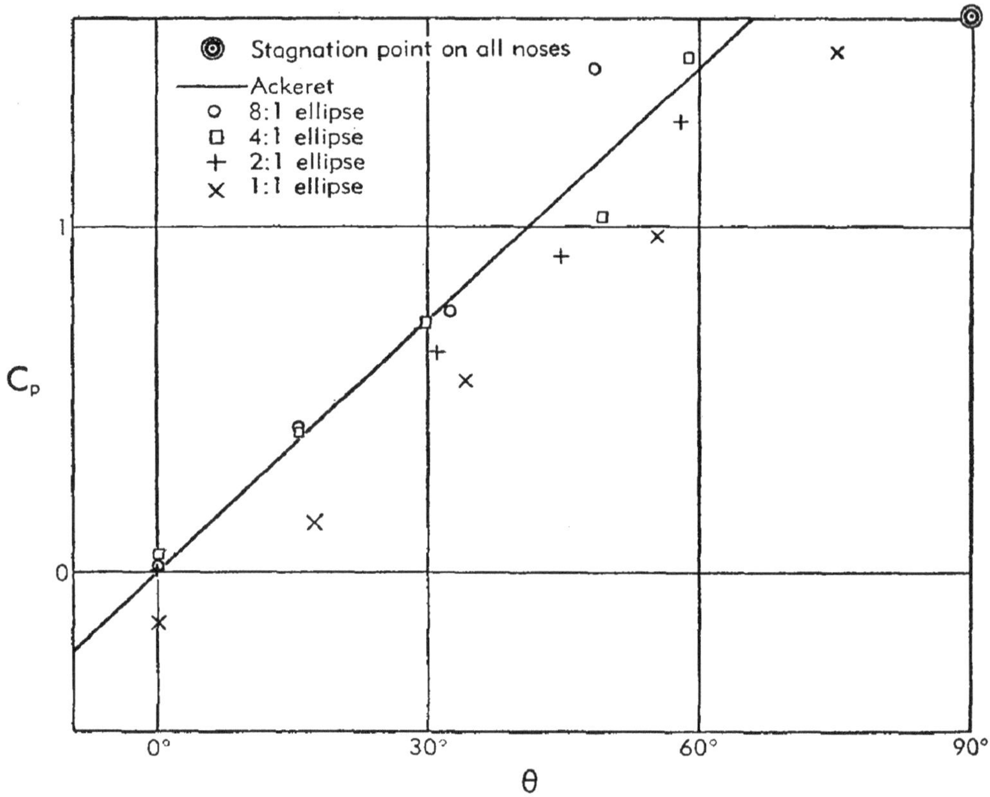

Fig. E,5e. Pressure coefficient on a blunt nose, plotted against flow direction.

respect to thickness is least. This is so for the elliptic noses of larger axis ratio, for which the rise in θ from 0° to 45°, for example, occurs in a larger region (viewed head-on) than the subsequent rise to 90°.

It is significant that the drag coefficient rises with Mach number, in the interval considered, for the smaller axis ratios. For very large axis ratios (nearly sharp noses) it must *fall* with Mach number, as Ackeret's formula indicates. But for very blunt noses, where most of the drag is probably associated with entropy change at the nearly normal part of the shock, the drag coefficient would be expected to rise, since the strength of this part of the shock increases with Mach number, and with it the entropy change increases rapidly.

To obtain the whole pressure drag at zero lift on a blunt-nosed airfoil, one must add the nose drag, just discussed, to the drag predicted by Busemann's approximate simple wave theory applied over the rest of the airfoil, and if necessary include also the drag due to the additional pressure distribution illustrated in Fig. E,5c. The latter is obtained by multiplying the said distribution of C_p by the flow direction θ, integrating along the chord, and multiplying by $2/c$. The resulting contribution, when significant, is negative, and due to the dip at the beginning of each

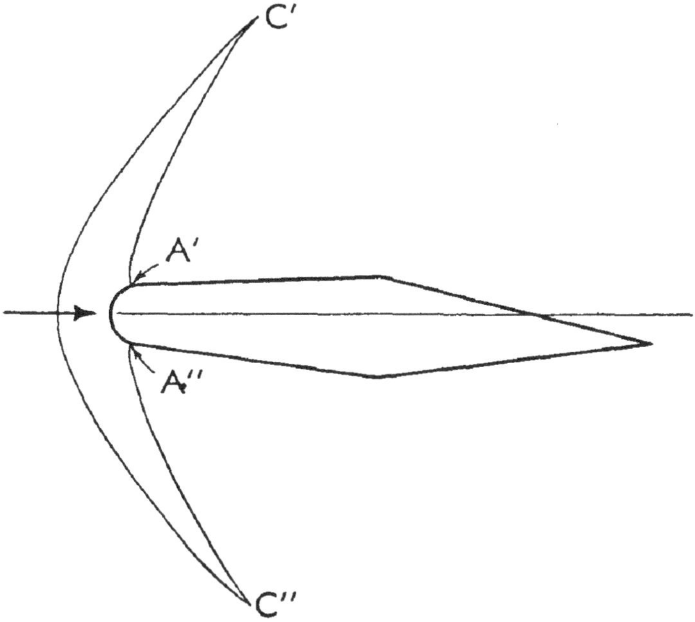

Fig. E,5f. Supersonic flow past a round-nosed airfoil at angle of attack.

of the curves with axis ratio $\leqq 1$ in Fig. E,5c. On the other hand, this will be partly balanced by a smaller negative pressure coefficient acting over a slightly larger frontal area at the rear, and probably for most practical purposes it may be ignored. Then effectively *the drag is increased by the blunting of the leading edge exactly as it would be if the pressure coefficient over the leading edge thickness were raised from its value for a perfectly sharp edge to that given by the ordinates in* Fig. E,5d.

There have not been such systematic measurements of the effect of leading edge thickness on lift, and on the extra drag due to angle of attack, but it is generally believed that these effects are not very important. There is a theoretical argument indicating this for the case of a semicircular leading edge. For then the flow pattern at angle of attack must be as in Fig. E,5f, where a very large leading edge thickness has been chosen for the sake of clarity. The flow to the left of $A'C'$ and $A''C''$ is identical with that to the left of AC, and of its reflection in the axis of symmetry, in Fig. E,5b. For the nose shape is invariant under rotation, and the fact that the surfaces downstream of A and A' are not identical cannot make a difference upstream of AC and $A'C'$. Hence no resultant lift, or additional drag, arises from this region, and therefore it is clear (from Eq. 3-22 and 3-23, which hold over the rest of the surface) that, if a factor of the form $1 + \tau/c$ is negligible, both these quantities are unchanged by making the leading edge blunt. Thus the main effect of blunting the leading edge is to increase the drag at zero lift as already analyzed in Fig. E,5e.

Finally, some brief mention must be made of the shape of the front shock. Its shape in the subsonic region is discussed elsewhere (Sec. H) both theoretically and experimentally, but it is desirable here to discuss its behavior farther out and the connection of this behavior with the theory of Art. 4.

It is obvious that the Friedrichs theory (Art. 4) applies to the shock above C. Indeed the streamline through C may be regarded as a thin sharp-nosed airfoil producing a shock exactly as in Art. 4, but more fundamentally one can regard the shock shape above C as deducible from Mach lines in Friedrichs' form (4-1), where, however, $Y(\theta)$ is related to the surface shape less directly than when the shock is attached. For the blunt flat plate, for which the Mach lines concur to good approximation in a region around AD, the asymptotic formulas may be expected to hold fairly well already at, say, ten leading edge thicknesses from the body, by the author's geometrical argument (Art. 4) leading to the parabolic asymptotic form. For example, the strength $(\Delta p/p)$ of the shock, multiplied by the square root of distance $\sqrt{y}$ from the axis of symmetry, should be nearly uniform, but would not be expected to equal the value of $2\gamma \sqrt{[y_p/\gamma + 1]}$ given by the first approximation of Art. 4, since the relation of $Y(\theta)$ to body shape is different. Indeed, with y_p replaced by $\frac{1}{2}\tau$, this approximation to $(\Delta p/p) \sqrt{y}$ for large y is $1.3 \sqrt{\tau}$, while the observed values are between two and three times as much. To within an error of ± 0.2, the values of $(\Delta p/p) \sqrt{y/\tau}$ for values of y/τ exceeding 10, for both $M_\infty = 1.6$ and 1.79, are as in the following table. We see that a round figure for blunt noses in general is 3.

Axis ratio of nose	0	$\frac{1}{2}$	1	2	4	8
$\lim_{y\to\infty} \Delta p/p \sqrt{y/\tau}$	3.4	3.2	3.0	2.8	2.6	2.4

But when the surface behind the blunted nose makes a positive angle with the incident stream, then large curvature of the surface ceases at some point D' ahead of D. This earlier flattening out is signaled along the Mach line through D', and the shock is altered from that described in the last paragraph after encountering this Mach line. It decays more slowly after this encounter, and indeed to any straight portion of surface behind D' corresponds a straight portion of shock of uniform strength.

On the other hand, if the angle of attack exceeds the inclination of the upper surface behind the blunt leading edge, then the shock is precisely as described two paragraphs above. For the Mach lines from the upper surface behind the leading edge then interact only with the rear shock. For such a flow, if the leading edge had been sharp, the flow over the upper surface would have been headed by a Prandtl-Meyer expansion. But in reality this is preceded by a shock whose shape must be exactly as in the NPL experiments on blunt-nosed plates. In particular its strength is given as

$$\frac{\Delta p}{p} \cong 3\sqrt{\frac{\tau}{y}} \tag{5-1}$$

when $y/\tau > 10$.

Bardsley [*65*] shows that even for the smallest leading edge thicknesses obtainable, this simple consideration is sufficient to account for the shocks which have always been observed to precede the Prandtl-Meyer expansion at the leading edge of a supersonic airfoil at high angle of attack. For example Eq. 5-1 predicts that, for a leading edge thickness of 20 microns, the shock strength at a distance 5 cm from the leading edge measured perpendicular to the stream, should be 0.06. (Shocks of such a strength are easily visible on schlieren photographs.) This value is, within the experimental error, the strength calculated by Bardsley from the measured shock inclination under such conditions. Bardsley also gives calculations leading to the rejection of the alternative theory that the shock in question is produced by boundary layer growth.

To sum up the present article, it has been demonstrated very briefly that, although the theories of Art. 3 and 4 do not apply directly to blunt-nosed airfoils, nevertheless suitable semiempirical modifications of these theories can in principle be constructed to give as much information about every aspect of the flow as we have in the simpler problems of sharp-nosed airfoils. But the carrying out of this program in full detail may perhaps have to wait until an even more systematic series of experiments than those described above has been performed.

E,6. Supersonic Projectile Theory: Complete Flow Pattern. In Art. 6 and 7 we seek information, of the kind obtained for airfoil flow in Art. 3, 4, and 5, in the analogous problem of axisymmetric flow past projectiles, and also in some problems where flow is not strictly axisymmetric but the projectile surface everywhere makes a fairly small angle with the stream. Even the axisymmetric flow problem is a more complicated one than that of airfoil flow, largely because the solution on small perturbation theory (D,15–D,18) is more complicated, which in turn is due to the famous fact that a cylindrical pulse of sound, unlike a plane or spherical pulse, leaves a "tail" behind it (cf. C,12).

In consequence, fewer strictly logical inferences are possible. But this is partly compensated by the possibility of arguments by analogy from the two-dimensional case. For example, it will be assumed that, at least for sharp-nosed projectiles with attached shocks and a completely supersonic flow pattern, the flow is given to excellent approximation by the potential flow, at the entropy of the undisturbed stream, which satisfies the boundary condition at the surface of the projectile—always provided that this potential flow is cut off by the front shock (which is to be determined by the conditions governing shocks with uniform flow ahead of them) to avoid the appearance of a region in which there are three values for each physical quantity, and that similarly, where necessary, a rear shock or shocks appear. The assumption here is by analogy with the "simple wave theory" of Art. 3, whose accuracy in predicting surface pressures was amply demonstrated in Fig. E,3f, and whose accuracy in predicting shock positions was carefully checked in Art. 4. The errors, being due to those differences between the behavior of shocks and of acoustic discontinuities which were collected in the last sentence of point (7) of Art. 1, should be no greater in the present problem since the shocks, which still have supersonic flow on both sides of them, can be no stronger than before.

Such analogies must of course be used intelligently. There is no reason to suppose that any of the flow quantities are constant along Mach lines, or that these are straight (the first point is not even true on the small perturbation theory). For these Mach lines are meridian sections of stationary sound waves of conical, or nearly conical, shape. There is no reason why they should remain exact conical waves, which have not the self-preserving property of plane waves. Again, it should not be assumed that the pressure at a point depends only on the flow direction, or that the flow is uniform behind the rear shock, even on this simplified theory, or indeed even on linear theory, owing to the "tail" property mentioned above.

Thus, although it is at least true that the essential problem is the determination of a potential flow, it is clear from the points in the last paragraph that a method like Busemann's (Art. 3) for proceeding to the second approximation to the surface pressures will not be available in axisymmetric supersonic flow. Therefore a straightforward approach by successive approximations is necessary, as for the subsonic flow problems of Art. 2, and especially as in the Hantzsche-Wendt method. But such a method is much more laborious in axisymmetric problems. Actually, as Broderick [*34*] found, and as will be reported briefly in Art. 7, it becomes simpler when only the pressures close to the projectile surface are sought, but it is still almost too complicated to be practically useful, although the situation has recently been improved by Van Dyke [*54*]. Further, even if this second approximation could be found away from the body surface, it would give one no information about the position of the shocks, which are among the most interesting features of the flow field. To see this, e.g. for the front shock in axisymmetric flow, it is sufficient to observe that the equation for the second approximation to the disturbance potential (which is derived from Eq. 2-14) has the same characteristic curves as the equation for the first approximation, namely, the straight lines whose equations in cylindrical polar coordinates are $x \pm (M_\infty^2 - 1)^{\frac{1}{2}}r = \text{const}$. Hence on the second approximation the flow is still uniform ahead of the downstream-pointing characteristic, given by $x - (M_\infty^2 - 1)^{\frac{1}{2}}r = \text{const}$, through the nose of the projectile, whereas in reality the front shock must lie ahead of this line. Actually, as is hardly surprising after this consideration, the process of successive approximation diverges near any shock, as foreshadowed in Art. 1. A similar difficulty arises at points whose distance r from the axis of the projectile is large. Owing to the continual separation, as r increases, of the approximate (straight) Mach lines from the true ones, the second approximation may be as bad as the first for large r; and indeed for sufficiently large r the process again diverges.

However, there is a quite different approach by which all this latter information (concerning the approximate strength and position of shocks, and the flow at large distances from the axis, which the second approximation as well as the first fails to give) may be obtained by a fairly direct and physically natural device from the form of the first approximation alone. This approach is due to Whitham, who has improved on his previous work [*32*] in a later paper [*66*]. The theory gives all the information analogous to that resulting from Friedrichs' theory of two-dimensional flow (Art. 4), but of course only to a first, not to a second, approximation.

Since the Whitham theory of the complete flow pattern is simpler than the Broderick theory of the surface pressures, it is here given at once, leaving the Broderick theory to Art. 7, and thus reversing the order followed in Art. 3 and 4 for the corresponding two-dimensional problem. This order also reflects the greater practical importance, in the present problem, of a knowledge of the complete flow pattern, which satisfies the assumptions of the theory at *all* distances from the projectile (while the flow past wings, even of large aspect ratio, is not nearly two-dimensional at large distances, as mentioned at the beginning of Art. 4). In fact the conclusions of this article may reasonably be expected to give the correct order of magnitude of the disturbances, at sufficiently large distances, for a projectile of any shape whatever, because the theory at least works in the right number of dimensions. In other words some knowledge of what should be observed at ground level, when any supersonic projectile passes overhead, must emerge.

The Whitham theory. Whitham's theory [*66*] will be described for axisymmetric flow, and afterward we shall show that it holds without modification in more general flows. The cardinal assumption on which it rests is as follows. The small perturbation theory gives a correct first approximation throughout the flow, *provided that the value which it predicts for any physical quantity, at a given distance from the axis on the straight (approximate) Mach line,* $x - (M_\infty^2 - 1)^{\frac{1}{2}}r$ = const, *pointing downstream from a given point on the projectile surface, is interpreted as the value, at that distance from the axis, on the exact Mach line which points downstream from the said point.* In other words the ineffectiveness, near the shocks and at large distances from the axis, of the linear theory is due to its incorrect placing of the Mach lines, not to incorrect prediction of the variation of physical quantities along them.

The following arguments support Whitham's assumption:

1. It is true, word for word, in the two-dimensional flow past airfoils. Applied there it would give the first approximation to the Friedrichs theory, which has been copiously checked in Art. 4.

2. It takes into account the convection of sound, and the variations in its speed (Art. 1), for the features of the problem through which they might be expected to act most effectively, namely for the positions of waves, not the state of the fluid thereon.

3. It includes the statement that the small perturbation theory gives a correct first approximation to the surface pressures, which is supported by strong evidence, some given in VII, B and some in Art. 7 below.

4. The conclusions are not in contradiction with any known facts, and several of them agree with results obtained previously by quite different methods, as is mentioned below in connection with each such conclusion.

The practical use of the assumption depends on the fact that, if the linearized solution is known, then a second approximation to the Mach lines may be deduced. If this second approximation is used, instead of the exact Mach lines mentioned in Whitham's assumption, the approximation should still be reasonable. The spread of the Mach lines at large distances (which causes the decay of the shocks) will be given, not accurately, but to an approximation probably adequate in practice.

To find this second approximation to the Mach lines, let the velocity components parallel and perpendicular to the axis be $q_\infty(1 + u)$ and $q_\infty v$ respectively, so that u and v are nondimensional disturbance velocities. (Then the pressure is given, on small perturbation theory, as $p_\infty - \rho_\infty q_\infty^2(u + \frac{1}{2}v^2)$, where the $\frac{1}{2}v^2$ need only be included near the axis, a point which has been mentioned in Sec. C and D and will be more fully substantiated by the nonlinear theory in Art. 7.) The Mach directions at any point satisfy

$$\frac{dx}{dr} = \cot(\pm\beta + \theta) \tag{6-1}$$

where, as in Art. 3, β is the local Mach angle and θ the direction of the stream, measured in the positive sense from the axis of symmetry. For the Mach direction pointing outward and downstream the plus sign must be taken in Eq. 6-1. For this Mach direction, Eq. 6-1 may be expanded, using Bernoulli's equation, in the form

$$\frac{dx}{dr} - (M_\infty^2 - 1)^{\frac{1}{2}} = \frac{(\gamma + 1)M_\infty^4}{2(M_\infty^2 - 1)^{\frac{1}{2}}} u - M_\infty^2[v + (M_\infty^2 - 1)^{\frac{1}{2}}u] \tag{6-2}$$

if the squares of disturbances are neglected. (This result is more complicated than the analogous result (4-4) for a simple wave, in which to a first approximation $v + (M_\infty^2 - 1)^{\frac{1}{2}}u = 0$.)

Now the values for u and v given by small perturbation theory (D,16), expressed as functions of r and of the variable $y = x - (M_\infty^2 - 1)^{\frac{1}{2}}r$,

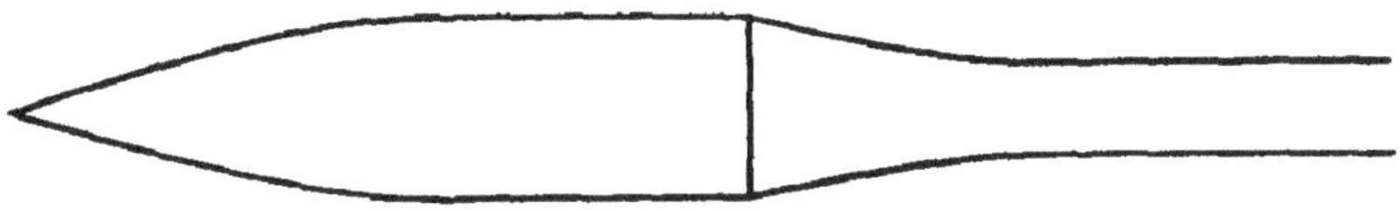

Fig. E,6a. Body-wake combination.

which determines on which straight (approximate) Mach line a point lies, are

$$u = -\int_0^y \frac{f'(t)dt}{(y - t)^{\frac{1}{2}}[y - t + 2(M_\infty^2 - 1)^{\frac{1}{2}}r]^{\frac{1}{2}}}$$

$$v = \frac{1}{r}\int_0^y \frac{[y - t + (M_\infty^2 - 1)^{\frac{1}{2}}r]f'(t)dt}{(y - t)^{\frac{1}{2}}[y - t + 2(M_\infty^2 - 1)^{\frac{1}{2}}r]^{\frac{1}{2}}} \tag{6-3}$$

The function $f(t)$ in these integrals depends on the body shape, and if the flow behind the body is to be determined the "body shape" should be understood to include also the effective shape of the wake (a sort of "mean" shape, in the sense of turbulence theory). In this sense a typical "body shape" is as illustrated in Fig. E,6a.

For a slender pointed body, the slope and curvature of whose surface vary only gradually, the function $f(t)$ is equal to $S'(t)/2\pi$, where $S(t)$ is the cross-sectional area of the body (or the effective cross-sectional area of the wake) at a distance t (measured along the axis) from the nose. But if, for example, $S'(t)$ (and hence the slope of the surface) is discontinuous anywhere, the function $f(t)$ lags behind, changing to the new value continuously over a distance of a radius or two; this makes a significant difference in Eq. 6-3, as Lighthill [*35*] showed (Art. 7), since in the region of rapid change $f'(t)$ is relatively large. Where the slope or curvature varies rapidly one may correct $f(t)$ suitably by replacing the said rapid variation by a number of small discontinuities of slope, and applying the author's theory to each. Finally, if the body is *not* slender, expressions of the form of Eq. 6-3 may be too inaccurate near the body because the disturbances are not small; but nevertheless at distances from the axis exceeding some suitable multiple of the radius of the projectile, in other words outside a certain circular cylinder enclosing the body, the disturbances will be small enough, and here at least equations of the form of Eq. 6-3 are valid. The function f in these equations may still sometimes be estimated, e.g. from experimental data, using the full small perturbation theory as a partial guide; whenever this function is known the theory below gives the complete flow pattern outside the cylinder mentioned.

Now Whitham's assumption, italicized above, implies that Eq. 6-3 give a valid first approximation (for small perturbations) *everywhere*, provided that in them y stands not for $x - (M_\infty^2 - 1)^{\frac{1}{2}}r$ itself, but for a quantity constant along any downstream-pointing Mach line and equal to $x - (M_\infty^2 - 1)^{\frac{1}{2}}r$ at the surface. Henceforth y will be understood in this sense. *To a close approximation it is where the Mach line, produced, meets the axis* (measured downstream from the nose).

Now we can obtain the relation between x, r, and y to a second approximation (the first being $x - (M_\infty^2 - 1)^{\frac{1}{2}}r = y$) by integrating Eq. 6-2, which holds all along any downstream-pointing characteristic (along which y is constant), starting from the initial value given by $x - (M_\infty^2 - 1)^{\frac{1}{2}}r = y$. This gives

$$x - (M_\infty^2 - 1)^{\frac{1}{2}}r = y + \int_R^r \left\{ \frac{(\gamma + 1)M_\infty^4}{2(M_\infty^2 - 1)^{\frac{1}{2}}} u - M_\infty^2[v + (M_\infty^2 - 1)^{\frac{1}{2}}u] \right\} dr \quad (6\text{-}4)$$

where the integration is performed at constant y. Here R signifies the radius of the projectile where the Mach line specified by y springs from it, i.e. where $x - (M_\infty^2 - 1)^{\frac{1}{2}}r = y$. (But for nonslender bodies Whitham's assumption can be applied only to the flow outside a certain cylinder enclosing the body, and then R may be taken to be the radius of this cylinder.)

By Eq. 6-3, Eq. 6-4 may be explicitly integrated to give

$$\begin{aligned} x - (M_\infty^2 - 1)^{\frac{1}{2}}r &= y \\ &+ \frac{(\gamma + 1)M_\infty^4}{2(M_\infty^2 - 1)} \int_0^y f'(t) \frac{[y - t + 2(M_\infty^2 - 1)^{\frac{1}{2}}R]^{\frac{1}{2}} - [y - t + 2(M_\infty^2 - 1)^{\frac{1}{2}}r]^{\frac{1}{2}}}{(y - t)^{\frac{1}{2}}} dt \\ &- 2M_\infty^2 \int_0^y f'(t) \ln \left\{ \frac{[y - t + 2(M_\infty^2 - 1)^{\frac{1}{2}}r]^{\frac{1}{2}} - (y - t)^{\frac{1}{2}}}{[y - t + 2(M_\infty^2 - 1)^{\frac{1}{2}}R]^{\frac{1}{2}} - (y - t)^{\frac{1}{2}}} \frac{R^{\frac{1}{2}}}{r^{\frac{1}{2}}} \right\} dt \end{aligned} \quad (6\text{-}5)$$

This is the second approximation to the equation of the Mach line specified by y. However, it is never used in the Whitham theory in this very complicated form, but always in the approximate form

$$x - (M_\infty^2 - 1)^{\frac{1}{2}}r = y - kF(y)r^{\frac{1}{2}} \quad (6\text{-}6)$$

where

$$k = 2^{-\frac{1}{2}}(\gamma + 1)M_\infty^4(M_\infty^2 - 1)^{-\frac{3}{4}}, \qquad F(y) = \int_0^y \frac{f'(t)dt}{(y - t)^{\frac{1}{2}}} \tag{6-7}$$

This is obtained from Eq. 6-5 by assuming that $(M_\infty^2 - 1)^{\frac{1}{2}}r$ is large compared with y, so that the form of the characteristic is considered at a point considerably farther downstream from its point of origin on the surface than the said point of origin is from the nose. The justification for making this approximation everywhere is that, where $(M_\infty^2 - 1)^{\frac{1}{2}}r/y$ is not large, the integrals in Eq. 6-5 are unimportant compared with the main term y, which is obvious because the range of integration is y but the integrand is of the same small order of magnitude as the disturbances. Hence these integrals can be approximated as $-kF(y)r^{\frac{1}{2}}$, because this also is unimportant when $(M_\infty^2 - 1)^{\frac{1}{2}}r/y$ is not large. Actually it will appear below that, wherever the small perturbation theory in its ordinary form is inadequate, including all regions near shocks, the condition that $(M_\infty^2 - 1)^{\frac{1}{2}}r/y$ be large is amply satisfied. But in fact the approximation (6-6) seems to be surprisingly good everywhere.

For any *constant* y, Eq. 6-6 is the approximate equation of a Mach line,[2] which shows that to this approximation the Mach lines are *parabolas*, with axes in the Mach direction of the undisturbed stream. (Oddly enough this is the same shape as was found, asymptotically, for the *shocks* in two-dimensional airfoil theory, in which, however, the Mach lines are straight.)

Of essential importance in the theory is the coefficient $kF(y)$ of $r^{\frac{1}{2}}$ in Eq. 6-6, which determines the curvature of the parabolic Mach lines. This is split up in Eq. 6-7 into a factor k depending only on the Mach number, in a way illustrated in Fig. E,6b, and a function $F(y)$ which is derived from the fundamental function $f(t)$ in a manner independent of Mach number. Thus for smooth slender bodies it depends on body shape alone. Mathematically F is the $(\frac{1}{2})$th derivative of f, multiplied by $\sqrt{\pi}$.

Now whereas for smooth slender bodies $F(y)$ is

$$\frac{1}{2\pi}\int_0^y \frac{S''(t)dt}{(y - t)^{\frac{1}{2}}} \tag{6-8}$$

an additional term arises when the surface has a discontinuity of slope, say at $x = t_1$, where the projectile radius is R_1, and $S'(X)$ changes by $(\Delta S')$. By a method similar to that used by Lighthill [*35*] to find the additional terms in the surface pressures (Art. 7), one can show that the additional term in $F(y)$ is

$$\left(\frac{\Delta S'}{2\pi}\right)\left[\frac{2}{(M_\infty^2 - 1)^{\frac{1}{2}}R_1}\right]^{\frac{1}{2}} h\left(\frac{y - t_1}{(M_\infty^2 - 1)^{\frac{1}{2}}R_1}\right) \tag{6-9}$$

[2] Only, of course, up to where it encounters a shock.

where the function $h(z)$ is plotted as a plain line in Fig. E,6c. It is zero for z less than -1, where it rises discontinuously to unity, and then falls off steadily, following closely the curve $(2z)^{-\frac{1}{2}}$ (dashed line) for $z > 3$. When this is so, the sum of Eq. 6-8 and 6-9 becomes effectively

$$\frac{1}{2\pi}\int_0^y \frac{dS'(t)}{(y-t)^{\frac{1}{2}}} \tag{6-10}$$

which is to be understood as including a term $(\Delta S'/2\pi)(y - t_1)^{-\frac{1}{2}}$ owing to the jump in $S'(t)$ at $t = t_1$. (This latter term could be written in the

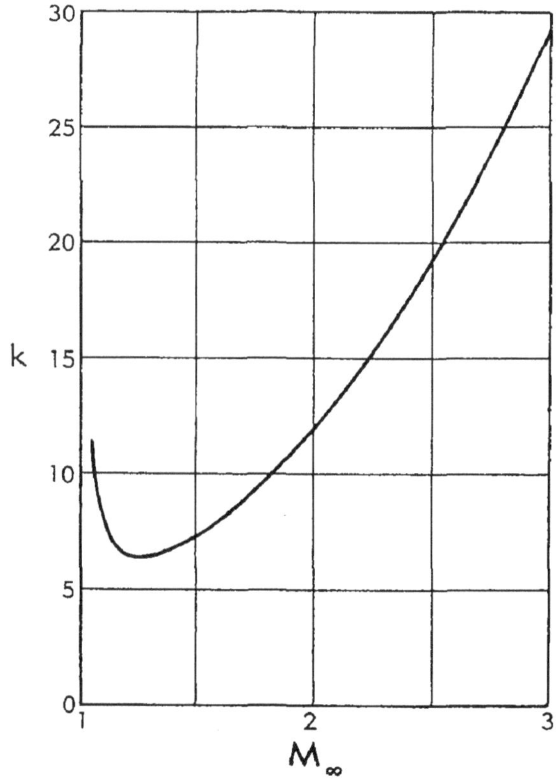

Fig. E,6b. k as a function of M_∞.

form of Eq. 6-9 with $h(z)$ given by the dashed line in Fig. E,6c. It is worth remarking that the total area between the two curves is zero.)

The discontinuity in $F(y)$, by an amount equal to the product of the first two factors in Eq. 6-9, takes place on the Mach lines given by $y = t_1 - (M_\infty^2 - 1)^{\frac{1}{2}}R_1$ through the point (t_1, R_1) where the slope of the surface changes. There are several such Mach lines although y has only one value, because (in Eq. 6-6 for the Mach lines) $F(y)$ may have any value between its two extremes. The Mach lines are a system of parabolas which concur to form a fan, just as they do in the Prandtl-Meyer expansion around a corner. Indeed the range of slopes dx/dr of these Mach lines, by Eq. 6-6, is

$$-\frac{1}{2}\,k(\Delta F)R_1^{-\frac{1}{2}} = -\frac{1}{2}\,k\left(\frac{\Delta S'}{2\pi}\right)2^{\frac{1}{2}}(M_\infty^2 - 1)^{-\frac{1}{4}}R_1^{-1} \qquad (6\text{-}11)$$

and this is $\frac{1}{2}(\gamma + 1)M_\infty^4(M_\infty^2 - 1)^{-1}$ times the change in flow direction which gives welcome support to the theory, since it agrees with the first approximation to the two-dimensional flow around a corner.[3] The fan of Mach lines has physical reality only if $\Delta S' < 0$; otherwise of course there

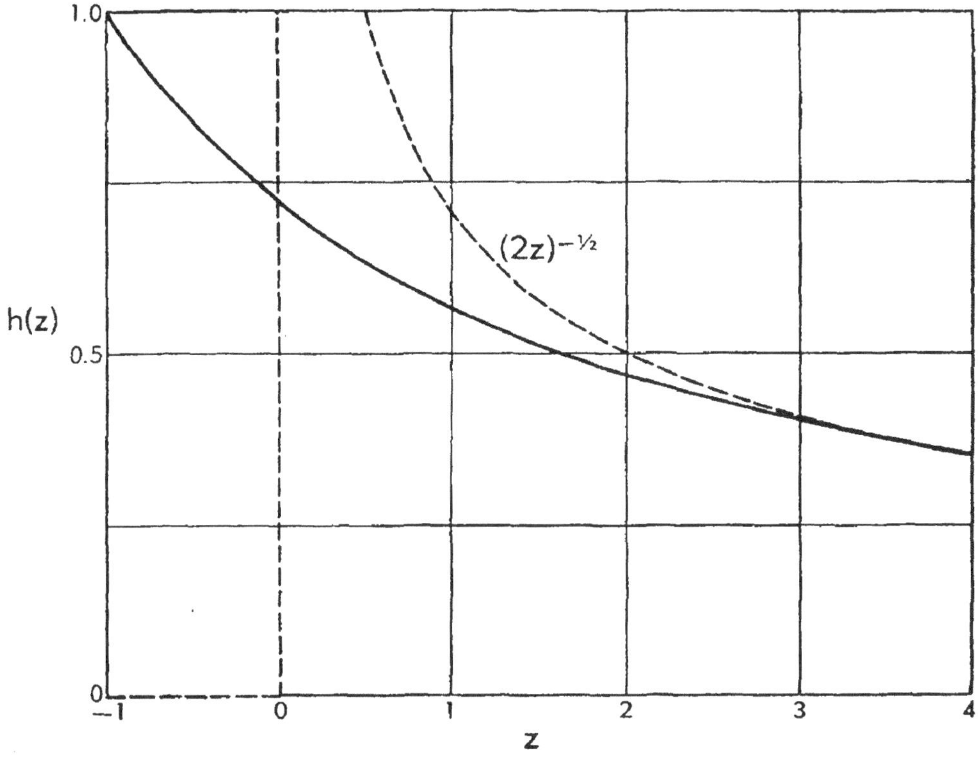

Fig. E,6c. The function $h(z)$.

is an attached shock; but its form can still be predicted from the form of $F(y)$, as will be seen below.

When there are several discontinuities of slope, say t_1, t_2, . . . , the expression (6-9) corresponding to each must be added to Eq. 6-8 to give $F(y)$. As indicated before, one can find $F(y)$ for a body whose slope or curvature changes too rapidly for Eq. 6-8 to be reliable, by simulating the region of rapid change by several (or a continuous distribution of) discontinuities of slope.

Using these considerations, the fundamental function $F(y)$ has been

[3] It is obvious that the immediate change at the surface must take place exactly as in two-dimensional flow, since the distances involved may be considered as infinitely small compared with the distance from the axis.

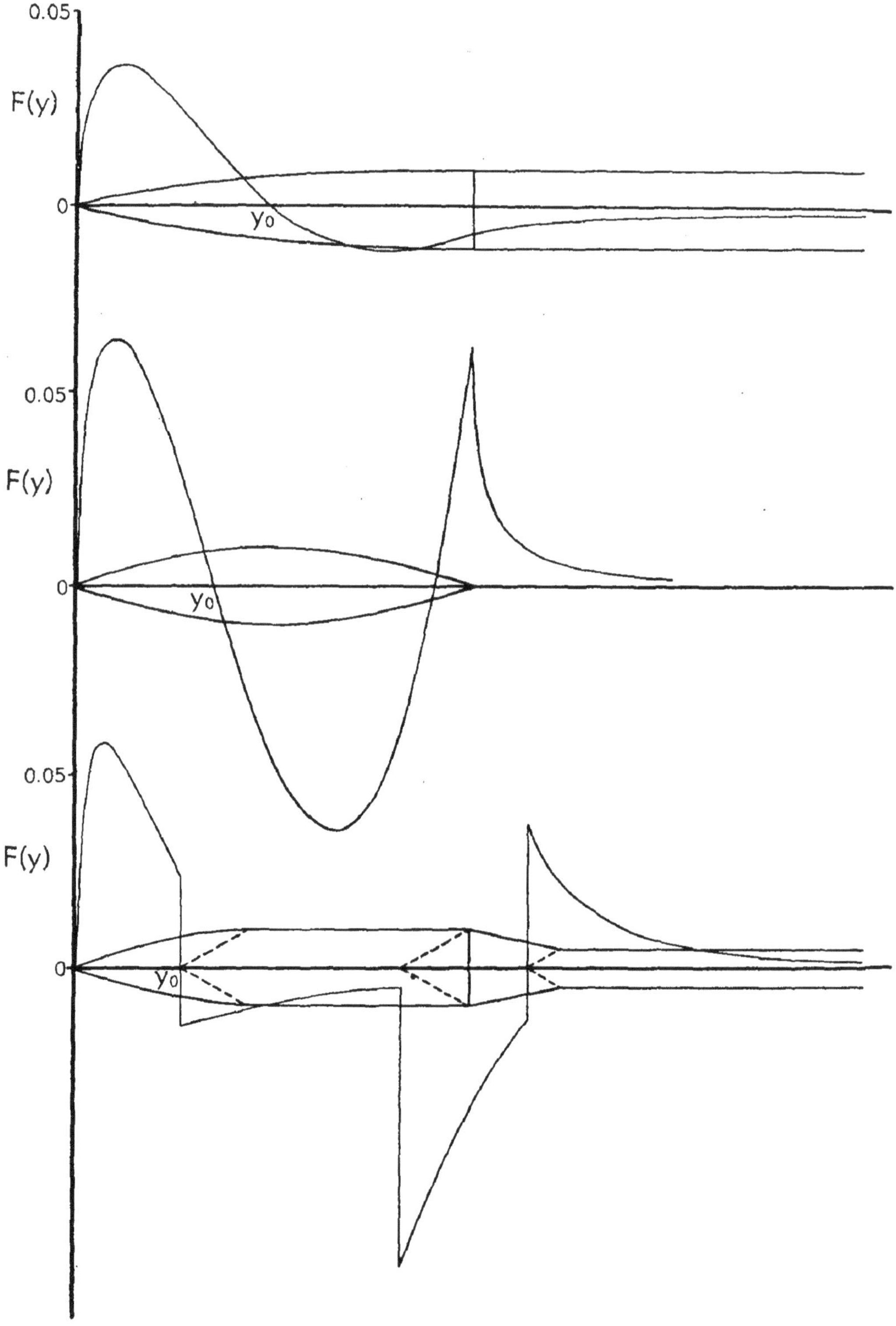

Fig. E,6d. The function $F(y)$ for three different body shapes.

calculated for three different body shapes, which are shown in Fig. E,6d with a graph of $F(y)$ superimposed on each. (In each case the length of the body is taken arbitrarily as unity.) The first is a shell shape ending in a cylindrical portion, with no allowance made for wake thinning. The second is the extreme case of a body pointed at both ends, which corresponds most closely to the airfoils of Art. 3 and 4; however, the linear theory (D,16; cf. Fig. D,16b) indicates for such bodies a very rapid compression ahead of the rear point, so that almost certainly the boundary layer separates, giving a true "mainstream" more as in Fig. E,6a. (On the other hand, if the boundary layer were controlled, the "trailing edge" might probably be a stagnation point, in a region of subsonic flow.) The third shape, which comes far closer to conditions met with in ordinary ballistics, incorporates three discontinuities in slope, one where the "5/10 caliber ogival head" is joined to the cylindrical portion, one where the wake leaves the cylindrical portion at the typical angle of 12°, and one where the wake straightens out. It would be slightly more realistic to make this straightening out rapid but continuous; then the jump in $F(y)$ would be similar. For this body both f and F depend on Mach number, even on linear theory; the calculations were done for $M_\infty = 2$. Fig. E,6d gives a general impression of the behavior of $F(y)$ for given maximum diameter; for variations of the maximum diameter, F is proportional to its square.

Notice that initially $F(y)$ is always positive. For that portion of the surface which is approximately conical (with semiangle δ, at the nose) $f(t)$ is approximately $\delta^2 t$, and so

$$F(y) \cong 2\delta^2 y^{\frac{1}{2}} \tag{6-12}$$

While $F(y)$ remains positive the corresponding parabolic Mach lines (through points on the axis a distance y from the nose) are, as Eq. 6-6 shows, convex to the oncoming stream. Where $F(y)$ is first zero, say for $y = y_0$, there is a straight Mach line, and those farther downstream are concave to the oncoming stream. In some cases there may be a further portion with $F(y) > 0$ again.

Now, when the condition that $(M_\infty^2 - 1)^{\frac{1}{2}} r/y$ is large (which was used to deduce Eq. 6-6) holds, i.e. on a given Mach line far downstream of its point of origin compared with the distance of the latter from the nose, a notable simplification to Eq. 6-3 for the disturbance velocities also results, giving

$$v = -(M_\infty^2 - 1)^{\frac{1}{2}} u = 2^{-\frac{1}{2}}(M_\infty^2 - 1)^{\frac{1}{4}} \frac{F(y)}{r^{\frac{1}{2}}} \tag{6-13}$$

Thus on any Mach line the disturbance velocity is ultimately perpendicular to the Mach line and proportional to the inverse square root of the distance from the axis. This is what would be expected of a conical sound wave, assuming that at large distances it leaves less and less tail behind it, and that its wave energy (proportional to the radius and the square of the amplitude) therefore becomes constant.

The information so far obtained, insofar as it relates to the potential flow field far from the axis, was deduced by two quite different methods in Whitham's first paper [32]. First he showed that the equation of continuity may be approximated with some appearance of reason, for large r, by the equation

$$2\left[\frac{\partial u}{\partial r} + (M_\infty^2 - 1)^{\frac{1}{2}}\frac{\partial u}{\partial x}\right] + \frac{u}{r} + \frac{(\gamma + 1)M_\infty^4}{(M_\infty^2 - 1)^{\frac{1}{2}}}u\frac{\partial u}{\partial x} = 0 \qquad (6\text{-}14)$$

of which the exact solution happens to be given by Eq. 6-6 and 6-13, with $F(y)$ an arbitrary function, which Whitham assumes to depend on the body shape. This solution at large distances from the axis may, for slender bodies, be compared with the linear theory (in some intermediate region where both are approximately valid) to determine $F(y)$ as in Eq. 6-7. Because he feels that the approximations leading to Eq. 6-14 are lacking in rigor, he then assumes a very general series expansion for u and v for large r, of the kind described at the end of Art. 1, and determines Eq. 6-6 and 6-13 as the first terms, respectively, in these series, by substituting the series in the exact equations of potential flow. Carried to one further approximation, the general behavior of supersonic axisymmetric potential flow at large distances from the axis is found to be given by

$$\begin{aligned} x - (M_\infty^2 - 1)^{\frac{1}{2}}r &= y - kF(y)r^{\frac{1}{2}} + k_1F^2(y)\ln r + \cdots \\ u &= -2^{-\frac{1}{2}}(M_\infty^2 - 1)^{-\frac{1}{4}}F(y)r^{-\frac{1}{2}} + k_2F^2(y)r^{-1} + \cdots \qquad (6\text{-}15) \\ v &= -(M_\infty^2 - 1)^{\frac{1}{2}}u - \frac{(\gamma + 1)M_\infty^4}{4(M_\infty^2 - 1)}F^2(y)r^{-1} + \cdots \end{aligned}$$

where k_1 and k_2 (like k) are specified functions of Mach number only, and the curves y = const are now *exact* Mach lines. Again the function $F(y)$ can be determined, for slender bodies, by comparison with linear theory. This systematic approach affords strong confirmation of the results obtained by the physical arguments given above. However, these latter arguments by themselves are sufficient to determine the shocks, as will be seen; it is therefore merely additional confirmation that in Whitham's earlier paper [32] the new terms, in $F^2(y)$, in Eq. 6-15 were not found to have any influence on the shock positions at large distances.

Before going on to find where the shocks cut off the potential flow field, it must be remarked that the conclusions obtained so far, and hence also what are to follow, are true not only for axisymmetric flows, but also, to a first approximation, for any flow about projectiles (not necessarily of revolution) which are slender in Ward's sense [43], i.e. make everywhere a small angle with the incident stream. For then (Art. 7) the flow, in regions where $(M_\infty^2 - 1)^{\frac{1}{2}}r/[x - (M_\infty^2 - 1)^{\frac{1}{2}}r]$ is not small, *is* approximately axisymmetric. For, if it is Fourier-analyzed with respect to the angular cylindrical polar coordinate θ, the coefficients of $\cos n\theta$ and $\sin n\theta$ are smaller than the axisymmetric term (independent of θ) by a factor of order the nth power of the fineness ratio, which is only canceled out near the body, where the fraction just mentioned *is* small and the coefficient of $\cos n\theta$ and $\sin n\theta$ behaves like r^{-n}. Further, the said axisymmetric flow is given by Eq. 6-3, with $f(t)$ related to the distribution of cross-sectional area downstream of the pointed nose exactly as for axisymmetric flow.

Thus, except near the projectile, the flow is roughly axisymmetric, and the Mach lines and shocks are where they would be for a symmetrically placed projectile of revolution with the same distribution of cross-sectional area. In particular, yaw makes little difference to the shocks on a slender projectile. They are yawed by only a small fraction of the projectile yaw. This is borne out by experiment and by more exact calculations on flow past yawing cones [37]. Little further reference to the case of nonaxisymmetric flow will be made in this article.

Determination of the front shock. The front shock will be determined first. This must lie ahead of the straight (approximate) Mach line through the nose, in the region where the left-hand side of Eq. 6-5 is negative. But y is essentially positive. Hence the integrals in Eq. 6-5 must be negative and of absolute magnitude exceeding y. This shows at once that the condition that $(M_\infty^2 - 1)^{\frac{1}{2}}r/y$ be large, on which Eq. 6-6 is (strictly speaking) based, must be amply satisfied. The latter equation is now used to determine the shock much as in the Friedrichs theory. Since (Art. 4) to a first approximation the direction of a shock bisects that of the Mach directions ahead of and behind it, the slope dx/dr of the shock must lie halfway between $(M_\infty^2 - 1)^{\frac{1}{2}}$ and the slope of the Mach line immediately behind it. Hence, by Eq. 6-6, along the shock,

$$\frac{d}{dr}\left[y - kF(y)r^{\frac{1}{2}}\right] = \frac{1}{2}\left[-\frac{1}{2}k\frac{F(y)}{r^{\frac{1}{2}}}\right] \tag{6-16}$$

The solution of this differential equation (which becomes linear when regarded as an equation for $r^{\frac{1}{2}}$ in terms of y), with the initial condition $r = 0$ when $y = 0$, is

$$r = \frac{4}{k^2}\frac{\left[\int_0^y F(t)dt\right]^2}{[F(y)]^4} \tag{6-17}$$

Thus the front shock is described parametrically by Eq. 6-6 and 6-17 (cf. Eq. 4-1 and 4-7 in the Friedrichs theory, which, it should however be remembered, goes to a second approximation).

Near the nose, or more precisely for that portion of the shock intersected by Mach lines springing from the portion of surface which is approximately conical, Eq. 6-12 holds, so that Eq. 6-6 and 6-17 become

$$x - (M_\infty^2 - 1)^{\frac{1}{2}} r = y - 2k\delta^2 (yr)^{\frac{1}{2}}, \qquad r = \frac{4}{9k^2\delta^4}\, y \tag{6-18}$$

whence

$$x = r[(M_\infty^2 - 1)^{\frac{1}{2}} - \tfrac{9}{4}k^2\delta^4] \tag{6-19}$$

Thus this forward portion of the shock is effectively straight and at an angle

$$(\tfrac{9}{4}k^2\delta^4)M_\infty^2 = \tfrac{9}{8}(\gamma + 1)^2 M_\infty^6 (M_\infty^2 - 1)^{-\frac{3}{2}}\delta^4 \tag{6-20}$$

to the undisturbed Mach direction. The strength of the shock in this region is therefore

$$\frac{\Delta p}{p} = \frac{3}{2}\gamma(\gamma + 1) M_\infty^6 (M_\infty^2 - 1)^{-1}\delta^4 \tag{6-21}$$

Notice that, by Eq. 6-18, this strength is maintained to a distance r from the axis such that the projectile remains practically conical over a distance $\frac{9}{4} k^2\delta^4 r$ from the nose; even if the latter distance be only small, the former distance may be quite large owing to the factor δ^4 multiplying r.

The correctness to a first approximation of Eq. 6-21 for the shock strength at the nose is checked by the analysis on the *exact* adiabatic theory of the axisymmetric flow past a cone of semiangle δ, which Taylor and Maccoll [*38*] first showed to be governed by ordinary differential equations only (H,16–19). The present author's proof [*36*] that Eq. 6-21 is, in this case, an approximation to the shock strength with an error at most of order δ^5 for small δ, is completely rigorous. (The true error, however, is more probably of about order δ^6.) Thus the theory is certainly correct for small enough δ. But how small need δ be for the approximation to be useful? This is answered by a comparison with numerical solutions of the ordinary differential equations mentioned above, which were computed by Kopal [*37*] at the Massachusetts Institute of Technology, and which, it should be mentioned, agree almost perfectly with experimental results. The comparison is made, on the angle between the shock and the undisturbed Mach direction, given approximately in Eq. 6-20, in Fig. E,6e. It is seen that, although Eq. 6-20 predicts a rather spectacularly weak shock compared with that produced by a symmetrically placed *wedge* of semiangle δ (for which the corresponding formula is $\frac{1}{4}(\gamma + 1)M_\infty^2(M_\infty^2 - 1)^{-1}\delta$), nevertheless it overestimates the true shock strength as δ increases. The formula is satisfactory for $\delta < 10°$ (indeed for $\delta = 5°$ the curves would be indistinguishable) which is a rather larger range than for corresponding formulas in Art. 3 and 4. But the error is about 50 per cent for $\delta = 15°$. It must be stressed that, to a much greater extent for projectiles of revolution than for airfoils, *the range of practical shapes extends far outside the range in which any theories assuming small disturbances are useful;* in fact nose semiangles of 30° are common, and where this is so the flow can only be determined by numerical methods (Sec. H). On the other hand, future projectiles may be more slender, since the wave drag for given total volume varies (D,18) as the $\frac{8}{3}$ power of the fineness ratio. Notice that there is no

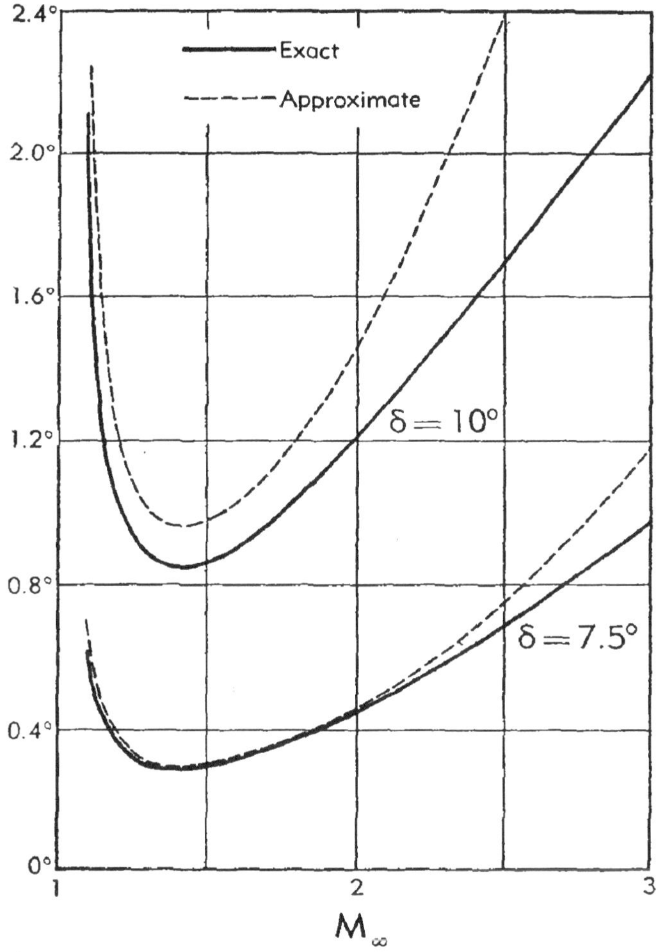

Fig. E,6e. Angle between shock and undisturbed Mach cone in axisymmetric flow at Mach number M_∞ past a solid cone of semiangle δ.

need to possess Eq. 6-20 and 6-21 for the cone, since accurate tables are available. Their essential purpose is to make possible the above comparison with exact adiabatic theory. Note also, as pointed out after Eq. 6-3, that the Whitham theory, even when inaccurate near the nose, is still correct outside a cylinder of sufficient radius.

Returning to the formula (6-17) for the distance from the axis at which the Mach line, emanating from the axis at a distance y from the nose, encounters the shock, notice that $F(y)$ becomes an increasingly larger multiple of $r^{-\frac{1}{4}}$, but that the coefficient cannot increase without limit because (Fig. E,6d) the maximum of the integral is

$$\int_0^{y_0} F(t)dt.$$

For very large r, $F(y)$ therefore becomes very small, and so $y \to y_0$. Thus all Mach lines ahead of the straight Mach line $y = y_0$ intersect the front shock, but the straight Mach line continues all the way to infinity. For very large r, with $y \cong y_0$, the integral in Eq. 6-17 has reached its maximum and combining Eq. 6-17 with Eq. 6-6,

$$x - (M_\infty^2 - 1)^{\frac{1}{2}} r \cong y_0 - \left[2k \int_0^{y_0} F(t)dt\right]^{\frac{1}{2}} r^{\frac{1}{4}} \tag{6-22}$$

Thus the front shock stands ahead of the straight Mach line $y = y_0$ by an amount asymptotically proportional to the *fourth* root of the distance from the axis (and, as indicated in the first sentence of this paragraph, by rather less than Eq. 6-22 gives for smaller r); this increases much less rapidly than the corresponding quantity for two-dimensional airfoil flow, which is proportional to the inverse *square* root of the distance (Art. 4).

The shock strength is deduced, either from the angle between the shock and the undisturbed Mach lines (obtained by differentiating Eq. 6-22), or from the expressions (6-13) for the disturbance velocities behind it, as

$$\frac{\Delta p}{p} \sim \gamma \left(\frac{2^{\frac{1}{2}}}{\gamma + 1}\right)^{\frac{1}{2}} (M_\infty^2 - 1)^{\frac{1}{8}} \left[\int_0^{y_0} F(t)dt\right]^{\frac{1}{2}} r^{-\frac{3}{4}} \tag{6-23}$$

for large r. For similar bodies, of different lengths l and fineness ratios ϵ, this expression for the shock strength at "miss-distance" r (from the path of the projectile) is proportional to $(r/l)^{-\frac{3}{4}}$ and to ϵ, while depending little on Mach number. It would be reasonable, if the shock strength at a large distance from a projectile not of revolution, and not even slender, had to be estimated, to hold to this result but with fineness ratio interpreted as (frontal area)$^{\frac{1}{2}}$/length. Using the curves in Fig. E,6c to estimate the integral in Eq. 6-23 in terms of frontal area and length, the rough order-of-magnitude formula for the shock strength far from the line of flight,

$$\frac{\Delta p}{p} \cong \epsilon \left(\frac{l}{r}\right)^{\frac{3}{4}} \tag{6-24}$$

would in such cases be used.

The asymptotic proportionality of the strength of ballistic shocks to the inverse $\frac{3}{4}$ power of the miss-distance was first observed experimentally, and also explained theoretically, by DuMond, Cohen, Panofsky, and Deeds [*39*]. For the bullets and shells used by them the $\frac{3}{4}$ power law was closely verified for miss-distances exceeding about 1,000 projectile diameters, while the rate of fall of strength with distance was rather more rapid closer to the projectile (as the above theory indicates). Two theoretical arguments explaining the law were given by these authors, of which the more convincing is an energy method which relates the rate of loss of wave energy between the shocks to the rate of gain of entropy of the fluid. It is very satisfactory that this gives the same conclusion as Whitham obtained by his quite different approach (which does not need to consider entropy variation); of course the reader should no longer regard such a fact as paradoxical, after the full discussion of the analogous two-dimensional problem in Art. 4.

Determination of the rear shock. We pass to the determination of the rear shock. This will not normally be attached, since the body-wake combination is not expected to have a discontinuity in slope concave to the stream. Rather the rear shock will originate in the midst of the flow, as a result of the running together of Mach lines, in the manner described in Art. 3. Photographs show much branching at the base of such shocks, as in the delta of a great river. This is incidental, being due to irregularities propagated along the Mach lines. The present theory applies to the form of the shock only where it becomes a recognizable single branch.

The flow is nonuniform both ahead of and behind the rear shock. If, at any point of it, the downstream-pointing Mach lines ahead of and behind it are those with the values y_1 and y_2 (respectively) for y, then, since the slope of the shock (to a first approximation) bisects that of these two Mach lines, Eq. 6-6 gives

$$\frac{d}{dr}[y - kF(y)r^{\frac{1}{2}}] = \frac{1}{2}\left[-\frac{1}{2}kF(y_1)r^{-\frac{1}{2}} - \frac{1}{2}kF(y_2)r^{-\frac{1}{2}}\right] \tag{6-25}$$

along the shock. Here the subscript on y on the left-hand side is immaterial because

$$y_1 - kF(y_1)r^{\frac{1}{2}} = y_2 - kF(y_2)r^{\frac{1}{2}} \tag{6-26}$$

since the two Mach lines meet on the shock.

Writing Eq. 6-26 as

$$r^{\frac{1}{2}} = \frac{y_2 - y_1}{k[F(y_2) - F(y_1)]} \tag{6-27}$$

and differentiating this, both r and dr may be eliminated from Eq. 6-25, giving an equation which may be put in the form

$$F(y_2)dy_2 - F(y_1)dy_1 = \tfrac{1}{2}d\{(y_2 - y_1)[F(y_1) + F(y_2)]\} \tag{6-28}$$

This form was ingeniously selected by Whitham to permit direct integration as

$$\int_{y_1}^{y_2} F(t)dt = \tfrac{1}{2}(y_2 - y_1)[F(y_1) + F(y_2)] \tag{6-29}$$

where no additive constant appears since no infinity in F can be supposed where $y_1 = y_2$ (i.e. where the shock originates).

Eq. 6-29 has a simple geometric interpretation which makes it easy, in practice, to evaluate the relation between Mach lines y_1 and y_2 which intersect on the rear shock. The left-hand side is the area under the curve $F(t)$ for $y_1 < t < y_2$. The right-hand side is the area under the straight segment joining the points y_1, $F(y_1)$ and y_2, $F(y_2)$. Hence for Eq. 6-29 to be satisfied, there must be parts of the graph of $F(y)$ lying both above and below the straight segment, and the two lobes which lie between the straight segment and the curve must have equal area. This is illustrated in Fig. E,6f, where the lobes are shaded. Further, by Eq. 6-27, the *slope* of the said straight segment is equal to $k^{-1}r^{-\frac{1}{2}}$, i.e. it is proportional to

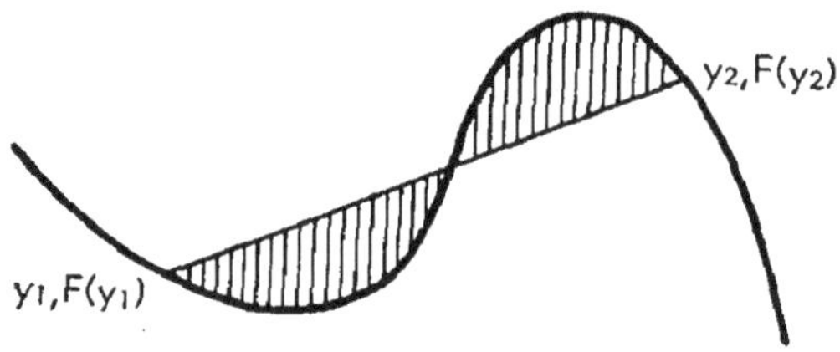

Fig. E,6f. The relation between Whitham's y_1 and y_2.

the inverse square root of the distance from the axis of the point on the shock where the two Mach lines meet. This shows that this slope must be positive; also, it gives the value of r at the corresponding point on the shock. From Eq. 6-6, the x coordinate follows, and, by varying the straight segment so that the lobes remain of equal area, the whole rear shock can be plotted.

A rear shock will *originate* on a Mach line specified by a value $y = y_i$ for which the slope of the graph of $F(y)$ is positive and locally a maximum, so that r is least (i.e. the point on the shock nearest the axis is being considered). When $F(y)$ is a smooth curve, y_i must be a point of inflection. This is illustrated in Fig. E,6g, where successive straight segments joining y_1, $F(y_1)$ to y_2, $F(y_2)$ and cutting off lobes of equal area are also shown, the function $F(y)$ being chosen as the simplest of those graphed in Fig. E,6d. As the slope decreases (so that we go farther out along the rear shock) it is clear that $y_1 \to y_0$ and $y_2 \to \infty$. Thus all the Mach lines behind the straight Mach line $y = y_0$ meet the rear shock: Those with $y_0 < y < y_i$ meet it in front, and those with $y_i < y$ meet it behind. The general properties of the shock system will be inferred from Fig. E,6g. Any slight alterations resulting from functions $F(y)$ of more complicated shape will be discussed later.

One may suppose that $F(y) \to 0$ as $y \to \infty$, if one takes the effective wake to be ultimately cylindrical. Then the behavior of the rear shock at very large distances from the axis, corresponding to y_1 near y_0 and large y_2, is given by approximating Eq. 6-27 and Eq. 6-29 as

$$r^{\frac{1}{2}} = \frac{y_2 - y_1}{-kF(y_1)}, \qquad \int_{y_0}^{\infty} F(t)dt = \frac{1}{2}(y_2 - y_1)F(y_1) \tag{6-30}$$

Substituting for $y_2 - y_1$ in the first equation from the second, and taking the positive square root (noticing from Fig. E,6d that $F(y_1) < 0$) we get

$$-F(y_1) = \left[\frac{-2\int_{y_0}^{\infty} F(t)dt}{kr^{\frac{1}{2}}}\right]^{\frac{1}{2}} \tag{6-31}$$

whence by Eq. 6-6 the asymptotic form of the rear shock is

$$x - (M_\infty^2 - 1)^{\frac{1}{2}}r \cong y_0 + [-2k\int_{y_0}^{\infty} F(t)dt]^{\frac{1}{2}}r^{\frac{1}{4}} \tag{6-32}$$

This asymptotic form should be compared with that of the front shock Eq. 6-22. It is seen at once that the rear shock stands behind the

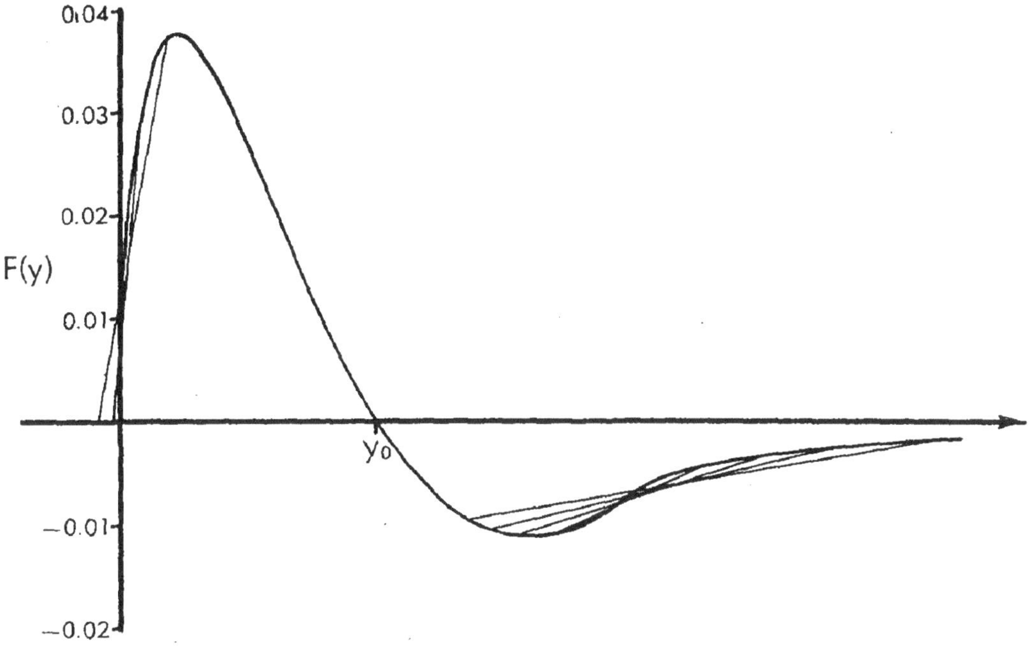

Fig. E,6g. Geometrical construction for determining shocks on projectiles.

straight Mach line $y = y_0$ by an amount proportional to $r^{\frac{1}{4}}$, just as the front shock so stands ahead of it. But, in fact, *the constants of proportionality are the same* in Eq. 6-22 and 6-32. This is because it may be shown that, because $F(y)$ as defined in Eq. 6-7 is a sort of derivative, the integral

$$\int_0^{\infty} F(t)dt$$

is zero (assuming that $f(0) = 0$ and $f = O(y^{-1})$ as $y \to \infty$). Thus at very large distances the front and rear shocks are of equal strength. Now the experiments [*39*] support this conclusion even at miss-distances of about 100 projectile diameters, for which the $-\frac{3}{4}$ power law has not yet come into operation. That this should be so is explained by evaluating the shock paths to the next approximation at large miss-distances, on which they are

$$x - (M_\infty^2 - 1)^{\frac{1}{2}} r \cong y_0 \mp \left[2k \int_0^{y_0} F(t)dt \right]^{\frac{1}{2}} \left[r^{\frac{1}{4}} + \frac{r^{-\frac{1}{4}}}{2kF'(y_0)} \right] \qquad (6\text{-}33)$$

showing that, as observed, the front and rear shock strengths should equalize out before they begin to satisfy the $-\frac{3}{4}$ power law accurately. The *length* of the N wave is, at large distances from the axis, twice the second term in Eq. 6-33. The experiments confirmed this by showing that the slope of length against distance on logarithmic scales is close to 0.25, but always less if anything.

It may be remarked that the arguments used to determine the rear shock apply equally to the front shock, for which, however, the Mach lines ahead are straight, so that effectively $F(y) = 0$. It is still true that straight segments cutting off lobes of equal area (see the left portion of Fig. E,6g) join values of y corresponding to Mach lines meeting at a point on the shock, whose distance from the axis is proportional to the inverse square of the slope of the segment. It is because $F = 0$ for the left-hand value of y that this geometrical relation between r and y has for the front shock a relatively simple analytical form (6-17). But in practice the geometrical relation may be easier for purposes of calculation.

We next consider the pressure variations measured by a suitable instrument as a supersonic projectile passes, missing it by a distance r large compared with the length of the projectile. By Eq. 6-13 the pressure is given by

$$p - p_\infty = \rho_\infty q_\infty^2 2^{-\frac{1}{2}} (M_\infty^2 - 1)^{-\frac{1}{4}} F(y) r^{-\frac{1}{2}} \qquad (6\text{-}34)$$

and it is the variation of this with x for constant r in which we are interested, where x and y are related by Eq. 6-6. It is seen at once that the pressure gradient $\partial p/\partial x$ (which when multiplied by q_∞ gives the time rate of change of the observed pressure) is

$$\frac{\partial p}{\partial x} = \rho_\infty q_\infty^2 2^{-\frac{1}{2}} (M_\infty^2 - 1)^{-\frac{1}{4}} \frac{F'(y) r^{-\frac{1}{2}}}{1 - kF'(y) r^{\frac{1}{2}}} \qquad (6\text{-}35)$$

The denominator in Eq. 6-35 marks a difference between the linear and Whitham theories by expressing the change in pressure gradient due to the spreading out or bunching in of the Mach lines. It never becomes negative (this would involve encountering a "limit line") as shocks always appear before this can happen.

Now, between the shocks at large distances from the axis, y is close to y_0, and so the pressure gradient (6-35) is practically constant, giving an "N wave" form of pressure curve (Fig. E,4a). Further, when r is much larger than $[kF'(y_0)]^{-2}$, only the second term in the denominator matters. When this is so, Eq. 6-35 becomes (using the expression (6-7) for k)

$$\frac{\partial p}{\partial x} = -\rho_\infty q_\infty^2 \frac{(M_\infty^2 - 1)^{\frac{1}{2}}}{(\gamma + 1) M_\infty^4} \frac{1}{r} \qquad (6\text{-}36)$$

This ultimate slope of the N is independent of body shape and inversely proportional to miss-distance. The result is very similar to that obtaining in two-dimensional airfoil theory, for which $\partial p/\partial x$ is exactly twice as much (as follows from Eq. 4-13); actually it is the one point for which there is such quantitative similarity. Eq. 6-36 means that at 15°C an observer will record a time rate of pressure fall (after the initial jump at the shock) equal to about

$$0.2\,\frac{(M_\infty^2 - 1)^{\frac{1}{2}}}{M_\infty}\quad \frac{\text{atmospheres per millisecond}}{\text{miss-distance in meters}} \tag{6-37}$$

In Eq. 6-37 the second factor is not far from 1 if the projectile speed is well supersonic. Hence an instrument that could measure this time rate of pressure fall between the shocks would give at once the miss-distance of a supersonic projectile to a good approximation even if its Mach number were not known. (Notice that Eq. 6-37 is true independently of any approximation for slender bodies, provided that the miss-distance is large enough, since the function $F(y)$ does not appear therein. Notice also that, since Eq. 6-36 is asymptotically true both for projectiles of revolution and, within a factor of 2, for the very extreme case of two-dimensional flow, it may be expected to be true to good approximation for projectiles of any shape.)

We pass to the consideration of the pressure behind the rear shock. Here $F'(y)$ is always considerably smaller than $k^{-1}r^{-\frac{1}{2}}$, since the geometry of Fig. E,6f shows that the straight segments (of slope $k^{-1}r^{-\frac{1}{2}}$) always intersect the curve from below. Hence the argument leading to Eq. 6-36 is completely falsified. In fact the pressure tails off rapidly behind the rear shock. By considering Eq. 6-7 for large y one sees that

$$F(y) \sim -f_0 y^{-\frac{1}{2}}, \quad \text{where } f_0 = \tfrac{1}{2}\int_0^\infty f(t)dt \tag{6-38}$$

For slender bodies f_0 is the effective cross-sectional area of the wake, divided by 4π. Using the approximation (6-38) one may calculate the pressure immediately behind the rear shock, from Eq. 6-34 and 6-30, as

$$p - p_\infty = -\rho_\infty q_\infty^2 2^{-\frac{1}{2}}(M^2 - 1)^{-\frac{1}{4}} f_0 \left[\frac{2}{k}\int_0^{y_0} F(t)dt\right]^{-\frac{1}{4}} r^{-\frac{7}{8}} \tag{6-39}$$

for large r. This is essentially negative, and of a slightly smaller order of magnitude than the value ($\sim r^{-\frac{3}{4}}$) immediately ahead of the shock. Also, since, for constant r, $dx/dy = 1 - kF'(y)r^{\frac{1}{2}}$ does not, as remarked above, depart greatly from 1, the subsequent behavior of the pressure as x increases is similar to its behavior as y increases, namely, a subsidence to zero from below, in proportionality (6-38) to $y^{-\frac{1}{2}}$.

To illustrate the great difference between the flow pattern near the body and that at large distances, the pressure is graphed against x for two fixed values of r in Fig. E,6h. The lower curve corresponds to a value of r soon after the appearance of the rear shock, which is still much weaker than the front shock. There is still a region of compression behind the front shock, and a sizeable tail behind the rear shock (indeed the predictions of linear theory are still very roughly correct in this region). The upper curve corresponds to a very large value of r, for which the asymptotic formulas are valid. The N wave dominates this pressure curve, and the tail is reduced in relation to it.

Additional shocks. We must now consider what modifications to the above theory are necessary when, as usually in practice (Fig. E,6d), the graph of $F(y)$ is more complicated than that used for illustration in

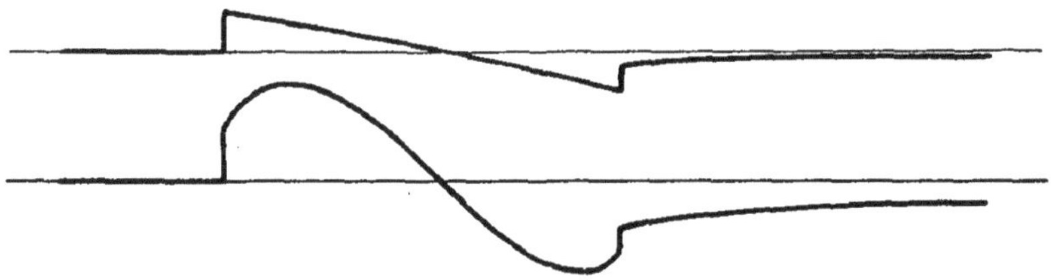

Fig. E,6h. Pressure signatures, far from and near a projectile, with time scales adjusted to make shocks coincide.

Fig. E,6g. First consider the effect of a "kink" in the body surface, producing a kink in the graph of $F(y)$ (Fig. E,6i) so that it has two points of inflection with positive slope. Then a shock will start from points corresponding to each of the two points of inflection. A typical point on the main rear shock will be represented by the straight segment BC. At first sight it would appear that the Mach line corresponding to B also meets the kink-produced shock, at a point represented by the straight segment AB. But the value of r corresponding to AB exceeds that corresponding to BC, and so there is no real encounter of the Mach line with the kink-produced shock before the former is cut off by the main rear shock. On the other hand the Mach line corresponding to E is cut off by the kink-produced shock, at a point represented by DE, before it can meet the rear shock at a point represented by EF. Between E and B there must be a point Y which is on a complete straight segment XYZ, such that both XY and YZ cut off lobes of equal area. Then both XY and YZ correspond to the same point, which is clearly where the two shocks run together. Above this they unite to form a single shock, whose progress may be followed by the use of single straight segments X_1Z_1 (with X_1 to the left of X, Z_1 to the right of Z) such that the total area of curve on both sides is the same. (However, it should be remarked that one can also draw kinks which produce no shock, because the rear shock cuts off *all* the associated Mach lines before they can run together; this means geometrically that no line such as DEF exists.)

Next consider a discontinuous function $F(y)$, such as the third curve in Fig. E,6d, corresponding to a shell with (somewhat idealized) wake thinning. The straight Mach line $y = y_0$, $F(y) = 0$, is seen to originate in the middle of the fan emanating from the shoulder. Behind this there are three places where the slope of the curve has locally a positive maximum. Of these the most important, connected with the origin of the main rear shock, is the third discontinuity, where the slope of the curve is $+\infty$. Obviously a series of straight segments can be drawn, crossing this vertical line and cutting off lobes of the curve with equal area on both sides, with slopes from $+\infty$ downward. They correspond to a shock,

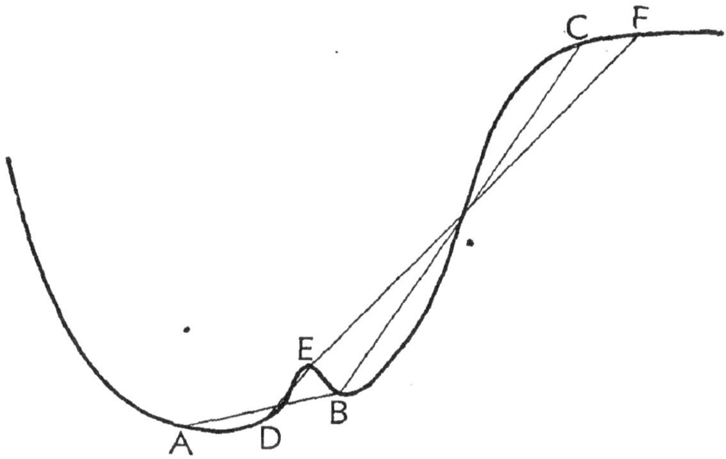

Fig. E,6i. Construction for determining bifurcated rear shock.

attached to the wake at the point where it straightens out. (If a wake shape which straightened out more gradually had been used, then $F(y)$ would have a steep but not vertical maximum of slope in this neighborhood and the shock would then be detached.) There is little reason to doubt the accuracy of this prediction of an attached shock in a corner, since the imaginary fan of Mach lines which it replaced was predicted correctly (see above).

But immediately behind each of the other discontinuities in $F(y)$ (where the slope of the curve is $-\infty$) there is a local positive maximum of slope which may correspond to a shock forming on the rear Mach line of the Prandtl-Meyer fan associated with the expansion around shoulder and rim respectively. In the case of the shoulder such a shock obviously forms, though with a value about 0.3 for the slope $k^{-1}r^{-\frac{1}{2}}$, which for $M_\infty = 2$ ($k = 12$) corresponds to a distance r from the axis of 7.7 projectile lengths. It gradually runs into the rear shock, exactly as discussed above for kink-produced shocks. But actually no shock forms on the rear Mach line emanating from the rim, because it is cut off by the rear shock first; careful measurement shows in fact that no such line as *DEF* (Fig. E,6i) can exist here.

The front part of the complete flow pattern in a meridian plane, as calculated by Whitham using his geometrical method, is illustrated in Fig. E,6j. The shocks are shown as heavy lines, and certain Mach lines are also shown, namely, in front and rear of each fan and also the straight Mach line in the middle of the forward fan.[4] The general pattern is of the form familiar from bullet photographs. Actually the strength of the front shock must be overestimated (see Fig. E,6d; actually the nose semi-angle is about 20°), but if a reliable estimate were required for practical

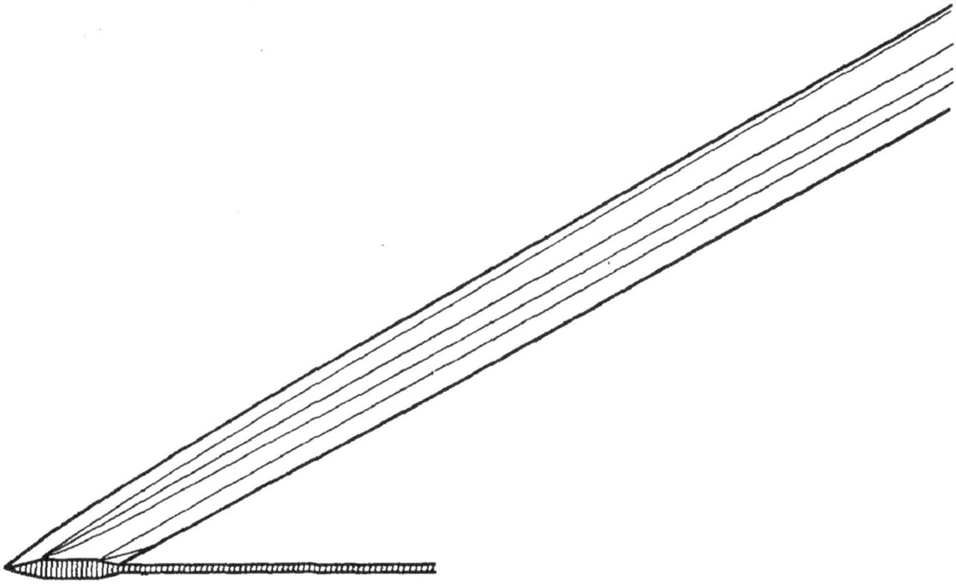

Fig. E,6j. Theoretically determined pattern of flow past a shell.

purposes, this error could be allowed for by adjusting $F(y)$ near the nose so that the deduced nose shock strength was as given by the numerical calculations for the cone. Similarly, accuracy could be improved near the base of the rear shock, making it detached by giving $F(y)$ a rapid but not discontinuous rise where the wake straightens out.

Drag related to downstream flow. Finally, the relation between the wave drag of the projectile and the flow behind it will be considered briefly, eking out, from analogies with the two-dimensional theory (Art. 4), our less precise knowledge of the flow. First we relate the drag to the flux of wave energy (or, what is essentially the same, to the flux of additional longitudinal momentum) across a plane far downstream and perpendicular to the axis. As in Art. 4, the wave energy of the N wave (being proportional to the wave length times the radius, i.e. to $r^{5/4}$, multiplied by the square of the amplitude, i.e. by $r^{-3/2}$) tends to zero far downstream (like $r^{-1/4}$), and the same is true, but more so (like $r^{-1/2}$), of the tail.

[4] The shock which forms on the rear Mach line of the forward fan is just off the picture; this hardly matters since it is not a phenomenon of any consequence.

Only the term Ts in the wave energy contributes, in the limit, to the total flux, and we have (cf. Eq. 4-43)

$$\text{Drag} = \int_0^\infty \rho_\infty Ts\, 2\pi r dr \tag{6-40}$$

where only the square of the specific entropy is neglected.

Since specific entropy is conserved for any fluid element, except where it increases at the shocks by an amount related to the deflection by Eq. 3-20, the integral in Eq. 6-40 may be evaluated to a first approximation by taking it over both shocks, with s signifying the *gain* in specific entropy across each. Thus

$$\begin{aligned} \text{Drag} &= \int_{\text{both shocks}} \rho_\infty T c_p \frac{\gamma^2 - 1}{12} \frac{M_\infty^6}{(M_\infty^2 - 1)^{\frac{3}{2}}} (\Delta v)^3 2\pi r dr \\ &= \rho_\infty q_\infty^2 \frac{\gamma + 1}{12} \frac{M_\infty^4}{(M_\infty^2 - 1)^{\frac{3}{2}}} 2\pi \left\{ \int_{\text{front shock}} 2^{-\frac{3}{2}} (M_\infty^2 - 1)^{\frac{3}{4}} \frac{F^3(y)}{r^{\frac{3}{2}}} r dr \right. \\ &\quad \left. + \int_{\text{rear shock}} 2^{-\frac{3}{2}} (M_\infty^2 - 1)^{\frac{3}{4}} \frac{[F(y_2) - F(y_1)]^3}{r^{\frac{3}{2}}} r dr \right\} \end{aligned} \tag{6-41}$$

where the nondimensional velocity v has been substituted from Eq. 6-13. By using Eq. 6-17, and the expression (6-7) for k, the contribution from the front shock is easily evaluated as

$$\rho_\infty q_\infty^2 \pi \int_0^{y_0} F^2(y) dy \tag{6-42}$$

By using Eq. 6-27, and also (in a sufficiently cunning manner) Eq. 6-28, the contribution from the rear shock can actually be put into the same form, but with the limits y_0 and ∞. Hence, combining these,

$$\text{Drag} = \rho_\infty q_\infty^2 \pi \int_0^\infty F^2(y) dy \tag{6-43}$$

If $F(y)$ is substituted from Eq. 6-7, Eq. 6-43 reduces to Kármán's familiar form (D, 18; Eq. 18-9)

$$\rho_\infty q_\infty^2 \pi \int_0^\infty \int_0^\infty f'(x) f'(y) \ln \frac{1}{|x - y|} dx dy \tag{6-44}$$

This is yet another check on the correctness of Whitham's theory.

Other satisfactory points are (1) the simple form of the drag when expressed in terms of the function $F(y)$ which is fundamental in the theory, and (2) the fact that the part of the integral (6-43) shown in Eq. 6-42 corresponds to drag associated with degradation of energy at the front shock, while the rest is similarly associated with the rear shock.

When the function $F(y)$ has been defined in terms of a distribution of cross-sectional area which decreases in the wake to an ultimate value less than the base area of the projectile, the drag (6-43) includes more than what is *normally* called the wave drag, namely, the resultant of the normal forces on the head associated with the difference of the pressure from its undisturbed value. For it includes, in addition, the suction forces on the thinning wake. The resultant of these forces is clearly that part of the base drag which acts outside the cylinder enclosing the ultimate wake. The remainder of the base drag, which may reasonably be called "form drag," and also the skin friction drag, are associated with flux of reduced longitudinal momentum in the turbulent wake itself, due essentially to *shear* effects. It would seem reasonable to use "wave drag" (except where there is danger of confusion with established usage) to signify the drag due to the appearance of the shocks (as a result of *compression* effects), a drag whose influence on the fluid behind the projectile is spread out very widely indeed and correspondingly thinly.

Of course the pressure drag on the head *can* be expressed in the form (6-42), provided only that $F(y)$ signifies its value for the same projectile with *unthinned* cylindrical wake. It is an interesting consequence that the difference between the expressions (6-42) for thinned and unthinned wake is a known proportion of the base drag. Since the base drag can be calculated independently from the angle through which the main stream turns at the rim (which must already have been used in calculating $F(y)$ for the shell with thinned wake), it appears that a check is here possible which may in further developments of the theory permit still more accurate values of $F(y)$, and so of the flow pattern to be inferred.

If the drag is compared with the wave energy flux across a plane not so far downstream, the contribution from entropy is smaller because the whole of either shock does not lie ahead of the plane. The deficiency is made up from the wave energy in the N wave and tail. Since, as remarked above, this is asymptotically proportional to $r^{-1/4}$, it falls off at a rate proportional to $r^{-5/4}$. This is balanced by the rate of entropy gain (proportional to the cube of the shock strength, and so to $r^{-9/4}$, but multiplied by $2\pi r$ to allow for axisymmetry). The details are omitted.

Some final remarks are needed on the mass flow behind the projectile, since (Art. 4) positive work is done by the pressure of the atmosphere in increasing this anywhere, and negative work in decreasing it. It is easiest to calculate the first approximation to the mass flow in the N wave and tail across a *cylinder* of radius r, coaxial with the projectile. This, by the expression (6-13) for v, is proportional to the integral of $F(y)$ with respect to x for constant r, taken between the shocks and behind the rear shock. By Eq. 6-6 this is

$$\int F(y)[1 - kF'(y)r^{1/4}]dy \tag{6-45}$$

Now Eq. 6-45 would be zero if the limits were 0 and ∞. Hence it may be replaced, with the sign changed, by the same integral taken over the intervals previously excluded from the range of integration, i.e. from zero to the value of y on the front shock, and from y_1 to y_2 in the notation used for the rear shock. That the integral over each of these intervals vanishes follows trivially from Eq. 6-17 for the front shock and Eq. 6-27 and 6-29 for the rear shock.

To the first order then, the mass flow is due to Whitham's approximate velocity field (6-13) and its total is zero. Note that to achieve this for large r it is essential that the N wave becomes balanced. Indeed the mass outflow through a cylinder of very large radius r, in the front half of the N wave, due to a disturbance velocity v proportional to $r^{-\frac{3}{4}}$ acting across an area proportional to $r \cdot r^{\frac{1}{4}}$, is *increasingly large* as the cylinder increases in radius. It is balanced however by the mass inflow in the rear half. (There is nothing extraordinary in this, which is equally true of any half wavelength in a harmonic train of spherical sound waves.) We have no second approximation to the flow, but we can be sure that, as in the two-dimensional case, there is, to a second approximation, a net mass outflow far from the axis to balance the reduced mass outflow behind the projectile due to increased entropy. One may also expect that pressure waves produced by reflection from the front shock and transmission through the rear shock play a part in this balance.

E,7. Supersonic Projectile Theory: Surface Pressures. In this article, as foreshadowed in Art. 6, we discuss fully the errors in the linear theory of the surface pressures on a supersonic projectile, and briefly describe the methods of Broderick [*34*] and Van Dyke [*54*] for proceeding to a second approximation for the pressure on projectiles of revolution in axisymmetric flow, and the author's analogous work [*40*] on the effect of yaw.

In the early work on the small perturbation theory of supersonic flow, it was gradually realized that, in its application to axisymmetric flow past projectiles, there were two peculiar features, as follows: (1) The axial and radial disturbance velocities u and v, given by

$$u = -\int_0^{x-(M_\infty-1)^{\frac{1}{2}}r} \frac{f'(t)dt}{[(x-t)^2-(M_\infty^2-1)r^2]^{\frac{1}{2}}}$$
$$v = \frac{1}{r}\int_0^{x-(M_\infty-1)^{\frac{1}{2}}r} \frac{(x-t)f'(t)dt}{[(x-t)^2-(M_\infty^2-1)r^2]^{\frac{1}{2}}} \tag{7-1}$$

(cf. Eq. 6-3), are *not of the same order of magnitude* near the projectile surface. In fact Eq. 7-1 shows that $-v/u$ is a sort of average of $(x-t)/r$ for $0 < t < x - (M_\infty^2 - 1)^{\frac{1}{2}}r$; this average would be expected to be large on the surface of a slender projectile. Indeed the square of v is normally comparable with u near the surface, and hence the question arises *whether the use of a linear theory is justifiable, since it predicts u by neglecting* v^2. (2) While on the one hand the calculation of the fundamental function

$f(t)$, by satisfying exactly the boundary condition at the surface, is a very lengthy process indeed, it is not at all obvious under what conditions the approximation $f(t) = S'(t)/2\pi$, which is obtained from the boundary condition at the surface by neglecting consistently the squares of the disturbances, is justified.

These two questions were tackled by the author [*42*] in 1945, by working out the order of magnitude of the next approximation to the flow (which Broderick and Van Dyke were later to calculate explicitly). First he showed that the errors in predictions of u and v near the surface by use of the linear equation of motion take the form of a factor differing from unity by not more than the order of magnitude of u itself. (Work equivalent to this investigation will be presented below.) But, as a consequence, the surface pressure distribution must be calculated from the equation

$$C_p = -2u - v^2 \tag{7-2}$$

not the equation $C_p = -2u$ which is usual in the linear theory, because as stated above v^2 is comparable with u in magnitude near the surface.

Secondly, the author showed that the approximation $f(t) = S'(t)/2\pi$ will not introduce errors of larger order of magnitude than those already implicit in the linear equation of motion, provided that the body is *smooth* as well as slender.[5] Thus not only must the slope of the surface to the incident stream be small, but changes in it must be spread evenly over the length of the projectile, so that the curvature, too, is consistently small; and similarly the curvature must vary only gradually (actually, this condition is not essential if the pressure distribution is only required for calculating the drag). Under these rather stringent conditions of smoothness the boundary condition $v/(1 + u) = R'(x)$ on the surface $r = R(x)$, which may be approximated as $v = R'(x)$ since factors like $1 + u$ have already been neglected, becomes by a further approximation

$$f(x) = R(x)R'(x) = \frac{S'(x)}{2\pi} \tag{7-3}$$

This is because the expansion of v (see Eq. 7-1) for small r is

$$v = \frac{f(x)}{r} + \frac{1}{2}(M_\infty^2 - 1)r \int_0^x \left[\ln \frac{(M_\infty^2 - 1)^{\frac{1}{2}} r}{2(x - t)} - \frac{1}{2}\right] df''(t) + \cdots \tag{7-4}$$

which indicates that the approximation (7-3) will be tolerable if on the surface $r^2 f''(x)$ is small compared with $f(x)$, which is clearly the case under the conditions stated. (But notice that the error increases with Mach number.) The expansion (7-4) is most simply obtained from Heaviside's operational calculus, in which v is $pK_1(pr)f(x)$, if p is the operator

[5] The nomenclature here introduced, which divides the class of "slender" bodies according to whether or not they are "smooth," is more suggestive than the one common at present, which takes "slender" to imply "smooth."

$(M_\infty^2 - 1)^{\frac{1}{2}}\partial/\partial x$ and K_1 is a modified Bessel function of the second kind. Eq. 7-4 then corresponds to the expansion of the Bessel function for small pr [*44*, p. 80] with the operational forms interpreted.

It is important to notice that, since for smooth bodies the approximation (7-3) to the axial distribution of source strength $f(x)$ has a proportional error of order the *square* of the fineness ratio, this source strength is best taken as the value of $S'(x)/2\pi$ at the cross section *through* the source itself, *not* as the value of $S'(x)/2\pi$ at the cross section farther downstream where the influence of the source first emerges into the flow (which would clearly differ from it by a proportional error of order the first power of the fineness ratio).

By expanding u as v was expanded in Eq. 7-4, and using Eq. 7-2 and 7-3, we obtain the first order approximation to the pressure distribution on smooth slender pointed projectiles in axisymmetric flow in the form

$$C_p = \frac{1}{\pi} S''(x) \ln \frac{2x}{(M_\infty^2 - 1)^{\frac{1}{2}} R(x)} - \frac{1}{\pi} \int_0^x \frac{S''(x) - S''(t)}{x - t} dt - R'^2(x) \quad (7\text{-}5)$$

The second and third terms are of order ϵ^2 (where ϵ is the fineness ratio). The mathematical order of magnitude of the first is strictly $\epsilon^2 \ln \epsilon^{-1}$; in practice, however, this is not very different from ϵ^2.

Flow past slender bodies which are not smooth. When, as in many practical problems, the projectile surface is not smooth, the function $f(t)$ must be found by some other means, of which the numerical solution of the boundary condition as an integral equation, which Kármán and Moore originally used [*41*], is a possibility. Another method is that introduced by the present author [*35*] in 1948. He showed that, as a result of a discontinuity in surface slope at $x = x_1$, where $R(x) = R_1$, by an amount $\Delta R'$, the pressure coefficient deduced from linear theory, with $f'(t) = S''(t)/2\pi$, and no account taken of the discontinuity, must be augmented by an amount

$$\frac{2(\Delta R')}{(M_\infty^2 - 1)^{\frac{1}{2}}} U\left(\frac{x - x_1}{(M_\infty^2 - 1)^{\frac{1}{2}} R_1}\right) \quad (7\text{-}6)$$

where $U(z)$ is a function which is zero for $z < 0$, unity for $z = 0$, and falls thence asymptotically to zero as $z \to \infty$, as shown in Fig. E,7a (where the function z^{-1} is shown as a broken line for comparison). This fact enables any regions of large rate of change of slope or curvature to be treated by approximating to them by means of a number of small changes of slope. Indeed such an approach can be applied all along the surface, and the resulting sum allowed to approach, in the limit of a very large number of very small changes of slope, an integral. This gives

$$C_p = \frac{1}{\pi}\int_0^x U\left(\frac{x-t}{(M_\infty^2-1)^{\frac{1}{2}}R(t)}\right)\frac{dS'(t)}{(M_\infty^2-1)^{\frac{1}{2}}R(t)} - R'^2(x) \qquad (7\text{-}7)$$

as an approximate solution of Kármán's integral equation which is equally valid for any slender body, whether smooth or not. It is also correct for the external flow about any slender ducted body of revolution, provided that the integral is understood to include a contribution of the

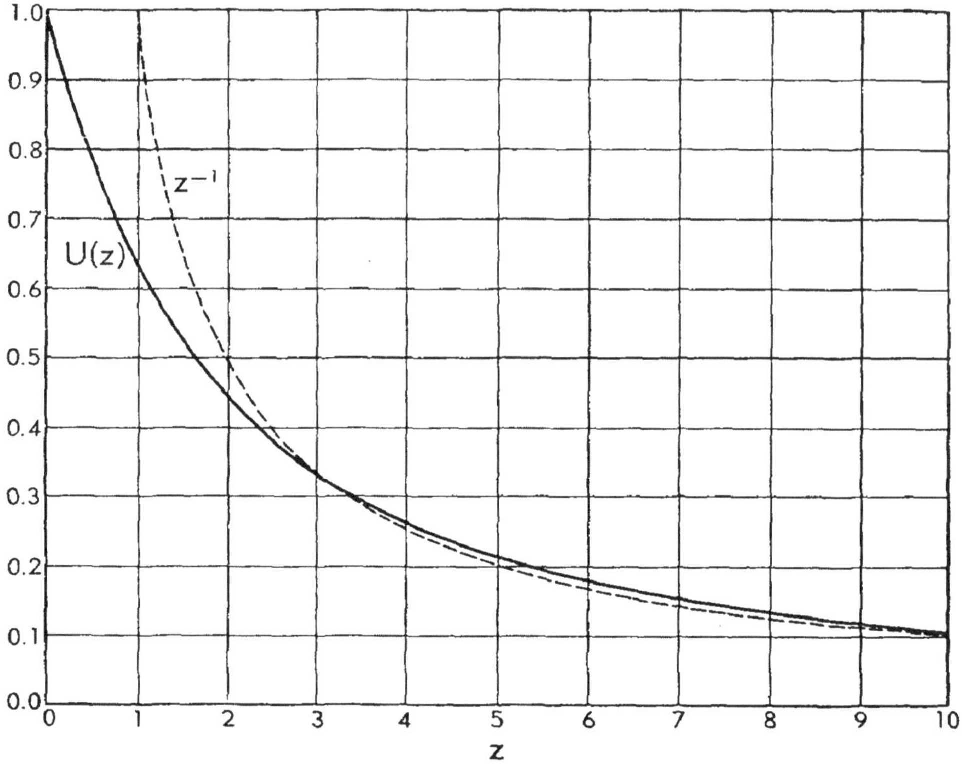

Fig. E,7a. The function $U(z)$.

form of Eq. 7-6, not only from places where the slope is discontinuous, but also from the lip, if there $S'(x) \neq 0$. This extreme versatility makes the formula (7-7) very useful. It is relatively easy to compute numerically, because the integrand is without singularities.

A brief explanation of the theory just stated will now be given. $U(z)$ is calculated as the function such that, if Eq. 7-1 are regarded as defining *for some constant value of* r (by elimination of f) u in terms of v, then

$$u(x, r) = -\int_0^x U\left(\frac{x-t}{(M_\infty^2-1)^{\frac{1}{2}}r}\right)\frac{dv(t, r)}{(M_\infty^2-1)^{\frac{1}{2}}} \qquad (7\text{-}8)$$

This function $U(z)$ was computed by the British Admiralty [42] by solving the integral equations when v is a unit function. But of course the integral equations (7-1) must in fact be solved not at constant r, but with r related to x by the equation $r = R(x)$ of the body surface. However, if the formula (7-8) be rewritten

$$u = -\int_0^x U\left(\frac{x - t}{(M_\infty^2 - 1)^{\frac{1}{2}}R(t)}\right)\frac{dS'(t)}{2\pi R(t)(M_\infty^2 - 1)^{\frac{1}{2}}} \tag{7-9}$$

two points are clear concerning the predicted contribution to u from any small change in rv ($= S'(t)/2\pi$ on the boundary): (1) It is approximately correct for small $x - t$, such that $R(x)$ has changed little from $R(t)$ (i.e. in this region r is approximately constant), by the definition of $U(z)$. (2) It is also approximately correct when $z = (x - t)/(M_\infty^2 - 1)^{\frac{1}{2}}R(t)$ is large enough for the approximation $U(z) \cong z^{-1}$ to be adequate (see Fig. E,7a for this), for then the addition to u due to the small change $dS'(t)$ has become, by Eq. 7-9,

$$-\frac{1}{x - t}\frac{dS'(t)}{2\pi} \tag{7-10}$$

This agrees with the first of Eq. 7-1 if $f(t)$ has adjusted itself fairly rapidly to the change in $S'(t)/2\pi$ at t, and that this must be so is clear from the second equation, as approximated, when $x - t$ is large compared with $(M_\infty^2 - 1)^{\frac{1}{2}}r$, by putting the change in f equal to the change in $rv = S'(x)/2\pi$.

Now these two regions will practically overlap for slender bodies.[6] Hence Eq. 7-9 is true not (as the author supposed [42] in 1945) only for ducted bodies of nearly constant cross section, but also for quite arbitrary slender bodies. Putting the reasoning of the last paragraph into words, this is because, though the pressure change due to a given change in slope is calculated by a formula deduced by assuming the cross-sectional area thereafter approximately constant, the said pressure change is actually *independent* of cross-sectional area sufficiently far downstream for the cross-sectional area to have altered significantly.

The formula (7-7) for C_p now follows at once from Eq. 7-9 and 7-2. The fact that for smooth bodies it agrees with Eq. 7-5, to within the order of approximation on which both are based, can be deduced without too much difficulty from the fact that

$$\int_0^x U(z)dz \sim \ln(2x) \text{ as } x \to \infty$$

Calculated examples are compared with experiment in Fig. E,7b. The upper figure represents the pressure distribution on a projectile consisting of a conical head of 10° semiangle, followed by a cylindrical

[6] Since the latter may be regarded as starting at $z = 2$, they overlap provided that $R(t + 2(M_\infty^2 - 1)^{\frac{1}{2}}R(t))$ is close to $R(t)$, which is true if $2(M_\infty^2 - 1)^{\frac{1}{2}}R'(t)$ is small. But even if they do not quite overlap, the approximation can hardly diverge far from the truth between them.

portion. In unpublished wind tunnel experiments by Cope at the NPL the measured pressures were clustered closely about the plain curve. The initial position of this curve is as calculated by Kopal [37] on exact adiabatic theory. The slight falls in pressure before the corner and

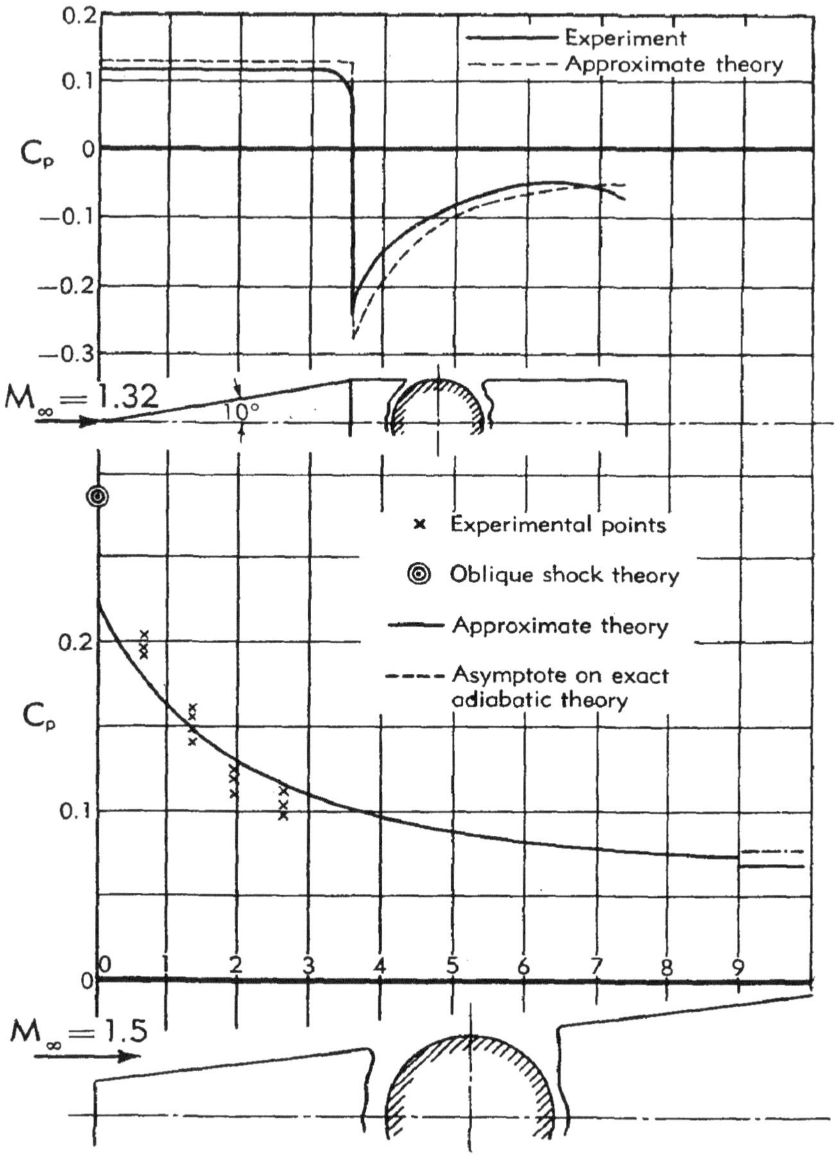

Fig. E,7b. Surface pressures in two axisymmetric flows.

before the base are boundary layer effects of familiar type. The discontinuity in the plain curve at the shoulder was calculated from the theory of the Prandtl-Meyer expansion, since the distance of the first experimental point behind it was as much as one third of the projectile radius. The dashed line is calculated on the author's theory and is in reasonable agreement. The lower figure represents the pressure distribution on the outside of a duct of conical shape, with semiangle 7.5°. The curve represents Eq. 7-7, calculated with $S(x) = \pi(1 + x \tan 7.5°)^2$ (measuring distances in multiples of the radius of the mouth). The value at the lip, $2 \tan 7.5°/\sqrt{M_\infty^2 - 1} - \tan^2 7.5°$, agrees to the first order with Ackeret's prediction of the pressure due to two-dimensional turning of the stream, but is of course an underestimate of the true pressure coefficient, which by oblique shock theory (Fig. E,3f) is as given by the double-ringed point. The crosses represent NPL measurements. The asymptote of the curve is represented by a plain line. The true asymptote of the pressure distribution, at least if boundary layer be neglected, is presumably the value for a solid cone, shown as a dashed line. In this lower figure the approximate theory is reasonably good except near the lip. Here the flow is nearly two-dimensional, which causes the disturbance pressure to be higher, and hence the error (of order its square) to be correspondingly greater. (The comparable, though smaller, error in the sudden fall of pressure in the upper figure is shared equally between the pressures ahead and behind, and so reduced in importance.) This shows how useful it would be if the theory could be improved so as to be uniformly correct to order ϵ^2, even in regions of pseudo-two-dimensional flow (where this would necessitate taking into account the nonlinearity of the equation of motion), so as to give (for example) the Busemann value (Art. 3) for the pressure change at a discontinuity of slope.

The author [*35*] was more concerned with total pressure drag than with pressure distribution. He showed that, of discontinuities, only those in slope affect the pressure drag significantly, and gave a fairly simple formula for evaluating it. (A discontinuity in *curvature* certainly alters the pressure distribution from that predicted by smooth body theory, but not so as to affect the drag to the order of approximation used.)

It should be emphasized in passing that no theory nearly so complete applies to the *internal* flow in slender ducts. Here, obviously, the linear theory applies only if the radius remains nearly constant; but even when this is so its predictions are difficult to interpret. Ward [*45*] showed that there are logarithmic infinities propagated along certain Mach lines in the linear theory, and it is still not understood what really happens in their neighborhood. (The theory in Lighthill's 1945 paper is altogether wrong.)

The theory of flow past bodies of revolution which are not smooth has been extended to a second approximation by Van Dyke [*54*], whose work therefore supersedes in point of accuracy Broderick's second order theory for smooth bodies [*34*], referred to below. Only a simplified account of Van Dyke's theory can be given here. If the perturbation potential is

$\varphi_1 + \varphi_2$ to a second approximation, where φ_1 is the first approximation, then φ_2 must be considered to satisfy not the ordinary Hantzsche-Wendt equation deduced from Eq. 2-14 but the more accurate equation

$$\frac{\partial^2 \varphi_2}{\partial r^2} + \frac{1}{r}\frac{\partial \varphi_2}{\partial r} - (M_\infty^2 - 1)\frac{\partial^2 \varphi_2}{\partial x^2}$$
$$= \frac{\partial}{\partial x}\left[\left(1 + \frac{\gamma - 1}{2} M_\infty^2\right)\left(\frac{\partial \varphi_1}{\partial x}\right)^2 + \left(\frac{\partial \varphi_1}{\partial r}\right)^2\right] + M_\infty^2 \left(\frac{\partial \varphi_1}{\partial r}\right)^2 \frac{\partial^2 \varphi_1}{\partial r^2} \quad (7\text{-}11)$$

The cubic term on the extreme right has to be included because $\partial\varphi_1/\partial r$ and $\partial^2\varphi_1/\partial r^2$ are, as has been seen, considerably larger than $\partial\varphi_1/\partial x$ and $\partial^2\varphi_1/\partial x^2$ near the body.

Van Dyke's great advance, which rendered possible the solution of Eq. 7-11, was the discovery of a particular integral $\varphi_2 = \varphi_2^{(p)}$ of the equation. Here

$$\varphi_2^{(p)} = M_\infty^2 \varphi_1 \frac{\partial \varphi_1}{\partial x} + \frac{(\gamma + 1)M_\infty^4}{2(M_\infty^2 - 1)} r \frac{\partial \varphi_1}{\partial r}\frac{\partial \varphi_1}{\partial x} - \frac{1}{4} M_\infty^2 r \left(\frac{\partial \varphi_1}{\partial r}\right)^3 \quad (7\text{-}12)$$

The general solution takes the form of the sum of $\varphi_2^{(p)}$ with an arbitrary solution of the homogeneous equation which vanishes ahead of the undisturbed Mach cone. Thus

$$\varphi_2 = \varphi_2^{(p)} - \int_0^{x-(M_\infty^2-1)^{\frac{1}{2}}r} \frac{h'(t)dt}{[(x-t)^2 - (M_\infty^2 - 1)r^2]^{\frac{1}{2}}} \quad (7\text{-}13)$$

for some $h(t)$.

Van Dyke obtains φ_1 by solving the exact boundary condition numerically as an integral equation (following von Kármán and Moore [*41*]) and then solves numerically the boundary condition for φ_2 as an integral equation for $h(t)$. It is not possible to describe here his technique for numerical solution, which is rather complicated. But it should be mentioned that the divergences between theory and experiment observed in cases like those illustrated in Fig. E,7b (with the exception of those noted as boundary layer effects) are removed by the improvement in the theory. It is also particularly satisfactory that the theory continues to give excellent agreement with exact calculations (by the method of characteristics) even at a Mach number exceeding 3.

Nonaxisymmetric flows. To return to the questions (1) and (2) put at the beginning of this article, it is seen that the Van Dyke theory goes further than was suggested in those questions, which had already been answered in the pages preceding its mention. Two similar questions arise in the study of yawed flow past projectiles of revolution, and in Ward's theory [*43*] of flow past smooth slender bodies of more general cross section. In these cases, which will now be considered, a theory as complete as that of Van Dyke for axisymmetric flow does not at present exist. It is necessary to consider extensions of the preceding theories.

For the yawed flow the answer to both questions is obtainable along similar lines. Thus, for the theory with discontinuities in slope the reader is referred to the author's paper [35]; the over-all effect of these on the excess pressures due to yaw is found to be less than their effect on the pressures at zero yaw. Next, for smooth bodies at yaw the order of the errors can be deduced from his second approximation [40]. The problem is, however, a special case of Ward's more general theory [43], which applies to bodies (of any cross section) which are both smooth and slender in all directions perpendicular to the stream. (One can visualize how the extension of the Ward theory to general slender body shapes which are not smooth would be made, but it would probably be too complicated for practical use.) Ward gave a first order theory and also considered the order of magnitude of higher order terms.

A study of Broderick's second approximation [34] to axisymmetric flow, the author's work on smooth bodies at yaw [40] and Ward's estimate of second order terms in the general case [43] makes it clear that the true character of all these theories of flow over smooth bodies emerges clearly only from Ward's general approach. The first two methods perhaps attain their answers most expeditiously, but the work is still very long, and some of the answers are too complicated for extensive use. But the improvement of Ward's arguments, which will now be given, obtains in principle the second approximation for flow over an arbitrary smooth and slender body, and each stage in the argument illuminates the physics of the flow and the relative values of the approximations employed. Therefore the essential steps in the argument will here be carefully described. The results are not however calculated explicitly, but are taken direct from the above two papers [34,40], except that the most complicated term in Broderick's theory has been simplified in the light of Van Dyke's ideas.

The basis of the Ward theory is that, near the surface of a smooth slender pointed body, variation parallel to the axis is gradual compared with variation perpendicular to it (cf. C, 9 and D, 19–D, 24). Thus (working to the first order), if we write the linear equation for the disturbance potential φ, in cylindrical coordinates (r, θ, x) with the stream in the x direction (so that the full velocity potential is $q_\infty[x + \varphi]$), as

$$\frac{\partial^2 \varphi}{\partial r^2} + \frac{1}{r}\frac{\partial \varphi}{\partial r} + \frac{1}{r^2}\frac{\partial^2 \varphi}{\partial \theta^2} = (M_\infty^2 - 1)\frac{\partial^2 \varphi}{\partial x^2} \tag{7-14}$$

then near the body the left-hand side of the equation is dominant, and effectively the solution near the body is a plane harmonic in any plane perpendicular to the stream. The boundary condition is obvious from Fig. E,7c, in which two cross sections (normal to the stream) at stations a distance dx apart are shown superposed. Here dn, at any point on either, is the perpendicular distance between them. Expressing that there is no flow component normal to the surface, we see that in either plane section

$$\frac{\partial \varphi}{\partial n} = \left(1 + \frac{\partial \varphi}{\partial x}\right) \frac{dn}{dx} \tag{7-15}$$

which to a linear approximation becomes

$$\frac{\partial \varphi}{\partial n} = \frac{dn}{dx} \tag{7-16}$$

However, the determination of the appropriate plane harmonic describing the flow in a plane section near the body surface, and satisfying

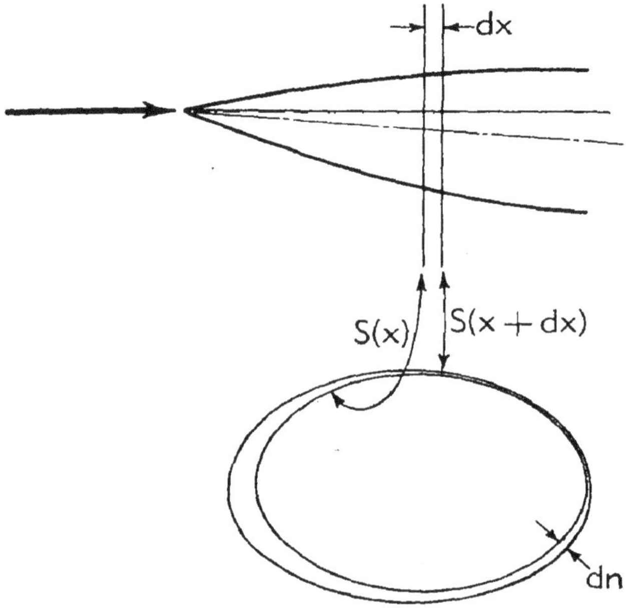

Fig. E,7c. Two superimposed adjacent cross sections of a smooth slender body.

the boundary condition (7-16), is not unique. It is obvious, for example, that any "constant" term (i.e. one depending on x alone) may be added without disturbing the conditions. Nor is it clear that the harmonic must be one which farther from the body has vanishingly small gradients, because, of course, the harmonic does not actually represent the potential out there, where the right-hand side of Eq. 7-14 is of equal importance with the left-hand side.

In fact the solution can be rendered unique only by continuing it away from the body surface into a solution of the full equation (7-14), and expressing the condition that the disturbance potential so obtained vanishes upstream of the Mach cone from the nose. But luckily this process can be done once for all in the general case. For if one considers the potential expanded in a Fourier series in θ, away from the axis, then the term independent of θ must be given by the axial source distribution potential

$$-\int_0^{x-(M_\infty^2-1)^{\frac{1}{2}}r} \frac{f(t)dt}{[(x-t)^2-(M_\infty^2-1)^2r^2]^{\frac{1}{2}}} \tag{7-17}$$

(whose derivatives with respect to x and r were given in Eq. 7-1), and the other terms are given by a similar axial distribution of multipoles, such as

$$\cos n\theta \int_0^{x-(M_\infty^2-1)^{\frac{1}{2}}r} \frac{F_n(t)dt}{[(x-t)^2-(M_\infty^2-1)r^2]^{\frac{1}{2}}} \left(\left\{\frac{x-t+[(x-t)^2-(M_\infty^2-1)r^2]^{\frac{1}{2}}}{(M_\infty^2-1)^{\frac{1}{2}}r}\right\}^n + \left\{\frac{(M_\infty^2-1)^{\frac{1}{2}}r}{x-t+[(x-t)^2-(M_\infty^2-1)r^2]^{\frac{1}{2}}}\right\}^n\right) \tag{7-18}$$

(These are obtained most simply by Heaviside's operational theory; for example Eq. 7-18 is $2K_n(pr)F_n(x)\cos n\theta$, where, as above, p stands for $(M_\infty^2-1)^{\frac{1}{2}}\partial/\partial x$; this satisfies Eq. 7-14; it also represents a potential zero ahead of the Mach cone through the nose, because the function $K_n(z)$ is like $e^{-z}\sqrt{\pi/2z}$ for large z, and, for example, $e^{-pr}F_n(x)$ signifies $F_n(x-(M_\infty^2-1)^{\frac{1}{2}}r)$.) Now for small r (i.e. in the region near the body, where the potential is expected to take the form of a plane harmonic) Eq. 7-17 becomes approximately

$$f(x)\ln r + \int_0^x \ln\left[\frac{(M_\infty^2-1)^{\frac{1}{2}}}{2(x-t)}\right] f'(t)dt \tag{7-19}$$

(this again is best obtained by expanding $-K_0(pr)f(x)$ for small r) and Eq. 7-18 becomes approximately

$$\frac{2^n}{(M_\infty^2-1)^{\frac{1}{2}n}}\frac{\cos n\theta}{r^n}\int_0^x (x-t)^{n-1}F_n(t)dt \tag{7-20}$$

Hence the whole potential near the surface is a combination of harmonics like Eq. 7-20, which vanish at infinity, with a part (7-19) involving a constant term and a term proportional to $\ln r$. Further these last two terms are *related;* the constant term is expressible in terms of the coefficient of $\ln r$.

Since, then, the plane harmonic required to satisfy the boundary condition (7-16) has zero gradient at infinity, it is specified uniquely by this boundary condition except for the possible addition of a function of x only. But the latter function is specified in terms of the coefficient of $\ln r$, as has just been seen. Also the coefficient of $\ln r$, by Gauss's theorem, is

$$\frac{1}{2\pi}\oint \frac{\partial\varphi}{\partial n}ds = \frac{1}{2\pi}\oint \frac{dn}{dx}ds = \frac{S'(x)}{2\pi} \tag{7-21}$$

Here the integral is taken around the cross section, and the geometry of Fig. E,7c makes it clear that the second integral is equal (as stated) to the rate of change $S'(x)$ of cross-sectional area (normal to the stream) with x. Hence the true form of the axisymmetric part of the first order disturbance potential (i.e. the term independent of θ in its Fourier analysis), by Eq. 7-19, is

$$\frac{1}{2\pi}\int_0^x \ln\frac{(M_\infty^2 - 1)^{\frac{1}{2}}r}{2(x - t)} S''(t)dt \tag{7-22}$$

This result for the axisymmetric part of the potential is thus independent of whether the projectile is one of revolution.

It should be added that if the boundary condition (7-16) is to be satisfied at the surface, by a sum of Eq. 7-22 with terms like Eq. 7-20, it must doubtless be roughly the case that the terms are of the same order of magnitude for values of r comparable with the values at the body surface. Thus the coefficient $F_n(t)$ in Eq. 7-18 must be smaller than that of Eq. 7-17, i.e. $f(t) = S'(t)/2\pi$, by an order of magnitude the nth power of the body radius measured from the axis. This proves what was stated in Art. 6, that, for given x, at distances from the axis large compared with the body radius for that x, these nonaxisymmetric terms are relatively unimportant.

Nonaxisymmetric flows to a second approximation. Having stated Ward's linear theory (for applications of which see [*43*]), we now investigate the second approximation to the flow. Near the body surface this is governed by an equation of which the left-hand side, embodying the largest terms, is as in Eq. 7-14, and of which the right-hand side contains those terms (out of the full equation of potential flow) next in importance; these include, besides the linear term in Eq. 7-14, nonlinear terms. The said equation is, in fact,

$$\frac{\partial^2\varphi}{\partial r^2} + \frac{1}{r}\frac{\partial\varphi}{\partial r} + \frac{1}{r^2}\frac{\partial^2\varphi}{\partial\theta^2} = -\frac{\partial^2\varphi}{\partial x^2} + \frac{1}{2}M_\infty^2\left(\frac{\partial}{\partial x} + \frac{\partial\varphi}{\partial r}\frac{\partial}{\partial r} + \frac{1}{r^2}\frac{\partial\varphi}{\partial\theta}\frac{\partial}{\partial\theta}\right)\left[2\frac{\partial\varphi}{\partial x} + \left(\frac{\partial\varphi}{\partial r}\right)^2 + \frac{1}{r^2}\left(\frac{\partial\varphi}{\partial\theta}\right)^2\right] \tag{7-23}$$

The selection from the full equation of the terms retained in Eq. 7-23 is based on the principle that axial variation (represented by $\partial/\partial x$) should be slower than normal variation ($\partial/\partial r$ and $r^{-1}\partial/\partial\theta$) by a factor of the order of the normal disturbance velocities $\partial\varphi/\partial r$ and $r^{-1}\partial\varphi/\partial\theta$. In particular squares of these latter velocities are comparable with $\partial\varphi/\partial x$. Notice that the convection of sound has to be rather fully taken into account. On the other hand, terms due to variation in the speed of sound are negligible compared with those retained in Eq. 7-23, a typical such term being $(\gamma - 1)M_\infty^4(\partial\varphi/\partial x)(\partial^2\varphi/\partial x^2)$.

To obtain the second approximation $\varphi = \varphi_1 + \varphi_2$, where φ_1 is the first approximation near the body as calculated by Ward (roughly, φ_1 is of the order of the square of the fineness ratio, and φ_2 may be expected to be of the order of its fourth power) we may solve Eq. 7-23 with φ replaced on the right-hand side by φ_1. A particular integral is found fairly easily by complex-variable methods. If $\zeta = re^{i\theta}$, then φ_1, as a plane harmonic, is the real part of a regular function $w(\zeta)$. Eq. 7-23 may be written, using this, as

$$\frac{\partial^2\varphi}{\partial\zeta\partial\bar{\zeta}} = \frac{1}{8}(M_\infty^2 - 1)\frac{\partial^2}{\partial x^2}(w + \bar{w}) + \frac{1}{8}M_\infty^2\left[2\frac{dw}{d\zeta}\frac{d}{d\bar{\zeta}}\left(\frac{\partial\bar{w}}{\partial x}\right)\right.$$

$$\left. + 2\frac{d\bar{w}}{d\bar{\zeta}}\frac{d}{d\zeta}\left(\frac{\partial w}{\partial x}\right) + \left(\frac{dw}{d\zeta}\right)^2\frac{d^2\bar{w}}{d\bar{\zeta}^2} + \left(\frac{d\bar{w}}{d\bar{\zeta}}\right)^2\frac{d^2w}{d\zeta^2}\right] \tag{7-24}$$

of which a particular integral is the real part of

$$\frac{1}{4}(M_\infty^2 - 1)\bar{\zeta}\int\frac{\partial^2w}{\partial x^2}d\zeta + \frac{1}{2}M_\infty^2 w\frac{\partial\bar{w}}{\partial x} + \frac{1}{4}M_\infty^2\frac{d\bar{w}}{d\bar{\zeta}}\int\left(\frac{dw}{d\zeta}\right)^2 d\zeta \tag{7-25}$$

(The physical significance of the terms in Eq. 7-25 will be discussed later.)

To this particular integral must be added a complementary function, i.e. a solution of Eq. 7-23 with right-hand side replaced by zero, in other words, a plane harmonic. This must be chosen to make $\varphi_1 + \varphi_2$ a one-valued function of r and θ satisfying, to within the approximation employed, the more exact boundary condition (7-15). Since φ_1 already satisfies Eq. 7-16, we must have in fact

$$\frac{\partial\varphi_2}{\partial n} = \frac{\partial\varphi_1}{\partial x}\frac{dn}{dx} \tag{7-26}$$

But it is clear that, again, φ_2 is so far indeterminate, at least to the extent of an arbitrary function of x alone. For Eq. 7-26 concerns only the derivatives $\partial\varphi_2/\partial r$ and $r^{-1}\partial\varphi_2/\partial\theta$. The specification of the part of φ_2 independent of r and θ can only be obtained, as was that of φ_1, by extending the solution outside the region near the surface, this time to a second approximation, and expressing the condition that the extended value of φ_2 vanishes ahead of the Mach cone from the nose. This process ultimately yields a relation between the coefficients of $\ln r$ and 1 in φ_2, whence the latter coefficient is determined.

Broderick carried out this process [34] for axisymmetric flow, and actually the result must be precisely the same in the general case of Ward's slender bodies as can be shown from the fact that, away from the axis, the flow *is* axisymmetric to a first approximation; but this part of his work exceeds by a large factor, both in length and in complexity of the final answer, all the rest of the present investigation. Now, however, it can be greatly shortened in both respects, by use of Van Dyke's form (7-13) of second approximation to the potential. For smooth bodies the form of this near the body is obtained by replacing φ_1 by Eq. 7-22 and expanding the integral as in Eq. 7-19; it is

$$M_\infty^2\left[f(x)\ln r + g(x) + \frac{(\gamma+1)M_\infty^2}{2(M_\infty^2-1)}f(x)\right][f'(x)\ln r + g'(x)]$$
$$-\frac{1}{4}M_\infty^2\frac{f^3(x)}{r^2} + h(x)\ln r + \int_0^x \ln\left[\frac{(M_\infty^2-1)^{\frac{1}{2}}}{2(x-t)}\right]h'(t)dt \quad (7\text{-}27)$$

Here $f(x) = S'(x)/2\pi$, as always in smooth body theory, and Eq. 7-22 has been written $f(x)\ln r + g(x)$. Of course the second term φ_2 in the approximation for smooth bodies must contain not only Eq. 7-27 but also the second term in the expansion near the body of the full linearized solution (7-17). However, this does not affect the coefficients of ln r and 1, which must be related as in Eq. 7-27. Thus, if

$$M_\infty^2(fg)' + \frac{(\gamma+1)M_\infty^4}{2(M_\infty^2-1)}ff' \quad \text{and} \quad M^2gg' + \frac{(\gamma+1)M_\infty^4}{2(M_\infty^2-1)}fg' \quad (7\text{-}28)$$

respectively be subtracted from these coefficients, the remainders are related as are the coefficients of ln r and 1 in Eq. 7-19.

This relationship is still complicated, but Broderick's [*34*] was so much more so that it could not be evaluated except for the single case of flow past a cone (see below). However, it should be noted that the complexity of the term independent of r and θ in φ_2 does not affect all aspects of the investigation, since for many purposes (e.g. evaluating lifts and moments) only the other terms in φ_2 are needed.

There is one other indeterminacy in φ_2, that is to say one other plane harmonic, with zero normal derivative on the boundary, of which any multiple could (without violating the conditions on the flow near the body) be added to φ_2, so that the particular multiple needs to be determined by a special device. This is the harmonic function which behaves far from the body like $r\cos(\theta+\delta)$, for some δ, and so represents the two-dimensional incompressible flow of a uniform stream past the cross section. But the determination of the appropriate multiple of *this* for incorporation in φ_2 is no more difficult than Ward's determination of the term independent of r and θ in φ_1. This is because, if a term of order $E(x)r\cos(\theta+\delta)$ is to be significant, even to the second approximation, near the body, then $E(x)$ must be of the order of the cube of the fineness ratio, and this (considered *not* near the body) is of the order of the coefficients of $\cos(\theta+\delta)$ in the *first* approximation φ_1. Hence $E(x)$ may be studied simply on the linear theory. The solutions of the linear equation of motion which are proportional to $\cos(\theta+\delta)$, and vanish ahead of the Mach cone from the nose, are of the form

$$\frac{\cos(\theta+\delta)}{r}\int_0^{x-(M_\infty^2-1)^{\frac{1}{2}}r}\frac{(x-t)B'(t)dt}{[(x-t)^2-(M_\infty^2-1)r^2]^{\frac{1}{2}}} \quad (7\text{-}29)$$

for some $B(x)$. But the expansion of Eq. 7-29 near the body is (cf. Eq. 7-4)

$$\cos(\theta+\delta)\left\{\frac{B(x)}{r} + \frac{1}{2}(M_\infty^2-1)r\int_0^x\left[\ln\frac{(M_\infty^2-1)^{\frac{1}{2}}r}{2(x-t)}-\frac{1}{2}\right]dB''(t)+\cdots\right\} \tag{7-30}$$

which shows that the coefficient $E(x)$ of $r\cos(\theta+\delta)$ in φ_2 is obtained at once from that of $r^{-1}\cos(\theta+\delta)$ in φ_1 (which has here been called $B(x)$) in the form

$$E(x)=\frac{1}{2}(M_\infty^2-1)\int_0^x\left[\ln\frac{(M_\infty^2-1)^{\frac{1}{2}}}{2(x-t)}-\frac{1}{2}\right]dB''(y) \tag{7-31}$$

On the other hand the fact that, by expression (7-30), the coefficient of

$$(r\ln r)\cos(\theta+\delta) \tag{7-32}$$

in φ_2 must be $\frac{1}{2}(M_\infty^2-1)B''(x)$, is merely an independent check on the work; indeed it is clear that the part of the first term in Eq. 7-25 corresponding to the term $B(x)e^{-i\delta}\zeta^{-1}$ in $w(\zeta)$ will, when a corresponding term with $\zeta e^{2i\delta}$ for $\bar{\zeta}$ has been added to it to make it single-valued, have just this behavior for large r.

There are no further indeterminacies in φ_2 to be rendered determinate by such devices. For the only single-valued harmonic functions with zero normal derivative on the boundary are those behaving at large distances like $r^n\cos(n\theta+\delta)$, for some δ, where n is a non-negative integer. Such a function could have a place in φ_2 only if multiplied by a coefficient of order the $(4-n)$th power of the fineness ratio. But Ward's theory shows (see the paragraph following Eq. 7-22 above) that even the first approximation φ_1 has, as coefficient of $\cos(n\theta+\delta)$ when r is not small, a term at most of order the $(2+n)$th power of the fineness ratio. Hence such a harmonic occurs in φ_2 only if $2+n\leqq 4-n$ or $n\leqq 1$. (If there were any question of going to still higher approximations to φ, which there is not, the said function would occur from φ_{n+1} onward.)

The whole process for determining φ_2 having now been specified, we may draw general conclusions from its form. Conclusions regarding its order of magnitude (for use in a critique of the first order theory) may reasonably be drawn from Eq. 7-25, since the operations to be performed on its various terms to obtain φ_2 should not greatly change their order of magnitude, and since the additional term $E(x)r\cos(\theta+\delta)$ is, by Eq. 7-31, of similar order to the first term in Eq. 7-25. Regarding the *relative* orders of magnitude of the three terms in Eq. 7-25, observe that the last two, which represent corrections to the linear theory due to the convection of sound, are for smooth bodies just as important as the first term, which represents corrections due to departures, near the body, of the linear equation of motion (7-14) from Ward's simplified form neglecting the right-hand side. Thus *for smooth bodies, if the linear equation of motion be used at all, the full smooth body approximation can be used without serious additional loss of accuracy.* (Further numerical evidence for this is given below, Fig. E,7d.)

Eq. 7-25 shows clearly that this conclusion is independent of Mach number. The error resulting from both approximations varies as M_∞^2. This is because, on the one hand, variation parallel to the axis is less gradual for larger M_∞, since the Mach lines make a smaller angle with the stream, and, on the other hand, the variations in fluid velocity become a larger fraction of the speed of sound as M_∞ increases.

The actual orders of magnitude of the three terms relative to φ_1 itself are

$$M_\infty^2 \frac{r^2}{x^2}, \qquad M_\infty^2 \frac{\varphi_1}{x}, \qquad M_\infty^2 \frac{\varphi_1^2}{r^2} \tag{7-33}$$

if smoothness is taken to mean that $\partial/\partial x$ is of order x^{-1}. Since φ_1/x is of the order of the square of the fineness ratio ϵ, it is seen that roughly the error due to each term (near the body) represents a factor $M_\infty^2\epsilon^2$. It is clear from this that the linear theory can be used, for practical body shapes, only at moderate supersonic speeds. This conclusion has been drawn by several writers, from different kinds of argument.

Broderick's theories and their significance. The specialization of the above theory to the axisymmetric and yawed flows past projectiles of revolution will now be indicated very briefly indeed, and the results discussed. In the axisymmetric case φ_1 consists only of the term (7-19) independent of θ, so $w(\zeta)$ is of the form $f(x) \ln \zeta + g(x)$, where as always in smooth body theory $f(x) = S'(x)/2\pi$, while $g(x)$ stands for expression (7-22) with r omitted. Hence the real part of Eq. 7-25 is easily calculated as a sum of multiples of $r^2 \ln r$, r^2, $\ln^2 r$, $\ln r$, 1, and r^{-2} by specific functions of x. (At least this is so after a term in θ^2 has been removed, by subtracting a suitable multiple of the real part of $\ln^2 \zeta$, in order to make the result single-valued.) To satisfy the boundary condition (7-26) on $r = R(x)$ a new multiple of $\ln r$ has to be added. The terms in φ_2 other than the term independent of r are thus rapidly calculated as

$$\begin{aligned}\tfrac{1}{4}(M_\infty^2 - 1)[f''r^2 \ln r + (g'' - f'')r^2] + M_\infty^2 ff' \ln^2 r + [(f' \ln R + g')RR' \\ - \tfrac{1}{4}(M_\infty^2 - 1)(2f'' \ln R + 2g'' - f'')R^2 - 2M_\infty^2 ff' \ln R \\ - \tfrac{1}{2}M_\infty^2 f^3 R^{-2}] \ln r - \tfrac{1}{4}M_\infty^2 f^3 r^{-2}\end{aligned} \tag{7-34}$$

The remaining term is then given by the rule incorporating expressions (7-28).

Broderick's method [*34*] of obtaining these terms was conceptually different. He considered the potential everywhere as expanded in powers of the fineness ratio ϵ, for a series of body shapes $r = \epsilon R(x)$. He had to consider the coefficients of ϵ^2, ϵ^4, and ϵ^6 in this expansion, and also, as the terms in $\ln R$ in Eq. 7-34 indicate, the coefficient of $\epsilon^4 \ln \epsilon^{-1}$. He pointed out that terms in $\epsilon^6 \ln \epsilon^{-1}$ and $\epsilon^6 \ln^2 \epsilon^{-1}$ exist, and also investigated terms in still higher powers of ϵ. He expanded the equation for each of these coefficients in ascending powers of r and $\ln r$, to determine their form for small r. Finally he picked out the terms which are of order ϵ^2 or $\epsilon^2 \ln \epsilon^{-1}$ near the body, namely, Eq. 7-22 (the first approximation for smooth slender bodies), and then the terms which are of order ϵ^4 or $\epsilon^4 \ln \epsilon^{-1}$ or $\epsilon^4 \ln^2 \epsilon^{-1}$ near the body, namely (apart from the term independent of r), Eq. 7-34. He had also to prove that no contribution arises from coefficients of other powers of ϵ and $\ln \epsilon^{-1}$. There is little doubt that the present method, in which the equation for the potential is approximated in a different manner, appropriate to flow conditions near the body, and the relevant terms emerge as the whole of the first and second approximations respectively, is superior for obtaining Eq. 7-34.

The pressure coefficient is calculated to the order of approximation required as

$$C_p = -2\frac{\partial \varphi_1}{\partial x} - \left(\frac{\partial \varphi_1}{\partial r}\right)^2 - \left(\frac{\partial \varphi_1}{\partial x}\right)^2 - 2\frac{\partial \varphi_2}{\partial x} - 2\frac{\partial \varphi_1}{\partial r}\frac{\partial \varphi_2}{\partial r} + \frac{1}{4}M_\infty^2\left[2\frac{\partial \varphi_1}{\partial x} + \left(\frac{\partial \varphi_1}{\partial r}\right)^2\right] \tag{7-35}$$

of which the first two terms constitute the first approximation and are of orders $\epsilon^2 \ln \epsilon^{-1}$ and ϵ^2 near the body, while the remaining terms are of orders $\epsilon^4 \ln^2 \epsilon^{-1}$, $\epsilon^4 \ln \epsilon^{-1}$, and ϵ^4 near the body. They are easily calculable from Eq. 7-22 and 7-34, except for the term depending on x alone arising from the fourth term of Eq. 7-35, which arises from that in φ_2. It is not here given, but it may be mentioned that it is of order ϵ^4.

For the flow past a cone a numerical comparison of the various terms is possible. If ϵ is the tangent of the semivertical angle, Eq. 7-35 becomes

$$C_p = -\epsilon^2 + 2\epsilon^2 \ln \frac{2}{(M_\infty^2 - 1)^{\frac{1}{2}}\epsilon} + 3(M_\infty^2 - 1)\epsilon^4\left[\ln \frac{2}{(M_\infty^2 - 1)^{\frac{1}{2}}\epsilon}\right]^2 - (5M_\infty^2 - 1)\epsilon^4 \ln \frac{2}{(M_\infty^2 - 1)^{\frac{1}{2}}\epsilon} + \left[\frac{13}{4}M_\infty^2 + \frac{1}{2} + \frac{(\gamma + 1)M_\infty^4}{M_\infty^2 - 1}\right]\epsilon^4 \tag{7-36}$$

of which the first two terms constitute the first order approximation. The term involving γ comes from Broderick's evaluation of the term independent of r in φ_2. In Fig. E,7d the pressure coefficient on a cone is plotted against Mach number for semiangles of 10° and 15°. Here the full lines represent Kopal's numerical calculation [37] of the exact adiabatic solution, the dashed lines represent the second order approximation (7-36), the dotted lines represent the first order approximation (the first

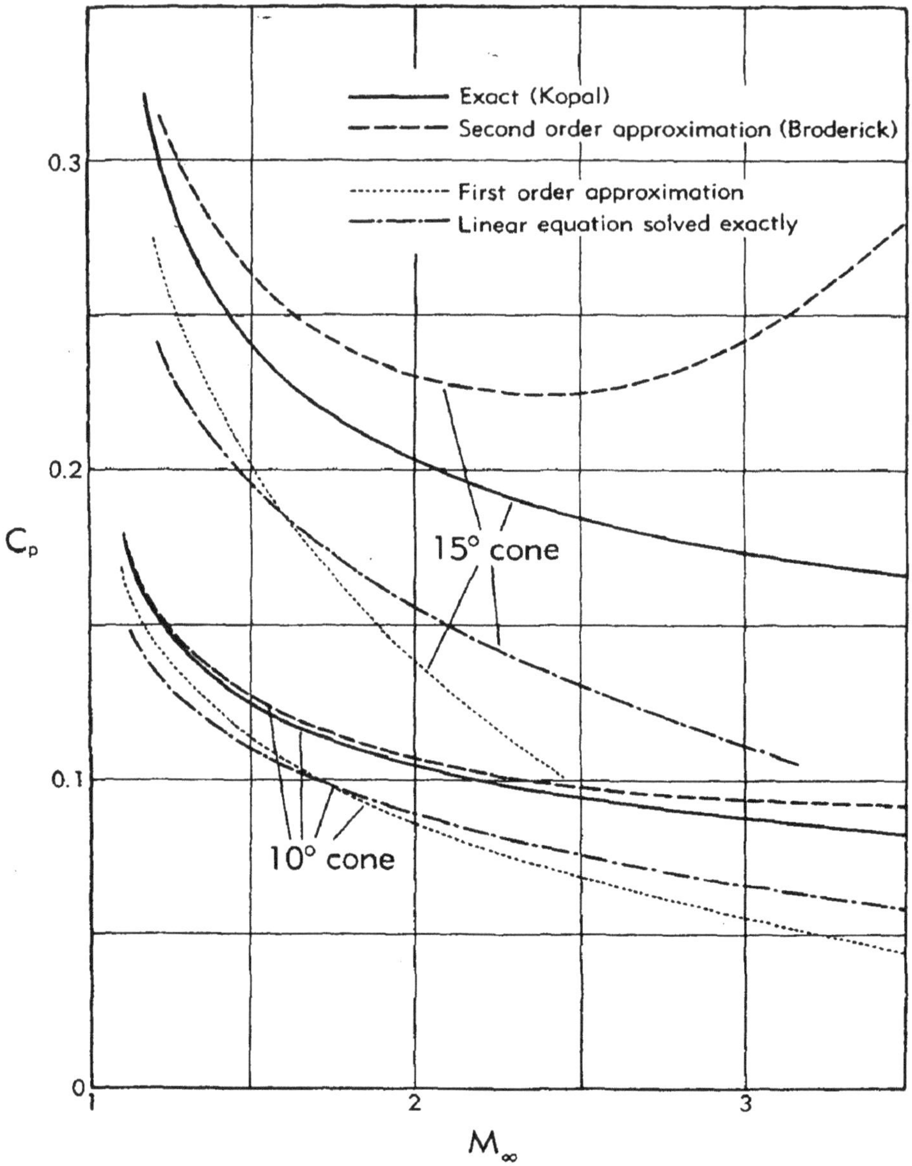

Fig. E,7d. Pressure coefficient on the surface of a cone in axisymmetric flow at Mach number M_∞.

two terms of Eq. 7-36), while the dash-dotted line is the curve obtained by using the linearized equation for the velocity field and satisfying the boundary condition at the surface exactly, and also deducing the pressure coefficient from the exact form of Bernoulli's equation. It is seen that, as predicted above from the orders of magnitude of the terms, there is, in the case of a cone, hardly any appreciable advantage, if the linear equation is being used at all, in not adopting the full smooth body approximation. For a semiangle of 5° the exact and Broderick's curves are indistinguishable in the range of M_∞ used in Fig. E,7d. It is seen that the Broderick theory is still excellent for 10°, but is beginning to be a serious overestimate for 15°. The errors in all the theories increase markedly with Mach number, as predicted above. (They are really all meaningless for semiangle 15° and $M_\infty = \csc 15° = 3.86$, since the cone then coincides with the undisturbed Mach cone.)

In Broderick's second paper in the journal cited [34] he extended further his solution for the flow past a cone to include the flow near the shock. In this way he was able to prove rigorously that his value (7-36) for the surface pressure was correct if terms of order $\epsilon^6 \ln^3 \epsilon^{-1}$ were neglected, even if full account be taken of the properties of the shock. This was hardly obvious (except by analogy with the two-dimensional flow) since the second approximation to the disturbance potential is of the same order of magnitude as the shock strength. Broderick's proof, therefore, sets the whole theory (not simply in the special case of the cone) on a firmer footing. To be precise, he showed that his series for φ in powers of ϵ converges only in a certain region stopping short of the undisturbed Mach cone through the nose. But he obtained a second series expansion in powers of ϵ, whose coefficients are functions of the ratio of distance from the shock to distance of the Mach cone from the shock. This expansion converged in a region including the shock and overlapping the former region, whereas it satisfied the full Rankine-Hugoniot condition at the shock. By comparison of the series in their common region of convergence he was able to determine all remaining arbitrary constants that were needed to obtain the full second approximation to the surface pressures.

Broderick's detailed conclusions regarding flow near the shock are easily understood in the light of Whitham's subsequent theory (Art. 6). He finds that in this region $v = \partial\varphi/\partial r$ (for example) is given, to a first approximation, in the notation of Art. 6, by

$$v \cong 2^{\frac{1}{2}}(M_\infty^2 - 1)^{\frac{1}{4}}\delta^2 \left\{\left[\frac{x}{r} - (M_\infty^2 - 1)^{\frac{1}{2}} + k^2\delta^4\right]^{\frac{1}{2}} + k\delta^2\right\} \qquad (7\text{-}37)$$

having obtained certain of the constants herein by a comparison (involving careful consideration of error terms) with the small perturbation theory solution

$$v \cong 2^{\frac{1}{2}}(M_\infty^2 - 1)^{\frac{1}{4}}\delta^2 \left[\frac{x}{r} - (M_\infty^2 - 1)^{\frac{1}{2}}\right]^{\frac{1}{2}} \qquad (7\text{-}38)$$

On the other hand Eq. 7-37 agrees perfectly with Whitham's (6-13) if $F(y)$ is given by Eq. 6-12 and if this value is also used to solve Eq. 6-6 as a quadratic in $y^{\frac{1}{2}}$. The result, like Eq. 6-12, must be true all along the shock for the cone, and also, it may be mentioned, along the extensive initial stretch of it for any projectile.

But there is yet another interpretation of Eq. 7-37, which is important in the general theory (Art. 8) of the shock, and of the second approximation to the surface pressures, in "conical flows." In this interpretation, the first term in braces is regarded as identical with the linear theory solution (7-38), except that the value of x/r is modified to become $(x/r) + k^2\delta^4$, so that, as it were, the "undisturbed Mach cone" (where the solution is singular) is shifted forward, and indeed *beyond* where the solution is cut off by the front shock. This "doctoring" of the linear solution is correct (as Eq. 7-37 shows) near the shock, while elsewhere, where $x/r - (M_\infty^2 - 1)^{\frac{1}{2}}$ is not small, the error is of order δ^6 and so negligible even to a second approximation. On the other hand the second term in braces in Eq. 7-37 is regarded as due to the second approximation φ_2, which has

$$\frac{\partial \varphi_2}{\partial r} = 2^{\frac{1}{2}}(M_\infty^2 - 1)^{\frac{1}{4}} k\delta^4 \tag{7-39}$$

on the undisturbed Mach cone $x/r = (M_\infty^2 - 1)^{\frac{1}{2}}$. Eq. 7-39 is in fact true, trivially, for the "natural" second approximate solution, whose equation away from the body surface is not Eq. 7-23 or 7-11 but follows from Eq. 2-14 in the form

$$\frac{\partial^2 \varphi_2}{\partial r^2} + \frac{1}{r}\frac{\partial \varphi_2}{\partial r} - (M_\infty^2 - 1)\frac{\partial^2 \varphi_2}{\partial x^2} = \frac{\partial}{\partial x}\left[\left(1 + \frac{\gamma - 1}{2} M_\infty^2\right)\left(\frac{\partial \varphi_1}{\partial x}\right)^2 + \left(\frac{\partial \varphi_1}{\partial r}\right)^2\right] \tag{7-40}$$

If φ_2 is taken as zero (because assumed continuous) on the Mach cone $x/r = (M_\infty^2 - 1)^{\frac{1}{2}}$, the first and third terms on the left vanish thereon; the right-hand side is easily calculated as r^{-1} times that of Eq. 7-39, whose truth therefore follows. It is a good approximation not only on the Mach cone but also anywhere sufficiently near to it, and hence Eq. 7-37 follows.

This gives a new principle (analogous to but essentially different from Whitham's of Art. 6) for obtaining a flow field which is everywhere correct to a second approximation, from knowledge of the second approximate flow field calculated by the straightforward process of successive approximation. The principle is as follows: *The second approximate flow field is uniformly valid if distance downstream from the undisturbed Mach cone is always interpreted therein as distance downstream from the limit cone.*

In the case of flow past a cone the limit cone referred to is given by $x/r = (M_\infty^2 - 1)^{\frac{1}{2}} - k^2\delta^4$. We are justified in calling this cone, on which the flow field is singular, a limit cone (by analogy with limit line in two dimensions) because it is in fact the envelope of the Mach lines (6-6), as the ordinary process of differentiation with respect to y and elimination of y shows.

This principle could probably, in theory, be applied to determine the shocks on general (nonconical) projectiles, but it is very unsatisfactory for this purpose (compared with Whitham's principle), since it involves considering the second approximation. However, it is the principle which must be applied in the general theory of conical fields (Art. 8), and so it is important to notice that in general physical soundness and plausibility it is the equal of Whitham's. It states in fact that the square root singularities in the first and second approximate solutions correspond to a real singularity in the theoretical potential flow, i.e. a limit line; notice that near the limit line the velocities necessarily vary like the square root of distance from it. But this singularity is really slightly ahead of where it is predicted to be on these approximate solutions, and therefore the flow field must be shifted forward so that the predicted singularity comes in the right place. This makes a big difference to the behavior in the neighborhood of the shock, but elsewhere the shift in the coordinates by a multiple of δ^4 produces a negligible effect to the order of approximation used. The shock has to be determined as a surface between the limit cone and the undisturbed Mach cone, such that the deflection behind the shock (predicted on the above principle) is consistent with the deviation in shock slope from the Mach value.

The yawed flow past projectiles of revolution. Having plumbed the significance of Broderick's work on the cone, we pass to the present author's theory [*40*] of the yawed flow past projectiles of revolution. He uses an expansion in powers of the angle of yaw α, and takes the axis of the body, rather than the stream direction, as the axis of his cylindrical coordinates; but a few remarks are desirable on how his conclusions could be deduced from the smooth slender body theory above, in which the axis of coordinates is in the direction of the incident stream.

This approach simplifies the treatment of the equations of motion but complicates the boundary conditions. To a first approximation (only) the cross section of the projectile by a plane perpendicular to the stream is a circle of radius $R(x)$. Assuming this it is not hard to see that the distance normal to the stream between adjacent cross sections is given by

$$\frac{dn}{dx} = R'(x) + \alpha \cos \theta \tag{7-41}$$

provided that (r, θ) are taken temporarily *not* as cylindrical coordinates but as polar coordinates in the plane of the cross section with pole at the center of the circle, with the line $\theta = 0$ in the direction in which the projectile is yawed. But when we go to a second approximation, it is, strictly speaking, necessary to include perturbation terms due to the fact that the cross section is really an ellipse of axis ratio $\cos \alpha$. These are not actually important in practice, but can be calculated by considering how far the condition on $\partial\varphi_1/\partial n$ has *not* been satisfied on the ellipse (owing to being satisfied exactly on the circle), and then correcting

$\partial\varphi_2/\partial n$ by just this amount, which may be done, for simplicity, to within the accuracy of the second approximation, on the circle itself.

Because of Eq. 7-41 and 7-16 there is an additional term

$$-\alpha R^2(x)r^{-1}\cos\theta$$

in φ_1 due to yaw, and so an additional term $-\alpha R^2(x)\zeta^{-1}$ in $w(\zeta)$. This produces terms in the particular integral (7-25), additional to those in Broderick's solution, in α, α^2, and α^3, which include (among others) terms due to interaction of the yaw and nonyaw terms in φ_1. It should also be noticed that the *origin* of ζ varies from cross section to cross section (since it is always on the axis of the projectile, which is yawed to the x axis), and that therefore, in writing down the first and second terms of Eq. 7-25, terms due to the dependence of ζ on x (in such a way that $\partial\zeta/\partial x = -\alpha$) have to be included. Harmonics (with zero gradient at infinity) must then be added to make φ_2 satisfy the boundary condition (7-26) (with perhaps an additional term, see above, due to the ellipticity of the true cross section) on the circular boundary. This modifies the coefficients of α, α^2, and α^3. The term independent of r and θ is then modified, as discussed in the general theory. After this the term independent of α in φ_2 becomes exactly Broderick's term. Finally, the term in $r\cos\theta$ must be modified, by subtraction of a multiple of the harmonic $(r + r^{-1})\cos\theta$ representing incompressible flow past the circle, so that the coefficient $E(x)$ of $r\cos\theta$ takes the form (7-31) with $B(x) = -\alpha R^2(x)$.

Thus, whereas φ_1 contains terms only in 1 and α, φ_2 contains (besides Broderick's term in 1) terms in α, α^2, and α^3. These were all[7] obtained by the author [*40*], in his system of axes, though his method of presentation hardly made it clear that they form part of the same second approximation. He calculated the lift force on the part of the projectile given by $0 < x < l$ as

$$L = \rho_\infty q_\infty^2 S(l)\alpha \left\{1 + \frac{7}{6}(M_\infty^2 - 1)\alpha^2 + \frac{2M_\infty^2 - 1}{\pi}\left[S''(l)\ln\frac{(M_\infty^2 - 1)^{\frac{1}{2}}R(l)}{2l} + \int_0^l \frac{S''(l) - S''(y)}{l - y}\,dy\right] + \frac{M_\infty^2}{\pi}S''(l) + (M_\infty^2 - 2)R'^2(l)\right\} + 3(M_\infty^2 - 1)D_0\alpha \quad (7\text{-}42)$$

where D_0 is the drag at zero yaw on that part of the projectile (still ignoring the effect of the boundary layer), given by

$$D_0 = \frac{\rho_\infty q_\infty^2}{2\pi}\left[\int_0^l S''(x)dx \int_0^x S''(y)\ln\frac{(M_\infty^2 - 1)^{\frac{1}{2}}R(l)}{2(x - y)}\,dy - S'(l)\int_0^l S''(y)\ln\frac{(M_\infty^2 - 1)^{\frac{1}{2}}R(l)}{l - y}\,dy\right] \quad (7\text{-}43)$$

[7] Except the term in α^2 independent of r and θ; this is not needed in calculating lifts and moments; in any case it is clear in principle, from the general account, how it could be obtained.

Here the lift L signifies force normal to the stream, which is less by an amount $D_0\alpha$ than the force N normal to the axis of the projectile. To obtain the part of the *pressure distribution* which produces a force normal to the axis of the projectile, namely, the coefficient of $(-\cos\theta)$ when the pressure is Fourier analyzed, write x for l in $N = L + D_0\alpha$, differentiate with respect to x to obtain the normal force on the projectile between planes x and $x + dx$ normal to the axis, and divide by $\pi R(x)$. This may be used to determine the yawing moment [*40*].

It is seen from Eq. 7-42 that only small deviations of the lift curve from linearity can be attributable to the potential flow, since the proportional error in taking Eq. 7-42 as a multiple of α is about $\frac{7}{6}(M_\infty^2 - 1)\alpha^2$, which for large yaws like 10° is still only ten per cent for $M_\infty = 2$. We may therefore confine our attention to the predicted lift curve *slope*.[8]

A comparison with the exact adiabatic flow is possible for the case of a cone. Kopal [*37*] gives values of a coefficient K_N, which is, according to his equations for it (though his definition is rather loosely expressed),

$$K_N = \frac{1}{4R^2(l)\rho_\infty q_\infty^2}\left(\frac{dN}{d\alpha}\right)_{\alpha=0} \tag{7-44}$$

On the author's theory

$$K_N = \frac{1}{4}\pi\left[1 + M_\infty^2\delta^2 \ln\frac{(M_\infty^2 - 1)^{\frac{1}{2}}\delta}{2} + \left(\frac{3}{2}M_\infty^2 - 1\right)\delta^2\right] \tag{7-45}$$

where δ is the semiangle. In Fig. E,7e the plain lines indicate the exact values of K_N, the dashed lines denote his approximation, the first order approximation $K_N = \frac{1}{4}\pi$ is shown by a dotted line, and the dash-dotted line represents Tsien's prediction [*50*], based on a more exact solution of the linearized equation of motion. The author's theory is almost perfect for 5° semiangle, is good for $M_\infty \leqq 2$ with 10° semiangle, and is beginning to be a serious overestimate for 15° semiangle. On the other hand no advantage emerged from Tsien's taking a purely linear theory beyond the first approximation.[9]

For a body shape with a cylindrical portion at the rear, $R'(l)$ and $S''(l)$ are zero and Eq. 7-42 takes the simpler form

$$\left(\frac{dL}{d\alpha}\right)_{\alpha=0} = \rho_\infty q_\infty^2 S(l)\left[1 - \frac{2M_\infty^2 - 1}{\pi}\int_0^{l_1}\frac{S''(y)dy}{l - y}\right] + 3(M_\infty^2 - 1)D_0\alpha \tag{7-46}$$

where $x = l_1$ is the shoulder and the cylindrical portion is given by $l_1 < x < l$. Notice that *both* terms additional to the first approximation

[8] In fact there is very considerable departure from linearity for larger angles of attack, in the direction of higher lift. But this is due to boundary layer separation (see note 10).

[9] But for better approximations, in which the nonlinearity of the equation of motion for the cross flow is still not included, see Van Dyke [*54*].

$\rho_\infty q_\infty^2 S(l)$ for lift curve slope are positive. In the case of the integral this is because, though the integral of $S''(y)$ between the limits would be zero, the positive values of $S''(y)$ near the nose are weighted less by the factor $(l - y)^{-1}$ than the negative values farther back. This conclusion, that the lift curve slope always exceeds $\rho_\infty q_\infty^2$ times the base area, is in agreement

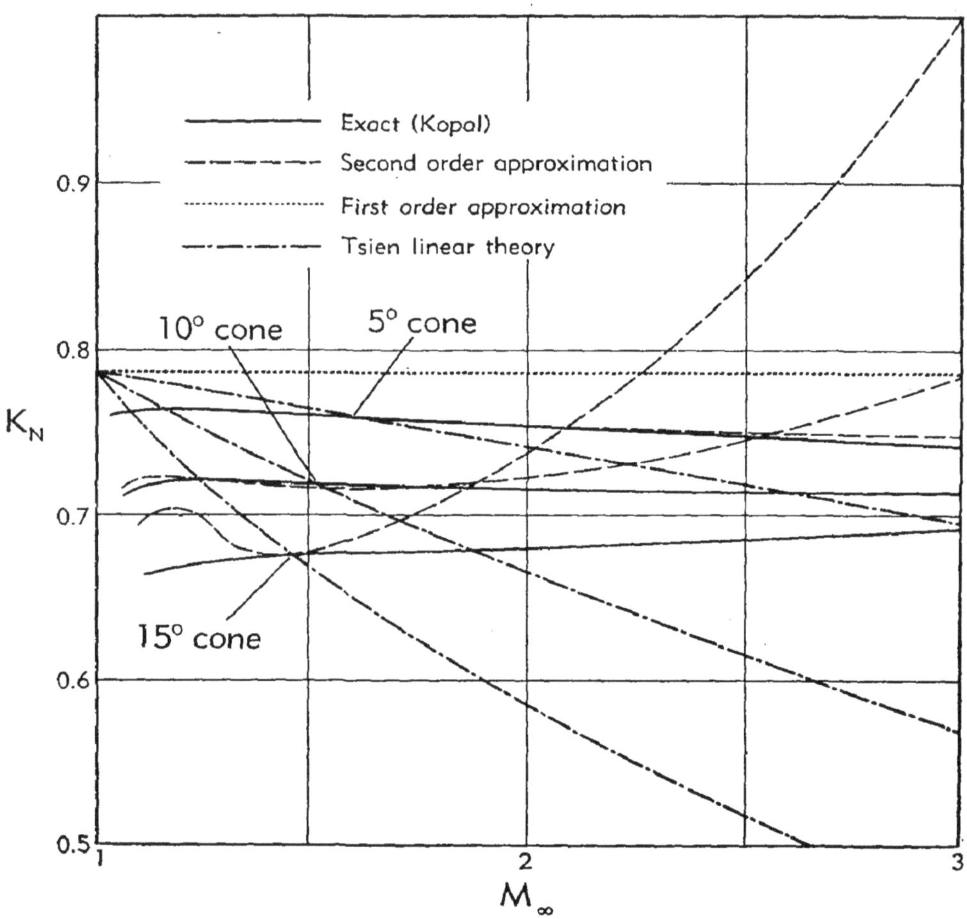

Fig. E,7e. Normal force coefficient K_N for a cone in slightly yawed flow at Mach number M_∞.

with experimentally determined lifts on projectiles, even though these contain an extra negative term due to the component normal to the stream of the skin friction and the base drag. It is also found, as the formula predicts, that the lift curve slope, divided by $\rho_\infty q_\infty^2$, increases with Mach number.

From a more quantitative point of view, take $D_0 = 0.8\rho_\infty q_\infty^2 S^2(l)/l_1^2$, (a typical value, about 25 per cent in excess of its theoretical minimum for given $S(l)$ and l_1), and estimate the integral in Eq. 7-46, by expanding $(l - y)^{-1}$ as $l^{-1} + yl^{-2}$, as $-S(l)/l^2$. This gives the approximation

$$\left(\frac{dL}{d\alpha}\right)_{\alpha=0} = \rho_\infty q_\infty^2 S(l)\left[1 + \frac{2M_\infty^2 - 1}{\pi}\frac{S(l)}{l^2} + 2.4(M_\infty^2 - 1)\frac{S(l)}{l_1^2}\right] \quad (7\text{-}47)$$

where $S(l)$ is the cross-sectional area of the cylindrical portion, whose length is $l - l_1$ out of a total length l. Clearly, the third term is more important than the second.

For a shell with 7.4-caliber radius ogival head, and total length 4.9 calibers, on which experiments have been made, $l = 9.8R(l)$ and $l_1 = R(l)\sqrt{4 \times 7.4 - 1}$. Eq. 7-47 then gives for the "cross wind force coefficient" commonly used in ballistics

$$\begin{aligned} f_L = \frac{1}{\rho_\infty q_\infty^2 R^2(l)}\left(\frac{dL}{d\alpha}\right)_{\alpha=0} &= \pi\left\{1 + \frac{2M_\infty^2 - 1}{\pi}\frac{\pi R^2(l)}{[9.8R(l)]^2}\right. \\ &\qquad \left. + 2.4\,(M_\infty^2 - 1)\frac{\pi R^2(l)}{28.6R^2(l)}\right\} \quad (7\text{-}48) \\ &= 3.17 + 0.89(M^2 - 1) \end{aligned}$$

This (which neglects the negative component of skin friction and base drags) is 3.79 and 4.56 for $M_\infty = 1.3$ and 1.6 respectively, whereas the values measured by Cope and Thurston in wind tunnel tests (unpublished) were 3.70 and 3.90 respectively. (The error in the theory would be expected to increase noticeably with M_∞, even for such low values, since the nose semiangle is as high as 21°, which is comparable with the Mach angle 39° when $M_\infty = 1.6$.)

On the cylindrical portion of a projectile the part of the pressure distribution proportional to $\cos\theta$ is very small. For it is

$$\begin{aligned} -\frac{1}{\pi R(l)}\frac{\partial}{\partial l}\left(\frac{dN}{d\alpha}\right)_{\alpha=0} &= -\frac{1}{\pi R(l)}\frac{\partial}{\partial l}\left[\left(\frac{dL}{d\alpha}\right)_{\alpha=0} + D_0\right] \\ &= -\rho_\infty q_\infty^2 S(l)\frac{2M_\infty^2 - 1}{\pi^2 R(l)}\int_0^{l_1}\frac{S''(y)dy}{(l-y)^2} \quad (7.49) \end{aligned}$$

by Eq. 7-46. This is at most a second order term, and would be about $\rho_\infty q_\infty^2(4M_\infty^2 - 2)[R(l)/l]^3$ near the base, or less than one per cent of $\rho_\infty q_\infty^2$ in typical cases. The only *first* order terms remaining in the pressure are

$$\rho_\infty q_\infty^2\left[\frac{1}{2\pi}\int_0^{l_1}\frac{S''(t)}{l-t}\,dt + \alpha^2\left(-\frac{1}{2} + \cos 2\theta\right)\right] \quad (7\text{-}50)$$

These represent a two-peaked variation about a negative mean. Near the nose, however, the term in α, representing a one-peaked variation (like $\cos\theta$), will normally exceed that in α^2 in importance, and the mean, represented by the axisymmetrical pressure distribution, will be positive. The transition from the one form to the others is well exemplified in the pressure plotting measurements of Cope and Thurston, which are shown in Fig. E,7f. The distance between peak and trough at the base is $0.13p_\infty$,

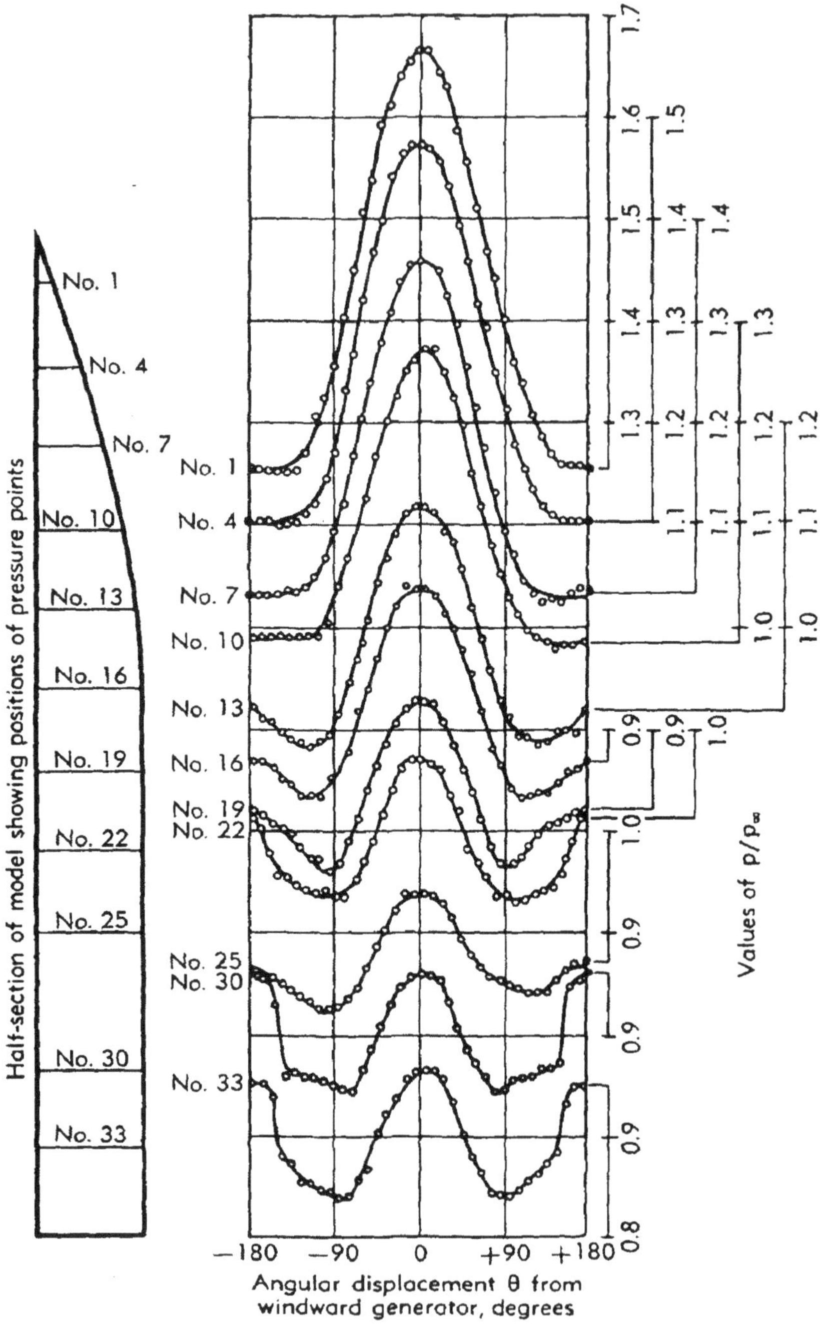

Fig. E,7f. Pressure distribution on a shell of six-caliber head radius in flow at Mach number 1.24, yawed to the shell by 10°.

which is precisely the value $2\rho_\infty q_\infty^2\alpha^2$ predicted by Eq. 7-44. Note that the term in α^2 in Eq. 7-50 is that associated with the incompressible potential flow of a stream with velocity $q_\infty\alpha$ normal to the cylinder. It is amusing that, among the various boundary layer control devices which permit the incompressible potential flow normal to a cylinder to be achieved in practice, its superposition on a supersonic stream parallel to the axis of the cylinder is among the most effective![10] It should also be noticed that, if the shell is spinning, this cross flow produces a Magnus effect tending to yaw the shell about an axis at 90° to that of the existing yaw.

E,8. Supersonic Three-dimensional Wing Theory. The extension of the ideas of the foregoing articles to fully three-dimensional problems, that is to problems where no approximate dependence on only two space variables can be assumed, even in part of the flow pattern, has not yet been achieved. In particular, in three-dimensional wing theory, the surface pressure distribution has not been found to a better approximation than the linear theory gives, nor has much of the complete flow pattern been elucidated. It may be thought that the investigation must necessarily be too complicated to be worthwhile; but this is not so. For example if, by arduous calculations in special cases, one were to find that a value of C_p on linear theory signifies a true value considerably nearer to the value $C_p + (C_2/C_1^2)C_p^2$, as Eq. 3-21 shows is true in two dimensions, then this could be used to improve greatly our knowledge of aerodynamic forces. Again, in experiments on shocks produced by wedges in supersonic streams, it would be desirable to know whether any departures from two-dimensional shape in the wedge are fully reproduced in the shock, or whether they disappear far from the leading edge. Here we make no attempt to predict whether advances in these fields will become possible from the application of improvements in the physical approaches which are principally used in this section, or from formal power series expansions, combined with the use of the author's mathematical technique (for removing their divergences) near shocks, etc.

At present the only nonlinear studies of three-dimensional wing theory have been made in relation to the regions of "conical flow" (Busemann's "kegelige Strömungsfelder" [*46*]) which, for many wing shapes, consti-

[10] The reason, of course, is that the boundary layer is swept off at the rear before it has time to separate. The cross flow, which has velocity $U\alpha$, takes a time approximately $\frac{1}{2}R(l)/U\alpha$ to separate [*55*, p. 185], and therefore will not do so if αl is less than about $\frac{1}{2}R(l)$. However, for longer projectiles, or higher angles of attack, the ordinary gradual separation process for low speed flow past a circular cylinder occurs in the cross flow. The resulting suction on the leeward side near the base adds to the lift (of which the remainder comes from the nose) and decreases the destabilizing yawing moment. But there is some experimental evidence that unfavorable complications of this regime occur when $U\alpha$ reaches about half the speed of sound, so that (Art. 2) the shock stall for the cross flow is attained.

tute the front portion of the flow pattern. (But since in these regions the flow quantities are independent of distance from the apex, the problems are not "fully three-dimensional" in the sense of the last paragraph.) It has been seen in Sec. D that on linear theory, for many triangular and quadrilateral planforms with symmetrical sections, the whole contribution to the surface pressures due to angle of attack can be obtained by evaluating a conical field, or superposing two or three such fields. However, where nonlinear effects are considered, the division of the problem into the effect of thickness and of angle of attack is not permissible, nor is the superposition of solutions. Hence the nonlinear theory of conical fields can be applied only to those parts of a flow field which are fully conical, which means that those parts of the boundary which influence the field must be generated by straight lines through a single apex. Thus the influence of angle of attack on the flow past flat plane wings, with planforms of the triangular or quadrilateral shapes alluded to above, can be so treated. Also the effect of thickness can be taken into account, *but only over the front portion of the wing*, if similar sections at different spanwise stations, arranged so as to meet this condition, for example wedge sections with uniform wedge angle, are used in that portion.

The study of the subject was begun by the present author [*33*] who used his general technique [*31*] to determine the strength of the shock (of conical shape, nearly but not quite right circular) in an arbitrary conical field, given only the solution to the problem on linear theory. This would appear to be an investigation of only theoretical interest, but when F. K. Moore [*47*] went on to determine the surface pressures to the second approximation in a special case, he found that the solution was not unique unless a certain boundary condition, derived from the author's theory, was applied on the Mach cone from the apex. When previously a simpler boundary condition was applied (in Moore's Ph.D. thesis), the resulting incorrect second order solution for the surface pressures differed far more widely from the linear theory than does the true solution. Moore's theory has since been amplified by Tan [*48*], who computes the surface pressures in a wider range of cases.

These three references constitute the whole existing literature on the subject of the present article, although further work is in progress, particularly in France on the use of an electric analogy to calculate the second approximate flow field, and in England on the problem of improving the linear theory for delta wings with rounded leading edge (inside the Mach cone from the apex) at angle of attack, in order to deduce a finite pressure distribution around the leading edge. This problem is similar to one encountered in subsonic flow (Art. 2), since the flow normal to the leading edge is subsonic, and it can perhaps be treated by similar methods.

The cone field theory will here be discussed only for cases when the wing lies entirely within the Mach cone from the apex. The reader is referred to [*33*] for a discussion of the shock strength in the general case, and in particular of the interesting shock behavior, near the tip of a rectangular wing at angle of attack, where the Prandtl-Meyer fan of expansion waves on the upper surface interacts with the tip shock, pulling it in toward the axis in a most peculiar corrugation. (See also [*57,58*].)

With this restriction, the author's conclusions will here be deduced by a simpler physical argument, based on the principle enunciated near the end of Art. 7. (Of course this is the more convincing for being supported by his original mathematical argument.) They will then be used in deriving the Moore-Tan theory of the surface pressures.

It is the general conclusion of this section that an adiabatic flow field can, even behind shocks, be evaluated to good approximation by solving the equations of potential motion. But it has been seen that the complete solution of these equations has in general singular regions, at each point of which it predicts three different values for each flow quantity. These regions must in practice be traversed by a shock, ahead of which the flow quantities take one of the three values, and behind which they take one of the others. For the two-dimensional flow past a sharp-nosed airfoil, the singular region associated with the front shock is of wedge shape, being bounded by a Mach line associated with the flow deflection at the nose on the upper surface, and by another associated with the undisturbed stream. For axisymmetric flow past a cone the region is bounded by the undisturbed Mach cone and by a cone of larger angle, which may be called the limit cone (being the upstream limit of disturbances in the potential flow) and which is an envelope of Mach lines. Such an envelope (not conical) also exists for any supersonic flow past slender projectiles, and may be obtained in the usual way by eliminating y from Eq. 6-6 and from the same equation differentiated with respect to y.

Similarly, it may be supposed that the exact solution, for potential flow, to our conical field problem has such a singular region, bounded behind by the undisturbed Mach cone from the apex, and ahead by a limit cone (not, of course, necessarily right circular), and that the shock cone must be placed in this region. The flow (and hence finally the shock position) is then deduced by applying word for word the principle stated in italics on p. 470, and expressing the condition that, since the limit cone is the forward boundary of the potential flow field, the component of velocity normal to it must equal the local speed of sound (for the limit cone, if not a stationary wave front itself, is at least an envelope of such wave fronts).

The first step is to write down the equations for the first and second approximate solutions. The first approximation φ_1 to the disturbance potential satisfies the linearized equation, while the second approximation, by Eq. 2-14, satisfies

$$\frac{\partial^2\varphi_2}{\partial r^2}+\frac{1}{r}\frac{\partial\varphi_2}{\partial r}+\frac{1}{r^2}\frac{\partial^2\varphi_2}{\partial\theta^2}-(M_\infty^2-1)\frac{\partial^2\varphi_2}{\partial x^2}$$

$$=M_\infty^2\frac{\partial}{\partial x}\left[\left(1+\frac{\gamma-1}{2}M_\infty^2\right)\left(\frac{\partial\varphi_1}{\partial x}\right)^2+\left(\frac{\partial\varphi_1}{\partial r}\right)^2+\frac{1}{r^2}\left(\frac{\partial\varphi_1}{\partial\theta}\right)^2\right] \quad (8\text{-}1)$$

in cylindrical coordinates. The solution of the linearized equation for φ_1 is most easily accomplished in terms of a variable

$$s=\cosh^{-1}\left(\frac{x}{(M_\infty^2-1)^{\frac{1}{2}}r}\right) \quad (8\text{-}2)$$

for then, in a cone-field problem, in which the disturbance velocities $u=\partial\varphi/\partial x$, $v=\partial\varphi/\partial y$, and $w=\partial\varphi/\partial z$ can depend only on nondimensional variables like s and θ, they are actually harmonic functions of s, θ, i.e. solutions of $\partial^2u/\partial s^2+\partial^2u/\partial\theta^2=0$. The determination of these to satisfy the boundary conditions and vanish on the undisturbed Mach cone $s=0$ is (D, 13) a fairly easy matter. The behavior of these velocities near the cone is given by equations like

$$\frac{\partial\varphi_1}{\partial x}\sim -A(\theta)\frac{s}{\sqrt{2}}\sim -A(\theta)\sqrt{\frac{x}{(M_\infty^2-1)^{\frac{1}{2}}r}-1} \quad (8\text{-}3)$$

where the function $A(\theta)$ is positive if the pressure increases (on linear theory) as the fluid passes the undisturbed Mach cone, and negative if it decreases. The forms for φ_1 and $\partial\varphi_1/\partial r$ near the Mach cone follow at once, whence the right-hand side of Eq. 8-1 becomes, in the limit as $x\to(M_\infty^2-1)^{\frac{1}{2}}r$,

$$M_\infty^2\frac{\partial}{\partial x}\left\{\left(1+\frac{\gamma-1}{2}M_\infty^2\right)A^2(\theta)\left[\frac{x}{(M_\infty^2-1)^{\frac{1}{2}}r}-1\right]\right.$$

$$\left.+(M_\infty^2-1)A^2(\theta)\left[\frac{x}{(M_\infty^2-1)^{\frac{1}{2}}r}-1\right]\right\}=\frac{(\gamma+1)M_\infty^4}{2(M_\infty^2-1)^{\frac{1}{2}}}\frac{A^2(\theta)}{r} \quad (8\text{-}4)$$

Now if φ_2 is taken continuous, and hence zero, on the undisturbed Mach cone $x=(M_\infty^2-1)^{\frac{1}{2}}r$, then both the third term, and the sum of the first and fourth terms, of the left-hand side of Eq. 8-1 vanish on the cone. Hence by Eq. 8-4

$$\frac{\partial\varphi_2}{\partial r}=\frac{(\gamma+1)M_\infty^4}{2(M_\infty^2-1)^{\frac{1}{2}}}A^2(\theta),\qquad \frac{\partial\varphi_2}{\partial\theta}=0 \quad (8\text{-}5)$$

on the cone, and the other nonvanishing component of velocity $\partial\varphi_2/\partial x$ follows in the form

$$\frac{\partial\varphi_2}{\partial x}=-\frac{r}{x}\frac{\partial\varphi_2}{\partial r}=-\frac{(\gamma+1)M_\infty^4}{2(M_\infty^2-1)}A^2(\theta) \quad (8\text{-}6)$$

from the fact that the field is conical.

The basic principle of the present method now tells us that Eq. 8-5 and 8-6 represent, not the velocity on the undisturbed Mach cone, but the velocity on the limit cone. The equation of the limit cone must take the general form

$$x = (M_\infty^2 - 1)^{\frac{1}{2}} r[1 - G(\theta)] \tag{8-7}$$

where $G(\theta)$ is small and positive. Expressing the condition that the component of velocity normal to this surface (calculated from the disturbance velocities (8-5) and (8-6)) is equal to the local speed of sound (calculated from Bernoulli's equation), and neglecting the squares of $G(\theta)$, $G'(\theta)$, φ_x, and φ_r, we obtain

$$G(\theta) = \frac{(\gamma + 1)^2 M_\infty^8}{4(M_\infty^2 - 1)^2} A^2(\theta) \tag{8-8}$$

Eq. 8-7 and 8-8 determine the limit cone as differing from the undisturbed Mach cone by an angle of the order of the square of the disturbances in the main part of the flow.

The approximation to $\partial\varphi/\partial x$ near the undisturbed Mach cone is given as the sum of Eq. 8-3 and 8-6, two terms which are of the same order of magnitude when the quantity under the square root in Eq. 8-3 is of the order of Eq. 8-8. But according to the basic principle, this expression would be correct only if the distance x were given a new origin, so that the curve $x - (M_\infty^2 - 1)^{\frac{1}{2}} r = 0$ became the limit cone. Thus, if no such alteration in the meaning of x is made, the true form of $\partial\varphi/\partial x$ in this neighborhood satisfies

$$\frac{\partial\varphi}{\partial x} \cong -A(\theta)\sqrt{\frac{x}{(M_\infty^2 - 1)^{\frac{1}{2}} r} - 1 + G(\theta)} - \frac{(\gamma + 1)M_\infty^4}{2(M_\infty^2 - 1)} A^2(\theta) \tag{8-9}$$

To the same order $\partial\varphi/\partial r = -(M_\infty^2 - 1)^{\frac{1}{2}}\partial\varphi/\partial x$.

The next term $\partial\varphi_3/\partial x$ would of course be infinite on $r = 1$ in successive approximation theory, but it is essential to the principle used that this is no longer so when the change in the equations due to shifting the origin of x (which begins to affect the φ_n for $n = 3$) is taken into account. The author [33] proves that this is in fact so. Hence Eq. 8-9 represents the terms of highest order in $\partial\varphi/\partial x$ near $x = (M_\infty^2 - 1)^{\frac{1}{2}} r$.

The shock must also be a cone, with equation say

$$x = (M_\infty^2 - 1)^{\frac{1}{2}} r[1 - H(\theta)] \tag{8-10}$$

The longitudinal disturbance velocity (8-9) behind it is related to the angle which it makes with the undisturbed stream. Expressing this relation, with terms of smaller order than $A^2(\theta)$ neglected, we have

$$\frac{4(M_\infty^2 - 1)}{(\gamma + 1)M_\infty^4} H(\theta) = A(\theta)\sqrt{G(\theta) - H(\theta)} + \frac{(\gamma + 1)M_\infty^4}{2(M_\infty^2 - 1)} A^2(\theta) \tag{8-11}$$

By suitable squaring, Eq. 8-11 can evidently be reduced to a quadratic equation in $H(\theta)$. But of the roots of this equation, only one is a solution of Eq. 8-11; the other satisfies the same equation with the sign changed in front of the square root. Actually, one of the said two roots is $H(\theta) = 0$. By Eq. 8-8 this is a solution of Eq. 8-11 when $A(\theta) < 0$. The other root, appropriate when $A(\theta) > 0$, is

$$H(\theta) = \frac{3}{4} G(\theta) = \frac{3}{16} \frac{(\gamma + 1)M_\infty^8}{(M_\infty^2 - 1)^2} A^2(\theta) \tag{8-12}$$

In Eq. 8-10 and 8-12 we have expressed the shock position, to the second order in the disturbances, in terms of the behavior (see Eq. 8-3) of the linearized solution to the cone field problem near the undisturbed Mach cone. But this expression has been shown to be the correct one only when $A(\theta) > 0$, i.e. when on the linear theory the pressure increases as the fluid passes the undisturbed Mach cone. When this is not so $H(\theta)$ is zero and the shock strength is at most of the order of the cube of the disturbances (the author [*33*] gives reasons for supposing it not to be exactly zero). The correspondence of this mathematical point with the physical fact that shock discontinuities are necessarily compressive is most interesting.

The pressure gradient along a streamline just behind the shock is easily calculated from Eq. 8-9 in each of the two cases. When $A(\theta) > 0$, so that Eq. 8-12 holds, it is given by

$$x \frac{\partial C_p}{\partial x} = \frac{2(M_\infty^2 - 1)}{(\gamma + 1)M_\infty^4} \tag{8-13}$$

which is not small, though pressure gradients of this magnitude act only over a small distance behind the shock. This thin region of rapid compression (not of course large enough to bring viscosity into play) "reinforces" the shock (itself a region of almost infinitely rapid compression). But when $A(\theta) < 0$, and so the shock is much weaker, the pressure gradient just behind it is given by

$$x \frac{\partial C_p}{\partial x} = - \frac{(M_\infty^2 - 1)}{(\gamma + 1)M_\infty^4} \tag{8-14}$$

and falls to half this value (for example) when $x/(M - 1_\infty^2)^{\frac{1}{2}}r$ is about $1 - 3G(\theta)$. The pressure drop in this region would probably be the most noticeable phenomenon on an interferometer record, as it exceeds the pressure rise at the shock. It is the analogue of the Prandtl-Meyer expansion in two-dimensional flow, in which also the pressure gradient is not small even though the total expansion be small.

When $A(\theta) > 0$ the strength of the shock is given, to the approximation considered, by

$$\frac{\Delta p}{p} = \frac{3}{4} \gamma(\gamma + 1) \frac{M_\infty^6}{M_\infty^2 - 1} A^2(\theta) \tag{8-15}$$

The author [*33*] gave two examples of Eq. 8-15. In one the obstacle was slender in all directions perpendicular to the stream. In this case $A(\theta) = (\Omega/\pi)\sqrt{2}$, in terms of the solid angle Ω at the apex. The resulting value for the strength agrees with that which follows from Whitham's later theory of Art. 6. In the second example the obstacle was a flat delta wing at angle of attack α. In the unyawed case, when the leading edges make an angle $\frac{1}{2}\pi - \tau$ with the incident stream, the shock strength below the wing is a multiple of α^2 varying with M_∞ like $M_\infty^6/(M_\infty^2 - 1)^2$ and with θ like $\sin^2\theta[1 - (M_\infty^2 - 1)\cot^2\tau\cos^2\theta]^{-3}$; above the wing it is of smaller order than α^2. There is one maximum of shock strength, in the middle, when $\tau > \tan^{-1}\sqrt{3(M_\infty^2 - 1)}$. For example when $M_\infty = \sqrt{2}$ and $\tau = 67.5°$ this maximum shock strength is only 0.026 for $\alpha = 10°$. But when $\tau < \tan^{-1}\sqrt{3(M_\infty^2 - 1)}$ the strength rises to a maximum on two symmetrically placed generators with $\cos^2\theta = \frac{1}{2}[3 - (M_\infty^2 - 1)^{-1}\tan^2\tau]$, while the central generator is a local minimum of strength. For example when $M_\infty = \sqrt{2}$, $\tau = 55°$, and $\alpha = 10°$ the maximum strength is 0.07; and this continues to increase as the leading edge approaches the undisturbed Mach cone.

We pass now to the calculation of the full second approximation to the flow field, especially on the body surface. The boundary condition appropriate on the Mach cone is shown by the above theory to be the very natural one obtained by using Eq. 8-1 for the second approximation also in the limit as $x \to (M_\infty^2 - 1)^{\frac{1}{2}}r$. The velocity vector on the Mach cone is in fact given by Eq. 8-5 and 8-6. In terms of the velocities $u_2 = \partial\varphi_2/\partial x$, $v_2 = \partial\varphi_2/\partial y$, $w_2 = \partial\varphi_2/\partial z$ which are found most convenient for solving the equations, this boundary condition becomes (using the definition (8-3) of $A(\theta)$)

$$u_2 = -\frac{(\gamma + 1)M_\infty^4}{M_\infty^2 - 1}\left(\frac{\partial u_1}{\partial s}\right)^2_{s=0}, \qquad v_2 = \frac{(\gamma + 1)M_\infty^4}{(M_\infty^2 - 1)^{\frac{1}{2}}}\left(\frac{\partial u_1}{\partial s}\right)^2_{s=0}\cos\theta,$$

$$w_2 = \frac{(\gamma + 1)M_\infty^4}{(M_\infty^2 - 1)^{\frac{1}{2}}}\left(\frac{\partial u_1}{\partial s}\right)^2_{s=0}\sin\theta \qquad (8\text{-}16)$$

But since the homogeneous part of Eq. 8-1, for the second approximation to the potential, is the same as the equation for the first approximation, the same is true of the equations for the velocities. But in the variables s (Eq. 8-2) and θ, Laplace's equation is satisfied by u_1, v_1, and w_1, and therefore equations of Poisson's type are satisfied by u_2, v_2, and w_2. Moore [*47*] worked out these equations, and also the relations which must subsist between u_2, v_2, and w_2 in the s, θ plane due to irrotationality. By working with complex variables he then obtains a particular integral for u_2 (rather as Eq. 7-22 was obtained). To this he has merely to add a solution of the ordinary linear cone-field equations, so that the whole satisfies the boundary conditions.

When the wing lies approximately in the plane $z = 0$, the boundary conditions thereon can be approximated (by using Taylor's theorem) in the form of a statement of the values of $\partial\varphi_1/\partial z = w_1$ and $\partial\varphi_2/\partial z = w_2$ on the plane $z = 0$. Using the irrotationality conditions we may then infer the boundary conditions on u_1 and u_2 on $z = 0$, and the kind of singularities which they may be permitted to satisfy at the leading edge. The second approximation may then be obtained uniquely using finally that Eq. 8-16 holds on $s = 0$. (The detailed process of calculation is not here discussed, as it is still in such an early stage that great improvements are certain to occur later on.) From u_2 and the first approximation the pressure coefficient follows in the form

$$C_p = -2u_1 - [2u_2 - (M_\infty^2 - 1)u_1^2 + v_1^2 + w_1^2] \tag{8-17}$$

Moore [47] and Tan [48] both considered, as an example of the method, the unyawed flow past a delta wing, whose airfoil section is a wedge of semiangle ϵ, at zero angle of attack. The cross section of the wing by a plane normal to the stream is thus a rhombus, whose center is the foot of the perpendicular from the apex to the said plane. If the planform makes an angle 2λ at the apex, the diagonals of this rhombus are in the ratio $\tan \epsilon : \tan \lambda$.

Moore considered only a single case, but Tan calculated the pressure distribution for a series of values for M_∞, λ, and ϵ. He took $M_\infty = 1.2$, $\sqrt{2}$ $(= 1.414)$, and 2, and used four different values for the parameter $(M_\infty^2 - 1)^{\frac{1}{2}} \tan \lambda$, which is the ratio of wing span to diameter of Mach cone on any plane normal to the stream. Since the wing is inside the Mach cone this parameter must be less than 1, and the four values used for it were 0.3651, 0.5505, 0.8, and 0.9535.

Tan's graphs of pressure distribution show quite clearly that the relation of the second approximation to the first remains closely similar for different M_∞ if the parameter $(M_\infty^2 - 1)^{\frac{1}{2}} \tan \lambda$ is kept constant. To be precise, at points whose distances from the axis of the Mach cone are a given fraction of its local radius, the ratio of the two approximations to the pressure coefficient changes little with Mach number. To illustrate the general behavior we may therefore confine ourselves to the single value $\sqrt{2}$ for M_∞.

For this value of M_∞, for $\epsilon = 3°$, and for the four values of

$$(M_\infty^2 - 1)^{\frac{1}{2}} \tan \lambda$$

listed above (which for $M_\infty = \sqrt{2}$ are simply $\tan \lambda$ and correspond to $\lambda = 20°4'$, $28°50'$, $38°39'$, and $43°38'$) the first and second approximations to the distribution of pressure coefficient C_p along the surface are illustrated in Fig. E,8. In this figure we work in a plane normal to the incident stream, which cuts the Mach cone in a circle whose radius is taken as the unit of length. Then the plane cuts the wing in four alternative sections as illustrated (confining ourselves by symmetry to the right-hand half of each) at the bottom of the figure. The two approximations to C_p are graphed for each wing until just before the tip. At the tip C_p takes in reality a fairly large value. This may be estimated, by assuming that

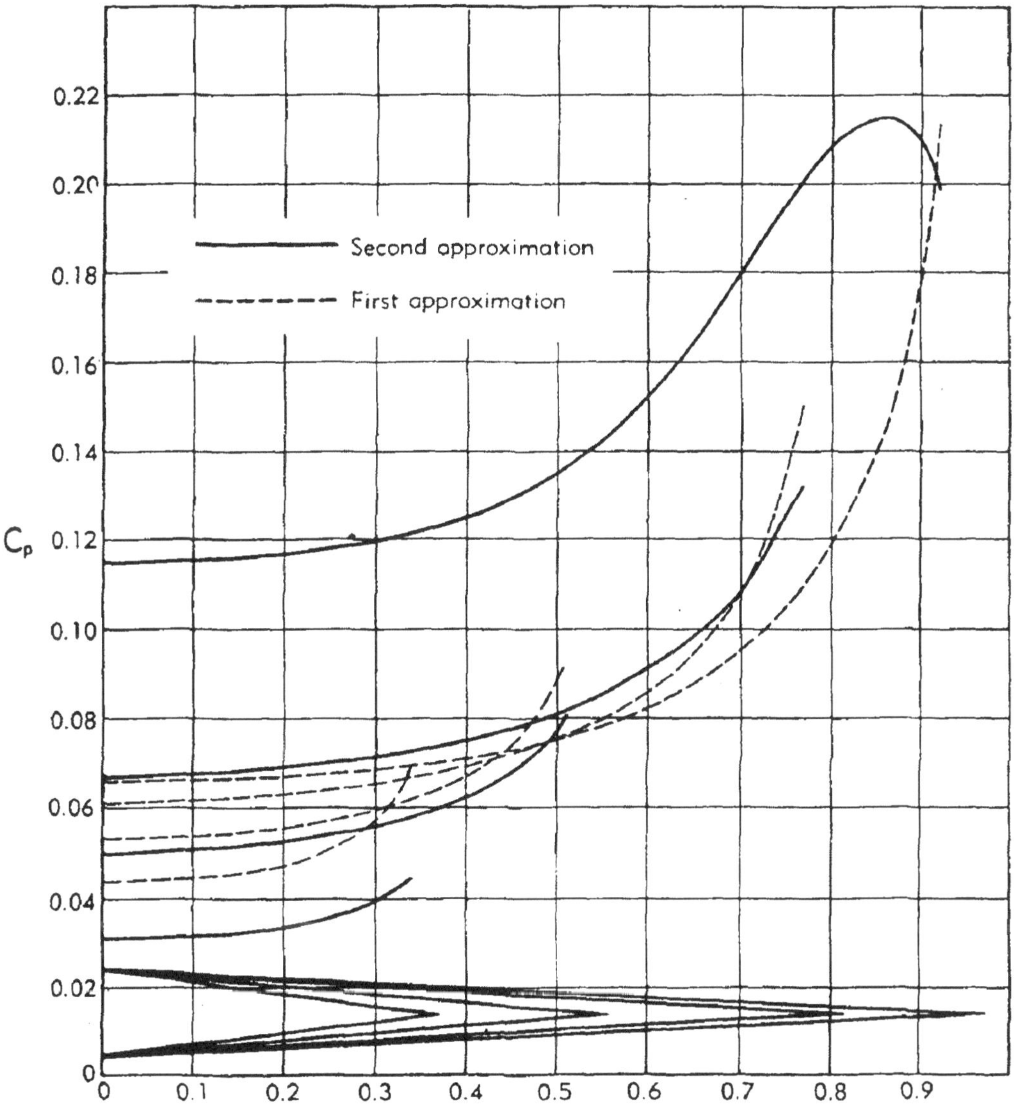

Fig. E,8. Pressure distribution on four delta wings, each with uniform wedge airfoil section of semiangle 3°, at Mach number $\sqrt{2}$.

the component of fluid velocity normal to the leading edge is reduced to zero there (by analogy with the subsonic flow past a wedge), while the component along the edge is as in the incident stream, as 0.125, 0.26, 0.47, and 0.60 in the four cases respectively. But, of course, no method of successive approximation could be expected to give the correct value here; in fact both approximations give integrable infinities for C_p at the tip. (There is no need to give the corresponding curves for other values of ϵ, the semiangle of the wedge airfoil section, since the first approximation to C_p is simply proportional to ϵ, and the difference between the two approximations is proportional to ϵ^2.)

We see from the figure that for $\epsilon = 3°$ the accuracy of the first approximation to the pressure distribution (as inferred from its difference from the second approximation) would be tolerable for practical purposes only in an intermediate range of $(M_\infty^2 - 1)^{\frac{1}{2}} \tan \lambda$, including 0.5505 and 0.8 but not 0.3651 and 0.9535. It is not in the least surprising that the error increases as $(M_\infty^2 - 1)^{\frac{1}{2}} \tan \lambda$ approaches 1, i.e. as the leading edge approaches the Mach cone (so that the flow normal to it is transonic) and this deficiency is far from remediable in the present state of knowledge. It is much odder, however, that the error should increase as λ decreases, so that when the planform has apex angle $2\lambda = 40°8'$ the error is no longer tolerable. It might be thought that this is due to the body becoming slender in all directions normal to the stream, as in Ward's theory (Art. 4), so that the terms in v_1^2 and w_1^2 in the expression (8-17) for C_p need to be taken into account even on a first approximation; but in fact λ is still much greater than ϵ, the terms in the squares of velocities in Eq. 8-17 are unimportant, and the difference u_2 in the axial perturbation velocities is just as considerable as the difference in the pressures.

This being so, it is very hard to believe that the linear theory should be in error as much as Fig. E,8 suggests, considering that the wing surface makes everywhere an angle less than 3° with the stream. It is probable, in fact, that the second approximation may in this case be farther from the truth than the first, since in its deduction from the first the singularity at the leading edges may play a dominant part, accentuated by the proximity of the leading edges for smaller λ and by the method of Taylor expansion to obtain boundary conditions on the plane $z = 0$. Thus we may, as indicated so often in this section, have run into difficulties by too readily using a formal expansion in series instead of proceeding from one approximation to another by physical reasoning.

It may nevertheless be the case that the theory is adequate in the intermediate range of $(M_\infty^2 - 1)^{\frac{1}{2}} \tan \lambda$, and if so it is very interesting that, as Tan remarks (interpolating between the curves for 0.5505 and for 0.8 in Fig. E,8), the first and second approximations to C_p are almost indistinguishable for $(M_\infty^2 - 1)^{\frac{1}{2}} \tan \lambda = 0.65$. This property is reproduced for $M_\infty = 1.2$ and for $M_\infty = 2$, and so may be supposed independent of Mach number. Whether it is also independent of airfoil section, so that the linear theory is always specially good for delta wings spanning two thirds of the Mach cone from the apex, is of course not known.

In future calculations on delta wings inside the Mach cone, airfoils with rounded leading edges should be used, since these are in practice essential to prevent stalling at low angles of attack. An example would be the right elliptical cone, for which the linear theory was calculated by Squire [*49*] and by Laporte and Bartels [*51*]. Boundary conditions should preferably be satisfied on the surface itself as in the Hantzsche-Wendt method, or if they are satisfied on the plane of symmetry a special technique should be applied to treat conditions near the leading edge (Art. 2).

There is little in Fig. E,8 to support the suggestion, made at the beginning of this article, that the two-dimensional result, "a given value of C_p on linear theory signifies a true value nearer to $C_p + (C_2/C_1^2)C_p^2$," might be true in some degree of generality. It is correct only for values of $(M_\infty^2 - 1)^{\frac{1}{2}} \tan \lambda$ around 0.8. However, although it is shown to be false, for a delta wing with leading edges markedly "subsonic," in the front portion where the field is conical, it might be expected to apply with greater accuracy to the pressure *changes* as the flow expands (more nearly two-dimensionally) around the rear portion of the wing (compare the expansion round the corner in Fig. E,7b).

If matters such as this are to be decided, the extension of the theories described in Sec. E to "fully three-dimensional" problems is necessary.

E,9. Cited References.

1. Janzen, O. Beitrag zu einer Theorie der stationären Strömung kompressibler Flüssigkeiten. *Phys. Zeits. 14*, 639 (1913).
2. Rayleigh, Lord. On the flow of compressible fluid past an obstacle. *Phil. Mag. 32*, 1 (1916).
3. Eser, F. Zur Strömung kompressibler Flüssigkeiten um feste Körper mit Unterschallgeschwindigkeit. *Luftfahrtforschung 20*, 220 (1943).
4. Imai, I. On the flow of a compressible fluid past a circular cylinder. *Proc. Phys. Math. Soc. Japan 20*, 636 (1938).
5. Tamada, K. On the flow of a compressible fluid past a sphere. *Proc. Phys. Math. Soc. Japan 21*, 743 (1939).
6. Kaplan, C. Compressible flow about symmetrical Joukowski profiles, A theoretical study of the moment on a body in a compressible fluid, *and* On the use of residue theory for treating the subsonic flow of a compressible fluid. *NACA Reports 621*, 1938; *671*, 1939; and *728*, 1942.
7. Goldstein, S., and Lighthill, M. J. Two-dimensional compressible flow past a solid body in unlimited fluid or symmetrically placed in a channel. *Phil. Mag. 35*, 549 (1944).
8. Tomotika, S., and Umemoto, H. On the subsonic flow of a compressible fluid past a symmetrical Joukowski aerofoil. *Rep. Aero. Res. Inst. Tokyo 16*, 35 (1941).
9. Taylor, G. I., and Sharman, C. F. A mechanical method for solving problems of flow in compressible fluids. *Proc. Roy. Soc. A121*, 194 (1928).
10. Görtler, H. Gasströmungen mit Übergang von Unterschall- zu Überschallgeschwindigkeiten. *ZAMM 20*, 254 (1940).
11. Hantzsche, W., and Wendt, H. Der Kompressibilitätseinfluss für dünne wenig gekrümmte Profile bei Unterschallgeschwindigkeit. *ZAMM 22*, 72 (1942).
12. Kaplan, C. Effect of compressibility at high subsonic velocities on the lifting force acting on an elliptic cylinder. *NACA Technical Note 1118*. 1946.
13. Glauert, H. *Elements of Aerofoil and Airscrew Theory*. Cambridge Univ. Press, 1926.
14. Theodorsen, T. Theory of wing sections of arbitrary shape, *NACA Report 411*. 1931.
15. Hantzsche, W. Die Prandtl-Glauertsche Näherung als Grundlage für ein Iterations-

verfahren zur Berechnung kompressibler Unterschallströmungen. *ZAMM 23*, 185 (1943).
16. Tsien, H. S. Two-dimensional subsonic flow of compressible fluids. *J. Aero. Sci. 6*, 399 (1939).
17. Prandtl, L. Neue Untersuchungen über die strömende Bewegung der Gase und Dämpfe. *Phys. Zeits. 8*, 23 (1907).
18. Ackeret, J. Luftkräfte auf Flügel die mit grösserer als Schallgeschwindigkeit bewegt werden. *Z. Flugtech. u. Motorluftschiffahrt 16*, 72 (1925).
19. Friedrichs, K. O. Formation and decay of shock waves. *Comm. Pure and Appl. Math. 1*, 211 (1948).
20. Busemann, A. Aerodynamischer Auftrieb bei Überschallgeschwindigkeit. *Luftfahrtforschung 12*, 210 (1935).
21. Epstein, P. S. On the air resistance of projectiles. *Proc. Nat. Acad. Sci. 17*, 532 (1931).
22. Donov, A. A plane wing with sharp edges in a supersonic stream. *Bull. Acad. Sci. URSS. Sér. Math. 1939*, 603 (1939).
23. Thomas, T. Y. The determination of pressure on curved bodies behind shocks. *Comm. Pure and Appl. Math 3*, 103 (1950).
24. Chandrasekhar, S. On the decay of plane shock waves. *Ballistics Research Lab. Report 423*. 1943.
25. Lighthill, M. J. The conditions behind the trailing edge of the supersonic aerofoil. *ARC Reports and Memoranda 1930*. 1944.
26. Martin, M. H. A problem in the propagation of shock. *Quart. Appl. Math. 4*, 330 (1947).
27. Meyer, R. E. The method of characteristics for problems of compressible flow involving two independent variables. *Quart. J. Mech. and Appl. Math. 1*, 196–219 (1948).
28. Lighthill, M. J. The flow behind a stationary shock. *Phil. Mag. 40*, 214 (1949).
29. Kahane, A., and Lees, L. The flow at the rear of a two-dimensional supersonic airfoil. *J. Aero. Sci. 15*, 167–170 (1948).
30. Lighthill, M. J. The energy distribution behind decaying shocks. I. Plane waves. *Phil. Mag. 41*, 1101 (1950).
31. Lighthill, M. J. A technique for rendering approximate solutions to physical problems uniformly valid. *Phil. Mag. 40*, 1179 (1949).
32. Whitham, G. B. The behaviour of supersonic flow past a body of revolution, far from the axis. *Proc. Roy. Soc. A201*, 89 (1950).
33. Lighthill, M. J. The shock strength in supersonic "conical fields." *Phil. Mag. 40*, 1202 (1949).
34. Broderick, J. B. Supersonic flow round pointed bodies of revolution, *and* Supersonic flow past a semi-infinite cone, *Quart. J. Mech. and Appl. Math. 2*, 98–120; 121–128 (1949).
35. Lighthill, M. J. Supersonic flow past slender bodies of revolution the slope of whose meridian section is discontinuous. *Quart. J. Mech. and Appl. Math. 1*, 90 (1948).
36. Lighthill, M. J. The position of the shock-wave in certain aerodynamic problems. *Quart. J. Mech. and Appl. Math. 1*, 309 (1948).
37. Kopal, Z. *Tables of Supersonic Flow Around Cones*, and *Tables of Supersonic Flow Around Yawing Cones*, Mass. Inst. of Tech., 1947.
38. Taylor, G. I., and Maccoll, J. W. The air pressure on a cone moving at high speeds. *Proc. Roy. Soc. A139*, 278 (1933).
39. DuMond, J. W. M., Cohen, E. R., Panofsky, W. K. H., and Deeds, E. A determination of the wave forms and laws of propagation and dissipation of ballistic shock waves. *J. Acoust. Soc. Amer. 18*, 97 (1946).
40. Lighthill, M. J. Supersonic flow past slender pointed bodies of revolution at yaw. *Quart. J. Math. and Appl. Math. 1*, 76 (1948).
41. von Kármán, Th., and Moore, N. B. Resistance of slender bodies moving with supersonic velocities, with special reference to projectiles. *Trans. ASME 54*, 303–310 (1932).
42. Lighthill, M. J. Supersonic flow past bodies of revolution. *ARC Reports and Memoranda 2003*. 1945.

43. Ward, G. N. Supersonic flow past slender pointed bodies. *Quart. J. Mech. and Appl. Math. 2*, 75 (1949).
44. Watson, G. N. *Theory of Bessel Functions*, 2nd ed. Cambridge Univ. Press, 1944.
45. Ward, G. N. The approximate external and internal flow past a quasi-cylindrical tube moving at supersonic speeds. *Quart. J. Mech. and Appl. Math. 1*, 225 (1948).
46. Busemann, A. Infinitesimale kegelige Überschallströmung. *Luftfahrtforschung 20*, 105 (1943).
47. Moore, F. K. Second approximation to supersonic conical flows. *J. Aero. Sci. 17*, 328 (1950).
48. Tan, H. S. *Second Approximation to Conical Flows.* Cornell Univ. Grad. School of Aero. Engrg., Ithaca, 1950. Also *WADC Technical Report 52-277*, Wright Air Dev. Center, Dayton, O., 1952.
49. Squire, H. B. Theory of the flow over a particular wing in a supersonic stream. *Roy. Air. Est. Rept. 2184.* Farnborough, 1947.
50. Tsien, H. S. Supersonic flow over an inclined body of revolution. *J. Aero. Sci. 5*, 480 (1938).
51. Laporte, O., and Bartels, R. C. F. An investigation of the exact solutions of the linearized equations for the flow past conical bodies. *Bumblebee Series Report 75.* University of Michigan, Ann Arbor, 1948.
52. Kaplan, C. The flow of a compressible fluid past a curved surface. *NACA Report 768.* 1943.
53. Kaplan, C. On the particular integrals of the Prandtl-Busemann iteration equations for the flow of a compressible fluid. *NACA Technical Note 2159.* 1950.
54. Van Dyke, M. First- and second-order theory of supersonic flow past bodies of revolution. *J. Aero. Sci. 18*, 161 (1951).
55. Goldstein, S. *Modern Developments in Fluid Dynamics.* Oxford Univ. Press, 1938.
56. Lighthill, M. J. A new approach to thin aerofoil theory. *Aero. Quart. 3*, 193 (1951).
57. Legras, M. J. Application de la méthode de Lighthill à un écoulement plan supersonique. *C.R. Acad. Sci. Paris 233*, 1005 (1951).
58. Legras, M. J. Écoulement conique au voisinage d'un point de jonction. *C.R. Acad. Sci. Paris 234*, 181 (1952).
59. Tan, H. S. *On Mach Reflection and Strength of a Reflected Shock.* Cornell Univ. Grad School of Aero. Engrg., Ithaca. Also *WADC Technical Report 52-163*, Wright Air Dev. Center, Dayton, O., 1952.
60. Miles, J. W. On virtual mass and transient motion in subsonic compressible flow. *Quart. J. Mech. and Appl. Math. 4*, 388 (1951).
61. Longhorn, A. L. The unsteady, subsonic motion of a sphere in a compressible inviscid fluid. *Quart. J. Mech. and Appl. Math. 5*, 64 (1952).
62. Longhorn, A. L. Subsonic compressible flow past bluff bodies. To be published in *Aero. Quart.*
63. Holder, D. W., and Chinneck, A. Observations of the flow past elliptic-nosed two-dimensional cylinders and bodies of revolution in supersonic airstreams. *ARC Report 14216* (unpublished). To be published in *Aero. Quart.*
64. Mair, W. A., and Bardsley, O. Separation of the boundary layer at a slightly blunt edge in supersonic flow. *Phil. Mag. 43*, 344 (1952).
65. Bardsley, O. The conditions at a sharp leading edge in supersonic flow. *Phil. Mag. 42*, 255 (1951).
66. Whitham, G. B. The flow pattern of a supersonic projectile. *Comm. on Pure and Appl. Math. 5*, 301 (1952).
67. Eggers, A. J., and Syvertson, C. A. Inviscid flow about airfoils at high supersonic speeds. *NACA Technical Note 2646*, 1952.

SECTION F

PLANE SUBSONIC AND TRANSONIC POTENTIAL FLOWS

Y. H. KUO
W. R. SEARS

F,1. Introduction. In classical hydrodynamics we take as a model a fluid that is both inviscid and incompressible. If the flow is assumed to be irrotational, as is frequently justified on physical grounds, a mathematical theory of hydrodynamics is built up upon the branch of mathematical physics known as potential theory. Fundamental questions of the existence and uniqueness of solutions can be answered with great generality in this framework. Once the hypothesis of incompressibility is rejected, however, the differential equations of the problem are nonlinear. Thus, high speed aerodynamics has a distinctive character, as has already been pointed out in Sec. A, and is no longer a branch of classical potential theory.

To meet this situation, a number of approximate techniques have been developed, and useful results obtained; these are the methods described in Sec. C, D, and E of this volume. In general, however, these do not suffice to answer questions such as existence and uniqueness of solutions. Moreover, they usually involve approximations of unknown magnitudes, so that their accuracy cannot be assessed, and it is desirable to compare their predictions with exact solutions, if such can be found. In the special case of steady plane flow, these objectives can be realized to some extent by the introduction of the *hodograph* method, i.e. the use of velocity components as independent variables. This is the subject of the present section.

In a broad sense, the hodograph method is not new in fluid mechanics. Prior to its application in high speed aerodynamics, Helmholtz *[58]* solved a number of problems involving plane walls and constant pressure surfaces (free boundaries), using the velocity plane to simplify the boundary-value problems. This method was subsequently perfected and has become a standard technique in dealing with this type of problem in hydrodynamics. The importance of this device as a means to linearize the equation of a certain gas flow was first recognized by Riemann *[1]* in his famous investigation of wave propagation.

Finally, at the end of the nineteenth century, Molenbroeck [*59*] and Chaplygin [*2*] gave essentially the method used in this section, and thus succeeded in reducing the steady potential flow of a gas to a linear problem in the hodograph plane.

Between the problem of jets considered by Molenbroeck and Chaplygin and the typical aerodynamic problem of flow past a cylindrical obstacle, however, there is an essential difference. The hodograph method is ideally suited to describe the free surface boundary condition that occurs in the jet problems described above. In the problem of flow past a cylinder, however, it linearizes the problem but complicates the boundary conditions, for specified physical conditions in the flow plane correspond to unknown boundaries in the velocity plane. What is even more serious, the boundary is dependent upon a parameter of the flow, such as free stream Mach number. The advantage gained by the transformation is therefore compensated, to a great extent, by the loss of simplicity in the boundary-value problem. Nevertheless, when the exact shape of boundary is not essential, exact solutions for particular boundaries are obtainable, and the salient features of a nonlinear motion can be studied.

In this section, we shall deal with isentropic plane potential flows that are either subsonic everywhere or partly subsonic and partly supersonic; the latter may be called mixed flows or transonic potential flows. Those involving shock waves or computed by methods other than the hodograph transformation will be excluded; these are discussed in Sec. G and H. Inasmuch as purely subsonic flows do not yield the most interesting new phenomena, our main concern will be the mixed subsonic-supersonic flows.

As mentioned above, a solution constructed in the hodograph plane cannot usually be made to satisfy a prescribed boundary condition in the flow plane for flow past a surface. Instead, a guess has to be made, in the hodograph plane, as to the form of solution which, when transformed, will yield a useful flow pattern. There are two such procedures, due, respectively, to Chaplygin and Cherry. In the case of a closed body, Chaplygin's method proceeds by analogy with an incompressible flow, namely, the solution in the limit of vanishing Mach number reduces to a known flow in the flow plane (Art. 5–8). On the other hand, when such an analogy is not possible, a solution has to be selected possessing the desired singular behavior in the whole velocity plane (Art. 9). In principle, Cherry's method is the same as Chaplygin's, as both involve first the determination of the analytic behavior of the solution, but Cherry's approach is indirect, as will be seen in Art. 9 below, and hence more difficult but also more general.

In the closing articles of this section, we discuss the apparent nonappearance of steady, mixed, potential flows in experiment, and some of the proposed explanations of this dilemma.

In the space available in this volume, it has been found impossible to reproduce in complete detail even the most important works using these methods. We are also forced to omit entirely a number of interesting investigations and merely to list them in an appended bibliography. In spite of this, the amount of mathematical analysis included in the section may seem objectionable to some readers (while it will certainly be considered insufficient by others). It has been our intention to point out the main features of the methods without unnecessary mathematical elaboration, but it must be agreed that many of these main features are mathematical in nature, and therefore cannot be discussed without equations and formulas. One compromise has been to collect in an appendix (Art. 16) a number of lemmas and formulas pertaining to the hypergeometric functions which are used occasionally throughout the section.

F,2. The Hodograph Transformation. Let u, v be the velocity components in the directions of the Cartesian coordinates x, y, respectively, and let ρ be the density of the fluid. If the flow is steady, isentropic, and irrotational, there exist both velocity potential $\phi(x, y)$ and stream function $\psi(x, y)$, defined by

$$u = \frac{\partial \phi}{\partial x}, \qquad v = \frac{\partial \phi}{\partial y}$$

and

$$\rho u = \rho^0 \frac{\partial \psi}{\partial y}, \qquad \rho v = -\rho^0 \frac{\partial \psi}{\partial x} \tag{2-1}$$

where ρ^0 is the stagnation density. By the elimination of u and v, we have as one fundamental system for two-dimensional flow,

$$\begin{aligned} \frac{\partial \phi}{\partial x} &= \frac{\rho^0}{\rho} \frac{\partial \psi}{\partial y} \\ \frac{\partial \phi}{\partial y} &= -\frac{\rho^0}{\rho} \frac{\partial \psi}{\partial x} \end{aligned} \tag{2-2}$$

In the case of isentropic potential flow, pressure p and density ρ are related by the equation of state

$$p = k\rho^\gamma \tag{2-3}$$

where k is a constant and γ the ratio of specific heats. In this case the equation of motion for steady flow can be integrated to give ρ in terms of the speed q $(= \sqrt{u^2 + v^2})$:

$$\rho = \rho^0 \left[1 - \frac{\gamma - 1}{2} \frac{q^2}{(a^0)^2}\right]^{\frac{1}{\gamma - 1}} \tag{2-4}$$

where a^0 is the speed of sound $a = \sqrt{\gamma p/\rho}$ at stagnation conditions.

By means of Eq. 2-4, the elimination of the stream function ψ from Eq. 2-2 yields a differential equation for ϕ (cf. III,A,9, Eq. 9-11):

$$(a^2 - \phi_x^2)\phi_{xx} - 2\phi_x\phi_y\phi_{xy} + (a^2 - \phi_y^2)\phi_{yy} = 0 \tag{2-5}$$

where the subscripts denote partial differentiation and the speed of sound a is

$$a^2 = (a^0)^2 - \frac{\gamma - 1}{2} q^2 = (a^0)^2 - \frac{\gamma - 1}{2} (\phi_x^2 + \phi_y^2) \tag{2-6}$$

Eq. 2-5 is a second order, nonlinear, differential equation with coefficients involving the first derivatives of the unknown function. For flows in which q is everywhere less than the speed of sound a, the differential equation is of *elliptic* type. For flows in which q is everywhere greater than a, the equation is of *hyperbolic* type. Since the coefficients are variable, conditions may occur where both elliptic and hyperbolic domains coexist and the flow therefore changes its character locally. For example, in the flow past an object placed in a subsonic stream of gas, the flow may locally become supersonic and then return again to subsonic. The possibility of the existence of such smooth, mixed, transonic flows is of both mathematical and practical importance.

In the hyperbolic domain, the differential equation (2-5) is associated with two families of real characteristics, described by the ordinary differential equation

$$(a^2 - u^2)dy^2 + 2uvdxdy + (a^2 - v^2)dx^2 = 0$$

or

$$(a^2 - u^2)dy = (-uv \pm a\sqrt{q^2 - a^2})dx \tag{2-7}$$

The significance of these characteristic curves in the hyperbolic region is discussed in several parts of the volume (see B, 2). Mathematically, it may be said the analyticity of the solution in a supersonic region may break down along these lines, instead of at isolated points as in the subsonic region. Physically, these lines of discontinuity represent weak disturbances or wave fronts whose orientation with respect to local flow direction was first given by E. Mach. They are the *Mach lines* frequently visible in schlieren photographs. Eq. 2-7 shows that Eq. 2-5 actually changes its type, as described above, along lines where $q = a = a^*$.

This brief discussion shows us that the transonic character of isentropic compressible flow is intimately related to the nonlinearity of the differential equation (2-5). In studying this type of flow, therefore, we cannot avoid this nonlinearity. On the other hand, we have no adequate technique to deal directly with such nonlinear equations. To get around this difficulty, the differential equation may be transformed to a linear form by a method first suggested by Riemann [*1*], in which the roles of dependent and independent variables are interchanged. In the following, we shall give two well-known methods, namely, the Chaplygin transformation and the Legendre transformation.

Chaplygin transformation. The Chaplygin transformation consists of a straightforward change of independent variables from the coordinates x, y to the hodograph coordinates q, θ, where q is the magnitude of the flow velocity vector and θ its inclination relative to the x axis. The functions ϕ and ψ are retained as dependent variables. The analysis given here is substantially that of Chaplygin's original paper [2]. A different derivation, essentially geometrical, has been given by von Kármán [3] and is reproduced in III,A,13; it is suggested that the reader compare these methods.

To effect this transformation, we start by writing

$$\begin{aligned} d\phi &= udx + vdy \\ \rho^0 d\psi &= -\rho vdx + \rho udy \end{aligned} \tag{2-8}$$

in accord with Eq. 2-1. If $\rho q \neq 0$, the inverse relations are

$$\begin{aligned} dx &= \frac{1}{q}\left[\cos\theta d\phi - \frac{\rho^0}{\rho}\sin\theta d\psi\right] \\ dy &= \frac{1}{q}\left[\sin\theta d\phi + \frac{\rho^0}{\rho}\cos\theta d\psi\right] \end{aligned} \tag{2-9}$$

where θ is the inclination of the velocity vector defined above. Now, if q and θ are taken as independent variables, and ϕ and ψ are now treated as functions of q and θ, we can write

$$\begin{aligned} d\phi &= \phi_q dq + \phi_\theta d\theta \\ d\psi &= \psi_q dq + \psi_\theta d\theta \end{aligned}$$

Substituting these expressions for $d\phi$ and $d\psi$ into Eq. 2-9 we have

$$\begin{aligned} dx &= \frac{1}{q}\left[\cos\theta\phi_q - \frac{\rho^0}{\rho}\sin\theta\psi_q\right]dq + \frac{1}{q}\left[\cos\theta\phi_\theta - \frac{\rho^0}{\rho}\sin\theta\psi_\theta\right]d\theta \\ dy &= \frac{1}{q}\left[\sin\theta\phi_q + \frac{\rho^0}{\rho}\cos\theta\psi_q\right]dq + \frac{1}{q}\left[\sin\theta\phi_\theta + \frac{\rho^0}{\rho}\cos\theta\psi_\theta\right]d\theta \end{aligned} \tag{2-10}$$

Since both dx and dy are total differentials, the condition of integrability requires that

$$\begin{aligned} \frac{\partial}{\partial\theta}\left[\frac{\cos\theta}{q}\phi_q - \frac{\rho^0}{\rho}\frac{\sin\theta}{q}\psi_q\right] &= \frac{\partial}{\partial q}\left[\frac{\cos\theta}{q}\phi_\theta - \frac{\rho^0}{\rho}\frac{\sin\theta}{q}\psi_\theta\right] \\ \frac{\partial}{\partial\theta}\left[\frac{\sin\theta}{q}\phi_q + \frac{\rho^0}{\rho}\frac{\cos\theta}{q}\psi_q\right] &= \frac{\partial}{\partial q}\left[\frac{\sin\theta}{q}\phi_\theta + \frac{\rho^0}{\rho}\frac{\cos\theta}{q}\psi_\theta\right] \end{aligned}$$

A simple reduction gives

$$\left(\frac{1}{q}\,\phi_\theta - \frac{\rho^0}{\rho}\,\psi_q\right)\cos\theta - \left[-\frac{1}{q}\,\phi_q + \frac{d}{dq}\left(\frac{\rho^0}{\rho q}\right)\psi_\theta\right]\sin\theta = 0$$

$$\left[-\frac{1}{q}\,\phi_q + \frac{d}{dq}\left(\frac{\rho^0}{\rho q}\right)\psi_\theta\right]\cos\theta + \left(\frac{1}{q}\,\phi_\theta - \frac{\rho^0}{\rho}\,\psi_q\right)\sin\theta = 0$$

For any θ, the vanishing of the determinant leads to

$$\begin{aligned} \phi_q &= -\frac{\rho^0}{\rho q}\,(1 - M^2)\psi_\theta \\ \frac{1}{q}\,\phi_\theta &= \frac{\rho^0}{\rho}\,\psi_q \end{aligned} \tag{2-11}$$

where we have made use of the relations

$$\frac{d}{dq}\left(\frac{\rho^0}{\rho q}\right) = -\frac{\rho^0}{\rho q^2}\,(1 - M^2) \tag{2-11a}$$

and $M = q/a$. The fundamental system Eq. 2-11 in the q, θ plane, i.e. the *hodograph plane,* is equivalent to the system Eq. 2-2 in the x, y plane. The essential difference is, however, that the transformed system Eq. 2-11 is linear. The linearization is valid as long as the Jacobian of the transformation, $J \equiv \partial(x, y)/\partial(q, \theta)$, is finite and nonvanishing.

Thus, we see that in the hodograph variable the equations are linear, and thus the principle of superposition of solutions is allowed. In other words, if particular solutions are obtained, linear combinations of such solutions are also solutions. By this principle, general solutions can be constructed. However, the fact that most of the physically interesting solutions are many-valued in the hodograph plane makes the actual construction of such solutions very difficult in practice.

Legendre transformation. To effect this transformation, the velocity components u, v are introduced as independent variables and a new function $\chi(u, v)$, related to the stream function, is introduced according to the following definition:

$$\chi = xu + yv - \phi, \qquad x = \chi_u, \qquad y = \chi_v \tag{2-12}$$

The differential equation corresponding to Eq. 2-5 is [62]

$$(a^2 - u^2)\chi_{vv} + 2uv\chi_{uv} + (a^2 - v^2)\chi_{uu} = 0 \tag{2-13}$$

The relation between χ and the stream function follows from Eq. 2-1 and 2-12; it is

$$\begin{aligned} \frac{\rho^0}{\rho}\,\psi_u &= v\chi_{uu} - u\chi_{uv} \\ \frac{\rho^0}{\rho}\,\psi_v &= v\chi_{uv} - u\chi_{vv} \end{aligned} \tag{2-14}$$

Eq. 2-13 and 2-14 form a complete system equivalent to the system Eq. 2-11. This transformation is also subject to the same condition as stated above for system Eq. 2-11.

Coordinate functions. When the stream function $\psi(q, \theta)$ is given in the hodograph plane, the streamlines in the flow plane, corresponding to constant values of ψ, can be calculated by integrating Eq. 2-10. One way to accomplish this purpose is to perform the partial integration of the equations as they stand, viz.

$$
\begin{aligned}
x &= \frac{1}{q}\int\left[\cos\theta\phi_\theta - \frac{\rho^0}{\rho}\sin\theta\psi_\theta\right]d\theta + \hat{x}(q)\\
y &= \frac{1}{q}\int\left[\sin\theta\phi_\theta + \frac{\rho^0}{\rho}\cos\psi_\theta\right]d\theta + \hat{y}(q)
\end{aligned}
\tag{2-15}
$$

The arbitrary functions $\hat{x}(q)$ and $\hat{y}(q)$ introduced as a result of the integration can be determined from

$$
\begin{aligned}
x_q &= \frac{1}{q}\left[\cos\theta\phi_q - \frac{\rho^0}{\rho}\sin\theta\psi_q\right]\\
y_q &= \frac{1}{q}\left[\sin\theta\phi_q + \frac{\rho^0}{\rho}\cos\theta\psi_q\right]
\end{aligned}
\tag{2-16}
$$

It will be clear later that, when ψ and hence also ϕ are expressed in the form of certain infinite series, the integration in Eq. 2-15 can easily be carried out.

In the case of the Legendre transformation, once χ is specified the coordinate functions can be obtained by differentiation. However, we have to integrate Eq. 2-14 to get ψ in order to define a relation between the velocity components, hence there is no obvious advantage to the Legendre transformation in our problem. If not otherwise stated, we shall follow the Chaplygin procedure.

F,3. The Particular Solutions of the Chaplygin Equation. In this article we shall obtain some particular solutions of the transformed linear equations (2-11), which are useful in the construction of practical flow patterns. Except for the simple solutions given in Eq. 3-8 and 3-10, below, these will involve the special functions known in mathematical literature as the *hypergeometric functions.*

By eliminating ϕ from the system (2-11), there results a differential equation satisfied by $\psi(q, \theta)$:

$$q^2\psi_{qq} + (1 + M^2)q\psi_q + (1 - M^2)\psi_{\theta\theta} = 0 \tag{3-1}$$

Since M depends only on q, this equation is linear and its solutions can be built up by adding together particular solutions. As particular solutions we can choose

$$\psi = q^{\nu}F_{\nu}(\tau)e^{i\nu\theta}, \qquad \tau = \frac{\gamma - 1}{2}\frac{q^2}{a^{02}}$$

where ν is a real number and the function $F_{\nu}(\tau)$ is a solution of the equation

$$\tau(1 - \tau)F_{\nu}'' + [\nu + 1 - (a_{\nu} + b_{\nu} + 1)\tau]F_{\nu}' - a_{\nu}b_{\nu}F_{\nu} = 0 \qquad (3\text{-}2)$$

where the parameters a_{ν} and b_{ν} are defined by

$$a_{\nu} + b_{\nu} = \nu - \beta, \qquad a_{\nu}b_{\nu} = -\frac{\beta\nu(\nu + 1)}{2} \qquad (3\text{-}3)$$

and

$$\beta = \frac{1}{\gamma - 1}$$

This is a special hypergeometric equation involving two arbitrary parameters, ν and γ. The two independent solutions are denoted by

$$F(a_{\nu}, b_{\nu}; \nu + 1; \tau), \qquad \tau^{-\nu}F(a_{\nu} - \nu, b_{\nu} - \nu; 1 - \nu; \tau) \qquad (3\text{-}4)$$

$F(a_{\nu}, b_{\nu}; \nu + 1; \tau)$ is known as the *hypergeometric function* [4, p. 281] and is defined by the power series

$$F(a_{\nu}, b_{\nu}; \nu + 1; \tau) = \frac{\Gamma(\nu + 1)}{\Gamma(a_{\nu})\Gamma(b_{\nu})} \sum_{0}^{\infty} \frac{\Gamma(a_{\nu} + m)\Gamma(b_{\nu} + m)}{\Gamma(\nu + 1 + m)} \frac{\tau^m}{m!} \qquad (3\text{-}5)$$

which is uniformly convergent in a domain $|\tau| < 1$.

The reader should note at this point the significance of the several parameters involved in each of these hypergeometric functions. The independent variable, for our work, is the variable τ, which, for a specified gas, is uniquely related to the local Mach number $\sqrt{2\beta\tau/(1 - \tau)}$. There are two parameters: the adiabatic exponent γ, which characterizes the gas, and the index ν, which will be given various values in order to construct solutions to satisfy given boundary conditions.

Now, as q increases from zero to the maximum possible value $q_{max} = \sqrt{2/(\gamma - 1)}a^0$, τ varies from zero to unity. The condition $q = a^*$, or $M = 1$, is given by $\tau^* = (\gamma - 1)/(\gamma + 1)$. Therefore, the integral as defined by the series (3-5) holds for the whole range of possible flow speeds.

The case $\tau = 0$ may be taken to represent either very low speed flow or the flow of an incompressible fluid. For this value of τ, $F_{\nu}(0) = 1$ for all $\nu > 0$; thus the particular solutions considered here reduce to the familiar particular solutions of the incompressible case. Thus, each of the particular solutions of Chaplygin's equation may be thought of as the analogue of a certain incompressible solution. This analogy was used by

Chaplygin and will be mentioned again in the detailed work of this section.

When the parameter ν is a positive integer, the second solution cannot be represented by a single power series. In this situation another independent solution is needed to replace the second of Eq. 3-4; it is given by [*4*, p. 201; *25*]

$$\tau^{-n}[-h_n\tau^n \ln \tau\, F(a_n, b_n; n+1; \tau) + \Psi_1(n)\tau^n F(a_n, b_n; n+1; \tau) + \tau^n G_n(\tau) + H_n(\tau)] \quad (3\text{-}6)$$

where the functions $G_n(\tau)$ and $H_n(\tau)$ are defined by

$$G_n(\tau) = \frac{\Gamma(n+1-b_n)}{\Gamma(a_n-n)\Gamma(n)} \sum_0^\infty \Psi_2(n, m) \frac{\Gamma(a_n+m)(-1)^m}{\Gamma(1-b_n-m)\Gamma(n+m+1)} \frac{\tau^m}{m!} \qquad |\tau| < 1$$

$$H_n(\tau) = \frac{\Gamma(n+1-b_n)}{\Gamma(a_n-n)\Gamma(n)} \sum_0^{n-1} \frac{\Gamma(a_n-n+m)\Gamma(n-m)}{\Gamma(n+1-b_n-m)} \frac{\tau^m}{m!}$$

$$\Psi_1(n) = h_n\left[\psi(n+1) + \psi(1) - \psi(a_n) - \psi(1-b_n) + \frac{d}{dn}\ln h_n\right]$$

$$\Psi_2(n, m) = \psi(1-b_n) - \psi(1-b_n-m) + \psi(a_n) - \psi(a_n+m) + \psi(m+1) - \psi(1) + \psi(n+m+1) - \psi(n+1)$$

$$h_n = \frac{\Gamma(a_n)\Gamma(n+1-b_n)}{\Gamma(a_n-n)\Gamma(1-b_n)\Gamma(n)\Gamma(n+1)}$$

$$\psi(z) = \frac{d}{dz}\ln\Gamma(z)$$

Here the constants a_n and b_n are chosen to be positive and negative, respectively, for $n > 1$.[1] Having the forms of the second solution defined as in Eq. 3-4 and 3-6, according to ν being nonintegral and integral, respectively, we can write the particular solutions of Eq. 3-1 as

$$q^\nu F_\nu(\tau)e^{-i\nu\theta}, \qquad q^{-\nu}F_{-\nu}(\tau)e^{i\nu\theta} \quad (3\text{-}7)$$

where

$$F_\nu(\tau) = F(a_\nu, b_\nu; \nu+1; \tau)$$

and $F_{-\nu}(\tau)$ denotes either $F(a_\nu - \nu, b_\nu - \nu; 1-\nu; \tau)$ (for nonintegral ν) or the expression within the brackets in Eq. 3-6 (for $\nu = n$, an integer).

Distinct from this class of solutions we have also for $\nu = 0$

$$\theta \quad \text{and} \quad \int^\tau (1-\tau)^\beta \frac{d\tau}{\tau} \quad (3\text{-}8)$$

The latter, in the limit of incompressible flow, reduces to 2 ln q.

[1] For $n = 1$, $F(1, -\beta; 2; \tau) = \frac{1}{(\beta+1)\tau}[1 - (1-\tau)^{\beta+1}]$ and $F_{-1}(\tau) = 1 + \frac{\beta}{2}\tau F_1(\tau)$.

These are our particular solutions for the stream function $\psi(q, \theta)$. We need now to find the corresponding particular velocity potentials $\phi(q, \theta)$. This can be done with the aid of the fundamental relations (2-11). We have

$$d\phi = -\frac{\rho^0}{\rho}(1 - M^2)\frac{1}{q}\psi_\theta dq + \frac{\rho^0}{\rho} q\psi_q d\theta$$

If $\psi(q, \theta)$ is given by the first of Eq. 3-7, then[2]

$$d\phi = i\frac{\rho^0}{\rho}(1 - M^2)\nu q^{\nu-1}F_\nu(\tau)e^{-i\nu\theta}dq + \frac{\rho^0}{\rho} q \frac{d}{dq} q^\nu F_\nu(\tau)e^{-i\nu\theta}d\theta$$

This can be integrated without difficulty with the aid of the differential equation (3-1), an alternative form of which is

$$\frac{\partial}{\partial q}\left(\frac{\rho^0}{\rho} q\psi_q\right) + \frac{\rho^0}{\rho q}(1 - M^2)\psi_{\theta\theta} = 0$$

or, in terms of the particular solution now under consideration,

$$\frac{\rho^0}{\rho}(1 - M^2)\nu q^{\nu-1}F_\nu(\tau) = \frac{1}{\nu}\frac{d}{dq}\left[\frac{\rho^0}{\rho} q \frac{d}{dq} q^\nu F_\nu(\tau)\right] = \frac{d}{dq}\left[\frac{\rho^0}{\rho} q^\nu F_\nu(\tau)\xi_\nu(\tau)\right]$$

where $\xi_\nu(\tau)$ denotes $2\left(\frac{\tau}{\nu}\right)\frac{d}{d\tau}\ln \tau^{\nu/2}F_\nu(\tau)$. The integral is

$$\phi = i\frac{\rho^0}{\rho} q^\nu F_\nu(\tau)\xi_\nu(\tau)e^{-i\nu\theta} + \text{const}$$

Therefore, the corresponding particular solutions for ϕ, corresponding to Eq. 3-7, are for $\nu \neq 0$

$$i\frac{\rho^0}{\rho} q^\nu F_\nu(\tau)\xi_\nu(\tau)e^{-i\nu\theta} \quad \text{and} \quad i\frac{\rho^0}{\rho} q^{-\nu}F_{-\nu}(\tau)\xi_{-\nu}(\tau)e^{i\nu\theta} \tag{3-9}$$

and, corresponding to Eq. 3-8, for $\nu = 0$

$$(1 - \tau)^{-\beta} - \frac{1}{2}\int^\tau (1 - \tau)^{-\beta}\frac{d\tau}{\tau} \quad \text{and} \quad \theta \tag{3-10}$$

F,4. Particular Solutions of the Equations Resulting from the Legendre Transformation. In analogous fashion, let us determine some particular solutions of the linear equation (2-13) which we obtained by the Legendre transformation. In the first place, from the relations (2-14) in polar coordinates, we have

$$\psi_\theta = -\frac{\rho^0}{\rho}(q\chi_q + \chi_{\theta\theta}) \tag{4-1}$$

[2] If $\psi(q, \theta)$ is given by the second of Eq. 3-7, we need only change ν to $-\nu$ in the subsequent formulas.

where $\chi(q, \theta)$ satisfies

$$q^2\chi_{qq} + (1 - M^2)q\chi_q + (1 - M^2)\chi_{\theta\theta} = 0 \tag{4-2}$$

If we assume again

$$\chi = q^\nu F(\tilde{a}_\nu, \tilde{b}_\nu; \nu + 1; \tau)e^{-i\nu\theta} \tag{4-3}$$

Eq. 4-2 reduces to

$$\tau(1 - \tau)\tilde{F}_\nu'' + [\nu + 1 - (\tilde{a}_\nu + \tilde{b}_\nu + 1)\tau]\tilde{F}_\nu' - \tilde{a}_\nu\tilde{b}_\nu\tilde{F}_\nu = 0 \tag{4-4}$$

where

$$\tilde{a}_\nu + \tilde{b}_\nu = \nu + \beta, \qquad \tilde{a}_\nu\tilde{b}_\nu = -\frac{\beta\nu(\nu - 1)}{2}$$

and

$$\tilde{F}_\nu(\tau) = F(\tilde{a}_\nu, \tilde{b}_\nu; \nu + 1; \tau) \tag{4-5}$$

Eq. 4-4 is again the hypergeometric equation (cf. Eq. 3-2), the only difference being that the parameters have been defined differently in terms of ν and γ. Thus, the particular functions $\chi(q, \theta)$ involve the same hypergeometric functions as we used in Art. 3, but with slightly different meanings of the parameters.

Let us now determine the forms of the particular stream functions $\psi(q, \theta)$ corresponding to these particular $\chi(q, \theta)$'s. Substituting $\chi(q, \theta)$ from Eq. 4-3 into Eq. 4-1, we have

$$\psi_\theta = -\frac{\rho^0}{\rho}[2\tau\tilde{F}_\nu' - \nu(\nu - 1)\tilde{F}_\nu]q^\nu e^{-i\nu\theta}$$

and by Eq. 4-4

$$\psi_\theta = -2(\gamma - 1)(1 - \tau)^{1+\beta}[\tau\tilde{F}_\nu'' + (\nu + 1)\tilde{F}_\nu']q^\nu e^{-i\nu\theta} \tag{4-6}$$

Following T. M. Cherry [5], we notice the identity

$$\tau\tilde{F}_\nu' + \nu\tilde{F}_\nu = \nu F(\tilde{a}_\nu, \tilde{b}_\nu; \nu; \tau)$$

By differentiation,

$$\tau\tilde{F}_\nu'' + (\nu + 1)\tilde{F}_\nu' = \tilde{a}_\nu\tilde{b}_\nu F(\tilde{a}_\nu + 1, \tilde{b}_\nu + 1; \nu + 1; \tau) \tag{4-7}$$

It is easily verified that the hypergeometric equation satisfied by $F(\tilde{a}_\nu + 1, \tilde{b}_\nu + 1; \nu + 1; \tau)$ can be transformed into Eq. 3-2 by the substitution

$$F(\tilde{a}_\nu + 1, \tilde{b}_\nu + 1; \nu + 1; \tau) = (1 - \tau)^{-1-\beta}F_\nu(\tau) \tag{4-8}$$

By means of Eq. 4-8 and 4-7, partial integration of Eq. 4-6 gives us

$$\psi(q, \theta) = i(\nu - 1)q^\nu F_\nu(\tau)e^{-i\nu\theta} \tag{4-9}$$

where the arbitrary function of q can only be the second of Eq. 3-8 and therefore has been omitted.

Corresponding to this particular solution (4-3), the coordinate functions are

$$x = q^{\nu-1}[2\tau\tilde{F}'_\nu(\tau)\cos\theta e^{-i\nu\theta} + \nu\tilde{F}_\nu(\tau)e^{-i(\nu-1)\theta}] + \text{const}$$
$$y = q^{\nu-1}[2\tau\tilde{F}'_\nu(\tau)\sin\theta e^{-i\nu\theta} - i\nu\tilde{F}_\nu(\tau)e^{-i(\nu-1)\theta}] + \text{const} \tag{4-10}$$

where the prime denotes differentiation with respect to τ. These functions $\tilde{F}_\nu(\tau)$ and $\tilde{F}'_\nu(\tau)$ can be expressed in terms of $F_\nu(\tau)$ and $F'_\nu(\tau)$ as follows [24,63]:

$$(1-\tau)^\beta\tilde{F}_\nu(\tau) = (\nu+1)F_\nu(\tau) + \frac{2\tau}{\nu}F'_\nu(\tau)$$
$$\frac{2\tau}{\nu}(1-\tau)^\beta\tilde{F}'_\nu(\tau) = (\nu-1)\left[\nu F_\nu(\tau) + \frac{2\tau}{\nu}F'_\nu(\tau)\right] \tag{4-11}$$

In like manner, if $\chi = q^{-\nu}\tilde{F}_{-\nu}(\tau)e^{+i\nu\theta}$, we find

$$\chi = +i(\nu+1)q^{-\nu}F_{-\nu}(\tau)e^{i\nu\theta} \tag{4-12}$$

Thus, it turns out that the particular stream functions corresponding to our elementary $\chi(q, \theta)$'s differ only in multiplicative factors from those selected in Art. 3 as elementary solutions of Chaplygin's equation. This identification is due to Cherry [5].

F,5. The Chaplygin-Kármán-Tsien Approximation. Before we proceed to treat the problem more exactly, let us deal first with an approximate method which has had considerable application in practice. It was first noted by Chaplygin [2] that, if we put $\gamma = -1$, the system (2-11) can be greatly simplified and its solutions are none other than those of the familiar Laplace equation. This can easily be established by the fact that, for $\gamma = -1$, the solutions of the hypergeometric equation (3-2) become[3]

$$F_\nu(\tau) = \left\{\frac{2}{1+[1+q^2/(a^0)^2]^{\frac{1}{2}}}\right\}^\nu,$$
$$q^{-2\nu}F_{-\nu}(\tau) = \left\{\frac{2q^2}{1+[1+q^2/(a^0)^2]^{\frac{1}{2}}}\right\}^{-\nu} \tag{5-1}$$

Thus, if we introduce as a new variable

$$q_i = \frac{2q}{1+[1+q^2/(a^0)^2]^{\frac{1}{2}}} \tag{5-2}$$

[3] This follows from the identity

$$F\left(\frac{\nu}{2}, \frac{\nu}{2}+\frac{1}{2}; \nu+1; -\frac{q^2}{(a^0)^2}\right) = 2^\nu\left[1+\sqrt{1+\frac{q^2}{(a^0)^2}}\right]^{-\nu}$$

(Forsyth, A. R. *A Treatise on Differential Equations*. 1885, p. 207.)

the particular integrals of Eq. 3-1 are

$$q_i^{\nu}e^{-i\nu\theta} \quad \text{and} \quad q_i^{-\nu}e^{i\nu\theta} \tag{5-3}$$

Consequently, the boundary-value problem of a compressible flow having a value of γ equal to -1 becomes, in a q_i, θ plane, identically that of an incompressible flow.

The question remains, however, whether this simplification has any meaning physically. We can look at this problem from two aspects. Following Chaplygin we can write the system (2-11) in the form

$$\begin{aligned} \phi_{q_i} &= K(\tau)\psi_\theta \\ \phi_\theta &= -\psi_{q_i} \end{aligned} \tag{5-4}$$

where

$$q_i = -\frac{1}{2}\int^{\tau} (1-\tau)^{\beta}\,\frac{d\tau}{\tau}$$

(which agrees with Eq. 5-2 in the case $\gamma = -1$) and $K(\tau) = (1 - \alpha^2\tau)/(1-\tau)^{\alpha}$, with $\alpha^2 = (\gamma + 1)/(\gamma - 1)$. The coefficient $K(\tau)$ varies very slowly with τ and is practically constant and close to unity if τ is small, i.e. for small Mach number. Therefore, the case $\gamma = -1$, for which $K = 1$, can be regarded as an approximation, to the order τ^2, for a purely subsonic flow.

On the other hand, it is seen that, for an isentropic flow, the pressure and density will be related by $p \sim \rho^{-1}$ when $\gamma = -1$. This can be interpreted as an approximation by a line tangent to the adiabatic curve in the pressure-volume diagram (Fig. F,5a). In a flow field, if pressure and density do not vary through a wide range, this linear law of state can give a fairly close approximation to the true situation governed by the adiabatic law. If the point of tangency is chosen to correspond to the free stream condition, the approximate equation of state given by von Kármán and Tsien [3,6] is obtained:

$$p - p_\infty = -\rho_\infty^2 a_\infty^2\left(\frac{1}{\rho} - \frac{1}{\rho_\infty}\right) \tag{5-5}$$

Here the subscript ∞ indicates the free stream condition. It is easy to verify that, at free stream conditions, this equation gives the exact slope, $(dp/d\rho)_\infty = a_\infty^2$.

With this linear equation of state, the integrated equation of motion, Eq. 2-6, becomes

$$a^2 - q^2 = (a^0)^2 \tag{5-6}$$

which, in conjunction with Eq. 5-5, gives

$$\frac{\rho^0}{\rho} = \left[1 + \frac{q^2}{(a^0)^2}\right]^{\frac{1}{2}} \tag{5-7}$$

Thus, we see from Eq. 5-6 that in this approximation the sonic speed can never be attained in the flow field.

In his original treatment [*2*], Chaplygin employed a different equation of state, since he chose the stagnation condition p^0, ρ^0 as the point of tangency to the adiabatic curve, rather than p_∞, ρ_∞. The Kármán-Tsien approximation seems better suited to most aeronautical applications.

It might be mentioned here that the Kármán-Tsien approximation can be adapted, at least formally, to the study of purely supersonic plane

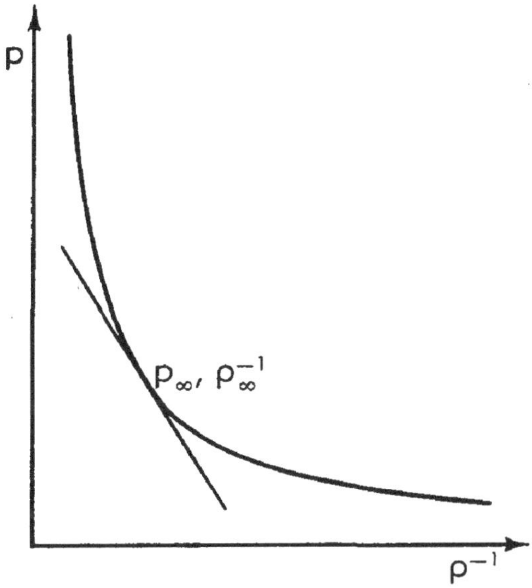

Fig. F,5a. Approximation to true isentropic pressure-volume curve.

flows [*64*]. The hodograph equation (5-4) assumes a simple hyperbolic form. The value of this procedure is difficult to assess, especially in comparison with other available techniques such as the numerical method of characteristics (Sec. G).

Transformation of incompressible to compressible flows. As a result of this simplification, we are able to show that, given an incompressible flow, a "similar" compressible flow, subsonic everywhere, can be constructed. Let $U_iW_i = \phi_i + i\psi_i$ be the given complex potential of an incompressible flow and U_i its undisturbed stream velocity. We have already seen, in connection with Eq. 5-3, that the functions

$$\phi = \phi_i(q_i, \theta), \qquad \psi = \psi_i(q_i, \theta) \tag{5-8}$$

represent the velocity potential and stream function of a compressible flow in terms of the quantity q_i defined in Eq. 5-2. Once ϕ and ψ are known, the coordinate functions (2-9) can be expressed as

$$dz = \frac{e^{i\theta}}{q}\left[d\phi_i + i\frac{\rho^0}{\rho}\,d\psi_i\right] \tag{5-9}$$

where $dz = dx + idy$. To eliminate q, we have, by Eq. 5-2,

$$q = \frac{q_i}{1 - q_i^2/4(a^0)^2}$$

whence

$$\frac{\rho^0}{\rho} = \frac{1 + q_i^2/4(a^0)^2}{1 - q_i^2/4(a^0)^2}$$

Substitution of q and ρ^0/ρ in Eq. 5-9 yields an expression for dz in terms of quantities of the q_i, θ plane only:

$$dz = \frac{1}{q_i e^{-i\theta}} \left\{ \left[1 - \frac{q_i^2}{4(a^0)^2} \right] d\phi_i + i \left[1 + \frac{q_i^2}{4(a^0)^2} \right] d\psi_i \right\} \tag{5-10}$$

Furthermore, define

$$w_i = q_i e^{-i\theta}, \qquad dz_i = \frac{U_i dW}{w_i}$$

so that dz_i is the increment of the position vector z_i in the original incompressible flow corresponding to increments $d\phi_i$, $d\psi_i$ of the velocity potential and stream function. We can now relate this increment dz_i to the analogous incremental vector dz of the compressible flow, which corresponds to the same increments $d\phi_i$, $d\psi_i$ of its velocity potential and stream function. We have, namely,

$$dz = dz_i - \lambda^2 \left(\frac{d\bar{W}_i}{d\bar{z}_i} \right)^2 d\bar{z}_i \tag{5-11}$$

where $\bar{W}_i$ and $\bar{z}_i$ are the complex conjugates of W_i and z_i, respectively, and λ stands for $M_\infty/(1 + \sqrt{1 - M_\infty^2})$, M_∞ being the free stream Mach number. Integration of Eq. 5-11 gives

$$z = z_i - \lambda^2 \int \left(\frac{d\bar{W}_i}{d\bar{z}_i} \right)^2 d\bar{z}_i \tag{5-12}$$

The constant of integration can be determined by proper choice of the origins of the two coordinate systems z and z_i. Eq. 5-12 can now be considered as a mapping function which transforms a given incompressible flow pattern into a compressible flow around a "similar" but different contour. Given the incompressible flow pattern, the second term of Eq. 5-12 is a correction to the complex coordinate of the compressible flow at which any given values ϕ, ψ are found. The similarity of the two patterns is closer as the Mach number M_∞ diminishes, until an identity occurs when $M_\infty = 0$.

A difficulty however is that, if the incompressible flow involves circulation, then the correction term in Eq. 5-12 is no longer single-valued. A closed boundary in the z_i plane will be mapped into an unclosed one in the z plane. Hence, Eq. 5-12 in the present form fails to deal with circulatory flow. This problem was attacked by several authors [7–11]. We shall briefly present Lin's method, which appears to be most simple and general.

Lin's suggestion is to interpret w_i more generally than was done by Tsien, as reproduced above; namely, we can choose

$$w_i = k(z_i)q_i e^{-i\theta} \tag{5-13}$$

and $dz_i = U_i dW_i/w_i$ as before, where $k(z_i)$ is another analytic function. Instead of Eq. 5-11, we then have, from Eq. 5-10,

$$dz = k(z_i)dz_i - \lambda^2 \left(\frac{\overline{dW_i}}{dz_i}\right)^2 \frac{\overline{dz_i}}{k(z_i)} \tag{5-14}$$

and, instead of Eq. 5-12,

$$z = \int k(z_i)dz_i - \lambda^2 \int \left(\frac{\overline{dW_i}}{dz_i}\right)^2 \frac{\overline{dz_i}}{k(z_i)} \tag{5-15}$$

We now impose two conditions on the function $k(z_i)$. The first is that Eq. 5-13 be analytic; this leads [7] to

$$|\tfrac{1}{2}W_i'(\zeta)| < |k(z_i)| < \infty \tag{5-16}$$

Secondly, to insure that a closed contour in the z_i plane will be mapped on a closed contour in the z plane, we have, from Eq. 5-15

$$\oint k(z_i)dz_i - \lambda^2 \oint \left(\frac{\overline{dW_i}}{dz_i}\right)^2 \frac{\overline{dz_i}}{k(z_i)} = 0 \tag{5-17}$$

Thus, given a function W_i which represents an incompressible flow with circulation, we can always construct a related compressible flow with circulation by choosing properly the auxiliary function $k(z_i)$, subject to Eq. 5-16 and 5-17. In Lin's paper, the problem is formulated in such a way as to place it within the scope of familiar incompressible fluid airfoil theory.

Pressure correction. When the flow about a certain body is determined, the pressure distribution over the surface can be calculated, although the numerical work may be somewhat tedious. The main interest in subsonic flow appears to be the effect of the compressibility of the fluid on the pressure distribution, under given geometrical conditions. Within the approximation permitted by the assumptions of the present theory, this has been achieved by von Kármán and Tsien.

Let C_p be the pressure coefficient at a point in the flow field, defined by

$$C_p = \frac{p - p_\infty}{\frac{1}{2}\rho_\infty U^2}$$

where p is the local pressure and p_∞, ρ_∞, and U denote, respectively, the free stream pressure, density, and velocity. By making use of Eq. 5-5 and 5-7, we find

$$C_p = \frac{2}{M_\infty^2}\left[1 - \sqrt{1 - M_\infty^2}\,\frac{1 + q_i^2/4(a^0)^2}{1 - q_i^2/4(a^0)^2}\right] \tag{5-18}$$

Now q_i is associated with a contour in z_i plane, while C_p is the pressure coefficient on a contour in the z plane. According to Eq. 5-12, these two contours are different. Consequently, the determination of C_p for a given contour involves the calculation of the incompressible flow about a different contour, which must be determined by use of the mapping function. Tsien has carried out this process for the case of compressible flow about an elliptical cylinder [*6*].

For practical engineering work, however, Tsien has suggested using a further approximation to avoid this lengthy procedure, namely, to neglect the difference between the two contours involved. If we ignore this difference, we can regard Eq. 5-18 as a formula relating the pressure coefficient C_p to the pressure coefficient on the same body in incompressible flow. The pressure coefficient in the incompressible flow would be

$$C_{p_{\text{inc}}} = 1 - \left(\frac{q_i}{U_i}\right)^2 \tag{5-19}$$

Eliminating q_i and after simple reduction, we obtain

$$C_p = C_{p_{\text{inc}}}\left[\sqrt{1 - M_\infty^2} + \frac{M_\infty^2}{1 + \sqrt{1 - M_\infty^2}}\,\frac{C_{p_{\text{inc}}}}{2}\right]^{-1} \tag{5-20}$$

This is the famous Kármán-Tsien pressure correction formula. It has frequently been used in engineering practice. Having the form $C_p = C_p(C_{p_{\text{inc}}}, M_\infty)$, it replaces the Prandtl-Glauert rule for plane flow (C,6)

It is clear that formula (5-20) becomes identical to the Prandtl-Glauert rule if the second term in the denominator is neglected, i.e. $C_{p_{\text{P.G.}}} = C_{p_{\text{inc}}}(1 - M_\infty^2)^{-\frac{1}{2}}$. Furthermore, the nature of the improvement afforded by Eq. 5-20 over the simpler first order result can easily be seen, viz.

$$C_p < C_{p_{\text{P.G.}}}$$

Thus, the Kármán-Tsien rule predicts pressure coefficients that are numerically greater in regions where $C_{p_{\text{inc}}}$ is negative and smaller where $C_{p_{\text{inc}}}$ is positive. Reference to the data of C,8 will confirm that this tendency is, qualitatively, in the right direction.

It seems clear that the validity of this result rests on the assumption that the distortion of the boundary due to compressibility is negligible. We can study this by taking $|d\overline{W}_i/d\bar{z}_i|$ in Eq. 5-12 to be $O(1)$, the integral $\int (d\overline{W}_i/d\bar{z}_i)^2 d\bar{z}_i$ must then be of the order of the body dimension. If τ and τ_i are the thickness ratios of the bodies in the z and z_i planes, respectively, a rough estimate shows that $\tau/\tau_i = 1 + O(M_\infty^2\tau_i)$. But we have shown

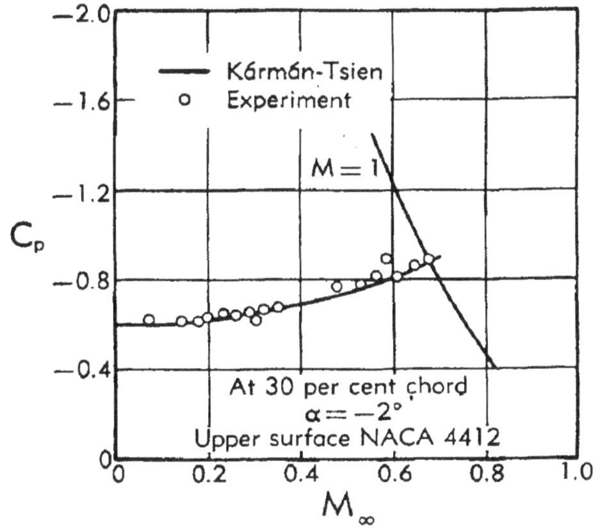

Fig. F,5b. Comparison of pressure correction formula with experiments [70] at different free stream Mach numbers M_∞, for NACA 4412. At 30 per cent chord, upper surface, $\alpha = -2°$.

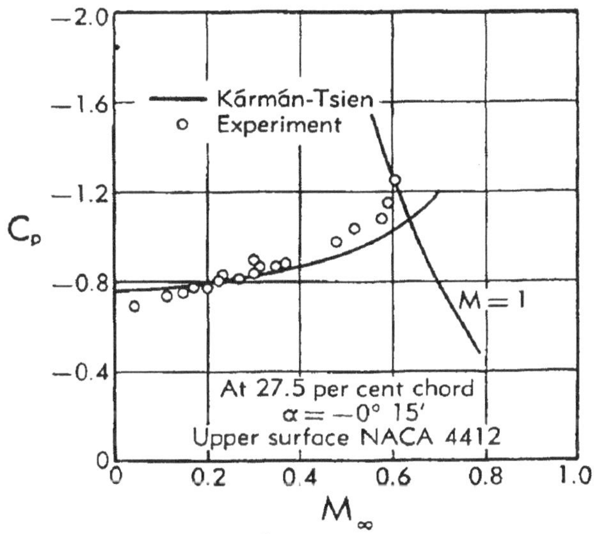

Fig. F,5c. Comparison of pressure-correction formula with experiments [70] at different free stream Mach numbers M_∞, for NACA 4412. At 27.5 per cent chord, upper surface, $\alpha = -0°15'$.

that Chaplygin's approximation holds only for small Mach numbers, and the idea of replacing the adiabatic curve by its tangent further precludes cases of large disturbance or thick bodies. Therefore, when both M_∞ and τ are restricted to be small, the distortion, being of $O(M_\infty^2\tau_i)$, can, of course, be neglected.

The merit of Kármán-Tsien formula, however, is not that it has a theoretically sound basis but rather that it is justified by its practical

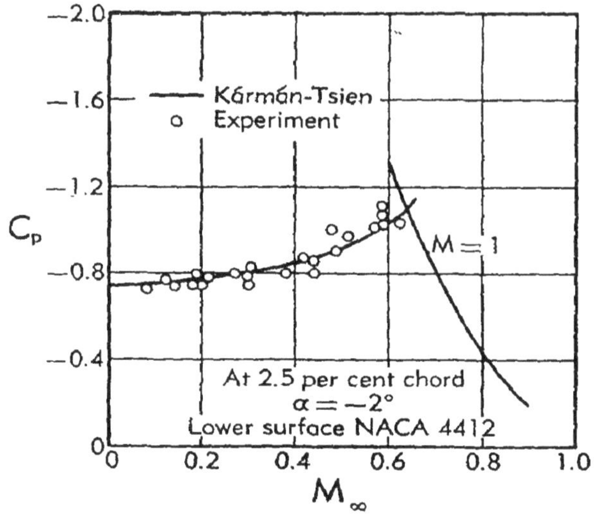

Fig. F,5d. Comparison of pressure-correction formula with experiments [70] at different free stream Mach numbers M_∞ for NACA 4412. At 25 per cent chord, lower surface, $\alpha = -2°$.

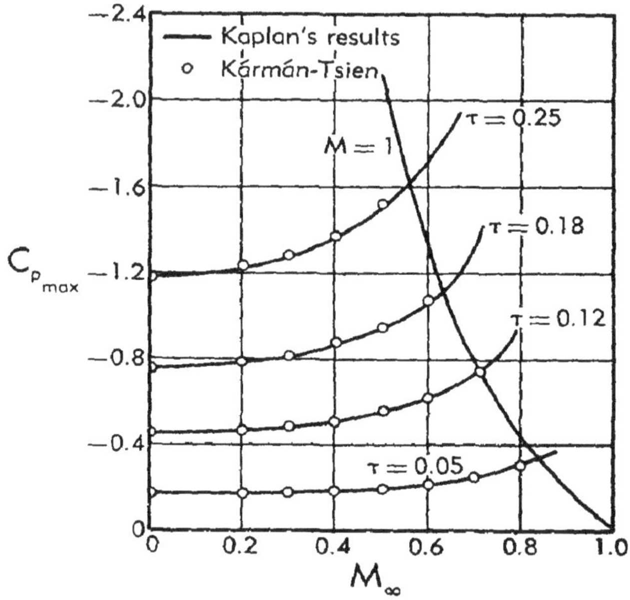

Fig. F,5e. Comparison of theoretical maximum pressure coefficient on a bump with pressure-correction formula at different free stream Mach numbers M_∞ and thickness ratios τ. (From [55].)

utility. In practice, we find close agreement between experimental results and theoretical predictions, even when the formula is applied at rather large subsonic Mach numbers. Examples are presented in Fig. F,5b to F,5e.

Tsien's tentative explanation [12] of this phenomenon can be summarized as follows. By examining Fig. F,5a, we notice that the linear equation of state generally *underestimates* the pressure. On the other hand, transformation of a body in the incompressible flow to that in compressible flow, as will be shown later, *increases* the thickness ratio of the body. Ignoring this increase is equivalent to *overestimating* the pressure in regions of reduced pressure. As the two effects work in opposite directions, they may counteract each other. A rigorous mathematical proof has not yet been offered. This subject is discussed in some detail in E,2.

Application to elliptic cylinder. As an example let us take W_i to be the complex potential for flow about an elliptic cylinder.

$$W_i = \zeta + \frac{b^2}{\zeta} \tag{5-21}$$

and

$$z_i = \zeta + \frac{1}{\zeta}$$

Substitution of this W_i into Eq. 5-12 gives us, after integration,

$$z = \zeta + \frac{1}{\zeta} - \lambda^2 \left[\bar{\zeta} + \frac{b^4}{\bar{\zeta}} + \frac{(b^2 - 1)^2}{2} \ln \frac{\bar{\zeta} - 1}{\bar{\zeta} + 1} \right]$$

Now, the body contour in the z_i plane is given by $\zeta = be^{i\theta}$ in the ζ plane; the parametric equations for the contour in the z plane are therefore

$$\begin{aligned} x = {} & \left(b + \frac{1}{b}\right) \cos\theta \\ & - \lambda^2 \left[b(b^2 + 1) \cos\theta + \frac{(b^2 - 1)^2}{4} \ln \frac{b^2 - 2b\cos\theta + 1}{b^2 + 2b\cos\theta + 1} \right] \\ y = {} & \left(b - \frac{1}{b}\right) \sin\theta - \lambda^2 \left[b(b^2 - 1) \sin\theta - \frac{(b^2 - 1)^2}{2} \tan^{-1} \frac{2b\sin\theta}{b^2 - 1} \right] \end{aligned} \tag{5-22}$$

The derived contour is again symmetrical (Fig. F,5f) but is not an ellipse except when M_∞ (and therefore λ) is zero. The deviation from an ellipse, due to the presence of the ln (. . .) and $\tan^{-1}$ (. . .) terms in Eq. 5-22, grows rapidly with Mach number; in fact, it is proportional to M_∞^2. As a measure of the change of the incompressible flow contour in going over to compressible flow, let us define a thickness parameter τ as the ratio of the minor to the major axis, namely

$$\tau = \tau_i \frac{1 - \lambda^2 \left[b^2 - \frac{1}{2} b(b^2 - 1) \tan^{-1} \frac{2b}{b^2 - 1} \right]}{1 - \lambda^2 \left[b^2 + \frac{b(b^2 - 1)^2}{2(b^2 + 1)} \ln \frac{b - 1}{b + 1} \right]} \tag{5-23}$$

where $\tau_i = (b^2 - 1)/(b^2 + 1)$.

As the parameter b ranges from 0 to ∞, the elliptical cylinder in the z_i plane ranges from a flat plate to a circular cylinder. Throughout this range of b, the ratio τ/τ_i is always greater than unity. A calculation, based on Eq. 5-23, for ellipses of small thickness ratio τ_i leads to the approximate relation

$$\tau = \tau_i\left[1 + \frac{\pi}{2}\frac{\lambda^2}{1-\lambda^2}\tau_i + O(\tau_i^2)\right] \tag{5-24}$$

This confirms our earlier estimate, which was $\tau/\tau_i = 1 + O(M_\infty^2\tau_i)$.

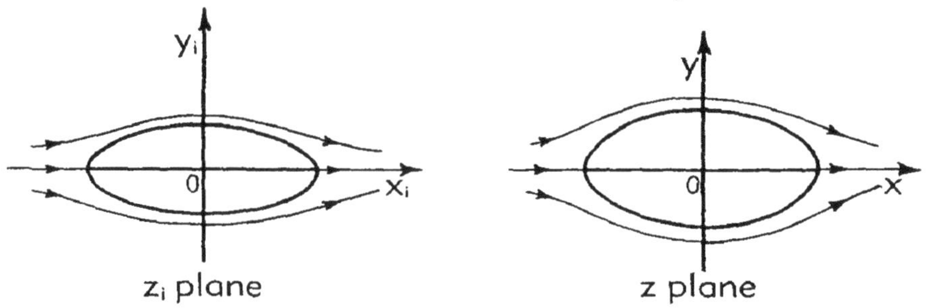

Fig. F,5f. Relationship between bodies in z_i and z planes.

F,6. Examples of Simple Flows. As was shown in the preceding section, the approximation $\gamma = -1$ makes it possible to transform the problem of subsonic compressible flow to one of incompressible flow. Clearly, if the problem is to be solved exactly, no such simplification generally exists. A quite different method has to be employed. To begin with, let us consider some simple flows such as sink, vortex, doublet, and so on, for which the solutions require no superposition.

Source or sink. Let the fluid be at rest at infinity and flow radially from or toward a center. When the steady state is established, the stream function must be a function of θ alone, that is

$$\psi = C\theta \tag{6-1}$$

where the constant C is arbitrary. Then, from the system (2-11)

$$\phi = -C\int\frac{\rho^0}{\rho q}(1 - M^2)dq + \text{const} \tag{6-2}$$

With this form of ψ, the coordinate functions can easily be obtained from Eq. 2-15:

$$x = C\frac{\rho^0}{\rho q}\cos\theta + \hat{x}(q)$$

$$y = C\frac{\rho^0}{\rho q}\sin\theta + \hat{y}(q)$$

since $\phi_\theta = 0$. Moreover, $\hat{x}(q)$ and $\hat{y}(q)$ must be constants, as a consequence of Eq. 2-16 and 2-11a. Let us choose the coordinate system such that $\hat{x}(q) = \hat{y}(q) = 0$; we have, then

$$x = C\frac{\rho^0}{\rho q}\cos\theta, \qquad y = C\frac{\rho^0}{\rho q}\sin\theta \tag{6-3}$$

As the streamlines are given by $\theta = \text{const}$, it is clear that these lines are radial. By elimination of θ, the equation of the lines of constant q, which are the equipotentials in this case, is found:

$$x^2 + y^2 = \frac{\rho^{02}C^2}{\rho^2 q^2} \tag{6-4}$$

These lines are concentric circles. If the constant C is so chosen that $\rho^{02}C^2 = (\rho^*)^2(a^*)^2$, asterisk indicating the value at $q = a = a^*$, then the equation of these circles can be written

$$x^2 + y^2 = \left(\frac{\rho^* a^*}{\rho q}\right)^2 \tag{6-5}$$

Now, the mass flow ρq is a function of q, which has a maximum value at $M = 1$, as is clear from Eq. 2-11a. The fact introduces an interesting phenomenon in the present problem, namely, the radius of the equipotentials (6-5) has a minimum value; our flow pattern is confined to the region outside of a certain circle at which sonic speed occurs.

If we differentiate Eq. 6-5 with respect to q along a streamline, viz.

$$\frac{d}{dq}(x^2 + y^2) = -2\left(\frac{\rho^* a^*}{\rho q}\right)^2 \frac{1 - M^2}{q}$$

we see that the radius of the circle decreases as q increases. But when the minimum circle is reached, at which $M = 1$, further increase of q increases the radius again, because $d(x^2 + y^2)/dq > 0$. Thus, we obtain another possible flow of the source or sink type, which has sonic speed at the minimum radius and increasing supersonic speeds at all greater radii, reaching maximum speed at infinity. (Naturally, for physical reasons we cut off this solution at some speed less than q_{max}, the speed corresponding to zero density, or, perhaps make an abrupt transition from supersonic flow through a shock.) The fact that these two different flows appear to be plotted on top of each other in the x, y plane here is of no significance. Given the boundary condition $M = 1$ at $x^2 + y^2 = 1$, we will have either one or the other of these flows, depending on the conditions specified at larger radii.

To be sure, the fact that these compressible source and sink flows possess a minimum radius where $dq/d\sqrt{x^2 + y^2}$ becomes infinite is, in this example, simply a manifestation of the familiar phenomenon of choking in a nozzle throat: The flow cannot progress inside of the minimum radius because the stream tube area does not increase there. Nevertheless, we shall find below that this is but a special example of a more general and profound phenomenon of plane isentropic compressible flow.

Vortex. For a vortex, we have, on the other hand,

$$\phi = C\theta \tag{6-6}$$

Then, system (2-11) gives

$$\psi = C \int \frac{\rho}{\rho^0} \frac{dq}{q} \tag{6-7}$$

Thus, in this case constant q lines are streamlines.

The coordinate functions, accordingly, are

$$x = C \frac{\sin \theta}{q} + \hat{x}(q)$$

$$y = -C \frac{\cos \theta}{q} + \hat{y}(q)$$

where $\hat{x}(q)$ and $\hat{y}(q)$ are again constant and can be taken equal to zero. The streamlines, by elimination of θ, are circles:

$$x^2 + y^2 = \frac{C^2}{q^2} \tag{6-8}$$

Thus, in contrast to the case of source flow, the speed varies monotonically with the radius, being subsonic at large radii and changing smoothly to supersonic at a specific radius. At a sufficiently small radius, q reaches q_{max}, a vacuum is theoretically created and, of course, no flow is possible. Therefore, in this case a core again exists, though the physical nature is entirely different from the previous case.

Doublet: flow around an edge. We know that the incompressible flow whose complex potential W_i has the form of a doublet in the hodograph plane, viz.

$$W_i = \frac{C}{w}$$

corresponds to flow about an edge in the z_i plane. The complex velocity for this flow in the z_i plane is given by the formula

$$w = \sqrt{\frac{C}{2z_i}}$$

where C is real and positive. The streamlines of this flow in both the flow plane and the hodograph plane are sketched in Fig. F,6a. If the fluid is compressible, the analogous flow pattern, in the analogy suggested by Chaplygin (Art. 3), would be described by the stream function

$$\psi = Cq^{-1}F_{-1}(\tau)e^{i\theta}$$

For the particular value $\nu = -1$, it can easily be verified that the hypergeometric equation (3-2) has the following solutions (cf. note 1):

$$F_{-1}(\tau) = \text{const} \quad \text{and} \quad F_{-1}(\tau) = \text{const } (1 - \tau)^{1+\beta}$$

The first of these corresponds to the second solution mentioned in Eq. 3-5. Let us follow Ringleb [13] and study the case F_{-1} = const,[4] i.e. we take for the stream function the simple form

$$\psi = C\frac{\cos\theta}{q} \tag{6-9}$$

where the constant C is again real. The streamlines in the hodograph plane are therefore the system of circles passing through the origin.

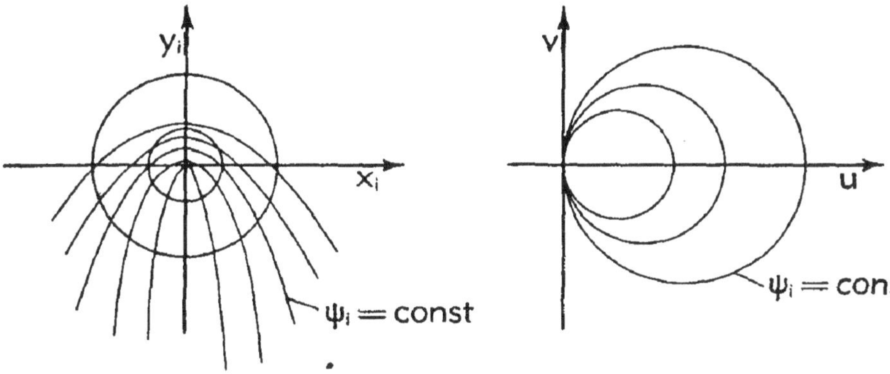

Fig. F,6a. Incompressible flow patterns around an edge in z_i and w planes.

Furthermore, by integration of Eq. 2-11,

$$\phi = -\frac{C\rho^0}{\rho q}\sin\theta \tag{6-10}$$

With these forms of ϕ and ψ, partial integration of Eq. 2-15, gives us

$$x = -\frac{C}{2q^2}(1-\tau)^{-\beta}\sin 2\theta + \hat{x}(q)$$

$$y = \frac{C}{2q^2}(1-\tau)^{-\beta}\cos 2\theta + \hat{y}(q)$$

where the arbitrary functions $\hat{x}(q)$ and $\hat{y}(q)$, by Eq. 2-16, satisfy

$$\hat{x}'(q) = 0, \qquad \hat{y}'(q) = -C\beta(1-\tau)^{-1-\beta}\frac{\tau}{q^3}$$

Integrating,

$$\hat{x}(q) = C_1 \quad \text{and} \quad \hat{y}(q) = -\frac{C}{4(a^0)^2}\int^{\tau}(1-\tau)^{-1-\beta}\frac{d\tau}{\tau} + C_2$$

where C_1 and C_2 are constants which can be taken to be zero. Finally, we

[4] The other solution has been studied by Temple and Yarwood [14].

have our coordinate functions

$$
\begin{aligned}
\frac{x}{l} &= -\frac{1}{4\beta}\frac{\sin 2\theta}{\tau(1-\tau)^{\beta}} \\
\frac{y}{l} &= \frac{1}{4\beta}\frac{\cos 2\theta}{\tau(1-\tau)^{\beta}} - \frac{1}{4}\int^{\tau}\tau^{-1}(1-\tau)^{-1-\beta}d\tau
\end{aligned}
\tag{6-11}
$$

if $C = l(a^0)^2$.

We note that, after elimination of θ, the constant q or constant τ lines are again circles:

$$\left(\frac{x}{l}\right)^2 + \left(\frac{y}{l} - y_0\right)^2 = R^2$$

where

$$y_0 = -\frac{1}{4}\int^{\tau}\tau^{-1}(1-\tau)^{-1-\beta}d\tau \quad \text{and} \quad R = \frac{1}{4\beta}\tau^{-1}(1-\tau)^{-\beta}$$

As the streamline hodographs are circles, namely

$$\tau^{\frac{1}{2}} = k^{\frac{1}{2}}\cos\theta \tag{6-12}$$

where k is a constant depending only on ψ, we can eliminate θ between Eq. 6-11 and Eq. 6-12. The resulting parametric equations for the stream-

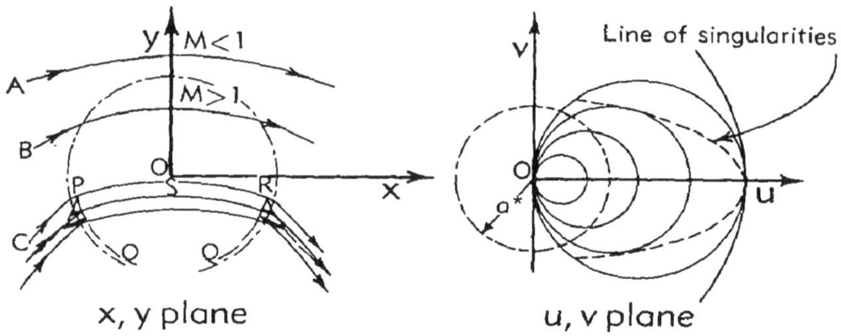

Fig. F,6b. The derived compressible flow patterns in x, y and u, v planes.

lines in the physical plane are

$$
\begin{aligned}
\frac{x}{l} &= -\frac{1}{2\beta}k^{-\frac{1}{2}}(1-\tau)^{-\beta}\left(\frac{1}{\tau} - \frac{1}{k}\right)^{\frac{1}{2}} \\
\frac{y}{l} &= -\frac{1}{2}\beta^{-1}(1-\tau)^{-\beta}\left(\frac{1}{2\tau} - \frac{1}{k}\right) - \frac{1}{4}\int^{\tau}\tau^{-1}(1-\tau)^{-1-\beta}d\tau
\end{aligned}
\tag{6-13}
$$

For various values of k, the streamlines are shown in Fig. F,6b.

Examining the flow pattern, we notice that the flow field is divided into subsonic and supersonic regions by the sonic circle $q = a^*$. In the subsonic region, which lies outside the circle, there is a striking resemblance between this flow and that of the incompressible flow around the edge (Fig. F,6a). However, in the supersonic region this similarity gradually disappears, and as the speed increases an entirely new phenomenon appears. First of all, the streamlines in the central region tend to bend more rapidly and hence become flatter than those of the incompressible case. This flattening produces two "throats" in each stream tube, at the two sonic lines, and between them the familiar expansion that is characteristic of supersonic ducts. As we move toward the origin, there occurs a first streamline (curve C in the figure) on which an infinite pressure gradient occurs. From there on, every streamline, instead of proceeding smoothly to the other side, makes four abrupt turns in the form of cusps. No streamlines enter the region $PSRQ$; consequently, we have again encountered an area in which no solution is possible.

Upon closer examination of the equations of the streamlines, it is easy to see the mathematical reasons why this happens. For if we differentiate these equations (6-13), the elements dx, dy, and $d\tau$ are related by

$$\frac{dx}{l} = \frac{1}{4\beta}\tau^{-2}(1-\tau)^{-1-\beta}\left(\frac{k}{\tau}-1\right)^{-\frac{1}{2}}\left[1-\alpha^2\tau+2\beta\frac{\tau^2}{k}\right]d\tau$$

$$\frac{dy}{l} = \frac{1}{4\beta}\tau^{-2}(1-\tau)^{-1-\beta}\left[1-\alpha^2\tau+2\beta\frac{\tau^2}{k}\right]d\tau$$

Thus, when a critical value of τ is reached whereby

$$1-\alpha^2\tau+2\beta\frac{\tau^2}{k}=0$$

dx and dy momentarily vanish for given $d\tau$, i.e. they become differentials of higher order. As a result all space derivatives such as $\partial/\partial x$ and $\partial/\partial y$ become locally indeterminate. The implication of this singular line PQR (Fig. F,6b) will be explored further in Art. 12.

From a different point of view, the example is not without practical interest. If two streamlines such as A and B are taken as channel walls, we find that the flow through this channel is partly subsonic and partly supersonic. It starts from rest at one end, accelerates to a supersonic speed, and then decelerates continuously, without shocks, back to subsonic speed again. Thus Ringleb's results emphasized the fact that sonic speed is not a theoretical barrier to smooth, isentropic, accelerating-decelerating flow, contrary to earlier beliefs (e.g. Taylor [*15*]). By suitable choice of wall shape according to these results, a supersonic Mach number as high as 2.5 can be reached, with subsequent deceleration, without occurrence of any abnormal effect. The fact that a mixed smooth transonic flow can be established theoretically is of considerable importance. This is one of the significant results of the hodograph method of attacking the nonlinear equations of gas flow. Its practical meaning, in terms of the design of ducts or airfoils, has not yet been completely established; flows of this type have apparently never been observed experimentally.

F,7. Two-dimensional Gas Jets. The first step toward the solution of the general problem was made by Chaplygin [2] in his study of the efflux of gas from a slit. The technique used to overcome, to a certain extent, the difficulty of boundary conditions is to imitate the analogous incompressible flow. For this purpose, we shall begin with an investigation of the incompressible problem.

The incompressible jet. Suppose the half-plane $x_i \leqq 0$ to be filled with incompressible fluid at a given pressure. If the plane $x_i = 0$ has an opening AA' symmetrical with respect to the x_i axis, the gas, when it

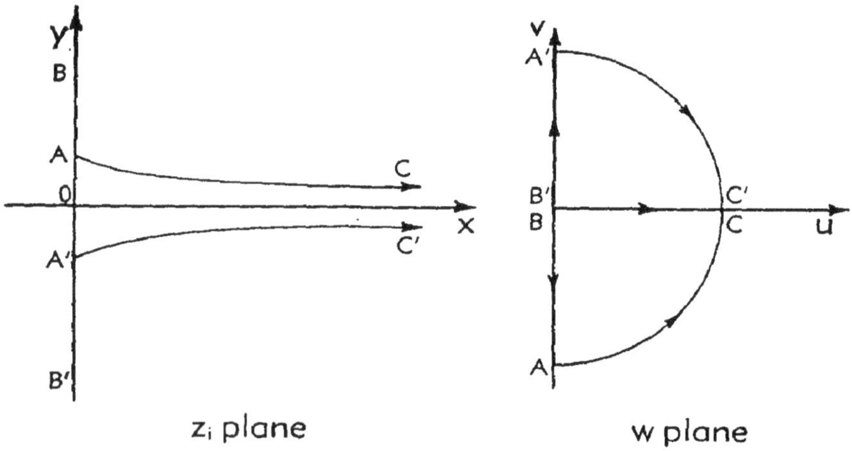

Fig. F,7a. Flow patterns in z_i and w planes.

rushes out through this slit, produces a jet bounded by two free boundaries AC and $A'C'$ on which pressure is constant (Fig. F,7a). The flow starts with zero velocity at infinity and converges to a parallel jet at positive infinity. The problem is to determine the shape of the free boundary and the velocity distribution inside the jet.

The first solution of this incompressible flow problem was given by Helmholtz by the hodograph method. To map conformally the flow on the w plane, we proceed as follows: Since the pressure of the free boundaries is constant, they correspond to a semicircle $ACC'A'$ in the w plane (Fig. F,7a), while the walls AB and $A'B'$ appear as segments of the v axis inside the circle. The whole flow field is therefore mapped into a region inside the semicircle. This region can be further mapped into half-plane R.P.$\{\zeta\} \geqq 0$ (Fig. F,7b) by the so-called Joukowski transformation

$$2\zeta = -w + \frac{U^2}{w} \tag{7-1}$$

and

$$w = qe^{-i\theta} = u - iv$$

where U is the velocity on the free boundary AC or $A'C'$.

To establish a relation between ζ and W_i, we note that on the wall and the free boundary the stream function ψ_i is constant. If the quantity of flow through the slit is Q, it is clear that $\psi_i = \frac{1}{2}Q$ on BAC and $\psi_i = -\frac{1}{2}Q$ on $B'A'C'$. The flow field in the W_i plane therefore lies in a

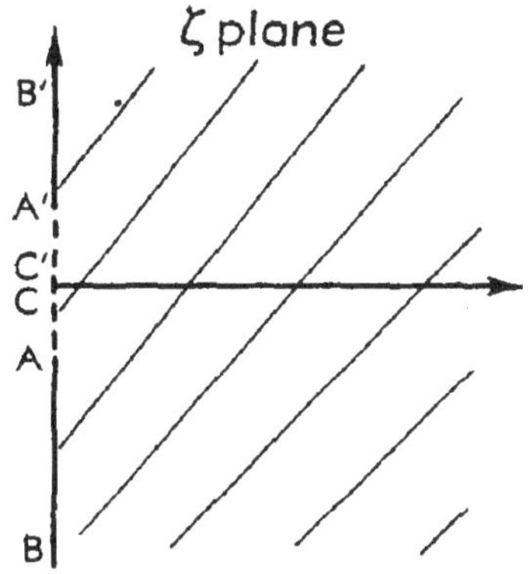

Fig. F,7b. Flow field in ζ plane.

strip (Fig. F,7c). The transformation of the half-plane into this strip can be effected by

$$W_i = -\frac{Q}{\pi} \ln \frac{\zeta}{U} \tag{7-2}$$

By substituting ζ from Eq. 7-1 in 7-2, we finally arrive at the expression for the complex potential W_i in the hodograph plane:

$$W_i = -\frac{Q}{\pi} \ln \left(\frac{U}{w} - \frac{w}{U}\right) \tag{7-3}$$

Study of this expression shows that the flow pattern in the w plane (considered, for the moment, as a flow plane itself, rather than the

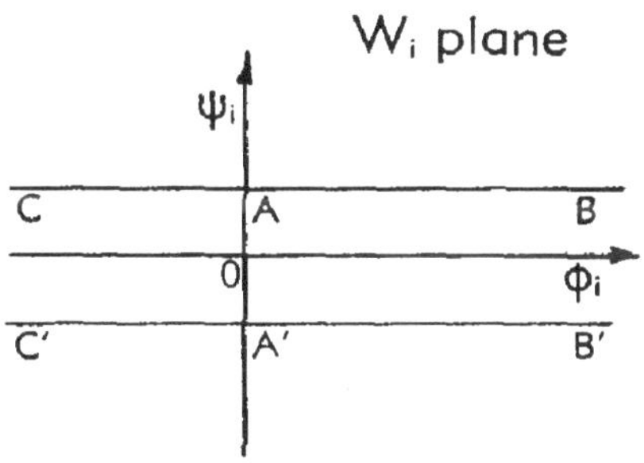

Fig. F,7c. Flow field in W_i plane.

hodograph of another flow pattern) consists of the superposition of a source at the origin and two sinks at $w = \pm U$. The function W_i is single-valued in the cut plane $-\pi < \theta < \pi$ and for $|w| < U$ possesses an expansion

$$W_i = \frac{Q}{\pi} \ln \frac{2w}{U} + \frac{Q}{\pi} \sum_1^{\infty} \frac{1}{n} \left(\frac{w}{U}\right)^{2n}, \qquad |w| < U \tag{7-3a}$$

Separating this into real and imaginary parts, we have

$$\left.\begin{aligned}\phi_i &= \frac{Q}{\pi}\ln\frac{2q}{U} + \frac{Q}{\pi}\sum_1^\infty \frac{1}{n}\left(\frac{q}{U}\right)^{2n}\cos 2n\theta \\ \psi_i &= -\frac{Q}{\pi}\theta - \frac{Q}{\pi}\sum_1^\infty \frac{1}{n}\left(\frac{q}{U}\right)^{2n}\sin 2n\theta\end{aligned}\right\} \quad q < U \quad (7\text{-}4)$$

The compressible jet. On the basis of Eq. 7-4, Chaplygin chose the form of solution for the compressible jet to be

$$\psi = -\frac{Q}{\pi}\theta - \frac{Q}{\pi}\sum_1^\infty \frac{1}{n}\left(\frac{q}{U}\right)^{2n}\frac{F_{2n}(\tau)}{F_{2n}(\tau_1)}\sin 2n\theta \tag{7-5}$$

in $-\pi < \theta < \pi$ and $q < U$ where τ_1 is the value of τ at $q = U$. It is seen that when a^0 is infinite ($\tau = \tau_1 = 0$) the solution reduces identically to the incompressible flow. In any case, on the free boundary $q = U$, $\psi = \psi_i(U, \theta) = \pm Q/2$ and the same value occurs on the walls $\theta = \pm\pi/2$, thus, formally Eq. 7-5 is a solution satisfying the boundary condition. To establish its convergence, we need only demonstrate the inequality: as $n \rightarrow \infty$

$$\left|\frac{1}{n}\left(\frac{q}{U}\right)^{2n}\frac{F_{2n}(\tau)}{F_{2n}(\tau_1)}\right| < \frac{C}{n}\left[\frac{T(\tau)q}{T_1U}\right]^{2n}$$

where T is the function defined in the appendix to this section, and T_1 denotes the value $T(\tau_1)$. This can easily be done, making use of the asymptotic properties of $F_n(\tau)$ discussed in the appendix (Eq. 16-3), provided that the flow is everywhere subsonic. Moreover, since $d(Tq)/dq > 0$, $Tq < T_1U$. Therefore, the dominant series converges absolutely if $Tq < T_1U$, and it follows that series (7-5) converges uniformly in any closed domain $Tq \leqq T_1U - \delta$, $\delta > 0$.

From Eq. 7-5 together with Eq. 3-9 and 3-10, the velocity potential ϕ is

$$\begin{aligned}\phi = {} & \frac{Q}{\pi}\left[(1-\tau)^{-\beta} - \frac{1}{2}\int^\tau \tau^{-1}(1-\tau)^{-\beta}d\tau\right] \\ & + \frac{Q}{\pi}(1-\tau)^{-\beta}\sum_1^\infty \frac{1}{n}\left(\frac{q}{U}\right)^{2n}\frac{F_{2n}(\tau)}{F_{2n}(\tau_1)}\xi_{2n}(\tau)\cos 2n\theta, \qquad q < U \qquad (7\text{-}6)\end{aligned}$$

This series converges in the same domain as Eq. 7-5, since $|\xi_{2n}(\tau)| = O(1)$ as $n \rightarrow \infty$ by Eq. 16-7, for subsonic flow.

At this point both ϕ and ψ are determined up to an arbitrary multiplier Q, proportional to the mass flow. The coordinate functions now can be integrated in the following manner: Upon substitution of Eq. 7-5 and 7-6 in Eq. 2-15, partial integration yields

$$x = \frac{Q}{\pi}\frac{(1-\tau)^{-\beta}}{q}\left\{\sum_1^\infty \left(\frac{q}{U}\right)^{2n} \frac{F_{2n}(\tau)}{F_{2n}(\tau_1)}\xi_{2n}(\tau)\left[\frac{\cos(2n+1)\theta}{2n+1} + \frac{\cos(2n-1)\theta}{2n-1}\right]\right.$$

$$\left. - \sum_1^\infty \left(\frac{q}{U}\right)^{2n} \frac{F_{2n}(\tau)}{F_{2n}(\tau_1)}\left[\frac{\cos(2n+1)\theta}{2n+1} - \frac{\cos(2n-1)\theta}{2n-1}\right] - \cos\theta\right\} + \hat{x}(q) \tag{7-7}$$

$$y = \frac{Q}{\pi}\frac{(1-\tau)^{-\beta}}{q}\left\{\sum_1^\infty \left(\frac{q}{U}\right)^{2n} \frac{F_{2n}(\tau)}{F_{2n}(\tau_1)}\xi_{2n}(\tau)\left[\frac{\sin(2n+1)\theta}{2n+1} - \frac{\sin(2n-1)\theta}{2n-1}\right]\right.$$

$$\left. - \sum_1^\infty \left(\frac{q}{U}\right)^{2n} \frac{F_{2n}(\tau)}{F_{2n}(\tau_1)}\left[\frac{\sin(2n+1)\theta}{2n+1} + \frac{\sin(2n-1)\theta}{2n-1}\right] - \sin\theta\right\} + \hat{y}(q)$$

Now $x = 0$ and $y = 0$ when $\theta = \pi/2$ and 0 respectively, hence, we must have

$$\hat{x} = \hat{y} = 0$$

Thus, the coordinate functions have the forms

$$x = \frac{Q}{\pi}\frac{(1-\tau)^{-\beta}}{q}\left\{\sum_1^\infty \left(\frac{q}{U}\right)^{2n} \frac{F_{2n}(\tau)}{F_{2n}(\tau_1)}[\xi_{2n}(\tau)C_{2n}^{(1)}(\theta) - C_{2n}^{(2)}(\theta)] - \cos\theta\right\} \tag{7-8}$$

$$y = \frac{Q}{\pi}\frac{(1-\tau)^{-\beta}}{q}\left\{\sum_1^\infty \left(\frac{q}{U}\right)^{2n} \frac{F_{2n}(\tau)}{F_{2n}(\tau_1)}[\xi_{2n}(\tau)S_{2n}^{(2)}(\theta) - S_{2n}^{(1)}(\theta)] - \sin\theta\right\}$$

where the coefficients $C_{2n}^{(1)}$, $C_{2n}^{(2)}$, $S_{2n}^{(1)}$, and $S_{2n}^{(2)}$ denote the four trigonometric functions of θ that appear in square brackets in Eq. 7-7, i.e.

$$C_{2n}^{(1)}(\theta) = \frac{\cos(2n+1)\theta}{2n+1} + \frac{\cos(2n-1)\theta}{2n-1}$$

$$C_{2n}^{(2)}(\theta) = \frac{\cos(2n+1)\theta}{2n+1} - \frac{\cos(2n-1)\theta}{2n-1}$$

$$S_{2n}^{(1)}(\theta) = \frac{\sin(2n+1)\theta}{2n+1} + \frac{\sin(2n-1)\theta}{2n-1}$$

$$S_{2n}^{(2)}(\theta) = \frac{\sin(2n+1)\theta}{2n+1} - \frac{\sin(2n-1)\theta}{2n-1}$$

On the jet boundary, $q = U$ and $\tau = \tau_1$, the parametric equations for the curve are

$$x = \frac{Q}{\pi}\frac{(1-\tau_1)^{-\beta}}{U}\left\{\sum_1^\infty [\xi_{2n}(\tau_1)C_{2n}^{(1)}(\theta) - C_{2n}^{(2)}(\theta)] - \cos\theta\right\}$$

$$y = \frac{Q}{\pi}\frac{(1-\tau_1)^{-\beta}}{U}\left\{\sum_1^\infty [\xi_{2n}(\tau_1)S_{2n}^{(2)}(\theta) - S_{2n}^{(1)}(\theta)] - \sin\theta\right\}$$

or

$$x = \frac{Q}{\pi}\frac{(1-\tau_1)^{-\beta}}{U}\sum_1^\infty \xi_{2n}(\tau_1)C_{2n}^{(1)}(\theta)$$

$$y = \frac{Q}{\pi}\frac{(1-\tau_1)^{-\beta}}{U}\left[\sum_1^\infty \xi_{2n}(\tau_1)S_{2n}^{(2)}(\theta) - \frac{\pi}{2}\right]$$

as, for $0 < \theta < \pi$, it is known that

$$\sum_1^\infty \frac{\sin(2n-1)\theta}{2n-1} = \frac{\pi}{4}$$

and

$$\sum_1^\infty C_{2n}^{(2)}(\theta) = -\cos\theta$$

Let $2b$ be the width of the jet at infinity, and let Q be defined as

$$Q = 2bU(1-\tau_1)^\beta.$$

We have

$$\frac{y}{b} = -\frac{2}{\pi}\left[\frac{\pi}{2} - \sum_1^\infty \xi_{2n}(\tau_1)S_{2n}^{(2)}(\theta)\right]$$

If the aperture of the slit in the wall is $2a$, then, for $\theta = \pi/2$, $y = -a$. The contraction ratio is therefore

$$\frac{b}{a} = \left[1 + \frac{8}{\pi}\sum_1^\infty (-1)^{n-1}\frac{n\xi_{2n}(\tau_1)}{4n^2-1}\right]^{-1} \tag{7-9}$$

This ratio was calculated by Chaplygin for a series of Mach numbers. He found that the ratio increases with the Mach number from the lowest value of 0.611 at zero Mach number to 0.74 when the maximum Mach number, i.e. the Mach number at infinity and along the free boundary, is close to 1 (Fig. F,7d).

When the pressure in the receiver is lower than the critical value corresponding to unit Mach number, the jet boundary is mapped on a circular arc in the hodograph plane with radius greater than sonic speed. The flow in the jet is then partly subsonic and partly supersonic, with the transition taking place along sonic curves projecting into the flow from points A and A'. At these two points the speed jumps from subsonic to supersonic in a discontinuous way. Consequently, the boundary hodograph curve is not closed, and the method used here to solve the problem

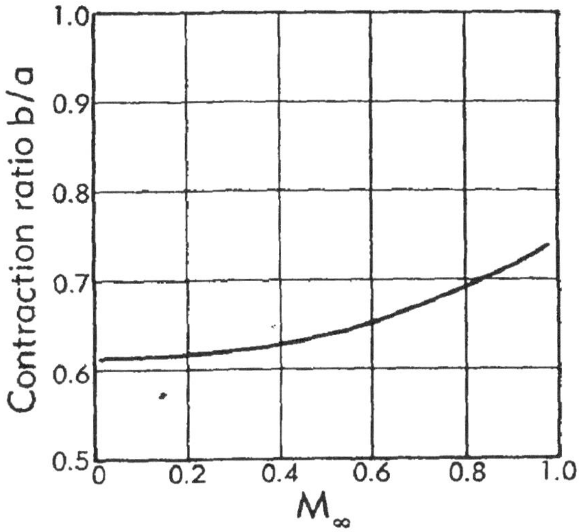

Fig. F,7d. Increase of contraction ratio with Mach number at infinity of the jet.

fails. For further development along this line, reference should be made to the work of Frankl [16] and Guderley [17] (cf. H,9).

F,8. Flow Around a Closed Body. As a generalization of Chaplygin's method, let us consider now the typical aeronautical problem of flow around a closed body. In the absence of circulation, the problem is to be solved subject to the following conditions: (1) the solution for the compressible flow should have the same analytic behavior as does the solution for a given incompressible flow; and (2) in the limiting case of vanishing Mach number, it must reduce to the incompressible solution. It appears that once more we must study the incompressible flow as our first step.

Conformal mapping of incompressible flow. For a simple example, we consider the flow about a circular cylinder. If the flow at infinity is represented by $w = 1$, then on the cylinder the well-known result is $|w| = 2 \cos \theta$. In going over from the z_i to the w plane, we observe, in the first place, that the infinity of the z_i plane corresponds to a point $w = 1$ in the w plane; the parts of the real axis outside the cylinder are mapped on the segment $v = 0, 0 \leqq u \leqq 1$; and the imaginary axis for $1 \leqq y_i < \infty$, on which v is again zero, is mapped on $1 \leqq u \leqq 2$ (see Fig. F,8a). Furthermore, the streamline $S_-P_+S_+$ in the z_i plane, according to the result just mentioned above, corresponds to the circle $S_-P_+S_+$ in the hodograph plane. Thus, the streamline $R_-P_+R_+$ appears as this circle together with the segment $v = 0$, $0 \leqq u \leqq 1$, on which both R_-S_- and S_+R_+ overlap. Since the highest velocity of the flow field is attained on the boundary, according to a general theorem for flows of this type, the whole upper half of the flow field must be mapped into the interior of the circle with

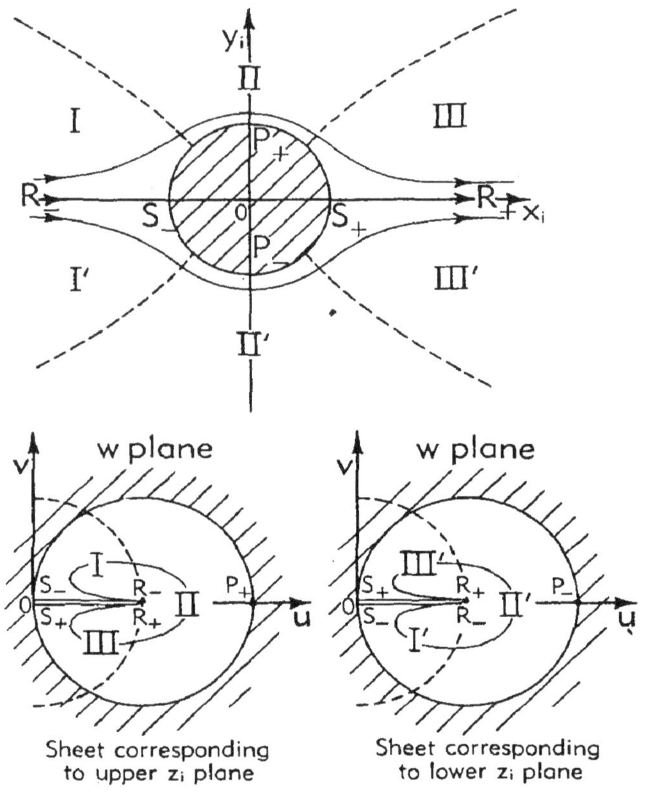

Fig. F,8a. Conformal mapping of incompressible flow from z_i into w plane.

center at $w = 1$. Proceeding to the lower half of the z_i plane, we have, by symmetry, the same mapping, as is also shown in Fig. F,8a.

Consequently, we see that the z_i plane, when transformed into the w plane, will be two-sheeted, i.e. will cover two w planes.

In Fig. F,8a six regions of the flow field, designated I, II, III, I′, II′, and III′ are indicated, their boundaries being the x_i axis and the lines along which $|w| = 1$. The locations of these regions in the two hodograph sheets are also shown; here, of course, their boundaries are the real axis and the unit circle.

The mapping can be considered one-valued if a cut is put in from $w = 0$ to $w = 1$, i.e. along R_-S_- and S_+R_+. This cut prevents our passing from the upper to the lower half of the flow field; thus, the w plane represents one or the other, but not both. We shall consider the upper half first, and obtain the flow in the lower by symmetry.

Even in the cut plane, it is not possible to write down a single expansion for W_i, in the form needed to carry out Chaplygin's method, which is valid throughout the region of interest. This is due to the presence of a singularity at the point $w = 1$, where all streamlines converge. Instead, it is necessary to use the following expansions in the various regions of the w plane:

$$
\begin{aligned}
{}_1W_i &= \sum_0^\infty A_n w^n, && && |w| < 1 \\
{}_2W_i &= i\sum_0^\infty B_n w^{-\nu}, && \nu = n - \tfrac{1}{2} && |w| > 1 \\
{}_3W_i &= -\sum_0^\infty A_n w^n, && && |w| < 1
\end{aligned}
\tag{8-1}
$$

where ${}_1W_i$, ${}_2W_i$, and ${}_3W_i$ are the proper expansions for W_i in the regions I, II, and III of Fig. F,8a, respectively. In the case of a circular cylinder, the constants A_n and B_n are

$$
\begin{aligned}
A_n &= -\frac{(n-1)\Gamma(n-\frac{1}{2})}{\Gamma(\frac{1}{2})\Gamma(n+1)} \\
B_n &= -\frac{(n+\frac{1}{2})\Gamma(n-\frac{1}{2})}{\Gamma(\frac{1}{2})\Gamma(n+1)}
\end{aligned}
$$

By symmetry, the complex potential in the other sheet (lower half of the flow field) is related to these expansions by the relations

$$
{}_3W_i' = {}_3W_i, \qquad {}_2W_i' = -{}_2W_i, \qquad {}_1W_i' = {}_1W_i \tag{8-2}
$$

Thus, when W_i is determined, W_i' is known.

A trial solution of compressible flow: region I. Following the principle of the preceding article, we can at once construct a solution for compressible flow about a distorted circular cylinder by applying Chaplygin's analogy to the expansions (8-1), if the flow is subsonic at infinity.

With considerable generality, we can select for our solution in region I the expansion

$$
W_1 = \sum_0^\infty A_n f_n(\tau_1) F_n(\tau) w^n, \qquad |w| < 1
$$

where W_1 has ψ_1 as its imaginary part, $F_n(\tau)$ is the hypergeometric function and $f_n(\tau_1)$ is a function of n and τ_1, τ_1 being the value of τ corresponding to the free stream condition.

To satisfy the conditions (1) and (2) stated at the beginning of this article, $f_n(\tau_1)$ must be chosen so that $f_n(0) = 1$ (since $F_n(0) = 1$); and for large n, since $F_n(\tau)$ is asymptotic to a multiple of T^n for $\alpha^2\tau < 1$ by Eq. 16-3, $f_n(\tau_1)$ must be asymptotic to a multiple of T_1^{-n} if W_1 and ${}_1W_1$ are to have the same analytic behavior at $\tau = \tau_1$. Clearly, to meet these requirements, $f_n(\tau_1)$ may either be $F_n(\tau_1)^{-1}$ (as in the example of our preceding article) or T_1^{-n}. Both of these forms have been used, but the latter, as pointed out by both Lighthill [*18*] and Cherry [*5*], is a more suitable choice if the boundary involved is closed. Accordingly, we take as our trial solution in region I the expansion

$$W_1 = \sum_0^\infty A_n \mathfrak{F}_n(\tau) w^n, \qquad |w| < 1 \tag{8-3}$$

where $\mathfrak{F}_n(\tau) = F_n(\tau)/T_1^n$.

Formally, we now have a solution satisfying both conditions (1) and (2). In order that it actually be a solution valid in the region indicated, we have to show that it converges. In the first place, we have the inequality

$$\left|\sum_0^\infty A_n \mathfrak{F}_n(\tau) w^n\right| \leqq \sum_0^\infty |A_n \mathfrak{F}_n w^n|$$

But from Eq. 16-3, $\mathfrak{F}_n(\tau) < C(T/T_1)^n$ if $\alpha^2\tau < 1$ and n is large. It follows, therefore, that

$$|A_n \mathfrak{F}_n(\tau) w^n| < C|A_n| \left(\frac{Tq}{T_1}\right)^n$$

the dominant series is then convergent absolutely if $Tq < T_1$. Consequently, series (8-3) converges uniformly and absolutely for any closed domain $Tq \leqq T_1 - \delta$, $\delta > 0$. Furthermore, as Tq is zero when $q = 0$, is T_1 when $q = 1$, and is monotonic in between, the series has exactly the same radius of convergence as Eq. 8-1, i.e. $|w| < 1$.

Continuation into region II. The solution, however, is still incomplete. In order to have a complete solution for all the regions, the question of analytic continuation arises. We obviously cannot write down an expansion for region II, for example, without making sure that it is the correct continuation of Eq. 8-3. This problem has been attacked by several authors [*5,18–22*] by different techniques—not all of which are correct. The method of Lighthill [*18*] and Cherry [*5*] is presented here.

The flow within region I is purely subsonic; it was discovered by Lighthill and Cherry that in this case $\mathfrak{F}_n(\tau)$ possesses the following representation (cf. Eq. 16-6):

$$\mathfrak{F}_n(\tau) = \left(\frac{T}{T_1}\right)^n \left[f(\tau) - \sum_1^\infty \frac{h_m \tau^m T^m F_m(\tau)}{n + m}\right] \tag{8-4}$$

By virtue of Eq. 8-4, expansion (8-3) can be rewritten[5] as

$$W_1 = f(\tau)_1W_1\left(\frac{T}{T_1}w\right) - \sum_0^\infty A_n\left(\frac{T}{T_1}w\right)^n \sum_2^\infty \frac{h_m\tau^m T^m F_m(\tau)}{n+m}$$

Since both series in this solution are absolutely convergent in region I, we may interchange the order of summation:

$$W_1 = f(\tau)_1W_1\left(\frac{T}{T_1}w\right) - \sum_2^\infty h_m\tau^m T^m F_m(\tau)\left(\frac{T}{T_1}w\right)^{-m} \sum_0^\infty \frac{A_n\left(\frac{T}{T_1}w\right)^{n+m}}{n+m}$$

But, in view of Eq. 8-1, it is easily verified that

$$\sum_0^\infty \frac{A\left(\frac{T}{T_1}w\right)^{n+m}}{n+m} = \int_0^\zeta \zeta^{m-1}{}_1W_1 d\zeta$$

with $\zeta = Tw/T_1$. It follows that

$$W_1 = f(\tau)_1W_1(\zeta) - \sum_2^\infty h_m(\tau_1\tau)^{\frac{m}{2}}F_m(\tau)e^{im\theta}\int_0^\zeta \zeta^{m-1}{}_1W_1 d\zeta \qquad (8\text{-}5)$$

Now, although we have restricted ourselves to region I ($|w| < 1$, i.e. $|\zeta| < 1$), Eq. 8-5 is actually in a form that constitutes a continuation into a larger region. For in region I, ${}_1W_1$ converges to W_1, which is an analytic function throughout a larger region, while the infinite series in Eq. 8-5 can also be shown to possess a larger radius of convergence. To prove this, we need only note that the integral appearing there (when we read once more W_1 for ${}_1W_1(\zeta)$) is bounded for all ζ, provided that the path of integration avoids the singular point $\zeta = w = 1$, where W_1 itself is infinite, i.e.

$$\left|\int_0^\zeta \zeta^{m-1}{}_1W_1 d\zeta\right| \leq \frac{C}{m}|\zeta|^m$$

Therefore the series in Eq. 8-5 is dominated by $\sum h_m\tau^m T^m F_m(\tau)/m$ which converges for all $\alpha^2\tau < 1$; hence, it is uniformly and absolutely convergent in any closed region where the flow is subsonic. Thus Lighthill and Cherry have succeeded in Eq. 8-5 in continuing W beyond region I into all subsonic and supersonic regions of the flow excepting, of course, the point $w = 1$.

[5] It is interesting to note that, in the special case $\gamma = -1$ considered in an earlier article, this solution becomes the one used by von Kármán and Tsien. For in that case $f(\tau) = 1$ and all the h_n's vanish (see Art. 16); thus $W_1 = W_1(q_1e^{-i\theta})$.

For simplicity, in going from region I to region II, let us perform the integration in Eq. 8-5 along the following path:

$$\int_0^\zeta = \int_0^{1-} + \oint + \int_{1+}^\zeta$$

where the second contribution along the semicircle about the branch point vanishes, and ζ is restricted to values corresponding to $\tau(\zeta)$ in $\tau_1 < \tau(\zeta) < 1$. Now in region II, according to Eq. 8-1

$$\int_{1+}^\zeta \zeta^{m-1}{}_2W_i d\zeta = i\sum_0^\infty \frac{B_n}{-\nu + m}\zeta^{-\nu+m} - i\sum_0^\infty \frac{B_n}{-\nu + m}$$

Substituting in Eq. 8-5 and again interchanging the order of summation, we finally arrive at a solution in region II, $|w| > 1$ and $\tau_1 < \tau < 1$:

$$W_2 = i\sum_0^\infty B_n \mathfrak{F}_{-\nu}(\tau)w^{-\nu} - \sum_2^\infty h_m g_m(\sqrt{\tau_1}\, T_1)^m \psi_m(\tau)e^{im\theta} \tag{8-6}$$

where

$$g_m = \int^1 \zeta^{m-1}{}_1W_i d\zeta - i\sum_0^\infty \frac{B_n}{-\nu + m} \text{ and } \psi_m(\tau) = \tau^{\frac{m}{2}}F_m(\tau)$$

In the present example

$${}_1W_i = -(1-\zeta)^{\frac{1}{2}} - (1-\zeta)^{-\frac{1}{2}}$$

whence

$$\int_0^1 \zeta^{m-1}{}_1W_i d\zeta = -\frac{(m+1)\Gamma(m)\Gamma(\frac{1}{2})}{\Gamma(m+\frac{3}{2})}$$

and

$$\sum_0^\infty \frac{B_n}{-\nu + m} = \frac{1}{m+\frac{1}{2}}\frac{\Gamma(-m+\frac{1}{2})\Gamma(\frac{3}{2})}{\Gamma(-m+1)} + \frac{1}{m-\frac{1}{2}}\frac{\Gamma(-m+\frac{3}{2})\Gamma(\frac{1}{2})}{\Gamma(-m+1)}$$

As m is a positive integer, $1/\Gamma(-m+1) \equiv 0$; therefore, g_m is real and equal to

$$g_m = -\frac{(m+1)\Gamma(m)\Gamma(\frac{1}{2})}{\Gamma(m+\frac{3}{2})} \tag{8-7}$$

It is interesting to note that the continued solution (8-6), which we have derived by continuation of Eq. 8-3, differs in form from its "parent solution," i.e. ${}_2W_i$ in Eq. 8-1, by the presence of an additional series. This series vanishes with vanishing τ, as it must, but destroys the symmetry of the required solution.

Modified solution and continuation. As the contour in the limit of incompressible flow is a circular cylinder, the solution of the compressible flow must at least possess fore-and-aft symmetry, which our trial solution fails to exhibit. We need, therefore, to revise our solution to permit supersonic local conditions.

Again we follow Lighthill [*18*] and Cherry [*5*] to accomplish this. Their method is to replace Eq. 8-3 by a different form of solution for region I; in fact, they add to Eq. 8-3 a series of exactly the form that gives trouble in region II, i.e.

$$W_1 = \sum_0^\infty A_n \mathfrak{F}_n(\tau) w^n + \sum_2^\infty h_n g_n(\sqrt{\tau_1}\, T_1)^n \psi_n(\tau) e^{in\theta} \tag{8-8}$$

Since the second series converges in $\tau_1, < \tau < 1$ only the first series has to be continued. By repeating the above process, we find the solution in region II:

$$W_2 = i \sum_0^\infty B_n \mathfrak{F}_{-\nu}(\tau) w^{-\nu}, \qquad |w| > 1 \tag{8-9}$$

In the subsonic range, by Eq. 16-3 the series (8-9) is dominated by

$$\sum_0^\infty \left| B_n \left(\frac{Tw}{T_1}\right)^{-\nu} \right|$$

which, according to Eq. 8-1, converges if $Tq > T_1$, which we have seen to be true. Again, in the supersonic range, the dominant is

$$\sum_0^\infty \left| B_n \left(\frac{r}{\sqrt{\tau_1}\, T_1}\right)^{-\nu} \right|$$

which converges if $\sqrt{\tau_1}\, T_1 < r$ (see Eq. 16-4). This inequality is found to be true in connection with the proof of Eq. 16-6. Consequently, the continued series (8-9) converges uniformly and absolutely in any closed domain in $1 < |w| < q_{\max}$.

To complete our solution, we have only to extend it into region III. If the path of integration for the integral in Eq. 8-5 is taken from $\zeta = 0$, around $w = 1$, back to $\zeta = 0$, and thence to an arbitrary point in III, we have

$$\begin{aligned} \int_0^\zeta \zeta^{m-1} W_i d\zeta &= \int_0^{1-} \zeta^{m-1}{}_1 W_i d\zeta + \int_{1-}^0 \zeta^{m-1}{}_3 W_i d\zeta + \int_0^\zeta \zeta^{m-1}{}_3 W_i d\zeta \\ &= -2g_m + \int_0^\zeta \zeta^{m-1}{}_3 W_i d\zeta \end{aligned} \tag{8-10}$$

by virtue of Eq. 8-1, where again the contribution from the circle about $\zeta = 1$ is zero. Substituting Eq. 8-10 in the continued solution, corresponding to Eq. 8-5, the solution valid in region III is

$$W_3 = -\sum_0^\infty A_n \mathfrak{F}_n(\tau) w^n - \sum_2^\infty h_n g_n(\sqrt{\tau_1}\, T_1)^n \psi_n(\tau) e^{in\theta} \tag{8-11}$$

The three expressions (8-8), (8-9), and (8-11) taken together make up the solution in the first complete Riemann sheet and, moreover, satisfy the conditions (1) and (2) at the beginning of this article. This final solution has another important advantage over our "trial solution" discussed above, namely, that it represents a flow symmetric about the y axis, while the solution based on Eq. 8-3, upon continuation into region III, is found to be asymmetric. Thus, even for purely subsonic flow patterns, the "trial solution" was inappropriate for the present problem.

By symmetry, the solution in the second Riemann sheet, representing the lower half of the flow plane, can also be given as in Eq. 8-2, namely

$$W_3' = W_3, \qquad W_2' = -W_2, \qquad W_1' = W_1 \tag{8-12}$$

Thus, it can be stated that, given an incompressible flow without circulation around a body, the Lighthill-Cherry method with proper conditions, such as symmetry, closed body, etc., leads to a unique solution for compressible flow [*18*]. The extension to the case of a body with an isolated infinite curvature, e.g. the trailing edge of an airfoil, can easily be carried out if the formula (16-6) is modified to include negative integers; this has been done by Kuo [*23*].

Coordinate functions. To complete our solution, we must calculate the coordinate functions relating x and y to the hodograph variables. The stream function is found by taking the imaginary parts of Eq. 8-8, 8-9, and 8-10 for the three regions of the plane. The methods of Art. 3 then lead, without difficulty, to the following expressions for the velocity potential:

$$\phi_1 = (1-\tau)^{-\beta}\sum_0^\infty \tilde{A}_n q^n \mathfrak{F}_n(\tau)\xi_n(\tau)\cos n\theta, \qquad 1 > q$$

$$\phi_2 = (1-\tau)^{-\beta}\sum_0^\infty B_n q^{-\nu}\mathfrak{F}_{-\nu}(\tau)\xi_{-\nu}(\tau)\sin \nu\theta, \qquad 1 < q < q_{\max} \tag{8-13}$$

$$\phi_3 = -(1-\tau)^{-\beta}\sum_0^\infty \tilde{A}_n q^n \mathfrak{F}_n(\tau)\xi_n(\tau)\cos n\theta, \qquad 1 > q$$

where $\tilde{A}_n = A_n - h_n g_n(\sqrt{\tau_1}\,T_1)^{2n}$.

Both ψ and ϕ being known, the integration of Eq. 2-15 and 2-16 can be performed. The results are as follows: if the stagnation point is taken to be $x_1 = +x_0$ and $y = 0$, we have, for $q < 1$, $\theta > 0$,

$$x_1 = +x_0 + \frac{(1-\tau)^{-\beta}}{2q}\sum_2^\infty n\tilde{A}_n q^n \mathfrak{F}_n(\tau)[\xi_n(\tau)C_n^{(1)}(\theta) - C_n^{(2)}(\theta)]$$

$$y_1 = \frac{(1-\tau)^{-\beta}}{2q}\sum_2^\infty n\tilde{A}_n q^n \mathfrak{F}_n(\tau)[\xi_n(\tau)S_n^{(2)}(\theta) - S_n^{(1)}(\theta)] \tag{8-14}$$

where the coefficients $C_n^{(1)}$, $C_n^{(2)}$, $S_n^{(1)}$, and $S_n^{(2)}$ are the trigonometric functions defined in connection with Eq. 7-8. Similarly, in $1 < q < q_{max}$ and $-\pi/2 < \theta < \pi/2$

$$x_2 = \frac{(1-\tau)^{-\beta}}{2q} \sum_0^\infty \nu B_n q^{-\nu} \mathfrak{F}_{-\nu}(\tau)[\xi_{-\nu}(\tau) S_\nu^{(1)}(\theta) - S_\nu^{(2)}(\theta)]$$

$$y_2 = y_0 - \frac{(1-\tau)^{-\beta}}{2q} \sum_0^\infty \nu B_n q^{-\nu} \mathfrak{F}_{-\nu}(\tau)[\xi_{-\nu}(\tau) C_\nu^{(2)}(\theta) - C_\nu^{(1)}(\theta)] \tag{8-15}$$

and in $q < 1$, $\theta < 0$

$$x_3 = -x_1 \qquad y_3 = -y_1 \tag{8-16}$$

The constant in x_2 has been determined so that $x_2 = 0$ when $\theta = 0$; the constants x_0 and y_0 will be determined by the condition that, on $q = 1$ and $\theta = \theta_0$ on the body, i.e. $\psi(1, \theta_0) = 0$, we have

$$x_1 = x_2, \qquad y_1 = y_2 \tag{8-17}$$

By symmetry, the coordinate functions of the second branch (lower part of the flow field) are given by

$$x_3' = x_3, \qquad y_3' = y_3$$
$$x_2' = -x_2, \qquad y_2' = -y_2 \tag{8-18}$$
$$x_1' = x_1, \qquad y_1' = y_1$$

Results. The theory outlined in this article has been applied by Cherry [*24*] to the flow about a blunt symmetric cylinder. The case studied by Cherry is the one that reduces to a circular cylinder when the Mach number is zero. His results are reproduced in Fig. F,8b, where several streamlines are plotted for flow at a stream Mach number of 0.5098. The cylinder is very nearly circular at this Mach number. The values of τ noted in the figure indicate various values of the speed, sonic speed corresponding to $\tau = 0.1684$. A local supersonic region appears with a maximum value of τ that indicates a local Mach number of about 1.39. In the same paper, Cherry has extended the theory to handle a cylinder that is symmetric fore and aft but has camber, and has again calculated and plotted the streamlines and local speeds.

Method of Tsien and Kuo. There is another method [*19,20*] that is mathematically simple and should be briefly mentioned. We return to a discussion of the form to be assumed for the stream function in region I and its continuation into the other regions. If, as in Eq. 8-3, the solution inside the unit circle is

$$\psi^{(i)} = -\sum_1^\infty \bar{A}_n \mathfrak{F}_n(\tau) q^n \sin n\theta, \qquad 0 < \theta < 2\pi, \tag{8-19}$$

the solution for $q > 1$ and $0 < \theta < 2\pi$ can be assumed to be

$$\psi^{(0)} = \sum_{0}^{\infty} B_n q^{-\nu} \mathfrak{F}_{-\nu}(\tau) \cos \nu\theta + \sum_{1}^{\infty} \tilde{C}_n q^{-n} \mathfrak{F}_{-n}(\tau) \sin n\theta \tag{8-20}$$

In order that $\psi^{(0)}$ be the analytic continuation of $\psi^{(i)}$, they must coincide on $q = 1$, $0 < \theta < 2\pi$, i.e.

$$\begin{aligned} \psi^{(i)}(1, \theta) &= \psi^{(0)}(1, \theta) \\ \psi_q^{(i)}(1, \theta) &= \psi_q^{(0)}(1, \theta) \end{aligned} \tag{8-21}$$

If the B_n are supposed to be given by Eq. 8-1, then the constants $\tilde{A}_n$ and $\tilde{C}_n$ are determined from Eq. 8-21 by Fourier's theorem. Multiplying

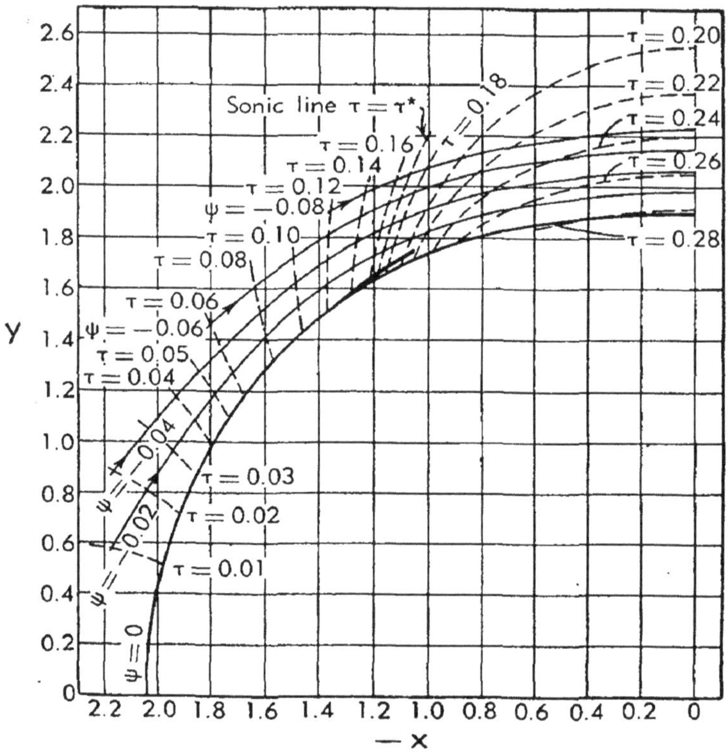

Fig. F,8b. Streamlines and constant-speed lines for flow at $M_\infty = 0.5098$ past a symmetrical cylinder, derived from incompressible flow around a circular cylinder [24]. $\gamma = 1.405$.

both sides by $\sin n\theta$ and integrating from $\theta = 0$ to $\theta = 2\pi$, we obtain

$$\begin{aligned} \tilde{A}_n \mathfrak{F}_n(\tau_1) - \tilde{C}_n \mathfrak{F}_{-n}(\tau_1) &= \frac{1}{\pi} \sum_{0}^{\infty} B_m \mathfrak{F}_{-\mu}(\tau_1) I_{n\mu} \\ n\tilde{A}_n \mathfrak{F}_n(\tau_1)\xi_n(\tau_1) - n\tilde{C}_n \mathfrak{F}_{-n}(\tau_1)\xi_{-n}(\tau_1) &= \frac{1}{\pi} \sum_{0}^{\infty} B_m \mathfrak{F}_{-\mu}(\tau_1)\xi_{-\mu}(\tau_1) I_{n\mu} \end{aligned} \tag{8-22}$$

where $I_{n\mu} = (n + \mu)^{-1} + (n - \mu)^{-1}$ and $\mu = m + \frac{1}{2}$. Now, by means of Eq. 16-6 together with Eq. 8-1, the first sum of Eq. 8-22 is shown to be

$$A_n\mathfrak{F}_n(\tau_1) - h_n g_n(\tau_1 T_1)^n F_n(\tau_1) \tag{8-23}$$

and the second, by similar process, is

$$nA_n\mathfrak{F}_n(\tau_1)\xi_n(\tau_1) - nh_n g_n(\tau_1 T_1^2)^n \mathfrak{F}_n(\tau_1)\xi_n(\tau_1) \tag{8-24}$$

When these are substituted back in Eq. 8-22, the algebraic equations yield the solution

$$\begin{aligned} \tilde{A}_n &= A_n - h_n g_n(\tau_1 T_1^2)^n \\ \tilde{C}_n &= 0 \end{aligned} \tag{8-25}$$

It is seen that the solution represented by Eq. 8-19 and 8-20 agree respectively with Eq. 8-8 and 8-9. In their original treatment, however, Tsien and Kuo [*19,20*] chose Chaplygin's form, viz. $F_n(\tau)/F_n(\tau_1)$ instead of $\mathfrak{F}_n(\tau)$. As a result, corresponding to the second terms in Eq. 8-23 and 8-24, they have infinite series, and for any nonvanishing Mach number the coefficient C_n is not zero. The mere presence of this additional series in Eq. 8-20 indicates that the analytic behavior of ψ and ψ_i is not the same. Consequently, their solutions do not have the character required by conditions (1) at the beginning of this article. Tsien and Kuo's results cannot, in fact, be considered correct. It appears, nevertheless, that the error is not numerically great, especially in the most interesting (transonic) regions, and that their results can be accepted as approximations. This is true of the interesting velocity distribution over a symmetric cylinder, calculated by Kuo [*20*], which exhibits smooth acceleration and deceleration through the sonic speed, just as do Cherry's results described above.

Conclusion. To summarize, the problem of constructing a solution along the line of Chaplygin for the flow about a closed body can be adequately dealt with so long as no circulation is present. However, when circulation is present, none of the proposed techniques is completely satisfactory. Cherry [*25*] has added circulation to the case mentioned above, where the body in the incompressible fluid is a circular cylinder; the results seem to be so involved as to be impractical for numerical application. More general problems involving circulation and obtained by modification of incompressible solutions have been treated by Lighthill [*18*]. Again the results are too complicated for utility.

Finally, we should call the reader's attention to Bergman's method [*26,27*] of constructing a solution for compressible flow by "integral-operator" methods. Due to the form of the solution, it does not appear suitable for practical computations (except perhaps by machine methods, as suggested by Bergman). Cherry [*28*] has shown that Bergman's solutions can be converted to Chaplygin's form; hence, the details of his technique will not be reproduced here.

F,9. Two-dimensional Laval Nozzle. The flow through a convergent-divergent symmetric channel can be either symmetric or asymmetric. The symmetric flow starts from rest at negative infinity, accelerates smoothly to a maximum speed in the throat, and then decelerates gradually in the divergent section to rest at positive infinity. For a given channel, the flow may either be entirely subsonic or may have local supersonic regions attached to the walls at the throat. This type of flow was first studied approximately by Taylor [*29*]. It can be treated exactly by the methods dealt with in the previous articles.

Distinct from this is the asymmetric flow, which, in the same channel, is subsonic in the convergent section but supersonic in the divergent one. Physically, it occurs when the pressure ratio is sufficiently increased beyond that of a subsonic symmetric flow. First, a shock wave appears downstream of the throat and then, as the pressure ratio is increased further, the shock moves downstream leaving the flow behind it as described here. From the mathematical standpoint, it is a new type of problem, and, furthermore, it is characteristically different from the one involving flow about a closed body. First of all, we note that, when the flow is mapped onto the hodograph plane, the streamlines cross one another in the supersonic region and, therefore, the hydrodynamic functions must be many-valued. Secondly, in a hyperbolic region, singularities, if they occur, cannot be isolated points but must be curves in the region. Consequently, the problem of analytic continuation must be expected to be very tricky.

Before we discuss the actual task of constructing a solution, let us determine the nature of the singularities involved in this type of flow. This can be accomplished by investigating the solution in the neighborhood of the sonic region. We know that, although the functions $\phi(q, \theta)$ and $\psi(q, \theta)$ are many-valued, the inverse functions $q(\phi, \psi)$ and $\theta(\phi, \psi)$ are, nevertheless, single-valued. Thus, in the neighborhood of the sonic region in the center of the throat, we can assume the expansions

$$\begin{aligned} \theta &= f_1\psi + \tfrac{1}{6}f_3\psi^3 + \cdots \\ \bar{q} &= f_0 + \tfrac{1}{2}f_2\psi^2 + \cdots \end{aligned} \tag{9-1}$$

where $f_0, f_1, f_2, f_3, \ldots$ are functions of ϕ, and $\bar{q}$ denotes $q - 1$ normalized with respect to sonic speed. For this form of solution, we have to invert the system (2-11). By approximating the coefficients in the sonic region by linear functions of $\bar{q}$, we find

$$\begin{aligned} \theta_\phi &= a\bar{q}_\psi \\ \theta_\psi &= 2b\bar{q}\bar{q}_\phi \end{aligned} \tag{9-2}$$

where $a = \left(\frac{2}{\gamma + 1}\right)^\beta$ and $b = \left(\frac{\gamma + 1}{2}\right)^{1+\beta}$. Upon substitution of Eq. 9-1 in Eq. 9-2, there result

$$
\begin{aligned}
f_1 &= 2bf_0f_0' \\
f_2 &= a^{-1}f_1' \\
f_3 &= 2b(f_0f_2' + f_0'f_2) \\
&\cdots\cdots\cdots\cdots
\end{aligned}
$$

or

$$
\begin{aligned}
f_1 &= 2bf_0f_0' \\
f_2 &= 2ba^{-1}(f_0f_0'' + f_0'^2) \\
f_3 &= 4b^2a^{-1}(f_0'^3 + 4f_0f_0'f_0'' + f_0^2f_0''') \\
&\cdots\cdots\cdots\cdots\cdots\cdots
\end{aligned}
\tag{9-3}
$$

Thus, the problem is uniquely determined if the axial velocity distribution $f_0(\phi)$ is specified. In the neighborhood of the origin, let us assume that f_0' is constant and equal to $k > 0$; θ can then be expressed as

$$\theta = 2bk\psi[\bar{q} - \tfrac{2}{3}k^2\psi^2 + \cdots] \tag{9-4}$$

Consider the region near the origin, wherein the approximation represented by the first two terms of Eq. 9-4 is valid. In the subsonic region $\bar{q} < 0$, according to this approximation, constant θ corresponds to only one real value of ψ. However, when $\bar{q} > 0$, i.e. in the supersonic region, constant θ corresponds to three real values of ψ. For $\theta = 0$, we have, for example, $\psi = 0$ and $\psi = \pm k^{-1}(3\bar{q}/2)^{\frac{1}{2}}$. This leads us to conclude that, in the supersonic region, $\psi(q, \theta)$ will be given by a Riemann surface of three sheets, joined at the branch lines

$$\theta_\psi = 0$$

Therefore, the branch lines, or envelopes of streamlines in the hodograph plane, are found to be

$$\theta = \pm \frac{2^{\frac{3}{2}}}{3}\left(\frac{\gamma + 1}{2}\right)^{\frac{1}{2}} \bar{q}^{\frac{3}{2}} \tag{9-5}$$

These results were first discovered by Lighthill [*65*].

It is interesting to note that this singular curve in the hodograph plane is independent of the velocity distribution $f_0(\phi)$ and, therefore, independent of the wall shape. As a result, the branch lines are fixed universal curves in the hodograph plane; they exist under all boundary conditions. It can, furthermore, be shown that each will belong to one family of characteristics of Eq. 3-1 (cf. Art. 12). Since the three surfaces overlap and are joined across two branch lines in two different planes, there must be four characteristics at the origin O; Fig. F,9a is a sketch showing the three surfaces and the branch lines.

Cherry's method of solution. From the previous discussions, we notice that in the case of asymmetric channel flows Chaplygin's analogy between incompressible and compressible solutions breaks down. We have shown, however, that the singular behavior of the solution for this type of flow is independent of the conditions at the boundary. Therefore, any possible solution must possess the following properties:

1. It is single-valued in the subsonic region and triple-valued in the supersonic region.
2. It has two branch lines Ω^+ and Ω^-.
3. It is finite and has finite partial derivatives.

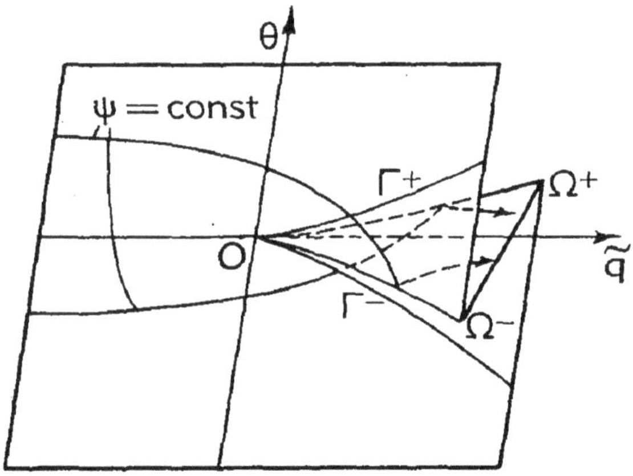

Fig. F,9a. Riemann surface of ψ in q, θ plane.

To construct such a solution for the Legendre potential, it is observed by Cherry [*30*] that the hypergeometric and Bessel functions have similar properties, as shown in Art. 16. If the solution is

$$\sum_0^\infty C_n q^n \tilde{F}_n(\tau) e^{in\theta}$$

it can be transformed into a new series involving Bessel functions by means of Eq. 16-9 and 16-11. By a suitable choice of the C_n, part of the resulting series will be found to be summable.

We begin by considering a solution

$$\sum_0^\infty \frac{(n/e)^n \chi_n(\tau) e^{-in\theta}}{r^n \Gamma(n+1)} + \sum_1^\infty \frac{\hbar_n r^n \Gamma(n) \chi_n(\tau) e^{in\theta}}{(n/e)^n} \tag{9-6}$$

where $\chi_n(\tau)$ denotes $\tau^{n/2}\tilde{F}_n(\tau)$ and r, $\hbar_n$ are defined respectively in Eq. 16-4 and 16-10. These series, according to Eq. 16-9 and Eq. 16-11, can be transformed into

$$Q(\tau)\left[1 + 2\sum_1^\infty J_n(nt)\cos n\theta\right]$$
$$- \sum_\mu f_\mu \left[\frac{1}{\mu} + \sum_1^\infty J_n(nt)\left(\frac{e^{-in\theta}}{\mu - n} + \frac{e^{in\theta}}{\mu + n}\right)\right] \quad (9\text{-}7)$$

where the summation $\sum_\mu$ is taken over the roots of $J_\mu(\mu t) = 0$ and $f_\mu(\tau)$ is defined in Eq. 16-11, provided rearrangement of the double series is allowed. The rearrangement can be shown to be permissible for $\alpha^2\tau < 1$ or $t < 1$. (The variable t is defined in terms of τ in Art. 16; see Eq. 16-4a and 16-11.)

The solution then is a superposition of Kapteyn series and, on account of the Bessel functions, considerable simplification results. Call

$$K(t, \theta) = 1 + 2\sum_1^\infty J_n(nt)\cos n\theta \quad (9\text{-}8)$$

$$L_\mu(t, \theta) = \frac{1}{\mu} + \sum_1^\infty J_n(nt)\left(\frac{e^{-in\theta}}{\mu - n} + \frac{e^{in\theta}}{\mu + n}\right) \quad (9\text{-}9)$$

We can then write Eq. 9-8 and 9-9, when $0 < t < 1$, as

$$QK + \sum_\mu f_\mu L_\mu \quad (9\text{-}10)$$

The series $K(t, \theta)$ is a well-known Kapteyn series and has the sum

$$K(t, \theta) = 1 + 2\sum_1^\infty J_n(nt)\cos n\theta = \frac{1}{1 - t\cos\xi} \quad (9\text{-}11)$$

where ξ is the root of

$$\xi - t\sin\xi = \theta \quad (9\text{-}12)$$

This is the celebrated Kepler equation. These relations (9-11) and (9-12) make it remarkably simple to verify that our solution has the correct behavior. In the first place, the Kapteyn series (9-11) provides a means of analytic continuation beyond $t = 1$, i.e. into the supersonic region, and the Kepler equation serves to determine the character of the solution. For $t \leqq 1$, the solution $\xi(t, \theta)$ is single-valued, but for $t > 1$ is triple-valued as in the neighborhood of $\xi = 0$ and $t = 1$,

$$\theta = \xi - t(\xi - \tfrac{1}{6}\xi^3) + \cdots$$

which is the same as Eq. 9-4 if ξ is replaced by ψ. Furthermore, the branch lines are

$$\theta_\xi = 1 - t\cos\xi = 0$$

i.e. $t = \sec \xi$. With this value of ξ, we have from Eq. 9-12 the equation of the branch lines

$$\theta = \pm(\tan^{-1}\sqrt{t^2 - 1} - \sqrt{t^2 - 1}) \tag{9-13}$$

But according to Eq. 16-10, the right-hand side of Eq. 9-13 is just $\pm\omega$, where ω is the quantity defined in Eq. 16-4a; i.e., the curves $\theta = \pm\omega$ are characteristic curves (epicycloids) passing through the sonic point on the axis in the hodograph plane. This verifies that K has the required singular behavior, except that it becomes infinite on the branch lines, in violation of (3) above. Consequently, Eq. 9-7 is a possible solution for the present problem. To establish its validity, we need, however, to

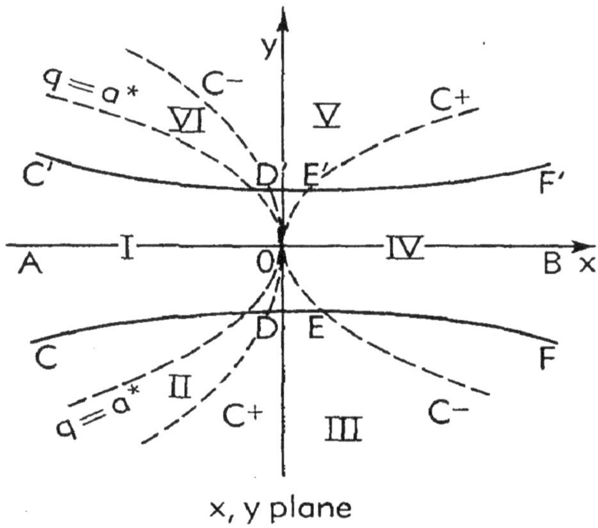

Fig. F,9b. Flow in a nozzle in x, y plane.

continue it into the supersonic region, verify the singular behavior of L_μ, and remove the infinities on the branch lines.

Analytic continuation of K and L_μ. Corresponding regions in the physical and hodograph planes are numbered as shown in Fig. F,9b and F,9c. The functions valid in the respective regions will be numbered accordingly, such as χ_{I}, χ_{II}, etc. In order ultimately to obtain the complete, continued solution in the form of a definite integral, we return to the series form for K in Eq. 9-11. Thus, both K and L_μ must be continued. To illustrate the principle employed, only K will be briefly considered here.

Since the Kapteyn series is convergent in $0 < t < 1$ and $-\pi \leqq \theta \leqq \pi$, then in region I we have, by definition,

$$K = 1 + 2\sum_{1}^{\infty} J_n(nt)\cos n\theta = K_{\mathrm{I}} \tag{9-14}$$

To continue K we consider in the half-plane R.P.$\{\nu\} > 0$ the integral

$$\oint \frac{e^{i\nu(\pi-\theta)}}{\sin \nu\pi} J_\nu(\nu t)d\nu$$

along a closed path. By choosing the path to be an appropriate semicircle in the half-plane R.P.$\{\nu\} > 0$, it can be shown by Cauchy's theorem that

$$\sum_1^\infty J_n(nt)e^{-in\theta} = \frac{1}{2i}\int_{\delta+\infty i}^{\delta-\infty i} \frac{e^{i\nu(\pi-\theta)}}{\sin \nu\pi} J_\nu(\nu t)d\nu \tag{9-15}$$

where the series represents the contributions from the simple poles $\nu = 1, 2, \ldots$ and the path of integration crosses the real axis between

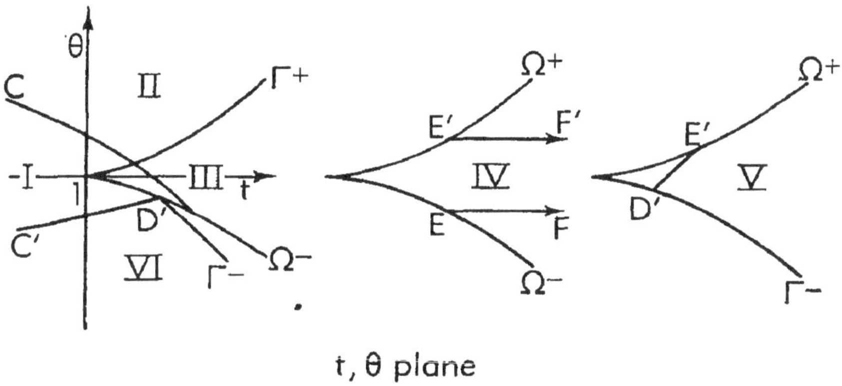

Fig. F,9c. Riemann surface in t, θ plane.

0 and 1, i.e. $0 < \delta < 1$. But, by the properties of $J_\nu(\nu t)$ near $t = 1$, we can verify that this integral actually converges uniformly in a small region G, surrounding $t = 1$, of a complex t plane, provided θ is also suitably restricted in a complex θ plane. Specifically, it converges uniformly for $|t - 1| < r$ and $\eta(r) <$ R.P.$\{\theta\} < 2\pi - \eta(r)$ where $\eta \to 0$ as $r \to 0$.

Thus, in Eq. 9-15 we have a form that constitutes a continuation of the series from I into a small region G around $t = 1$. The next step is to identify this, along a part of the real axis in G, with a form that permits continuation throughout II. This is done, for $t \geqq 1$, by distorting the path of integration of Eq. 9-15 to lie along the rays arg $\nu = \pm\pi$, avoiding the points $\nu = 0, -1, -2, \ldots$ by indentations. We obtain

$$\sum_1^\infty J_n(nt)e^{-in\theta} = \int_0^\infty J_{-\nu}(\nu t)e^{-i\nu(\pi-\theta)}d\nu - 1 - \sum_1^\infty J_n(nt)e^{in\theta}$$

or

$$1 + 2\sum_1^\infty J_n(nt)\cos n\theta = \int_0^\infty J_{-\nu}(\nu t)e^{-i\nu(\pi-\theta)}d\nu \tag{9-16}$$

The integral in Eq. 9-16 converges in $t \geqq 1$ and $\omega < \theta < 2\pi - \omega$, where $\theta = \pm\omega$ again defines two characteristics. In consequence, we have succeeded in continuing K through G into region II. The continued form is given by Eq. 9-16, i.e.

$$K_{\text{II}} = \int_0^\infty J_{-\nu}(\nu t)e^{-i\nu(\pi-\theta)}d\nu \tag{9-17}$$

For the continuation into III, it is most convenient to consider fixed $t > 1$ and K_{II} an analytic function of θ. From the relations between the Bessel and Hankel functions, it can be shown that the integral is regular in the half-plane I.P.$\{\theta\} > 0$, is singular at $\theta = -\omega$ and $\theta = 2\pi + \omega$, and can be continued between these points into the half plane I.P.$\{\theta\} < 0$. In particular, Eq. 9-17 is valid in region III, where $t \geqq 1$, $-\omega < \theta < \omega$.

Now, the singularity of the integral (9-17) at $\theta = -\omega$ in the θ plane can be shown to be a simple branch point. In order to obtain a function K_{IV} different from K_{III}, as required by condition (1) above, it is necessary to choose the other branch in passing over from $\theta > -\omega$ to $\theta < -\omega$, viz. [30, p. 560]

$$K_{\text{IV}} = -\int_0^\infty J_\nu(\nu t)e^{i\nu\theta}d\nu \tag{9-18}$$

This branch diverges at $\theta = \pm\omega$ and is symmetric in θ; thus $\theta = \omega$ is again a simple branch point, and by a process similar to the above it is found that

$$K_{\text{V}} = \int_0^\infty J_{-\nu}(\nu t)e^{-i\nu(\pi+\theta)}d\nu \tag{9-19}$$

By comparison of Eq. 9-17 and 9-19, the symmetry of the solution is verified.

In the case of L_μ, the key formula analogous to Eq. 9-15 is

$$\sum_1^\infty \frac{J_n(nt)e^{-in\theta}}{n-\mu} + \frac{\pi J_\mu(\mu t)e^{i\mu(\pi-\theta)}}{\sin \mu\pi} = \int_{\infty e^{\pi i}}^{(0-)} \frac{J_\nu(\nu t)e^{i\nu(\pi-\theta)}}{(\nu-\mu)\sin \nu\pi}\,d\nu \tag{9-20}$$

Here we assume, for the moment, that μ is complex and I.P.$\{\mu\} > 0$. After the continuation, however, it will be put equal to a zero of $J_\mu(\mu t)$, as in our definition of L_μ. By carrying out the similar steps, we can deduce that L_μ has the same character as K and has, in the various regions, the following continuations:

$$\begin{aligned}
(L_\mu)_{\text{I}} &= \sum_0^\infty \frac{J_n(nt)e^{in\theta}}{n+\mu} + \sum_1^\infty \frac{J_n(nt)e^{-in\theta}}{-n+\mu} = (L_\mu)_{\text{II}} \\
(L_\mu)_{\text{II}} &= \int_0^\infty \frac{J_{-\nu}(\nu t)e^{-i\nu(\pi-\theta)}}{\nu+\mu}\,d\nu = (L_\mu)_{\text{III}} \\
(L_\mu)_{\text{IV}} &= \int_0^\infty \frac{J_\nu(\nu t)e^{-i\nu\theta}}{\nu-\mu}\,d\nu \\
(L_\mu)_{\text{V}} &= \int_0^\infty \frac{J_{-\nu}(\nu t)e^{i\nu(\pi+\theta)}}{\nu+\mu}\,d\nu
\end{aligned} \tag{9-21}$$

Analytic continuation of χ. We have shown that both K and L_μ satisfy the conditions (1) and (2) and, therefore, Eq. 9-10 can be accepted as a possible solution. Let this solution be denoted by χ. Then in region I, i.e. $0 < t < 1$ and θ real, we have by definition

$$\chi_{\mathrm{I}} = QK - \sum_\mu f_\mu (L_\mu)_{\mathrm{I}}$$

Since in region I rearrangement of the double series is allowed, by applying Eq. 16-9 and 16-11, the series can be reconverted into Eq. 9-6, namely

$$\chi_{\mathrm{I}} = \sum_0^\infty \frac{(n/e)^n \chi_n(\tau) e^{-in\theta}}{r^n \Gamma(n+1)} + \sum_1^\infty \frac{\bar{h}_n r^n \Gamma(n) \chi_n(\tau) e^{in\theta}}{(n/e)^n} \tag{9-22}$$

The continuation of χ_{I} to region II, according to Eq. 9-16 and 9-21, can be expressed in the same form as Eq. 9-22, namely

$$\chi_{\mathrm{II}} = \chi_{\mathrm{I}} \tag{9-23}$$

provided $t \geqq 1$ and $\omega < \theta < 2\pi - \omega$.

In region III, taking the proper forms of K and L_μ, we have, on the other hand

$$\chi_{\mathrm{III}} = Q \int_0^\infty J_{-\nu}(\nu t) e^{-i\nu(\pi-\theta)} d\nu - \sum_\mu f_\mu \int_0^\infty \frac{J_{-\nu}(\nu t) e^{-i\nu(\pi-\theta)}}{\nu + \mu} d\nu$$

By an interchange of the summation and integral signs, it follows that

$$\chi_{\mathrm{III}} = \int_0^\infty J_{-\nu}(\nu t) e^{-i\nu(\pi-\theta)} \left(Q - \sum_\mu \frac{f_\mu}{\mu + \nu} \right) d\nu$$

From Eq. 16-9 we have finally

$$\chi_{\mathrm{III}} = \int_0^\infty \frac{\Gamma(\nu) r^\nu \sin \nu\pi \chi_{-\nu}(\tau) e^{-i\nu(\pi-\theta)}}{\pi(\nu/e)^\nu} d\nu \tag{9-24}$$

Similarly, in regions IV and V the continuations are respectively

$$\chi_{\mathrm{IV}} = -\int_0^\infty \frac{(\nu/e)^\nu \chi_\nu(\tau) e^{-i\nu\theta}}{r^\nu \Gamma(\nu+1)} d\nu \tag{9-25}$$

and

$$\chi_{\mathrm{V}} = \int_0^\infty \frac{\Gamma(\nu) r^\nu \sin \nu\pi \chi_{-\nu}(\tau) e^{i\nu(\pi+\theta)}}{\pi(\nu/e)^\nu} d\nu \tag{9-26}$$

Having verified that the solution made up by Eq. 9-22 to 9-26 satisfies the conditions (1) and (2), we notice, however, that according to Eq. 9-11 it is infinite on the branch lines Ω^- and Ω^+ and hence violates condition

(3). In order to remove this infinity, we can integrate the solution twice with respect to θ, because partial integration of a solution with respect to θ also yields a solution. The arbitrary functions introduced by integration must vanish if the branch character is to be preserved. Once this is done, the condition (3) will clearly be satisfied. As this final step involves no new principle, its details are omitted here.

For details of Cherry's analysis, the reader must refer to his original paper [30]. To date, no numerical examples of the flows obtained by this theory are available.

F,10. Approximate Solutions for Transonic Flows: Channel Flows. In the case of flow past boundaries with small curvature, the flow field may be expected to be nearly sonic everywhere, except possibly near stagnation points, if it is near sonic at infinity. For this type of flow, the problem can be considerably simplified.

We begin by making approximations to the fundamental system of Eq. 2-11. In this article it is most convenient to redefine the stream function, replacing ρ^0 by ρ^*, i.e. $\rho u = \rho^*\psi_y$ and $\rho v = -\rho^*\psi_x$; then the fundamental relations are

$$\begin{aligned} \phi_q &= -\frac{\rho^*}{\rho q}(1 - M^2)\psi_\theta \\ \frac{1}{q}\phi_\theta &= \frac{\rho^*}{\rho}\psi_q \end{aligned} \qquad (10\text{-}1)$$

where, as usual, the asterisk denotes sonic conditions. Let us define a new variable

$$\bar{q} = \int_{a^*}^{q} \frac{\rho}{\rho^*}\frac{dq}{q}$$

Now we are assuming $q = a^* + q'$, where q' is much smaller than a^*; under this assumption, we obtain the approximate relations

$$1 - M^2 \cong (\gamma + 1)\frac{q'}{a^*}$$

$$\bar{q} \cong \frac{q'}{a^*}$$

whereupon Eq. 10-1 become, approximately,

$$\begin{aligned} \phi_{\bar{q}} &= (\gamma + 1)\bar{q}\psi_\theta \\ \phi_\theta &= \psi_{\bar{q}} \end{aligned} \qquad (10\text{-}2)$$

The approximate differential equation for the stream function then becomes

$$\psi_{\bar{q}\bar{q}} - (\gamma + 1)\bar{q}\psi_{\theta\theta} = 0 \qquad (10\text{-}3)$$

It is seen that this equation is elliptic if $\bar{q} < 0$, i.e. in the subsonic region, and hyperbolic if $\bar{q} > 0$, i.e. in the supersonic region. This is known as Tricomi's equation [*31*]; it is simple, and yet it possesses all the important features of the exact equation in the transonic region. It is, in fact, the equation obtained by applying the hodograph transformation to the Kármán-Guderley transonic equation discussed in A,7, D,32, and D,33.

A number of interesting solutions has been obtained from the Tricomi equation, both for smooth flow past special boundaries and for supersonic flows involving shock waves. In the former category are the investigation of von Kármán and Fabri of flow past a corrugated wall [*32*] and the studies by Tomotika and Tamada [*33*] regarding transonic flow about certain cylinders. In the second category, involving weak shock waves, are the calculations of Guderley and Yoshihara [*34,35*], Cole [*36*], and Vincenti [*37*]. Flows of this type are discussed in some detail in H, 6–H,13.

In the present article, we shall present an application of this approximation to flow through convergent-divergent channels. This is also based on the work of Tomotika and Tamada [*38*].

Tomotika-Tamada solution for channel flows. We begin by interchanging the roles of the dependent and independent variables; we transform Eq. 10-2 into

$$\begin{aligned} \theta_\phi &= \bar{q}_\psi \\ \theta_\psi &= (\gamma + 1)\bar{q}\bar{q}_\phi \end{aligned} \tag{10-4}$$

(cf. Eq. 9-2). Elimination of θ yields

$$\frac{\partial^2 \bar{q}}{\partial \psi^2} - \frac{\gamma + 1}{2} \frac{\partial^2(\bar{q}^2)}{\partial \phi^2} = 0 \tag{10-5}$$

This equation is nonlinear and hence cannot generally be solved in any simple manner. However, as Tomotika and Tamada [*38*] have shown, a special group of solutions can easily be discovered. In the following, we shall outline their derivation of the one that gives convergent-divergent channel flow. Let the form of solution be

$$\frac{\gamma + 1}{2} \bar{q} = Z(\zeta) + 2\psi^2$$

where (10-6)

$$\zeta = \phi + \psi^2$$

With $\bar{q}$ written in the form of Eq. 10-6 together with Eq. 10-8, the second of Eq. 10-4 gives, upon integration,

$$(\gamma + 1)\theta = 4\psi[Z(\zeta) + 2\phi + \tfrac{2}{3}\psi^2] \tag{10-7}$$

by noticing that

$$\frac{\partial}{\partial \psi}(\psi Z) = Z + 2\psi^2 \frac{dZ}{d\zeta}$$

where the arbitrary function of ϕ is chosen to be zero. In view of the striking similarity between this form and Eq. 9-4, it is expected that the flow represented by Eq. 10-7 in the x, y plane must be of the Laval nozzle type. The function $Z(\zeta)$ serves to describe the axial velocity distribution.

If $\bar{q}$ in this form is substituted in Eq. 10-5, there results the following ordinary differential equation for Z:

$$\frac{d}{d\zeta}\left(Z\frac{dZ}{d\zeta}\right) - \frac{dZ}{d\zeta} - 2 = 0$$

or

$$Z\frac{dZ}{d\zeta} - Z = 2\zeta \tag{10-8}$$

where the constant of integration has been absorbed in ζ, as both ϕ and ψ are arbitrary in this regard. To integrate Eq. 10-8 let us introduce a new variable η defined by

$$\frac{d\zeta}{d\eta} = Z \tag{10-9}$$

Then

$$\frac{d^2\zeta}{d\eta^2} = \frac{dZ}{d\zeta}\frac{d\zeta}{d\eta} = Z\frac{dZ}{d\zeta}$$

As a result, instead of Eq. 10-8 we have the linear equation

$$\frac{d^2\zeta}{d\eta^2} - \frac{d\zeta}{d\eta} - 2\zeta = 0$$

of which the solution is

$$\zeta = c_1e^{2\eta} + c_2e^{-\eta} \tag{10-10}$$

Thus, according to Eq. 10-9, the function Z in terms of the variable η is

$$Z = 2c_1e^{2\eta} - c_2e^{-\eta} \tag{10-11}$$

Elimination of η between Eq. 10-10 and 10-11 leads to an integral

$$(Z + \zeta)(Z - 2\zeta)^2 = 2c^3 \tag{10-12}$$

where c is a constant.

To understand the nature of the flow, we shall discuss only the behavior of $Z(\zeta)$, leaving out the numerical details, which can be found in the work of Tomotika and Tamada. First, we note that for $c = 0$, $Z(\zeta)$ reduces to two straight lines through the origin, viz. $Z = -\zeta$ and $Z = 2\zeta$. Secondly, if c^3 is considered as a function of the two variables Z and ζ, it evidently changes sign when Z and ζ both change sign, and hence is antisymmetric in this sense. The integral curves $Z(\zeta)$ will lie in two distinct groups according as $c < 0$ and $c > 0$. These curves are shown in Fig. F,10a for the values $c = 0$, ± 0.05 and ± 0.10. There are four branches A, B, A', and B' with $Z = -\zeta$ and $Z = 2\zeta$ as asymptotes.

By Eq. 10-6, when $Z(\zeta)$ is positive, $\bar{q}$ is positive and the flow is supersonic. Branch A' (see Fig. F,10a) therefore corresponds to purely supersonic flow and is of very little practical interest. B and B', on

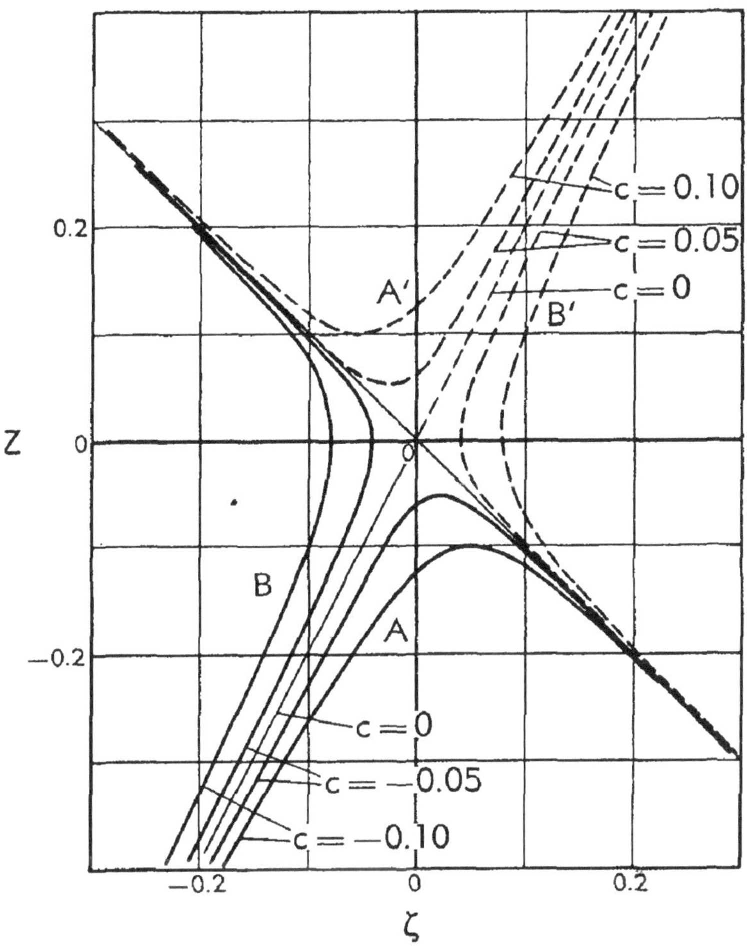

Fig. F,10a. Families of integral curves $Z(\zeta)$ [38].

the other hand, have singularities $dZ/d\zeta = \infty$ on $Z = 0$. However, $\bar{q}_\psi = [4/(\gamma + 1)][Z'(\zeta) + 2]\psi$ and $\bar{q}_\phi = [2/(\gamma + 1)]Z'(\zeta)$. Infinite $\bar{q}_\psi$ and $\bar{q}_\phi$ imply that the Jacobian $J \equiv \partial(x, y)/\partial(q, \theta)$ is zero; for

$$J^{-1} = \frac{\partial(q, \theta)}{\partial(\phi, \psi)} \frac{\partial(\phi, \psi)}{\partial(x, y)} \cong (a^*)^3[(\gamma + 1)\bar{q}\bar{q}_\phi^2 - \bar{q}_\psi^2]$$

in the present approximation. A solution with this type of singularity is physically impossible (cf. Art. 12) and hence must also be rejected.

On branch A, we have $Z(\zeta)$ everywhere negative and the axial velocity therefore subsonic. If the stream function is increased in either the positive or negative direction, $\bar{q}$ increases from its axial value and attains the largest value on the wall, i.e. at $\psi = \pm\psi_0$, say. Where $-Z(\zeta)$ is small, the value attained by $\bar{q}$ at the wall may be positive; i.e., the flow may be supersonic near the wall. We note that this occurs only near $\zeta = 0$, which means that supersonic flow can appear only in the neighborhood of the throat. This type of flow is shown in Fig. F,10b, for the particular case $c = -0.10$. This will be designated Taylor's type of flow [29]; it is the sort of flow called "symmetric" in Art. 9.

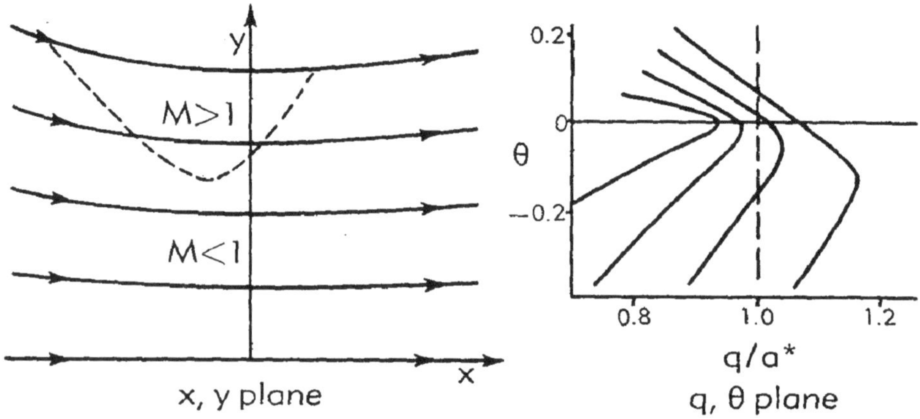

Fig. F,10b. Flow patterns in x, y and q, θ planes for $c = 0.10$, Taylor's type [38].

The limiting case of this family of solutions is the case $c = 0$, i.e.

$$\begin{aligned} Z &= -\zeta, \qquad \zeta > 0 \\ &= 2\zeta, \qquad \zeta < 0 \end{aligned}$$

The supersonic region then extends all the way to the axis at one point, viz. $\phi = 0, \psi = 0$. This solution has a discontinuity in $Z'(\zeta)$ at $\zeta = Z = 0$, i.e. along the curve $\phi + \psi^2 = 0$ in the flow plane. Since this singular curve lies in the supersonic region ($\bar{q} > 0$) and is a characteristic of Eq. 10-5, it must correspond to a Mach wave in the physical plane (cf. Art. 12).

On this singular line the velocity is continuous, but the curvature of the streamlines is discontinuous. This curvature is $qd\theta/d\phi = 4q\psi[Z'(\zeta) + 2]$; thus a discontinuity in $Z'(\zeta)$ causes a jump in its value in crossing the singular line from $\zeta > 0$ to $\zeta < 0$. If a slight discontinuity of wall curvature is allowed, this limiting solution may still represent a feasible flow (Fig. F,10c).

On the other hand, if the whole branch $Z = -\zeta$ for both positive and negative ζ is considered, the axial velocity is purely subsonic on one side, $\zeta > 0$, and supersonic on the other, $\zeta < 0$. This evidently corresponds to Meyer's type of flow, i.e. Laval nozzle flow (Fig. F,10d). In this case again there is a singularity on a characteristic. But the discontinuity in $(dy/dx)_{\bar{q}}$ and $d\theta/d\phi$ does not occur, and the flow is a physically possible flow, of a type for which an exact solution has already been given in the preceding article.

These solutions, although approximate, are interesting in several aspects. In the first place, from the practical point of view the calculation

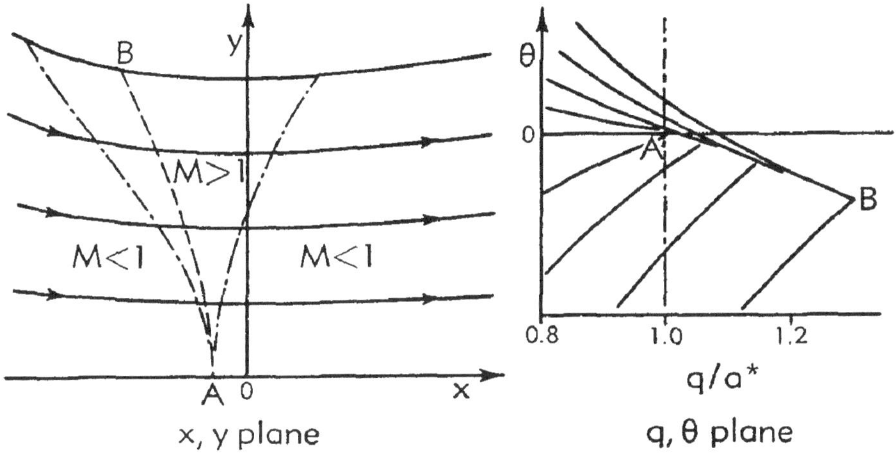

Fig. F,10c. Flow patterns in x, y and q, θ planes for $c = 0$, the last of the family $c < 0$ [38].

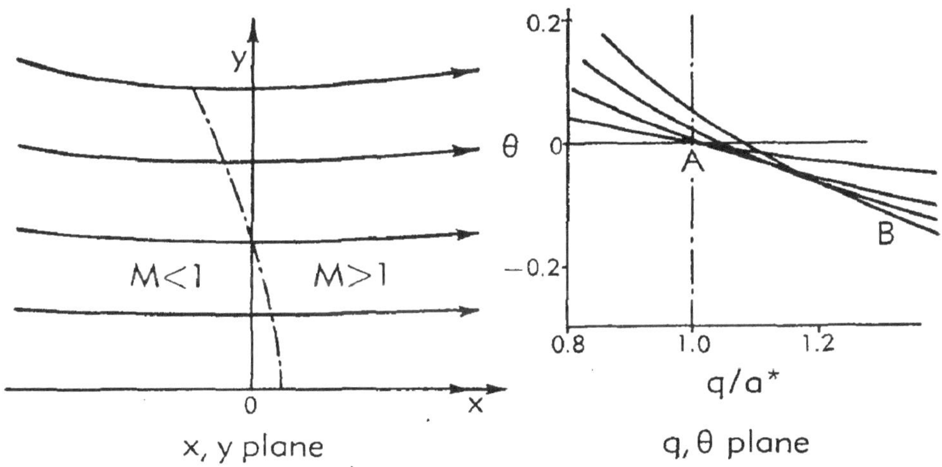

Fig. F,10d. Flow patterns in x, y and q, θ planes for $c = 0$, Meyer's type [38].

is simple and applies with sufficient accuracy in the important throat region. Secondly, they unify this type of channel flow in one family of solutions and clearly show why power series fail to deal with the limiting case of Taylor's type of flow (Fig. F,10c), namely, because of the appearance of discontinuities in the derivatives of velocities. The simplification afforded by this method has its origin in the fact that the solution is single-valued in the ϕ, ψ plane, although it may become many-valued in the q, θ plane. This many-valuedness, it will be recalled, is essentially the reason why the solution given in Art. 9 has such a complex nature.

F,11. Approximate Solution for Transonic Flow Past an Airfoil. Since the approximation made in the previous case consists in a simplification of the coefficients of the fundamental system (2-11), the region of validity of the solutions is, of course, restricted. To extend the region

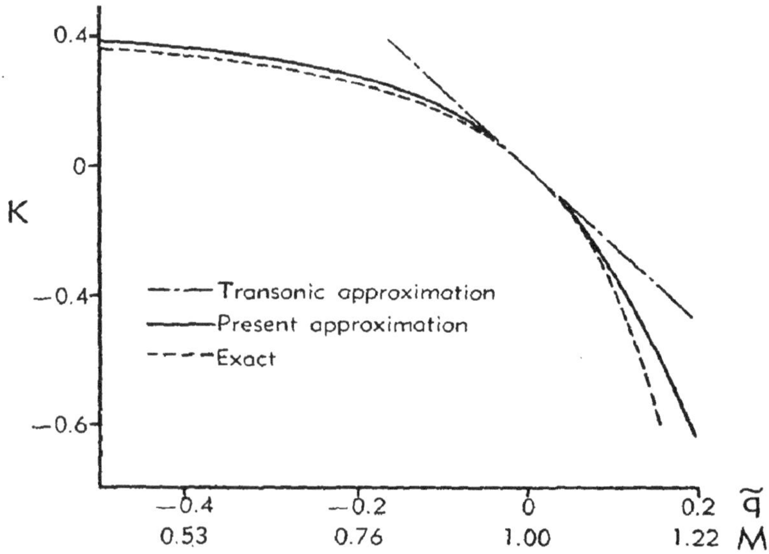

Fig. F,11a. Approximations to the exact function $K(\tilde{q})$.

of validity, but without complication of the mathematical problem, an alternative approximation was proposed by Tomotika and Tamada [39].

Let us again introduce the variable

$$\tilde{q} = \int_{a^*}^{q} (\rho/\rho^*)dq/q$$

The system (10-1), without approximation, becomes

$$\begin{aligned} \phi_{\tilde{q}} &= -K(q)\psi_\theta \\ \phi_\theta &= \psi_{\tilde{q}} \end{aligned} \tag{11-1}$$

where $K(q)$ is defined by

$$K(q) = \left(\frac{\rho^*}{\rho}\right)^2 (1 - M^2)$$

The new approximation proposed by Tomotika and Tamada is

$$K(\tilde{q}) = \left(\frac{2}{\gamma + 1}\right)^{2\beta} (1 - e^{2k\tilde{q}}) \tag{11-2}$$

with

$$k = \left(\frac{\gamma + 1}{2}\right)^{\alpha^2}, \quad \alpha^2 = \frac{\gamma + 1}{\gamma - 1}$$

The differential equation for the stream function ψ then becomes

$$\psi_{\bar{q}\bar{q}} + \left(\frac{2}{\gamma + 1}\right)^{2\beta} (1 - e^{2k\bar{q}})\psi_{\theta\theta} = 0 \tag{11-3}$$

Before we proceed further, let us examine this approximation. Since both the exact $K(\bar{q})$ and the approximate one defined by Eq. 11-2 are

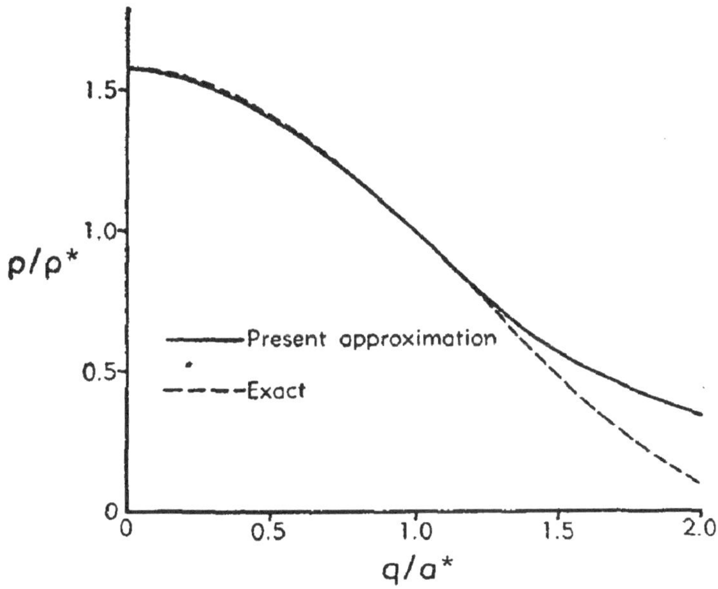

Fig. F,11b. Approximation to the exact density $\rho(\bar{q})$.

functions of $\bar{q}$, they can be plotted on one graph, as in Fig. F,11a. (The linear approximation to K represented by our previous approximation, Eq. 10-4, is also plotted for comparison.) As the present approximation to K is chosen to have the same value and tangent at both the stagnation point $\bar{q} = -\infty$ and the sonic point $\bar{q} = 0$, relatively good agreement is secured in the whole subsonic range. Thus, for any transonic flow with moderate local supersonic speeds, the approximation can be expected to be useful.

Accepting Eq. 11-2 to be valid, we can now determine the corresponding density law $\rho(\bar{q})$ implied by the relations

$$K = \left(\frac{\rho^*}{\rho}\right)^2 (1 - M^2) = \left(\frac{2}{\gamma + 1}\right)^{2\beta} (1 - e^{2k\bar{q}})$$

Making use of the definition of $\bar{q}$, we find a differential equation for ρq which has been integrated numerically by Tomotika and Tamada. Their result is shown in Fig. F,11b. By comparison, in the range $0 < q < 1.2a^*$, the difference between the exact and approximate laws is indeed negligible.

Method of constructing solutions. We now introduce new variables

$$\sigma = e^{k\tilde{q}}, \qquad \tilde{\theta} = \left(\frac{\gamma + 1}{2}\right)^{1+3\beta} \theta \tag{11-4}$$

by which the differential equation (11-3) becomes

$$\sigma^2 \psi_{\sigma\sigma} + \sigma \psi_\sigma + (1 - \sigma^2)\psi_{\tilde{\theta}\tilde{\theta}} = 0 \tag{11-5}$$

By definition, $\sigma = 0$ corresponds to $\tilde{q} = -\infty$ (stagnation) and $\sigma = 1$ to $\tilde{q} = 0$. Thus, in subsonic flow σ lies between 0 and 1, and in supersonic flow $\sigma > 1$; consequently, σ can be interpreted as some kind of Mach number.

The particular integrals of Eq. 11-5 are well known, and can be expressed as

$$J_\nu(\nu\sigma)e^{-i\nu\theta} \qquad \text{and} \qquad J_{-\nu}(\nu\sigma)e^{i\nu\tilde{\theta}} \tag{11-6}$$

It is of interest to compare this approximate solution with the exact one given by Cherry and reproduced in Art. 9. The similarity of these two solutions demonstrates again the degree of accuracy of the new approximation.

A general symmetric solution valid in the neighborhood of the stagnation point has the form

$$W = \sum_0^\infty A_n J_n(n\sigma)e^{-in\tilde{\theta}}, \qquad \sigma < 1 \tag{11-7}$$

Since $\tilde{\theta}$ is a constant nonintegral multiple of θ according to Eq. 11-4, the expansion (11-7) involves a very complicated singularity at the origin of the $\tilde{q}$, θ plane. Nevertheless, in a cut plane the change of variable from θ to $\tilde{\theta}$ is merely a distortion of the hodograph plane, and, if discontinuity of tangents is admitted at the stagnation point, such as a sharp leading or trailing edge, useful solutions can be found.

Now for Bessel's function $J_\nu(\nu\sigma)$, there is an integral representation [*40*, p. 20]

$$J_n(n\sigma) = \frac{1}{2\pi i}\int^{(0+)} \left[z^{-1}e^{\frac{\sigma}{2}(z-z^{-1})}\right]^n \frac{dz}{z}$$

where the path is taken around the origin in a positive sense. Substituting it in Eq. 11-7 and rearranging, we have

$$W = \frac{1}{2\pi i}\int^{(0+)} \sum_0^\infty A_n \left[z^{-1}e^{\frac{\sigma}{2}(z-z^{-1})-i\tilde{\theta}}\right]^n \frac{dz}{z} \tag{11-8}$$

provided the interchange of the order of integration and summation is justified. This is justified if the series is absolutely convergent, as can easily be shown for $\sigma > 1$.

Special solution. A case of special interest is $A_n = \lambda^{-n}$ where λ is a positive constant, as yet undetermined. It will develop, in what follows, that flows of interesting nature can be derived by proper selection of λ. For this form of A_n, the solution (11-8) becomes

$$W = \frac{1}{2\pi i}\int^{(0+)} \frac{1}{1 - e^Z}\frac{dz}{z}$$

with (11-9)

$$Z(z) = \frac{\sigma}{2}(z - z^{-1}) - \ln z - i\tilde{\theta} - \ln \lambda$$

This integral can be evaluated by calculating the residues at the roots of $Z(z) = 0$ inside the contour, since the contribution at the origin is zero. Let us fix our attention on the solution obtained by including only one root in the contour, say $z = z_0$. If this is a root, $Z(z)$ can be written in the form

$$Z(z) = \left[\frac{\sigma}{2}\left(1 + \frac{1}{z_0 z}\right) - \frac{\ln z - \ln z_0}{z - z_0}\right](z - z_0)$$

The residue of $z^{-1}(1 - e^Z)^{-1}$ is therefore

$$\left[-\frac{\sigma}{2}(z_0 + z_0^{-1}) + 1\right]^{-1}$$

Writing $z_0 = e^{-i\Theta}$ where Θ is complex, the integral (11-9) yields

$$W = \frac{1}{1 - \sigma\cos\Theta}$$

and from the second of Eq. 11-9 (11-10)

$$\tilde{\theta} - i\ln\lambda = \Theta - \sigma\sin\Theta$$

Eq. 11–10 are formally the same as we obtained in Eq. 9-11 and 9-12, except that complex values are involved here. The elimination of Θ gives W as a function of σ and $\tilde{\theta}$. From our analysis of Eq. 9-11 and 9-12, we deduce immediately that the function $\Theta(\sigma, \tilde{\theta})$ has branch lines given by

$$\tilde{\theta}_\Theta = 1 - \sigma\cos\Theta = 0$$

or, in terms of $\tilde{\theta}$,

$$\tilde{\theta} = \pm\cos^{-1}\frac{1}{\sigma} \mp \sqrt{\sigma^2 - 1} + i\ln\lambda \qquad (11\text{-}11)$$

There are two possibilities of finding interesting flow patterns here: namely, the branch lines may occur in the supersonic ($\sigma > 1$) or subsonic ($\sigma < 1$) regions. Consider first the supersonic case; here Eq. 11-11 leads to

$$\ln\lambda = 0$$

and

$$\tilde{\theta} = \pm \cos^{-1}\frac{1}{\sigma} \mp \sqrt{\sigma^2 - 1} = \pm\omega(\sigma) \tag{11-12}$$

The last equation is the equation of two characteristics of an approximate differential equation (11-5). Thus, by choosing $\lambda = 1$ in this solution, one obtains a flow of the type (Laval nozzle flow) discussed in Art. 9.

The other possibility is of more interest to us in this article. For $\sigma < 1$, Eq. 11-11 should be rewritten as

$$\tilde{\theta} = i\cosh^{-1}\frac{1}{\sigma} - i\sqrt{1-\sigma^2} + i\ln\lambda$$

whence

$$\begin{aligned}\tilde{\theta} &= 0 \\ \ln\lambda &= \pm\sqrt{1-\sigma^2} \mp \cosh^{-1}\frac{1}{\sigma}\end{aligned} \tag{11-13}$$

We see that the branch line in this case has degenerated into a single point. On the basis of our experience in Art. 8 with mixed flow past closed bodies, we immediately identify this branch point with flow conditions at infinity; i.e. we write

$$\tilde{\theta}_\infty = 0 \quad \text{and} \quad \ln\lambda = \pm\sqrt{1-\sigma_\infty^2} \mp \cosh^{-1}\frac{1}{\sigma_\infty}$$

This means that, by proper selection of the constant λ for any given subsonic stream Mach number, we obtain here a flow, parallel and horizontal at infinity, of the same character as the exact solutions treated in Art. 8.

The integral of this solution of Eq. 11-10 with respect to $\tilde{\theta}$ is also a solution of the differential equation (11-5); Tomotika and Tamada find that it is desirable to include it in order to set up an airfoil type flow. The inclusion of the integral does not affect the singular behavior at infinity, which was determined above. The integral is

$$\begin{aligned}\int \frac{d\tilde{\theta}}{1-\sigma\cos\Theta} = \int \frac{1}{1-\sigma\cos\Theta}\,\frac{d\Theta}{d\tilde{\theta}/d\Theta} &= \Theta + \text{const} \\ &= \tilde{\theta} + \sigma\sin\Theta + \text{const}\end{aligned} \tag{11-14}$$

according to Eq. 11-10. But $\tilde{\theta}$ is clearly a particular solution of the differential equation (11-5), so it can be omitted from Eq. 11-14. We therefore take as our complete solution

$$\psi = \text{I.P.}\left\{\frac{1}{1-\sigma\cos\Theta} + iK\sigma\sin\Theta\right\} \tag{11-15}$$

with $\tilde{\theta} - i\ln\lambda = \Theta - \sigma\sin\Theta$ where K is arbitrary.

Flow past a symmetric airfoil. To produce the transonic potential flow about a certain symmetric airfoil profile, we begin by choosing K

to make the leading edge a stagnation point; i.e. we choose K so as to place the point $\sigma = 0 = \psi$ on the real axis in the finite region. An appropriate choice turns out to be $K = (1 - \lambda^2)/(1 + \lambda^2)$.

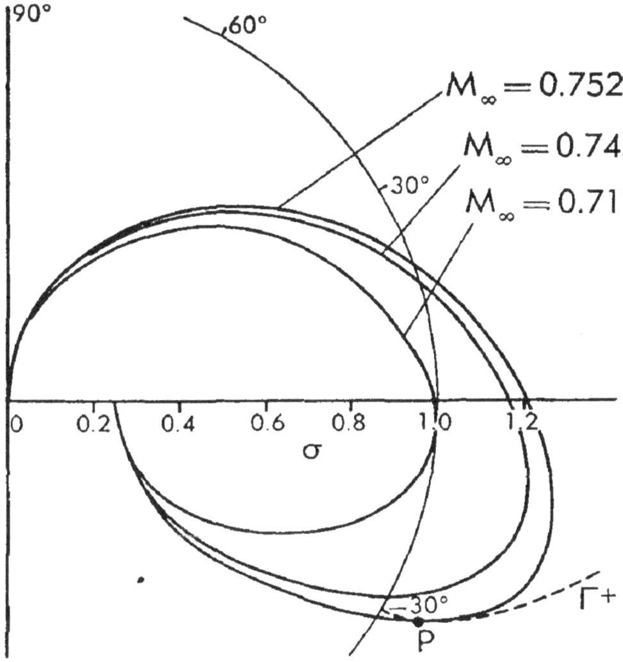

Fig. F,11c. Zero streamlines in the hodograph plane for free stream Mach numbers 0.717, 0.745, and 0.752 [39].

In terms of the variables σ and $\tilde{\theta}$, our fundamental set (11-1) reads

$$\begin{aligned} \phi_\sigma &= \left(\frac{2}{\gamma + 1}\right)^\beta \left(\sigma - \frac{1}{\sigma}\right) \psi_{\tilde{\theta}} \\ \phi_\theta &= \left(\frac{2}{\gamma + 1}\right)^\beta \sigma \psi_\sigma \end{aligned} \tag{11-16}$$

Upon integration of these, the velocity potential corresponding to Eq. 11-15 is found to be

$$\phi = \left(\frac{2}{\gamma + 1}\right)^\beta \sigma \, \text{I.P.} \left\{ \frac{\sin \Theta}{1 - \sigma \cos \Theta} - iK \left(\frac{\sigma}{2} + \cos \Theta \right) \right\} \tag{11-17}$$

Finally, the transformation from the $(\sigma, \tilde{\theta})$ to the (x, y) plane has been carried out by Tomotika and Tamada for three different Mach numbers M_∞. The relation between M_∞ and σ_∞ is

$$M_\infty = \left[1 + \left(\frac{\rho_\infty}{\rho^*}\right)^2 \left(\frac{2}{\gamma + 1}\right)^{2\beta} (\sigma_\infty^2 - 1) \right] \tag{11-18}$$

The resulting streamlines $\psi = 0$, together with velocity distributions, are shown in Fig. F,11c and F,11d. The calculated airfoil sections are quite similar to a 10 per cent thick symmetric Joukowski section, except that the leading edge is not rounded—as was already predicted in connection with our discussion of the approximation earlier in this article.

It is an interesting fact that the airfoil profiles for the three Mach numbers of Fig. F,11d are almost identical; the differences cannot be

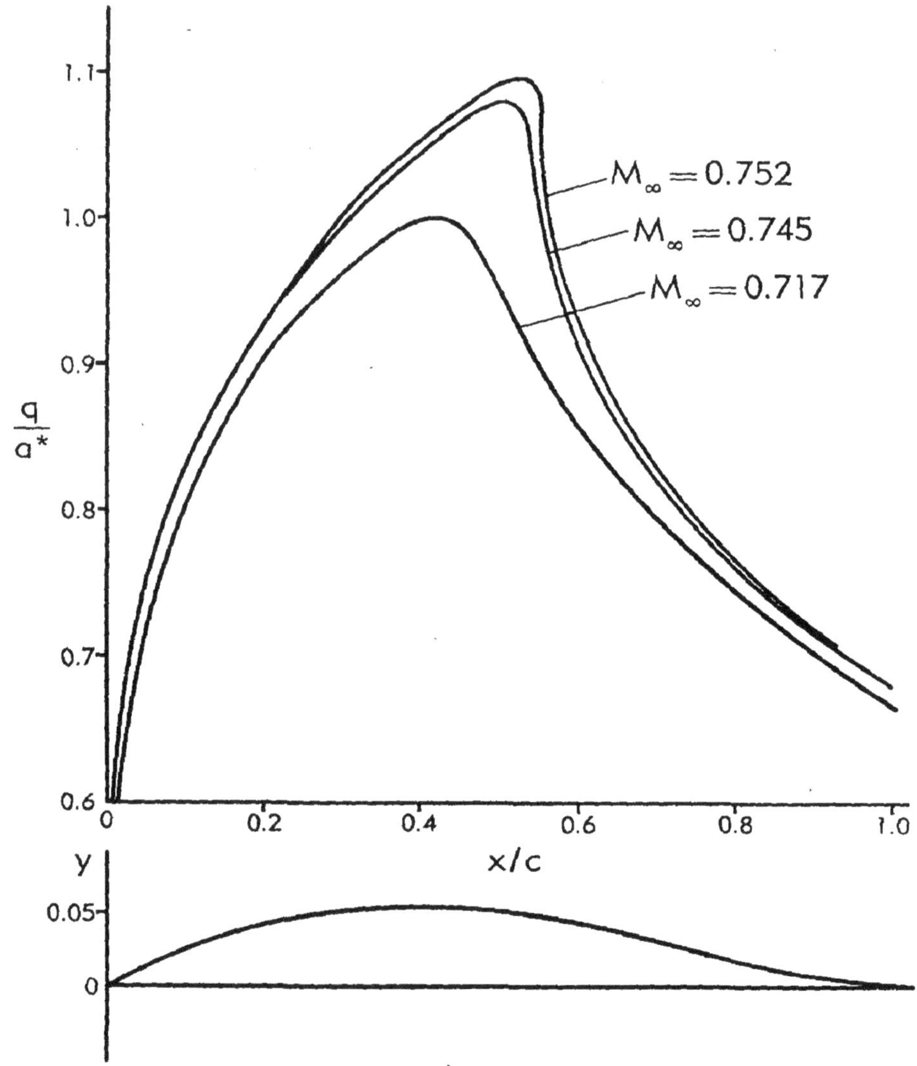

Fig. F,11d. Airfoil sections and surface q distribution at Mach numbers 0.717, 0.745, and 0.752 [39].

plotted in the figure. Thus, from the Tomotika-Tamada results, we obtain a close approximation to the flow at three Mach numbers about a fixed profile. The lowest of these Mach numbers produces entirely subsonic flow; the second, mixed flow with a considerable supersonic region and a maximum local speed of about $1.10a^*$. The highest Mach number produces still greater local speeds and exhibits the phenomenon of the singularity $J = 0$ at one point of the upper surface contour, and, of course, at a symmetrically located point of the lower surface. The significance of this type of singularity is discussed in the next article.

F,12. Singularities of the Hodograph Transformation. Let u and v be the velocity components of a flow in the x, y plane. Consider (dx, dy), an element in the x, y plane, and (du, dv) an element in the u, v plane. If they correspond to each other, they must be related by

$$du = u_x dx + u_y dy$$
$$dv = v_x dx + v_y dy$$

These equations define du and dv uniquely if the functions $u(x, y)$ and $v(x, y)$ are known and are differentiable, and dx and dy are given. From a different point of view, we can consider the system as a transformation. If the transformation is locally one-to-one, the transformation Jacobian J $(= \partial(x, y)/\partial(u, v))$ does not vanish nor become infinite. When these conditions are violated, the hodograph transformation fails and singularities occur. If this Jacobian vanishes identically throughout any region, u and v are no longer independent of each other. The flow in this case is particularly simple and is known as Prandtl-Meyer flow or simple wave flow (see E,3 and G,5). In the following, we will confine our attention to two distinct types of singularities, which are of special interest in connection with the mixed subsonic-supersonic potential flows. These are called, respectively, "limiting lines" [*42*] and "branch lines" [*65*].

Limiting lines: $J = 0$. When J becomes zero, viz.

$$J = \frac{\partial(x, y)}{\partial(q, \theta)} = -\left(\frac{\rho^0}{\rho}\right)^2 \frac{1}{q}\left[\psi_q^2 - \left(\frac{1}{a^2} - \frac{1}{q^2}\right)\psi_\theta^2\right] = 0 \qquad (12\text{-}1)$$

we must have, since $\rho^2 q$ cannot become infinite,

$$q^2\psi_q^2 - (M^2 - 1)\psi_\theta^2 = 0 \qquad (12\text{-}2)$$

For subsonic flow, this requires

$$q\psi_q = \psi_\theta = 0$$

This means that in a subsonic field, the singularities are isolated. The singularity may be either a stagnation point or a point where the streamline has infinite curvature. To show this, we note first of all, from the system Eq. 2-11,

$$q\phi_q = \phi_\theta = 0$$

Now the first partial derivatives of x and y are given in terms of the partial derivatives of ϕ and ψ by Eq. 2-15 and 2-16. It follows, then, that at the singular point, if the partial derivatives of x and y are finite, q must be zero. If q does not vanish, then the partial derivatives must be zero and in the flow plane the velocity gradient or the acceleration of the fluid particle must be infinite. As the acceleration is proportional to the curvature of the streamline, we see that $J = 0$ occurs at such points as the sharp trailing edge of an airfoil, as well as at stagnation points. This possibility was first noticed by Kuo [23] and later confirmed by Cherry. [71]

When $M > 1$, ψ_q and ψ_θ may either vanish identically in a region or satisfy the differential relation (12-2) along certain lines. Since $\psi = \text{const}$ is a trivial solution, we must consider the second alternative; we have the solutions

$$\begin{aligned} &\text{on } \Lambda^+ \qquad q\psi_q - \sqrt{M^2 - 1}\,\psi_\theta = 0 \\ &\text{on } \Lambda^- \qquad q\psi_q + \sqrt{M^2 - 1}\,\psi_\theta = 0 \end{aligned} \tag{12-3}$$

It can be shown that on each line J has a simple root, because on Λ^+

$$J_\theta = -\left(\frac{\rho^0}{\rho}\right)^2 \frac{1}{q^3} (q\psi_q + \sqrt{M^2 - 1}\,\psi_\theta)(q\psi_{q\theta} - \sqrt{M^2 - 1}\,\psi_{\theta\theta}) \neq 0$$

In the vicinity of these lines, a small neighborhood in the x, y plane corresponds to two different regions in the q, θ plane. This is seen clearly from the transformation (2-10), which on a streamline reduces to

$$dx = -\frac{\rho^0 \cos\theta}{\rho q^2 \psi_\theta} [(q\psi_q + \sqrt{M^2 - 1}\,\psi_\theta)(q\psi_q - \sqrt{M^2 - 1}\,\psi_\theta)]dq$$

$$dy = -\frac{\rho^0 \sin\theta}{\rho q^2 \psi_\theta} [(q\psi_q + \sqrt{M^2 - 1}\,\psi_\theta)(q\psi_q - \sqrt{M^2 - 1}\,\psi_\theta)]dq$$

Since J has simple zero, dx and dy both change their signs upon proceeding from one side of Λ^+ to the other, but without affecting the slope of the streamline. Thus, the streamline must have a cusp there, and as a result the stream function becomes many-valued, as shown in Fig. F,12a, the branches coming together along the singular line or "limiting line" L^+, which is the image in the flow plane of Λ^+. The immediate consequence of this singular line is that all space derivatives, such as pressure gradient, curvature of the streamline, and so on, are infinite there. Hence, once this singularity occurs, the solution will no longer be physically possible and must be rejected.

Now, let us investigate the geometric relationship between the Mach waves in the x, y plane and the characteristic hodographs in the q, θ plane. We know, from Eq. 3-1, that the characteristic hodographs are the curves that satisfy

$$q^2 d\theta^2 - (M^2 - 1)dq^2 = 0$$

or,

$$\begin{aligned} &\text{on } \Gamma^+ \qquad q\,d\theta - \sqrt{M^2 - 1}\,dq = 0 \\ &\text{on } \Gamma^- \qquad q\,d\theta + \sqrt{M^2 - 1}\,dq = 0 \end{aligned} \tag{12-4}$$

These equations can be integrated to give two families of fixed curves Γ^+ and Γ^- in the q, θ plane:

$$\Gamma^+ \qquad \theta - \omega = \text{const}$$

$$\Gamma^- \qquad \theta + \omega = \text{const}$$

where ω is defined in Eq. 16-4. It is easy to verify (G,3) that these curves are epicycloids, all radiating from the sonic circle and converging

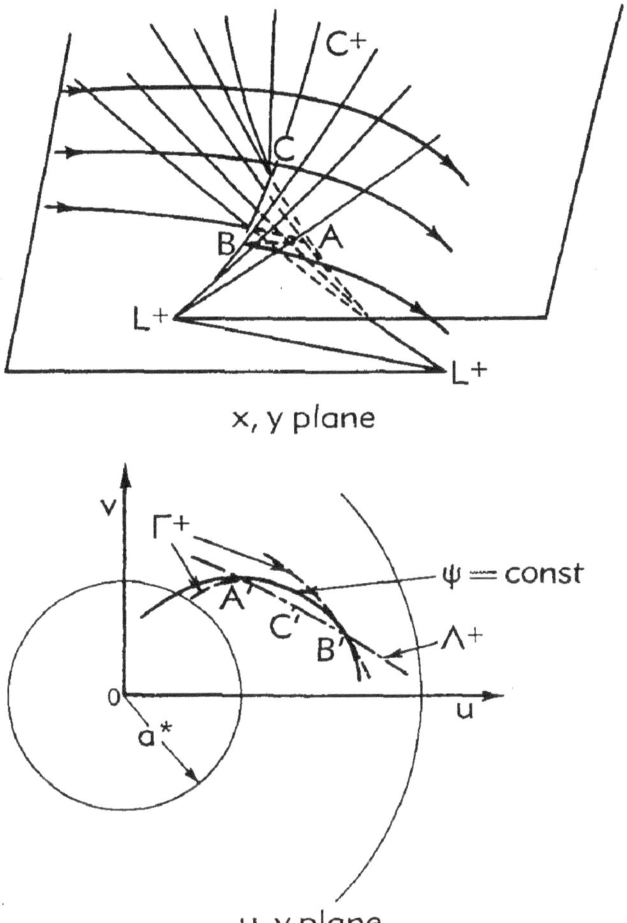

Fig. F,12a. The geometrical relations between the limiting line L^+ and the limiting hodograph Λ^+ (dashed Mach lines indicating lower or folded sheets).

asymptotically to the limit circle, $q = q_{max}$ or $M = \infty$. The corresponding lines in the physical plane are characteristics of Eq. 2-7, namely,

$$\text{on } C^+ \qquad \frac{dy}{dx} = \tan(\theta + \mu)$$

$$\text{on } C^- \qquad \frac{dy}{dx} = \tan(\theta - \mu) \tag{12-5}$$

with

$$\mu = \sin^{-1} \frac{1}{M}$$

Theodore von Kármán [3] first noticed that the singular line Λ^+ or Λ^- is the locus of the points of tangency of the streamline hodographs with one family of characteristic hodographs Γ^- or Γ^+. For along every streamline

$$\left(\frac{d\theta}{dq}\right)_\psi = -\frac{\psi_q}{\psi_\theta}$$

On the line Λ^+

$$q\psi_q - \sqrt{M^2 - 1}\,\psi_\theta = 0$$

Thus, whenever a streamline intersects the line Λ^+, this derivative must have the value

$$\left(\frac{d\theta}{dq}\right)_\psi = -\frac{\sqrt{M^2 - 1}}{q} = \left(\frac{d\theta}{dq}\right)_{\Gamma^-} \tag{12-6}$$

according to Eq. 12-4. Similarly, on Λ^-

$$\left(\frac{d\theta}{dq}\right)_\psi = \frac{\sqrt{M^2 - 1}}{q} = \left(\frac{d\theta}{dq}\right)_{\Gamma^+} \tag{12-7}$$

Each streamline generally exhibits two such points of tangency, i.e. intersects the line Λ^+, for example, in two points, such as A' and B' in Fig. F,12a. At the image points A and B in the flow plane, the streamline forms cusps. When the streamline just touches Λ^+ in the hodograph plane, such as at C', the streamline in the flow plane has no cusp but the limiting line L^+ does.

Next let us transform the Mach waves C^+ and C^- into the q, θ plane by means of Eq. 2-7 and 2-10; there result, respectively,

$$\begin{aligned} &\text{on } C^+ \qquad (\sqrt{M^2 - 1}\,dq - q\,d\theta)(q\psi_q - \sqrt{M^2 - 1}\,\psi_\theta) = 0 \\ &\text{on } C^- \qquad (\sqrt{M^2 - 1}\,dq + q\,d\theta)(q\psi_q + \sqrt{M^2 - 1}\,\psi_\theta) = 0 \end{aligned} \tag{12-8}$$

The Mach waves C^+ and C^- thus have as images the characteristics Γ^+ and Γ^- for $J \neq 0$. As $J = 0$ defines two lines Λ^+ and Λ^- that nowhere coincide with Γ^+ and Γ^-, Mach wave C^+ can touch the limiting line L^+ only at one point.

To see this, the behavior of one family of Mach waves is studied in the neighborhood of Λ^+. By eliminating $q\,d\theta = -\sqrt{M^2 - 1}\,dq$ from Eq. 2-10, we can define a small element of C^- by

$$\begin{aligned} dx &= -\frac{\rho^0}{\rho q^2}(q\psi_q - \sqrt{M^2 - 1}\,\psi_\theta)(\sqrt{M^2 - 1}\cos\theta + \sin\theta)dq \\ dy &= -\frac{\rho^0}{\rho q^2}(q\psi_q - \sqrt{M^2 - 1}\,\psi_\theta)(\sqrt{M^2 - 1}\sin\theta - \cos\theta)dq \end{aligned} \tag{12-9}$$

In the neighborhood of Λ^+, dx and dy vanish and, consequently, along any member of C^- the space derivatives of physical quantities, such as q or θ, will be infinite. This means that the characteristic curves of the other family C^+ are densely piled up on one another at such a point; i.e., they form an envelope L^+ that corresponds to Λ^+ in the hodograph plane.

Thus, we arrive at the conclusion that the singular lines Λ^+ and Λ^- correspond respectively to the limiting lines L^+ and L^-, and that on one of these limiting lines, L^+ say, the Mach waves C^+ form an envelope, while the streamlines and the Mach lines C^- have cusps (Fig. F,12a). Similarly, on the line L^- the Mach waves C^- form an envelope while the streamlines and the waves C^+ have cusps.

We have seen several examples of the appearance of limiting lines in the present section. The first example was the behavior of our source-or-sink flow (Art. 6) at the sonic circle. To be sure, this singularity was recognized as the familiar "choking" phenomenon of a nozzle. Nevertheless, it is easy to verify that $J = 0$ at $M = 1$ in this case, for $\phi_q = 0$ and ψ_θ is a constant, so that J is proportional to $M^2 - 1$. This is confirmed by the behavior of the streamlines, which reverse their directions at the sonic circle. Consequently, in this simple flow pattern, the sonic circle is a limiting line, and the anomalous behavior there can immediately be attributed to our specification of the streamline pattern. In more general cases, as we shall see, the explanation for the appearance of a limiting line in the calculated flow pattern is not so obvious.

Our second encounter with limiting lines was in the flow around an edge (Ringleb's case) in Art. 6. Again, it is easy to verify from the equations of that article that the Jacobian J vanishes along PQR (Fig. F,6b). The appearance of cusps in the streamlines and the folding of the flow plane have already been noted; this relatively simple case might be taken, in fact, as typical of the phenomenon discussed here (cf. Fig. F,6b and F,12a).

Again, in the Tomotika-Tamada results (Fig. F,11c and F,11d), the phenomenon of vanishing J was noted at a single point, where the hodograph of the streamline $\psi = 0$ became tangent to a characteristic. Fig. F,11d confirms that the pressure gradient becomes infinite at such a point. The same phenomenon appears in a flow calculated by Cherry [*24*] and mentioned above in Art. 8, namely, the flow about a slightly cambered cylinder. This flow pattern is reproduced in Fig. F,12b.

Branch lines. $J = \infty$. Let us consider the possibility of the occurrence of a different sort of singularity, namely, $J = \infty$. In subsonic flow, as we see in Eq. 12-1, this might occur at a stagnation point or at a point where

$$q^2\psi_q^2 + (1 - M^2)\psi_\theta^2 = \infty$$

It is easier to understand this condition if we compute the reciprocal of J, viz.,

$$J^{-1} = \frac{\partial(q, \theta)}{\partial(\phi, \psi)} \frac{\partial(\phi, \psi)}{\partial(x, y)} = -\left(\frac{\rho}{\rho^0}\right)^2 q \left[q_\psi^2 + \left(\frac{\rho^0}{\rho}\right)^2 (1 - M^2) q_\phi^2\right]$$

For this to vanish, except at a stagnation point, implies $q_\phi = q_\psi = 0$ and therefore $\theta_\phi = \theta_\psi = 0$. These relations can define only an isolated singularity [*41*] or a trivial flow.

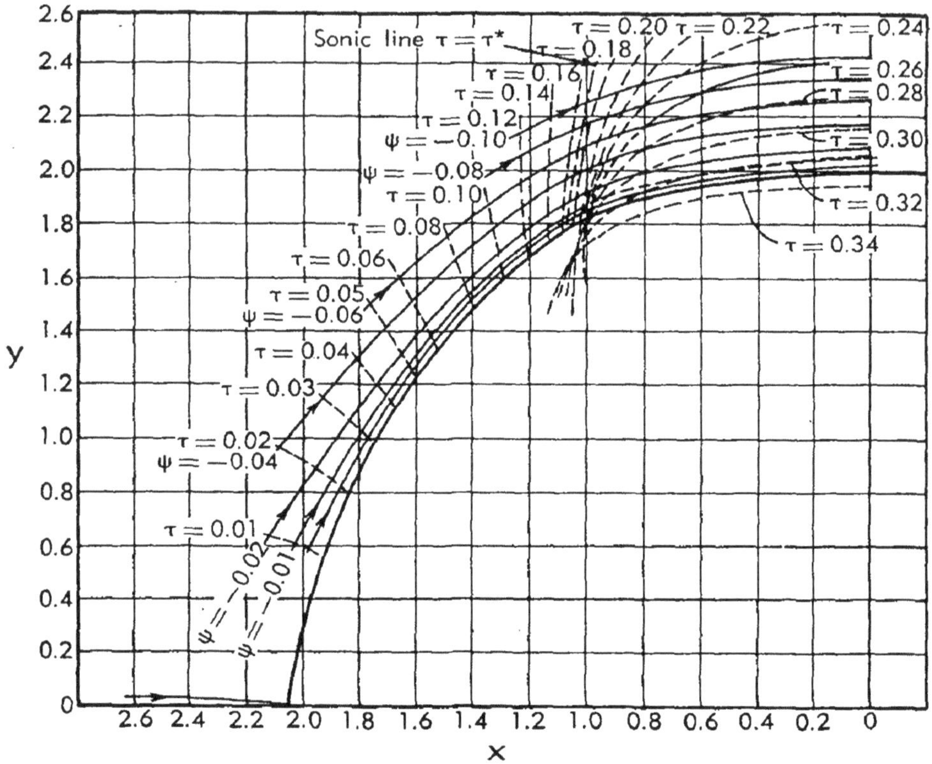

Fig. F,12b. Streamlines and constant-speed lines for flow at $M_\infty = 0.5098$ past a slightly cambered cylinder [*24*]. Note first appearance of limiting line at the boundary (near $x = 1.0$). $\gamma = 1.405$.

On the other hand, when $M > 1$, we have the possibility of singular curves in the flow; namely, we have the following solutions:

$$\begin{aligned} &\text{on } \Omega^+ && q\psi_q - \sqrt{M^2 - 1}\,\psi_\theta = \infty \\ & && q\psi_q + \sqrt{M^2 - 1}\,\psi_\theta \quad \text{is finite} \\ &\text{on } \Omega^- && q\psi_q + \sqrt{M^2 - 1}\,\psi_\theta = \infty \\ & && q\psi_q - \sqrt{M^2 - 1}\,\psi_\theta \quad \text{is finite} \end{aligned} \tag{12-10}$$

provided $\rho \neq 0$. In either case, both ψ_q and ψ_θ must be infinite but the ratio ψ_q/ψ_θ is finite.

To show that the singular lines Ω^+ and Ω^- are, in fact, characteristics

Γ^+ and Γ^- respectively, we consider a line

$$q\psi_q - \sqrt{M^2 - 1}\,\psi_\theta = \text{const}$$

By taking the total derivative along this line, we obtain

$$\left[q\psi_{qq} + \psi_q - \sqrt{M^2 - 1}\,\psi_{\theta q} - \frac{M^2\left(1 + \frac{\gamma - 1}{2}M^2\right)}{q\sqrt{M^2 - 1}}\right] + [q\psi_{q\theta} - \sqrt{M^2 - 1}\,\psi_{\theta\theta}]\frac{d\theta}{dq} = 0$$

By means of the differential (3-1), we simplify this expression and obtain for the derivative $q\,d\theta/dq$ along such a line

$$q\frac{d\theta}{dq} = \sqrt{M^2 - 1}\left\{1 - \frac{M^2[2\sqrt{M^2 - 1}\,(q\psi_q - \sqrt{M^2 - 1}\,\psi_\theta) + (\gamma + 1)M^2\psi_\theta]}{2(M^2 - 1)(\sqrt{M^2 - 1}\,\psi_{\theta\theta} - q\psi_{q\theta})}\right\}$$

But on the line Ω^+, both ψ_q/ψ_θ and $\psi_{\theta\theta}/\psi_{q\theta}$ are finite, while $\psi_\theta/\psi_{q\theta}$ is zero; we have, therefore

$$\left(\frac{d\theta}{dq}\right)_{\Omega^+} = \frac{\sqrt{M^2 - 1}}{q} = \left(\frac{d\theta}{dq}\right)_{\Gamma^+} \tag{12-11}$$

Similarly, we have

$$\left(\frac{d\theta}{dq}\right)_{\Omega^-} = -\frac{\sqrt{M^2 - 1}}{q} = \left(\frac{d\theta}{dq}\right)_{\Gamma^-} \tag{12-12}$$

Moreover, the streamline hodographs become tangent to these singular characteristic curves. This is clear from the fact that $q\psi_q + \sqrt{M^2 - 1}\,\psi_\theta$ remains finite along Ω^+, for example, while both ψ_q and ψ_θ become infinite. This can only mean that the ratio ψ_q/ψ_θ has the limiting value $-\sqrt{M^2 - 1}/q$; but this is just to say that

$$\left(\frac{d\theta}{dq}\right)_\psi = -\frac{\psi_q}{\psi_\theta} = \left(\frac{d\theta}{dq}\right)_{\Omega^+}$$

and similarly for Ω^-. The streamlines do not coincide with characteristics except in Prandtl-Meyer flow; thus, the singular characteristics Ω^+ and Ω^- must be envelopes of streamlines. This situation is sketched, for example, in Fig. F,10d.

Now consider the nature of these singular characteristics in the flow plane. According to Eq. 12-9 and 12-10, the elements dx, dy remain finite even when the characteristic C^- (for example) is the singular member of the family. Thus, infinite curvature of streamlines and infinite pressure gradient do not occur at singular curves of this type, and solutions possessing this kind of singularity are physically acceptable. Flows through convergent-divergent nozzles, such as we described in Art. 9 and 10, are typical examples. The significance of such singular curves is mathematical, since they imply the multivaluedness of $\psi(q, \theta)$ that makes the boundary-value problems difficult, as in Art. 9.

A much fuller discussion of these singularities, and others, has been given in a paper by Craggs [*41*], to which the interested reader is referred.

Significance of the limiting line. We have already seen examples of flow patterns calculated by the hodograph technique that exhibit the phenomenon of the limiting line. This situation appears to be quite general, so that whenever a family of flows is controlled by a parameter such as a stream Mach number, M_∞, there is the danger that sufficiently large values of the parameter will cause these singular curves to appear. As has already been pointed out above, the behavior of the streamlines in such a case is physically inacceptable and the flow pattern must be rejected. Tollmien [*42*] has proved that, if a limiting line appears, it is impossible to continue the smooth isentropic flow pattern across it.

When limiting lines were first encountered, for example in the paper by Ringleb [*13*], the phenomenon stimulated considerable interest in the scientific world. It was interpreted as a sign that irrotational, isentropic, plane flow had become impossible, and that other effects in the flow, such as viscosity and heat conduction, had become predominant. The name "shock wave" was even applied to the limiting line. Following this line of reasoning, the appearance of limiting lines was held directly responsible for the breakdown of potential flow, and the stream Mach number M_∞ at which the phenomenon first appeared was called the "upper critical Mach number."

But we must remember that, when we calculate a family of related flows about closed bodies using the present hodograph methods, the streamlines alter in shape as M_∞ varies, and this is also true of the boundary streamline. In Ringleb's problem, the effect is observed if any two streamlines are considered as channel walls; as the boundary streamlines are selected closer to the high speed regions of flow (Fig. F,6b), their shape is changed. The fundamental question therefore remains: Is there a critical Mach number at which the limiting line first appears, for flow about a given smooth boundary? In the earlier interpretation, this possibility was implied.

A qualified answer to this question has been given by Friedrichs [*43*], who showed that, assuming the existence of a family of analytic[6] coordinate functions $x(u, v)$ and $y(u, v)$ dependent on a parameter M_∞ and satisfying smooth boundary conditions (which may vary with M_∞), the

[6] More recently, Morawetz and Kolodner [*66*] have proved Friedrichs' theorem assuming only that $x(u, v)$ and $y(u, v)$ have continuous derivatives.

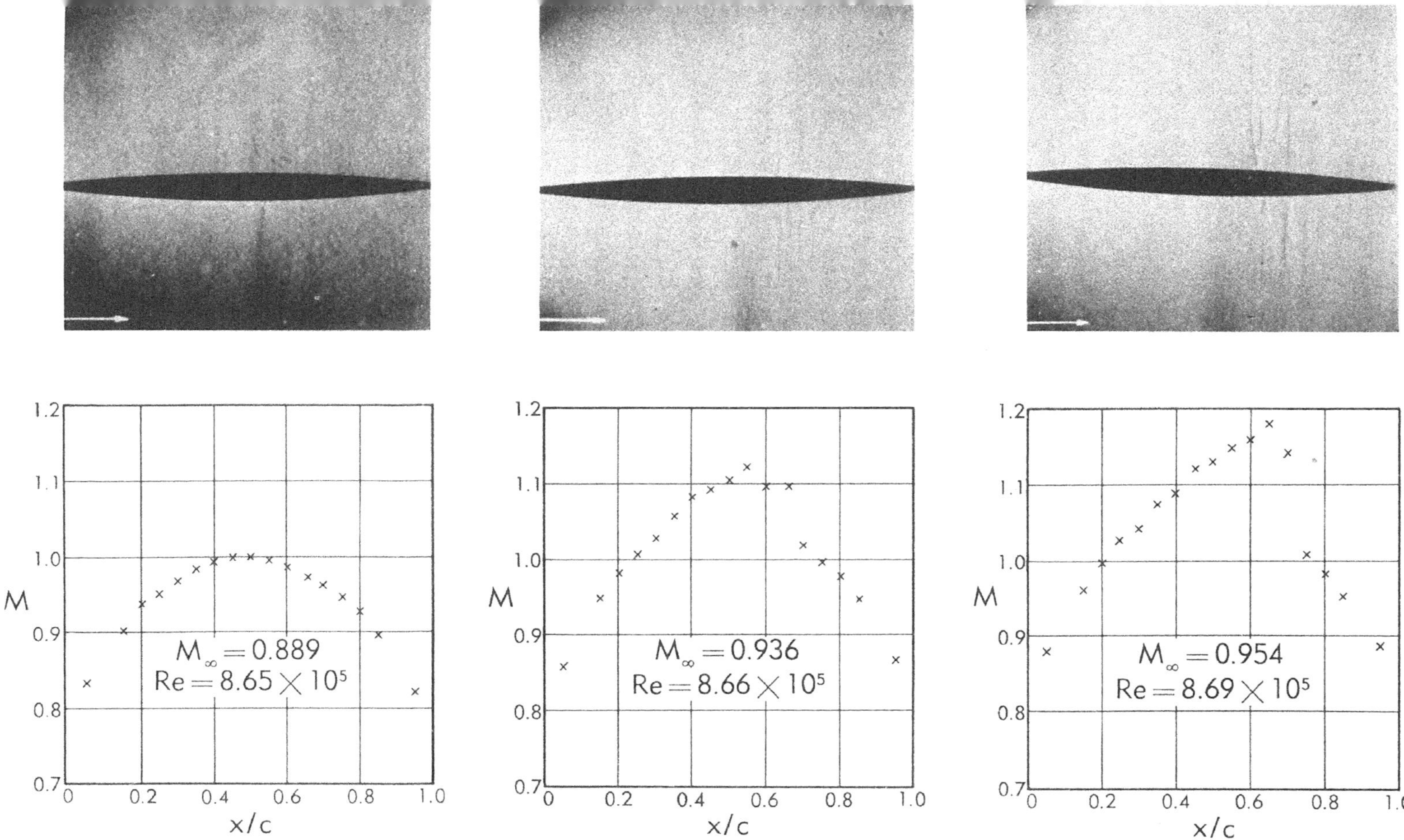

Plate F,13a, b, and c. Measured Mach number distributions over a biconvex circular arc airfoil at three free stream Mach numbers M_∞ and Reynolds numbers Re [47].

limiting line cannot occur. His argument runs substantially as follows: Assume $-J > 0$ everywhere for sufficiently small M_∞. (Note that $-J$ is always positive in subsonic flow.) Then, if the limiting line is to appear in the flow at some values of M_∞, say M_c, it follows from continuity of J that $J = 0$ somewhere at $M_\infty = M_c$. He proves this to be impossible. The proof has been extended by Manwell [*44*], who concludes that at $M_\infty = M_c$, J cannot vanish in any region nor along any line. Similar conclusions were previously reached by Nikolsky and Taganoff [*45*] by a different approach.

This does not completely answer the question posed above, because the existence of the analytic functions $x(u, v)$, $y(u, v)$ has not been proved. To be sure, their existence may be assumed to be quite independent of the vanishing of J, but there is no conclusive evidence of the existence or nonexistence of such functions, dependent upon a parameter M_∞, for flows past a prescribed boundary streamline.

It does not seem possible to proceed further along this line, since the use of the hodograph equations, to date, appears to eliminate the possibility of satisfying prescribed physical boundary conditions. It must be concluded that the identification of the limiting line as the cause of breakdown of potential flow should be viewed with suspicion, at least.

F,13. Conflicts between Theory and Experiment. Both the exact and the approximate solutions of this section have established beyond doubt the theoretical existence of continuous, steady, mixed transonic flows past solid boundaries. Other approximate methods of calculation, such as those discussed in Sec. E and in [*46*], have also predicted flows of this type. Theoretically, then, the fluid particle in such a flow should be able to decelerate isentropically to a subsonic speed after it has been accelerated isentropically to a supersonic state. Were this proved possible in practice, an aircraft might fly in the transonic regime without any greater drag than that caused by skin friction. In the case of the 10 per cent thick symmetric airfoil calculated by Tomotika and Tamada (Fig. F,11d), a free stream Mach number as high as 0.75 might be attained with substantial supersonic local speeds, the flow remaining smooth and free of shock waves.

Wind tunnel experiments, however, contradict these theoretical predictions, for when the sonic speed is exceeded anywhere on a solid boundary, the flow either becomes unsteady or involves shock waves, or both. Typically, the deceleration from supersonic speed takes place in shock waves rather than by isentropic compression. Apparently, the sequence of events as the Mach number is gradually increased is as follows: Immediately after the occurrence of local sonic speed at the surface, highly unsteady disturbances begin to appear in the flow field near the surface, downstream of the minimum pressure point. As the

Mach number is gradually raised, the traveling disturbances become proportionately stronger, although the pressure distribution over the surface may still show no peculiarities, except that far downstream there is some distortion due to the thickening of the boundary layer. But when the Mach number is high enough, the disturbances suddenly coalesce to form a stationary shock. The shock, involving as it does an irreversible thermodynamic process, causes drag; moreover, it changes the pressure distribution appreciably, and may produce boundary layer separation and all of the attendant effects.

H. W. Liepmann and his associates [47] have studied this type of flow over a six per cent, biconvex, circular arc airfoil in a 2 × 20 in. transonic wind tunnel. For each free stream Mach number, the local Mach number distribution over the surface was measured. Local supersonic speeds were reached when the free stream Mach number was slightly below 0.889. At $M_\infty = 0.889$, the highest local Mach number appeared to be 1.01. The Mach number distribution still maintained its low speed character, as can be seen in Plate F,13a. Yet the disturbances are already visible in the accompanying schlieren photograph. When the free stream Mach number was raised to 0.936, the disturbances were more clearly seen in the schlieren picture and the Mach number distribution also showed some irregularities at about 60 per cent chord, possibly due to boundary layer transition (Plate F,13b). A stationary shock was definitely formed at a slightly higher Mach number, viz. 0.954, as shown by the Mach number distribution curve (Plate F,13c).

In this series of experiments, steady laminar boundary layer was maintained over most of the airfoil, and transition most likely occurred in each case near the trailing edge.

In any region of boundary layer transition and turbulence, unsteady disturbances are excited, and these propagate into the fluid in all directions. All except those traveling against the main flow quickly disappear, due to their relatively high speed of propagation. Those traveling against the stream have a tendency to accumulate, particularly in the transonic region. If there is no mechanism to dissipate the energy of these waves, they have sufficient time to grow and eventually to destroy the smooth flow. There is strong experimental evidence that smooth, plane, transonic flows involving local supersonic regions are not stable to small disturbances, and that the flow pattern becomes stable only when stationary shocks are formed. It seems certain that phenomena similar to those described here do occur in channel flow that is substantially one-dimensional, but it is not completely clear how far this description applies to more general classes of flows. This problem is discussed further in the next article.

As an alternative to the instability theory to explain the conflict between experiment and potential flow theory, it has been proposed that the error lies in the neglect of viscosity in our theory. For example, the difference between theory and experiment might be due to the presence of the boundary layer. In support of this idea, it has been pointed out that the smooth deceleration can be thought of as the result of effects carried rearward by the family of Mach waves originating farther forward, as sketched in Fig. F,13a, and that a distortion to the contour, as might be caused by a boundary layer, would suffice to cause an envelope

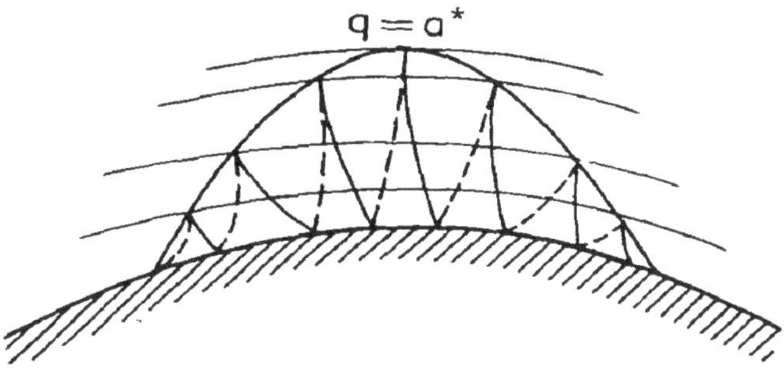

Fig. F,13a. Structure of local supersonic region.

of waves, as in Fig. F,13b. Such a discontinuity, of course, marks the breakdown of potential flow, and the pattern must be replaced by one involving a shock. But this would imply that a slightly different body, producing the Mach waves of Plate F,13a outside of its boundary layer, would permit smooth potential flow. This proposal hardly seems adequate to explain the complete absence of flows of this type in experiments.

Moreover, if this explanation is correct, it should be possible to achieve shockless mixed flows experimentally with the aid of boundary

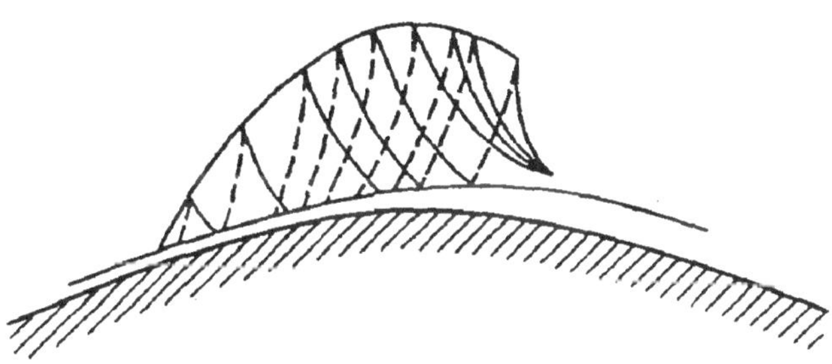

Fig. F,13b. The distortion of transonic flow by a boundary layer.

layer removal. It seems important to decide whether or not the inviscid potential flow itself has any peculiarities that would make it impossible.

A third proposal has been advanced to explain the discrepancy; namely, the smooth potential flows exist, even theoretically, only for certain special boundary shapes. If these are, in fact, special flows without neighboring solutions for slightly different boundaries, their nonappearance in nature would be explained, for real boundaries are always slightly different from ideal ones. The boundary layer would also constitute a deviation from the ideal special shape; hence, this proposal is not completely unrelated to the preceding one.

This suggestion, like the instability theory, has a parallel in channel theory, where it can easily be shown that, due to multiple reflections from channel walls, any small geometrical deviation from an ideal transonic diffuser shape must lead to large alterations of the flow pattern. The theory of nonexistence of neighboring solutions is discussed briefly in Art. 15.

We have already pointed out that there would be practical advantages to achieving smooth transonic deceleration in practice. Furthermore, it is

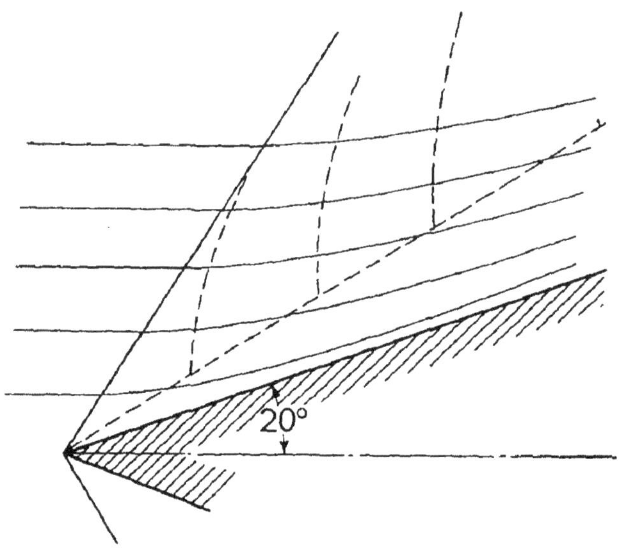

Fig. F,13c. Smooth transonic flow over a cone.

important to remember that such steady decelerations through the sonic speed, without shock waves, do occur in experiment in several cases wherein the flow is not plane. The simplest case is that of flow past a yawed infinite cylinder; this is discussed, within the limitations of linearized theory in C,4. The reader can easily verify that the same principle of solution by superposition of plane and axial flows is valid without linearization, and that extensive supersonic regions are possible without appearance of any shocks. This is confirmed by experiment.

Another famous example of isentropic compression from supersonic to subsonic speeds is afforded by the right circular cone solutions discussed in H,17–H,19; in certain ranges of Mach number and cone angle, there is supersonic flow behind the conical shock wave that is subsequently decelerated isentropically to subsonic speeds (Fig. F,13c). The existence of this flow has been well confirmed by experiment. To be sure, the exactly conical flow postulated in H,17–H,19 cannot be achieved in experiment because of the finite dimensions of the models tested. Nevertheless, the results [*68,69*] clearly show that for a given cone angle there is a range of Mach numbers for which there is no appreciable deviation from conical conditions near the tip of the cone, and the shock-free deceleration discussed here is observed. Consequently, this case of cone flow stands as evidence that isentropic irrotational deceleration through the sonic speed is not always impossible or unstable.

F,14. Theory of Stability of Transonic Flows. An approximate theory of the propagation of one-dimensional unsteady disturbances in a transonic decelerating channel flow has been given by Kantrowitz [*48*] and is discussed in III,C,7. In briefest outline, Kantrowitz's method is as follows: The histories of expansion and compression disturbances (pulses) originating downstream of a linearly decelerating sonic region are traced by means of the equations for one-dimensional unsteady isentropic flow. It is found that such disturbances are deformed as they propagate upstream, and that the disturbances tend to collect in the sonic region. Actually, the deformation of the pulse shape causes weak shock waves to form, either at head front or rear of the pulse. Expansion pulses, which terminate in shock waves at their rear, are subsequently "consumed" by their own shock waves, while compression pulses, which have shock fronts, grow in strength as their shock fronts progress into the supersonic region. Ultimately, in a real channel, they would destroy the supersonic flow, converting it into subsonic flow.

Although this is an approximate theory, since the isentropic process of deformation and propagation and nonisentropic shock processes are treated separately, it probably provides a correct explanation of the instability of channel flows. A more elaborate treatment of similar questions has been given by Meyer [*49*], who arrives at substantially the same conclusions.

It is desirable to carry out analogous studies relating to two-dimensional flows involving isentropic acceleration and deceleration. There is an obvious difference between such flows and the quasi-one-dimensional channel flows treated by Kantrowitz; viz., the propagation of disturbances in the curved nonuniform flow field must be essentially more complex than that of his one-dimensional pulses in a channel. It is surely possible that the lateral propagation in such nonuniform field might provide just the dissipative mechanism required to stabilize the flow under some circumstances.

Such a study has been made by Kuo [*50*] for thin cylinders. Let the velocity potential of the flow be

$$\phi(x, y, t) = \Phi(x, y) + \varphi(x, y, t) \tag{14-1}$$

where $\varphi(x, y, t)$ denotes the potential of the small transient disturbance superimposed on the steady transonic flow $\Phi(x, y)$. Now, substitution of

Eq. 14-1 into the differential equation in III,A, Eq. 9-8 for isentropic, irrotational, unsteady flow, even neglecting second order terms in $\varphi(x, y, t)$, results in a linear partial differential equation with variable coefficients whose solution cannot readily be expressed in a form convenient for the present investigation. Moreover, this linearization is questionable, since the processes under study may be expected to exhibit typically nonlinear behavior in the neighborhood of the sonic speed.

To avoid these difficulties, we restrict ourselves to such cases that the steady flow itself, represented by $\Phi(x, y)$, differs only slightly from a parallel stream, and therefore a perturbation potential $\Phi'(x, y)$ satisfies the Kármán-Guderley nonlinear approximation (A,9 and D,32,33), at least in the region close to the sonic lines. This assumption determines the relative orders of magnitude of the derivatives of $\Phi'(x, y)$ in the differential equation, in terms of the body thickness parameter τ. Kuo also draws analogous conclusions regarding the derivatives of $\varphi(x, y, t)$, and extends them to the time derivatives. Since the sonic region is most critical, and since the interesting disturbances travel slowly in this region, he assumes that the disturbances take on the character typical of sonic flow; i.e. their space derivatives have the same relative orders of magnitude as those of $\Phi'(x, y)$. This leads to a conclusion about the time derivatives; it turns out that there is only one possibility here that gives a reasonable result for the nonstationary flow. The differential equation can now be simplified greatly by neglecting higher order terms, and the result is

$$(\gamma + 1)(\Phi'_x + \varphi_x)\varphi_{xx} + 2\varphi_{xt} + (\gamma + 1)\Phi'_{xx}\varphi_x = \varphi_{yy} - (\gamma + 1)\Phi'_{xx}\varphi_t - \varphi_{tt} \quad (14\text{-}2)$$

where now Φ' and φ and also the coordinates x, y, and t have been made dimensionless in terms of a^* and a characteristic length l.

It is likely that the disturbances present in such a flow have their origin in a boundary layer or wake farther downstream. Since the flow is nearly parallel and close to sonic, such disturbances will be approximately plane waves propagating upstream when they arrive in the critical area. Furthermore, the process of steepening, in Riemann's sense, described just above, will long since have converted the disturbances to the characteristic triangular pulses considered by Kantrowitz. This is the type of disturbance considered by Kuo, i.e. a triangular pulse propagating from the rear, with a shock at its head (compression pulse) or tail (expansion pulse), and with velocity dropping off linearly to zero behind or before the shock. The boundary condition at the solid wall requires that the perturbation velocity always be parallel to the wall contour, which is approximated as parabolic in the sonic region.

The simplifications made are sufficient to permit a solution $\varphi(x, y, t)$ for the sonic region, provided Φ'_x is approximated as varying linearly with both x and y. The solution satisfies Eq. 14-2 and the wall boundary condition, and has the form required for a triangular pulse. It contains two arbitrary constants, which can be replaced in terms of the initial form and strength of the pulse. The pulse strength drops off linearly with distance normal to the surface. The solution permits the history of a pulse to be followed as it propagates into the sonic region. It is found that expansion pulses always decay to sound waves, whereas compression pulses decay as they propagate in accelerating sonic flow but grow in decelerating sonic flow.

As a consequence, Kuo concludes that smooth transonic plane flow is dynamically unstable to upstream-moving compression pulses.

Since this result is the same as Kantrowitz's, mentioned above, the question arises whether Kuo has succeeded in introducing the typically two-dimensional character of the flow, or whether his approximations have actually been equivalent to reducing the problem essentially to the one-dimensional one. The latter possibility seems to exist, both in the thin-body approximations, which emphasize the streamwise derivatives, and in the very special kind of pulse shape treated. Surely, the theory is a rough one and involves a combination of mathematical and physical approximations whose impact on the final conclusions is difficult to assess. It is our suggestion that this investigation be considered as an initial step only, and that the question of stability of highly curved transonic potential flows be recognized as still open to question.

F,15. Theory of Nonexistence of Smooth Transonic Flows. The suggestion has already been made above that the smooth transonic flows calculated by the hodograph technique may be exceptional cases pertaining to particular ideal boundary shapes, and that they have no neighboring solutions for slightly different boundaries as would necessarily exist in nature.[7] In favor of this theory it is argued that the effect on the flow pattern of a small alteration of the boundary is very much different in supersonic than in subsonic flow.

Consider, for example, the case sketched in Fig. F,15a, which is taken from a paper by Busemann [*51*]. It is supposed that there exists a smooth transonic flow over the original smooth boundary, and the flow past a slightly altered boundary, having a symmetric depression at the center, is to be studied. Since the local flow is supersonic, the effects of this alteration, in the new steady flow pattern, will be confined to an "alley" leading to the sonic curve, from which point it is reflected back to the body, and so forth, as sketched. But according to this process it finally

[7] This "theory of nonexistence' may even be an alternative explanation of a situation that is unstable in the sense of the preceding article; that is, the two phenomena may go hand in hand, as they do in the case of the decelerating transonic nozzle (see below).

affects the nearly sonic flow toward the rear, and we know that the pressure, density, and speed increments will be magnified in such a region. Thus a small change of the boundary shape must produce a large change in the flow pattern: i.e. there is no neighboring solution. Substantially the same argument has been put forth by Frankl [52].

All this is quite analogous to the case of decelerating channel flow. It is easy to draw up a channel for smooth deceleration of supersonic flow,

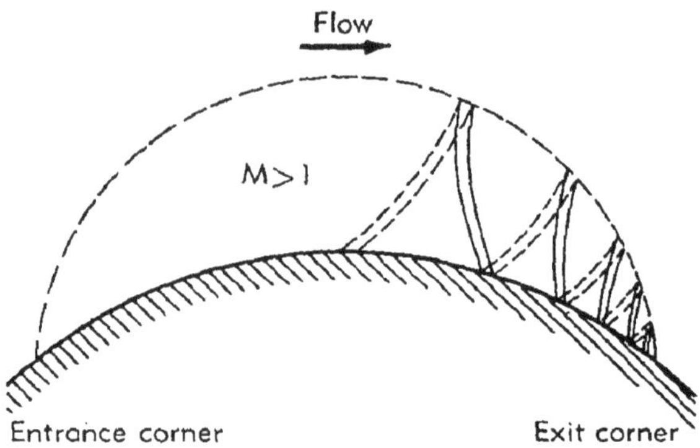

Fig. F,15a. Sketch showing the unsymmetric disturbance pattern.

through sonic, to subsonic; in fact, the channel is simply a Laval nozzle in reverse (Fig. F,15b). But the flow is unreal because there is no neighboring solution. A slight change of the contour at *A* produces violent effects at *B*. The real flow is the one involving a shock wave.

Another parallel is afforded by a familiar argument in statistical thermodynamics. An ordinary motion of the molecules of a gas, with increasing entropy, would produce decreasing entropy if followed in

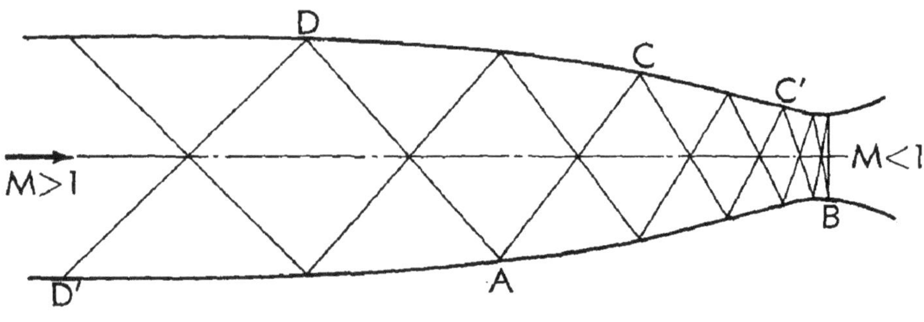

Fig. F,15b. Sketch showing reversed Laval nozzle for deceleration of supersonic flow to subsonic.

reverse; but again, the reversed motion is unreal because there is no neighboring solution.

Thus, Busemann's explanation is certainly very plausible. It would be indisputable if the perturbation at the sonic curve did not actually affect the whole flow pattern through the subsonic flow, but of course it does. To carry along the channel analogy, the trouble at *B* (Fig. F,15b) can be avoided by a small alteration at C or C', or by an alteration upstream, as at D or D', which would cancel the effect at A. It is possible that the integrated effects of the disturbances produced in the subsonic flow where the "alleys" reach the sonic curve are just sufficient to modify the upstream part of the flow and to cancel the large effect of the boundary change.

Thus the correctness of Busemann's and Frankl's conjectures can only be ascertained by more detailed calculation. An attempt in this direction has been provided by Busemann and Guderley [*53*]. They consider a

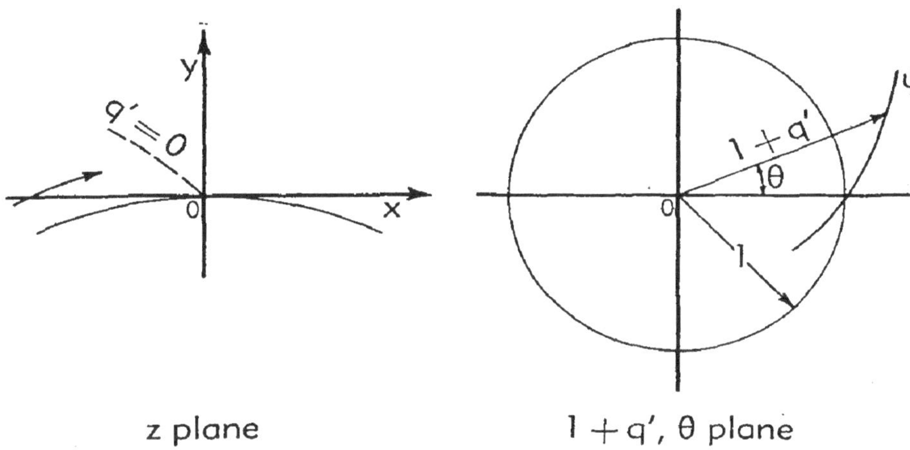

Fig. F,15c. Diagram showing the relationship between the x, y and q', θ planes.

known plane smooth transonic flow, described in the hodograph plane by a transformed potential function $\chi^{(1)}(q, \theta)$. Assuming that the flow speed is everywhere near sonic, it is convenient to use the coordinates q', θ where q' denotes $q/a^* - 1$. The approximate form assumed by Eq. 4-2 is then

$$\chi_{q'q'} - (\gamma + 1)q'\chi_{q'} - (\gamma + 1)q'\chi_{\theta\theta} = 0 \tag{15-1}$$

It is assumed that $\chi^{(1)}(q', \theta)$ satisfies this equation.

Let the coordinates be orientated so that $q' = 0$ and $\theta = 0$ correspond to $x = 0$ and $y = 0$ on the surface of the body (Fig. F,15c). If the basic flow is perturbed by an infinitesimal amount $\chi^{(2)}(q', \theta)$, the velocity at any point in the field will be altered by an amount $(\Delta q', \Delta\theta)$. The corresponding displacement of any point in the physical plane originally specified by q' and θ is calculated by means of the coordinate functions (2-12)

$$\Delta x = \chi^{(1)}_{q'q'}\Delta q' + (\chi^{(1)}_{q'\theta} - \chi^{(1)}_{\theta})\Delta\theta + \chi^{(2)}_{q'}$$
$$\Delta y = (\chi^{(1)}_{q'\theta} - \chi^{(1)}_{\theta})\Delta q' + (\chi^{(1)}_{q'} + \chi^{(1)}_{\theta\theta})\Delta\theta + \chi^{(2)}_{\theta}$$

where both q' and θ can be assumed small. But, as we are fixing our attentions on a fixed point in the physical plane, the displacement must be zero, and the disturbance velocity at the corresponding point can be determined in terms of $\chi^{(1)}$ and $\chi^{(2)}$, namely

$$\begin{aligned} \Delta q' &= D^{-1}[-(\chi^{(1)}_{\theta\theta} + \chi^{(1)}_{q'})\chi^{(2)}_{q'} + (\chi^{(1)}_{q'\theta} - \chi^{(1)}_{\theta})\chi^{(2)}_{\theta}] \\ \Delta\theta &= D^{-1}[(\chi^{(1)}_{q'\theta} - \chi^{(1)}_{\theta})\chi^{(2)}_{q'} - \chi^{(1)}_{q'q'}\chi^{(2)}_{\theta}] \end{aligned} \qquad (15\text{-}2)$$

where

$$D = \chi^{(1)}_{q'q'}(\chi^{(1)}_{\theta\theta} + \chi^{(1)}_{q'}) - (\chi^{(1)}_{q\theta} - \chi^{(1)}_{\theta})^2$$

In a small neighborhood of the origin, the potential $\chi^{(1)}$ is assumed to be of the form

$$\chi^{(1)} = -Rq'\theta + R\epsilon\theta^2 + \frac{\gamma+1}{6} R\epsilon q'^3 + \cdots \qquad (15\text{-}3)$$

where R is the arbitrary radius of curvature at the sonic point and ϵ is a constant. Upon substitution of this form in Eq. 15-2, there result, approximately

$$\begin{aligned} R\Delta q' &= \epsilon\chi^{(2)}_{q'} + \chi^{(2)}_{\theta} \\ R\Delta\theta &= \chi^{(2)}_{q'} + (\gamma + 1)\epsilon q'\chi^{(2)}_{\theta} \end{aligned} \qquad (15\text{-}4)$$

with

$$D \cong -R^2$$

Furthermore, from Eq. 15-3 and the coordinate functions, we obtain as boundary condition an approximate relation between q' and θ for points on the surface near the sonic point (origin)

$$q' \cong \epsilon\theta, \qquad |q'| \ll 1 \qquad (15\text{-}5)$$

Having thus established a relationship between the change of velocity and the perturbation potential Eq. 15-4, Busemann and Guderley formulate, besides more general ones, two simple boundary-value problems in conjunction with Eq. 15-5. First, if the boundary in the neighborhood of the sonic line is rigid, then when $q' = \xi_0\epsilon\theta$

(a) $$\Delta\theta = 0 \qquad (15\text{-}6)$$

where ξ_0 is a constant.
Secondly, if the boundary is soft, like a boundary layer, then when $q' = \xi_0\epsilon\theta$

(b) $$\Delta q' = 0 \qquad (15\text{-}7)$$

The problem remaining is to find a solution $\chi^{(2)}(q', \theta)$ satisfying Eq. 15-1 and condition 15-6, and corresponding to some small alteration of the boundary shape in the supersonic region. Busemann and Guderley claim that it is impossible to find any such $\chi^{(2)}(q', \theta)$ that is finite at the origin, in any case where a disturbance persists downstream of the last boundary alteration (i.e. excluding the "cancellation" possibility mentioned in connection with Fig. F,15b above). To demonstrate this, they assume their asymptotic solution in the form

$$\chi^{(2)} = \theta^n f(\xi) \sin \tilde{\theta} + \frac{\gamma + 1}{3} \theta^{n-\frac{3}{2}}\epsilon^{-\frac{3}{2}}q'^{2}f(\xi) \cos \tilde{\theta} + \frac{4n + 1}{6} \theta^{n-\frac{1}{2}}\epsilon^{\frac{1}{2}}q'f'(\xi) \cos \tilde{\theta} \tag{15-8a}$$

where $f(\xi)$ is a solution of the equation

$$f''(\xi) + (\gamma + 1)\xi f(\xi) = 0 \tag{15-8b}$$

with $f(-\infty) = 0$, $\xi = q'/\epsilon\theta$, and $\tilde{\theta} = 2/\epsilon^{\frac{1}{2}}\theta^{\frac{1}{2}}$; thus the boundary corresponds to $\xi = \xi_0$. The constant n is to be determined by boundary condition (a) or (b). To satisfy these conditions they select $n = -\frac{1}{4}$ in both cases and specify for the values of $f(\xi)$ at the boundary either (a) $f'(\xi_0) = 0$ or (b) $f(\xi_0) = 0$. In either case, the disturbance velocity $\Delta q'$ becomes infinite at the origin, and it is concluded that any small change of flow pattern in the supersonic region must involve a strong singularity where the sonic line meets the body surface.

Subsequently, Guderley [*54*] continued the analysis, using a series solution in place of Eq. 15-8a, attempting to provide a solution valid over a greater range of speed. The analysis is much more complex, but the same asymptotic singularity appears at the sonic point and his conclusions are therefore unchanged.

In criticism of Busemann's and Guderley's theory, it may be argued that they have not proved beyond doubt that nonsingular solutions of the boundary-value problem of the perturbed flow do not exist. The asymptotic form of Eq. 15-8a, as well as Guderley's asymptotic series, seems very special. There are indications that families of finite solutions that satisfy Busemann's and Guderley's problem, at least locally, can be found. It is also disturbing that their solution $\chi^{(2)}(q', \theta)$ is singular at both accelerating and decelerating sonic points. Since we do not expect any large disturbances at the upstream accelerating sonic point, nor observe any such disturbances in schlieren photographs, it seems possible that the whole solution of Eq. 15-8a should be rejected as physically inadmissible, removing the trouble at the downstream point as well. Busemann appeals to viscous effects to resolve this difficulty.

From a broader viewpoint, it is difficult to reconcile any theory of nonexistence of solutions with the fact that numerous continuous transonic solutions have been calculated for quite arbitrary boundary shapes by approximate methods. There are, for example, the results of Taylor [*29*] and Görtler [*67*] for the hyperbolic nozzle, Kaplan's results [*55*] for flow over a bump (see also E,2), and Emmons' numerical work by the relaxation method [*56*]. All of these show smooth transonic flows with acceleration and subsequent deceleration through the speed of sound, and in all three cases the boundary shapes were predetermined for all

Mach numbers. Although these are only approximate methods, it seems very likely that nonexistence of exact solutions would cause extreme difficulties in the calculations.

Another attack on the question of nonexistence of solutions for arbitrary boundaries is being made by Kaplan [57], who works directly with the nonlinear equation in the physical plane and seeks solutions in series form. His aim, it appears, is not to provide additional approximate flow patterns but to determine the convergence of the solutions in the transonic regime, and thus to answer the questions of this article.

The conclusion to be drawn at this writing appears to be the following: The argument presented by Busemann (Fig. F,15a) is surely plausible from the physical viewpoint. Nevertheless, his conjecture has not been convincingly proved in a mathematical way, possibly because the formulation of the problem has been inadequate.

F,16. Properties of the Hypergeometric Functions. For use in the present section, certain analytic properties of $F_\nu(\tau)$ and $\bar{F}_\nu(\tau)$ are required. For easy reference they are presented as a series of lemmas, each followed by a demonstration, which may be skipped over by readers who are not concerned with mathematical detail.

(i) *If* $\nu > 0$ and $0 \leqq \tau < \tau^*$, $F_\nu(\tau) > 0$ [2,60].

Let $\psi_\nu(\tau) = \tau^{\nu/2}F_\nu(\tau)$ and $d\eta = \tau^{-1}(1-\tau)^\beta d\tau$; from Eq. 3-1,

$$\frac{d^2\psi_\nu}{d\eta^2} = \nu^2 \frac{1-\alpha^2\tau}{4(1-\tau)^{\alpha^2}}\psi_\nu$$

If $\tau < \tau^*$, $d^2\psi_\nu/d\eta^2$ and ψ_ν are of the same sign. But in the neighborhood of $\tau = 0$, $\psi_\nu > 0$ and $\psi_\nu(0) = 0$ as $\nu > 0$; hence $d^2\psi_\nu/d\eta^2 > 0$. Therefore, for small τ the slope $d\psi_\nu/d\eta$ or $d\psi_\nu/d\tau$ will continue to increase until $\tau = \tau^*$ is reached. Since $\tau^{\nu/2} > 0$, it follows that $F_\nu(\tau) > 0$.

(ii) *In* $|\tau| < 1$, *$F_\nu(\tau)$ is an analytic function of ν, with simple poles at $\nu = -n$, n being a positive integer greater than 1* [5,60].

From Eq. 3-5, $F_\nu(\tau)$ is an analytic function of τ and ν if $|\tau| < 1$ and $\nu \neq -n$. When $\nu = -n$ and $n > 1$, $F_\nu(\tau)$ can be written as

$$F_\nu(\tau) = F_\nu^{(1)}(\tau) + \frac{1}{\nu+n}F_\nu^{(2)}(\tau)$$

where

$$F_\nu^{(1)}(\tau) = 1 + \frac{a_\nu b_\nu}{\nu+1}\tau + \cdots + \frac{a_\nu(a_\nu+1)\cdots(a_\nu+n-2)b_\nu(b_\nu+1)\cdots(b_\nu+n-2)}{(\nu+1)(\nu+2)\cdots(\nu+n-1)}\frac{\tau^{n-1}}{(n-1)!}$$

$$F_\nu^{(2)}(\tau) = \frac{a_\nu(a_\nu+1)\cdots(a_\nu+n-1)b_\nu(b_\nu+1)\cdots(b_\nu+n-1)}{(\nu+1)(\nu+2)\cdots(\nu+n-1)n!}\tau^n \left[1 + \frac{(a_\nu+n)(b_\nu+n)}{(\nu+n+1)(n+1)}\tau + \cdots\right]$$

This shows that at $\nu = -n$, both $F_\nu^{(1)}$ and $F_\nu^{(2)}$ remain regular. Therefore, the points $\nu = -n$ are simple poles of $F_\nu(\tau)$. The residue is evidently

$$\lim_{\nu \to -n} (\nu + n)F_\nu(\tau) = F_n^{(2)}(\tau)$$

where

$$F_n^{(2)}(\tau) = \frac{a_{-n}(a_{-n}+1)\cdots(a_{-n}+n-1)b_{-n}(b_{-n}+1)\cdots(b_{-n}+n-1)}{(-1)^n(n-1)!n!} \tau^n F(a_{-n}+n, b_{-n}+n; n+1; \tau)$$

Since

$$(a_{-n}+n) + (b_{-n}+n) = a_n + b_n$$

and

$$(a_{-n}+n)(b_{-n}+n) = a_n b_n$$

we can take

$$a_{-n} + n = a_n, \qquad b_{-n} + n = b_n$$

As a result, we have

$$F_n^{(2)}(\tau) = -h_n \tau^n F_n(\tau) \tag{16-1}$$

where, since $b_n \lessdot 0$,

$$h_n = \frac{\Gamma(a_n)\Gamma(n+1-b_n)}{\Gamma(a_n-n)\Gamma(1-b_n)(n-1)!n!} > 0 \tag{16-2}$$

It is noted that when $n = 1$, $a_1 = 1$, $h_1(\gamma) \equiv 0$. Furthermore, when $\gamma = -1$ (a case discussed in Art. 5), $a_n = (n+1)/2$ and $b_n = -n/2$; then h_n will involve factors $\sin[\pi(n+1)/2] \sin(\pi n/2)$. Hence $h_n \equiv 0$ in this special case.

(iii) *If the negative integers in the ν plane are excluded, then as $|\nu| \to \infty$ [19,60,63]*

$$\left.\begin{aligned} F_\nu(\tau) &= f(\tau)[T(\tau)]^\nu \left[1 + O\left(\frac{1}{\nu}\right)\right], \\ F_{-\nu}(\tau) &= f(\tau)[T(\tau)]^{-\nu}\left[1 + O\left(\frac{1}{\nu}\right)\right] \end{aligned}\right\} \quad \tau < \tau^* \tag{16-3}$$

where $q^{-2\nu}F_{-\nu}(\tau)$ denotes the second solution and

$$\begin{aligned} f(\tau) &= (1-\tau)^{\alpha^2/4}(1-\alpha^2\tau)^{-\frac{1}{4}} \\ T(\tau) &= \frac{2[\alpha(1-\tau)^{\frac{1}{2}} + (1-\alpha^2\tau)^{\frac{1}{2}}]^\alpha}{(1+\alpha)^\alpha[(1-\tau)^{\frac{1}{2}} + (1-\alpha^2\tau)^{\frac{1}{2}}]} \\ \alpha^2 &= \frac{\gamma+1}{\gamma-1} \end{aligned}$$

Also, $\tau^* < \tau < 1$

$$F_\nu(\tau) = f(\tau)[T(\tau)]^\nu \cos\left(\nu\omega - \frac{\pi}{4}\right)\left[1 + O\left(\frac{1}{\nu}\right)\right]$$

$$F_{-\nu}(\tau) = \frac{1}{2} f(\tau)[T(\tau)]^{-\nu}\left[\cos\left(\nu\omega + \frac{\pi}{4}\right) + \cot \nu\pi \sin\left(\nu\omega + \frac{\pi}{4}\right)\right]\left[1 + O\left(\frac{1}{\nu}\right)\right] \tag{16-4}$$

where

$$\begin{aligned} f(\tau) &= 2(1 - \tau)^{\alpha^2/4}(\alpha^2\tau - 1)^{-\frac{1}{4}} \\ T(\tau) &= r\tau^{-\frac{1}{2}} \\ r &= 2(\alpha - 1)^{\frac{\alpha - 1}{2}}(\alpha + 1)^{-\frac{\alpha+1}{2}} \\ \omega(\tau) &= \alpha \tan^{-1}\frac{1}{\alpha}\sqrt{\frac{\alpha^2\tau - 1}{1 - \tau}} - \tan^{-1}\sqrt{\frac{\alpha^2\tau - 1}{1 - \tau}} \end{aligned} \tag{16-4a}$$

Let $F_\nu(\tau) = \tau^{-\frac{\nu+1}{2}}(1 - \tau)^{\frac{1}{2(\gamma+1)}}g_\nu(\tau)$; then Eq. 3-2 reduces to

$$g_\nu'' - \left[\nu^2\sigma(\tau) - \sigma + \frac{2\gamma - 1}{4(\gamma - 1)(1 - \tau)^2}\right]g_\nu = 0 \tag{16-5}$$

where

$$\sigma(\tau) = \frac{1 - \alpha^2\tau}{4\tau^2(1 - \tau)}$$

When $|\nu|$ is large, by excluding the negative real axis, we deduce from Eq. 16-5 that

$$g_\nu = \sigma^{-\frac{1}{4}}e^{\nu\zeta(\tau)}\left[1 + \frac{g_1}{\nu} + \cdots\right], \qquad \alpha^2\tau < 1$$

where $\zeta = \int^\tau \sigma^{\frac{1}{2}} d\tau$. Evaluating the integral ζ and assuming, for the dominating terms, $g_\nu(0) = 1$ we obtain Eq. 16-3.

When $\tau^* < \tau < 1$, ζ becomes a purely imaginary quantity and the form of representation (16-3) fails. In that case, the asymptotic solutions for both regions $\tau < \tau^*$ and $\tau^* < \tau < 1$ are discontinuous at $\tau = \tau^*$. The method of continuation of the solution from $\tau < \tau^*$ to $\tau > \tau^*$ is generally quite elaborate. In the present problem, $\sigma(\tau)$ has a simple zero at $\tau = \tau^*$, and a technique originally due to Jeffreys proves to be convenient. According to this, Eq. 16-5, for large ν, is solved exactly at $\tau = \tau^*$ and hence reduces to the well-known Stokes equation with solution involving Bessel functions of one-third order. By carrying out this procedure, we find a phase constant of $-\pi/4$ and also a multiplicative factor of 2 in going over from $\tau < \tau^*$ to $\tau > \tau^*$. For details the reader is referred to the Jeffreys' book [*61*, p. 488].

(iv) *If* $0 < \tau < \tau^*$ *and* $\nu \neq -n$, *n being integer* [*5,60*]

$$[T(\tau)]^{-\nu}F_\nu(\tau) = f(\tau) - \sum_2^\infty \frac{h_m\tau^m T^m F_m(\tau)}{\nu + m} \tag{16-6}$$

From Lemma (ii), $F_\nu(\tau)$ has simple poles at $\nu = -n$, and from Lemma (iii), $|T^{-\nu}F_\nu(\tau)|$ is bounded on $|\nu| = n + \frac{1}{2}$ as $n \to \infty$. Then, by applying Mittag-Leffler's theorem,

$$T^{-\nu}F_\nu(\tau) = F_0(\tau) + \sum_1^\infty R_m \left(\frac{1}{\nu + m} - \frac{1}{m}\right)$$

where R_m is the residue of $T^{-\nu}F_\nu(\tau)$ at $\nu = -m$ and, since $|T^{-\nu}F_\nu(\tau)| \to f(\tau)$, we have

$$f(\tau) = F_0(\tau) - \sum_1^\infty \frac{R_m}{m}$$

Eliminating $F_0(\tau)$ and R_m by Eq. 16-1, we have Eq. 16-6.

To show that the series in Eq. 16-6 converges, we deduce from Eq. 16-2, for large m, that

$$|h_m| < M \left[\frac{(1+\alpha)^{\frac{\alpha+1}{2}}}{2(\alpha - 1)^{\frac{\alpha-1}{2}}}\right]^{2m}$$

by means of the properties of $\Gamma(z)$ at large z, where M is a constant. The series is therefore dominated by

$$\sum_1^\infty \left[\frac{(\alpha+1)^{\frac{\alpha+1}{2}}}{2(\alpha - 1)^{\frac{\alpha-1}{2}}}\tau^{\frac{1}{2}}T(\tau)\right]^{2m}$$

and therefore converges if $r^{-1}\tau^{\frac{1}{2}}T < 1$. It can be verified that the radius of convergence is the speed of sound $\tau = \tau^*$, for $\tau T^2 = r^2$ when $\tau = \tau^*$ (cf. Eq. 16-4).

(v) *If $0 < \tau < \tau^*$ and $\nu > 0$ and large* [20]

$$\xi_\nu(\tau) = \sqrt{\frac{1 - \alpha^2\tau}{1 - \tau}} + O\left(\frac{1}{\nu}\right) \tag{16-7}$$

This follows directly from Eq. 16-3.

(vi) *When $0 < \tau < \tau^*$, the values of $\nu(\tau)$ for which $F_\nu(\tau) = 0$ can be written $\nu_1, \nu_2, \nu_3, \ldots$, where $-(n+1) < \nu_n < -n$* [60].

For $0 < \tau < \tau^*$, $F_\nu(\tau) > 0$ from Lemma (i). Furthermore, by Lemma (ii), $F_{-1}(\tau) = 1 + \tau F_1\ (1, -\beta; 2; \tau)$, which is positive. But, we have, when $\nu + 2 > 0$,

$$F_\nu(\tau) \to -\infty \text{ as } \nu + 2 \to 0$$

according to Eq. 16-1. Hence $F_{\nu_2}(\tau) = 0$ and $-2 < \nu_2 < -1$. As soon as ν passes over to the other side of the pole $\nu = -2$, $F_{-2+\delta}(\tau) > 0$, $\delta > 0$, and $F_{-3+\delta}(\tau) < 0$; therefore $-3 < V_3 < -2$ and so on. From Eq. 16-6, when $|\nu_n| \to \infty$, we have, approximately

$$0 = T^{-\nu_n} F_\nu(\tau) \cong f(\tau) - \frac{h_n \tau^n T^n F_n(\tau)}{\nu_n + n} - \cdots$$

Therefore as $|\nu_n| \to \infty$

$$\nu_n \to -n + O(\epsilon^n) \tag{16-8}$$

where

$$\epsilon = \tau^{-2}{}_\tau T^2 < 1$$

(vii) *For $0 \leqq \tau < 1$ and $|\arg \nu| \leqq \pi$ [30]*

$$\frac{(\nu/e)^\nu \chi_\nu(\tau)}{\tau^\nu \Gamma(\nu + 1)} = J_\nu(\nu t) \left[Q(\tau) + \sum_\mu \frac{f_\mu(\tau)}{\nu - \mu} \right] \tag{16-9}$$

where $\chi_\nu(\tau) = \tau^{\frac{\nu}{2}} \bar{F}_\nu(\tau)$ and $Q(\tau) = (1 - \tau)^{\frac{1}{2} - \frac{\alpha^2}{4}} \left(\frac{t^2 - 1}{\alpha^2 \tau - 1} \right)^{\frac{1}{4}}$. Here t is a continuous function of τ defined by

$$\omega' \equiv \sqrt{t^2 - 1} - \tan^{-1} \sqrt{t^2 - 1} = \omega(\tau)$$

and

$$f_\mu(\tau) = \frac{(\mu/e)^\mu \chi_\mu(\tau)}{\tau^n} \left[\frac{\partial}{\partial \nu} \Gamma(\nu + 1) J_\nu(\nu t) \right]^{-1}_{\nu = \mu} \tag{16-10}$$

The summation in Eq. 16-9 is over the zeros of $J_\mu(\mu t)$.

In the first place, from Eq. 3-16 and 3-17 we can conclude that $\bar{F}_\nu(\tau)$ behaves exactly the same as $F_\nu(\tau)$. Moreover, for $\tau > 1$ and large $|\nu|$ with $|\arg \nu| \leqq \pi$, $J_\nu(\nu t)$ possesses the representation [40]

$$J_\nu(\nu t) = \left(\frac{2}{\nu \pi \sqrt{t^2 - 1}} \right)^{\frac{1}{2}} \left[\cos \left(\nu \omega' - \frac{\pi}{4} \right) + O\left(\frac{1}{\nu} \right) \right]$$

and, similar to Eq. 16-4, we have, for $\alpha^2 \tau > 1$

$$\chi_\nu(\tau) = \frac{2(1 - \tau)^{\frac{1}{2} - \frac{\alpha^2}{4}}}{(\alpha^2 \tau - 1)^{\frac{1}{4}}} \tau^\nu \cos \left(\nu \omega - \frac{\pi}{4} \right) \left[1 + O\left(\frac{1}{\nu} \right) \right]$$

These expressions together with the asymptotic expansions of $\Gamma(\nu + 1)$ and Eq. 16-10 leads to

$$\frac{(\nu/e)^\nu \chi_\nu(\tau)}{\tau^\nu \Gamma(\nu + 1) J_\nu(\nu t)} = Q(\tau) + O\left(\frac{1}{\nu} \right)$$

which can be easily verified from the asymptotic properties of $J_\nu(\nu t)$, $\Gamma(\nu + 1)$, and $\chi_\nu(\tau)$. Now both $\chi_\nu(\tau)$ and $\Gamma(\nu + 1) J_\nu(\nu t)$ have simple poles at $\nu = -n$, the poles of $(\nu/e)^\nu \chi_\nu(\tau)/r^\nu \Gamma(\nu + 1)J_\nu(\nu t)$ are therefore the zeros of $\Gamma(\nu + 1)J_\nu(\nu t)$. Consequently, the residue is

$$(\nu/e)^\mu \chi_\mu(\tau) r^{-\mu} \left[\frac{\partial}{\partial \nu} \Gamma(\nu + 1) J_\nu(\nu t)\right]^{-1}_{\nu=\mu}$$

Then, from Mittag-Leffler's theorem, we immediately obtain Eq. 16-9.

Once this is established for $t > 1$ and $\alpha^2 \tau > 1$, it is obvious that it is equally true for $t < 1$ or $\alpha^2 \tau < 1$. In that case, the only change lies in the fact that the asymptotic representations of $\chi_\nu(\tau)$ and $J_\nu(\nu t)$ are exponential and ω', ω become purely imaginary.

(*viii*) *For* $0 \leqq \tau < 1$ *and* $\nu \to -n$, *n being a positive integer* [30],

$$\frac{\bar{h}_n r^n \Gamma(n) \chi_n(\tau)}{(n/e)^n} = J_n(nt)\left(Q - \sum_\mu \frac{f_\mu}{\mu + n}\right) \tag{16-11}$$

where

$$\bar{h}_n = \frac{n+1}{n-1} h_n$$

Replacing ν by $\nu e^{\pi i}$, and by virtue of $J_{-\nu}(ze^{\pi i}) = e^{-\nu\pi i} J_{-\nu}(z)$ we have

$$\frac{\Gamma(\nu) \sin \nu\pi \chi_{-\nu}(\tau)}{\pi r^{-\nu}(\nu/e)^\nu} = J_{-\nu}(\nu t)\left(Q - \sum_\mu \frac{f_\mu}{\mu + \nu}\right) \tag{16-12}$$

Now $\chi_{-\nu}(\tau)$ has simple poles at $\nu = n$ at which its residues are $\bar{h}_n \chi_n(\tau)$ and $J_{-n}(z) = (-1)^n J_n(z)$. Therefore, as $\nu \to n$, Eq. 16-12 reduces to Eq. 16-11.

F,17. Cited References and Bibliography.

Cited References

1. Riemann, B. Über die Fortpflanzung ebener Luftwellen von endlicher Schwingungsweite. *Abhandlungen d. Gesellschaft d. Wissenschaften zu Göttingen, Math.-Phys. Klasse 8*, 43–65 (1858–1859).
2. Chaplygin, S. A. On gas jets. *Sci. Annals Imp. Univ. of Moscow, Phys.-Math. Div. 21*, Moscow (1904). Also *NACA Technical Memorandum 1063*. 1944.
3. von Kármán, Th. Compressibility effects in aerodynamics. *J. Aero. Sci. 8*, 337–356 (1941).
4. Whittaker, E. T., and Watson, G. N. *A Course of Modern Analysis*. Cambridge Univ. Press, 1940.
5. Cherry, T. M. Flow of a compressible fluid about a cylinder. *Proc. Roy. Soc. A192*, 45–79 (1947).
6. Tsien, H. S. Two-dimensional subsonic flow of compressible fluids. *J. Aero. Sci. 6*, 399–407 (1939).
7. Lin, C. C. On an extension of the von Kármán-Tsien method to two-dimensional subsonic flows with circulation around closed profiles. *Quart. Appl. Math. 4*, 291–297 (1946–1947).
8. Bers, L. On a method of constructing two-dimensional subsonic compressible flows around closed profiles. *NACA Technical Note 969*. 1945.
9. Gelbart, A. On subsonic compressible flows by a method of correspondence. *NACA Technical Note 1170*. 1947.

10. Christianovich, S. A., and Yuriev, I. M. Subsonic gas flow past a wing profile. *Prikladnaya Matematika i Mekhanika 11* (1947). Also *NACA Technical Memorandum 1250*. 1950.
11. Germain, P. Fluides compressibles. Étude directe du cas simplifié de Chaplygin. *C.R. Acad. Sci. Paris 223*, 532–534 (1946).
12. Tsien, H. S. Notes on Aerodynamics of Compressible Fluids. Mass. Inst. of Tech., Cambridge, 1948.
13. Ringleb, F. Exakte Lösungen der Differentialgleichungen einer adiabatischen Gasströmung, *ZAMM 20*, 185–189 (1940).
14. Temple, G., and Yarwood, J. Compressible flow in a convergent-divergent nozzle. *ARC Reports and Memoranda 2077*. 1947.
15. Taylor, G. I., and Sherman, C. F. A mechanical method for solving problems of flow in compressible fluids. *Proc. Roy. Soc. A121*, 194–217 (1928).
16. Frankl, F. On the problems of Chaplygin for mixed subsonic and supersonic flows *Bull. Acad. Sci. URSS 9*, 121–143 (1945). Also *NACA Technical Memorandum 1155*. 1947.
17. Guderley, G. Considerations of the structure of mixed subsonic-supersonic flow pattern. *Wright Field Report F-TR-2168-ND*. 1947.
18. Lighthill, M. J. Hodograph transformation in transonic flow. III. Flow round a body. *Proc. Roy. Soc. A191*, 352–369 (1947).
19. Tsien, H. S., and Kuo, Y. H. Two-dimensional mixed subsonic and supersonic flow of a compressible fluid and the upper critical Mach number. *NACA Technical Note 995*. 1946.
20. Kuo, Y. H. Two-dimensional irrotational transonic flows. *NACA Technical Note 1445*. 1948.
21. Guderley, G. On the compressible potential flow over a closed body. *Wright Field Report F-TR-2176-ND*. 1947.
22. Bergman, S. On two-dimensional flows of compressible fluids. *NACA Technical Note 972*. 1945.
23. Kuo, Y. H. Two-dimensional transonic flow past airfoils. *NACA Technical Note 2356*. 1951.
24. Cherry, T. M. Numerical solutions for compressible flow past a cylinder. *Council for Sci. and Ind. Res. Report A48*. Melbourne, 1948. Or *Proc. Roy. Soc. A196*, 32–36 (1949).
25 Cherry, T. M. Flow of a compressible fluid about a cylinder. *Proc. Roy. Soc. A196*. 1–31 (1949).
26. Bergman, S. Methods for determination and computation of flow patterns of a compressible fluid. *NACA Technical Note 1018*. 1946.
27. Bergman, S., and Greenstone, L. Numerical determination by use of special computational devices of an integral operator in the theory of compressible fluids, *J. Math. and Phys. 26*, 1–9 (1947).
28. Cherry, T. M. Relation between Bergman's and Chaplygin's methods of solving the hodograph equations. *Quart. Appl. Math. 9*, 91–94 (1951).
29. Taylor, G. I. The flow of air at high speeds past curved surfaces. *ARC Reports and Memoranda 1381*. 1930.
30. Cherry, T. M. Exact solutions for flow of a perfect gas in a two-dimensional Laval nozzle. *Proc. Roy. Soc. A203*, 551–571 (1950).
31. Tricomi, F. Sulle equazioni lineari derivate parziali del secondo ordine, di tipo misto, parts 1–7. *Memorie Reale Accad. Nazl. Lincei, classe sci. fis. mat. e nat. (ser. 5) 14*, 133–247 (1923). Also *Brown Univ. Transl. A9-T-26*.
32. von Kármán, Th. and Fabri, J. Écoulement transsonique à deux dimensions le long d'une paroi ondulée. *C.R. Acad. Sci. Paris 231*, 1271–1274 (1950).
33. Tomotika, S. and Tamada, K. Studies on two-dimensional transonic flows of a compressible fluid. II. *Quart. Appl. Math. 8*, 127–136 (1950).
34. Guderley, G. and Yoshihara, H. The flow over a wedge profile at Mach number 1. *J. Aero. Sci. 17*, 723–735 (1950).
35. Guderley, G. Two-dimensional flow patterns with a free-stream Mach number close to one. *Wright Field Report F-TR-6343*. 1951.

36. Cole, J. D. Drag of a finite wedge at high subsonic speeds. *J. Math. and Phys. 30,* 79–93 (1951).
37. Vincenti, W. G., and Wagoner, C. B. Transonic flow past a wedge profile with detached bow wave—details of analysis. *NACA Technical Note 2588.* 1951.
38. Tomotika, S., and Tamada, K. Studies on two-dimensional transonic flows of compressible fluid. I. *Quart. Appl. Math. 7,* 381–397 (1950).
39. Tomotika, S., and Tamada, K. Studies on two-dimensional transonic flows of compressible fluids. III. *Quart. Appl. Math. 9,* 129–147 (1951).
40. Watson, G. N. *A Treatise on the Theory of Bessel Functions.* Cambridge Univ. Press, 1945.
41. Craggs, J. W. The breakdown of the hodograph transformation for irrotational compressible fluid flow in two dimensions. *Proc. Camb. Phil. Soc. 44,* 360–379 (1948).
42. Tollmien, W. Grenzlinien adiabatischen Potentialströmungen. *ZAMM 21,* 140–152 (1941).
43. Friedrichs, K. O. On the non-occurrence of a limiting line in transonic flow. *Comm. on Pure and Appl. Math. 1,* 287–299 (1948).
44. Manwell, A. R. Note on hodograph transformation. *Quart. Appl. Math. 10,* 177–184 (1952).
45. Nikolsky, A. A., and Taganoff, G. I. Flow of a gas in a local supersonic zone and some conditions for the breakdown of potential flow. *Prikladnaya Matematika i Mekhanika 10,* 481–502 (1946). Also *Brown Univ. Trans. A9-T-17.* 1948.
46. Sears, W. R. Transonic potential flow of a compressible fluid. *J. Appl. Phys. 21,* 771–778 (1950).
47. Liepmann, H. W., Ashkenas, H., and Cole, J. D. Experiments in transonic flow. *Wright Field Technical Report 5667.* 1950.
48. Kantrowitz, A. The formation and stability of normal shock waves in channel flows. *NACA Technical Note 1225.* 1947.
49. Meyer, R. E. Waves of finite amplitude in ducts. *Quart J. Mech. Appl. Math. 5,* 257–291 (1952).
50. Kuo, Y. H. On the stability of two-dimensional smooth transonic flows. *J. Aero. Sci. 18,* 1–6 (1951).
51. Busemann, A. The drag problem at high subsonic speeds. *J. Aero. Sci. 16,* 337–344 (1949).
52. Frankl, F. I. On the formation of shock waves in subsonic flows with local supersonic velocities. *Prikladnaya Matematika i Mekhanika 11,* 199–202 (1947). Also *Brown Univ. Trans. F-TS-1223-IA.* 1949.
53. Busemann, A., and Guderley, G. The problem of drag at high subsonic speeds. *MOS Reports and Translations 184.* 1947.
54. Guderley, G. On the transition from a transonic potential flow to a flow with shocks. *Wright Field Report F-TR-2160-ND.* 1947.
55. Kaplan, C. The flow of a compressible fluid past a curved surface. *NACA Report 768.* 1943.
56. Emmons, H. W. The theoretical flow of a frictionless, adiabatic, perfect gas inside of a two-dimensional hyperbolic nozzle. *NACA Technical Note 1003* (1946).
57. Kaplan, C. On a solution of the nonlinear differential equation for transonic flow past a wave-shaped wall. *NACA Technical Note 2383.* 1951.
58. von Helmholtz, H. On discontinuous movements of fluids. *Phil. Mag. 36,* 337–346 (1868). Also *Monatsber. preussischen Akad. Wiss. Berlin,* 125 (1868).
59. Molenbroeck, P. Über einige Bewegungen eines Gases bei Annahme eines Geschwindigkeitspotentials. *Archiv. Math. und Phys., Grunert Hoppe, 9,* 157–195 (1890).
60. Lighthill, M. J. The hodograph transformation in transonic flow. II. *Proc. Roy. Soc. A191,* 341–351 (1947).
61. Jeffreys, H., and Jeffreys, B. S. *Methods of Mathematical Physics.* Cambridge Univ. Press, 1946.
62. Bateman, H. Notes on a differential equation which occurs in the two-dimensional motion of a compressible fluid and the associated variational problem. *Proc. Roy. Soc. A125,* 598–618 (1929).

63. Cherry, T. M. Asymptotic expansions for the hypergeometric functions occurring in gas-flow theory. *Proc. Roy. Soc. A202*, 507–522 (1950).
64. Coburn, N. The Kármán-Tsien pressure-volume relation in the two-dimensional supersonic flow of compressible fluids. *Quart. Appl. Math. 3*, 106–116 (1945).
65. Lighthill, M. J. The hodograph transformation in transonic flow. I. Symmetric channels. *Proc. Roy. Soc. A191*, 323–341 (1947).
66. Morawetz, C. S., and Kolodner, I. On the non-occurrence of the limiting lines in transonic flow. *Comm. on Pure and Appl. Math. 6* (1953).
67. Görtler, H. Zum Übergang von Unterschall zu Überschallgeschwindigkeiten in Düsen. *ZAMM 19*, 325–337 (1939).
68. Taylor, G. I. and Maccoll, J. W. The air pressure on a cone moving at high speeds. II. *Proc. Roy. Soc. A139*, 298–311 (1933).
69. Solomon, G. E. Transonic flow past cone-cylinders. *Guggenheim Aeronautical Laboratory, California Institute of Technology*. 1953. Also *NACA Technical Note 3213*. 1954.
70. Stack, J., Lindsey, W. F., and Littell, R. E. The compressibility burble and the effect of compressibility on pressure and force acting on an airfoil. *NACA Report 646*. 1939.
71. Cherry, T. M. A transformation of the hodograph equation and the determination of certain fluid motions. *Trans. Roy. Soc. 245*, 583–624 (1953).

Supplementary Bibliography

Mathematical papers.

Bers, L., and Gelbart, A. On a class of differential equations in mechanics of continua. *Quart. Appl. Math. 1*, 168–188 (1943).

Bers, L. On the continuation of a potential gas flow across the sonic line. *NACA Technical Note 2058*. 1950.

Frankl, F. I. On the Cauchy problem for equations of mixed elliptic-hyperbolic type with initial data on the parabolic line. *Bull. Acad. Sci. URSS 18*, 195–245 (1945).

Frankl, F. I. Asymptotic resolution of Chaplygin's functions. *Trans. Acad. Sci. URSS 58*. 1947. Also *Wright Field Technical Report F-TS-1212-IA*. 1949.

Bergman, S. Two-dimensional subsonic flows of a compressible fluid and their singularities. *Trans. Amer. Math. Soc. 62*, 452–498 (1947).

Weinstein, A. Discontinuous integrals and generalized potential theory. *Trans. Amer. Math. Soc. 63*, 342–354 (1948).

Cherry, T. M. Uniform asymptotic formulae for functions with transition points. *Trans. Amer. Math. Soc. 68*, 224–257 (1950).

Carrier, G. F., and Ehlers, F. E. On some singular solutions of the Tricomi equation. *Quart. Appl. Math. 6*, 331–334 (1948).

Goldstein, S., and Lighthill, M. J. A note on the hodograph transformation for the two-dimensional vortex flow of an incompressible fluid. *Quart. J. Mech. and Appl. Math. 3*, 297–302 (1950).

Loewner, C. A transformation theory of the partial differential equations of gas dynamics. *NACA Technical Note 2065*. 1950.

General theory.

Tollmien, W. The direct hodograph method in the theory of the flow of compressible fluids. *H. Reissner Anniv. Vol., Contributions to Appl. Mech.* Edwards Bros., 1949.

Bergman, S. A formula for the stream function of a certain flow. *Proc. Nat. Acad. Washington 29*, 276–281 (1943).

Bergman, S. On two-dimensional flows of compressible fluids. *NACA Technical Note 972*. 1945.

Goldstein, S., Lighthill, M. J., and Craggs, J. W. On the hodograph transformation for high-speed flow. I. A flow without circulation. *Quart. J. Mech. and Appl. Math. 1*, 344–357 (1948).

Lighthill, M. J. On the hodograph transformation for high-speed flow. II. A flow with circulation. *Quart. J. Mech. and Appl. Math. 1*, 442–450 (1948).

Levey, H. C. High-speed flow of a gas past an approximately elliptic cylinder. *Proc. Camb. Phil. Soc. 46*, 479–491 (1950).
Chang, C. C. General consideration of problems in compressible flow using the hodograph method. *NACA Technical Note 2582*. 1952.
Mackie, A. G., and Pack, D. C. Transonic flow past finite wedges. *Proc. Camb. Phil. Soc. 48*, 178–187 (1951).
Ovsiannikov, L. V. Gas flow with straight transition line. *Prikladnaya Matematika i Mekhanika 13* (1949). Also *NACA Technical Memorandum 1295*. 1951.

Approximate subsonic solutions.

Riabouchinsky, D. Mouvement d'un fluide compressible autour d'un obstacle. *C.R. Acad. Sci. Paris 194*, 1215 (1932).
Demtchenko, B. Sur les mouvements lents des fluides compressibles. *C.R. Acad. Sci. Paris 194*, 1218–1222 (1932).
Busemann, A. Die Expansionsberichtigung der Kontraktionsziffer von Blenden. *Forschungsbericht 4*, 186–187 (1933).
Busemann, A. Hodographenmethode der Gasdynamik. *ZAMM 12*, 73–79 (1937).
Tamada, K. Application of the hodograph method to the flow of a compressible fluid past a circular cylinder. *Proc. Phys. Math. Soc. Japan 22*, 207–219 (1940).
Suzuki, H. On the subsonic flow of a compressible fluid past a circular cylinder. *Proc. Phys. Math. Soc. Japan 23*, 337 (1941).
Tomotika, S., and Tamada, K. Application of the hodograph method to the flow of a compressible fluid past an elliptic cylinder. *Proc. Phys. Math. Soc. Japan 23*, 958 (1941).
Bers, L. On the circulatory subsonic flow of a compressible fluid past a circular cylinder. *NACA Technical Note 970*. 1945.
Bers, L. Velocity distribution on wing sections of arbitrary shape in compressible potential flow. I. Symmetric flows obeying simplified density-speed relation. *NACA Report 1006*. 1946.
Bers, L. Velocity distribution on wing sections of arbitrary shape in compressible potential flow. III. Circulatory flows obeying the simplified density-speed relation. *NACA Technical Note 2056*. 1950.
Bartnoff, S., and Gelbart, A. On subsonic compressible flows by a method of correspondence. II. Application of methods to studies of flow with circulation about a circular cylinder. *NACA Technical Note 1171*. 1947.
Gelbart, A., and Resch, A. A method of computing subsonic flows around given airfoils. *NACA Technical Note 2057*. 1950.
Leray, J. Fluides compressibles. Application a l'aile portante d'envergure infinie de la méthode approchée de Tchapliguine. *J. Math. Pures Appl. 28*, 181–191 (1949).
Bugaenko, G. A. On the problem of gas flow over an infinite cascade using Chaplygin's approximation. *Prikladnaya Matematik i Mekhanika 13* (1949). Also *NACA Technical Memorandum 1298*. 1951.
Clauser, F. H. Two-dimensional compressible flows having arbitrary specified pressure distributions for gases with gamma equal to minus one. *Symposium on Theoretical Compressible Flow, 28 June 1949*. Nav. Ord. Lab. Report 1132, 1–33 (1950).
Bers, L. Velocity distribution on wing sections of arbitrary shape in compressible potential flow. II. Subsonic symmetric adiabatic flows. *NACA Technical Note 1012*. 1946.
Garrick, I. E., and Kaplan, C. On the flow of a compressible fluid by hodograph method. I. Unification and extension of present-day results. *NACA Technical Report 789*. 1944.
Temple, G., and Yarwood, J. The approximate solution of the hodograph equations for compressible flow. *Armament Research Establishment Report S.M.E. 3201*. 1942.
Imai, I. Application of the W.K.B. method to the flow of a compressible fluid, I. *J. Math. and Phys. 28*, 173–182 (1949).
Imai, I., and Hasimoto, H. Application of the W.K.B. method to the flow of a compressible fluid. II. *J. Math. and Phys. 28*, 205–214 (1950).

Craggs, J. W. The application of the hodograph method to problems of subsonic flow in two dimensions. *ARC Reports and Memoranda 2273 (9801)*. 1949.

Transonic approximations.

Falkovich, S. V. On the theory of Laval nozzle. *Prikladnaya Matematika i Mekhanika 10* (1946). Also *NACA Technical Memorandum 1212*. 1949.

Falkovich, S. V. On a class of Laval nozzles. *Prikladnaya Matematika i Mekhanika 11* (1947). Also *Wright Field Technical Report F-TS-1223-IA*. 1949.

Germain, P. Application de l'approximation homographique a l'étude des écoulements transsoniques. *C.R. Acad. Sci. Paris 232*, 1811–1813 (1951).

Frankl, F. I. On a family of particular solutions of the equation of Darboux-Tricomi and their application to the approximate calculation of the critical velocity in a given plane-parallel Laval nozzle. *Doklady Akad. Nauk SSSR 56*, 683–686 (1947).

Ehlers, F. E. Methods of linearization in compressible flow. II. Hodograph method (Supplement 1, The transonic flow through a deLaval nozzle). *Wright Field Technical Report 102-AC49/5-100*. 1949.

Kiebel, I. A. Some stydies of the flow of a gas in the region of transition through velocity of sound. *Izvestia Akad. Nauk SSSR 3*, 253–259 (1947). Also *NACA Technical Memorandum 1252*. 1950.

Special solutions.

Daymond, S. D. Some solutions of the hodograph equations governing the two-dimensional flow of a compressible fluid. *Quart. J. Math. 16*, 78–85 (1945).

Williams, J. The two-dimensional irrotational flow of a compressible fluid in the acute region made by two rectilinear walls. *Quart. J. Math. 20*, 129–134 (1949).

Kraft, H., and Dibble, C. G. Some two-dimensional, adiabatic compressible flow patterns. *J. Aero. Sci. 11*, 283–298 (1944).

Saito, O. On some exact solutions of the two-dimensional steady flow of compressible fluid. *J. Phys. Math. Soc. Japan 6*, 248–254 (1951).

Craggs, J. W. The compressible flow corresponding to a line doublet. *Quart. Appl. Math. 10*, 88–93 (1952).

Tables of the hypergeometric functions.

Ferguson, D. F., and Lighthill, M. J. The hodograph transformation in transonic flow. IV. Tables. *Proc. Roy. Soc. A192*, 135–142 (1947).

Huckel, V. Tables of hypergeometric functions for use in compressible flow theory. *NACA Technical Note 1716*. 1948.

Chang, C. C., and O'Brien, V. Some exact solutions of two-dimensional flow of compressible fluid with hodograph method. *NACA Technical Note 2885*. 1953.

Munk, M. M., and Rawling, G. Tables of Chaplygin functions. *Nav. Ord. Lab. Memorandum 10816*. 1950.

Cherry, T. M. Tables and approximate formulae for hypergeometric functions, of high order, occurring in gas-flow theory. *Proc. Roy. Soc. A217*, 222–234 (1935).

SECTION G

THE METHOD OF CHARACTERISTICS

ANTONIO FERRI

CHAPTER 1. INTRODUCTORY CONSIDERATIONS

G,1. Introduction. In Sec. A and B of this volume, one of the fundamental rules of supersonic aerodynamics, "the rule of forbidden signals," is presented from the physical and mathematical points of view, respectively. In this section, the physical and mathematical results of Sec. A and B are used with the purpose of obtaining methods of analysis of supersonic flow fields related to practical problems.

It is shown in Sec. A that any stationary point source of infinitesimal disturbance placed in a uniform supersonic stream produces action only on points that lie on or within the Mach cone extending downstream from the point source. Conversely, pressure and velocity at an arbitrary point of the stream can be influenced only by disturbances acting at points that lie on or within a fore Mach cone of the same apex angle extending upstream from the point considered.

In a nonuniform supersonic flow field, where the Mach angle and stream direction are not constant, the rule of forbidden signals still applies. However, the concept of the geometric Mach cone must be substituted by the concept of an aerodynamic Mach conoid. The conoid has its apex at the point source of the infinitesimal disturbance, and starts tangent to the local Mach cone at the point considered and is the envelope of all the local Mach cones having their apexes at the surface of the conoid. For example, assume point P of Fig. G,1a to be a point in a two-dimensional stream on the x, y plane. The curves PA or PB, intersections with the plane x, y of the downstream Mach conoid from P, are given by the envelopes of the tangents t inclined at the local Mach angle μ with respect to the local stream direction, $\mathbf{q}$. The intersections PC or PD of the fore Mach conoid, with the x, y plane, are defined in a similar manner.

Sec. B of this volume shows that the motion of a supersonic stream is governed by a system of first order partial differential equations of the hyperbolic type having characteristic solutions, and the concept of characteristic surfaces has also been introduced. Characteristic surfaces are surfaces that divide two adjacent regions described by expressions which are analytically different (for example, one may be at a uniform state while in the other the state is not uniform). In general, characteristic surfaces exist when the transition from one region to another involves discontinuities of some derivatives.

When the system of partial differential equations considered governs a supersonic steady motion having constant entropy and constant total enthalpy (enthalpy at rest), the characteristic surfaces are coincident

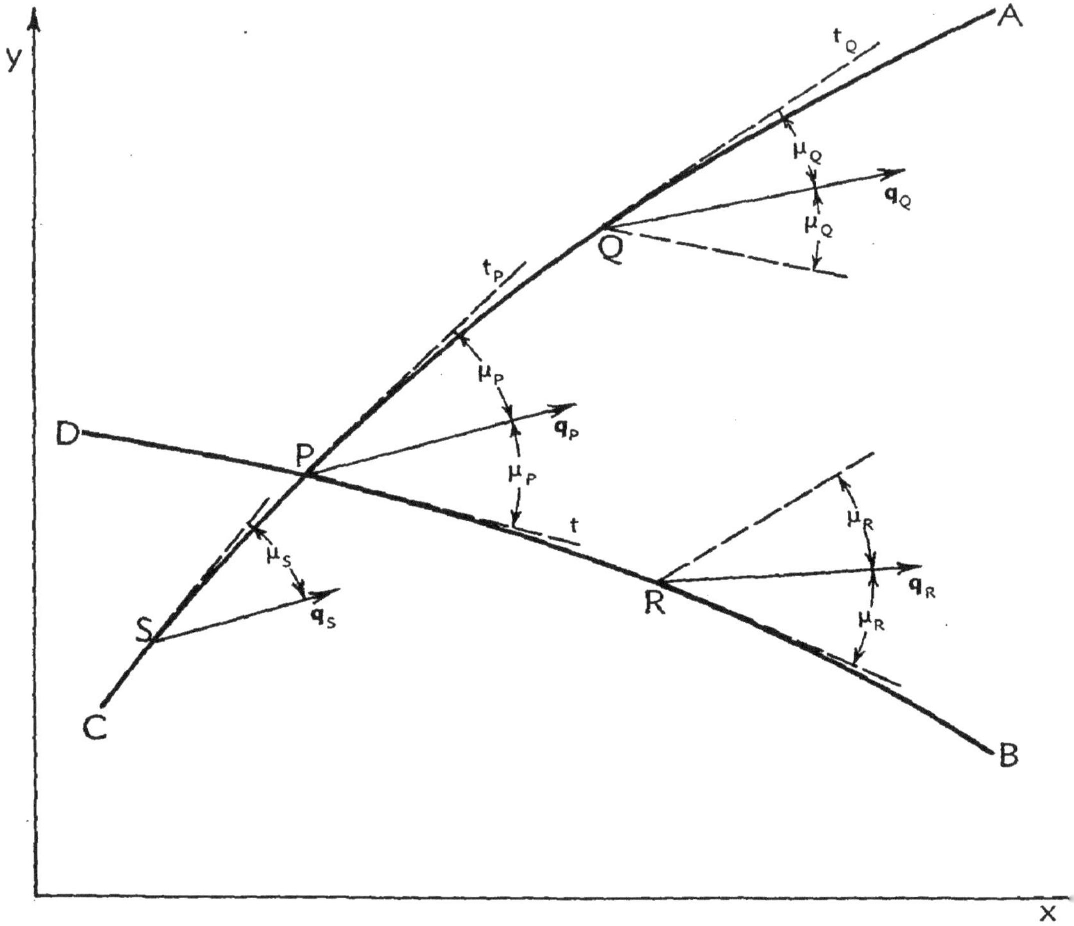

Fig. G,1a. Aerodynamic Mach conoid in a nonuniform supersonic flow field.

with the envelope of Mach conoids having vertices at the points where the disturbances are introduced in the flow. These characteristic surfaces divide the region of the flow where action is produced from the region of the flow that ignores the presence of the disturbances. For example, if infinitesimal disturbances are produced in a supersonic stream along the line AB (Fig. G,1b), the disturbances produce action on and within a region limited by the envelopes of all the Mach conoids having their vertices along the line AB, and the envelope $AHCFEDGB$ is coincident with the characteristic surfaces $BDGHCA$ and $BDEFCA$ of the system of differential equations which governs the motion. If the entropy and total enthalpy are not constant in the entire flow field (neglecting heat transmission and viscosity), the total enthalpy is constant along each streamline and the entropy remains constant along each streamline between shock waves, and the flow is rotational flow. In this case, the system of partial differential equations governing the supersonic motion still has characteristic surfaces which are coincident with the envelopes of Mach

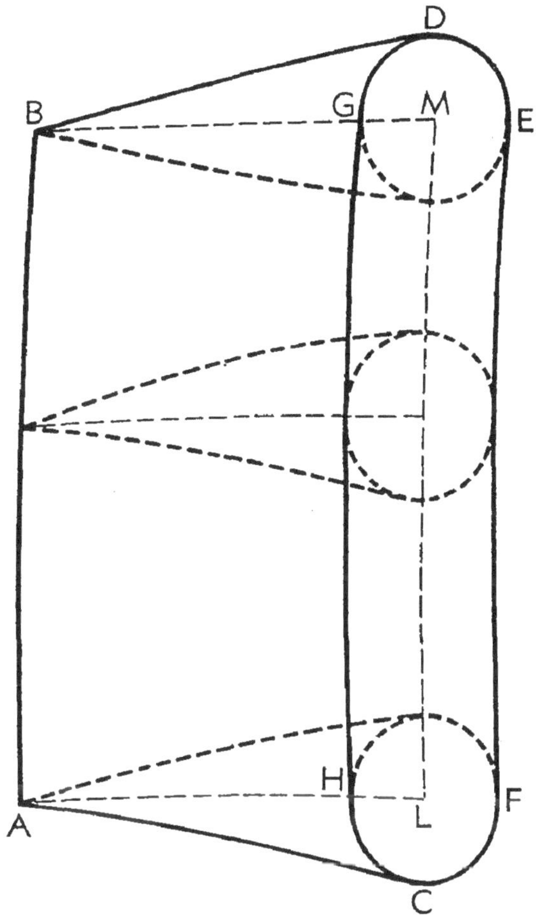

Fig. G,1b. Characteristic surfaces in a supersonic flow field.

conoids. However, another surface *ABML* must also be considered. This surface consists of the envelope of streamlines passing through each point on the line *AB*, where the disturbances are introduced.

If viscosity and heat conductivity cannot be neglected in the region of the flow considered, the rule of forbidden signals no longer applies and the system of partial differential equations differs from those discussed previously. For this condition characteristic surfaces no longer exist and an analysis by the method of characteristics cannot be used.

The physical and mathematical properties of the flow fields to which the method of characteristics is applied in this section can be briefly summarized as follows: Unless otherwise specified, steady flow fields are considered and external forces are neglected. The flow fields considered are divided into two groups: one in which the density can be expressed as a direct function of the pressure alone and is independent of the position coordinates (flow fields for which entropy and total enthalpy can be assumed constant). For this group the flow can be considered potential flow. The second group consists of flow fields in which the density is a function of the pressure and the position coordinates, such as flow fields in which entropy and total enthalpy gradients exist; then the flow is rotational. The effects of viscosity and heat conductivity are neglected for all cases considered; therefore, total enthalpy and entropy are assumed constant along the streamlines between discontinuities.

For the first group of flow fields considered the equation of the conservation of energy and the expression which relates the density to the pressure define pressure and density as functions of velocity. Therefore, the entire flow field is determined when the direction and intensity of the velocity are determined at every point of the flow field. The problem is therefore defined by a system of two differential equations, for two-dimensional or axially symmetric flow conditions, or by a system of three differential equations for the general three-dimensional case. For this group of flow fields, the motion is analyzed along particular lines obtained from the intersections of the characteristic surfaces given by the envelopes of the Mach conoids, with planes properly chosen in order to simplify the analysis.

For the second group of flow fields, the relation between pressure and density must also be determined for each point of the flow field; therefore, another system of differential equations is required in order to determine the flow field. In this case the analysis of motion is performed by moving simultaneously along the intersections of the characteristic surfaces given by the envelopes of the Mach conoids with reference planes, and along the streamlines passing through each point of the flow field considered.

G,2. The Equations of Characteristics for Two-dimensional and Axially Symmetric Flow. Consider a two-dimensional and an axially symmetric flow field. Assume for the two-dimensional case a Cartesian coordinate system x and y parallel to the plane in which the flow is considered two-dimensional. For the axially symmetric flow field, assume a cylindrical coordinate system x, y, and ω having the x axis coincident with the axis of symmetry of the flow. If p and ρ indicate pressure and density, and u and v are the velocity components along the x and y axes respectively, then Euler's equations are

$$u \frac{\partial u}{\partial x} + v \frac{\partial u}{\partial y} = -\frac{1}{\rho} \frac{\partial p}{\partial x}$$
$$u \frac{\partial v}{\partial x} + v \frac{\partial v}{\partial y} = -\frac{1}{\rho} \frac{\partial p}{\partial y} \tag{2-1}$$

and the equation of continuity is

$$\frac{\partial(\rho u)}{\partial x} + \frac{\partial(\rho v)}{\partial y} + \frac{\epsilon(\rho v)}{y} = 0 \tag{2-2}$$

where $\epsilon = 0$ for two-dimensional flow and $\epsilon = 1$ for axially symmetric flow. The total enthalpy h^0 is constant along each streamline, and the equation

$$u \frac{\partial h^0}{\partial x} + v \frac{\partial h^0}{\partial y} = 0 \tag{2-3}$$

is valid everywhere; while for the entropy s the equation

$$u \frac{\partial s}{\partial x} + v \frac{\partial s}{\partial y} = 0 \tag{2-4}$$

is valid everywhere except for passage through a shock wave.

The velocity of sound a for the general case depends on the direction of the line along which it is computed. Along the streamline

$$a = \left(\sqrt{\frac{\partial p}{\partial \rho}}\right)_{s=\text{const}} = \sqrt{\frac{\gamma p}{\rho}} = \sqrt{\gamma \mathcal{R} T} \tag{2-5}$$

The relation between pressure and density in an ideal gas having c_p and c_v constant is given by

$$\frac{p}{\rho^\gamma} = \frac{p_0}{\rho_0^\gamma} e^{\frac{s-s_0}{c_v}} = C e^{\frac{s}{c_v}} \tag{2-6}$$

where p_0, ρ_0, and s_0 are the initial conditions and $\gamma = c_p/c_v$.

Because of Eq. 2-3 and 2-4, the initial conditions and therefore the function C remain constant along each streamline.

From Eq. 2-5 and 2-6 the following relations are obtained.

$$\frac{\partial p}{\partial x} = \rho \frac{a^2}{c_p} \frac{\partial s}{\partial x} + a^2 \frac{\partial \rho}{\partial x} + \rho \frac{a^2}{\gamma} \frac{\partial \ln C}{\partial x} \tag{2-7}$$

$$\frac{\partial p}{\partial y} = \rho \frac{a^2}{c_p} \frac{\partial s}{\partial y} + a^2 \frac{\partial p}{\partial y} + \rho \frac{a^2}{\gamma} \frac{\partial \ln C}{\partial y} \tag{2-8}$$

The relation between entropy, total enthalpy, and vorticity component normal to the streamline in a gas flow is given by (D,32, Eq. 32-8; also III,A,11)

$$-T \frac{\partial s}{\partial n} + \frac{\partial h^0}{\partial n} = (\mathbf{q} \times \text{curl } \mathbf{q}) \cdot \mathbf{n} \tag{2-9}$$

where **n** is the direction normal to the streamline.

For two-dimensional or axially symmetric flow fields

$$(\mathbf{q} \times \text{curl } \mathbf{q}) \cdot \mathbf{n} = -q \left(\frac{\partial v}{\partial x} - \frac{\partial u}{\partial y} \right) \tag{2-10}$$

therefore

$$\frac{\partial v}{\partial x} - \frac{\partial u}{\partial y} = -\frac{1}{q} \frac{\partial h^0}{\partial n} + \frac{a^2}{\gamma \mathcal{R} q} \frac{\partial s}{\partial n} \tag{2-11}$$

By combining Eq. 2-1 to 2-5, 2.7 and 2.8, the following expression is obtained

$$\frac{\partial u}{\partial x} \left(1 - \frac{u^2}{a^2}\right) + \frac{\partial v}{\partial y} \left(1 - \frac{v^2}{a^2}\right) - \frac{uv}{a^2} \left(\frac{\partial u}{\partial y} + \frac{\partial v}{\partial x} \right) + \epsilon \frac{v}{y} = 0 \tag{2-12}$$

or because of Eq. 2-11

$$\frac{\partial u}{\partial x} \left(1 - \frac{u^2}{a^2}\right) + \frac{\partial v}{\partial y} \left(1 - \frac{v^2}{a^2}\right) - \frac{2uv}{a^2} \frac{\partial u}{\partial y} + \epsilon \frac{v}{y} = \frac{uv}{\gamma \mathcal{R} q} \frac{\partial s}{\partial n} - \frac{vu}{a^2 q} \frac{\partial h^0}{\partial n} \tag{2-13}$$

If $\partial s/\partial n$ and $\partial h^0/\partial n$ are zero, $\partial v/\partial x$ is equal to $\partial u/\partial y$ (Eq. 2-11), therefore a potential function ϕ exists which is defined by

$$\frac{\partial \phi}{\partial x} = u, \qquad \frac{\partial \phi}{\partial y} = v \tag{2-14}$$

and the right-hand terms of Eq. 2-13 become equal to zero.

Consider a line in the x, y plane passing through a point of the flow in which the tangent t at P is inclined at an angle $\tan^{-1} \lambda$ with respect to the x axis. The line can be expressed analytically as

$$\frac{dy}{dx} = \lambda \tag{2-15}$$

The variations of the velocity components u and v along the line are expressed in the form

$$\begin{aligned} du &= \frac{\partial u}{\partial x} dx + \frac{\partial u}{\partial y} dy = \left(\frac{\partial u}{\partial x} + \lambda \frac{\partial u}{\partial y} \right) dx \\ dv &= \frac{\partial v}{\partial x} dx + \frac{\partial v}{\partial y} dy = \left(\frac{\partial v}{\partial x} + \lambda \frac{\partial v}{\partial y} \right) dx \end{aligned} \tag{2-16}$$

Substituting Eq. 2-16 in Eq. 2-13 there results

$$H \frac{du}{dx} + (2K - H\lambda) \frac{dv}{dx} + \frac{\partial v}{\partial y} (H\lambda^2 - 2K\lambda + L) + (H\lambda - K)A + \epsilon \frac{v}{y} = 0 \tag{2-17}$$

where

$$H = 1 - \frac{u^2}{a^2}$$

$$K = -\frac{uv}{a^2}$$

$$L = 1 - \frac{v^2}{a^2} \tag{2-18}$$

$$A = \frac{a^2}{\gamma \mathcal{R} q}\frac{\partial s}{\partial n} - \frac{1}{q}\frac{\partial h^0}{\partial n}$$

Now consider two particular lines L_1 and L_2 passing through P, whose tangents λ are defined by the equation

$$H\lambda^2 - 2K\lambda + L = 0 \tag{2-19}$$

or

$$\lambda_{1,2} = \left(\frac{K}{H} \mp \frac{1}{H}\sqrt{K^2 - HL}\right)$$

Then

$$2K - H\lambda_1 = \lambda_2 H$$

$$2K - H\lambda_2 = \lambda_1 H \tag{2-20}$$

For these two lines defined by

$$\frac{dy}{dx} = \lambda_1 \tag{2-21a}$$

and

$$\frac{dy}{dx} = \lambda_2 \tag{2-21b}$$

Eq. 2-17 becomes

$$du + \lambda_2 dv + \left[A\left(\lambda_1 - \frac{K}{H}\right) + \frac{v}{y}\frac{\epsilon}{H}\right]dx = 0 \tag{2-22a}$$

and

$$du + \lambda_1 dv + \left[A\left(\lambda_2 - \frac{K}{H}\right) + \frac{v}{y}\frac{\epsilon}{H}\right]dx = 0 \tag{2-22b}$$

where Eq. 2-22a is valid along the line (2-21a), and Eq. 2-22b is valid along the line (2-21b). If θ is the inclination of the velocity vector, with respect to the x axis, and μ is the local Mach angle

$$\frac{u}{q} = \cos\theta$$

$$\frac{v}{q} = \sin\theta \tag{2-23}$$

$$\frac{q}{a} = \frac{1}{\sin\mu}$$

From Eq. 2-19 the following expressions are obtained by means of Eq. 2-23

$$\lambda_1 = \tan(\theta + \mu) \tag{2-24a}$$

$$\lambda_2 = \tan(\theta - \mu) \tag{2-24b}$$

Therefore, the simplified expressions (2-22a) and (2-22b) are valid along two lines whose tangents are inclined at the Mach angle with respect to the direction of the local velocity. These two lines, called characteristic lines, are the intersections of characteristic surfaces with the x, y plane.

The characteristic surfaces are cylindrical surfaces for two-dimensional flow and surfaces of revolution for axially symmetric flow. The lines defined by Eq. 2-24a will be called characteristic lines of the first family, and the lines defined by Eq. 2-24b characteristic lines of the second family. These lines are real only if $K^2 - HL$ in Eq. 2-19 is greater than zero, which corresponds to the condition $q/a > 1$. The application of the method of characteristics to two-dimensional supersonic potential flow was presented for the first time in [*1*,*2*], and to supersonic axially symmetric potential flow in [*3*,*4*,*5*]. For the application of the method of characteristics to rotational flow see references [*6–14*].

Eq. 2-22 are given in terms of the velocity components u and v, total enthalpy h^0, and the entropy s; however, they can be transformed in equivalent expressions containing other flow properties. The transformation which considers the quantities p and θ as independent variables is of interest because it simplifies the form of the differential equations. For this transformation Eq. 2-22 becomes [*10*]

$$\frac{\gamma + 1}{2\gamma^2}\frac{dp}{p} \pm \frac{d\theta}{\cos\mu \sin\mu} + \frac{\epsilon \sin\theta}{\cos\mu \cos(\theta \pm \mu)}\frac{dx}{y} = 0 \tag{2-25}$$

where the plus sign refers to the first family (2-24a), and the minus sign to the second family (2-24b). The coefficient ϵ is zero for two-dimensional flow and equal to 1 for axially symmetric flow.

Eq. 2-25 has the same form for potential or rotational flow, and can be used in the general case. Eq. 2-25 is simpler than Eq. 2-22; however, the simplification is partly formal because when rotational flow is considered Eq. 2-3 and 2-4 are still required along the streamlines in order to obtain a relation between pressure and velocity at every point of the flow.

In Eq. 2-22, the term A defined by the last equation of Eq. 2-18 becomes zero for irrotational flow; therefore, in this case, motion can be analyzed by moving only along the Mach lines.

For two-dimensional flow, ϵ is zero and, when A is zero, Eq. 2-22, contain only the velocity components and not the coordinates x and y; therefore, the equations can be solved analytically in the hodograph plane. This solution cannot be found for the axially symmetric flow because the coordinate y appears in the equations.

CHAPTER 2. TWO-DIMENSIONAL POTENTIAL FLOW

G,3. The Equations of Characteristics for Two-dimensional Potential Flow. When the entropy and total enthalpy are constant in the entire flow field considered, Eq. 2-22a along the characteristic line of the first family defined by

$$\lambda_1 = \left(\frac{dy}{dx}\right)_1 = \tan\,(\theta + \mu) \tag{3-1}$$

becomes

$$du + \lambda_2 dv = 0 \tag{3-2}$$

and Eq. 2-22b along the characteristic line of the second family defined by

$$\lambda_2 = \left(\frac{dy}{dx}\right)_2 = \tan\,(\theta - \mu) \tag{3-3}$$

becomes

$$du + \lambda_1 dv = 0 \tag{3-4}$$

By using polar components q and θ defined by

$$\begin{aligned} u &= q\cos\theta \\ v &= q\sin\theta \end{aligned} \tag{3-5}$$

since

$$M = \frac{1}{\sin\mu}$$

and

$$\begin{aligned} du &= dq\cos\theta - (q\sin\theta)d\theta \\ dv &= dq\sin\theta + (q\cos\theta)d\theta \end{aligned}$$

Eq. 3-2 and 3-4 become

$$\frac{dq}{d\theta}\sqrt{1 - \frac{1}{M^2}} = \pm\frac{q}{M} \tag{3-6}$$

or

$$\frac{dq}{q} = \pm\,(\tan\mu)d\theta \tag{3-7}$$

where the plus sign in Eq. 3-6 and 3-7 corresponds to Eq. 3-2 and the minus sign to Eq. 3-4. However

$$\frac{dM}{M} = \frac{dq}{q} - \frac{da}{a} = \frac{dq}{q} - \frac{\gamma - 1}{2}\frac{d\rho}{\rho} = \frac{dq}{q}\left(1 + \frac{\gamma - 1}{2}M^2\right)$$

hence

$$\pm d\theta = \frac{dM}{M\left(1 + \frac{\gamma - 1}{2}M^2\right)}\sqrt{M^2 - 1} \tag{3-8}$$

Eq. 3-8 can be written in the form

$$\pm d\theta = dM \left[\frac{\gamma + 1}{2} \frac{M}{\sqrt{M^2 - 1}\left(1 + \frac{\gamma - 1}{2} M^2\right)} \right] - \frac{dM}{M\sqrt{M^2 - 1}} \tag{3-9}$$

which becomes, upon integrating,

$$\mp \theta = \cos^{-1} \frac{1}{M} + \frac{1}{2}\sqrt{\frac{\gamma + 1}{\gamma - 1}} \cos^{-1}\left(1 - \frac{\gamma + 1}{1 + \frac{\gamma - 1}{2} M^2}\right) + \text{const} \tag{3-10}$$

or by changing the value of the constant

$$\mp \theta = \cos^{-1} \frac{1}{M} - \sqrt{\frac{\gamma + 1}{\gamma - 1}} \tan^{-1} \sqrt{(M^2 - 1)\left(\frac{\gamma - 1}{\gamma + 1}\right)} + \text{const} \tag{3-11}$$

or

$$\mp 2\theta = \sqrt{\frac{\gamma + 1}{\gamma - 1}} \sin^{-1} \frac{\frac{\gamma - 1}{2} M^2 - \gamma}{\frac{\gamma - 1}{2} M^2 + 1} + \sin^{-1}\left(\frac{2}{M^2} - 1\right) + \text{const} \tag{3-12}$$

Eq. 3-11 gives the relation between the direction θ of the streamline, and the value of Mach number along the characteristic of the first family (minus sign) and along the characteristic of the second family (plus sign). In Eq. 3-9 and 3-11 no position coordinates appear, indicating that, when an infinitesimal disturbance $d\theta$ is produced in a supersonic two-dimensional stream, the disturbance is transmitted along the Mach line with an intensity that does not change in going from the source along the Mach line to infinity [*1*].

Eq. 3-12 can also be expressed in terms of velocity or pressure. In place of Mach number the ratio between local velocity and critical velocity can be used. The critical velocity is defined as the velocity of sound at Mach number equal to 1, and is given by the expression

$$(q^*)^2 = \frac{2\gamma}{\gamma + 1} \frac{p^0}{\rho^0} = \frac{\gamma - 1}{\gamma + 1} q_{max}^2 \tag{3-13}$$

where q_{max} is the limiting velocity corresponding to an expansion into the vacuum, p^0 and ρ^0 are stagnation pressure and density respectively. Critical velocity and limiting velocity are functions only of the total enthalpy and independent of the type of transformation occurring in the flow.

Since

$$\frac{q_{max}^2}{q^2} = 1 + \frac{2}{\gamma - 1} \frac{1}{M^2} \tag{3-14}$$

Eq. 3-12 becomes

$$\mp 2\theta = \sqrt{\frac{\gamma+1}{\gamma-1}} \sin^{-1}\left[(\gamma-1)\frac{q^2}{(q^*)^2} - \gamma\right] + \sin^{-1}\left[(\gamma+1)\frac{(q^*)^2}{q^2} - \gamma\right] + \text{const} \quad (3\text{-}15)$$

In terms of pressure, Eq. 3-10 becomes

$$\mp\theta = \cos^{-1}\frac{1}{\sqrt{\frac{2}{\gamma-1}\left[\left(\frac{p^0}{p}\right)^{\frac{\gamma-1}{\gamma}} - 1\right]}} + \frac{1}{2}\sqrt{\frac{\gamma+1}{\gamma-1}}\cos^{-1}\left[1 - (\gamma+1)\left(\frac{p}{p^0}\right)^{\frac{\gamma-1}{\gamma}}\right] + \text{const} \quad (3\text{-}16)$$

G,4. The Characteristic Lines in the Hodograph Plane. Eq. 3-10 to 3-15 relate the direction of velocity to its intensity along the characteristic lines for two-dimensional potential flow, and permit a graphical representation in the hodograph plane [*1*,*2*].

The curve $\theta = f(q)$ defined by Eq. 3-15 is real only for

$$-1 \leqq (\gamma-1)\frac{q^2}{(q^*)^2} - \gamma \leqq 1$$

and

$$-1 \leqq (\gamma+1)\frac{(q^*)^2}{q^2} - \gamma \leqq 1$$

Therefore, q must vary between two values $q^* \leqq q \leqq q_{max}$ defined by Eq. 3-13, and the hodograph diagram will be contained between two circles, one of radius q and the other of radius q_{max}. For $q = q^*$ Eq. 3-15 becomes

$$\mp 2\theta^* = +\sqrt{\frac{\gamma+1}{\gamma-1}}\frac{3\pi}{2} + \frac{\pi}{2} + \text{const} \quad (4\text{-}1)$$

Therefore, for each sign of Eq. 3-15 an infinite number of curves can be obtained. These curves are all equal and obtainable one from the other by a rotation having as center the origin of the coordinates and magnitude equal to one half of the difference of the constants used in Eq. 3-15. The constant defines by means of Eq. 4-1 the position of the point of each curve on the circle $q = q^*$.

Eq. 3-15 represents the equation of an epicycloid in polar coordinates; therefore, the hodograph diagram can be constructed in the following manner (Fig. G,4a). Two concentric circles of radius q^* and q_{max} are drawn. The value of θ^* which fixes the point G is determined by the value of the constant chosen in Eq. 3-15, and is given by Eq. 4-1. The circle of diameter

$$q_{max} - q^* = \left(\sqrt{\frac{\gamma+1}{\gamma-1}} - 1\right)q^*$$

is drawn tangent to the circle of radius q^*. This circle is rolled around the circle of radius q^*, and the trajectory of the point G on this circle describes a curve which represents the hodograph diagram defined by Eq. 3-15. The two possible directions of rotation of the circle are related to the different signs of Eq. 3-15. Through every point B of the hodograph

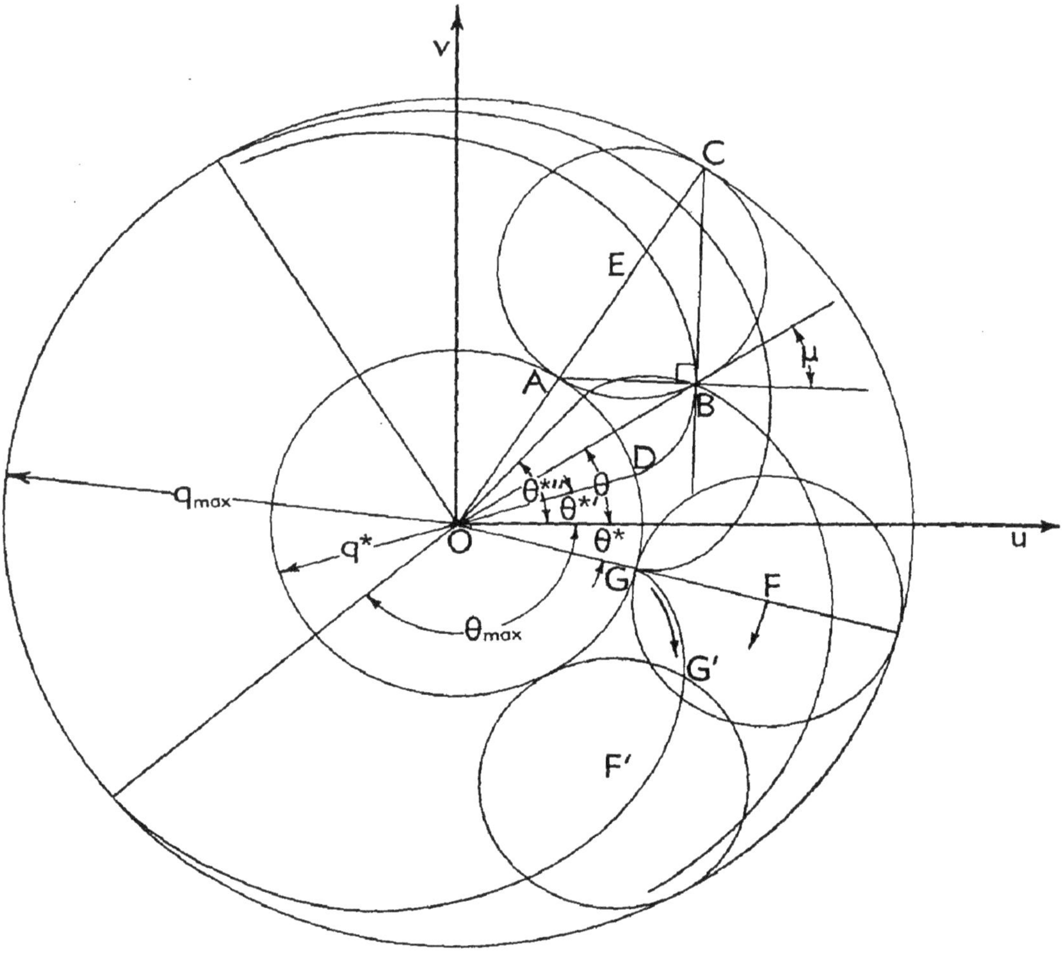

Fig. G,4a. The construction of the characteristic polar diagram in the hodograph plane.

plane, two epicycloids pass, one for each family of characteristics. The two epicycloids are symmetric with respect to the line θ = const passing through B, because from Eq. 3-15 is obtained

$$\theta^{*\prime\prime} - \theta = \theta - \theta^{*\prime} = \frac{1}{2}\left\{\sqrt{\frac{\gamma+1}{\gamma-1}}\frac{3\pi}{2} + \frac{\pi}{2} - \sqrt{\frac{\gamma+1}{\gamma-1}}\sin^{-1}\left[(\gamma-1)\frac{q^2}{(q^*)^2} - \gamma\right] - \sin^{-1}\left[(\gamma+1)\frac{(q^*)^2}{q^2} - \gamma\right]\right\} \quad (4\text{-}2)$$

The point on the circle $q = q_{max}$ of each epicycloid is defined by Eq.

3-15 and Eq. 4-1

$$\mp 2[\theta_{max} - \theta^*] = \sqrt{\frac{\gamma+1}{\gamma-1}} \sin^{-1}\left[(\gamma-1)\frac{q_{max}^2}{(q^*)^2} - \gamma\right] + \sin^{-1}\left[(\gamma+1)\frac{(q^*)^2}{q_{max}^2} - \gamma\right] - \frac{\pi}{2} - \sqrt{\frac{\gamma+1}{\gamma-1}}\frac{3\pi}{2} \quad (4\text{-}3)$$

Since q_{max}/q^* is given by Eq. 3-13, there results

$$\theta_{max} - \theta^* = \pm\frac{\pi}{2}\left(\sqrt{\frac{\gamma+1}{\gamma-1}} - 1\right) \quad (4\text{-}4)$$

For $\gamma = 1.40$ the maximum variation of direction along a characteristic line expanding from the sonic velocity to the limiting velocity is $\pm 130.45°$.

The tangent BC to the epicycloid BD at the point B must be normal to the radius of curvature at B and therefore must be normal to the line BA passing through the center of rotation A (Fig. G,4a).

The relationship between characteristic lines in the physical plane and epicycloids in the hodograph plane can be obtained from Eq. 3-1 to 3-4. From Eq. 3-2 and 3-3

$$\left(\frac{du}{dv}\right)_1 = -\lambda_2 = -\left(\frac{dy}{dx}\right)_2$$

or

$$\left(\frac{dv}{du}\right)_1\left(\frac{dy}{dx}\right)_2 = -1 \quad (4\text{-}5)$$

and from Eq. 3-4 and 3-1

$$\left(\frac{dv}{du}\right)_2\left(\frac{dy}{dx}\right)_1 = -1 \quad (4\text{-}6)$$

The tangent $t_1 = (dy/dx)_1$, to the characteristic line C_1 of the first family at a point B (Fig. G,4b) in the physical plane, is perpendicular to the tangent $(dv/du)_2$ to the epicycloid e_2 representing the characteristic line of the second family at the point corresponding to B in the hodograph plane. Conversely, the tangent $t_2 = (dy/dx)_2$, to the characteristic line C_2 of the second family at point B in the physical plane, is perpendicular to the tangent $(dv/du)_1$ to the epicycloid e_1 representing the characteristic line of the first family at the point corresponding to B in the hodograph plane. In Fig. G,4b, the epicycloid e_1 represents the characteristic line of the first family, and the normal $n_1 = A_1B$ must be parallel to the tangent to the characteristic line C_2 of the second family in the physical plane at point B. Therefore n_1 is inclined at the Mach angle μ_B with respect to the direction of velocity OB. The normal n_2 to the epicycloid e_2 at point B must be parallel to the tangent t_1 to the characteristic line C_1 at B, and it is also inclined at μ_B with respect to the direction OB.

The constants in Eq. 3-15 can be used to distinguish the different epicycloids on the hodograph plane. Eq. 3-15 for the first characteristic

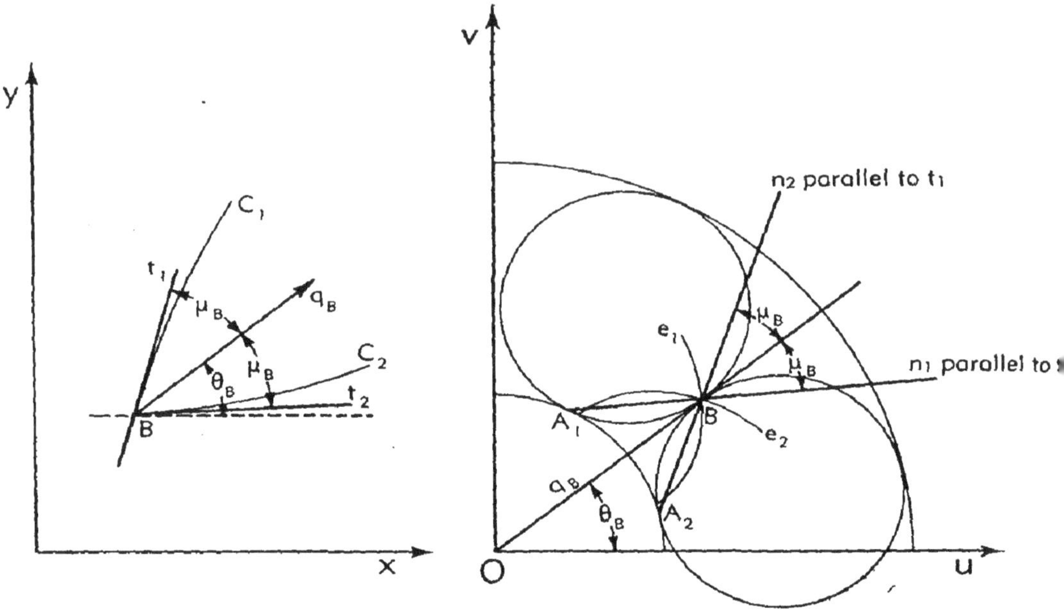

Fig. G,4b. Relation between tangents to the characteristic lines in the physical and hodograph planes.

line can be written in the form

$$-\theta = f(q) - C_1 \tag{4-7}$$

and the second characteristic line can be written in the form

$$+\theta = f(q) - C_2 \tag{4-8}$$

Then the direction θ = const corresponds to the condition

$$\theta = \frac{C_1 - C_2}{2} = \text{const} \tag{4-9}$$

while the circle q = const in the hodograph plane corresponds to the condition

$$f(q) = \frac{C_1 + C_2}{2} = \text{const} \tag{4-10}$$

The velocity vector at any point of the flow field can be defined by the constants of the two epicycloids representing the characteristic lines passing through the point considered. The sum of the constants defines the intensity of velocity and the difference of the constants defines the direction of velocity. Epicycloids which cross each other along a radius

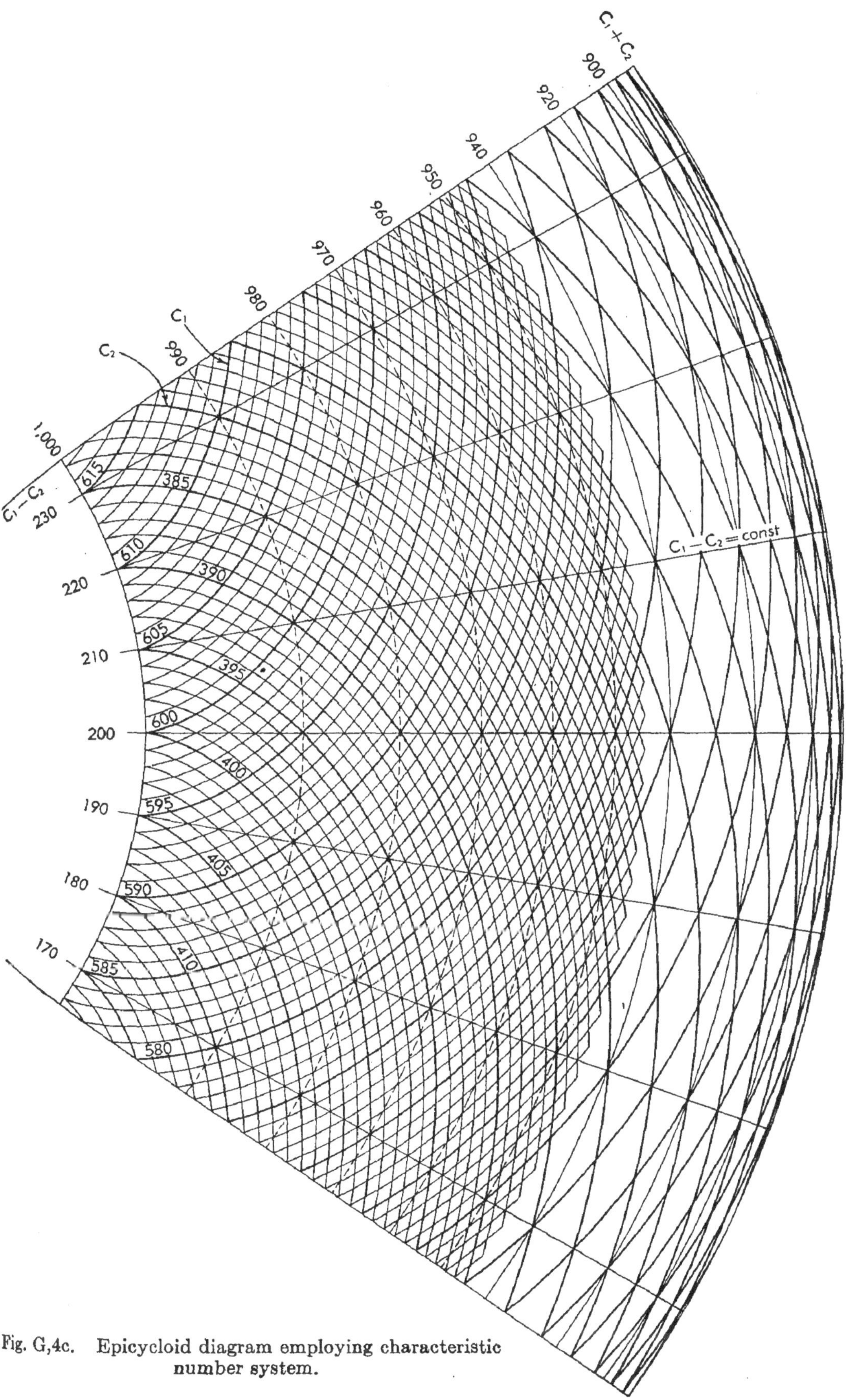

Fig. G,4c. Epicycloid diagram employing characteristic number system.

$\theta =$ const are distinguished by constants whose difference is constant (Eq. 4-9); while epicycloids which cross each other along a circle $q =$ const are distinguished by constants whose sum has a constant value (Eq. 4-10). The absolute value of the constants depends on the coordinate axis chosen for the value $\theta = 0$; therefore, the axis can be chosen in such a manner that the value of θ is positive in the entire flow field considered, and the values of the constants are whole integers. For example, assume that the circle $(C_1 + C_2)/2$ corresponding to $q = q^*$ is denoted by the number 1,000, and that the axis $\theta = 0$ is designated by the number 200

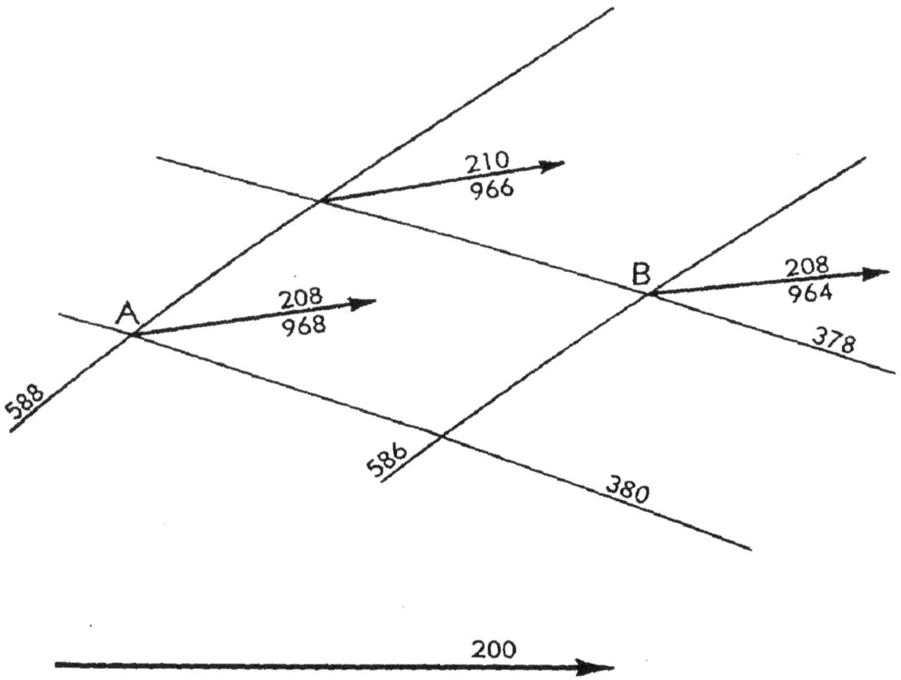

Fig. G,4d. Practical use of the characteristic number system.

(larger than the value of the maximum deviation, $\theta_{max} - \theta^* = 130.45°$); then a velocity vector 605-395 indicates a vector of direction 210 and intensity 1,000 or a vector turned 10° with respect to the 200 axis and of velocity $q = q^*$. The epicycloids at this point are denoted by the numbers 605 and 395. Similarly, the velocity vector of direction 190 and intensity 960 is defined by the two epicycloids 575, 385. An epicycloid diagram in which this number system is used is shown in Fig. G,4c.

This parametric system of numbers is practical for two-dimensional calculations in which the flow must be analyzed along characteristic lines of opposite families. Along each characteristic line, the number designating the characteristic remains constant; therefore, the velocity at each point of intersection of two characteristic lines is immediately determined from the numbers of the two characteristic lines. The value of the intensity of velocity can be obtained from the hodograph diagram,

Table G,4. Mach number and pressure ratio as functions of angle of expansion.

N	ν	M	$\frac{p}{p^*}$	μ	N	ν	M	$\frac{p}{p^*}$	μ
1000	0°	1.0000	1.00000	90°	946	54°	3.2263	.03669	18° 3′
999	1°	1.0813	.90724	67°37′	945	55°	3.2866	.03375	17°43′
998	2°	1.1327	.85100	61°59′	944	56°	3.3455	.03103	17°24′
997	3°	1.1764	.80486	58°13′	943	57°	3.4065	.02843	17° 4′
996	4°	1.2185	.76192	55° 9′	942	58°	3.4682	.02605	16°45′
995	5°	1.2577	.72329	52°39′	941	59°	3.5300	.02382	16°27′
994	6°	1.2938	.68896	50°37′	940	60°	3.5945	.02175	16° 9′
993	7°	1.3298	.65588	48°45′	939	61°	3.6606	.01984	15°51′
992	8°	1.3648	.62488	47° 6′	938	62°	3.7283	.01838	15°33′
991	9°	1.4000	.59484	45°35′	937	63°	3.7975	.01643	15°16′
990	10°	1.4350	.56608	44°10′	936	64°	3.8699	.01489	14°58′
989	11°	1.4690	.53921	42°54′	935	65°	3.9425	.01350	14°42′
988	12°	1.5032	.51325	41°42′	934	66°	4.0175	.01223	14°25′
987	13°	1.5368	.48875	40°36′	933	67°	4.0950	.01103	14° 8′
986	14°	1.5706	.46512	39°33′	932	68°	4.151	.00990	13°56′
985	15°	1.6046	.44233	38°33′	931	69°	4.256	.00889	13°35′
984	16°	1.6382	.42075	37°37′	930	70°	4.340	.00800	13°19′
983	17°	1.6721	.39995	36°44′	929	71°	4.427	.00720	13° 3′
982	18°	1.7064	.37981	35°53′	928	72°	4.517	.00646	12°47′
981	19°	1.7408	.36053	35° 4′	927	73°	4.610	.00577	12°32′
980	20°	1.7753	.34210	34°16′	926	74°	4.704	.00510	12°16′
979	21°	1.8098	.32454	33°32′	925	75°	4.802	.00450	12° 1′
978	22°	1.8446	.30768	32°49′	924	76°	4.905	.00400	11°46′
977	23°	1.8796	.29152	32° 8′	923	77°	5.010	.00354	11°31′
976	24°	1.9148	.27610	31°29′	922	78°	5.119	.00313	11°16′
975	25°	1.9503	.26136	30°50′	921	79°	5.231	.00277	11° 1′
974	26°	1.9860	.24727	30°14′	920	80°	5.348	.002413	10°47′
973	27°	2.0235	.23324	29°37′	919	81°	5.472	.002106	10°32′
972	28°	2.0600	.22034	29° 2′	918	82°	5.599	.001835	10°17′
971	29°	2.0968	.20805	28°29′	917	83°	5.730	.001598	10° 3′
970	30°	2.1320	.19690	27°58′	916	84°	5.865	.001383	9°49′
969	31°	2.1726	.18483	27°24′	915	85°	6.007	.001193	9°35′
968	32°	2.2105	.17430	26°54′	914	86°	6.152	.001023	9°21′
967	33°	2.2485	.16400	26°24′	913	87°	6.307	.000879	9° 7′
966	34°	2.2879	.15406	25°55′	912	88°	6.472	.000751	8°53′
965	35°	2.3290	.14480	25°25′	911	89°	6.642	.000638	8°39′
964	36°	2.3693	.13582	24°58′	910	90°	6.821	.000540	8°26′
963	37°	2.4107	.12735	24°30′	909	91°	7.008	.000455	8°12′
962	38°	2.4525	.11970	24° 4′	908	92°	7.202	.000381	7°59′
961	39°	2.4946	.11177	23°37′	907	93°	7.407	.000318	7°45′
960	40°	2.5373	.10454	23°12′	906	94°	7.622	.000265	7°32′
959	41°	2.5812	.09770	22°48′	905	95°	7.853	.000220	7°19′
958	42°	2.6256	.09117	22°23′	904	96°	8.092	.000182	7° 6′
957	43°	2.6712	.08500	21°59′	903	97°	8.343	.000149	6°53′
956	44°	2.7175	.07917	21°36′	902	98°	8.616	.0001194	6°40′
955	45°	2.7650	.07359	21°13′	901	99°	8.902	.0000965	6°27′
954	46°	2.8118	.06848	20°50′	900	100°	9.210	.0000770	6°13′
953	47°	2.8601	.06363	20°28′	899	101°	9.535	.0000614	6° 1′
952	48°	2.9112	.05895	20° 5′	898	102°	9.881	.0000481	5°48′
951	49°	2.9613	.05466	19°44′	897	103°	10.260	.0000377	5°36′
950	50°	3.0122	.05065	19°23′	896	104°	10.665	.0000294	5°22′
949	51°	3.0650	.04680	19° 2′	895	105°	11.088	.0000224	5°16′
948	52°	3.1188	.04310	18°42′	894	106°	11.552	.0000168	4°58′
947	53°	3.1735	.03985	18°22′					

or more conveniently from Table G,4. In the table N corresponds to $C_1 + C_2/2$, ν is the value $\pm(\theta - \theta^*)$, p/p^* the ratio between the local pressure and critical pressure, and μ the Mach angle.

For example, in Fig. G,4d, the point A is on the characteristic lines 588 and 380; therefore, the intensity of velocity is given by 968 and the direction by 208. The velocity at A is inclined $+8°$ with respect to the reference direction, 200, and the Mach number is 2.21 (corresponding to $N = 968$ in the table). The tangent to the line 588 at A is inclined

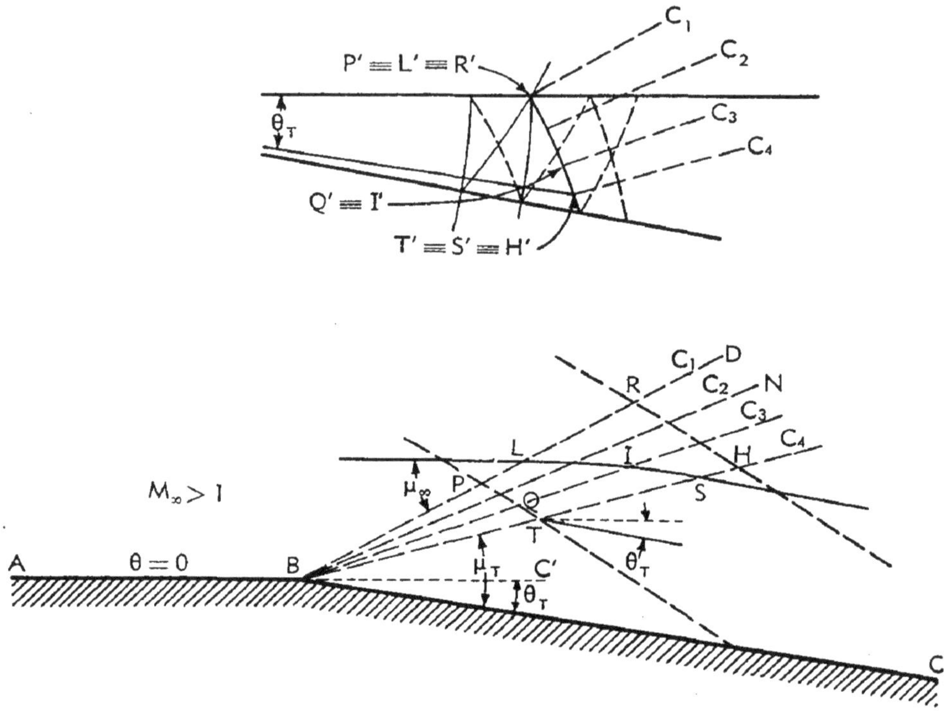

Fig. G,5a. The flow around a corner in the physical and hodograph planes.

$\theta + \mu = 8° + 26°54'$ with respect to the direction 200, and the tangent to the line 380 is inclined $\theta - \mu = 8° - 26°54'$ with respect to the same direction. Similarly, at B the velocity is defined by two numbers 586 and 378. The direction is 208 and the intensity is 964; therefore, $\theta_B = 8°$ and $M_B = 2.37$.

When increments of θ smaller than one degree are required in the calculations, 10,000 and 2,000 can be assumed as basic numbers in place of 1,000 and 200, so that each unit corresponds to one tenth of a degree.

G,5. Prandtl-Meyer Flow. Consider a uniform supersonic stream flowing tangent to the wall AB (Fig. G,5a) and assume that at the point B a change of stream direction from AB to BC is produced. Assume that the change of direction at B is produced gradually by infinitesimal rotations $d\theta$ of the wall BC' from the original direction AB to the final direction BC. The first disturbance produced at B is propagated in a region of the flow limited by the Mach line BD inclined at the Mach angle μ_∞ with respect to the undisturbed stream direction. Since the undisturbed flow is uniform, the Mach line BD is a straight line and the infinitesimal disturbance $d\theta$, transmitted along BD, produces a flow of constant velocity behind BD. Inasmuch as the flow behind BD is still uniform, the next infinitesimal disturbance $d\theta$ produced at B is transmitted along a Mach line BN more inclined than BD, but still straight. In a similar manner, all the infinitesimal disturbances are transmitted along Mach lines that are straight lines and originate at B.

Now consider a characteristic line PQT of the second family. Along PQT the direction θ changes and the variation of Mach number along PQT can be determined as a function of local direction by means of Eq. 3-11, where the plus sign must be assumed. If BT is the characteristic line which transmits the last disturbance $d\theta$, the direction θ_T along BT is given by the direction of BC; therefore, knowing the value θ_T, the Mach number in the zone BTC where the flow is uniform can be obtained with the use of Eq. 3-11. Knowing the Mach number M_T, the Mach angle μ_T is determined and the inclination of the line BT, equal to $\theta_T + \mu_T$, is obtained (θ_T in Fig. G,5a is negative). For each value θ_Q of θ between θ_P and θ_T (corresponding respectively to the directions AB and BC), the values M_Q and μ_Q can be determined from Eq. 3-11 and the direction BQ inclined at $\theta_Q + \mu_Q$ can be obtained. Since the velocity is constant along any characteristic line of the first family, the velocities at L and P or S and T are equal. As a result, the relation between θ and M (Eq. 3-11) determined along the characteristic line PQT also applies along any other line LIS drawn in the flow field, and can also be used along each streamline LS. In the hodograph plane all the streamlines LS, characteristic line PT, and all other lines that cross the characteristic lines BD and BT are represented by the same portion of one epicycloid, $P'T'$. If the value of θ_T increases, the Mach number along BC increases and the pressure decreases. The maximum value of θ_T possible for the condition of the stream still attached to the wall BC corresponds to the condition of zero pressure along BC. For this condition, the velocity along BC is equal to the limiting velocity. The value of $(\theta_T)_{max}$ is equal to the value of $(\theta_{max} - \nu)$, where ν is the deviation required to expand along a characteristic line of the flow from $M = 1$ to the value M_∞ of the undisturbed stream and is the value given in Table G,4. The maximum expansion θ_{max} for $\gamma = 1.40$ is equal to 130.5°. With the hypothesis of perfect continuous medium, the space behind the line from the corner inclined at $(\theta_{max} - \nu)$ with respect to the undisturbed stream direction cannot be reached by the flow. This result is related to the assumption of a continuous medium, which is not physically valid when the static pressure approaches zero.

In Fig. G,5b, the streamlines and the characteristic lines of the first and second families are shown for an expansion around the corner from $M = 1$ to a vacuum. The streamlines are shown as continuous lines, the characteristic lines of the first family as dashed lines, and the characteristic lines of the second family as dash-dot lines.

The flow around the corner can be analyzed without using the characteristic solution, as has already been noted in E,3. By means of simple

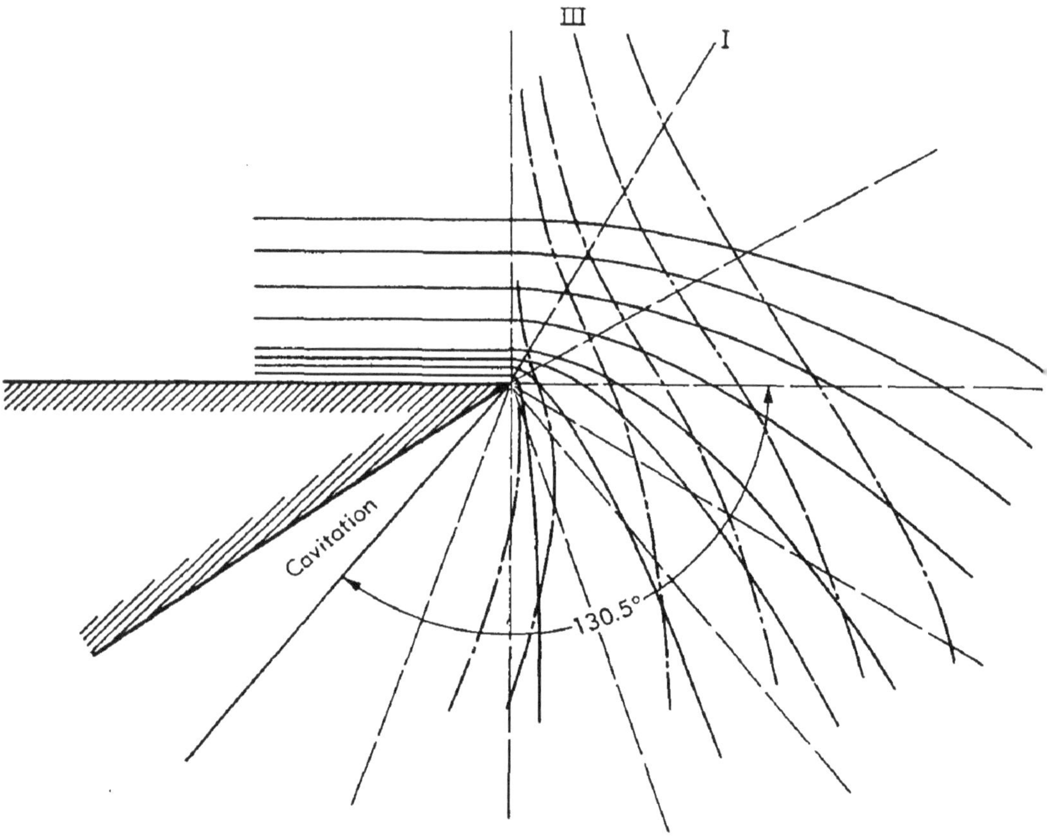

Fig. G,5b. The flow field around a corner. Expansion to zero pressure.

similitude considerations, it can be shown that the flow field has conical properties with respect to the apex at the corner (cf. D,13; also Art. 15 below); therefore, the velocity and pressure are constant along each radius BN originating at the corner B of Fig. G,5a.

Use polar coordinates r and ψ, and call p the static pressure, ρ the density, and v_r and v_n, respectively, the velocity components along the radius and in the normal direction to the radius (Fig. G,5c). Because conical flow exists

$$\frac{\partial p}{\partial r} = 0, \qquad \frac{\partial v_r}{\partial r} = 0, \qquad \frac{\partial v_n}{\partial r} = 0, \qquad \frac{\partial \rho}{\partial r} = 0 \tag{5-1}$$

The law of continuity is

$$\frac{\partial}{\partial r}(r\rho v_r) + \frac{\partial}{\partial \psi}(\rho v_n) = 0 \tag{5-2}$$

and Euler's equations are

$$v_r \frac{\partial v_r}{\partial r} + \frac{v_n}{r}\frac{\partial v_r}{\partial \psi} - \frac{v_n^2}{r} = -\frac{1}{\rho}\frac{\partial p}{\partial r} \tag{5-3}$$

$$v_r \frac{\partial v_n}{\partial r} + \frac{v_n}{r}\frac{\partial v_n}{\partial \psi} + \frac{v_r v_n}{r} = -\frac{1}{\rho r}\frac{\partial p}{\partial \psi} \tag{5-4}$$

Eq. 5-3 shows that for the conical flow considered

$$\frac{dv_r}{d\psi} = v_n \tag{5-5}$$

Then, combining Eq. 5-2 with Eq. 5-4, and using the conditions (5-1),

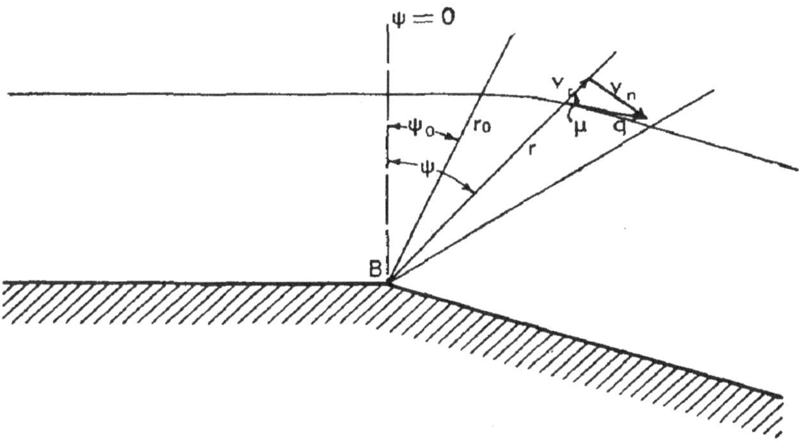

Fig. G,5c. The flow around a corner, considered as conical flow.

the following is obtained

$$\left(\frac{dv_n}{d\psi} + v_r\right)\left(1 - \frac{v_n^2}{a^2}\right) = 0 \tag{5-6}$$

However, from Eq. 5-4 there results

$$\left(\frac{dv_n}{d\psi} + v_r\right) v_n = -\frac{1}{\rho}\frac{dp}{d\psi} \tag{5-7}$$

Therefore, in the zone where $dp/d\psi \neq 0$ it must be that

$$1 - \frac{v_n^2}{a^2} = 0$$

or

$$v_n = a \tag{5-8}$$

Eq. 5-8 states that in the region where $dp/d\psi \neq 0$ each radial line starting from the corner B of Fig. G,5c is a Mach wave, as has been shown in the preceding analysis.

Using Eq. 5-8 the law of conservation of energy becomes

$$\frac{1}{2}(v_r^2 + v_n^2) + \frac{\gamma}{\gamma - 1}\frac{p}{\rho} = \frac{1}{2}\left(v_r^2 + \frac{\gamma + 1}{\gamma - 1}v_n^2\right) = \frac{1}{2}q_{max}^2 \qquad (5\text{-}9)$$

or from Eq. 5-5

$$\left(\frac{dv_r}{d\psi}\right)^2 = \frac{\gamma - 1}{\gamma + 1}(q_{max}^2 - v_r^2) \qquad (5\text{-}10)$$

Eq. 5-10 can be integrated and the following expression is obtained

$$v_r = q_{max} \sin \sqrt{\frac{\gamma - 1}{\gamma + 1}}\,\psi \qquad (5\text{-}11)$$

where $\psi = 0$ when $v_r = 0$ (Fig. G,5c). From Eq. 5-5

$$v_n = \sqrt{\frac{\gamma - 1}{\gamma + 1}}\,q_{max} \cos \sqrt{\frac{\gamma - 1}{\gamma + 1}}\,\psi \qquad (5\text{-}12)$$

If $\sqrt{\frac{\gamma - 1}{\gamma + 1}}\,\psi = \frac{\pi}{2}$ or $\psi = \sqrt{\frac{\gamma + 1}{\gamma - 1}}\frac{\pi}{2} = 130.45° + \frac{\pi}{2}$, then $v_n = 0$ and $v_r = q_{max}$. From Eq. 5-11 and 5-12 there results

$$q^2 = q_{max}^2\left[1 - \frac{1}{\gamma + 1}\left(1 + \cos 2\sqrt{\frac{\gamma - 1}{\gamma + 1}}\,\psi\right)\right] \qquad (5\text{-}13)$$

Because the component v_n of the total velocity is equal to a (Eq. 5-8)

$$\frac{v_n}{v_r} = \tan \mu \qquad (5\text{-}14)$$

and the inclination of the streamline is (Fig. G,5c)

$$-\theta = \mu + \psi - \frac{\pi}{2} \qquad (5\text{-}15)$$

(On Fig. G,5c θ is negative)

The equation of any streamline is

$$\frac{1}{r}\frac{dr}{d\psi} = \cot \mu = \sqrt{\frac{\gamma + 1}{\gamma - 1}} \tan \sqrt{\frac{\gamma - 1}{\gamma + 1}}\,\psi \qquad (5\text{-}16)$$

Eq. 5-16, after integration, gives

$$r = r_0\left[\frac{\cos \sqrt{\frac{\gamma - 1}{\gamma + 1}}\,\psi_0}{\cos \sqrt{\frac{\gamma - 1}{\gamma + 1}}\,\psi}\right]^{\frac{\gamma+1}{\gamma-1}} \qquad (5\text{-}17)$$

where r_0 and ψ_0 correspond to the initial conditions along the Mach wave from the corner for the undisturbed stream (Fig. G,5c).

G,6. Flow Along a Curved Surface. The considerations developed for the flow around a sharp corner can be extended to the flow along a two-dimensional curved surface having a uniform velocity field along an initial characteristic line. Consider a supersonic stream flowing tangent to a curved surface *ABC* (Fig. G,6a). Assume that the flow is originally uniform and parallel to the flat surface *AB*. At *B* the surface starts to curve and an infinitesimal disturbance is produced which is transmitted along the characteristic line *BD* from *B*. The flow behind *BD*

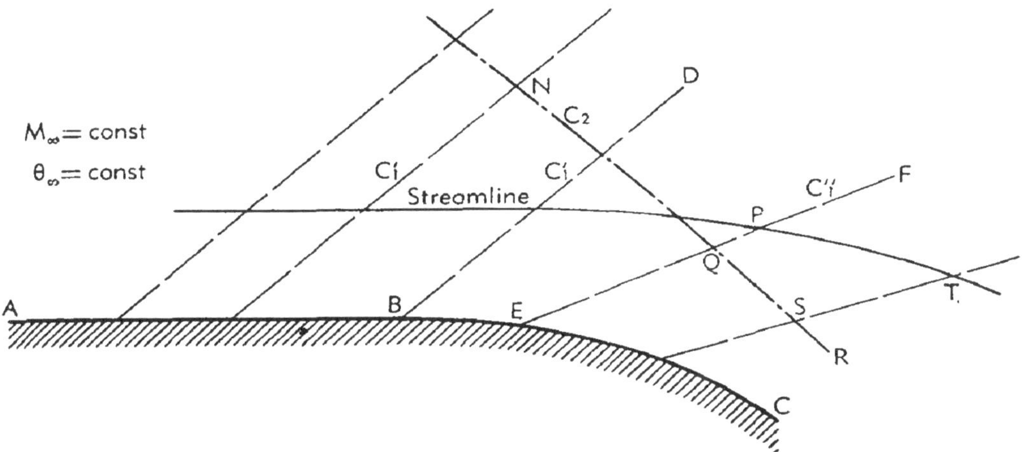

Fig. G,6a. The flow along a curved surface.

is still uniform; therefore, the same considerations can be applied to any successive characteristic line which transmits, in the flow field, the infinitesimal disturbances produced by the change in direction of the boundary streamline, *BC*. All the characteristic lines of the same family are straight lines and lines of constant velocity.

Along the characteristic line *NR* of the other family, a variation of Mach number occurs, and Eq. 3-11, which relates the local value of *M* to the stream direction, can be used. Eq. 3-11 is valid only along the line *NR*. However, for this particular case, the flow along any characteristic line *EF* is constant; therefore, the flow direction at *P* and *Q*, or *S* and *T*, is the same. Consequently, Eq. 3-11 can be used along any line *PT* that crosses the straight characteristic lines and in particular along a streamline. Since the boundary streamline *ABC* gives the stream direction, Eq. 3-11 permits the determination of the entire flow field.

The possibility of using Eq. 3-11 along lines other than characteristic lines, for which the equation has been derived, is general for all the problems distinguished by the property that all characteristic lines of one family are straight lines. This property exists when the initial supersonic flow is uniform and the boundary conditions are such that disturbances are transmitted in the flow only along one family of waves. These problems are often called simple wave problems (E,3).

These problems can be analyzed by means of Eq. 3-11 or by means of an epicycloid diagram which gives the velocity variation along the

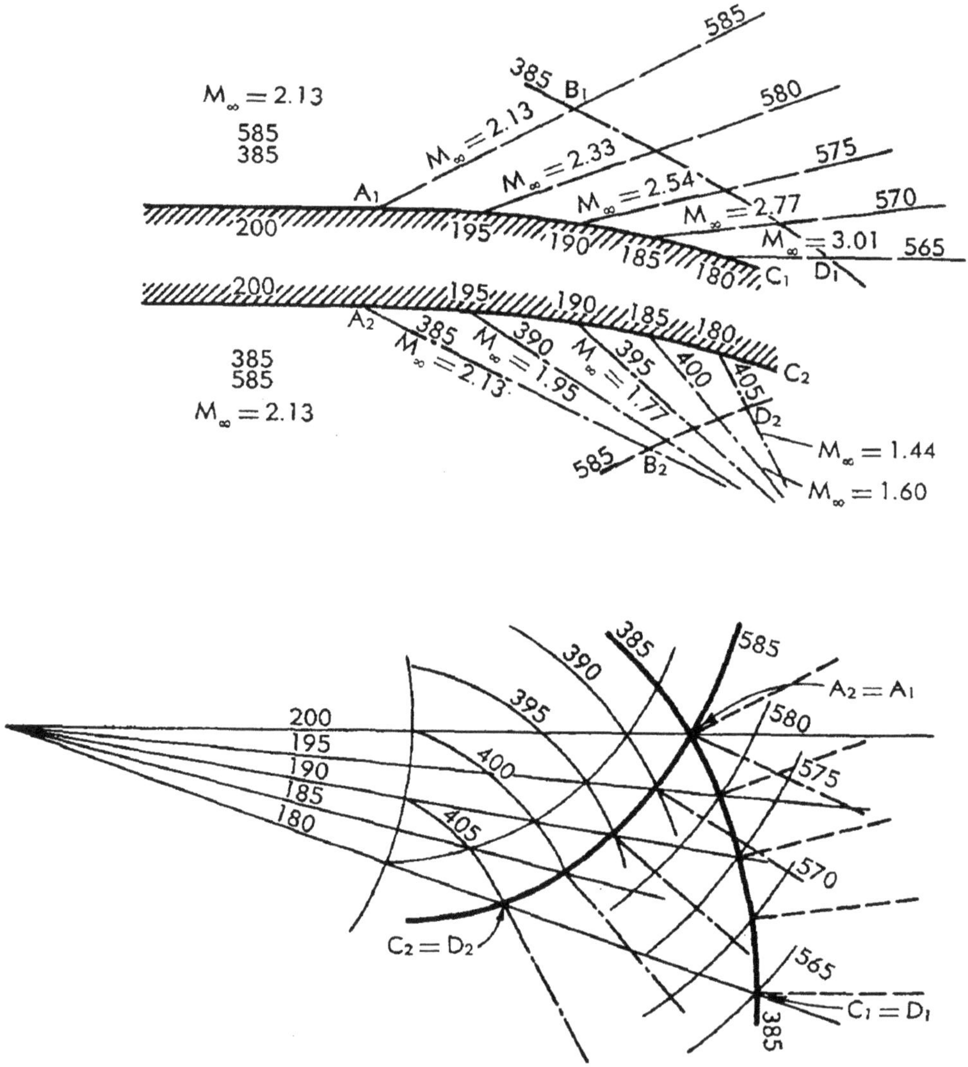

Fig. G,6b. The flow field along concave and convex surfaces in the physical plane (top) and in the hodograph plane (bottom).

characteristic line *NR*. This epicycloid diagram represents in the hodograph plane, in a unique curve, all the characteristic lines which belong to the family other than that of the straight characteristic lines, all the streamlines, and every other curve crossing the straight characteristic lines.

The proper choice of sign in Eq. 3-11 or of the epicycloid representing the streamline can be made by recalling that the flow along the streamline is equal to the flow along the characteristic lines which are not straight. For example, consider the flow fields of concave and convex surfaces having the same contour, as shown in the top part of Fig. G,6b. For the surface A_1C_1, the characteristics of the first family are the straight characteristic lines. Therefore, the flow variations occur along the characteristic lines of the second family B_1D_1, and the variations of flow along A_1C_1 are given by Eq. 3-11, with the plus sign. In the hodograph plane (Fig. G,6b, bottom) the streamline A_1C_1 is represented by an epicycloid of the second family A_1C_1. The straight characteristic lines in the physical plane are parallel to the normals to the epicycloid A_1C_1, in the hodograph plane.

For the surface A_2C_2, the characteristic lines of the second family are the straight characteristic lines. The flow variations along A_2C_2 are equal to the flow variations along B_2D_2 and are given by Eq. 3-11 with the minus sign. In the hodograph plane, the streamline is represented by an epicycloid of the first family A_2C_2 and the normals to A_2C_2 are parallel to the straight characteristic lines in the physical plane. Along the surface A_2C_2 the flow velocity decreases and the envelope of compression waves produced along A_2C_2 tends to produce a shock wave (see E,3). The Mach number at each point of the flow field can be obtained from the hodograph plane, or from table G,4 as a function of the number $N = (C_1 + C_2)/2$, or from Eq. 3-11.

G,7. The Linearized Theory Derived from the Characteristics Theory. In Sec. C the linearized theory for two-dimensional flow is derived. The same results obtained there can be derived by introducing in the characteristics theory the simplifications used by the linearized theory. Consider a two-dimensional flow field having the undisturbed stream direction parallel to the x axis. In the approximation of the linearized theory

$$(M_\infty^2 - 1)\frac{\partial u}{\partial x} = \frac{\partial v}{\partial y} \tag{7-1}$$

where M_∞ is the undisturbed Mach number. When the analysis performed in Art. 2 is applied to Eq. 7-1, since

$$H = -(M_\infty^2 - 1), \qquad L = 1, \qquad K = 0 \tag{7-2}$$

there results

$$\lambda_{1,2} = \pm \frac{1}{\sqrt{M_\infty^2 - 1}} = \pm \tan \mu_\infty \tag{7-3}$$

In the approximation of linearized theory the characteristic lines are all parallel and inclined at the free stream Mach angle to the undisturbed velocity. Consequently, the envelope shocks discussed in connection with the flow along surface A_2C_2 of Fig. G,6b are not found in this approximation.

In this approximation, if v' and u' are the disturbance velocities and U the free stream velocity, then

$$\tan\theta = \frac{v}{u} = \frac{v'}{U + u'} = \frac{v'}{U} \tag{7-4}$$

Therefore, by means of Eq. 3-2 and 3-3, Eq. 3-7 becomes

$$\frac{dq}{U} = \pm d\theta \tan\mu_\infty \tag{7-5}$$

or

$$\frac{\Delta p}{\frac{1}{2}\rho_\infty U^2} = \pm\frac{2\Delta\theta}{\sqrt{M_\infty^2 - 1}} \tag{7-6}$$

Eq. (7-6) was obtained by Ackeret for linearized two-dimensional flow.

G,8. Numerical and Graphical Methods for General Two-dimensional Potential Flow Problems. In the preceding articles, "simple wave" problems are analyzed which involve variation of flow properties along only one family of Mach lines passing through each point of the flow, while the flow remains constant along the other family of Mach lines. For these problems, one of the two equations (3-11) can be used along each streamline and the flow properties at every point are immediately determined from the boundary conditions.

In many practical problems the boundary conditions are such that changes of flow properties occur along both families of Mach lines. In these cases, both equations (3-11) must be used, and only a step-by-step numerical evaluation of the problem is possible. The basic concept of the numerical step-by-step procedure is the following:

Assume that the flow properties are known along a line AF (Fig. G,8a, left). Through each point A, B, C, D, E, F of the line AF two Mach lines pass, one of the first family ($a_1, b_1, c_1, d_1, e_1, f_1$) and one of the second family ($a_2, b_2, c_2, d_2, e_2, f_2$). Since the flow properties at the points of the line AF are known, the constants of Eq. 3-11 for each characteristic line $a_1 \ldots f_1, a_2 \ldots f_2$, or the characteristic numbers at each point of AF, are also known. Consequently, from Eq. 3-11 applied along two Mach lines of opposite families, as for example e_1 and b_2 of Fig. G,8a, two expressions between θ and M are obtained which give the value of M and θ at the intersection point Q as functions of θ and M at the points B and E. In this way, the velocity vector at each point lying within the characteristic lines a_1a_2, f_1f_2 can be determined. Although the velocity vector at the intersection points is determined in the hodograph plane, the position of the corresponding point in the physical plane is unknown because both families of characteristic lines are curved. The determination of the position of the intersection point in the physical plane requires a step-by-step construction of the characteristic lines obtained as an envelope of the tangents at each point. The direction of the tangent at each point of the physical plane is known because the directions of tangents to the characteristic lines are given by the normals to the epicycloids of the opposite family of the characteristic considered, or by Eq. 3-1 and 3-3.

The step-by-step construction of the characteristic lines can be performed by assuming that for small increments the characteristic curve

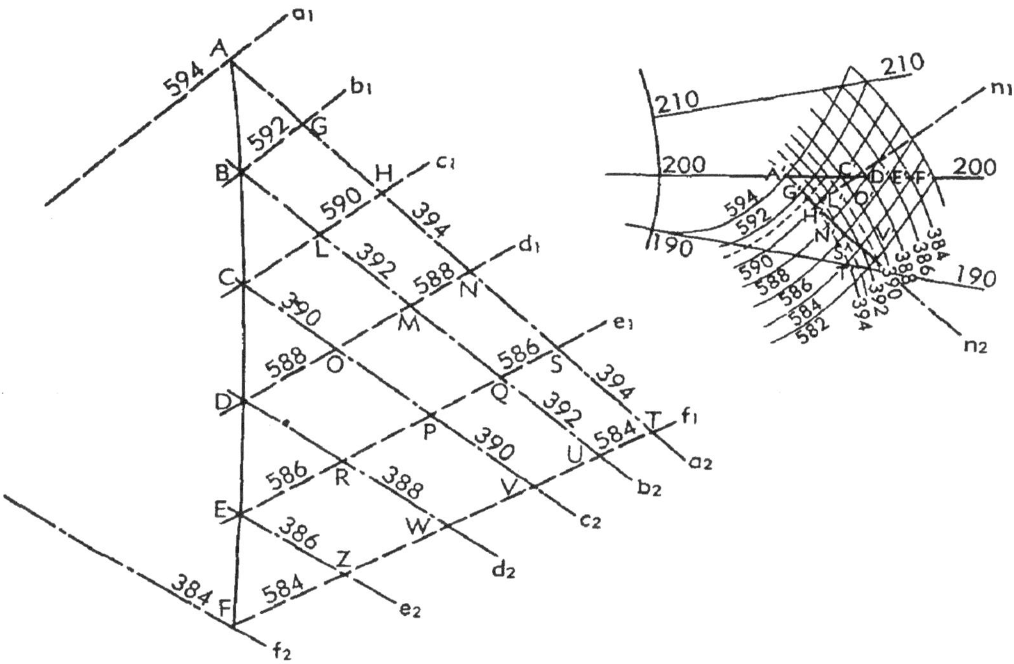

Fig. G,8a. Determination of the flow field for general two-dimensional potential flow problems. Lattice procedure. Left, physical plane; right, hodograph plane.

between the two end points can be represented by a straight line. This line is inclined with respect to the reference axis at an angle $\tan^{-1} \lambda$, equal to the average between the two values at the end points. (The values of $\tan^{-1} \lambda$ at the end points are known because θ and M are known (Eq. 3-1 and 3-3).)

In practical applications the characteristic net can be obtained by means of graphical procedures or by using the characteristic numbers and tabulated values (Table G,4).

The flow field at the points A, B, C, D, E, F defines at each point the two characteristic numbers. In the example of Fig. G,8a (left) the Mach number at C is 1.775, and the streamline is parallel to the reference axis; therefore, the sum of the two characteristic numbers at C must be 980 (see Table G,4) and the differences between the number of the characteristic line of the first family and the number of the characteristic line of the second family must be 200. The numbers at C are 590, 390.

The characteristic numbers at each point G, L, and O of the net are given by the numbers of the two crossing characteristic lines. For example, at L (Fig. G,8a, left) the characteristic numbers are 590, 392; the Mach number corresponds to 982 and it is 1.706; the direction corresponds

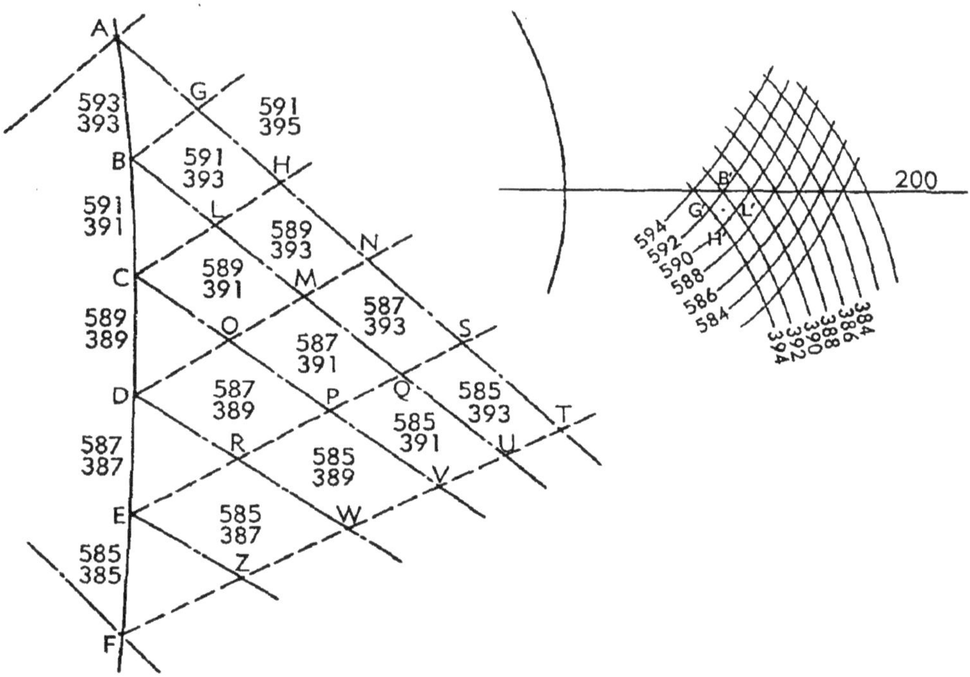

Fig. G,8b. Determination of the flow field for general two-dimensional potential flow problems. Mesh procedure.

to 198 and it is $\theta_L = -2°$. The Mach angle μ_C at C is 34°16′ and at H is 35°53′; therefore, the line CL is defined by

$$2\,\frac{y_L - y_C}{x_L - x_C} = \tan\,(\theta_C + \mu_C) + \tan\,(\theta_L + \mu_L) = \tan 34°16' + \tan 33°53' \qquad (8\text{-}1)$$

Similarly, line BL is defined by

$$2\,\frac{y_L - y_B}{x_L - x_B} = \tan\,(\theta_B - \mu_B) + \tan\,(\theta_L - \mu_L) = -\tan 35°53' - \tan 37°53' \qquad (8\text{-}2)$$

Eq. 8-1 and 8-2 define the position of point L in the physical plane. The construction of the whole characteristic net proceeds in a similar way. If a graphical procedure is used, any line HL of Fig. G,8a (left) is parallel to the normal n_1 to the epicycloid of the second family at the mid-point between H' and L' in the hodograph diagram (Fig. G,8a, right). Similarly, any line *HG* is parallel to the normal n_2 to the epicycloid of the first family at the mid-point between H' and G'.

In the preceding analysis, the procedure used to determine the flow at each point *G*, *L*, etc. is known as the lattice procedure. In many practical analyses, it is assumed that the flow field can be divided into many small regions limited by two adjacent pairs of characteristic lines, and the average values of the flow properties in each mesh is determined. This procedure is called the mesh procedure. In this case, two numbers are assigned to each mesh, which define the average value of the velocity vector in the mesh, and in passing from one mesh to the adjacent mesh one of the two numbers must change. In this scheme, the assumption is implied that the gradual variation of flow properties existing in the flow field is substituted by a finite number of variations, of finite strength, transmitted along lines which divide the adjacent meshes. These lines are drawn by assuming for their inclination the average value between the corresponding values in the two meshes. In Fig. G,8b, the same flow field analyzed in Fig. G,8a with the lattice procedure is analyzed with the mesh procedure. The mesh *GBLH* is defined by the numbers 591 and 393. The region *AB* is defined by the numbers 593 and 393. The numbers that define the line *BG* are 592 and 393; therefore, the line *BG* is inclined at $\theta + \mu = -1° + 38°33'$. As is shown from the analysis of Fig. G,8a or G,8b, all the region of flow between the characteristic lines a_1a_2 and f_1f_2 is defined by the boundary conditions given along the line *AF*.

G,9. Boundary Conditions: Solid Boundary. When the Mach waves reach a region where boundary conditions are prescribed, the waves are reflected as waves of the opposite family. The intensity and sign of the reflected waves depend on the type of boundary prescribed and can be determined by considerations similar to those presented previously.

Consider first a solid boundary. The solid boundary corresponds to a streamline; therefore, the stream direction at the boundary is given. Assume that a family of waves produced from a curved surface *AN* meets a solid boundary *EF* as shown in Fig. G,9a. The change in direction from *A* to *N* is transmitted along the waves of the first family. If the solid boundary is represented by a straight wall *EF*, when the waves from *AN* meet the wall *EF* they are reflected as waves of the second family. Along the characteristic line *CD* the stream direction changes from θ_A to θ_B; therefore, if no reflected waves were produced along *CL*, the streamline from *C* to L_1 would turn an angle of $\theta_B - \theta_A$. However, the streamline from *C* must remain straight; therefore, the reflected waves produced along *CL* must produce a deviation of opposite sign from the deviation produced by the waves generated along *AB*. The waves are of the opposite family; therefore, the variation of Mach number across the reflected waves has the same sign as the variation across the incoming wave. If the deviation AB produces an expansion, the reflected waves between CH and LM also produce an expansion. Reflection of incoming waves at a straight wall produces waves of the opposite family having the same intensity of deflection and the same sign.

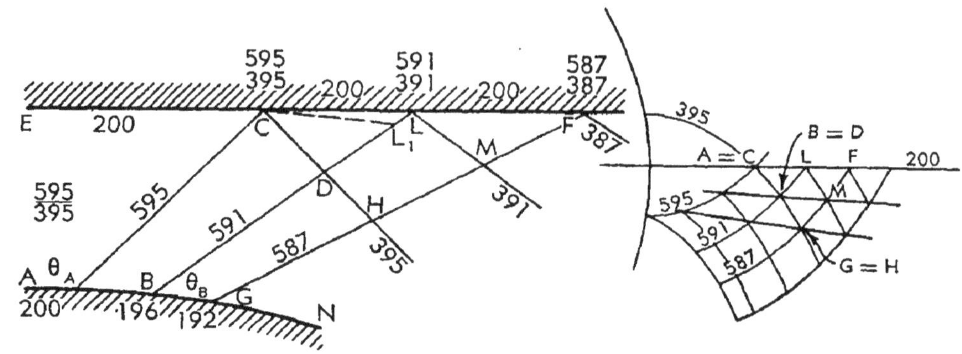

Fig. G,9a. Wave reflections at a solid straight wall.

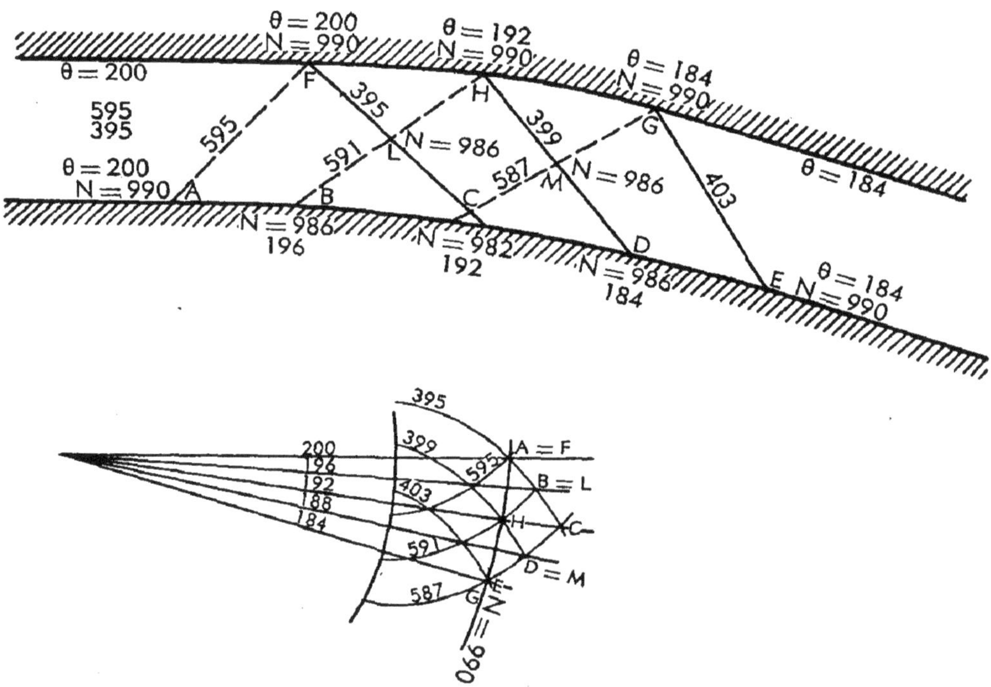

Fig. G,9b. Wave reflections at a curved wall.

If the rigid boundary EF is a curved surface, the reflection can be considered as obtained by superimposing on the waves reflected by a flat wall the waves originated locally by a curved wall; therefore, the reflections can be of the same or opposite sign depending on the curvature of the wall.

The reflection of the waves can be analyzed easily by means of the characteristic numbers because the difference of the characteristic numbers of the two characteristics which meet at the boundary is fixed at the boundary by its direction; therefore, from the incoming characteristic line the number of the reflected characteristic line can be determined. In Fig. G,9a, the wall is straight; therefore, the difference of the two characteristic numbers must be constant. The number of the reflected characteristic wave is obtained by subtracting the direction of the wall from the number of the incoming characteristic wave. The number N, sum of the two characteristic numbers, decreases moving downstream along the wall; therefore, the velocity increases. The wave pattern can be seen easily in the hodograph plane. In Fig. G,9b, the reflection from a curved wall is shown. The curvature chosen for the wall is such that the velocity along the curved wall remains constant; therefore, incoming expansion waves are reflected as compression waves producing opposite variation of velocity and variation of direction of the same sign and equal intensity.

G,10. Boundary Conditions: Free Boundary. The boundary conditions can be given by fixing the pressure, instead of direction, along a

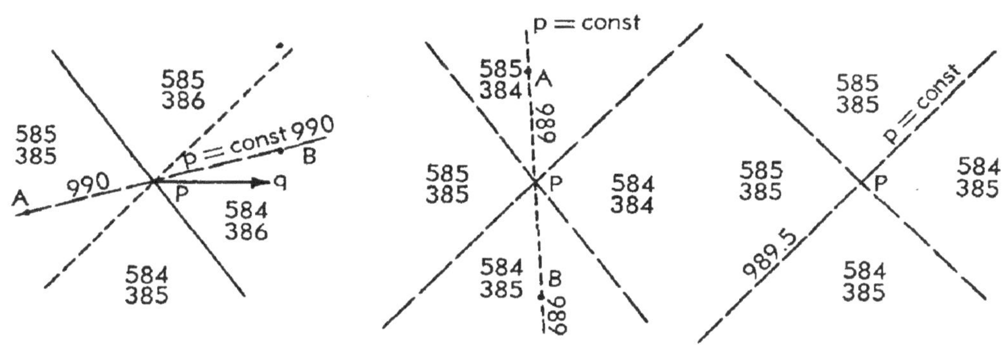

Fig. G,10a. Different types of wave reflections at a constant pressure boundary line.

boundary, as occurs when a supersonic jet exhausts into constant pressure ambient conditions or into another stream.

Since in isentropic transformation the pressure is a single-valued function of the Mach number, when the pressure is given along the boundary the boundary Mach number distribution is immediately determined.

Consider first the case in which the pressure is maintained constant along the given boundary. Each wave meeting the boundary is reflected as a wave of the opposite family. This reflected wave at the boundary can be of the same or opposite sign depending on whether the line of constant pressure falls inside the internal or external angle of the characteristic lines passing at the points of the boundary. Consider two Mach lines crossing at a point P of the flow (Fig. G,10a). The internal angle is the angle bisected by the streamline and the external angle is the angle bisected by the potential line. When the line of constant pressure falls inside the internal angle (Fig. G,10a, left), the reflected wave must be of opposite sign to the sign of the incoming wave because the pressure at the points A and B on the opposite side of the crossing point must be

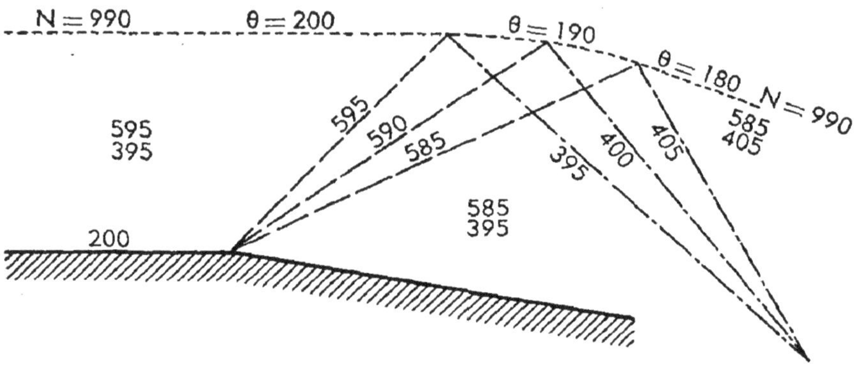

Fig. G,10b. Wave reflections at a free boundary.

equal. When the line of constant pressure falls inside the external angle (Fig. G,10a, center) incoming and reflected waves must be of the same sign, for should they be of opposite sign, the pressure at the points A and B would not be the same.

When the line of constant pressure coincides with a characteristic line, the flow becomes a single wave flow with one family of waves having zero strength (Fig. G,10a, right).

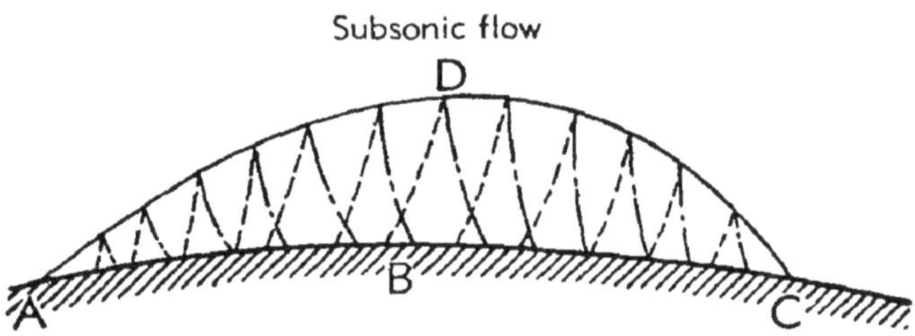

Fig. G,10c. Wave reflections at the sonic line. Subsonic stream with localized supersonic region.

Along the line of constant pressure the Mach number is constant; therefore, the sum N of the two characteristic numbers corresponding to the two characteristic lines crossing each other along the constant pressure line remains constant. The analysis of the reflected waves in all cases can easily be performed by means of the characteristic numbers system. In Fig. G,10b, the reflections of expansion waves from a free boundary are shown. The free boundary is a streamline; therefore, the reflected waves are of opposite sign from the incoming waves.

An important case of reflection of waves from a constant pressure boundary is encountered in transonic phenomena. The line of constant pressure considered in the transonic phenomena is the sonic line. In subsonic flow having a localized supersonic region and smooth flow (no shock waves), the sonic line must intersect each streamline twice; therefore, a sonic line as shown in Fig. G,10c is expected. In Fig. G,10c the line *ABC* represents the solid boundary; therefore, it is a streamline, and line *ADC* is the sonic line that bounds the local supersonic region imbedded in the subsonic stream. Along *AB*, expansion waves are produced which

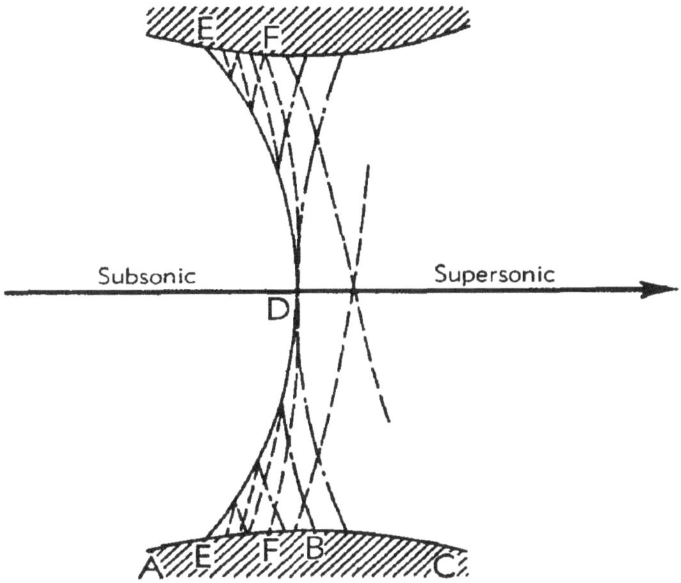

Fig. G,10d. Wave reflections at the sonic line. Transonic stream with supersonic downstream flow.

are reflected at *AD* as compression waves. The compression waves turn the stream in the same direction as the expansion waves and meet the solid wall downstream from the points where the expansion waves originate. These compression waves tend to neutralize the expansion waves that would be produced along *BC* and tend to converge and form shock waves. (Cf. F,13; especially Fig. F,13a and b.)

When a transonic stream having only supersonic downstream flow is considered, the sonic line intersects each streamline only once and appears as in Fig. G,10d in which the flow in a channel is analyzed. The expansion waves from the surface *ABC* are reflected at the sonic line as compression waves. The wave *FD* is the last wave produced along *EB* which meets the sonic line. The portion of the boundary *EF* can change the shape of the sonic line, thus affecting the subsonic part of the stream and changing the upstream flow. This part of the boundary conserves subsonic properties in spite of its local supersonic velocity. The flow downstream of the wave *FD* becomes truly supersonic. The point *F* is on the converging part of the channel.

CHAPTER 3. AXIALLY SYMMETRIC POTENTIAL FLOW

G,11. Numerical Methods. Consider a flow field having axial symmetry and assume a system of cylindrical coordinates whose x axis is coincident with the axis of symmetry. If the flow is potential flow, Eq. 2-22 becomes

$$du + \lambda_2 dv + \frac{v}{y} \frac{1}{\left(1 - \frac{u^2}{a^2}\right)} dx = 0 \tag{11-1}$$

along the characteristic line defined by

$$\lambda_1 = \left(\frac{dy}{dx}\right)_1 = \tan\,(\mu + \theta) \tag{11-2}$$

and

$$du + \lambda_1 dv + \frac{v}{y} \frac{1}{\left(1 - \frac{u^2}{a^2}\right)} dx = 0 \tag{11-3}$$

along the characteristic line

$$\lambda_2 = \left(\frac{dy}{dx}\right)_2 = \tan\,(\theta - \mu) \tag{11-4}$$

Using polar velocity components q and θ defined by Eq. 2-23, Eq. 11-1 and 11-3 become respectively

$$\cot \mu \frac{dq}{q} - d\theta - \frac{\sin \mu \sin \theta}{\cos\,(\theta + \mu)} \frac{dx}{y} = 0 \tag{11-5}$$

and

$$\cot \mu \frac{dq}{q} + d\theta - \frac{\sin \mu \sin \theta}{\cos\,(\theta - \mu)} \frac{dx}{y} = 0 \tag{11-6}$$

The relation between q and μ is given by

$$\frac{q_{max}^2}{q^2} = 1 + \frac{2}{\gamma - 1} \sin^2 \mu \tag{11-7}$$

Eq. 11-5 and 11-6 are similar to the equations used for two-dimensional flow; however, the presence of the point position coordinates x and y in the expressions makes it impossible to obtain a unique analytic relation between q and θ, as was obtained for the two-dimensional problem. For an axially symmetric flow field this relation can be determined only by a step-by-step numerical calculation simultaneous with the step-by-step calculation which gives the shape of the characteristic lines in the physical plane.

The procedure for the step-by-step calculation used can vary, to some extent, depending upon the form used for Eq. 11-5 and 11-6. Only a few typical methods are illustrated briefly here.

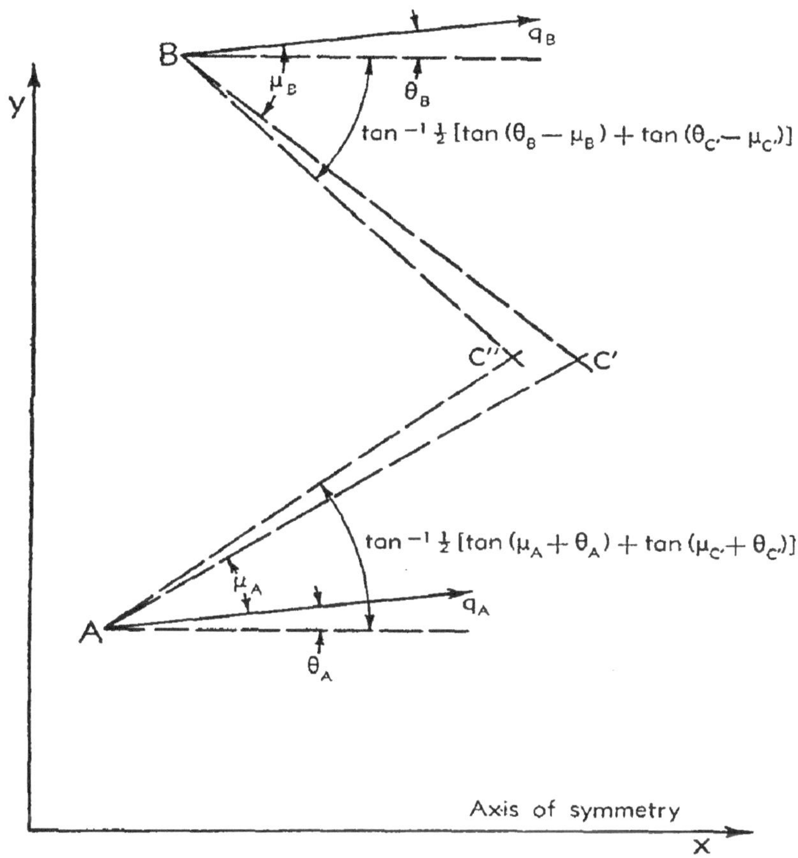

Fig. G,11. Determination of an axially symmetric flow field with a numerical step-by-step procedure.

A numerical procedure used often for the step-by-step calculation of the flow field is the following: Consider two points, A and B, in the flow field where the flow properties are known (Fig. G,11). From Eq. 11-2 and 11-4, the tangents to the characteristic lines of the first family at A, and of the second family at B, can be determined. The two tangents determine a point C' in the meridian plane. The points A and B are chosen close to one another and the lines AC' and BC' are assumed, in the first approximation, coincident with the characteristic lines between A and C' and B and C'. The variation of q as a function of θ and x along the characteristic lines AC' and BC' is obtained by substituting, for the differential equations (11-5) and (11-6), difference equations having coefficients that are constant along the steps AC' and BC' considered.

As a first approximation, the coefficients are assumed equal to the value corresponding to the point A along AC' and to the value corresponding to the point B along BC'. As a result, Eq. 11-5 along AC' becomes in the first approximation

$$\cot \mu_A \frac{q_{C'} - q_A}{q_A} - (\theta_{C'} - \theta_A) - \frac{\sin \mu_A \sin \theta_A}{y_A \cos (\theta_A + \mu_A)} (x_{C'} - x_A) = 0 \quad (11\text{-}8)$$

and Eq. 11-6 along BC' becomes

$$\cot \mu_B \frac{q_{C'} - q_B}{q_B} + (\theta_{C'} - \theta_B) - \frac{\sin \mu_B \sin \theta_B}{y_B \cos (\theta_B - \mu_B)} (x_{C'} - x_B) = 0 \quad (11\text{-}9)$$

From Eq. 11-8 and 11-9 $q_{C'}$ and $\theta_{C'}$ can be determined; in the first approximation

$$q_{C'} = \frac{1}{\dfrac{\cot \mu_A}{q_A} + \dfrac{\cot \mu_B}{q_B}} \left[\theta_B - \theta_A + \cot \mu_A + \cot \mu_B \right.$$
$$\left. + \frac{\sin \theta_A \sin \mu_A}{y_A \cos (\theta_A + \mu_A)} (x_{C'} - x_A) + \frac{\sin \mu_B \sin \theta_B}{y_B \cos (\theta_B - \mu_B)} (x_{C'} - x_B) \right] \quad (11\text{-}10)$$

and

$$\theta_{C'} = \theta_A + \frac{\cot \mu_A}{q_A} (q_{C'} - q_A) - \frac{\sin \mu_A \sin \theta_A}{y_A \cos (\theta_A + \mu_B)} (x_{C'} - x_A) \quad (11\text{-}11)$$

In this first approximation, since the coefficients of the differential equation have been assumed constant and equal to the values at the points A and B, terms of the order $(\Delta L)^2$ have been neglected (where ΔL is the distance between A and C' or B and C') as can be seen from the following considerations. The quantities q and θ of Eq. 11-5 and 11-6 can be expressed along AC' and BC' in series form as

$$\begin{aligned} q_C &= q_A + \left(\frac{dq}{dx}\right)_A dx + \left(\frac{d^2q}{dx^2}\right)_A \frac{(dx)^2}{2} + \cdots \\ \theta_C &= \theta_A + \left(\frac{d\theta}{dx}\right)_A dx + \left(\frac{d^2\theta}{dx^2}\right)_A \frac{(dx)^2}{2} + \cdots \end{aligned} \quad (11\text{-}12)$$

In Eq. 11-8 and 11-9 $q_{C'}$ and $\theta_{C'}$ have been obtained from the expression

$$q_{C'} = q_A + \left(\frac{dq}{dx}\right)_A dx$$

and

$$\theta_{C'} = \theta_A + \left(\frac{d\theta}{dx}\right)_A dx$$

Therefore, terms of the order of $(\Delta x)^2$ have been neglected in Eq. 11-8 and 11-9. However, terms of the order of $(\Delta L)^2$ can be retained if an iteration process is used. Knowing the values of $q_{C'}$ and $\theta_{C'}$ in the first approximation, the lines AC and BC, therefore, the position of C in the second approximation (point C'' of Fig. G,11), can be determined by assuming for the inclination of AC the average between the value of the tangent at A and the value of the tangent at C determined in the first approximation. Then

$$\frac{y_{C''} - y_A}{x_{C''} - x_A} = \frac{1}{2}[\tan(\mu_A + \theta_A) + \tan(\mu_{C'} + \theta_{C'})] \tag{11-13}$$

and similarly

$$\frac{y_{C''} - y_B}{x_{C''} - x_B} = \frac{1}{2}[\tan(\theta_B - \mu_B) + \tan(\theta_{C'} - \mu_{C'})] \tag{11-14}$$

The values of q and θ are obtained in the second approximation at the point C'' by assuming as coefficients of each term of Eq. 11-5 and 11-6, the averages between the values of the coefficients at A and C' for Eq. 11-5 and 11-6 respectively. Eq. 11-5 and 11-6 for the second approximation becomes

$$\frac{1}{2}\left(\frac{\cot\mu_{C'}}{q_{C'}} + \frac{\cot\mu_A}{q_A}\right)(q_{C''} - q_A) - (\theta_{C''} - \theta_A)$$
$$- \frac{1}{2}\left[\frac{\sin\mu_A \sin\theta_A}{y_A\cos(\theta_A + \mu_A)} + \frac{\sin\mu_{C'}\sin\theta_{C'}}{y_{C'}\cos(\theta_{C'} + \mu_{C'})}\right](x_{C''} - x_A) = 0 \tag{11-15}$$

and

$$\frac{1}{2}\left(\frac{\cot\mu_{C'}}{q_{C'}} + \frac{\cot\mu_B}{q_B}\right)(q_{C''} - q_B) + (\theta_{C''} - \theta_B)$$
$$- \frac{1}{2}\left[\frac{\sin\mu_B \sin\theta_B}{y_B\cos(\theta_B - \mu_B)} + \frac{\sin\mu_{C'}\sin\theta_{C'}}{y_{C'}\cos(\theta_{C'} - \mu_{C'})}\right](x_{C''} - x_B) = 0 \tag{11-16}$$

In a similar way, other approximations can be obtained. When the position of the point C does not change in successive iterations, terms of the order of $(\Delta L)^3$ or higher are the only ones neglected and all terms of the order of $(\Delta L)^2$ have been retained; indeed

$$\left(\frac{dq}{dx}\right)_C = \left(\frac{dq}{dx}\right)_A + \left(\frac{d^2q}{dx^2}\right)_A \Delta x \tag{11-17}$$

Therefore, from Eq. 11-12 and 11-17 there results

$$(q)_C = (q)_A + \frac{1}{2}\left[\left(\frac{dq}{dx}\right)_A + \left(\frac{dq}{dx}\right)_C\right]$$
$$+ \text{terms of order of } (\Delta x)^3 \text{ or higher} \tag{11-18}$$

Usually for small steps AC, BC the variations from C' to C'' are very small and the second approximation is sufficiently close to the final results; therefore, only the second iteration is determined.

In the analysis of two-dimensional problems, q_C and θ_C are determined exactly in the first step; therefore, the first numerical calculation considers also terms of the order of $(\Delta x)^2$ and iteration processes are not useful for improving precision of results.

G,12. Points on the Boundaries and Points on the Axis. Points on the boundaries can be handled by means of Eq. 11-8 or 11-9 with a

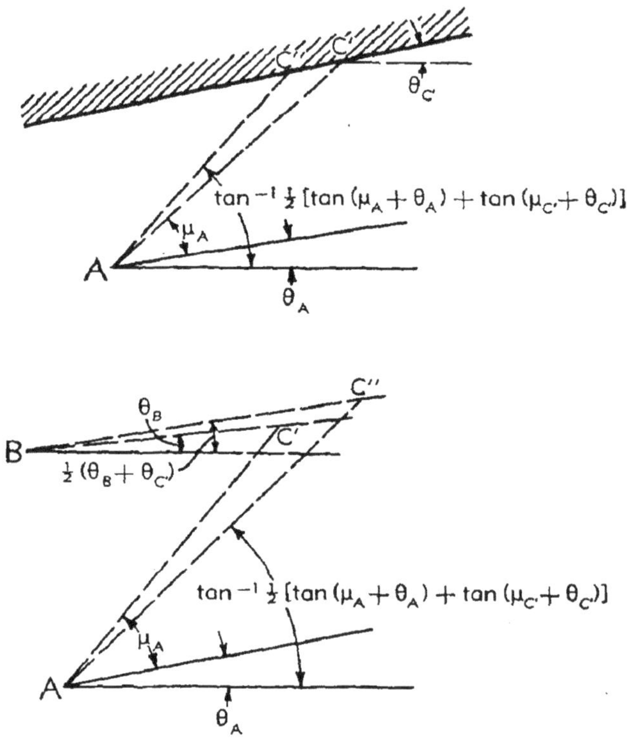

Fig. G,12a. Determination of points on the boundaries of supersonic axially symmetric flows. Top, solid boundary; bottom, free boundary.

procedure similar to the procedure used for the determination of flow properties at a general point. If the boundary is a solid boundary, the direction θ of the velocity vector at any point on the boundary is known. From a point A of the flow, the position of the point C' on the boundary and on the characteristic line, passing through A, is determined in the first approximation from the intersection of the tangent to the characteristic line at A and of the boundary line (Fig. G,12a, top). From Eq. 11-8 or 11-9, depending upon the family of the characteristic line used, the intensity of velocity at the point C' is determined ($\theta_{C'}$ is known). A second approximation can be obtained by determining the position of C and the value of q_C in a manner similar to that used for a general point.

If the boundary is a free boundary, the position of the point C (Fig. G,12a, bottom) is obtained in the first approximation by the intersection of the tangent to the streamline at a known point B of the free boundary, and of the tangent to the characteristic line at A which meets the boundary. At C', the value of q can be obtained directly from the boundary conditions which give the value of static pressure at C'; therefore, the value of θ_C can be obtained from either Eq. 11-8 or 11-9, corresponding to the family of the characteristic line AC. A second approximation for C can be obtained by determining a point C'' on the intersection of a line $C''B$ inclined at an angle equal to $\frac{1}{2}(\theta_B + \theta_{C'})$ with respect to the x axis, and of a line AC'' inclined at an angle equal to $\tan^{-1} \frac{1}{2}(\lambda_A + \lambda_{C'})$. Consequently, the flow at C'' is determined in the second approximation by means of Eq. 11-15 or 11-16.

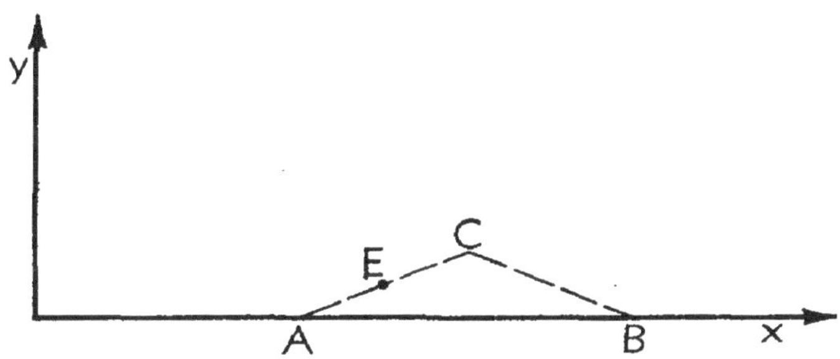

Fig. G,12b. Determination of points on the axis of symmetry, for supersonic axially symmetric flows.

In numerical calculations, particular care is required for the determination of flow properties at points in the neighborhood of the axis of symmetry, because near the axis, y and θ become very small, and the values of the last terms of Eq. 11-5 and 11-6 become very sensitive to small errors in the value of θ.

The discontinuity of curvature of the streamlines, produced by the Mach waves and carried by each wave, changes in amplitude moving along the wave because of the presence of the last term in Eq. 11-6 and increases moving along the Mach cone toward the axis [*12*].

At the apex of the Mach conoid on the axis of symmetry a singularity exists. On the axis, θ and y are both zero and the last term which contains $\sin \theta/y$ becomes indeterminate. However, the limit of the value of $\sin \theta/y$ for $y \to 0$ has a finite value and expressions (11-5) and (11-6) still have physical significance.

Consider a point C near the axis of symmetry, and call A and B the points on the axis and on the characteristic lines from C (Fig. G,12b). At C, Eq. 11-5 and 11-6 can be written in the following form:

$$\cot \mu_C \frac{(dq)_{CA}}{q_C} - (d\theta)_{CA} - \frac{\theta_C}{y_C} \frac{\sin \mu_C}{\cos \mu_C} (dx)_{CA} = 0 \tag{12-1}$$

$$\cot \mu_C \frac{(dq)_{BC}}{q_C} + (d\theta)_{BC} - \frac{\theta_C}{y_C} \frac{\sin \mu_C}{\cos \mu_C} (dx)_{BC} = 0 \tag{12-2}$$

or

$$\frac{\cot \mu_C}{q_C} [(dq)_{CA} + (dq)_{BC}] + [(d\theta)_{BC} - (d\theta)_{CA}] - \frac{\theta_C}{y_C} \tan \mu_C [(dx)_{CA} + (dx)_{BC}] = 0 \tag{12-3}$$

where the subscript CA indicates variations along the characteristic line CA and the subscript BC along the line BC.

If the value of $y_C \rightarrow 0$ then

$$\tan_{y\rightarrow 0} (\mu_C + \theta_C) = \tan \mu_C = \lim_{y\rightarrow 0} \frac{y_C}{(dx)_{CA}} = \lim_{y\rightarrow 0} \frac{y_C}{(dx)_{BC}}$$

$$\left(\frac{d\theta}{dy}\right)_C = \lim_{y\rightarrow 0} \left(\frac{\theta}{y}\right)_C$$

$$(dq)_{CA} + (dq)_{BC} = (dq)_{BA}$$

and

$$(d\theta)_{BC} - (d\theta)_{CA} = \frac{\tan \mu_C}{y_C} [(d\theta)_{BC}(dx)_{BC} - (d\theta)_{CA}(dx)_{CA}]$$

$$= -\left(\frac{\theta}{y}\right)_C \tan \mu_C [(dx)_{CA} + (dx)_{BC}]$$

There results

$$\left(\frac{\theta_C}{y_C}\right)_{y_C \rightarrow 0} = \frac{1}{2} \frac{\cot^2 \mu_C}{q_C} \left(\frac{\partial q}{\partial x}\right)_{\text{axis}} \tag{12-4}$$

For points of the axis, Eq. 11-5 and 11-6 become

$$\cot \mu \frac{dq}{q} \mp d\theta - \frac{1}{2} \cot \mu \frac{1}{q} \left(\frac{\partial q}{\partial x}\right)_{\text{axis}} dx = 0 \tag{12-5}$$

For numerical calculations, Eq. 12-5 can be transformed into an equation containing finite differences in the following way: Consider the characteristic line AC (Fig. G,12b) and the mid-point E between A and C. Then

$$q_C = q_A + \left(\frac{dq}{dx}\right)_A (\Delta x)_{AC} + \frac{1}{2} \left(\frac{d^2q}{dx^2}\right)_A (\Delta x)^2_{AC} = q_A + \left(\frac{dq}{dx}\right)_E (\Delta x)_{AC}$$

and

$$q_E = q_A + \left(\frac{dq}{dx}\right)_A \frac{(\Delta x)_{AC}}{2} + \frac{1}{8} \left(\frac{d^2q}{dx^2}\right)_A (\Delta x)^2_{AC}$$

$$= \frac{q_A + q_C}{2} - \frac{1}{8} \left(\frac{d^2q}{dx^2}\right)_A (\Delta x)^2_{AC} \tag{12-6}$$

$$\theta_C = \left(\frac{d\theta}{dx}\right)_A (\Delta x)_{AC} + \left(\frac{d^2\theta}{dx^2}\right)_A \frac{(\Delta x)^2_{AC}}{2} = \left(\frac{d\theta}{dx}\right)_E (\Delta x)_{AC}$$

and

$$\theta_E = \frac{1}{2}\left(\frac{d\theta}{dx}\right)_A (\Delta x)_{AC} + \frac{1}{8}\left(\frac{d^2\theta}{dx^2}\right)_A (\Delta x)^2_{AC} = \frac{\theta_C}{2} - \left(\frac{d^2\theta}{dx^2}\right)_A \frac{(\Delta x)^2_{AC}}{8} \tag{12-7}$$

At E, Eq. 11-5 gives

$$\cot \mu_E \frac{1}{q_E}\left(\frac{dq}{dx}\right)_E - \left(\frac{d\theta}{dx}\right)_E - \frac{\theta_E \sin \mu_E}{\cos (\mu_E + \theta_E)} \frac{1}{y_E} = 0 \tag{12-8}$$

but, when terms of the order of $(\Delta x)^2_{AC}$ are neglected

$$\frac{\cot \mu_E}{q_E} = \frac{1}{2}\left(\frac{\cot \mu_A}{q_A} + \frac{\cot \mu_C}{q_C}\right)$$

$$\frac{(\Delta x)_{AC}}{y_E} \frac{\sin \mu_E}{\cos (\mu_E + \theta_E)} = \frac{2 \sin \mu_E}{\sin (\mu_E + \theta_E)} = 2 \frac{\sin \mu_C + \sin \mu_A}{\sin (\mu_C + \theta_C) + \sin \mu_A}$$

and

$$\theta_E = \frac{\theta_C}{2}$$

Therefore, Eq. 12-8, maintaining terms of the order of $(\Delta x)^2$, can be written in the form

$$(q_C - q_A)\left(\frac{\cot \mu_A}{q_A} + \frac{\cot \mu_C}{q_C}\right) - \theta_C\left[1 + \frac{\sin \mu_C + \sin \mu_A}{\sin (\mu_C + \theta_C) + \sin \mu_A}\right] = 0 \tag{12-9}$$

Eq. 11-2 to 11-4 and 11-8 to 11-9 and the corresponding equations for the boundary conditions and for the points on the axis permit the determination of q and θ in the whole flow field as a function of the assigned boundary conditions. The determination can be obtained by a completely numerical procedure, which is more convenient when automatic computing machines are used, or by means of graphical and numerical procedure, where graphical methods are used in the determination of the position of the lattice points (points of intersection of the characteristic lines) in the physical plane. In this case Eq. 11-8 and 11-9 are more often used in a transformed form given in the next paragraph, which permits the use of tabulated values.

In place of determining the flow properties at the points of intersection of characteristic lines, the average flow properties in a region included between two pairs of characteristic lines of opposite family can be determined, as is discussed in Art. 8 for two-dimensional flow. This procedure, called mesh procedure, is often adopted when normal operations and graphical methods are used. In this procedure, for coefficients of the difference equations (11-8) and (11-9), the average values in each mesh are used; therefore, the mesh method permits only the determination of first order variations between adjacent meshes, and terms of the order of $(\Delta L)^2$ cannot be obtained by iteration processes. The practical use of the mesh procedure for axially symmetric problems is described in Art. 14, where graphical methods are presented.

G,13. Numerical Method Using Tabulated Values. The equations of characteristics can be transformed in order to introduce the characteristic number system used for the analysis of two-dimensional flow, as described in Art. 4. Eq. 11-5 and 11-6 can be written in the form

$$\cot\mu\frac{dq}{q} - d\theta = \sin\theta\sin\mu\frac{d\eta}{y} \tag{13-1}$$

$$\cot\mu\frac{dq}{q} + d\theta = \sin\theta\sin\mu\frac{d\xi}{y} \tag{13-2}$$

where $d\eta$ is the element along the characteristic line of the first family of which dx of Eq. 11-5 is the x component, and $d\xi$ is the element of the second family of which dx in Eq. 11-6 is the x component.

Eq. 13-1 and 13-2 differ from Eq. 3-7 only by the presence of the terms on the right side of the equations. For two-dimensional flow, the equations of characteristics have been expressed in the form of Eq. 4-7 and 4-8

$$C_1 = f(q) + \theta \tag{13-3}$$

$$C_2 = f(q) - \theta \tag{13-4}$$

where $f(q) = -\int_1^{q/q^*}\cot\mu(dq/q)$, or

$$\theta = \frac{C_1 - C_2}{2} \tag{13-5}$$

$$f(q) = \frac{C_1 + C_2}{2} \tag{13-6}$$

where C_1 and C_2 are the constants distinguishing each characteristic line, and half of each constant has been called the characteristic number. The system of numbers suggested for two-dimensional flow assumes the value of 1,000 for $q = q^*$ and the value of 200 for the direction $\theta = 0$, or in radians

$$\theta = \left(-200 + \frac{C_1 - C_2}{2}\right)\frac{\pi}{180} \tag{13-7}$$

$$f(q) = \left(1000 - \frac{C_1 + C_2}{2}\right)\frac{\pi}{180} \tag{13-8}$$

Assuming as new variables the constants C_1 and C_2 defined by Eq. 13-3 and 13-4, Eq. 13-1 and 13-2 become

$$\frac{dC_1}{d\eta} = -\frac{180}{\pi}\frac{\sin\theta\sin\mu}{y} \tag{13-9}$$

along the characteristic line of the first family and

$$\frac{dC_2}{d\xi} = -\frac{180}{\pi}\frac{\sin\theta\sin\mu}{y} \tag{13-10}$$

along the characteristic line of the second family [7].

Eq. 13-9 and 13-10 show that, while for two-dimensional flow the characteristic number remains constant along each characteristic line, in axially symmetric flow the characteristic number changes.

The use of Eq. 13-9 and 13-10 is simple. The variation of the characteristic number is determined by a finite difference method for each step along the characteristic line, and the values of the constants C_1 and C_2 are obtained at each point of the flow. Knowing the values of the constants, the value of θ can be obtained from Eq. 13-7 (in radians) and the value of M and μ from the value of $N = (C_1 + C_2)/2$ in Table G,4. Using Fig. G,11, the constants at the point C' are determined from the following equations in first approximation

$$(C_1)_{C'} - (C_1)_A = -\frac{180}{\pi}\frac{\sin\theta_A\sin\mu_A}{y_A}\frac{x_{C'} - x_A}{\cos(\mu_A + \theta_A)} \tag{13-11}$$

$$(C_2)_{C'} - (C_2)_B = -\frac{180}{\pi}\frac{\sin\mu_B\sin\theta_B}{y_B}\frac{x_{C'} - x_B}{\cos(\theta_B - \mu_B)} \tag{13-12}$$

Determine the position of C'' in the second approximation, as indicated before; then the constants at C'' are obtained from

$$\begin{aligned}(C_1)_{C''} - (C_1)_A &= \frac{1}{2}\left[\left(\frac{dC_1}{d\eta}\right)_A + \left(\frac{dC_1}{d\eta}\right)_{C'}\right]\Delta\eta \\ &= -\frac{180}{\pi}\left(\frac{\sin\theta_A\sin\mu_A}{y_A} + \frac{\sin\theta_{C'}\sin\mu_{C'}}{y_{C'}}\right) \\ &\qquad \left[\frac{x_{C''} - x_A}{\cos(\mu_A + \theta_A) + \cos(\mu_{C'} + \theta_{C'})}\right]\end{aligned} \tag{13-13}$$

$$\begin{aligned}(C_2)_{C''} - (C_2)_B &= \frac{1}{2}\left[\left(\frac{dC_2}{d\xi}\right)_B + \left(\frac{dC_2}{d\xi}\right)_{C'}\right]\Delta\xi \\ &= -\frac{180}{\pi}\left(\frac{\sin\theta_B\sin\mu_B}{y_B} + \frac{\sin\theta_{C'}\sin\mu_{C'}}{y_{C'}}\right) \\ &\qquad \left[\frac{x_{C''} - x_B}{\cos(\theta_B - \mu_B) + \cos(\theta_{C'} - \mu_{C'})}\right]\end{aligned} \tag{13-14}$$

The method requires a small amount of numerical calculation, especially if the distance $d\eta$ and $d\xi$ are determined graphically, because the coefficients of the terms on the right side of Eq. 13-11 and 13-12 are the same for both characteristic lines passing through a given point and the values of μ and M are determined by using tabulated values.

G,14. Graphical Methods. The method described in Art. 13 can be adapted to construct graphical solutions in the hodograph plane similar to the solutions obtained for the two-dimensional flow [*13,14*].

Consider in the physical plane as independent variables the projections f and g of the velocity vector on the two directions of the tangents to the characteristic lines η and ξ of the first and second families respectively (Fig. G,14a). The projections f and g at each point A of the flow field as functions of the polar velocity components are defined by

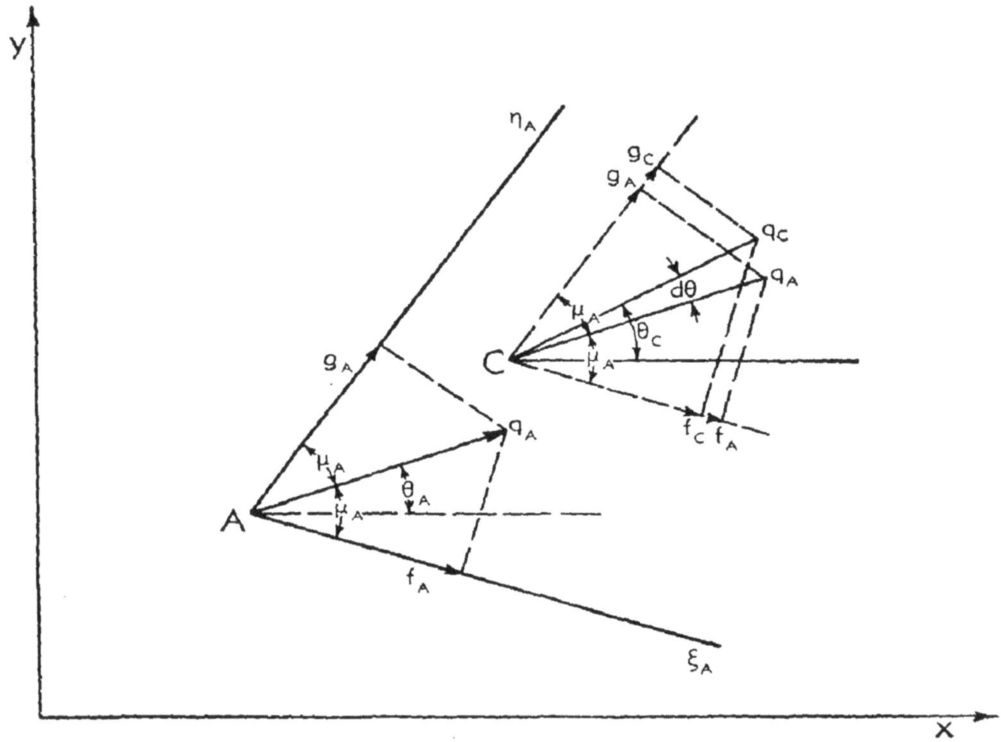

Fig. G,14a. Graphical solution of the characteristic equations for axially symmetric flows.

$$g_A = q_A \cos \mu_A \tag{14-1}$$

$$f_A = q_A \cos \mu_A \tag{14-2}$$

Assuming the coordinate axes η and ξ fixed at A, the projections of the velocity vector for any other point C of the flow field are (Fig. G,14a)

$$g_C = q_C \cos (\mu_A - d\theta) \tag{14-3}$$

$$f_C = q_C \cos (\mu_A + d\theta) \tag{14-4}$$

Therefore

$$(g_C - g_A) = dg_{AC} = dq_{AC} \cos \mu_A + q_A \sin \mu_A d\theta_{AC} \tag{14-5}$$

$$(f_C - f_A) = df_{AC} = dq_{AC} \cos \mu_A - q_A \sin \mu_A d\theta_{AC} \tag{14-6}$$

Substituting Eq. 14-5 and 14-6 in the equations of the characteristics (13-1) and (13-2), Eq. 13-1 along the characteristic line of the first family η becomes

$$\frac{1}{q}\,df = \frac{\sin^2\mu\,\sin\theta}{y}\,d\eta \tag{14-7}$$

where df is the variation of the component f of the velocity vector in the direction of the tangent to the characteristic line of the second family ξ

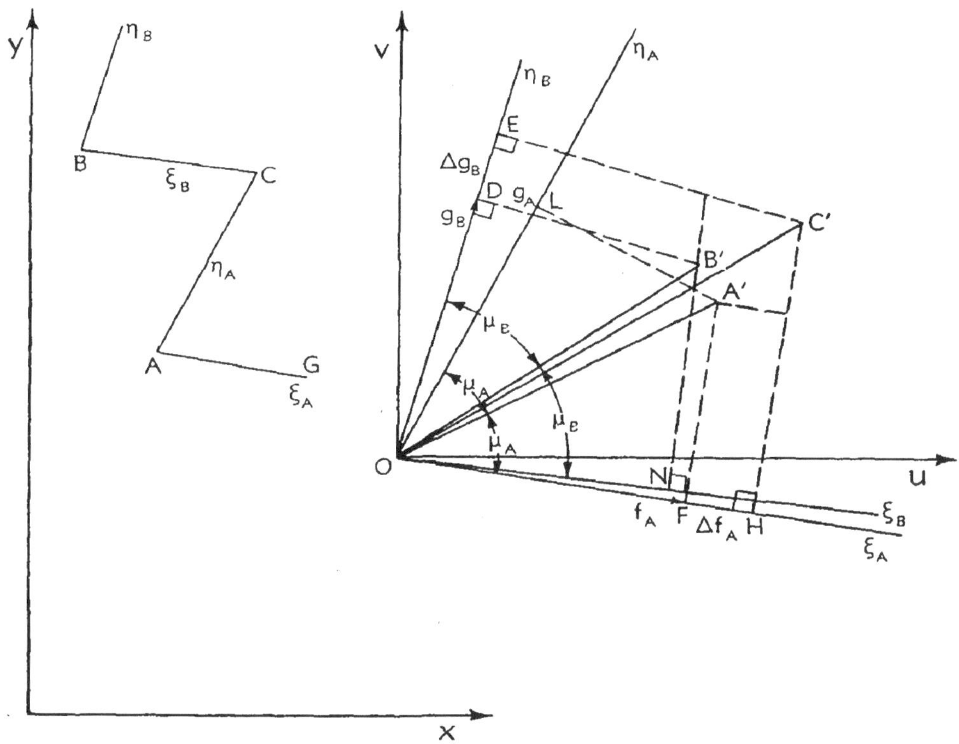

Fig. G,14b. The graphical solution of the characteristic equations for axially symmetric flows in the physical plane (left) and in the hodograph plane (right).

at the point considered. Eq. 13-2 along the characteristic line of the second family ξ becomes

$$\frac{1}{q}\,dg = \frac{\sin^2\mu\,\sin\theta}{y}\,d\xi \tag{14-8}$$

where dg is the variation of the component g of the velocity vector in the direction of the tangent to the characteristic line of the first family at the point considered.

Eq. 14-7 gives the variation of the component f_A at a point A of the physical plane moving along the line AC (Fig. G,14b, left), while Eq. 14-8 gives the variation of g_A moving along AG.

Consider now the point C at the intersection of the characteristic line η_A from A and ξ_B from B. The velocity vector at C represented in the hodograph plane (Fig. G,14b, right) by OC' must have a component g_C parallel to the direction η_B of intensity

$$g_C = g_B + \frac{\sin^2 \mu_B \sin \theta_B}{y_B} q_B(\xi_C - \xi_B)_{BC} \tag{14-9}$$

and a component f_C parallel to the direction ξ_A, of intensity

$$f_C = f_A + \frac{\sin^2 \mu_A \sin \theta_A}{y_A} q_A(\eta_C - \eta_A)_{AC} \tag{14-10}$$

If OF is the component f_A of the velocity vector OA' at A in the ξ_A direction (Fig. G,14b, right) and FH is the variation $\Delta f_A = f_C - f_A$, given by Eq. 14-10, OH gives the value f_C in the ξ_A direction, and the point C' must be on the normal to ξ_A at H.

If OD is the component g_B of the velocity vector OB' at B in the η_B direction and DE is the variation $\Delta g_B = g_C - g_B$ given by Eq. 14-9, OE represents the value g_C in the η_B direction, and the point C' must be on the normal to η_B at E. As a result the point C' is obtained on the intersection of the two normals to ξ_A at H and to η_B at E.

If higher approximation is required, the values of $(\Delta f)_A$ and $(\Delta g)_B$ can be obtained from an iteration process. Usually only the first approximation is used because of the graphical determination of OC'.

The determination of points on the boundary with this method is analogous to the determination used in numerical methods because either the direction of the vector OC' (solid boundary) or its intensity (free boundary) is given by the boundary conditions; therefore, only one component f_C or g_C is required in order to determine the position of the point C'.

For irrotational two-dimensional problems Δg_B and Δf_A are zero; therefore, the point C' lies on the normals from D to η_B and from F to ξ_A (Fig. G,14b, right). This property is expressed by Eq. 4-5 and 4-6. These equations state that the normal to the characteristic line of one family is tangent to the epicycloid representing the variation of velocity along the other family in the hodograph plane, and in the present considerations the epicycloid is approximated by its tangent at the point considered.

G,15. Conical Flow. Conical flow fields are analyzed in detail in Sec. H, where analytic expressions defining the flow field are obtained directly, using the conical properties of the flow. However, two-dimensional or axially symmetric conical flow can be analyzed also by means of the characteristics methods or as a particular case of two-dimensional or axially symmetric flow.

Consider a flow field whose velocity, pressure, and density are constant along each straight line originating from a point called the apex. This flow is defined as conical flow. The velocity at point B along the straight line OB from the apex O (Fig. G,15) is equal to the velocity at any other point A on the same straight line. Consider the characteristic line of the first family AC from A and the characteristic line of the second family BC from B. The velocity at C can be determined from the values of the velocity at A and B; therefore, the velocity along the line OC can be determined from the value along the line OB.

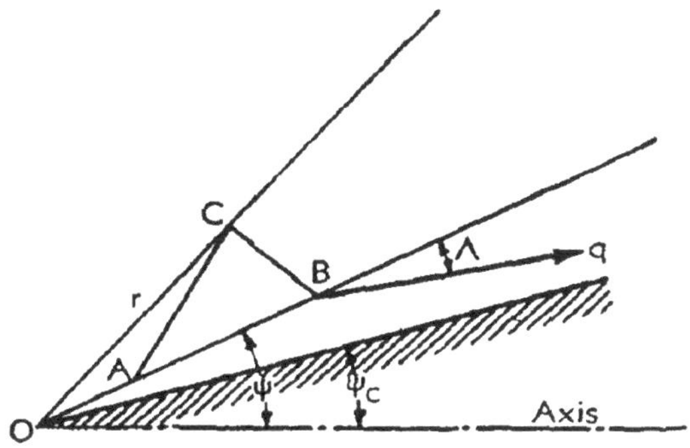

Fig. G,15. Conical flow field.

From Eq. 11-5, 11-6, and 3-7 the following expressions are obtained for the variation of velocity components from the line OB to the line OC.

$$2 \cot \mu_A \frac{(\Delta q)_{AC}}{q_A} = \epsilon \left[\frac{(\Delta x)_{AC}}{y_A \cos (\theta_A + \mu_A)} + \frac{(\Delta x)_{BC}}{y_B \cos (\theta_A - \mu_A)} \right] \times \sin \mu_A \sin \theta_A \quad (15\text{-}1)$$

$$2(\Delta\theta)_{AC} = \epsilon \left[\frac{(\Delta x)_{BC}}{y_B \cos (\theta_A - \mu_A)} - \frac{(\Delta x)_{AC}}{y_A \cos (\theta_A + \mu_A)} \right] \sin \mu_A \sin \theta_A \quad (15\text{-}2)$$

where $\epsilon = 0$ for two-dimensional flow and $\epsilon = 1$ for axially symmetric flow. If a two-dimensional conical flow is considered, Eq. 15-1 and 15-2 become equal to zero; therefore, for two-dimensional conical flow the flow along OB must be the same as the flow along OC. The flow field remains uniform (or changes in a discontinuous way) in the entire region where the characteristic line AC is different from the radius OA. (Therefore, the flow behind a shock wave produced by a two-dimensional wedge is uniform and constant in all the region between the shock wave and the wedge.) Variations of velocity components for two-dimensional conical flow are possible only upon crossing a line OC that is a characteristic, i.e. has the same inclination as AC. (Then O, A, C, and B are colinear.)

Eq. 15-1 and 15-2 can be transformed, for axially symmetric conical flow, into a form more convenient for practical use. Introduce polar coordinates r and ψ, and let q denote the magnitude of velocity vector and Λ its inclination with respect to the radius r at the point considered, defined positive as in Fig. G,15. Then Eq. 15-1 and 15-2, because of the conical property, becomes

$$\frac{dq}{d\psi} = q \sin \Lambda \cos \Lambda \left(\frac{1 - \tan \Lambda \cot \psi}{1 - \dfrac{\sin^2 \Lambda}{\sin^2 \mu}} \right) \tag{15-3}$$

$$\frac{d\Lambda}{d\psi} = 1 + \cos^2 \Lambda \left(\frac{1 - \tan \Lambda \cot \psi}{1 - \dfrac{\sin^2 \Lambda}{\sin^2 \mu}} \right) \tag{15-4}$$

Eq. 15-3 and 15-4 permit determination of all the flow field by means of a step-by-step integration. As initial values, the velocity q at the surface of the conical body $\psi = \psi_C$ or the velocity behind a conical shock wave are given. The angle Λ is defined as

$$\psi - \theta = \Lambda$$

Along the Mach wave from 0, $\psi = \mu$ and θ is zero; therefore, Λ is equal to μ, and Eq. 15-3 and 15-4 become indeterminate.

At the surface of the cone $\psi = \psi_C$ the angle Λ is zero; therefore, between ψ_C and $\psi = \mu$, Λ increases from zero to the value μ, and θ decreases from $\theta = \psi_C$ to zero.

G,16. Flow Around Corners in Axially Symmetric Flow. Consider an axially symmetric body having x as axis of symmetry and the line OAC as generatrix. At the point A of the boundary a discontinuous variation of cross section (corner) exists. Assume that the variation of direction of the generatrix OAC of the body in the meridian plane x, y at A is as indicated in Fig. G,16 and therefore has a sign such that no discontinuities of velocity components occur in any point of the flow field other than A. (A deviation such as AC' would produce a shock wave and the flow field would change in a discontinuous way across the shock wave.)

The boundary streamline AC undergoes at A a finite variation of direction $\Delta\theta$, which is transmitted in the region contained between the characteristic line AB corresponding to the flow conditions at the point A before the deviation $\Delta\theta$ and the characteristic line AD starting also at A but corresponding to the flow conditions at A after the variation of direction from θ_A to $\theta_A + \Delta\theta_A$.

In order to determine the flow field in the region ABD, consider the variation along a characteristic line PQ of opposite family from the characteristic lines AB and AD. Along PQ Eq. 11-6 is valid.

If the point P moves along the line BA toward A, the point Q moves along the characteristic line AE toward A, and the distance PQ decreases; and when P tends to A, PQ tends to zero.

Now in Eq. 11-6, the last term of the equation contains the expression $dx/\cos(\theta - \mu)$, which represents the infinitesimal distance PQ between

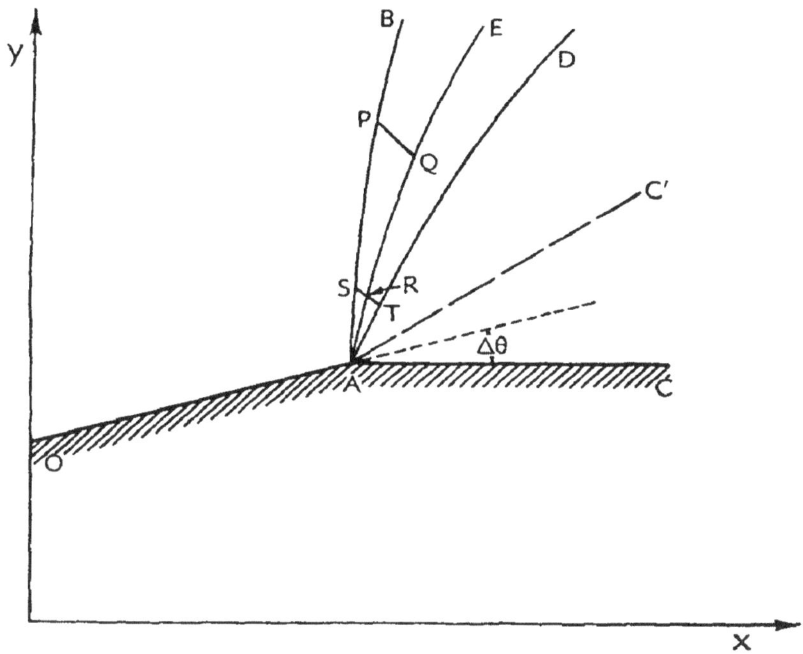

Fig. G,16. Flow around a corner in axially symmetric flow.

the two points P and Q along the characteristic line PQ; therefore, when $P \to A$, the last term of Eq. 11-6 tends to zero and at A Eq. 11-6 becomes

$$\cot\mu \frac{dq}{q} + d\theta = 0 \tag{16-1}$$

Eq. 16-1 is the equation obtained for two-dimensional flow; therefore' the expansion at A occurs around a two-dimensional corner. Along the streamline OAC the variation at A from OA to AC can be determined from Table G,4, and the tangents to the characteristic lines AB, AE, and AD at A can be obtained as for the two-dimensional case [4].

The flow in the region ABD can be determined with the procedure described for the general case. From the tangent to the characteristic line of the second family at S, and the tangent to the characteristic line at A, corresponding to a direction $(\theta_1 + \Delta\theta_R)$, the point R is obtained. The velocity vector at R is determined as described in Art. 11 and 13; then by a similar procedure all the points along the characteristic line ST are determined. The tangent AT to the last characteristic line from A is inclined at μ_{A_2} with respect to the line AC where μ_{A_2} corresponds to

N_{A_1} in Table G,4. The value of N_{A_2} is obtained by decreasing the number N_{A_1} corresponding to μ_{A_1} or M_{A_1} at A for the line AS by the deviation $(\Delta\theta)_A$. If a deviation having the sign of the deviation AC' of Fig. G,16 occurs at A, a shock wave is produced. The shock wave at the point A is equal to a shock wave of deviation $\Delta\theta$, and therefore at A the shock wave also has two-dimensional properties.

G,17. The Shape of the Characteristic Lines at Infinity. At a large distance from the axis of an axially symmetric supersonic flow field the velocity at any point differs only slightly from the undisturbed velocity U. Therefore, the velocity components in this region can be expressed in the form

$$u = U(1 + u') \tag{17-1}$$

$$v = Uv' \tag{17-2}$$

where u' and v' can be assumed to be small numbers. Neglecting terms of higher order, the inclination of the velocity can be expressed in the form

$$\tan\theta = \frac{v'}{1 + u'} \tag{17-3}$$

or

$$\theta = v' \tag{17-4}$$

From the equation of conservation of energy expressed in the form

$$\frac{2}{\gamma - 1}\sin^2\mu = \frac{q_{max}^2}{q^2} - 1$$

(where q_{max} is the limiting velocity) the following expression is obtained

$$\mu = \mu_\infty - \left(1 + \frac{\gamma - 1}{2} M_\infty^2\right)\tan\mu_\infty u' \tag{17-5}$$

where the subscript ∞ indicates quantities at undisturbed stream conditions.

Because v' and u' are small numbers with respect to unity, it can be expected that the inclination of the characteristic lines of the first family, $\tan(\theta + \mu)$, will not change appreciably from one characteristic line to another measured at the intersection with the same characteristic line of the second family. Therefore, the distance $\Delta\xi$ between two adjacent characteristic lines of the first family (Fig. G,17) measured along a characteristic line of the second family will not increase appreciably moving from a point A to another point C of the same family at a larger distance from the axis.

If the value of $\Delta\xi$ does not increase appreciably as y increases, the term $\sin\mu\sin\theta\,(d\xi/y)$ of Eq. 13-10 becomes small for large values of y

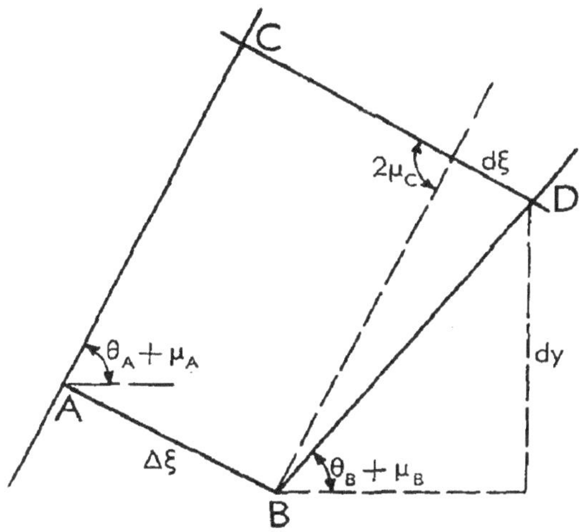

Fig. G,17. The length $\Delta\xi$ between characteristic lines of the first family as function of the distance from the axis.

and tends to zero where y tends to infinity, because $\sin\theta$ approaches zero at infinity and $\sin\mu$ approaches $\sin\mu_\infty$; therefore, the variation of the characteristic number C_2 decreases when y increases and tends to zero for $y \to \infty$. If

$$\lim_{y\to\infty}\left(\frac{dC_2}{d\xi}\right) = 0 \tag{17-6}$$

then Eq. 13-2 becomes

$$\cot\mu\,\frac{dq}{q} + d\theta = 0 \tag{17-7}$$

or neglecting terms of higher order

$$\theta = v' = -\cot\mu_\infty u' \tag{17-8}$$

In the region considered Eq. 13-1 can be expressed in the form

$$-2d\theta = \frac{\sin\theta\,\sin\mu}{\sin(\theta+\mu)}\,\frac{dy}{y} \tag{17-9}$$

or

$$\theta = \frac{K}{\sqrt{y}} \tag{17-10}$$

$$u' = -\frac{K\tan\mu_\infty}{\sqrt{y}} \tag{17-11}$$

The characteristic lines of the first family defined by

$$\frac{dx}{dy} = \cot(\theta+\mu) \tag{17-12}$$

become

$$\frac{dx}{dy} = \cot\mu_\infty + \frac{1}{2}u'M_\infty^4(\gamma+1)\tan\mu_\infty \tag{17-13}$$

or

$$x = \cot \mu_\infty y - [\tan^2 \mu_\infty K M^4(\gamma + 1)] y^{\frac{1}{2}} + \text{const} \tag{17-14}$$

The characteristic lines of the first family at a large distance from the axis are parabolic with $u' \sqrt{y}$ and $v' \sqrt{y}$ constant along them [*15*].

These results satisfy the assumption of Eq. 17-6 that the variation of C_2 at a large distance from the axis is small and tends to zero as y goes to infinity. Consider two characteristic lines of the first family AC and BD (Fig. G,17). The variation $d\xi$ of the length $\Delta\xi$ with y can be expressed in the form

$$\frac{d\xi}{dy} = \frac{\sin (\theta_A + \mu_A - \theta_B - \mu_B)}{\sin 2\mu_C \sin (\theta_B + \mu_B)} \tag{17-15}$$

But the values $\theta_B - \theta_A$ and $\mu_B - \mu_A$ are proportional to $1/\sqrt{y}$ (see Eq. 17-10, 17-11, and 17-5); therefore, $\Delta\xi$ increases in the first order proportionally to $\sqrt{y}$. Since in the expression

$$dC_2 = \frac{\sin \theta \sin \mu \, d\xi}{y} \tag{17-16}$$

$\sin \theta$ is proportional to $1/\sqrt{y}$, the value of dC_2 tends to zero as y approaches infinity in agreement with the assumptions made in the derivation of Eq. 17-10 and 17-11. (See E,6.)

CHAPTER 4. THE CHARACTERISTICS METHOD FOR TWO-DIMENSIONAL OR AXIALLY SYMMETRIC ROTATIONAL FLOW

G,18. Numerical and Graphical Methods. When the density is not a unique function of the pressure but depends also on the position coordinates (rotation exists in the flow), the analysis of the flow properties with the characteristics method must be performed by considering the characteristic surfaces which are the envelopes of Mach conoids and also the surfaces which are the envelopes of the streamlines.

Heat conductivity, heat addition, and viscosity are neglected and along each streamline the total enthalpy is considered constant for the flow field considered. Since, in addition, the entropy remains constant along each streamline between discontinuities, the density along a streamline is a single-valued function of the pressure.

The numerical methods used for the analysis of two-dimensional or axially symmetric flows are similar to the methods used for potential flows. However, the step-by-step procedure must be used in both the hodograph and physical planes in these cases.

The equation of motion, when the total enthalpy h^0 is constant in the flow, can be expressed along the characteristic lines in the following form

$$\cot \mu \frac{dq}{q} - d\theta - \epsilon \frac{\sin \mu \sin \theta}{\cos (\mu + \theta)} \frac{dx}{y} + \sin \mu \cos \mu \frac{ds}{\gamma \mathcal{R}} = 0 \quad (18\text{-}1)$$

along the characteristic line

$$\tan (\mu + \theta) = \frac{dy}{dx} \quad (18\text{-}2)$$

and

$$\cot \mu \frac{dq}{q} + d\theta - \epsilon \frac{\sin \mu \sin \theta}{\cos (\theta - \mu)} \frac{dx}{y} + \sin \mu \cos \mu \frac{ds}{\gamma \mathcal{R}} = 0 \quad (18\text{-}3)$$

along the line

$$\tan (\theta - \mu) = \frac{dy}{dx} \quad (18\text{-}4)$$

and

$$ds = 0 \quad (18\text{-}5)$$

along the streamline defined by

$$\frac{dy}{dx} = \tan \theta \quad (18\text{-}6)$$

where ϵ is equal to zero for two-dimensional flow and equal to one for axially symmetric flow.

If the total enthalpy h^0 is not constant in the flow, the equations along the characteristic lines (18-2) and (18-4) can be expressed more conveniently in the form of Eq. 2-25, and along the streamline (18-6) the equation

$$dh^0 = 0 \quad (18\text{-}7)$$

applies when heat addition does not occur.

The use of Eq. 18-2, 18-3, or 2-25 is similar to the use of Eq. 12-1 and 12-2. If the velocity vector, total enthalpy, and entropy are known at the two points A and B of Fig. G,18, the position of a point C' at the intersection of the tangents AC' to the characteristic line of the first family from A and the tangent BC' to the characteristic line of the second family from B can be obtained since the value of μ can be calculated at A and B.

Assuming that the tangents AC' and BC' can be substituted for the characteristic lines between AC' and BC', the first approximation to the variation of flow properties between A and C' and B and C' can be obtained from Eq. 18-2, 18-3, or 2-25 by assuming that the coefficients are constant and equal to the corresponding values at A and B. However, in the equations the variation of s and h^0 between A and C' or B and C' must be known in order to determine the variation of q and θ, or p and θ, between A and C' and B and C'. The assumption that h^0 and s are continuous functions in the region of the flow between the streamlines that pass through A and B, plus the fact that total enthalpy and the entropy remain constant along each streamline, leads to the equations

$$(\Delta s)_{AC'} = s_{C'} - s_A \tag{18-8}$$

$$(\Delta s)_{BC'} = -(s_B - s_{C'}) \tag{18-9}$$

and

$$s_{C'} = s_A + \left(\frac{ds}{dn}\right)_{AC'} (\Delta n)_{AC'}$$
$$= s_A + \left(\frac{ds}{dn}\right)_{AC'} (x_{C'} - x_A) \left[\frac{\sin \mu}{\cos (\theta + \mu)}\right]_{AC'} \tag{18-10}$$

$$s_{C'} = s_B + \left(\frac{ds}{dn}\right)_{BC'} (\Delta n)_{BC'}$$
$$= s_B - \left(\frac{ds}{dn}\right)_{BC'} (x_{C'} - x_B) \left[\frac{\sin \mu}{\cos (\theta - \mu)}\right]_{BC'} \tag{18-11}$$

where ds/dn is the gradient normal to the streamlines, and Δn is the

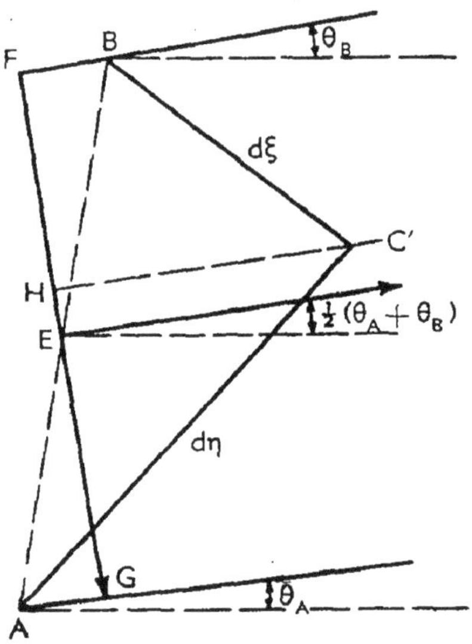

Fig. G,18. The use of the characteristic methods for rotational two-dimensional and axially symmetric flow.

normal distance between the streamline at A or B and the streamline passing at C'. Therefore, assuming that

$$\left(\frac{ds}{dn}\right)_{AC'} = \left(\frac{ds}{dn}\right)_{BC'} = \left(\frac{ds}{dn}\right)_{AB} \tag{18-12}$$

$s_{C'}$ can be determined by determining ds/dn from the expressions

$$\left(\frac{ds}{dn}\right)_{AB} = \frac{s_B - s_A}{\dfrac{x_{C'} - x_A}{[\cos (\mu + \theta)]_{AC'}} (\sin \mu)_{AC'} + \dfrac{(x_{C'} - x_B)}{[\cos (\theta - \mu)]_{BC'}} (\sin \mu)_{BC'}} \tag{18-13}$$

Similar considerations are valid for the determination of $(h^0)_{C'}$. The first approximation may then be obtained for $q_{C'}$ and $\theta_{C'}$ or $p_{C'}$ and $\theta_{C'}$ from Eq. 18-11, 18-14, and the following expressions

$$q_{C'} = \frac{1}{(\cot \mu_B/q_B) + (\cot \mu_A/q_A)} \Bigg\{ \theta_B - \theta_A + \cot \mu_A + \cot \mu_B + \epsilon \left[\frac{\sin \theta_A \sin \mu_A}{\cos (\theta + \mu)_A} \frac{x_{C'} - x_A}{y_A} + \frac{\sin \theta_B \sin \mu_B}{\cos (\theta - \mu)_B} \frac{x_{C'} - x_B}{y_B} \right] - \sin \mu_A \cos \mu_A \frac{s_{C'} - s_A}{\gamma \mathfrak{R}} - \sin \mu_B \cos \mu_B \frac{s_{C'} - s_B}{\gamma \mathfrak{R}} \Bigg\} \qquad (18\text{-}14)$$

and

$$\theta_{C'} = \theta_B - \frac{\cot \mu_B}{q_B} (q_{C'} - q_B) + \epsilon \frac{\sin \theta_B \sin \mu_B}{y_B \cos (\theta - \mu)_B} (x_{C'} - x_B) - \cos \mu_B \sin \mu_B \frac{s_{C'} - s_B}{\gamma \mathfrak{R}} \qquad (18\text{-}15)$$

A second approximation can be obtained by iteration methods in which the average values of the coefficients between A and C'' and B and C'' are assumed as is discussed in Art. 12.

The determination of points on the boundary does not present any more difficulty than that experienced in potential flow. Since the boundary represents a streamline, the value of the entropy on the boundary is constant, and therefore the value of Δs in Eq. 18-1 and 18-3 is known for the process of computing a boundary point from a point of the flow field.

The system described in Art. 13 can also be extended to the analysis of rotational flow and the amount of numerical work can be somewhat reduced by using tabulated values. Eq. 13-9 and 13-10, along the characteristic lines of first and second family for flow with variable entropy, become

$$\frac{dC_1}{d\eta} = -\left(\frac{180}{\pi} \frac{\sin \theta \sin \mu}{y}\right) \epsilon + \frac{1}{\gamma \mathfrak{R}} \frac{ds}{d\eta} \cos \mu \sin \mu \qquad (18\text{-}16)$$

and

$$\frac{dC_2}{d\xi} = -\left(\frac{180}{\pi} \frac{\sin \theta \sin \mu}{y}\right) \epsilon + \frac{1}{\gamma \mathfrak{R}} \frac{ds}{d\xi} \cos \mu \sin \mu \qquad (18\text{-}17)$$

The use of Eq. 18-16 and 18-17 is similar to the use of Eq. 13-9 and 13-10. The values of $ds/d\eta$ and $ds/d\xi$ can be obtained from Eq. 18-13. If a graphical-numerical procedure is used, they may be found by measuring $d\eta$ and $d\xi$ and dn in the characteristic diagram, or by constructing the streamline passing through point C.

When graphical methods are used for the analysis of rotational flow, Eq. 14-7 and 14-8 along the characteristic lines for the case of constant total enthalpy become

$$\frac{1}{q}\frac{df}{d\eta} = \epsilon\frac{\sin^2\mu\sin\theta}{y} - \frac{1}{\gamma\mathfrak{R}}\frac{ds}{d\eta}\cos\mu\sin^2\mu \tag{18-18}$$

and

$$\frac{1}{q}\frac{dg}{d\xi} = \epsilon\frac{\sin^2\mu\sin\theta}{y} - \frac{1}{\gamma\mathfrak{R}}\frac{ds}{d\xi}\cos\mu\sin^2\mu \tag{18-19}$$

The use of Eq. 18-18 and 18-19 is similar to the use of Eq. 14-7 and 14-8, and the only difference arises in the determination of the last terms, which involve the variation of entropy along the characteristic lines. This determination can be performed by means of Eq. 18-13, or graphically as shown in Fig. G,18 where the distance Δn corresponds to the segment GF normal to the streamline at E. This streamline through E halves the line AB and has a slope $\frac{1}{2}(\theta_A + \theta_B)$ with respect to the x axis. Then

$$(\Delta s)_{AC'} = \frac{s_B - s_A}{\overline{GF}} \times (\Delta n)_{AC'} = \frac{s_B - s_A}{\overline{GF}} \times \overline{GH} \tag{18-20}$$

G,19. Reflections from an Entropy Discontinuity or Enthalpy Discontinuity. Consider a supersonic flow field in which an entropy or enthalpy discontinuity occurs, produced, for example, by interaction of two shock waves or by mixing of two streams. The discontinuity is transmitted in the flow along a streamline and across the streamline a finite variation of velocity, Mach number, density, and stagnation pressure occurs. Any wave striking the discontinuity is partly transmitted and partly reflected at the discontinuity. Because such discontinuities are frequent in supersonic flow, an analysis of this phenomenon is of natural interest.

The flow conditions in the region of the discontinuity can be analyzed with the method of characteristics if the flow is supersonic on both sides of the discontinuity or if the static pressure on one side of the discontinuity is given (i.e. a jet discharging into constant pressure ambient conditions). The conditions of equilibrium require that both static pressure and direction of velocity be the same on both sides of the discontinuity. The total enthalpy h^0 and the entropy s on the two sides of the discontinuity line AF of Fig. G,19a are known; therefore, from the equation of conservation of energy written in the form

$$\left(\frac{q}{q_{max}}\right)^2 = 1 - \left(\frac{p}{p^0}\right)^{\frac{\gamma-1}{\gamma}} \tag{19-1}$$

a relation between the velocities q'/q'_{max} and q''/q''_{max} on opposite sides of the discontinuity for the same value of pressure p can be obtained as a function of $p^{0'}$ and $p^{0''}$.

Assume that the flow is two-dimensional and consider the numbers

C_1 and C_2 which distinguish each characteristic line of the first and second family. Call C_1' and C_2' the characteristic numbers on one side of the discontinuity and C_1'' and C_2'' the characteristic numbers on the other side of the discontinuity. Because the quantity $C_1 + C_2$ defines a function $f(q)$ of q, a relation between $C_1' + C_2'$ and $C_1'' + C_2''$ can be obtained from the condition of equal pressure on both sides of the discontinuity. Calling N' and N'' the values of $C_1 + C_2/2$ on the opposite sides of the

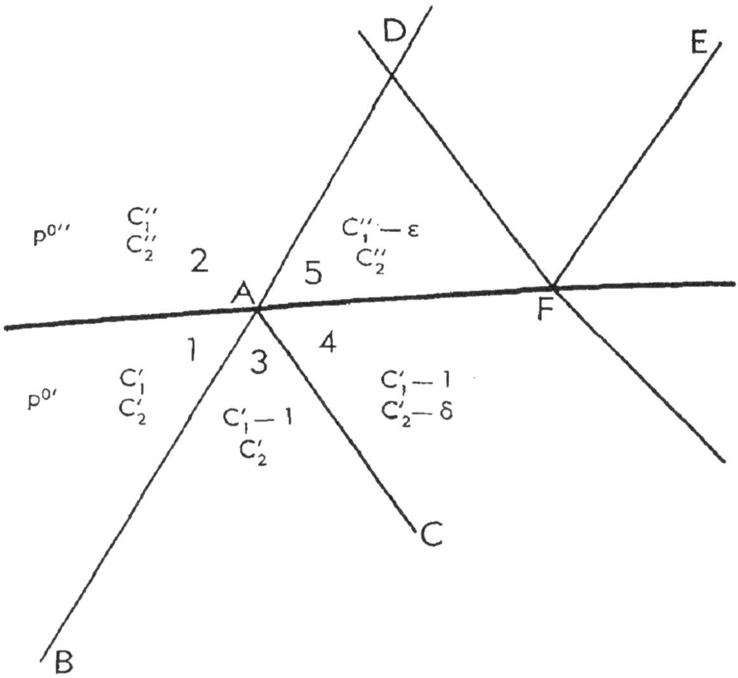

Fig. G,19a. Wave reflections at an entropy or enthalpy discontinuity in two-dimensional flow.

discontinuity, N'' is given as a function of N' by the expression

$$1 - \left(\frac{q''}{q''_{\max}}\right)^2 = \left[1 - \left(\frac{q'}{q'_{\max}}\right)^2\right]\left(\frac{p^{0'}}{p^{0''}}\right)^{\frac{\gamma-1}{\gamma}} \tag{19-2}$$

where p^0 represents the stagnation pressure. (From $q/q_{\max}$ the value of N can be determined from Table G,4.)

The other condition along the discontinuity is that the direction θ or the difference $C_1 - C_2$ must be the same on both sides of the discontinuity; therefore

$$C_1' - C_2' = C_1'' - C_2'' \tag{19-3}$$

From these two conditions and from the values of the two constants of the incoming characteristic lines the values of C_1' and C_2', or C_1'' and C_2'', at each point of discontinuity can be determined.

The actual calculations of the reflected waves can be performed by plotting $C_1'' + C_2''$ as a function of $C_1' + C_2'$.

Consider first a two-dimensional flow (Fig. G,19a) in which AB is a wave of strength equal to 1, hitting the discontinuity; AD is the transmitted wave and AC is the reflected wave. The meshes in the regions 1, 2, 3, 4, 5 are determined in the following way: In the region 1 the characteristic numbers are C_1' and C_2' while in 3 they are $C_1' - 1$ and C_2'. If δ is the intensity of the reflected wave in 4, the characteristic numbers are $C_1' - 1$, $C_2' - \delta$. Assume that ϵ is the intensity of the wave AD. Then

$$C_1' - 1 + \delta - C_2' = C_1'' - C_2'' - \epsilon \tag{19-4}$$

or

$$1 - \delta = \epsilon \tag{19-5}$$

Now, if α is the slope of the plot of $C_1'' + C_2''$ against $C_1' + C_2'$ evaluated at $C_1' + C_2'$ of region 1, then

$$\left[\frac{d(C_1'' + C_2'')}{d(C_1' + C_2')}\right]_1 = \tan\alpha = \frac{\epsilon}{1+\delta} = \frac{1-\delta}{1+\delta} \tag{19-6}$$

Therefore, knowing the value of α, the values of δ and ϵ can be determined.

$$\delta = \left(\frac{1 - \tan\alpha}{1 + \tan\alpha}\right) \tag{19-7}$$

From Eq. 19-7 it is evident that for the condition of $\tan\alpha = 1$ the reflected wave has zero intensity. If $\tan\alpha > 1$ the reflected wave has the opposite sign to that of the incoming wave (an expansion is reflected as compression) while it has the same sign if $\tan\alpha < 1$. By a similar process (which only varies in the inversion of the C indices) the intensity of the reflected wave EF can be determined.

The value of $C_1' + C_2'$, and therefore the value of M', corresponding to the condition $\tan\alpha = 1$ depends on the intensity of the discontinuity. For discontinuities of small intensity the relation between $(C_1' + C_2')$ and $(C_1'' + C_2'')$ can be obtained from the following relations. From Eq. 3-7, 4-7, and 4-8 or from the corresponding Eq. 13-8

$$\left[\frac{\partial(C_1 + C_2)}{\partial p^0}\right]_{p=\text{const}} = \cot\mu\,\frac{\partial q}{\partial p^0}\frac{1}{q} = \frac{\partial M}{\partial p^0}\,\frac{\sqrt{M^2 - 1}}{M\left(1 + \frac{\gamma - 1}{2}M^2\right)} \tag{19-8}$$

But

$$\left(\frac{\gamma - 1}{2}M^2 + 1\right) = \left(\frac{p^0}{p}\right)^{\frac{\gamma-1}{\gamma}}$$

and for a given value of the static pressure p

$$\frac{\partial M}{\partial p^0} = \frac{\left(1 + \frac{\gamma - 1}{2}M^2\right)}{M\gamma p^0} \tag{19-9}$$

Therefore

$$\left[\frac{\partial(C_1 + C_2)}{\partial \ln p^0}\right]_{p=\text{const}} = \frac{\sqrt{M^2 - 1}}{\gamma M^2} \tag{19-10}$$

For small variations of p^0 across the discontinuity Eq. 19-10 can be written in the form

$$C_1'' + C_2'' = C_2' + C_1' + \frac{\sqrt{M'^2 - 1}}{\gamma M'^2} \ln \frac{p^{0''}}{p^{0'}} \tag{19-11}$$

Eq. 19-11 permits the determination of $C_1'' + C_2''$ as a function of $C_1' + C_2'$ for small discontinuities. From Eq. 19-11 and 19-8, for a fixed value of $p^{0''}/p^{0'}$, there results

$$\frac{d(C_1'' + C_2'')}{d(C_2' + C_1')} = \tan\alpha = 1 + \frac{1 + \frac{\gamma - 1}{2} M'^2}{\gamma M'^2(M'^2 - 1)} (2 - M'^2) \ln \frac{p^{0''}}{p^{0'}} \tag{19-12}$$

Eq. 19-12 shows that $\tan\alpha = 1$ for $M' = \sqrt{2}$; therefore, for $M' = \sqrt{2}$ the reflected waves have zero intensity.

For $M'^2 > 2$, waves from a region of high $p^{0'}$ are partially reflected as waves of opposite sign (expansions as compressions and vice versa) at a discontinuity separating the regions of high $p^{0'}$ from that of a lower $p^{0''}$, while waves from the region of lower $p^{0''}$ are reflected at the discontinuity as waves of the same sign. The converse occurs for $M'^2 < 2$.

The same considerations are valid for axially symmetric flow. However, the determination of the flow field is slightly more involved because the values of the characteristic numbers C_1 and C_2 vary along the characteristic lines (Eq. 18-16 and 18-17) and therefore are not known a priori. Also in this case calculations are needed for the determination of the values of $(C_1'' + C_2'')$ as a function of $(C_1' + C_2')$ for the given value of $p^{0''}/p^{0'}$ across the discontinuity. Then along AB (Fig. G,19b) the variation of C_1 (Eq. 18-16) and along AC the variation of C_2'' (Eq. 18-17) can be determined; therefore, the values of C_1' and C_2' at A are obtained. But

$$(C_1' - C_2'')_A = (C_1'' - C_2')_A \tag{19-13}$$

and

$$(C_1'' + C_2'')_A = (C_1'' + C_2'')_B + (\tan\alpha)_B[(C_1' + C_2')_A - (C_1' + C_2')_B] \tag{19-14}$$

where $(C_1'' + C_1'')_B$ is the value corresponding to $(C_1' + C_2')_B$ for the condition of same static pressure and $(\tan\alpha)_B$ is defined by Eq. 19-6. Therefore, C_1', C_2', C_1'', and C_2'' at A can be determined.

The position of A is obtained in the first approximation from the tangent to the streamline at D and the tangent to the characteristic line of the first family at B. From A the tangent to the characteristic line of the second family at an interpolated point C between D and E is drawn. The values of C_1'' and C_2'' at C are obtained from linear interpolation between the corresponding values at D and E.

Because of the simplicity of the determination of the changes of flow properties across discontinuities for two-dimensional problems, flows having continuous distribution of vorticity are often analyzed by approximating the actual vortex distribution by an equivalent vortex distribution made up by a finite number of discontinuities of finite strength.

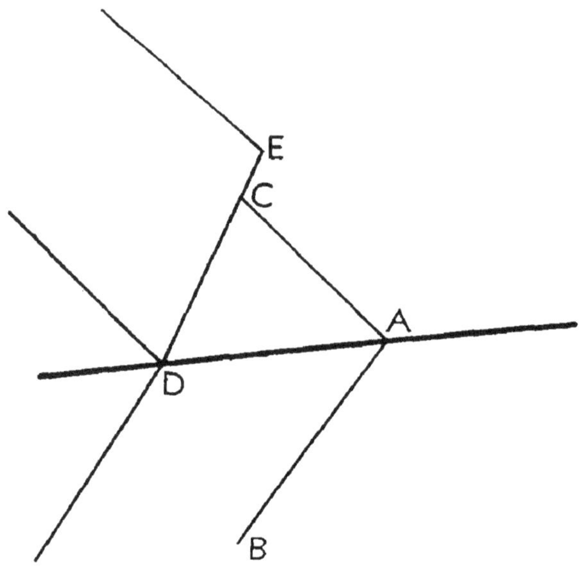

Fig. G,19b. Wave reflections at an entropy or enthalpy discontinuity in axially symmetric flow.

CHAPTER 5. CHARACTERISTICS METHODS FOR PROBLEMS IN THREE INDEPENDENT VARIABLES

G,20. The Equations of Motion Along Characteristic Surfaces. Consider a flow, having a constant total enthalpy, defined by a velocity vector which in Cartesian coordinates has components u, v, w, and by a vorticity vector $\mathbf{\Omega}$ of components ξ, η, ζ.

From Euler's equations, the continuity equation, and the equation of conservation of energy the following expressions can be obtained (D,32; also III,A)

$$\left(1 - \frac{u^2}{a^2}\right)\frac{\partial u}{\partial x} + \left(1 - \frac{v^2}{a^2}\right)\frac{\partial v}{\partial y} + \left(1 - \frac{w^2}{a^2}\right)\frac{\partial w}{\partial z} - \frac{uv}{a^2}\left(\frac{\partial u}{\partial y} + \frac{\partial v}{\partial x}\right) - \frac{uw}{a^2}\left(\frac{\partial u}{\partial z} + \frac{\partial w}{\partial x}\right) - \frac{vw}{a^2}\left(\frac{\partial v}{\partial z} + \frac{\partial w}{\partial y}\right) = 0 \quad (20\text{-}1)$$

$$\frac{\partial w}{\partial y} - \frac{\partial v}{\partial z} = \xi \tag{20-2}$$

$$\frac{\partial u}{\partial z} - \frac{\partial w}{\partial x} = \eta \tag{20-3}$$

$$\frac{\partial v}{\partial x} - \frac{\partial u}{\partial y} = \zeta \tag{20-4}$$

and along the streamline between discontinuities

$$a^2 = \left(\frac{\partial p}{\partial \rho}\right)_{s=\text{const}} = \gamma \frac{p}{\rho}$$

For irrotational flow ξ, η, ζ are zero in all the field. Moreover, for a flow having constant total enthalpy, the quantities ξ, η, ζ are related to the entropy gradients existing in the flow field and are defined by the following expressions (D,32, Eq. 32-9; also III,A,11)

$$-w\eta + v\zeta = -\frac{\partial s}{\partial x}\frac{a^2}{\gamma \mathcal{R}} \tag{20-5}$$

$$-u\zeta + w\xi = -\frac{\partial s}{\partial y}\frac{a^2}{\gamma \mathcal{R}} \tag{20-6}$$

$$-v\xi + u\eta = -\frac{\partial s}{\partial z}\frac{a^2}{\gamma \mathcal{R}} \tag{20-7}$$

and

$$\frac{\partial \xi}{\partial x} + \frac{\partial \eta}{\partial y} + \frac{\partial \zeta}{\partial z} = 0 \tag{20-8}$$

In order to find the equations of motion along characteristic conoids consider an orthogonal coordinate system t, n, N with origin at the point P under consideration and the t axis parallel to the velocity vector at P (Fig. G,20), and consider in the plane n, t a straight line L with slope expressed as

$$\tan \lambda = \left(\frac{dn}{dt}\right)_N \tag{20-9}$$

In this coordinate system, if v_t, v_n, v_N are the components of the velocity at any point along the three axes, and Ω_t, Ω_n, Ω_N are the components of the vorticity, Eq. 20-1 becomes

$$\left(1 - \frac{q^2}{a^2}\right)\frac{\partial v_t}{\partial t} + \frac{\partial v_n}{\partial n} + \frac{\partial v_N}{\partial N} = 0 \tag{20-10}$$

Eq. 20-6 becomes

$$-q\left(\frac{\partial v_n}{\partial t} - \frac{\partial v_t}{\partial n}\right) = -q\Omega_N = -\frac{\partial s}{\partial n}\frac{a^2}{\gamma \mathcal{R}} \tag{20-11}$$

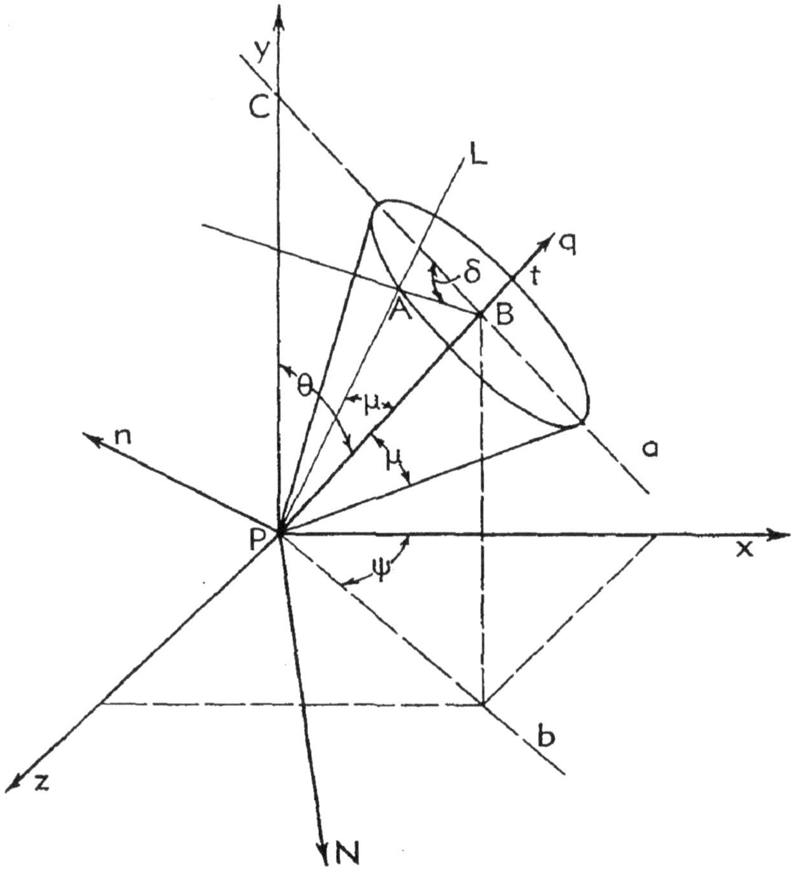

Fig. G,20. The equations of characteristics for three-dimensional rotational flow.

and Eq. 20-7 becomes

$$q\left(\frac{\partial v_t}{\partial N} - \frac{\partial v_N}{\partial t}\right) = -\frac{\partial s}{\partial N}\frac{a^2}{\gamma \mathcal{R}} = q\Omega_n \tag{20-12}$$

while

$$\left(\frac{\partial v_N}{\partial n} - \frac{\partial v_n}{\partial N}\right) = \Omega_t \tag{20-13}$$

and

$$\left(\frac{\partial \Omega_t}{\partial t} + \frac{\partial \Omega_n}{\partial n} + \frac{\partial \Omega_N}{\partial N}\right) = 0 \tag{20-14}$$

Along the line L defined by Eq. 20-9

$$dv_t = \left(\frac{\partial v_t}{\partial t} + \frac{\partial v_t}{\partial n}\frac{dn}{dt}\right) dt \tag{20-15}$$

and

$$dv_n = \left(\frac{\partial v_n}{\partial t} + \frac{\partial v_n}{\partial n}\frac{dn}{dt}\right) dt \tag{20-16}$$

therefore Eq. 20-10 becomes

$$\frac{dv_t}{dt}\left(1-\frac{q^2}{a^2}\right)-\lambda\frac{dv_n}{dt}\left(1-\frac{q^2}{a^2}\right)+\frac{dv_n}{\partial n}\left[1+\lambda^2\left(1-\frac{q^2}{a^2}\right)\right]+\frac{\partial v_N}{\partial N}$$

$$+\lambda\Omega_N\left(1-\frac{q^2}{a^2}\right)=0 \quad (20\text{-}17)$$

Therefore along each straight line L of slopes

$$\left(\frac{dn}{dt}\right)_N=\lambda_{1,2}=\pm\sqrt{\frac{1}{\frac{q^2}{a^2}-1}}=\pm\tan\mu \quad (20\text{-}18)$$

and

$$\left(\frac{dN}{dt}\right)_n=0, \qquad \left(\frac{dN}{dn}\right)_t=0$$

the following equations are valid

$$\cot\mu\frac{dv_t}{dt}\mp\frac{dv_n}{dt}\pm\Omega_N-\tan\mu\frac{dv_N}{dN}=0 \quad (20\text{-}19)$$

$$\frac{dv_N}{dt}=-\Omega_n\pm\tan\mu\Omega_t+\frac{dv_t}{dN}\pm\tan\mu\frac{dv_n}{dN} \quad (20\text{-}20)$$

The line L is real only when $q/a > 1$.

Eq. 20-19 and 20-20 are valid along each two lines λ_1 and λ_2 inclined at an angle $\pm\tan\mu$ to the direction of the local velocity in each plane t, n, containing the velocity vector q. However, the direction n is arbitrary; therefore, Eq. 20-19 and 20-20 are valid along each tangent L to the Mach cone from the point P considered, and along the normal N at P to the plane of the tangent L and the velocity vector q.

For practical use it is more convenient to refer Eq. 20-19 and 20-20 to a fixed coordinate system independent of the direction of the local velocity. If the velocity vector is defined by its magnitude q, and its directions θ, ψ in a spherical coordinate system (Fig. G,20), when the line L tangent to the Mach cone is in the plane BCP, the following equations hold

$$qd\theta = dv_n \quad (20\text{-}22)$$

$$q\sin\theta\, d\psi = dv_N \quad (20\text{-}23)$$

When the line L is a generic line tangent to the Mach cone in a plane L, t inclined at an angle δ to the plane t, y, then

$$dv_n = qd\theta\cos\delta + q\sin\theta\sin\delta d\psi \quad (20\text{-}24)$$

$$dv_N = -qd\theta\sin\delta + q\sin\theta\cos\delta d\psi \quad (20\text{-}25)$$

and Eq. 20-19 and 20-20 become

$$\frac{1}{q}\frac{dq}{dL} - \frac{d\theta}{dL}\tan\mu\cos\delta - \sin\theta\sin\delta\tan\mu\frac{d\psi}{dL} + \frac{\Omega_N}{q}\sin\mu + \frac{\sin^2\mu}{\cos\mu}\left(\sin\delta\frac{d\theta}{dN} - \sin\theta\cos\delta\frac{d\psi}{dN}\right) = 0 \quad (20\text{-}26)$$

and

$$\frac{1}{q}\frac{dq}{dN} + \tan\mu\frac{d\theta}{dN}\cos\delta + \tan\mu\sin\theta\sin\delta\frac{d\psi}{dN} - \frac{\Omega_n}{q} + \tan\mu\frac{\Omega_t}{q} + \frac{\sin\delta}{\cos\mu}\frac{d\theta}{dL} - \frac{\sin\theta\cos\delta}{\cos\mu}\frac{d\psi}{dL} = 0 \quad (20\text{-}27)$$

Combination of these yields

$$\frac{1}{q}\frac{dq}{dL}\frac{\sin\delta}{\sin\mu} - \frac{d\psi}{dL}\frac{\sin\theta}{\cos\mu} + \frac{1}{q}\frac{dq}{dN}\cos\delta + \tan\mu\frac{d\theta}{dN} = -\frac{1}{q}(\Omega_N\sin\delta + \Omega_n\cos\delta + \Omega_t\cos\delta\tan\mu) \quad (20\text{-}28)$$

and

$$\frac{1}{q}\frac{dq}{dL}\cos\delta - \frac{d\theta}{dL}\tan\mu - \frac{1}{q}\frac{dq}{dN}\sin\mu\sin\delta - \frac{d\psi}{dN}\tan\mu\sin\mu\sin\theta = -\frac{\sin\mu}{q}(\Omega_N\cos\delta + \Omega_n\sin\delta - \Omega_t\tan\mu\sin\delta) \quad (20\text{-}29)$$

where both signs of Eq. 20-19 and 20-20 are included in the definition of δ, N, and θ.

Eq. 20-28 and 20-29 give the variation of the velocity vector along any tangent to the Mach cone and along the normal N to the plane L, t and they become equal to the equations for two-dimensional flow when δ is zero and ψ is zero.

G,21. The Characteristic Surfaces. Consider a point P of the flow field and the Mach conoid $PABCD$ from P (Fig. G,21a). For each line PQ passing through the point P, two characteristic surfaces, $PQEA$ and $PCFQ$, can be obtained as the envelope of all the Mach conoids having their apexes on the line PQ and their generatrices inclined at the Mach angle to the local velocity.

Each Mach conoid $PABCD$ having its apex P on the line PQ is tangent to each characteristic surface along one generatrix L (PA or PC of Fig. G,21a) along which Eq. 20-28 and 20-29 can be applied. The lines PA or PC common to both the Mach conoids and to the characteristic surfaces are called bicharacteristic lines.

Because the line PQ passing through P is an arbitrary line and since an infinite number of such lines exist, an infinite number of characteristic surfaces $PQAE$ also exist.

When the line passing through P changes, the bicharacteristic line AP also changes and moves around the Mach conoid $PABCA$. (The corresponding value of δ in Eq. 20-28 and 20-29 changes accordingly.)

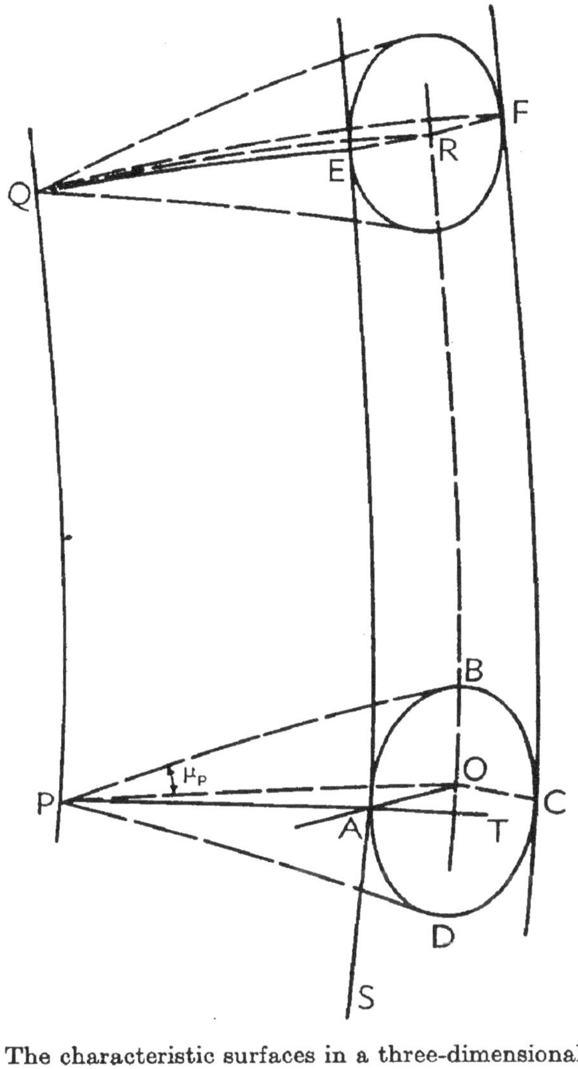

Fig. G,21a. The characteristic surfaces in a three-dimensional steady flow.

If the line PQ has a finite length, the two characteristic surfaces are connected at the ends by the Mach conoids whose apexes are at the ends of the line.

Consider now the plane tangent to the characteristic surface $PAEQ$ at the point A. In this plane two tangents can be considered, one (AT) tangent to the bicharacteristic line PA, and another (AS) normal to the bicharacteristic line (Fig. G,21b).

The plane ATS is tangent to the characteristic surface $PAQE$ of Fig. G,21a, and to the Mach cone from A; the line AS is perpendicular to a generatrix of the Mach cone from A, and lies in the plane ATS. Consequently, the line AS must be perpendicular to the plane containing the generatrix (which is the bicharacteristic line) and the axis of the Mach cone from A (which coincides with the velocity vector at A).

Along TA and AS Eq. 20-28 and 20-29 can be used. These equations define the variation of the velocity along a characteristic surface (TA

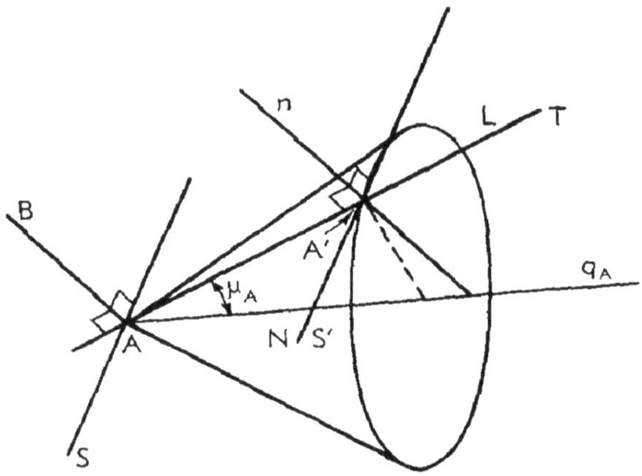

Fig. G,21b. The bicharacteristic line and the binormal.

corresponds to L and AS to N). The analytic expression for the characteristic surfaces can be determined from the consideration that the normal AB to the characteristic surface at any point A (Fig. G,21b) must be normal to the plane of AS, and AT, which is a plane tangent to the Mach cone at A. Therefore, the normal AB must be inclined at the complement of the Mach angle with respect to the velocity vector.

If the surface is expressed in the form

$$y = f(x, z) \tag{21-1}$$

at the characteristic surface the equation

$$\left(1 - \frac{u^2}{a^2}\right)(f'_x)^2 + 2\frac{uv}{a^2}f'_x + \left(1 - \frac{v^2}{a^2}\right) - \frac{2uw}{a^2}f'_xf'_z + \frac{2vw}{a^2}f'_z + \left(1 - \frac{w^2}{a^2}\right)(f'_z)^2 = 0 \tag{21-2}$$

must be satisfied; where f'_x, f'_z are the partial derivatives with respect to x and z of the expression (21-1). This can be shown from the following considerations. The direction cosines of the normal AB to the surface given by Eq. 21-1 at a given point are given by

$$\cos(AB, x) = \frac{f'_x}{\sqrt{1 + (f'_x)^2 + (f'_z)^2}} \tag{21-3}$$

$$\cos(AB, y) = -\frac{1}{\sqrt{1 + (f'_x)^2 + (f'_z)^2}} \tag{21-4}$$

$$\cos(AB, z) = \frac{f'_z}{\sqrt{1 + (f'_x)^2 + (f'_z)^2}} \tag{21-5}$$

and the direction cosines of the velocity vector are

$$\cos(x, q) = \frac{u}{q} \tag{21-6}$$

$$\cos(y, q) = \frac{v}{q} \tag{21-7}$$

$$\cos(z, q) = \frac{w}{q} \tag{21-8}$$

Therefore, if the line AB is inclined with respect to the velocity vector at an angle

$$\cos(AB, q) = \sin\mu \tag{21-9}$$

Eq. 21-2 must be valid.

At each point P of coordinates x_0, y_0, z_0 and velocity components u, v, w, the Mach conoid in the forward and downward direction must be tangent to the local Mach cone which is defined by the expression

$$\begin{aligned}(v^2 + w^2 - a^2)(x - x_0)^2 + (u^2 + w^2 - a^2)(y - y_0)^2 \\ + (u^2 + v^2 - a^2)(z - z_0)^2 - 2uv(x - x_0)(y - y_0) \\ - 2uw(x - x_0)(z - z_0) - 2vw(y - y_0)(z - z_0) = 0\end{aligned} \tag{21-10}$$

G,22. Numerical Methods of Characteristics for General Problems. By means of Eq. 20-28, 20-29, and 21-1 or 21-10 the position and the flow properties at the intersections of a hexahedral network can be determined if the flow properties at given boundary conditions are known.

The following procedure may be applied. Consider three points A, B, and C where the flow properties are given (Fig. G,22a). The characteristic surfaces containing the lines AB, BC, CA define the position of a point P. If the distances between the points A, B, and C are small, the planes tangent to the characteristic surfaces at the middle points between A and B, A and C, and B and C can be substituted for the characteristic surfaces ABP, ACP, BCP, and the point P can be assumed in first approximation to be defined by a tetrahedron having plane surfaces. Considering a cone inscribed in the tetrahedron at the points EFG, the bicharacteristic lines EP, FP, GP can be determined, and by means of Eq. 20-28 and 20-29 the velocity components at P obtained [16].

The determination of the flow field by means of a hexahedral network involves a large amount of numerical work and the introduction of the boundary conditions is difficult. Therefore, a different network, consisting of intersections of characteristic surfaces with planes parallel to the reference planes, is often more convenient [17,18]. The intersections of characteristic surfaces with reference planes are not bicharacteristics

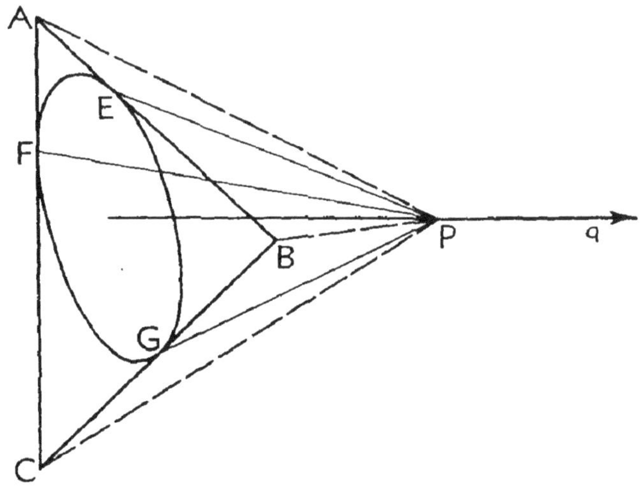

Fig. G,22a. The hexahedral characteristic network for three-dimensional supersonic flow analysis.

and therefore Eq. 20-28 and 20-29 must be transformed in order to be used in this different network.

Consider a characteristic surface $ABGC$ (Fig. G,22b) passing through the line AB in the plane $x = \text{const}$. If C is a point on the line σ_A at the intersection of the plane $z = z_A$ with the characteristic surface considered, Eq. 20-28 and 20-29 must be applied along the line LC, tangent to the fore Mach conoid from C, and along LA normal to the bicharacteristic line LC. However, by means of an axis transformation two equations expressing the variation of velocity components along the lines $\sigma_A = AC$ and $\tau_A = AB$, which are the intersections of the characteristic surface considered with the planes $z = z_A$ and $x = x_A$, can be obtained, and, therefore, one expression which relates the flow properties at C and at A can be obtained.

In this transformation the assumption is implied that the flow at C depends on the flow properties at A, which is outside of the Mach conoid from C, and therefore the rule of forbidden signals is not exactly satisfied. However, the assumption is consistent with the approximation used in the analysis where finite difference procedures are used. Consider, for example, the point P in Fig. G,22a. In order to determine the flow at P from the points F, E, and G, it is assumed that the flow properties at the points F, E, and G can be obtained by interpolation of the corresponding flow properties at two neighboring points A, B, and C respectively. This assumption is similar to the assumption that the flow at P can be

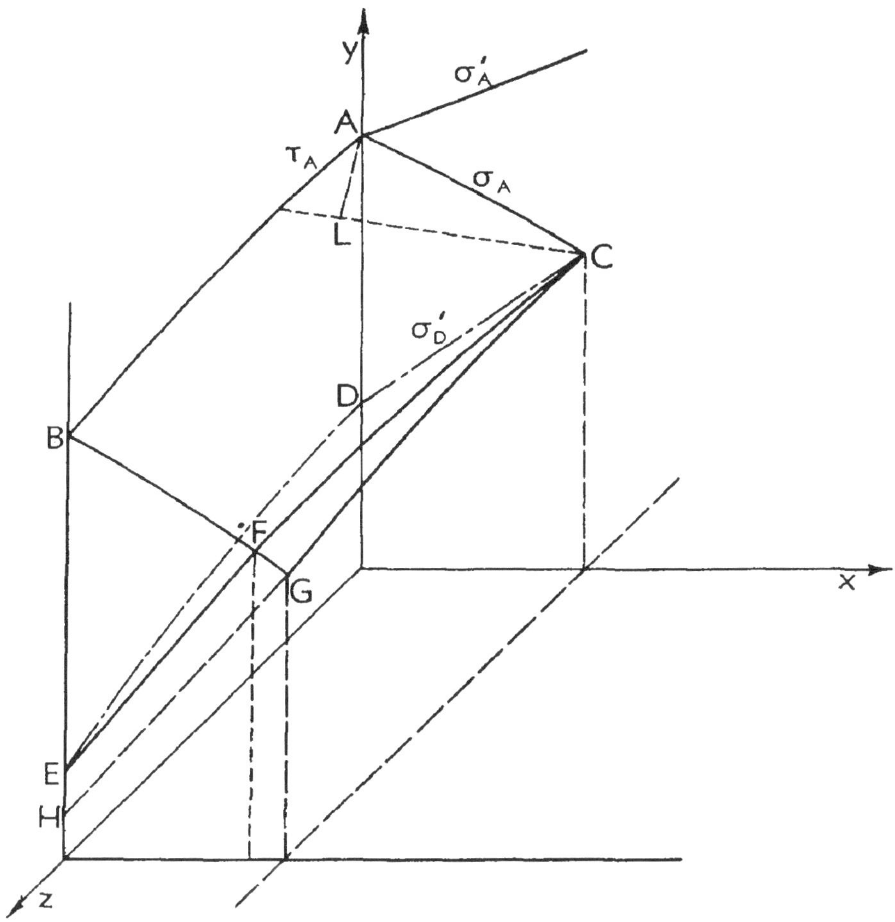

Fig. G,22b. The characteristic network for three-dimensional supersonic flow analysis using characteristic surfaces and planes parallel to reference planes.

obtained by a finite difference procedure, i.e. that the flow can be considered as continuous.

The transformation gives equations which relate the variations of velocity components u, v, w along the lines σ_A or σ'_A in the plane $z = z_A$ and along the line τ_A in the plane $x = x_A$ (Fig. G,22b). Therefore, the flow properties at a point C are calculated from the equations applied along AC, DC, and AB. The line σ'_A is the intersection with the plane $z = z_A$ of the other characteristic surface passing at the line BA.

The equations valid along the lines σ, σ', and τ can be obtained from

Eq. 20-28 and 20-29, or directly from Eq. 20-1. If λ_1 and λ_2 are the inclinations of the lines σ'_A and σ_A at A defined by

$$\left(\frac{dy}{dx}\right)_{\sigma_A} = \lambda_1 \tag{22-1}$$

$$\left(\frac{dy}{dx}\right)_{\sigma'_A} = \lambda_2 \tag{22-2}$$

Then

$$(du)_{\sigma_A,\sigma'_A} = \left(\frac{\partial u}{\partial x} + \lambda_{1,2}\frac{\partial u}{\partial y}\right) dx \tag{22-3}$$

$$(dv)_{\sigma_A,\sigma'_A} = \left(\frac{\partial v}{\partial x} + \lambda_{1,2}\frac{\partial v}{\partial y}\right) dx \tag{22-4}$$

$$(dw)_{\sigma_A,\sigma'_A} = \left(\frac{\partial w}{\partial x} + \lambda_{1,2}\frac{\partial w}{\partial y}\right) dx \tag{22-5}$$

If K_A is the inclination of the line τ_A in the plane $x = x_A$, defined by

$$\left(\frac{dy}{dz}\right)_{\tau_A} = K_A \tag{22-6}$$

Then

$$(du)_{\tau_A} = \left(\frac{\partial u}{\partial z} + K_A\frac{\partial u}{\partial y}\right) dz \tag{22-7}$$

$$(dv)_{\tau_A} = \left(\frac{\partial v}{\partial z} + K_A\frac{\partial v}{\partial y}\right) dz \tag{22-8}$$

$$(dw)_{\tau_A} = \left(\frac{\partial w}{\partial z} + K_A\frac{\partial w}{\partial y}\right) dz \tag{22-9}$$

where λ_1, λ_2, and K are related by Eq. 21-2 because $ABGC$ is a characteristic surface. Therefore

$$\left(1 - \frac{u^2}{a^2}\right)\lambda^2 + \frac{2uv}{a^2}\lambda - \frac{2uw}{a^2}\lambda K + \frac{2vw}{a^2}K + \left(1 - \frac{w^2}{a^2}\right)K^2 + \left(1 - \frac{v^2}{a^2}\right) = 0 \tag{22-10}$$

For each value of K_A at A along τ_A two values of λ (λ_1 and λ_2) along σ_A and σ'_A can be obtained.

By substituting Eq. 22-3 to 22-9, 20-2 to 20-4, and 22-10 into Eq. 20-1, the following two equivalent expressions are obtained

$$\left\{\left(1-\frac{u^2}{a^2}\right)\frac{du}{dx}-\left[\left(1-\frac{u^2}{a^2}\right)\lambda+\frac{2uv}{a^2}\right]\frac{dv}{dx}-\frac{2uw}{a^2}\frac{dw}{dx}\right\}_{\sigma_A,\sigma'_A}$$
$$+\left\{\frac{dv}{dz}\left[2\lambda\frac{uw}{a^2}-\left(1-\frac{w^2}{a^2}\right)K-\frac{2vw}{a^2}\right]+\left(1-\frac{w^2}{a^2}\right)\frac{dw}{dz}\right\}_{\tau_A}$$
$$+\zeta\left[\lambda\left(1-\frac{u^2}{a^2}\right)+\frac{uv}{a^2}\right]+\xi\left[\frac{2uw}{a^2}\lambda-\frac{vw}{a^2}-\left(1-\frac{w^2}{a^2}\right)K\right]$$
$$-\frac{uw}{a^2}\eta=0 \quad (22\text{-}11)$$

$$\left\{\left(1-\frac{u^2}{a^2}\right)\frac{du}{dx}-\frac{dv}{dx}\left[\left(1-\frac{u^2}{a^2}\right)\lambda+\frac{2uv}{a^2}-\frac{2uw}{a^2}K\right]\right\}_{\sigma_A,\sigma'_A}$$
$$+\left\{-\frac{2uw}{a^2}\frac{du}{dz}-\frac{dv}{dz}\left[\frac{2vw}{a^2}+\left(1-\frac{w^2}{a^2}\right)K\right]+\left(1-\frac{w^2}{a^2}\right)\frac{dw}{dz}\right\}_{\tau_A}$$
$$+\zeta\left[\lambda\left(1-\frac{u^2}{a^2}\right)+\frac{uv}{a^2}-\frac{2uw}{a^2}K\right]-\xi\left[\frac{vw}{a^2}+\left(1-\frac{w^2}{a^2}\right)K\right]$$
$$+\frac{uw}{a^2}\eta=0 \quad (22\text{-}12)$$

or by combining Eq. 22-11 and 22-12

$$\left(K\frac{dv}{dx}+\frac{dw}{dx}\right)_{\sigma_A,\sigma'_A}-\left(\frac{du}{dz}+\lambda\frac{dv}{dz}\right)_{\tau_A}-K\zeta-\xi\lambda+\eta=0 \quad (22\text{-}13)$$

Eq. 22-11, 22-12, and 22-13 permit the determination of the flow field by means of a step-by-step procedure.

Assume that the velocity components along the lines BA and DE and the shape of the lines BA and DE in the plane $x = x_A$ are known (Fig. G,22b). From Eq. 22-10 the value of λ_2 along σ_A can be determined, because K at A is known and the coefficients of the equations are also known. From the same equation the value of λ_1 along σ'_D at D can be determined. Now in a first approximation the tangents at A and D can be substituted for the lines σ_A and σ'_D and the position of the point C in the plane $z = z_A = z_D$ can be determined.

Along AB and DE, du/dz, dv/dz, dw/dz are known, therefore, by a method of finite differences applied along AC and DC, the value of u_C and v_C can be obtained from Eq. 22-12. In this equation $\lambda = \lambda_{2A}$ and $K = K_A$ along AC, while $\lambda = \lambda_{1D}$ and $K = K_D$ along DC. From Eq. 22-13 the value of w can be obtained at C.

In a similar way the velocity components at F can be determined. Since the point F obtained from B and E is not in the plane $x = x_C$, an extrapolated value at the point G along BF at $x = x_C$ must be determined by assuming

$$\left(\frac{du}{dx}\right)_{BG}=\left(\frac{du}{dx}\right)_{BF}, \text{ etc.} \quad (22\text{-}14)$$

From the values obtained in the first approximation a second approximation can be determined by iteration calculations. The point G can be determined by interpolating a point H on the line $x = x_B$, $z = z_B$ and then determining G from B and H (Fig. G,22b). In order to determine the vorticity at each point, the streamlines from C, G, etc. must be determined. The variation of ξ, η, and ζ can be obtained from Eq. 20-5 to 20-8.

G,23. Two-dimensional Unsteady Flow. The considerations presented for the analysis of steady supersonic problems in three independent variables can be extended with few modifications to the analysis of two-dimensional unsteady problems.

Consider, for example, an irrotational two-dimensional flow in nonsteady motion where

$$u = \frac{\partial \phi}{\partial x} \tag{23-1}$$

$$v = \frac{\partial \phi}{\partial y} \tag{23-2}$$

$$\frac{dp}{d\rho} = \gamma \frac{p}{\rho} = a^2 \tag{23-3}$$

$$\frac{u^2 + v^2}{2} + \int \frac{dp}{\rho} = -\frac{\partial \phi}{\partial t} = \left(\frac{u^2 + v^2}{2} + \frac{a^2}{\gamma - 1}\right) \tag{23-4}$$

From the continuity equation and Euler's equations the following expression can be obtained (III,A,9)

$$(u^2 - a^2)\frac{\partial^2 \phi}{\partial x^2} + (v^2 - a^2)\frac{\partial^2 \phi}{\partial y^2} + \frac{\partial^2 \phi}{\partial t^2} + 2v\frac{\partial^2 \phi}{\partial y \partial t} + 2u\frac{\partial^2 \phi}{\partial x \partial t} + 2uv\frac{\partial^2 \phi}{\partial x \partial y} = 0 \tag{23-5}$$

or

$$a^2\left(\frac{\partial u}{\partial x} + \frac{\partial v}{\partial y}\right) = -\left(u\frac{\partial a}{\partial x} + v\frac{\partial a}{\partial y} + \frac{\partial a}{\partial t}\right)\frac{2a}{\gamma - 1}$$

Eq. 23-5 is similar to Eq. 20-1 and has characteristic surfaces

$$t = f(x, y) \tag{23-6}$$

defined by

$$(a^2 - u^2)(f'_x)^2 + (a^2 - v^2)(f'_y)^2 - 1 + 2vf'_y + 2uf'_x - 2uvf'_xf'_y = 0 \tag{23-7}$$

where f'_x and f'_y are the partial derivatives of $f(x, y)$ with respect to x and y. The direction cosines of the normal to the characteristic surfaces are proportional to $f'(x)$, $f'(y)$, and -1.

Eq. 23-7 can be written in the form

$$a^2[(f'_x)^2 + (f'_y)^2] = (uf'_x + vf'_y - 1)^2 \tag{23-8}$$

and therefore can be expressed parametrically by

$$\frac{f'_x}{\cos\delta} = \frac{f'_y}{\sin\delta} = \frac{1}{u\cos\delta + v\sin\delta + a} \tag{23-9}$$

where δ is defined in Fig. G,23, and the corresponding bicharacteristic direction is

$$(u + a\cos\delta,\ v + a\sin\delta,\ 1)$$

The characteristic surfaces are real for any value of the velocity components. Consider the coordinate system x, y, t of Fig. G,23. At the time

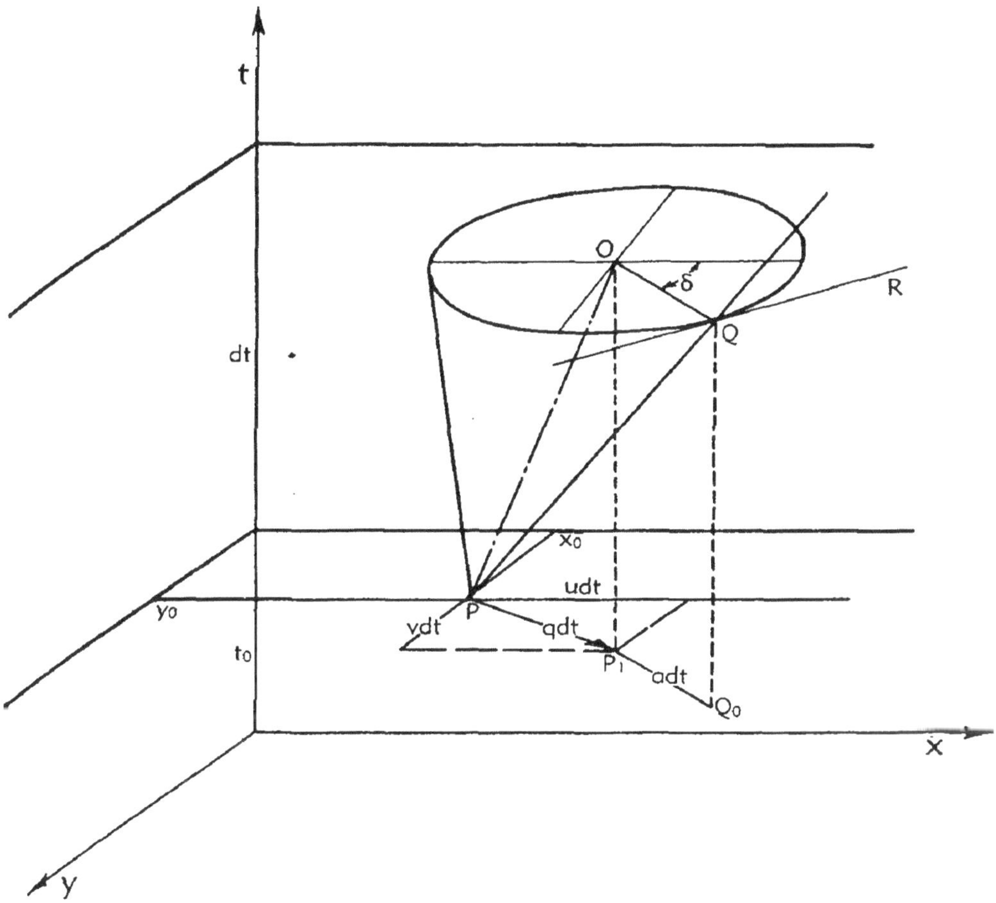

Fig. G,23. The Mach conoid in a two-dimensional unsteady flow.

t_0, a fluid particle is located at point P with coordinates x_0, y_0, t_0, and has a velocity q with components u and v. After a time dt the particle has moved to P_1 with coordinates $x = x_0 + udt$ and $y = y_0 + vdt$. An infinitesimal disturbance produced at the time t_0 at P, after an interval dt has propagated inside a circle of center $x_0 + udt$, $y_0 + vdt$ and radius adt, defined by the equation

$$(x - x_0 - udt)^2 + (y - y_0 - vdt)^2 = (adt)^2 \tag{23-10}$$

Therefore, the Mach cone from P is defined by the equation

$$(x - x_0)^2 + (y - y_0)^2 + (t - t_0)^2(u^2 + v^2 - a^2) - 2u(t - t_0)(x - x_0) - 2v(t - t_0)(y - y_0) = 0 \tag{23-11}$$

Consider now the line PQ defined by (Fig. G,23)

$$\left(\frac{dx}{adt}\right)_{PQ} = \frac{u}{a} + \cos \delta \tag{23-12}$$

$$\left(\frac{dy}{adt}\right)_{PQ} = \frac{v}{a} + \sin \delta \tag{23-13}$$

and the line QR normal to the line OQ in the plane $t =$ const. If OQ is inclined at a slope δ to the x axis

$$\left(\frac{dy}{dx}\right)_{OQ} = \tan \delta \tag{23-14}$$

$$\left(\frac{dt}{dy}\right)_{OQ} = \left(\frac{dt}{dx}\right)_{OQ} = 0 \tag{23-15}$$

Along the bicharacteristic line PQ Eq. 23-9 is valid; therefore, if $\partial/\partial l$ indicates the variation along PQ and $\partial/\partial n$ the variation along QR, from Eq. 23-5 and 23-8 the following expressions are obtained

$$(a + v \sin \delta)\left(\frac{\partial u}{\partial l} - \frac{\partial v}{\partial n}\right) - v \cos \delta \left(\frac{\partial v}{\partial l} + \frac{\partial u}{\partial n}\right) + \frac{2a}{\gamma - 1} \cos \delta \frac{\partial a}{\partial l} + \frac{2}{\gamma - 1}(v + a \sin \delta)\frac{\partial a}{\partial n} = 0 \tag{23-16}$$

$$\sin \delta \left(\frac{\partial u}{\partial l} - \frac{\partial v}{\partial n}\right) - \cos \delta \left(\frac{\partial v}{\partial l} + \frac{\partial u}{\partial n}\right) + \frac{2}{\gamma - 1}\frac{\partial a}{\partial n} = 0 \tag{23-17}$$

which give

$$\left(\frac{\partial u}{\partial l} - \frac{\partial v}{\partial n}\right) + \frac{2}{\gamma - 1}\left(\cos \delta \frac{\partial a}{\partial l} + \sin \delta \frac{\partial a}{\partial n}\right) = 0 \tag{23-18}$$

$$\left(\frac{\partial u}{\partial n} + \frac{\partial v}{\partial l}\right) + \frac{2}{\gamma - 1}\left(\sin \delta \frac{\partial a}{\partial l} - \cos \delta \frac{\partial a}{\partial n}\right) = 0 \tag{23-19}$$

Eq. 23-18 and 23-19 can be used for numerical computation along a hexahedral grid as discussed for steady three-dimensional flow. The same transformations used for steady phenomena can be used in order to simplify the practical applications [*16*].

If $v = 0$ and $\delta = 0$ or $\delta = \pi$

$$u \pm \frac{2a}{\gamma - 1} = \text{const}$$

along

$$\frac{dx}{dt} = u \pm a$$

which are the equations for nonsteady one-dimensional flow.

CHAPTER 6. LINEARIZED CHARACTERISTICS METHODS

G,24. Linearized Characteristic Equations. The use of the method of characteristics for the solution of supersonic flow problems requires numerical procedures which must be repeated for each set of boundary conditions. Therefore, the investigations by the characteristics methods of the effects on the flow field of changes of any parameter of the boundary conditions are difficult and lengthy and procedures which simplify such investigations with some sacrifice of precision in the results are valuable for practical applications.

A simplification which can be introduced into the procedure is based on the assumption that one of the velocity components, or the variation of the velocity components with a given parameter, can be considered small, so that terms of second or higher order in the quantities considered small can be neglected.

With this assumption, the flow field for any condition can be represented by the superposition of a linearized flow perturbation on a nonlinear basic flow field [*19*]. The linear flow field then represents the variation of the basic flow due to the changes in a given geometrical parameter or due to the presence of a small velocity component which was of zero magnitude in the basic flow.

Because of the linearization, the superposed flow field must be determined only once for all the range where the simplification is acceptable, and therefore the amount of numerical work is appreciably reduced.

Consider, for example, a flow field defined by a velocity vector $\mathbf{q}(u, v, w)$ whose components can be expressed in the form

$$u = u_0 + \sum_1^N a_m u_m \tag{24-1}$$

$$v = v_0 + \sum_1^N a_m v_m \tag{24-2}$$

$$w = \sum_1^N a_m w_m \tag{24-3}$$

and assume that each coefficient a_m is small, so that terms of the order of $a_n a_m$ or higher can be neglected.

In this flow the basic flow field is represented by a velocity vector $\mathbf{q}_0(u_0, v_0)$ and therefore is a two-dimensional flow if Cartesian coordinates are used, or an axially symmetric flow if cylindrical coordinates are used.

The linearized flow is represented by a summation of N three-dimensional flow fields, each of which is proportional to the corresponding coefficient a_m.

In the approximation accepted Eq. 20-1 becomes for Cartesian coordinates

$$\frac{\partial u_0}{\partial x}\left(1 - \frac{u_0^2}{a_0^2}\right) + \frac{\partial v_0}{\partial y}\left(1 - \frac{v_0^2}{a_0^2}\right) - \frac{u_0 v_0}{a_0^2}\left(\frac{\partial u_0}{\partial y} + \frac{\partial v_0}{\partial x}\right)$$
$$= -\sum a_m \left[\frac{\partial w_m}{\partial z} + \frac{\partial u_m}{\partial x}\left(1 - \frac{u_0^2}{a_0^2}\right) + \frac{\partial v_m}{\partial y}\left(1 - \frac{v_0^2}{a_0^2}\right)\right.$$
$$- \frac{u_0 v_0}{a_0^2}\left(\frac{\partial v_m}{\partial x} + \frac{\partial u_m}{\partial y}\right) - \frac{\partial u_0}{\partial x}\left(\frac{2u_0 u_m + u_0^2 A_m}{a_0^2}\right) - \frac{\partial v_0}{\partial y}\left(\frac{2v_0 v_m + v_0^2 A_m}{a_0^2}\right)$$
$$\left. - \left(\frac{\partial u_0}{\partial y} + \frac{\partial v_0}{\partial x}\right)\frac{u_0 v_0 A_m + u_0 v_m + u_m v_0}{a_0^2}\right] \tag{24-4}$$

where

$$A_m = \frac{\gamma - 1}{a_0^2}(u_0 u_m + v_0 v_m) \tag{24-5}$$

The terms on the left side of Eq. 24-4 are equal to zero and define the flow field $\mathbf{q}_0(u_0, v_0)$; therefore, the terms on the right side of the equation are also equal to zero and represent the superposed linearized flow. Thus, for each value of a_m

$$a_m \left[\frac{\partial w_m}{\partial z} + \frac{\partial u_m}{\partial x}\left(1 - \frac{u_0^2}{a_0^2}\right) + \frac{\partial v_m}{\partial y}\left(1 - \frac{v_0^2}{a_0^2}\right) - \frac{u_0 v_0}{a_0^2}\left(\frac{\partial v_m}{\partial x} + \frac{\partial u_m}{\partial y}\right)\right.$$
$$- \frac{\partial u_0}{\partial x}\left(\frac{2u_0 u_m + u_0^2 A_m}{a_0^2}\right) - \frac{\partial v_0}{\partial u}\left(\frac{2v_0 v_m + v_0^2 A_m}{a_0^2}\right)$$
$$\left. - \left(\frac{\partial u_0}{\partial y} + \frac{\partial v_0}{\partial x}\right)\frac{u_0 v_0 A_m + u_0 v_m + u_m v_0}{a_0^2}\right] = 0 \tag{24-6}$$

Eq. 24-6 represents a flow field whose characteristic surfaces do not change when the value of a_m changes and are coincident with the characteristic surfaces of the basic flow (Eq. 21-2).

Assume that an investigation is to be made for a given variation of boundary conditions, which can be represented by a variation of the velocity vector at a given number N of points of the basic characteristic net. The variation of the boundary conditions at each point m of the net produces a disturbance in the flow field which travels inside the Mach conoid from the point m considered, and can be determined by means of

Eq. 24-6, where the geometry of the conoid is known and given by the basic characteristic net, and only the variation of velocity components must be determined. Now once the variation in the flow field produced by a disturbance at the point m, due to a variation in the boundary conditions, has been determined for a given value of a_m, the effect of any other disturbance produced at m can be obtained immediately. This follows from the fact that the velocity components u_m, v_m, w_m remain constant in this approximation, while only the a_m varies with change in boundary conditions at m.

In a similar way the effect on the flow field of disturbances at each of the other points of the net for all possible variations in boundary conditions in the zone investigated can be determined with only one characteristic calculation. This method can be applied for example to analysis of interference and downwash effects or to the determination of flow for boundary conditions close to two-dimensional or axially symmetric ones. Three examples of possible applications are described in the following paragraphs.

G,25. The Linearized Characteristics Method for Axially Symmetric Flows. Consider an axially symmetric flow field about a body of revolution at zero angle of attack and assume that the effect of small variations of boundary conditions, such as small variation of body shape, is to be investigated. The analysis can be performed in two steps. First, the flow field around the body for the basic shape is determined and the shape of the shock wave and the characteristic net obtained. From the points of the boundary streamline other points $A \ldots B \ldots N \ldots$ inside the boundary line must also be determined (Fig. G,25) [*19*].

Any small variation of the boundary conditions can be represented by a superposed flow field, and Eq. 24-6 for axially symmetric flow fields becomes

$$a_m\left[\frac{v_m}{y} - \frac{\partial u_0}{\partial x}\left(\frac{2u_m u_0}{a_0^2} + \frac{u_0^2 A_m}{a_0^2}\right) - \frac{\partial v_0}{\partial y}\left(\frac{2v_m v_0 + v_0^2 A_m}{a_0^2}\right)\right.$$
$$- \frac{u_0 v_0 A_m + v_0 u_m + u_0 v_m}{a_0^2}\left(\frac{\partial u_0}{\partial y} + \frac{\partial v_0}{\partial x}\right) + \frac{\partial u_m}{\partial x}\left(1 - \frac{u_0^2}{a_0^2}\right)$$
$$\left. + \frac{\partial v_m}{\partial y}\left(1 - \frac{v_0^2}{a_0^2}\right) - \frac{u_0 v_0}{a_0^2}\left(\frac{\partial u_m}{\partial y} + \frac{\partial v_m}{\partial x}\right)\right] = 0 \quad (25\text{-}1)$$

The characteristic lines of Eq. 25-1 are the same as the characteristic lines of the basic flow field; therefore, the characteristic net does not change.

The variation of the disturbance velocity $\mathbf{q}_m$ of polar coordinates components q_m and θ_m along the characteristic line of the first and second families can be obtained from Eq. 25-1 in a similar way as for Eq. 18-1.

Considering only terms of the first order along the characteristic line of the first family

$$\frac{dy}{dx} = \lambda_1 = \tan(\mu_0 + \theta_0) \tag{25-2}$$

the following equation is valid

$$\frac{1}{q_0}\frac{dq_m}{dx} - \tan\mu_0 \frac{d\theta_m}{dx} + \frac{\sin^3\mu_0}{\cos(\theta_0+\mu_0)}\frac{1}{\gamma\mathcal{R}}\frac{ds_m}{dn_0} + \theta_m B_1 + \frac{q_m}{q_0}C_1 = 0 \tag{25-3}$$

and along the line of the second family

$$\frac{dy}{dx} = \lambda_2 = \tan(\theta_0 - \mu_0) \tag{25-4}$$

the following equation is valid

$$\frac{1}{q_0}\frac{dq_m}{dx} + \tan\mu_0 \frac{d\theta_m}{dx} - \frac{\sin^3\mu_0}{\cos(\theta_0-\mu_0)}\frac{1}{\gamma\mathcal{R}}\frac{ds_m}{dn_0} - \theta_m B_2 + \frac{q_m}{q_0}C_2 = 0 \tag{25-5}$$

when n_0 is the normal to the streamline for the basic flow and the coefficients B_1, C_1, B_2, C_2 are functions of the flow properties of the basic flow at each point of the characteristic net and at each point are given by the following relations:

$$B_1 = \frac{1}{\cos\mu_0\cos(\theta_0+\mu_0)}\left[\epsilon\frac{\sin\mu_0}{y}\sin(\theta_0-\mu_0) + \frac{\sin^4\mu_0}{\gamma\mathcal{R}}\frac{ds_0}{dn_0} - \frac{1}{q_0}\left(\frac{dq_0}{dx}\right)_{\lambda_2}\frac{\cos(\theta_0-\mu_0)}{\sin\mu_0}\right] \tag{25-6}$$

$$B_2 = \frac{1}{\cos\mu_0\cos(\theta_0-\mu_0)}\left[\epsilon\frac{\sin\mu_0}{y}\sin(\theta_0+\mu_0) - \frac{\sin^4\mu_0}{\gamma\mathcal{R}}\frac{ds_0}{dn_0} - \frac{1}{q_0}\left(\frac{dq_0}{dx}\right)_{\lambda_1}\frac{\cos(\theta_0+\mu_0)}{\sin\mu_0}\right] \tag{25-7}$$

$$C_1 = \frac{1}{q_0}\left(\frac{dq_0}{dx}\right)_{\lambda_1}\left(\tan^2\mu_0 + \frac{\gamma-1}{2\sin^2\mu_0\cos^2\mu_0}\right) - \left[\frac{2}{\gamma\mathcal{R}}\frac{ds_0}{dn_0}\frac{\sin^3\mu_0}{\cos(\theta_0+\mu_0)} - \frac{1}{q_0}\left(\frac{dq_0}{dx}\right)_{\lambda_2}\frac{\cos(\theta_0-\mu_0)}{\cos(\theta_0+\mu_0)}\frac{1}{\cos^2\mu_0}\right]\left(1+\frac{\gamma-1}{2\sin^2\mu_0}\right) \tag{25-8}$$

$$C_2 = \frac{1}{q_0}\left(\frac{dq_0}{dx}\right)_{\lambda_2}\left(\tan^2\mu_0 + \frac{\gamma-1}{2\sin^2\mu_0\cos^2\mu_0}\right) + \left[\frac{2}{\gamma\mathcal{R}}\frac{ds_0}{dn_0}\frac{\sin^3\mu_0}{\cos(\theta_0-\mu_0)} + \frac{1}{q_0}\left(\frac{dq_0}{dx}\right)_{\lambda_1}\frac{\cos(\mu_0+\theta_0)}{\cos(\theta_0-\mu_0)}\frac{1}{\cos^2\mu_0}\right]\left(1+\frac{\gamma-1}{2\sin^2\mu_0}\right) \tag{25-9}$$

The symbol $(dq_0/dx)_{\lambda_1}$ represents the total derivative along the characteristic of the first family and $(dq_0/dx)_{\lambda_2}$ represents the total derivative along the characteristic of the second family, and ϵ is equal to 1, for cylindrical coordinates.

The determination of values of q_m and θ_m at any point of the net is made by means of finite difference methods and requires evaluation of expressions of the type

$$(q_m)_S = a_S(q_m)_T + b_S(q_m)_H + c_S(\theta_m)_T + d_S(\theta_m)_H + e_S(s_m)_T + f_S(s_m)_H \quad (25\text{-}10)$$

where subscripts S, H, and T indicate values at points S, H, and T of Fig. G,25. Because the coefficients a, b, c, d, e, f of Eq. 25-10 are constant

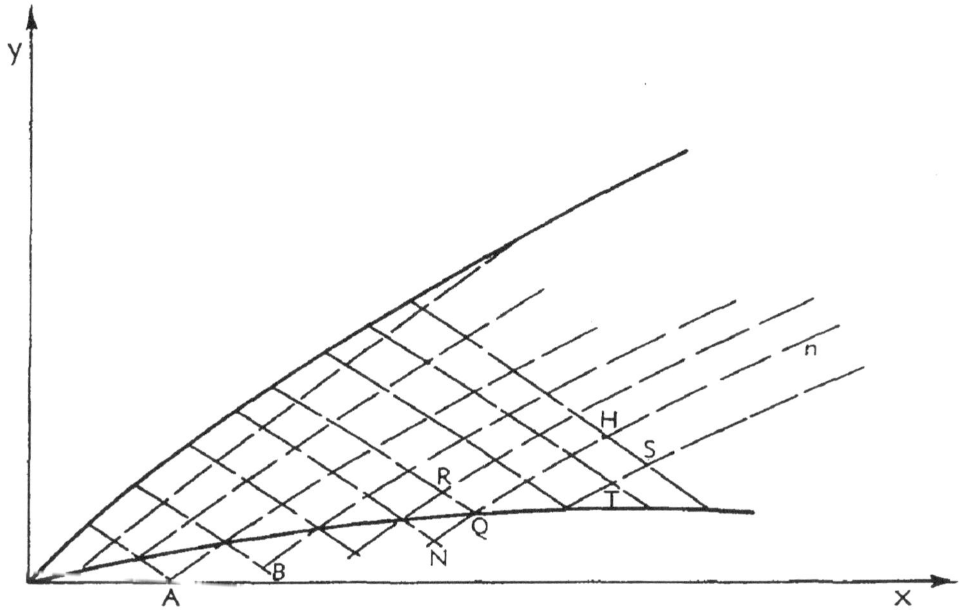

Fig. G,25. The characteristic net for the linearized characteristic system.

at each point, the effects of each variation of boundary conditions at points A, B, and N in the velocity components at H, T, and S can be determined rapidly, and the pressure distribution for each boundary streamline obtained.

The work required in the calculation of the flow field for the basic body and the determination of the coefficients $a, \ldots, f$ of Eq. 25-10 can be reduced to a minimum when conical bodies are assumed in the analysis as basic bodies. In this case the data on the properties of the basic flow are available in tabulated form (see H,17) and the coefficients $a, \ldots, f$ are function only of the polar coordinate ϕ and are constant along each radius of the conical flow.

For conical flow the coefficients B_1, B_2, C_1, and C_2 given by Eq. 25-6 to 25-9 become

$$B_1(dx)_{\lambda_1} = \frac{d\phi}{\cos\mu_0}\left[\frac{\sin\mu_0\sin(\theta_0-\mu_0)}{\sin\phi\sin(\theta_0+\mu_0-\phi)} + \frac{1}{q_0}\left(\frac{dq_0}{d\phi}\right)_{\psi}\frac{1}{\sin\mu_0}\frac{\sin(\phi-\theta_0+\mu_0)}{\sin(\theta_0+\mu_0-\phi)}\right] \quad (25\text{-}11)$$

$$B_2(dx)_{\lambda_2} = \frac{d\phi}{\cos\mu_0}\left[-\frac{\sin\mu_0\sin(\theta_0+\mu_0)}{\sin\phi\sin(\phi-\theta_0+\mu_0)} + \frac{1}{q_0}\left(\frac{dq_0}{d\phi}\right)_{\psi}\frac{1}{\sin\mu_0}\frac{\sin(\theta_0+\mu_0-\phi)}{\sin(\phi-\theta_0+\mu_0)}\right] \quad (25\text{-}12)$$

$$C_1(dx)_{\lambda_1} = \frac{1}{q_0}\left(\frac{dq_0}{d\phi}\right)_{\phi}\left[\frac{1}{\cos^2\mu_0}\left(\sin^2\mu_0+\frac{\gamma-1}{2\sin^2\mu_0}\right) - \frac{1}{\cos^2\mu_0}\left(1+\frac{\gamma-1}{2\sin^2\mu_0}\right)\frac{\sin(\phi-\theta_0+\mu_0)}{\sin(\theta_0+\mu_0-\phi)}\right] \quad (25\text{-}13)$$

$$C_2(dx)_{\lambda_2} = \frac{1}{q_0}\left(\frac{dq_0}{d\phi}\right)_{\phi}\left[\frac{1}{\cos^2\mu_0}\left(\sin^2\mu_0+\frac{\gamma-1}{2\sin^2\mu_0}\right) - \frac{1}{\cos^2\mu_0}\left(1+\frac{\gamma-1}{2\sin^2\mu_0}\right)\frac{\sin(\theta_0+\mu_0-\phi)}{\sin(\phi-\mu_0+\theta_0)}\right] \quad (25\text{-}14)$$

and

$$\frac{dq_0}{d\phi} = -\frac{v_{n_0}}{q_0}\left[\frac{v_{r_0}+v_{n_0}\cot\phi}{1-\dfrac{2v_{n_0}^2}{(\gamma-1)(1-q_0^2)}}\right] \quad (25\text{-}15)$$

where v_{n_0} and v_{r_0} are the components normal to the radius and in radial direction of the velocity $\mathbf{q}_0$ of the conical flow.

The application of the method to any body shape can be performed by choosing for different regions of the flow field along the body different conical bodies, in order to reduce the magnitude of the linearized quantities in each region considered.

The passage from a basic body to another, during the process of calculation can be performed by determining first the total components q and θ of the total flow along a characteristic line of the first family that starts at the body surface; then the flow along this characteristic line and the body surface downstream of the characteristic line are assumed as new boundary conditions, and the total flow which satisfies these boundary conditions is approximated by the new basic flow field and by a superposed linearized flow which can be determined by the method described before.

The best choice of the cone angle and position of the apex of the cone along the axis is dictated by the requirement that the differences between the flow at the given boundaries and the conical flow in the region used for the calculation must be small and of the same order in the entire region considered.

It is convenient that, for the position of the apex and the angle of the cone chosen, the conical body be contained inside the body analyzed in order to avoid the necessity of extending the available conical flow data to the region contained inside the conical body.

In place of conical fields, cylindrical fields can be used for the central part of the body.

For the analysis of the flow around the tail of the body conical solutions are again available (see H,20). Any streamline of the conical solution which transforms a uniform stream to another uniform stream at a lower Mach number can be used as the shape of the basic body for this part of the analysis.

G,26. Bodies of Revolution at Small Angles of Attack. For a system of cylindrical coordinates x, y, ω with velocity components

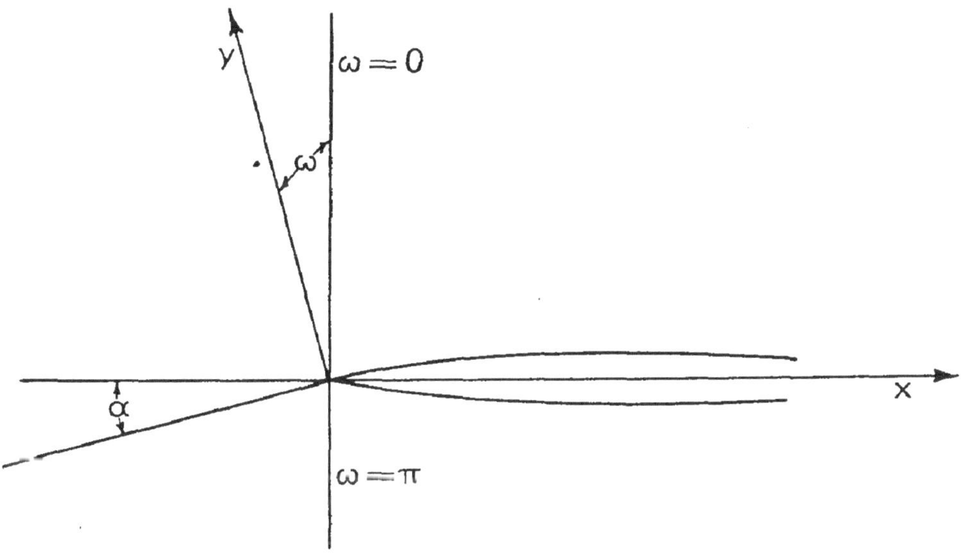

Fig. G,26. Cylindrical coordinate system x, y, ω for body of revolution at angle of attack.

u, v, and w, the equation of motion (22-10), obtained from Euler's equations and the continuity equation, is (Fig. G,26)

$$\frac{\partial u}{\partial x}\left(1-\frac{u^2}{a^2}\right)+\frac{\partial v}{\partial y}\left(1-\frac{v^2}{a^2}\right)+\frac{\partial w}{y\partial\omega}\left(1-\frac{w^2}{a^2}\right)+\frac{v}{y}-\frac{vu}{a^2}\left(\frac{\partial u}{\partial y}+\frac{\partial v}{\partial x}\right)$$
$$-\frac{uw}{a^2}\left(\frac{\partial w}{\partial x}+\frac{\partial u}{y\partial\omega}\right)-\frac{vw}{a^2}\left(\frac{\partial v}{y\partial\omega}+\frac{\partial w}{\partial y}\right)=0 \quad (26\text{-}1)$$

For small values of the component w, the velocity components can be expressed in the form

$$u = u_0 + \sum_{1}^{N} a_m u_m \cos m\omega \tag{26-2}$$

$$v = v_0 + \sum_{1}^{N} a_m v_m \cos m\omega \tag{26-3}$$

$$w = \sum_{1}^{N} a_m w_m \sin m\omega \tag{26-4}$$

where u_0 and v_0 are the velocity components in the meridian plane of an axially symmetric flow field and a_m, u_m, v_m, w_m must be determined in order to satisfy the boundary conditions at the body and at the shock.

If N is equal to 1, Eq. 26-2, 26-3, and 26-4 represent the flow field around a body of revolution at an angle of attack, and a_1 is proportional to the angle of attack. In this case Eq. 26-2, 26-3, and 26-4 satisfy the conditions at the body and at the shock surface [*20,21,22*] if terms of the order of α^2 are neglected. If α is substituted for a_1, Eq. 26-1, by means of Eq. 26-2, 26-3, and 26-4, becomes

$$\begin{aligned}
&\frac{\partial u_0}{\partial x}\left(1 - \frac{u_0^2}{a_0^2}\right) + \frac{\partial v_0}{\partial y}\left(1 - \frac{v_0^2}{a_0^2}\right) - \frac{u_0 v_0}{a_0^2}\left(\frac{\partial u_0}{\partial y} + \frac{\partial v_0}{\partial x}\right) + \frac{v_0}{y} \\
&\quad = -\alpha \cos\omega \left[\frac{\partial u_1}{\partial x}\left(1 - \frac{u_0^2}{a_0^2}\right) + \frac{\partial v_1}{\partial y}\left(1 - \frac{v_0^2}{a_0^2}\right) - \frac{u_0 v_0}{a_0^2}\left(\frac{\partial u_1}{\partial y} + \frac{\partial v_1}{\partial x}\right)\right. \\
&\quad + \frac{v_1}{y} + \frac{w_1}{y} - \frac{\partial u_0}{\partial x}\left(\frac{2u_0u_1 + u_0^2 A_1}{a_0^2}\right) - \frac{\partial v_0}{\partial y}\left(\frac{2v_0v_1 + v_0^2 A_1}{a_0^2}\right) \\
&\quad \left. - \left(\frac{\partial u_0}{\partial y} + \frac{\partial v_0}{\partial x}\right)\frac{u_0v_1 + u_1v_0 + A_1u_0v_0}{a_0^2}\right] = 0
\end{aligned} \tag{26-5}$$

where A_1 is defined by Eq. 24-5.

The left term of Eq. 26-5 is zero, from the condition for zero angle of attack, and therefore the right side is also zero.

In the form of Eq. 26-5 the characteristic surfaces are still surfaces of revolution and the calculations can be performed along the characteristic net used for the determination of the flow at zero angle of attack. The calculations are usually performed in the plane $\omega = \pi$, which contains the direction of the undisturbed velocity. The calculations are performed only once and are valid for all the range of angles of attack where the precision of the results given by the linearized solution is sufficient.

Assume polar components q_0, q_1 and θ_0, θ_1, in place of u_0, u_1 and v_0, v_1, given by

$$u_0 = q_0 \cos\theta_0 \tag{26-6}$$

$$v_0 = q_0 \sin\theta_0 \tag{26-7}$$

$$u_1 = q_1 \cos\theta_0 - \theta_1 q_0 \sin\theta_0 \tag{26-8}$$

$$v_1 = q_1 \sin\theta_0 + \theta_1 q_0 \cos\theta_0 \tag{26-9}$$

From the characteristic equations for the condition of zero angle of attack there is obtained

$$\frac{\partial q_0}{\partial x} = \frac{\lambda_2}{\lambda_2 - \lambda_1}\left(\frac{dq_0}{dx}\right)_{\lambda_1} - \frac{\lambda_1}{\lambda_2 - \lambda_1}\left(\frac{dq_0}{dx}\right)_{\lambda_2} \tag{26-10}$$

$$\frac{\partial q_0}{\partial y} = \frac{1}{\lambda_2 - \lambda_1}\left[\left(\frac{dq_0}{dx}\right)_{\lambda_2} - \left(\frac{dq_0}{dx}\right)_{\lambda_1}\right] \tag{26-11}$$

If n is the normal to the local tangent to the streamline in the meridian plane $\omega = \text{const}$

$$\frac{\partial s}{\partial n} = -\frac{\partial s}{\partial x}\frac{v}{q} + \frac{\partial s}{\partial y}\frac{u}{q} = \frac{\gamma \mathcal{R}}{a^2} q\left(\frac{\partial v}{\partial x} - \frac{\partial u}{\partial y}\right) \tag{26-12}$$

Therefore in the approximation accepted

$$\frac{\partial v_1}{\partial x} - \frac{\partial u_1}{\partial y} = \frac{a_0^2}{\gamma \mathcal{R} q_0}\left[\frac{\partial s_1}{\partial n_0} - \frac{\partial s_0}{\partial n_0}\left(\frac{q_1}{q_0} + A_1\right)\right] \tag{26-13}$$

where $\alpha s_1 \cos \omega$ is the increase of entropy for the condition of $\alpha \neq 0$ and s_0 the entropy at the same point for the condition of $\alpha = 0$. Thus the entropy at each point can be expressed in this approximation as

$$s = s_0 + \alpha s_1 \cos \omega \tag{26-14}$$

Eq. 26-14 is valid everywhere with the exception of the surface of the body, where a vortical layer of infinitesimal thickness exists and where the entropy remains constant and equal to $s_0 - \alpha s_1$. However the pressure distribution does not change across the discontinuity. (See H,22.) By means of Eq. 26-6 to 26-14, Eq. 26-5, if considered along the characteristic lines of the first family and second family (see Art. 18), gives

$$\frac{1}{q_0}\frac{dq_1}{dx} - \tan \mu_0 \frac{d\theta_1}{dx} + \frac{\sin^3 \mu_0}{\cos(\mu_0 + \theta_0)}\frac{1}{\gamma \mathcal{R}}\frac{ds_1}{dn_0} + \theta_1 B_1 + \frac{q_1}{q_0} C_1 - \frac{w_1}{q_0 y}\frac{\sin \mu_0 \tan \mu_0}{\cos(\mu_0 + \theta_0)} = 0 \tag{26-15}$$

along

$$\frac{dy}{dx} = \tan(\theta_0 + \mu_0) \tag{26-16}$$

and

$$\frac{1}{q_0}\frac{dq_1}{dx} + \tan \mu_0 \frac{d\theta_1}{dx} - \frac{\sin^3 \mu_0}{\cos(\theta_0 - \mu_0)}\frac{1}{\gamma \mathcal{R}}\frac{ds_1}{dn_0} - \theta_1 B_2 + \frac{q_1}{q_0} C_2 - \frac{w_1}{q_0 y}\frac{\sin \mu_0 \tan \mu_0}{\cos(\mu_0 - \theta_0)} = 0 \tag{26-17}$$

along

$$\frac{dy}{dx} = \tan(\theta_0 - \mu_0) \tag{26-18}$$

where B_1, B_2, C_1, and C_2 are defined by Eq. 25-6 to 25-9 with $\epsilon = 1$.

The velocity components q_1 and θ_1 can be determined at each point, if the value of w_1 is known, by a finite difference procedure applied to the characteristic net for $\alpha = 0$.

The value of w_1 can be obtained in the following way: from the relation between vorticity and entropy gradient (D,32, Eq. 32-9)

$$\frac{\partial s}{y\partial\omega}\frac{a^2}{\gamma\mathcal{R}} = v\left(\frac{\partial w}{\partial y} - \frac{\partial v}{y\partial\omega}\right) + \frac{wv}{y} - u\left(\frac{\partial u}{y\partial\omega} - \frac{\partial w}{\partial x}\right) \tag{26-19}$$

Therefore, in the approximation accepted, i.e. $N = 1$, $a_1 = \alpha$ in Eq. 26-2 to 26-4,

$$v_0\frac{\partial w_1}{\partial y} + u_0\frac{\partial w_1}{\partial x} = -\frac{v_0w_1 + v_0v_1 + u_0u_1}{y} + \frac{a_0^2}{\gamma\mathcal{R}}\frac{\partial s}{y\partial\omega}\frac{1}{\alpha\sin\omega} \tag{26-20}$$

But in the whole flow field Eq. 26-14 is valid; therefore, along the streamline l,

$$\frac{dw_1}{dl} = -\frac{v_0w_1 + v_0v_1 + u_0u_1}{yq_0} - \frac{\sin^2\mu_0}{y\gamma\mathcal{R}}q_0s_1 \tag{26-21}$$

or

$$\frac{1}{q_0}\frac{dw_1}{dl} = -\frac{1}{y}\left[\frac{q_1}{q_0} + \frac{w_1}{q_0}\sin\theta_0 + \sin^2\mu_0\frac{s_1}{\gamma\mathcal{R}}\right] \tag{26-22}$$

Eq. 26-22 permits the determination of q_1 at each point of the characteristic net.

Similar procedures can be applied for the determination of the flow fields about bodies having cross sections slightly different from a circular cross section. In this case more than one term of the summations in Eq. 26-2 to 26-4 must be used in order to satisfy the boundary conditions.

G,27. Quasi-two-dimensional Flow Fields. Consider three-dimensional flow fields which are not too different from two-dimensional flows, as for example the flows around two-dimensional bodies moving in a slightly nonuniform stream (analysis of the downwash or interference effects) or the flows around wings having the same planform as any wing which can be analyzed by means of two-dimensional flow theory, but having a small twist or thickness variation along the span.

In these flow fields the w component of the velocity can be considered small and the first terms of the summations of 24-1, 24-2, and 24-3, where a_1 is a small quantity, can be used with sufficient approximation [*19*].

Then the three-dimensional flow field can be represented by a two-dimensional flow field in each plane $z = $ const, plus a flow field with velocity components a_1u_1, a_1v_1, a_1w_1, which take into account the three-dimensional effects. Because the cross-flow component a_1w_1 is small, the coefficient a_1 is a small number, and therefore the terms in a_1^2 and higher powers can be neglected with respect to the linear terms, and the superposed flow field can be represented by Eq. 24-6 with $m = 1$.

Eq. 24-6 can be analyzed along characteristic lines of the basic flow in each plane z = const, and by expressing the disturbance velocity in polar coordinates a_1q_1, $a_1\theta_1$ in the plane z = const and a_1w_1 in the direction normal to the plane, the following equations are obtained

$$\frac{1}{q_0}\frac{dq_1}{dx} - \tan\mu_0 \frac{d\theta_1}{dx} + \frac{\sin^2\mu_0}{\gamma\mathfrak{R}}\frac{ds_1}{dx} + B_1\theta_1 + \frac{q_1}{q_0}C_1 = \frac{dw_1}{dz}\frac{1}{q_0}\frac{\tan\mu_0 \sin\mu_0}{\cos(\theta_0+\mu_0)} \tag{27-1}$$

along each characteristic line

$$\lambda_1 = \tan(\mu_0 + \theta_0) \tag{27-2}$$

in a plane z = const, and

$$\frac{1}{q_0}\frac{dq_1}{dx} + \tan\mu_0 \frac{d\theta_1}{dx} + \frac{\sin^2\mu_0}{\gamma\mathfrak{R}}\frac{ds_1}{dx} - B_2\theta_1 + \frac{q_1}{q_0}C_2 = \frac{dw_1}{dz}\frac{1}{q_0}\frac{\tan\mu_0 \sin\mu_0}{\cos(\theta_0-\mu_0)} \tag{27-3}$$

along the line

$$\lambda_2 = \tan(\theta_0 - \mu_0) \tag{27-4}$$

in a plane z = const.

Here q_0, θ_0, and μ_0 are quantities defined by the basic two-dimensional flow in the plane z = const considered, and B_1, B_2, C_1, and C_2 are quantities defined by Eq. 25-6 to 25-9 with $\epsilon = 0$.

The quantity w_1 can be determined by the relation

$$\frac{dw_1}{dl_0} = \frac{\partial q}{\partial z} + \frac{\partial s_1}{\partial z}\frac{a_0^2}{\gamma\mathfrak{R}q_0} \tag{27-5}$$

along the streamline l_0 of the basic two-dimensional flow.

The determination of the linearized three-dimensional flow field by means of Eq. 27-1 to 27-5 can be performed with the following procedure:

Some cross sections of the flow field are chosen in planes z = const. The flow in each plane is determined by means of the two-dimensional flow theory, and the basic characteristic net, the values of q_0, θ_0, and μ_0 and the coefficients B_1, B_2, C_1, and C_2 are obtained at each point of the net.

In this determination the effect of entropy gradient can be considered small, so that it can be incorporated in the linearized flow and potential flow theory can be used. Then s_1 in Eq. 27-1, 27-3, and 27-5 indicates the value of the entropy of the basic flow plus the value of the entropy of the linearized flow.

When the two-dimensional flow is a potential "simple wave" flow the coefficients B_1, B_2, C_1, and C_2 assume the simpler form

$$B_1 = -\frac{\cos(\theta_0-\mu_0)}{\cos\mu_0 \sin\mu_0 \cos(\theta_0+\mu_0)}\frac{1}{q_0}\left(\frac{dq_0}{dx}\right)_{\lambda_2} \tag{27-6}$$

$$B_2 = 0 \tag{27-7}$$

$$C_1 = \frac{1}{q_0}\left(\frac{dq_0}{dx}\right)_{\lambda_2} \frac{\cos(\theta_0 - \mu_0)}{\cos(\theta_0 + \mu_0)} \frac{1}{\cos^2 \mu_0}\left(1 + \frac{\gamma - 1}{2 \sin^2 \mu_0}\right) \tag{27-8}$$

$$C_2 = \frac{1}{q_0}\left(\frac{dq_0}{dx}\right)_{\lambda_2}\left(\tan^2 \mu_0 + \frac{\gamma - 1}{2 \sin^2 \mu_0 \cos^2 \mu_0}\right) \tag{27-9}$$

and

$$\frac{1}{q_0}\left(\frac{dq_0}{dx}\right)_{\lambda_2} = \frac{1}{q_0}\left(\frac{\partial q_0}{\partial x}\right)_{y=0}\left[1 - \frac{\cot(\theta_0 + \mu_0)}{\cot(\theta_0 - \mu_0)}\right] \left\{1 - \frac{y}{\sin^2(\theta_0 + \mu_0)}\left[\frac{\partial(\theta_0 + \mu_0)}{\partial x}\right]_{y=0}\right\}^{-1} \tag{27-10}$$

where

$$\left(\frac{\partial q_0}{\partial x}\right)_{y=0}, \quad \left[\frac{\partial(\theta_0 + \mu_0)}{\partial x}\right]_{y=0}$$

are the gradients along the axis and are constant along each characteristic line of the first family.

Then

$$q = q_0(x, y, z) + a_1 q_1(x, y, z) \tag{27-11}$$

$$\theta = \theta_0(x, y, z) + a_1 \theta_1(x, y, z) \tag{27-12}$$

$$w = a_1 w_1(x, y, z) \tag{27-13}$$

where q_0 and θ_0 are the potential flow solution in each plane $z =$ const and satisfy the boundary conditions in this plane; q_1, θ_1, and w_1 are the components of the linearized flow field due to the presence of w_1 (and of the entropy gradient); and a_1 is a coefficient, for example proportional to the twist or thickness distribution.

Because q_0 and θ_0 are functions of x, y, and z, while for the basic flow they have been determined from two-dimensional considerations, $\partial q_0/\partial z$ and $\partial \theta_0/\partial z$ are not zero; therefore, Eq. 27-5 becomes

$$\frac{\partial w_1}{\partial l_0} = \frac{\partial w_1}{\partial x}\frac{u_0}{q_0} + \frac{\partial w_1}{\partial y}\frac{v_0}{q_0} = \frac{dq_0}{\partial z} + \frac{\partial q_1}{\partial z} + \frac{\partial s_1}{\partial z}\frac{a_0^2}{\gamma \mathcal{R} q_0} \tag{27-14}$$

where l_0 is the streamline of the basic flow in a plane $z =$ const, $\partial q_0/\partial z$ the variation of the velocity for the basic flow, $\partial q_1/\partial z$ the variation of the linearized flow, and $\partial s_1/\partial z$ the total entropy gradient.

By means of Eq. 27-1 to 27-4 and 27-14, the whole flow field can be determined, using Eq. 27-1 to 27-4 along the basic characteristic net in the planes $z =$ const considered, and Eq. 27-14 in the direction normal to each plane, in order to determine the variation of w_1 along each streamline l_0. For numerical applications, finite difference equations must be obtained from the differential equations (27-1) to (27-4) and (27-14).

G,28. Cited References.

1. Prandtl, L., and Busemann, A. Näherungsverfahren zur zeichnerischen Bestimmung von ebenen Strömungen und Überschallgeschwindigkeiten. *Stodola-Festschrift*, Zürich, 1929.

2. Busemann, A. Gasdynamik. *Handbuch der Experimentalphysik 4*, 343–460 (1931). Leipzig.
3. Frankl, F. I. Supersonic flows with axial symmetry. *Izvestia Akademii RKKA 1*, 1934. Reported in Kisenko, M. T. Comparative results of tests on several different types of nozzles. *Central Aero-Hydrodynamical Institute Report 478*, Moscow, 1940. Translated in *NACA Technical Memorandum 1066*. 1944.
4. Ferrari, C. Campo aerodinamico a velocitá iperacustica attorno a un solido di rivoluzione a prora accuminata. *L'Aerotecnica 16*, 121–130 (1936). Also *Brown Univ. Transl. A9-T-18.*
5. Ferrari, C. Campi di corrente ipersonora attorno a solidi di rivoluzione. *L'Aerotecnica 17*, 507–518 (1937). Also *Brown Univ. Transl. A9-T-29.*
6. Tollmien, W. *Charakteristikentheorie.* Technische Hochschule Dresden, 1940, Archiv 44/2. And *Adiabatische Wirbelströmungen*. Technische Hochschule Dresden, 1940, Archiv 44/6. Also *Brown Univ. Trans. Summary 9043.*
7. Guderley, G. Die Charakteristikenmethode für ebene und achsensymmetrische Überschallströmungen. *Jahrb. der deut. Luftfahrtforschung 1940*, 522–535. Also *Brown Univ. Transl. Summary A-9099.*
8. Pfeiffer, R., and Meyer-Koenig, W. Die Charakteristikenmethode bei stationären ebenen und rotationssymmetrischen Gasströmungen. *Deut. Luftfahrtforschung, Forschungsbericht 1581*. March 1942. Also *Brown Univ. Transl. Summary A-9021.*
9. Temple, G. The method of characteristics in supersonic flow. *ARC Reports and Memoranda 2091*. 1944.
10. Ferri, A. Application of the method of characteristics to supersonic rotational flow. *NACA Report 841*. 1946.
11. Shapiro, A. H., and Edelman, G. M. Method of characteristics for two-dimensional supersonic flow—graphical and numerical procedures. *J. Appl. Mech. 14*, A154–162 (1947).
12. Meyer, R. E. The method of characteristics for problems of compressible flow involving two independent variables, Parts 1 and 2. *Quart. J. Mech. and Appl. Math. 1*, 196–219, 451–469 (1948).
13. Sauer, R. Charakteristikenverfahren für räumliche achsensymmetrische Überschallströmungen. *Deut. Luftfahrtforschung, Forschungsbericht 1269*, 1940. Translated in *NACA Tech. Memorandum 1133*, 1947. Also *Brown Univ. Transl. Summary A-9026.*
14. Sauer, R. Charakteristikenverfahren für räumliche achsensymmetrische Überschallströmungen. *Jahrb. der deut. Luftfahrtforschung 1941*, 178–181. Also *Brown Univ. Transl. Summary A-9109.*
15. Whitham, G. B. The behavior of supersonic flow past a body of revolution far from the axis. *Proc. Roy. Soc. A201*, 89–109 (1950).
16. Thornhill, C. K. The numerical method of characteristics for hyperbolic problems in three independent variables. *Armament Research Establishment Report N 29/48.*
17. Ferrari, C. Determinazione della pressione sopra solidi di rivoluzione a prora accuminata disposti in deriva in corrente di fluido compressible a velocitá ipersonora. *Atti Reale Accad. sci., Torino 72*. 1936.
18. Ferrari, C. Interference between wing and body at supersonic speeds—theory and numerical application. *J. Aero. Sci. 15*, 317–336 (1948).
19. Ferri, A. The linearized characteristics method and its application to practical nonlinear supersonic problems. *NACA Technical Note 2515*. 1951. Also *NACA Report 1102*. 1952.
20. Sauer, R. Überschallströmung um beliebig geformte Geschosspitzen unter kleinem Anstellwinkel. *Luftfahrtforschung 19*, 148–152 (1942). Also *Brown Univ. Transl. Summary 9110.*
21. Ferri, A. The method of characteristics for the determination of supersonic flow over bodies of revolution at small angles of attack. *NACA Technical Note 1809*. 1949.
22. Ferri, A. Supersonic flow around circular cones at angles of attack. *NACA Technical Note 2236*. 1950.

SECTION H

SUPERSONIC FLOWS WITH SHOCK WAVES

ANTONIO FERRI

H,1. Introductory Remarks. In III,C and III,D, the formation of shock waves and their physical properties are analyzed and discussed from a general point of view, and the relations determining the change of flow properties across shock waves are derived. In the present section the relation between boundary conditions and shape of the shock waves in a steady flow field are discussed.

Shock waves can be produced in steady flow by bodies traveling at supersonic speeds or at high subsonic speeds. In the latter case shock waves can occur only if a local supersonic region exists in the stream.

For supersonic free stream conditions the head shock waves which divide the free stream from the perturbed region are of particular interest from an analytic point of view. These shock waves, also called front shock waves, can be classified as attached shock waves or detached shock waves, depending upon whether or not there exists at least one point of contact between the shock and the solid boundary. When detached shock waves occur, a subsonic region exists in the disturbed flow field, and the transonic flow theory is required for the analysis of the problem.

For the case of attached shock waves either transonic or pure supersonic flow can be found in the disturbed flow field. If the contact between the shock wave and the boundary occurs only at one point, the flow in the immediate neighborhood of the point of contact is of the conical type. The relation between the shape of the conical shock tangent to the shock wave and the cone tangent to the body must be obtained from conical flow calculations. If the contact between the shock wave and the solid boundary occurs along a line, two-dimensional considerations can be applied locally in the plane normal to the tangent to the contact line. Direct relations can be found between the inclination of the tangent to the body and the tangent to the shock wave in this plane, from the shock wave equations. The points at the ends of such a contact line are equivalent to contact points and here the local flow at the point is of the conical type.

CHAPTER 1. TWO-DIMENSIONAL AND AXIALLY SYMMETRIC FLOW WITH SHOCK WAVES

H,2. Shape of the Shock Wave in Two-dimensional Flow and Axially Symmetric Flow. Consider a two-dimensional flow field in which the flow is everywhere supersonic and assume, at first, that the

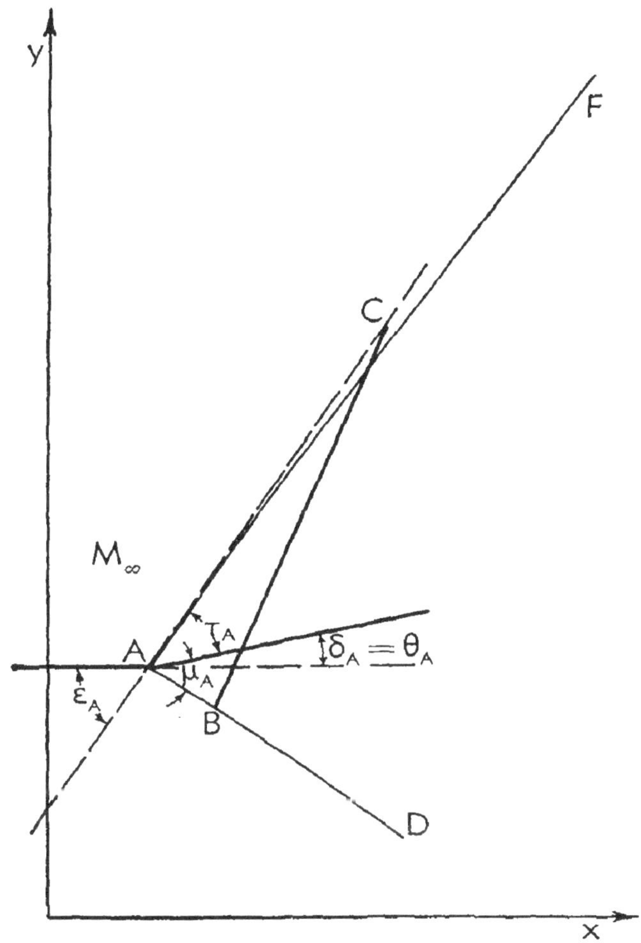

Fig. H,2a. Determination of the shock wave shape.

shock wave to be analyzed divides the undisturbed from the disturbed flow region. Let the velocity vector be defined by its magnitude q and its inclination θ to an axis; assume that the uniform flow in front of the shock wave moves parallel to this axis so that the free stream direction corresponds to $\theta = 0$. Let the free stream Mach number be M_∞. Consider a point A on the shock wave and the line ABD in Fig. H,2a which corresponds to a characteristic line of the family opposite to that of the shock wave at A. Assume that the velocity vector at A and along ABD has been determined from the free stream conditions and from the boundary conditions. The shape of the shock wave AF and the flow field in the region ADF are not yet known and must be determined.

In the region ADF the method of characteristics for rotational flow can be applied, while along the shock wave AF the equations of the shock wave must be valid. The flow in the region ADF is rotational because the shock wave AF is curved. From a point B close to A on the line AD the tangent to the characteristic line of the same family as the shock wave is drawn as shown in Fig. H,2a:

$$\left(\frac{dy}{dx}\right)_B = \tan\,(\theta + \mu)_B \tag{2-1}$$

From the point A the tangent to the shock wave at A is drawn. The tangent is obtained from the equation of the shock wave by remembering that at A the deviation across the shock must be equal to θ_A because θ of the free stream direction is zero. If δ is the deviation across the shock wave and ϵ is the local inclination of the front of the shock wave with respect to the direction of the flow in front of the shock wave, then (see III,E,2)

$$\frac{1}{\tan\,\delta_A} = \left(\frac{\gamma + 1}{2}\,\frac{M_\infty^2}{M_\infty^2 \sin^2 \epsilon_A - 1} - 1\right)\tan\,\epsilon_A \tag{2-2}$$

The point C at the intersection of the two tangents can be assumed in first approximation as a point on the shock wave and on the characteristic line drawn from B.

Along the line BC, Eq. 18-1 of Sec. G (for two-dimensional flow) is valid:

$$\cot\,\mu\,\frac{dq}{q} - d\theta + \sin\,\mu\,\cos\,\mu\,\frac{ds}{\gamma\mathcal{R}} = 0 \tag{2-3}$$

and at C the direction of the velocity vector is

$$\theta_C = \theta_B + \Delta\theta_B = \delta_C \tag{2-4}$$

But at C the equations of the shock wave are valid. For every deviation δ across the shock wave or for every inclination ϵ of the shock wave, all the conditions behind the shock wave are defined as functions of the conditions in front of the shock wave. These relations can be written in the following form (see III,E,2 or [*1*])

$$\tan\,\epsilon = \frac{2}{\gamma + 1}\tan\,\tau\left(\frac{1}{M_2^2 \sin^2 \tau} + \frac{\gamma - 1}{2}\right) \tag{2-5}$$

where M_2 is the Mach number behind the shock wave and τ the inclination of the shock wave with respect to the direction of the flow behind it. Therefore

$$\tau + \delta = \epsilon \tag{2-6}$$

The velocity q is usually referred to the limiting velocity q_{max} in such calculations, and when M_2 is known the value of q behind the shock can be obtained from the equation

$$\frac{q_{max}^2}{q^2} = 1 + \frac{2}{\gamma - 1} \cdot \frac{1}{M_2^2} \tag{2-7}$$

Eq. 2-2, 2-5, and 2-7 relate q and θ to the value of ϵ. The entropy increase Δs across the shock wave is given by

$$\Delta s = c_v \ln \left[\frac{2\gamma}{\gamma + 1} M_\infty^2 \sin^2 \epsilon - \frac{\gamma - 1}{\gamma + 1} \right] + c_p \ln \left[\frac{2}{(\gamma + 1) M_\infty^2 \sin^2 \epsilon} + \frac{\gamma - 1}{\gamma + 1} \right] \tag{2-8}$$

If the inclination ϵ of the shock wave changes, q, δ, and Δs across the shock wave also change. The rates of change are given by the expressions

$$\frac{1}{q} \frac{dq}{d\epsilon} = -\sin \tau \cos \tau \left[\frac{\tan \epsilon}{\tan \tau} + \frac{\tan \tau}{\tan \epsilon} - 2 \frac{\gamma - 1}{\gamma + 1} \right] \tag{2-9}$$

$$\frac{d\Delta s}{d\epsilon} = \gamma \mathcal{R} \frac{\cos^2 \tau}{\tan \epsilon} (\tan \epsilon - \tan \tau)^2 M_2^2 \tag{2-10}$$

and

$$-\frac{d\delta}{d\epsilon} = \frac{\sin 2\delta}{\sin 2\epsilon} - \sin^2 \delta \frac{(\gamma + 1) M_\infty^4 \sin^2 \epsilon}{(M_\infty^2 \sin^2 \epsilon - 1)^2} \tag{2-11}$$

or

$$\frac{d\delta}{d\epsilon} = \cos^2 \tau \left[\frac{4}{\gamma + 1} - \frac{(\tan \epsilon - \tan \tau)^2}{\tan \delta \tan \epsilon} \right]$$

Now the velocity q_C at C behind the shock wave must correspond to the velocity given by Eq. 2-2 and 2-5 for the deviation $\delta = \theta_C$, and can be obtained from tabulated values of q as a function of δ, when θ_C has been determined. By means of Eq. 2-4, q_C can be expressed in the form

$$(q)_C = (q)_{\theta_C} = (q)_{\theta_B} + \left(\frac{dq}{d\epsilon} \right)_{\theta_B} \left(\frac{d\epsilon}{d\delta} \right)_{\theta_B} \Delta\theta_B \tag{2-12}$$

where $(q)_{\theta_B}$, the velocity behind the shock wave, $dq/d\epsilon$, $d\epsilon/d\delta$ are determined for a deviation equal to θ_B, and $\Delta\theta_B$ is defined by Eq. 2-4.

In a similar way the variation of entropy for a deviation θ_C across the shock can be expressed as

$$(\Delta s)_C = (\Delta s)_{\theta_C} = (\Delta s)_{\theta_B} + \left(\frac{\partial \Delta s}{\partial \epsilon} \right)_{\theta_B} \left(\frac{\partial \epsilon}{\partial \delta} \right)_{\theta_B} \Delta\theta_B \tag{2-13}$$

By means of Eq. 2-12 and 2-13, Eq. 2-3 between B and C becomes

$$-\Delta\theta_B + \cot \mu_B \left[\left(\frac{q_{\theta_B}}{q_B} - 1 \right) + \left(\frac{dq}{d\epsilon} \frac{d\epsilon}{d\delta} \right)_{\theta_B} \frac{1}{q_B} \Delta\theta_B \right] + \frac{\sin \mu_B \cos \mu_B}{\gamma \mathcal{R}} \left[(\Delta s_{\theta_B} - \Delta s_B) + \left(\frac{d\Delta s}{d\epsilon} \frac{d\epsilon}{d\delta} \right)_{\theta_B} \Delta\theta_B \right] = 0 \tag{2-14}$$

A first approximation to the value of $\Delta\theta_B$ can be determined from Eq. 2-14. The values of ϵ_C and μ_C can be determined from the value of θ_C. A second approximation can then be obtained by determining the point C from the intersection of two straight lines, one inclined at an angle which is the average of the inclination of the tangents to the shock wave at A and C, and the other inclined at an angle which is the average of the inclination of the characteristic lines at B and C. Then an equation similar to Eq. 2-14 can be used along BC where, as coefficients in the equation, average values between the values at B and at C are used in place of the quantities at θ_B and B. In this way all of the shock wave can be determined [*2,3,4*]. The first point for the shock wave calculation is obtained at the contact point between the solid boundary and the shock wave. At this point the tangent to the shock wave is determined directly by the deviation produced by the solid boundary by means of Eq. 2-2 [*5*].

Because the point C is determined by the intersection of shock waves and characteristic lines of the same family inclined one with respect to the other by a small angle, the distances AC and BC are much longer than the distance AB; therefore, in practical calculations the point B must be very close to A. Usually the values of q_θ, Δs_θ, $dq/d\delta$, $d\Delta s/d\delta$ are tabulated as functions of $\theta = \delta$ for the free stream conditions, and are calculated by assuming that ϵ and not δ is the independent variable in Eq. 2-2 to 2-11.

Eq. 18-17 or 18-18 with $\epsilon = 0$ in Sec. G can be used in the calculations in place of Eq. 2-3. In this case the values of C_1 and C_2 are determined and tabulated for the free stream conditions as functions of ϵ in place of q_θ and δ.

When the flow field in front of the shock wave is not uniform, the calculations are performed in a similar manner, but are more involved because the Mach number in front of the shock wave must also be determined.

Consider a nonuniform flow field $NEFBL$ as in Fig. H,2b and assume that at A a shock wave exists. The flow along the characteristic lines FL and FN is known; the flow field in the region NFL and the shape of the shock wave AG must be determined.

The calculation is performed by first determining the conditions in front of the shock wave at each point A, C, and G of the shock wave. The following numerical procedure can be used: The presence of the shock wave at A is ignored, and from a point E along FG and from the point A in front of the shock a point H is obtained corresponding to conditions in front of the shock wave. Then, the tangent to the shock wave is drawn at A and, by assuming for a first approximation the tangent at A to be coincident with the shock wave, the point C along EH on the shock wave is obtained. By interpolating between the values at H and E the conditions in front of the shock wave at C can be obtained.

The flow along AL behind the shock wave is known and a point B can be determined by interpolating between A and L in a manner such that the tangent to the characteristic line of the first family from B passes through C. Then the flow conditions at C behind the shock can be obtained by applying Eq. 2-14 along BC. When θ and ϵ at C are known

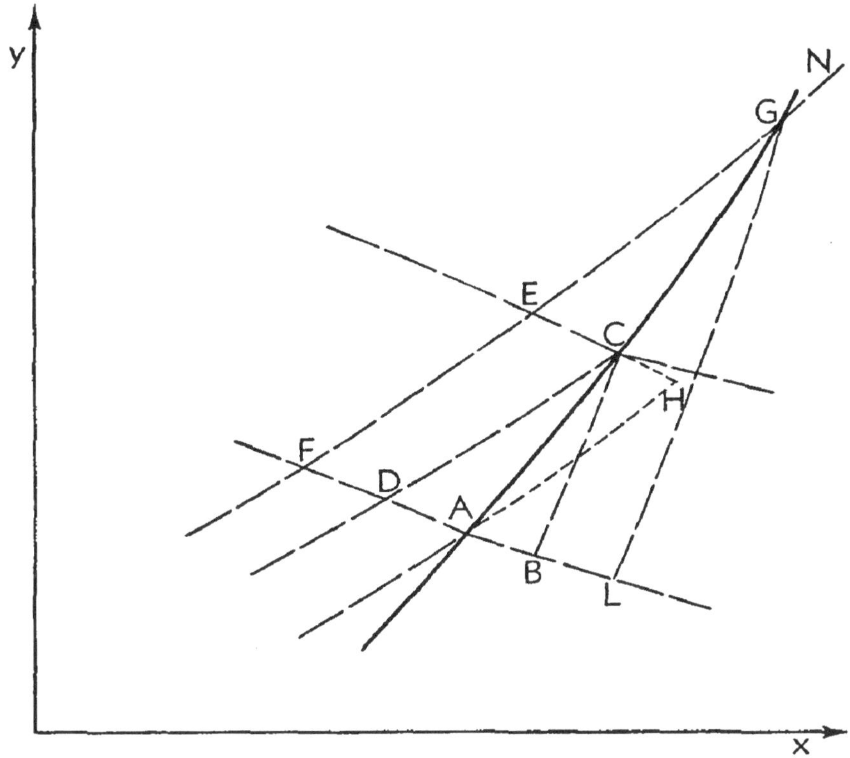

Fig. II,2b. Determination of the shock wave shape in a nonuniform flow field.

to a first approximation, a second approximation of the position and intensity of the shock wave at C can be obtained.

Other different procedures can be used for the choice of the point C on the shock wave. For example, the point C can be obtained from the intersection of the tangent to the shock wave at A with the characteristic line at a point B chosen arbitrarily along AL. Then, the flow in front of the shock wave at C must be obtained by determining the characteristic lines CE and CD passing through C, and by calculating the velocity at C from the velocities at E and D.

The Mach number at each point C is variable; therefore, Eq. 2-5 to 2-11 must be evaluated at each point.

The shape of the shock wave in axially symmetric flow can be determined by means of the shock wave equations and the characteristics theory as for two-dimensional problems, with only the difference that the equations for axially symmetric flow must be used along the characteristic line and therefore a term such as

$$-\frac{x_C - x_B}{y_B}\frac{\sin \mu_B \sin \theta_B}{\cos (\theta + \mu)_B}$$

must be added to the left side of Eq. 2-14.

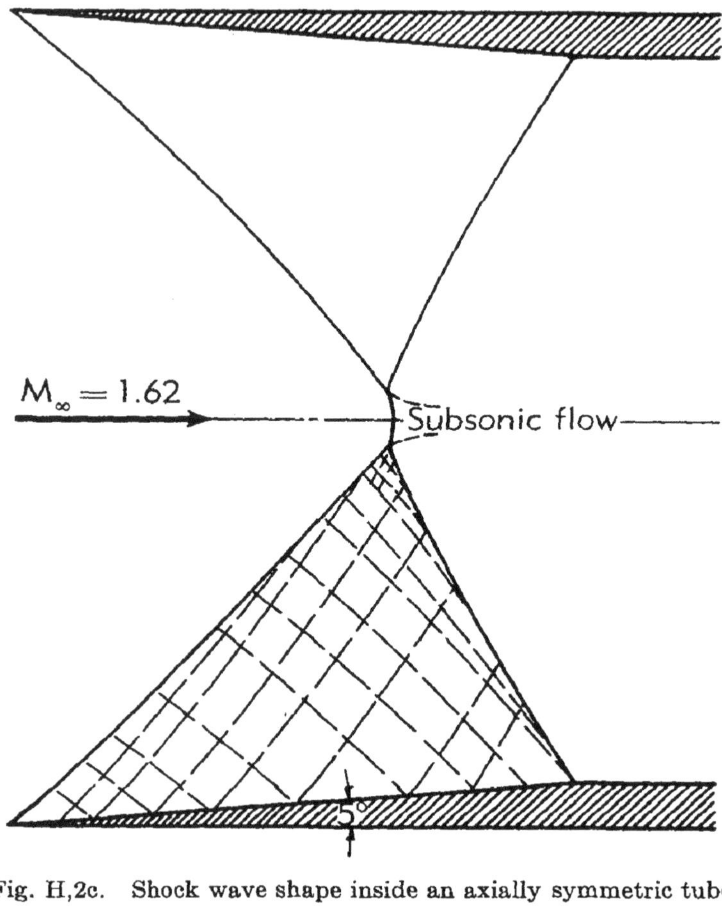

Fig. H,2c. Shock wave shape inside an axially symmetric tube.

In the neighborhood of the axis of symmetry, the shock wave can be considered to be a conical shock, and therefore may be treated by means of conical flow theory [*5*]. At the sharp nose of a body of revolution, for example, a conical flow solution has been found for the flow field (see Art. 19) and therefore no particular discussion is required here.

Thus, if the flow around a pointed-nosed body of revolution is considered, the inclination of the shock wave and the flow properties between the shock wave and the body near the nose can be obtained from the conical flow theory. The characteristic line of the first family can be drawn for the conical flow field; from the flow properties along this characteristic line and from the boundary conditions, the other part of the flow field and the remaining part of the shock wave can be determined.

The flow at the pointed tail of a body of revolution has not yet been fully investigated. In this case a conical flow solution which transforms the converging flow at the tail of the body to a flow parallel to the axis would be required in order to obtain a supersonic solution at the tail.

A brief analysis of the properties of the points on the axis of axially symmetric shock waves is of interest for flow inside tubes and for circular free jets.

Consider for example the flow field illustrated in Fig. H,2c. The shock at the lip of the body is of the two-dimensional type, and the tangent to the shock at the lip can be obtained from two-dimensional considerations. In this case the flow near the lip can be obtained by assuming a point of the shock near the lip on the two-dimensional shock and by determining, along the two characteristic lines from the point, the flow field along the body. Then, from the points on the body the other part of the shock wave can be obtained.

Moving from the lip toward the axis the strength of the shock increases as is required by the equations along the characteristic lines which go from the body toward the axis. The increase in strength is such that, in the neighborhood of the axis, subsonic flow exists behind the shock, and at the axis the shock is a normal shock [*4*].

The possibility of a supersonic flow solution at the axis for a shock wave moving downstream as shown in Fig. H,2c would require in the *neighborhood of the axis the possibility of existence of a fore conical* shock (having the apex downstream of any other point of the shock), which would be reflected at the axis as a downstream conical shock with the same apex. Such reflection of conical shocks is impossible as discussed in Art. 20; therefore, all the shock waves of the type shown in Fig. H,2c have at the axis a Mach reflection, as shown.

H,3. Intensity of the Reflected Waves Due to the Presence of a Curved Shock Wave. When shock waves of nonuniform strength exist in the flow field, entropy gradients are produced along these waves and in the flow behind them. The flow behind such shocks is rotational, and reflected waves like *EB* and *GC* of Fig. H,3 exist in the flow, originating at the shock wave or within the region *ADC* due to the entropy gradient existing there.

The analysis of rotational flow requires a step-by-step calculation in both physical and hodograph planes, and therefore is more involved than the analysis of potential flow, especially for the case of two-dimensional flow fields.

The effect of the presence of reflected waves on the flow field properties is usually small, and therefore is often neglected; therefore, two-dimensional flow fields such as the flow considered in Fig. H,3 can still be analyzed by means of simple wave theory with good approximation.

It is possible to show (see E,4) that both entropy distribution and pressure distribution along the shock are in error when simple wave theory is used in the analysis. However, the difference in pressure coefficient and entropy distribution introduced by the case of simple wave theory is of the order of δ^3 where δ is the deviation of the flow through the shock wave, and therefore the approximation of simple wave theory is good for the analysis of flows where only small stream deviations are produced.

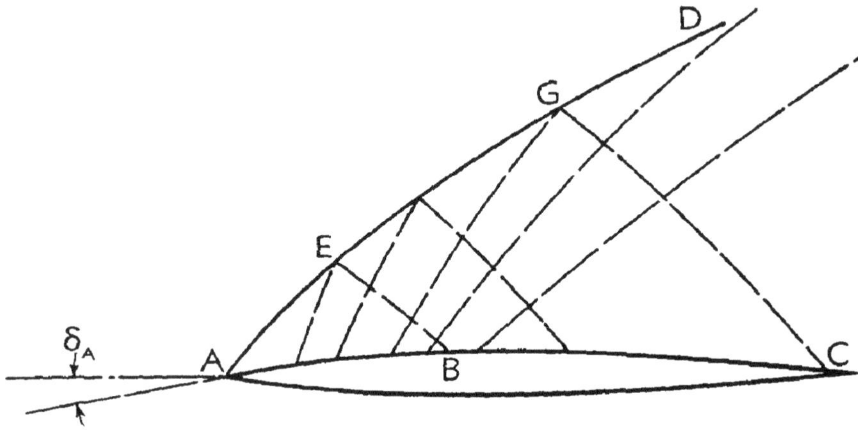

Fig. H,3. Reflected waves in rotational flow.

H,4. Relations Between the Curvature of the Head Shock Wave and the Curvature of the Streamlines. Consider in a two-dimensional flow a streamline L_P from a point P of a head shock wave S as shown in Fig. H,4. The curvature of the streamline L_P at P is related to the curvature of the shock wave at P because, if the shape of the shock wave is given in the neighborhood of P, the slope of the streamline at P is determined. The relation between curvature of the shock wave and curvature of the streamline can be determined analytically from the properties of the shock wave [*1*,*6*,*7*]. Because of the curvature of the shock wave, the flow behind the shock wave is rotational.

Assume a Cartesian coordinate system having the x axis tangent to the streamline at P. Then

$$(v)_P = 0 \tag{4-1}$$

If the streamline is defined by the expression

$$y = L(x) \tag{4-2}$$

then at P

$$(L')_P = 0 \tag{4-3}$$

$$(L'')_P = r^{-1} \tag{4-4}$$

where r is the radius of curvature of the streamline at P and the prime denotes differentiation with respect to x. Assume that the velocity components u and v in the flow region between the shock front and the streamline from P can be expressed by series of the form

$$u = \bar{u} + \bar{u}_x x + \bar{u}_y y + \tfrac{1}{2}\bar{u}_{xx}x^2 + \tfrac{1}{2}\bar{u}_{yy}y^2 + \bar{u}_{xy}xy \cdots \tag{4-5}$$

$$v = \bar{v}_x x + \bar{v}_y y + \tfrac{1}{2}\bar{v}_{xx}x^2 + \tfrac{1}{2}\bar{v}_{yy}y^2 + \bar{v}_{xy}xy \cdots \tag{4-6}$$

where the bars denote quantities evaluated at P and the subscripts denote partial differentiation. Eq. 4-5 and 4-6 can be used in a region of the flow

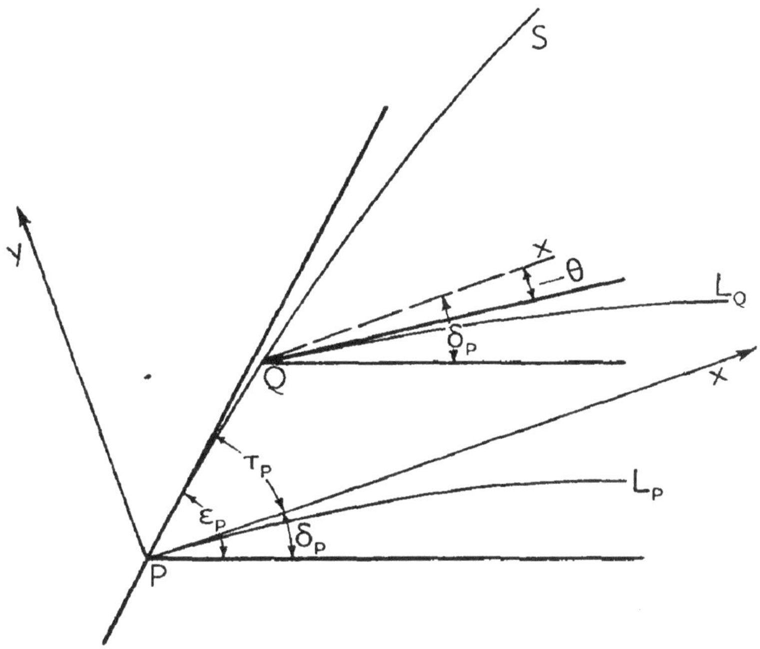

Fig. H,4. The flow field behind a curved shock wave.

without singularities. At any point the streamline is defined by

$$\frac{v}{u} = \left(\frac{dy}{dx}\right)_L, \text{ and at } P, \left(\frac{dy}{dx}\right)_P = 0 \tag{4-7}$$

Therefore, at P

$$\left(\frac{d^2y}{dx^2}\right)_L = \bar{L}'' = \left(\frac{1}{u} v_x\right)_P \tag{4-8}$$

If θ is the inclination of the streamline with respect to the x axis along the shock wave S, there is obtained

$$\frac{v}{u}\bigg| = \tan\theta \tag{4-9}$$

and $\theta = 0$ at P. The shock wave is defined by an equation

$$y = S(x) \tag{4-10}$$

and at P

$$(S')_P = \tan \tau_P = \tan (\epsilon - \delta)_P \tag{4-11}$$

Therefore

$$(S'')_P = \frac{1}{\cos^2 (\epsilon - \delta)} \left(\frac{d\epsilon}{dx}\right)_S \tag{4-12}$$

or

$$\left(\frac{d\epsilon}{dx}\right)_S = \cos^2 \tau_P (S'')_P \tag{4-13}$$

Along the shock wave then

$$\frac{1}{q_2}\left(\frac{dq_2}{dx}\right)_S = \frac{1}{q_2}\left(\frac{dq_2}{d\epsilon}\right)\left(\frac{d\epsilon}{dx}\right)_S$$

But $dq_2/d\epsilon$ is given by Eq. 2-9; therefore, at P

$$\frac{1}{q_2}\left(\frac{dq_2}{dx}\right)_{S_P} = -H \cos^4 \tau_P S''_P \tag{4-14}$$

where

$$H = \tan \tau_P \left[\frac{\tan \epsilon_P}{\tan \tau_P} + \frac{\tan \tau_P}{\tan \epsilon_P} - 2\frac{\gamma - 1}{\gamma + 1}\right] \tag{4-15}$$

At a point Q of the shock wave the deviation across the shock wave is

$$\epsilon - \tau = \delta + \theta$$

but

$$\left(\frac{d\theta}{dx}\right)_{S_P} = \left(\frac{d\theta}{d\epsilon}\right)_P \left(\frac{d\epsilon}{dx}\right)_S \tag{4-16}$$

and the value of $(d\theta/d\epsilon)_P$ at P is given by Eq. 2-11. Therefore, by means of Eq. 4-13, Eq. 4-16 becomes

$$\left(\frac{d\theta}{dx}\right)_{S_P} = \frac{4}{\gamma + 1} K \cos^4 \tau_P S''_P \tag{4-17}$$

where

$$K = 1 - \frac{\gamma + 1}{4} \frac{(\tan \epsilon_P - \tan \tau_P)^2}{\tan \epsilon_P \tan \delta_P} = \left(\frac{d\delta}{d\epsilon}\right)_P \frac{\gamma + 1}{4 \cos^2 \tau_P} \tag{4-18}$$

The velocity behind the shock wave is defined by

$$u^2 + v^2 = q_2^2 \tag{4-19}$$

$$\frac{v}{u} = \tan \theta \tag{4-20}$$

Therefore at P along the shock wave

$$\left(\frac{dq_2}{dx}\right)_{S_P} = (u_x)_P + S'(u_y)_P = (u_x + \tan \tau_P u_y)_P \tag{4-21}$$

and

$$\left(\frac{d\theta}{dx}\right)_{S_P} = \left(\frac{v_x}{u} + \frac{\tan \tau_P}{u} v_y\right)_P \tag{4-22}$$

The relation between the rotation and the entropy gradient in the flow gives (see D,32, Eq. 32-9; also III,A,11)

$$\frac{\partial v}{\partial x} - \frac{\partial u}{\partial y} = \frac{a_2^2}{q_2 \gamma \mathcal{R}} \frac{ds}{dn} \tag{4-23}$$

where n is the normal to the streamline.
Therefore

$$\left[\frac{v_x - u_y}{q_2}\right]_{S_P} = \frac{1}{\mathcal{R} M_2^2 \gamma} \left(\frac{\partial s}{\partial \epsilon}\right) \left(\frac{\partial \epsilon}{\partial x}\right)_{S_P} \left(\frac{\partial x}{\partial n}\right)_{S_P} \tag{4-24}$$

But at P, $\dfrac{\partial x}{\partial n} = \cot \tau_P$ and $\dfrac{\partial s}{\partial \epsilon}$ is given by Eq. 2-10; therefore, by means of Eq. 4-13 there results

$$\left(\frac{v_x}{u} - \frac{u_y}{u}\right)_{S_p} = \cos^4 \tau_P S'' J \tag{4-25}$$

where

$$J = \frac{(\tan \epsilon_P - \tan \tau_P)^2}{\tan \epsilon_P \tan \tau_P} \tag{4-26}$$

From Eq. 4-14 and 4-21 there results

$$\frac{u_x}{u} + \tan \tau_P \frac{u_y}{u} + H \cos^4 \tau_P S'' = 0 \tag{4-27}$$

and from Eq. 4-17 and 4-22, and 4-8

$$\frac{4}{\gamma + 1} K \cos^4 \tau_P S'' = L'' + \tan \tau_P \frac{v_y}{u} \tag{4-28}$$

But from the equation of motion (Sec. G, Eq. 2-13) at P where v is zero, there results

$$u_x (1 - M_2^2) + v_y = 0 \tag{4-29}$$

where M_2 is the Mach number at P behind the shock. Therefore

$$\frac{4}{\gamma + 1} K \cos^4 \tau_P S'' = L'' + \tan \tau_P (1 - M_2^2) \frac{u_x}{u} \tag{4-30}$$

From Eq. 4-25 and 4-8

$$J \cos^4 \tau_P S'' = L'' - \frac{u_y}{u} \tag{4-31}$$

Eq. 4-27, 4-30, and 4-31 relate u_x, u_y, and S'' to L'' at the point P, and give in explicit form, the following

$$\frac{S''}{L''} = \frac{\gamma + 1}{4 \cos^4 \tau_P} \frac{1 - \tan^2 \tau_P (M_2^2 - 1)}{K + \tan^2 \tau_P (M_2^2 - 1)} \tag{4-32}$$

$$\frac{u_x}{u} = -L'' \tan \tau_P \frac{K + 1}{K + \tan^2 \tau_P (M_2^2 - 1)} \tag{4-33}$$

$$\frac{u_y}{u} = -L'' \left[\frac{\gamma + 1}{4} J \frac{1 - \tan^2 \tau_P (M_2^2 - 1)}{K + \tan^2 \tau_P (M_2^2 - 1)} - 1\right] \tag{4-34}$$

Eq. 4-32 gives the relation between curvature of the shock wave and curvature of the streamline. The relation between the curvature of the streamline and the pressure gradient along the streamline can be obtained from Eq. 4-33 and 4-34.

Along the streamline

$$\frac{p}{p_2} = \left(\frac{T}{T_2}\right)^{\frac{\gamma}{\gamma-1}} = \left(\frac{q_{\max}^2 - q^2}{q_{\max}^2 - q_2^2}\right)^{\frac{\gamma}{\gamma-1}} \tag{4-35}$$

therefore

$$\frac{1}{p}\left(\frac{dp}{dx}\right)_P = -\gamma \frac{q_P^2}{a_P^2} \frac{1}{q_P} \frac{dq_P}{dx} = -\frac{1}{2} \frac{\rho_P}{p_P} \left(\frac{dq^2}{dx}\right)_P \tag{4-36}$$

But across the shock wave

$$\frac{\rho_P}{\rho_\infty} = \frac{\tan \epsilon}{\tan \tau}$$

$$\frac{q_P}{U} = \frac{\cos \epsilon}{\cos \tau} \tag{4-37}$$

and at P, q_x is equal to u_x, therefore Eq. 4-36 and 4-33 give

$$\frac{1}{\frac{1}{2}\rho_\infty U^2} \left(\frac{dp}{dx}\right)_L = \frac{\sin 2\epsilon_P}{\sin 2\tau_P} \frac{L''(K + 1) \tan \tau_P}{K + \tan^2 \tau_P (M_2^2 - 1)} \tag{4-38}$$

The curvature of the shock wave and the pressure gradient along the streamline are both proportional to the curvature of the streamline and the coefficients of proportionality are functions of the inclination of the streamline to the shock wave and of the local Mach number. (Values of the coefficients are tabulated in [7].)

In all the Eq. 4-32, 4-33, 4-34, and 4-38 the expression

$$A = K + \tan^2 \tau_P (M_2^2 - 1) \tag{4-39}$$

appears in the denominator of a fraction, where K is defined by Eq. 4-18. The quantity K becomes zero for the condition of δ_{max}, and is positive for a deviation δ less than δ_{max}. For the condition of δ_{max} the flow behind the shock wave is subsonic; therefore, A for this condition is negative. For δ corresponding to sonic velocity behind the shock wave ($\delta = \delta^*$ and $q_P = a_P = q^*$), K is positive and A is positive. Therefore, for a value of δ between δ^* and δ_{max}, A becomes equal to zero. For this condition the curvature of the streamline, L'', must be zero. Therefore, for this value of δ indicated by δ_C corresponding to $A = 0$, the streamline at the shock is straight even though the shock wave is curved ($S'' \neq 0$). This special point δ_C of the shock polar diagram located between δ^* and δ_{max} is often called the "Crocco point."

For conditions between δ_C and δ_{max}, A becomes negative and Eq. 4-32 would therefore indicate that the shock wave and streamline would have opposite curvature. Consider now a solid boundary having a sharp edge causing a deviation of δ between δ_C and δ_{max}. Eq. 4-32 would indicate that, in order to have an attached shock at the edge of the profile, the shock wave must have a concave shape if the profile is convex. This result, however, is a consequence of the analytic approach used in the development and is not physically correct. In the range between δ_C and δ_{max}, singularities occur in the flow, as is discussed in Art. 10, invalidating the use of Eq. 4-5 and 4-6 (see [8]).

H,5. The Curvature of the Head Shock Wave at Infinity and the Determination of the Head Shock Drag. Consider a two-dimensional shock wave having uniform flow ahead of the shock wave. If the subscript ∞ indicates quantities in front of the shock wave, and the subscript 2 indicates quantities behind the shock wave, then as illustrated in III,E,2,

$$\cot \delta = \left(\frac{\gamma + 1}{2} \frac{M_\infty^2}{M_\infty^2 \sin^2 \epsilon - 1} - 1\right) \tan \epsilon \tag{5-1}$$

and

$$\sin^2 \mu_2 = \frac{\left(1 + \frac{\gamma - 1}{2} M_\infty^2 \sin^2 \epsilon\right)\left(\gamma M_\infty^2 \sin^2 \epsilon - \frac{\gamma - 1}{2}\right)}{\left(\frac{\gamma + 1}{2}\right)^2 M_\infty^4 \sin^2 \epsilon - (M_\infty^2 \sin^2 \epsilon - 1)(\gamma M_\infty^2 \sin^2 \epsilon + 1)} \tag{5-2}$$

where μ is the Mach angle, δ the deviation across the shock, and ϵ the inclination of the shock with respect to the free stream direction.

At a large distance from the body the variations across the shock wave become small and the flow can therefore be considered as a simple wave flow (see Art. 3).

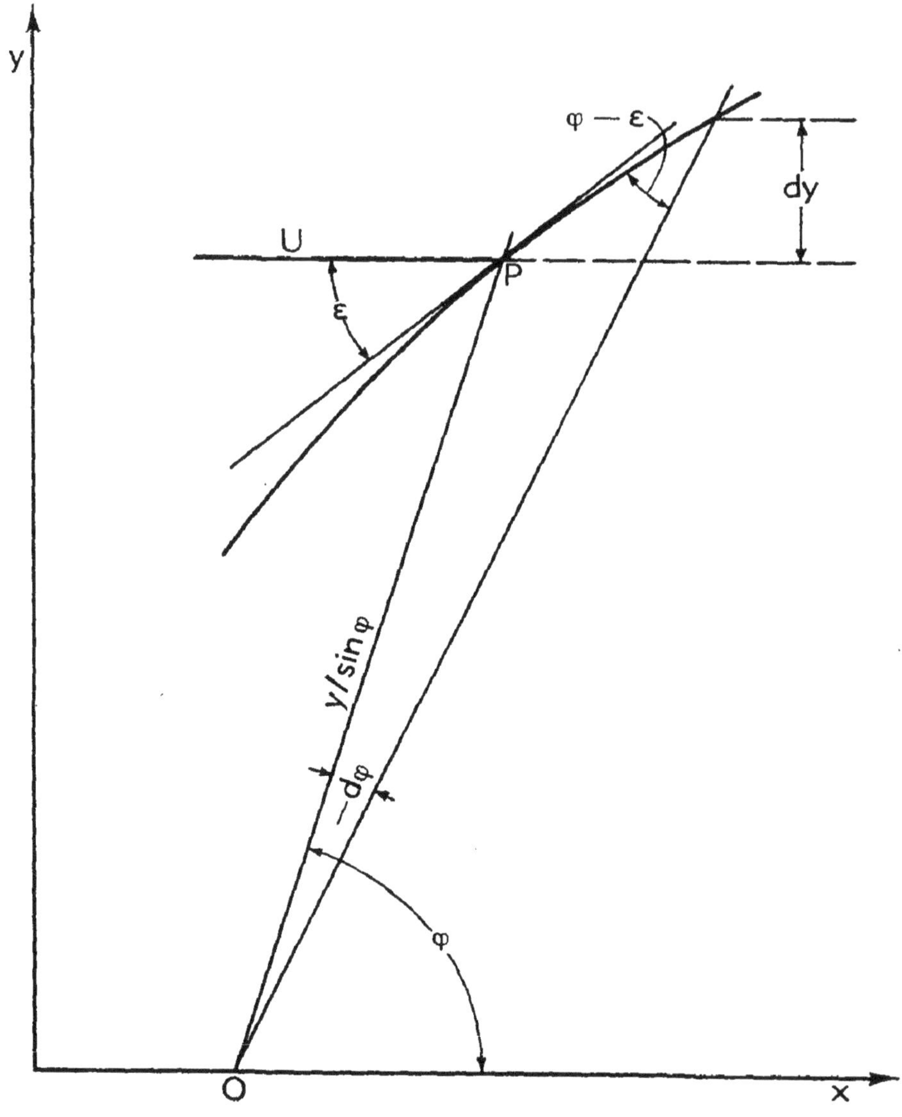

Fig. H,5a. The shape of the head shock wave at large distance from the body. Two-dimensional flow.

Let

$$M_\infty^2 \sin^2 \epsilon - 1 = \Delta \tag{5-3}$$

At a large distance from the body, Δ can be considered small, and the characteristic lines of the simple wave flow are straight. If the distance y between the chord of the body and the region considered is large and tends to infinity (Fig. H,5a), the chord of the body becomes small with respect to y and all the characteristic lines can be considered as originating at one point which to this scale represents the body chord. Then (Fig. H,5a) the equation of the shock wave in this region can be expressed in the form

$$dy = -y[\cot(\varphi - \epsilon) - \cot\varphi]d\varphi \tag{5-4}$$

$$\frac{dy}{y} = -\frac{\sin\epsilon}{\sin\varphi\sin(\varphi - \epsilon)}\,d\varphi \tag{5-5}$$

If δ is the deviation of the stream at P, the inclination of the line OP is

$$\tan\varphi = \tan(\mu_2 + \delta) \tag{5-6}$$

Therefore, Eq. 5-5 can be expressed by means of Eq. 5-1 and 5-2 as a function of only M_∞ and ϵ. By expressing ϵ as

$$\epsilon = \mu_\infty + \sigma \tag{5-7}$$

and by assuming that Δ is small relative to unity, Eq. 5-5 becomes, by means of Eq. 5-1, 5-2, and 5-3

$$\frac{dy}{y} = -2\frac{d\sigma}{\sigma} \tag{5-8}$$

or

$$\frac{\sigma}{\sigma_0} = \sqrt{\frac{y_0}{y}} \tag{5-9}$$

where the subscript 0 indicates quantities at some point where Δ is small, which are boundary values.

Eq. 5-9 shows that the shock angle approaches its asymptote inversely as the square root of the distance from the body. It follows also from Eq. 5-9 that, while the angle of the shock has an asymptote, the shock wave contour does not have an asymptote [9].

From Eq. 5-9 an approximate value of the head shock drag due to the increase of entropy across the part of the shock wave at large distance from the body can be determined. The drag due to the increase of entropy across the head shock wave can be obtained by integration of the momentum of the shock wave. If q_3 is the velocity behind the shock wave for the condition of static pressure downstream of the shock equal to free stream static pressure, the shock wave drag can be expressed as

$$D = \rho_\infty U^2 \int_{-\infty}^{+\infty}\left(1 - \frac{q_3}{U}\right)dy \tag{5-10}$$

If p^0 is the stagnation pressure

$$\frac{U^2 - q_3^2}{U^2} = \frac{2}{(\gamma - 1)M_\infty^2}\left[\left(\frac{p_\infty^0}{p_3^0}\right)^{\frac{\gamma-1}{\gamma}} - 1\right]$$

or

$$\frac{q_3^2}{U^2} = 1 - \frac{2}{(\gamma - 1)M_\infty^2}\left[\frac{\Delta s}{C_p} + \frac{1}{2}\left(\frac{\Delta s}{C_p}\right)^2 + \cdots\right] \tag{5-11}$$

where Δs denotes the increase of entropy across the shock wave or in terms of the free stream Mach number and the inclination of the shock wave, if terms of higher order are neglected

$$\frac{q_3^2}{U^2} = \left\{1 - \frac{2}{(\gamma - 1)M_\infty^2}\left[\frac{1}{M_\infty^2 \sin^2 \epsilon}\left(\frac{2\gamma}{\gamma + 1} M_\infty^2 \sin^2 \epsilon - \frac{\gamma - 1}{\gamma + 1}\right)^{1/\gamma}\right.\right.$$
$$\left.\left.\left(\frac{2}{\gamma + 1} + \frac{\gamma - 1}{\gamma + 1} M_\infty^2 \sin^2 \epsilon\right) - 1\right]\right\} \qquad (5\text{-}12)$$

The part of the drag due to the shock wave in the region near the body must be determined from the shape of the shock wave, obtained, for example, experimentally, while the part of the drag depending upon the increase in entropy across the shock at large distances from the body can be evaluated by means of Eq. 5-9.

Let y_0 be the ordinate of a point of the shock wave at a large distance from the body and ϵ_0 the inclination of the shock at this point. In this region Eq. 5-3 is valid; therefore, Eq. 5-12 becomes

$$1 - \frac{q_3}{U} = \frac{2}{3}\frac{\Delta^3}{(\gamma + 1)^2 M_\infty^2} \qquad (5\text{-}13)$$

and

$$\Delta = 2\sigma \cot \mu_\infty \qquad (5\text{-}14)$$

From Eq. 5-9

$$dy = -\frac{2y_0}{\sigma^3}\sigma_0^2 d\sigma \qquad (5\text{-}15)$$

Therefore, by means of Eq. 5-10, 5-13, 5-14, and 5-15 the drag due to the shock wave from y_0 to infinity becomes

$$\Delta D = \frac{64}{3}\frac{1}{2}\rho_\infty U^2 \frac{(M_\infty^2 - 1)^{\frac{3}{2}}}{M_\infty^2(\gamma + 1)^2} y_0 \sigma_0^3 \qquad (5\text{-}16)$$

It follows from Eq. 5-9 that $y_0\sigma_0^2$ is constant or

$$\sigma_0 \sim \frac{1}{\sqrt{y_0}} \qquad (5\text{-}17)$$

and for given free stream conditions, if y_0 changes

$$\Delta D \sim \frac{B}{\sqrt{y_0}} \qquad (5\text{-}18)$$

where B is an appropriate constant. Thus, the asymptotic value of the wave drag is approached very slowly and the part of the wave drag attributable to the portion of the head shock wave at a large distance from the two-dimensional body decreases slowly as y increases [*9*].

Consider now a body of revolution at zero angle of attack [*10*]. From G,17 the characteristic line at large distance from the axis can be expressed in the form (Sec. G, Eq. 17-14)

$$x = \cot \mu_\infty y - by^{\frac{1}{4}} + \text{const} \qquad (5\text{-}19)$$

where x and y are now cylindrical coordinates and

$$b = \tan^2 \mu_\infty M_\infty^4 K(\gamma + 1) \tag{5-20}$$

The inclination of the streamline and the local velocity components are given in Sec. G, Eq. 17-8, 17-10, and 17-11 and are

$$\theta = \frac{K}{\sqrt{y}} = \frac{v}{U} \tag{5-21}$$

$$\frac{u}{U} - 1 = u' = -\frac{K \tan \mu_\infty}{\sqrt{y}} \tag{5-22}$$

At large distances from the axis the head shock wave is weak and is inclined to the free stream direction approximately at the undisturbed Mach angle μ_∞. Therefore, the equation of the shock can be expressed in the form

$$x = y \cot \mu_\infty - cy^{n+1} \tag{5-23}$$

or

$$\frac{dy}{dx} = \frac{1}{\cot \mu_\infty - (n+1)cy^n} \tag{5-24}$$

where c and n are constants and $n < 0$ since the shock wave must decrease in intensity when y increases and $c > 0$ because the shock is ahead of the Mach wave.

Consider now a characteristic line passing through a point Q of coordinates X, Y which meets the head shock S at a point P (x_P, y_P) (Fig. H,5b). The equation of the characteristic line is (see Eq. 5-19)

$$x = y \cot \mu_\infty - by^{\frac{1}{2}} + X - \cot \mu_\infty Y + bY^{\frac{1}{2}} \tag{5-25}$$

The slope of the characteristic line is

$$\frac{dy}{dx} = \frac{1}{\cot \mu_\infty - \frac{1}{2}by^{-\frac{1}{2}}} \tag{5-26}$$

The angle ϵ of the shock wave at P can be expressed as in Eq. 5-7, therefore from Eq. 5-2

$$\sigma = c(n+1)y_P^n \sin^2 \mu_\infty \tag{5-27}$$

If the inclination of the characteristic line at P with respect to the x axis is expressed as

$$\mu_2 + \delta = \mu_\infty + \vartheta \tag{5-28}$$

then from Eq. 5-26 it results

$$\vartheta = \tfrac{1}{2}by_P^{-\frac{1}{2}} \sin^2 \mu_\infty \tag{5-29}$$

From Eq. 5-3 and 5-7 there results, to the first order in Δ,

$$\sigma = \frac{\tan \mu_\infty}{2} \Delta \tag{5-30}$$

while from Eq. 5-1 and 5-3 there results, to the same order,

$$\delta = \frac{2\Delta}{\gamma + 1} \sin \mu_\infty \cos \mu_\infty \tag{5-31}$$

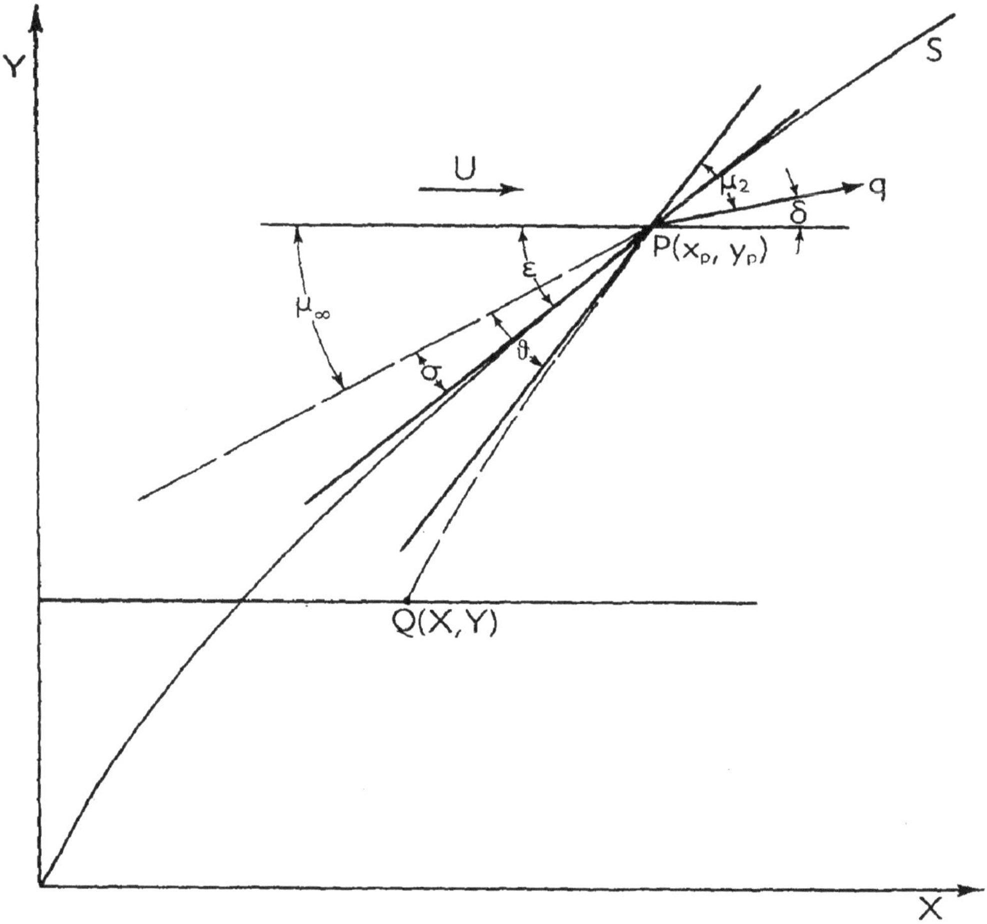

Fig. H,5b. The shape of the head shock wave at large distance from the body. Axially symmetric flow.

and from Eq. 5-2

$$\mu_2 = \mu_\infty + \tan \mu_\infty \left(\frac{\gamma - 1}{\gamma + 1} \Delta + \frac{2\Delta}{\gamma + 1} \sin^2 \mu_\infty \right) \tag{5-32}$$

Therefore

$$\vartheta = \mu_2 + \delta - \mu_\infty = \Delta \tan \mu_\infty = 2\sigma \tag{5-33}$$

Eq. 5-33 shows that to the first order in Δ the direction of the shock wave bisects the Mach directions of the flow in front of and behind the shock (cf. E,4).

From Eq. 5-24, 5-26, and 5-33,

$$b = 4(n + 1)cy_P^{n+\frac{1}{2}} \tag{5-34}$$

But point P is on the characteristic line and on the shock; therefore, from Eq. 5-23 and 5-25,

$$(4n + 3)cy_P^{n+1} - 4(n + 1)cy_P^{n+1} \left(\frac{Y}{y_P} \right)^{\frac{1}{2}} + \cot \mu_\infty Y - X = 0 \tag{5-35}$$

The second term of Eq. 5-35 is a function of Y/y_P and becomes smaller if y_P increases; therefore, if the coefficient $(4n + 3)$ of the first term is different from zero, the second term can be neglected for values of $y_P \gg Y$. Then

$$(4n + 3)cy_p^{n+1} = X - \cot \mu_\infty Y \tag{5-36}$$

However, σ in Eq. 5-7 is positive; therefore, in Eq. 5-27, $(n + 1)$ must be positive. Then, from Eq. 5-36, it results that, for a given value of Y, when y_P tends to infinity the value of X tends to infinity. This is impossible because any characteristic meeting the head shock must originate at the surface of the body. This can be seen, for example, by considering a closed body of finite length. Then all the characteristics which meet the head shock originate between the head shock and the tail shock wave.

If Eq. 5-36 is physically impossible, the first term of Eq. 5-35 must be of the same order as the second term for any value of y_P; therefore,

$$4n + 3 = 0 \tag{5-37}$$

or

$$n = -\tfrac{3}{4}$$

The shock at large distances from the axis is given by

$$x = y \cot \mu_\infty - cy^{\frac{1}{4}} \tag{5-38}$$

where c is a constant which depends upon the shape of the body. Then

$$\sigma = \frac{c}{4} y^{-\frac{3}{4}} \sin^2 \mu_\infty$$

or

$$\frac{\sigma}{\sigma_0} = \left(\frac{y_0}{y}\right)^{\frac{3}{4}} \tag{5-39}$$

A comparison of Eq. 5-9 and 5-39 indicates that the strength of the head shock for bodies of revolution decreases more rapidly than the head shock for two-dimensional bodies.

From Eq. 5-35

$$cy_P^{-\frac{3}{4}} Y^{\frac{1}{4}} = \cot \mu_\infty Y - X \tag{5-40}$$

and from Eq. 5-34

$$b = cy_P^{-\frac{3}{4}} \tag{5-41}$$

Therefore, along the characteristic line which meets the shock at P, (Eq. 5-20, 5-22, and 5-41)

$$u' = -\frac{cy_P^{-\frac{3}{4}}}{M_\infty^4(\gamma + 1) \tan \mu_\infty}$$

By means of Eq. 5-40 there results

$$u'(X, Y) = -\left(\cot \mu_\infty - \frac{X}{Y}\right) \frac{\cot \mu_\infty}{(\gamma + 1)M_\infty^4} \tag{5-42}$$

$$v'(X, Y) = \frac{\cot^2 \mu_\infty}{(\gamma + 1)M_\infty^4} \left(\cot \mu_\infty - \frac{X}{Y}\right) \tag{5-43}$$

Eq. 5-38, 5-42, and 5-43 permit determination of the shock drag along a streamline at a large distance from the axis (see also E,6).

CHAPTER 2. TRANSONIC FLOW

H,6. Simplified Basic Equations. Many purely supersonic or subsonic flow patterns are transformed into a mixed flow by a steady change of the initial boundary conditions as, for example, the undisturbed stream velocity or the shape of the solid boundary.

The complete solution of such problems is extremely difficult even in the approximation of nonviscous flow theory. The structure of the flow patterns and their properties can be determined without going into a detailed analysis if simplified hypotheses are accepted which permit the use of the transonic similarity law introduced by von Kármán [*11*]. With these hypotheses the quantitative analysis is also simplified (see Sec. D).

For a two-dimensional or axially symmetric flow having constant stagnation enthalpy the following equations are valid (see Sec. G, Eq. 2-12)

$$(a^2 - u^2)\frac{\partial u}{\partial x} - uv\left(\frac{\partial u}{\partial y} + \frac{\partial v}{\partial x}\right) + \frac{\partial v}{\partial y}(a^2 - v^2) = -\epsilon\frac{v}{y}a^2 \qquad (6\text{-}1)$$

where ϵ is zero for two-dimensional flows and is equal to 1 for flows having axial symmetry. Also

$$a^2 = \frac{\gamma + 1}{2}q^{*2} - \frac{\gamma - 1}{2}q^2 \qquad (6\text{-}2)$$

where q^* is the critical velocity (velocity of sound at $M = 1$) and a the local speed of sound. Furthermore (Sec. D, Eq. 32-9),

$$-\frac{\partial u}{\partial y} + \frac{\partial v}{\partial x} = \frac{T}{q}\frac{ds}{dn} = \frac{a^2}{\gamma \mathcal{R} q}\frac{ds}{dn} \qquad (6\text{-}3)$$

where ds/dn is the entropy gradient taken normal to the streamline.

Eq. 6-1, 6-2, and 6-3 become somewhat simpler if the similarity law for transonic problems is assumed valid.

Mathematically speaking, the similarity law can be expressed in the following form: Consider a family of flows, each of which is designated by a parameter Z, and assume that for small values of the parameter Z ($Z \ll 1$) the following relations are valid between flows of the same family

$$u(x, y) = q^*[1 + Z\alpha(\xi, \eta)] \qquad (6\text{-}4)$$

$$v(x, y) = q^*Z^{\frac{3}{2}}\beta(\xi, \eta)\sqrt{\gamma + 1} \qquad (6\text{-}5)$$

where

$$x = C\xi \qquad (6\text{-}6)$$

and

$$y = C\eta Z^{-\frac{1}{2}}(\gamma + 1)^{-\frac{1}{2}} \qquad (6\text{-}7)$$

The quantities $\alpha(\xi, \eta)$ and $\beta(\xi, \eta)$ are suitable functions of their arguments ξ and η, and C is a constant, for example, a reference length.

From Eq. 6-4 and 6-5 it can be seen that the velocities at infinity for two flows distinguished by the parameters Z_1 and Z_2 are

$$u_1(\infty) = q^*[1 + Z_1\alpha(\infty, \infty)] \tag{6-8}$$

$$u_2(\infty) = q^*[1 + Z_2\alpha(\infty, \infty)] \tag{6-9}$$

Therefore, the parameter Z which distinguishes the flow patterns is proportional to the difference between undisturbed velocity and critical velocity.

From Eq. 6-6 and 6-7 it results that corresponding points in the two flow patterns having the same value of ξ and η have the same x coordinate but have y coordinates which are stretched by a factor $[(\gamma + 1)Z]^{1/2}$, proportional to the square root of the difference between the undisturbed velocity and the critical velocity. The velocity components of the flow at the two corresponding points are different, as shown from Eq. 6-4 and 6-5. When Z decreases, u approaches q^* and v decreases by a factor $Z^{3/2}$; therefore, v decreases more rapidly than u. Consequently, the slope of the streamlines decreases, and for small values of Z the slope decreases approximately as $Z^{3/2}$.

By means of Eq. 6-4 to 6-9, from the flow over a given body at a certain Mach number close to 1, an infinite number of flows at different Mach numbers (still in the neighborhood of 1) can be obtained over bodies having different shapes. The shape of the bodies is not arbitrary, but must be derived from the shape of the original body according to Eq. 6-4 to 6-9. This procedure cannot be extrapolated to flows at a free stream Mach number of 1, because bodies of zero thickness would be obtained.

By means of Eq. 6-4 to 6-9, and by assuming that the flow field is described by the lowest order terms in Z, Eq. 6-2 and 6-1 become, respectively

$$a^2 - u^2 = -(\gamma + 1)Zq^{*2}\alpha \tag{6-10}$$

$$-\alpha\frac{\partial\alpha}{\partial\xi} + \frac{\partial\beta}{\partial\eta} = -\epsilon\frac{\beta}{\eta} \tag{6-11}$$

From Eq. 5-3, 5-11, and 5-13 it is seen that the variation of entropy is of the order of the cube of the difference $[(q_{n\infty}^2/a^2) - 1]$ (where $q_{n\infty}$ is the component of the stream velocity normal to the shock wave), or because of Eq. 6-10, of the order of Z^3, and terms of the order of Z^3 are neglected in these considerations. Thus Eq. 6-3 becomes

$$\frac{\partial\alpha}{\partial\eta} = \frac{\partial\beta}{\partial\xi} \tag{6-12}$$

Because of Eq. 6-12 the vorticity in the flow can be neglected, and the flow can be treated as potential flow.

Eq. 6-11 is independent of γ, Z, and M_∞, and therefore represents a family of flows.

The characteristic lines of Eq. 6-11 in the supersonic region of the flow field can be obtained as shown in G,2 and are given by

$$\frac{d\eta}{d\xi} = \pm\alpha^{-\frac{1}{2}} \tag{6-13}$$

The differential equations valid along the characteristic lines become [12]

$$\mp\alpha^{\frac{1}{2}}\frac{d\alpha}{d\eta} + \frac{d\beta}{d\eta} = -\epsilon\frac{\beta}{\eta} \tag{6-14}$$

where ϵ assumes the values 0 and 1, for two-dimensional and axially symmetric flow, respectively. If $\epsilon = 0$,

$$\beta = \pm\tfrac{2}{3}\alpha^{\frac{3}{2}} + \text{const} \tag{6-15}$$

Eq. 6-11 for two-dimensional flow, where $\epsilon = 0$, can be linearized by introducing hodograph variables. Assume as independent variables the quantities q and θ (where θ is referred to the x axis), which define the velocity vector in polar coordinates.

In the neighborhood of the critical velocity, from Eq. 6-4 and 6-5 it is found that

$$q = q^*(1 + Z\alpha) \tag{6-16a}$$

$$\theta = \frac{v}{u} = Z^{\frac{3}{2}}\beta\sqrt{\gamma + 1} \tag{6-16b}$$

or

$$\alpha = \frac{q - q^*}{Zq^*} \tag{6-17}$$

and

$$\beta = \frac{\theta}{Z^{\frac{3}{2}}\sqrt{\gamma + 1}} \tag{6-18}$$

The equation of the stream function in the hodograph plane in its complete form is (Chaplygin's equation, see Sec. F, Eq. 3-1).

$$q^2\psi_{qq} + \psi_q q\left(1 + \frac{q^2}{a^2}\right) + \left(1 - \frac{q^2}{a^2}\right)\psi_{\theta\theta} = 0 \tag{6-19}$$

and the relations in the physical plane in their complete form are (cf. Sec. F, Eq. 2-10, 2-11)

$$\frac{\partial x}{\partial q} = -\frac{1}{\rho q}\left[\left(1 - \frac{q^2}{a^2}\right)\psi_\theta\frac{\cos\theta}{q} + \psi_q\sin\theta\right] \tag{6-20}$$

$$\frac{\partial x}{\partial \theta} = \frac{1}{\rho q}(q\psi_q\cos\theta - \psi_\theta\sin\theta) \tag{6-21}$$

$$\frac{\partial y}{\partial q} = -\frac{1}{\rho q}\left[\left(1 - \frac{q^2}{a^2}\right)\psi_\theta\frac{\sin\theta}{q} - \psi_q\cos\theta\right] \tag{6-22}$$

$$\frac{\partial y}{\partial \theta} = \frac{1}{\rho q}\,(q\psi_q \sin\theta + \psi_\theta \cos\theta) \tag{6-23}$$

In the transonic flow approximation, where the parameter Z is small and terms of the order of Z^2 or higher are neglected, Eq. 6-19, by means of Eq. 6-17 and 6-18, becomes

$$\psi_{\alpha\alpha} - \alpha\psi_{\beta\beta} = 0 \tag{6-24}$$

Eq. 6-20 to 6-23 in the transonic approximation become

$$\frac{\partial x}{\partial \alpha} = \frac{1}{\rho^* q^*}\,\alpha\,\sqrt{\gamma + 1}\,Z^{\frac{1}{2}}\psi_\beta \tag{6-25}$$

$$\frac{\partial x}{\partial \beta} = \frac{1}{\rho^* q^*}\,\sqrt{\gamma + 1}\,Z^{\frac{1}{2}}\psi_\alpha \tag{6-26}$$

$$\frac{\partial y}{\partial \alpha} = \frac{1}{\rho^* q^*}\,\psi_\alpha \tag{6-27}$$

and

$$\frac{\partial y}{\partial \beta} = \frac{1}{\rho^* q^*}\,\psi_\beta \tag{6-28}$$

or

$$y = \frac{1}{\rho^* q^*}\,\psi \tag{6-29}$$

The symbol ρ^* indicates the density at the critical velocity q^* ($M = 1$). When the flow field $\alpha(\xi, \eta)$, $\beta(\xi, \eta)$ has been determined, any corresponding flow field $u(x, y)$, $v(x, y)$ represented by the parametric flow α, β can be obtained from the following considerations.

At infinity, Eq. 6-8 and 6-10 give

$$\frac{U - q^*}{q^*} = \frac{2}{\gamma + 1}\,(M_\infty - 1) = Z\alpha_\infty(\infty) \tag{6-30}$$

Therefore, Z must be proportional to $\dfrac{2}{\gamma + 1}\,(M_\infty - 1)$.

The boundary conditions at the surface of a two-dimensional body can be expressed in the form

$$\left(\frac{dy}{dx}\right)_L = \theta(x) = t\,h(x) \tag{6-31}$$

where $(dy/dx)_L$ is the slope of the boundary streamline and t is a dimensionless quantity proportional to a characteristic length in the y direction, for example, the maximum per cent thickness. Eq. 6-31 must be evaluated at the boundary, but, in this simplified solution, can be evaluated at the x axis. Eq. 6-18 indicates that, in order to satisfy the boundary conditions at the body, Z must be proportional to $t^{\frac{2}{3}}(\gamma + 1)^{-\frac{1}{3}}$. Therefore, the flow field α, β in the approximation of the similarity law represents all the flows u, v having the same distribution $h(\xi)$ of β at the ξ axis and a constant value of the expression k given by

$$k = \frac{M_\infty - 1}{[(\gamma + 1)t]^{\frac{2}{3}}} \tag{6-32}$$

Because Z is proportional to $t^{\frac{2}{3}}(\gamma + 1)^{-\frac{1}{3}}$ the relation (6-7) between y and η becomes

$$y = C[t(\gamma + 1)]^{-\frac{1}{3}}\eta \tag{6-33}$$

The pressure coefficient at each point of the flow is given by

$$C_p = -4\frac{M_\infty - 1}{\gamma + 1}\left(\frac{\alpha}{\alpha_1} - 1\right) = -\frac{t^{\frac{2}{3}}}{(\gamma + 1)^{\frac{1}{3}}}P(\xi, \eta, k) \tag{6-34}$$

the lift coefficient is given by

$$C_L = \frac{1}{c}\oint C_p dx = \frac{t^{\frac{2}{3}}}{(\gamma + 1)^{\frac{1}{3}}}L(k) \tag{6-35}$$

and the drag coefficient by

$$C_D = \frac{1}{c}\oint C_p \frac{dy}{dx}\,dx = \frac{t^{\frac{5}{3}}}{(\gamma + 1)^{\frac{1}{3}}}D(k) \tag{6-36}$$

All considerations in which the transonic similarity law is introduced apply strictly only when the local velocity is everywhere close to the critical velocity. The considerations cannot be used for the analysis of transonic flows over bodies producing detached shock waves at large supersonic velocity. However, qualitative information on the properties of such flow fields can still be obtained from this analysis.

H,7. The Shock Polar Diagram for Transonic Flow. The shock polar diagram in the hodograph plane is given by a curve defined by the following equation (see III,E,2)

$$v_2 = \pm\,(q_1 - u_2)\sqrt{\frac{u_2 - \dfrac{q^{*2}}{q_1}}{\dfrac{q^{*2}}{q_1} + \dfrac{2}{\gamma + 1}q_1 - u_2}} \tag{7-1}$$

where v_2 and u_2 are the velocity components behind the shock wave, and q_1 is the velocity in front of the shock wave, assumed parallel to the x axis.

In the physical plane the shock wave is defined by

$$\frac{dy}{dx} = \frac{q_1 - u_2}{v_2} \tag{7-2}$$

If the transonic similarity law is introduced in Eq. 7-1 and terms of the order of Z^2 or higher are neglected, Eq. 7-1 in the vicinity of the critical velocity q^* becomes

$$\beta_2 = \pm(\alpha_1 - \alpha_2)\sqrt{\frac{\alpha_1 + \alpha_2}{2}} \tag{7-3}$$

and

$$\frac{d\beta_2}{d\alpha_2} = \mp \frac{3\alpha_2 + \alpha_1}{\sqrt{8(\alpha_2 + \alpha_1)}} \tag{7-4}$$

In order to pass from the hodograph plane to the physical plane, the variations dx and dy along the shock wave in the physical plane are related to the variations $d\alpha$ and $d\beta$ along the shock wave in the hodograph plane by the expressions

$$dx = \frac{\partial x}{\partial \alpha} d\alpha + \frac{\partial x}{\partial \beta} d\beta \tag{7-5}$$

$$dy = \frac{\partial y}{\partial \alpha} d\alpha + \frac{\partial y}{\partial \beta} d\beta \tag{7-6}$$

Eq. 7-2 along the shock wave, by means of Eq. 7-5 and 7-6, 6-25, 6-26, 6-27, 6-28, and 7-4, becomes

$$\psi_\alpha \mp \psi_\beta \sqrt{\frac{\alpha_1 + \alpha_2}{2}} \frac{7\alpha_2 + \alpha_1}{5\alpha_2 + 3\alpha_1} = 0 \tag{7-7}$$

Along the streamline the following equation is valid

$$\psi_\alpha + \psi_\beta \frac{d\beta}{d\alpha} = 0 \tag{7-8}$$

Therefore, from Eq. 7-7 and 7-8 it is seen that the direction of the tangent to the streamlines along the shock polar is given by

$$\frac{d\beta}{d\alpha} = \mp \sqrt{\frac{\alpha_1 + \alpha_2}{2}} \frac{7\alpha_2 + \alpha_1}{5\alpha_2 + 3\alpha_1} \tag{7-9}$$

If the value of $d\beta/d\alpha$ is zero

$$\alpha_2 = -\tfrac{1}{7}\alpha_1 \tag{7-10}$$

The variation of β, and therefore of θ, along the streamline is zero at the shock wave; i.e., the curvature is zero; therefore, Eq. 7-10 defines the "Crocco point" (see Art. 4).

If θ behind the shock wave is zero, β is also zero; therefore, the normal shock is defined by

$$\alpha_1 = -\alpha_2 \tag{7-11}$$

The sonic velocity q^* is defined by $\alpha_2 = 0$ or (Eq. 7-3)

$$\beta_2 = \pm\, \alpha_1 \sqrt{\frac{\alpha_1}{2}} \tag{7-12}$$

The point of maximum deviation across the shock is defined by the condition $d\beta/d\alpha = 0$ in Eq. 7-4,

$$\alpha_1 = -3\alpha_2 \tag{7-13}$$

and

$$\beta_2 = \pm \frac{4}{3} \alpha_1 \sqrt{\frac{\alpha_1}{3}} \tag{7-14}$$

Eq. 7-9 gives the gradient along the streamlines at points of the shock wave. The condition $d\beta/d\alpha$ equal to infinity in Eq. 7-9 indicates the point of the shock wave where the streamline starts with zero velocity gradient.

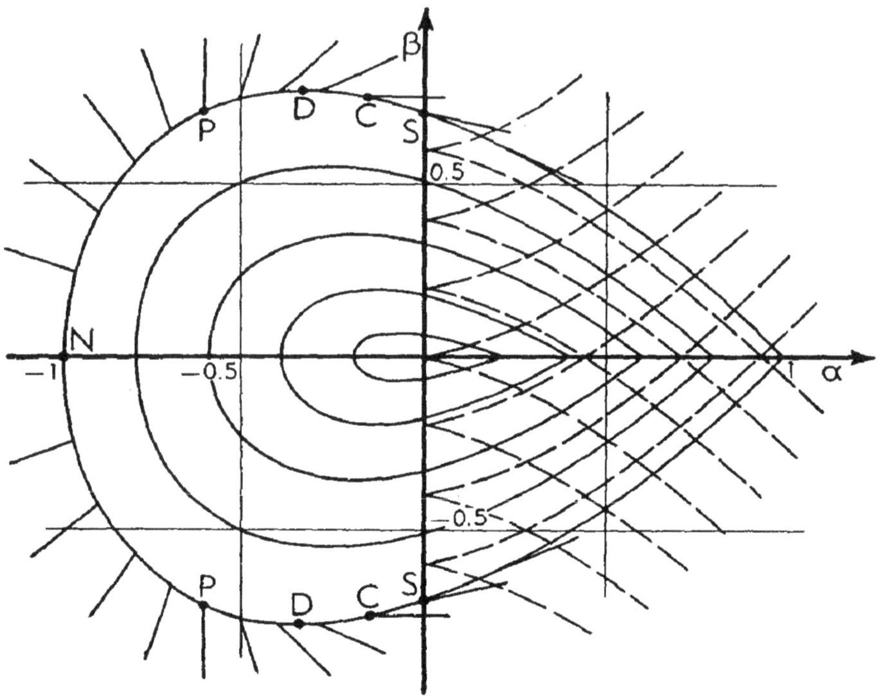

Fig. H,7. The shock polar diagram for transonic flow.

This condition corresponds to

$$\alpha_2 = -\tfrac{3}{5}\alpha_1 \tag{7-15}$$

$$\beta_2 = \frac{8}{5} \alpha_1 \sqrt{\frac{\alpha_1}{5}} \tag{7-16}$$

In Fig. H,7 the shock polar diagram for transonic flow is shown for different values of α_1. Point C corresponds to the "Crocco point" defined by Eq. 7-10, point N to the normal shock (Eq. 7-11), point S to the sonic velocity behind the shock wave (Eq. 7-12), the point of maximum deviation θ is indicated by D (Eq. 7-13 and 7-14), while the point corresponding to Eq. 7-15 and 7-16 is indicated by P.

In Fig. H,7 the tangents to the streamlines (Eq. 7-9) have been drawn for the shock polar diagram corresponding to $\alpha = 1$. (The shock polar diagram for transonic flow was introduced by Guderley and Busemann

[*8,12,13*]. In Fig. H,7 some characteristic lines defined by Eq. 6-15 have also been shown and are indicated by broken lines.

H,8. Boundary Conditions. By means of the hodograph transformation the problem of determining two-dimensional flow fields in the physical plane has been reduced to the problem of determining boundary conditions in the hodograph plane, and by means of the similarity law the differential equation for transonic problems has been reduced to

$$\psi_{\alpha\alpha} - \alpha\psi_{\beta\beta} = 0 \tag{8-1}$$

Mathematical proofs of the existence and uniqueness of the solutions of this problem are presented in [*14* and *15*], and a detailed discussion of this problem is given in [*8*]. Here the differences of required boundary

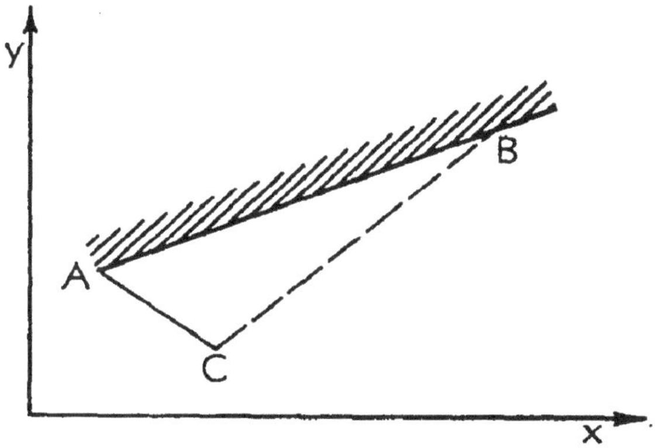

Fig. H,8a. Relations between boundary conditions in the physical and hodograph planes. Physical plane.

conditions for subsonic and supersonic problems are briefly discussed in order to obtain information for transonic problems. Eq. 8-1 is of the elliptic type if α is negative, while it is of the hyperbolic type if α is positive. When α is positive, the characteristic theory can be applied.

Consider a supersonic flow field (Fig. H,8a) where AB represents a solid boundary and AC a line in the flow along which the flow properties are known. If BC is a characteristic line from C, the whole flow field ABC is defined and can be determined by means of the characteristic theory; therefore, the boundary streamline AB and the flow along AC are sufficient boundary conditions to define the flow in ABC.

If these boundary conditions are represented in the hodograph plane (Fig. H,8b) the boundary streamline AB is represented by $A'B'$, and the line AC by $A'C'$. The line $B'C'$ is not an independent boundary condition but must coincide with the characteristic line from C'. Therefore, in supersonic flow the boundary conditions in the hodograph plane generally cannot be given along a closed line, but must be given along an open line, which is closed by characteristic lines.

In subsonic flow the boundary conditions must be different, because disturbances can travel upstream in the flow. If the flow along AC and BC is subsonic, the flow along the line BC is dependent not only upon the direction AB and the flow properties along AC, but also upon the flow outside of the region ABC. Therefore, the flow at ABC is defined

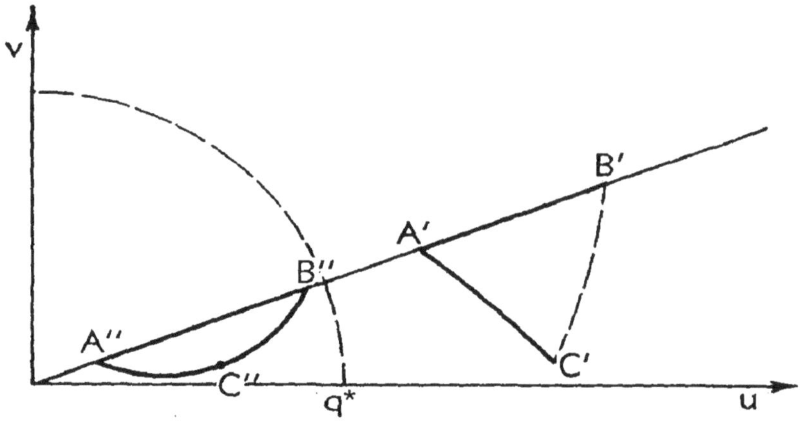

Fig. H,8b. Relations between boundary conditions in the physical and hodograph planes. Hodograph plane.

only when the flow properties along BC are also given. In the hodograph plane the flow in ABC is defined by boundary values along a closed line $A''B''C''$.

Consider now a transonic flow, as for example the upper half of a Laval nozzle as shown in Fig. H,8c. The line AB represents the sonic line and the line BC the last characteristic line that meets the sonic line.

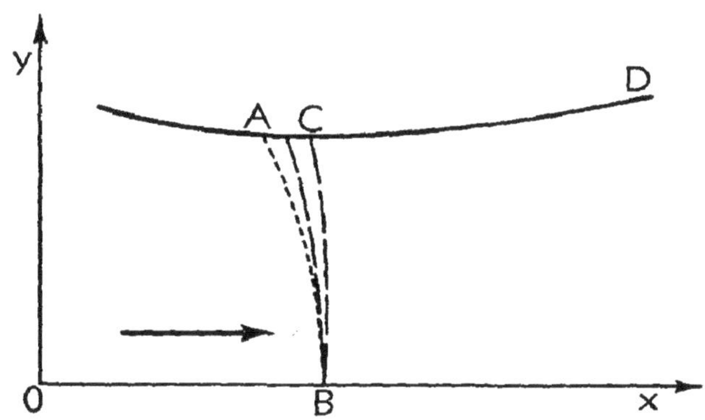

Fig. H,8c. Transonic flow at the minimum cross section of a Laval nozzle.

The supersonic flow is defined when the boundary streamline CD and the flow along BC are given. The flow in the region ABC has supersonic speed, and is physically determined if the shape of the sonic line and the streamline AC are given. However, any change of the streamline AC would change the shape of the sonic line and thus influence the subsonic

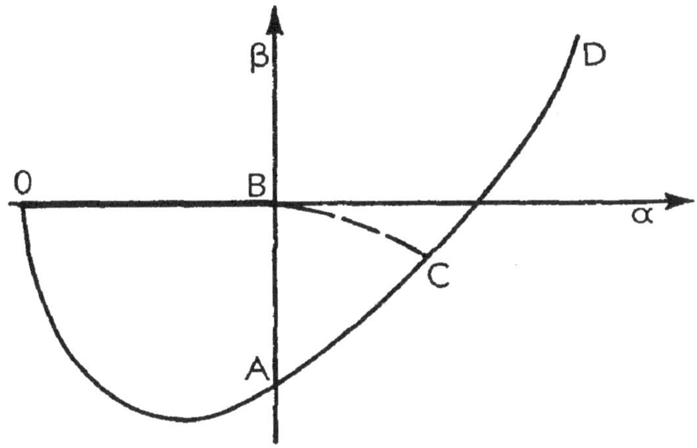

Fig. H,8d. Hodograph representation of the flow of Fig. H,8c.

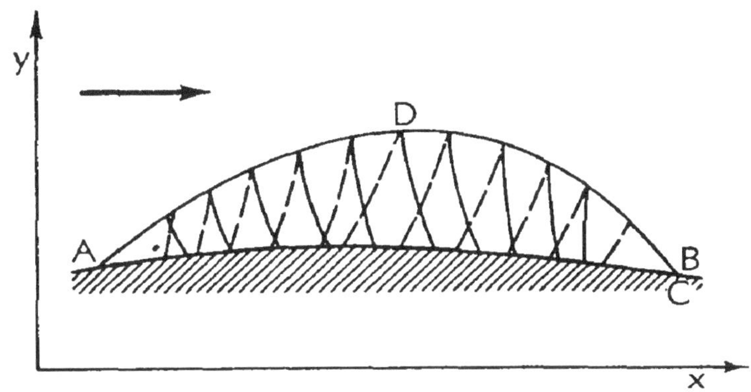

Fig. H,8e. Supersonic flow region imbedded in a subsonic flow.

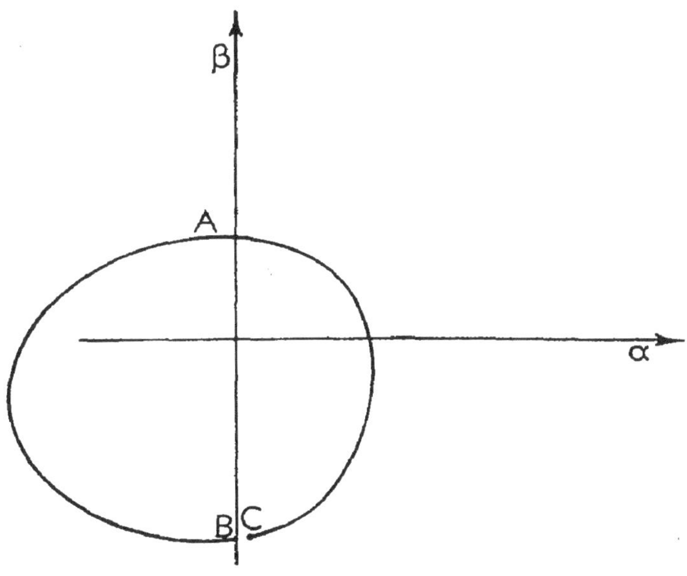

Fig. H,8f. Hodograph representation of the flow field of Fig. H,8e.

flow. The flow upstream of the sonic line AB is subsonic, and in this region the boundary conditions cannot be given on a closed line but in general must be given as shown in Fig. H,8d along an open line BOA. Therefore, in transonic flow, boundary conditions such as shown in Fig. H,8d seem sufficient, so long as no limiting lines are produced, and this boundary extends to the last Mach wave BC which still influences the sonic line.

Consider now a supersonic field imbedded in a subsonic flow illustrated in Fig. H,8e. In this case every point of the supersonic boundary AC influences the sonic line ADB and therefore influences the subsonic region of the flow field. In the hodograph diagram as shown in Fig. H,8f, the gap between B and C found in Fig. H,8d is not possible in this case. In this case a singularity is expected at the end of the supersonic region (point B), which represents the gap between B and C [8,13].

H,9. Transition from Subsonic to Supersonic Flow around Corners. Consider a free jet flowing from a large container (infinite

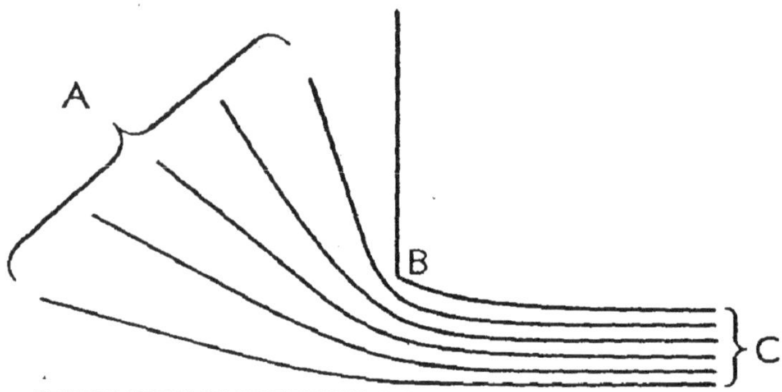

Fig. H,9a. Flow through a two-dimensional orifice. The ratio p_a/p^0 is higher than the initial value.

volume) at stagnation pressure p^0 through a two-dimensional orifice (Fig. H,9a) [8]. Let the outside ambient pressure be equal to p_a.

If the pressure ratio p_a/p^0 is higher than the critical value, the jet is subsonic. The velocity at some distance from the orifice inside the tank is zero (point A of Fig. H,9a) and outside the tank the velocity becomes constant (point C). Along the streamline BC the pressure and velocity are constant, and at C, which is far from the orifice, the stream is parallel to the axis of the jet. The hodograph representation of the flow field is as shown in Fig. H,9b (where coordinates u and v are used because flow far from sonic velocity is also considered).

If the pressure ratio p_a/p^0 is lower than the critical value, the jet is supersonic (Fig. H,9c and H,9d). The axis of the jet AC has constant direction and in the hodograph representation maps into the u axis. The rigid wall AB is represented by $A'B'$ and because the boundary of the jet BE is a streamline, having constant pressure and velocity, it maps into a segment of the circle of radius $A'D'_0$ defined by the ratio p_a/p^0.

Now consider the flow along the surface AB. At B the velocity must

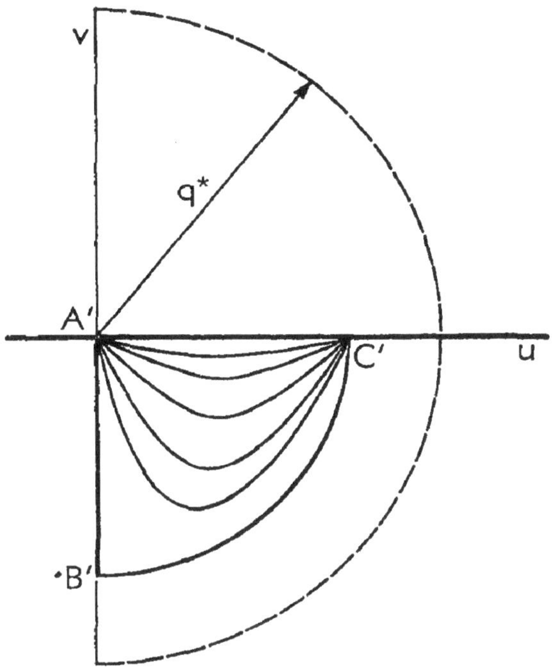

Fig. H,9b. Hodograph representation of the flow field of Fig. H,9a.

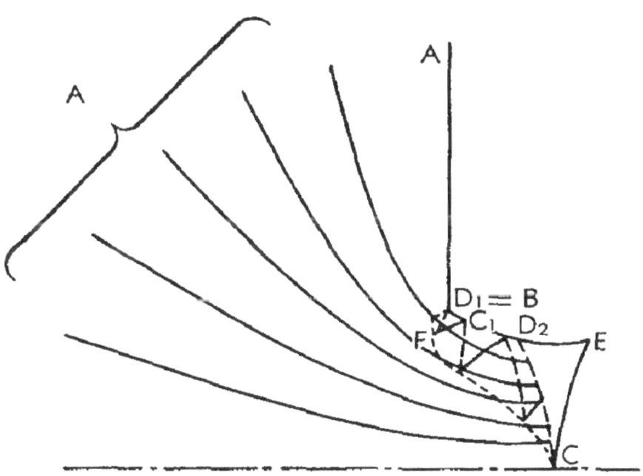

Fig. H,9c. Flow through a two-dimensional orifice. The ratio p_a/p^0 is lower than the critical value. (From [8].)

reach the intensity $A'D'_0$. However, the hodograph representation $A'D'_0$ would indicate a supersonic expansion from sonic velocity $A'B'$ to $A'D'_0$ without any change in direction, which requires the presence of incoming expansion waves meeting the wall AD and reflecting at the wall as expansion waves as illustrated in Fig. H,9d (top right). But, if the sonic line originates at B as in Fig. H,9d (top left) then the waves coming to the wall AB are the waves reflected at the sonic line. These waves can only be compression waves because going along any wave from BD to the sonic line the velocity decreases and the pressure increases. Consequently, the hodograph representation $A'B'D_0'$ is not correct. The correct representation is $A'B'D_1'$ where $B'D_1'$ is a characteristic line.

This representation indicates that along the wall AB sonic velocity is reached at corner B of Fig. H,9c and here an expansion around a

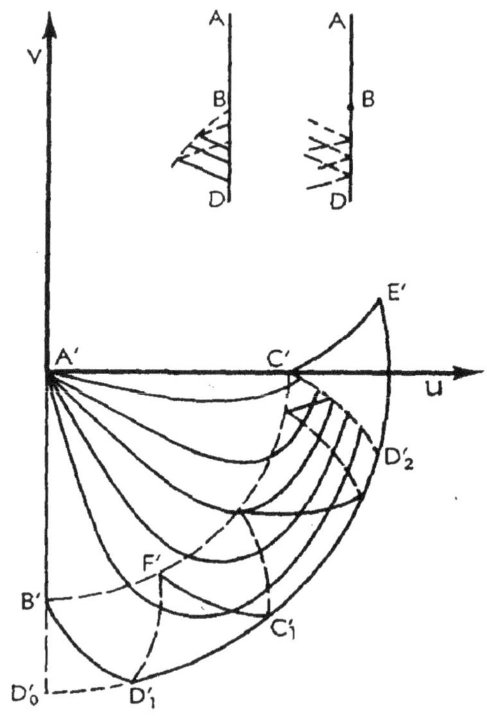

Fig. H,9. Bottom, the hodograph representation of the flow field of Fig. H,9c. (From [12].) Top right, waves required by the hodograph representation $A'D_0'$ of Fig. H,9d, bottom. Top left, waves reflected at the sonic line.

corner occurs (characteristic line $B'D_1'$) adjacent to the subsonic field. At B the sonic line and the expansion fan start [8].

The expansion waves originating at B are reflected at the sonic line BC as compression waves (wave D_1F is reflected as FC_1 of Fig. H,9c). The compression waves are reflected once again as expansion waves at the free boundary.

In Fig. H,9c expansion wave D_2C is the last expansion wave which meets the sonic line, and downstream of D_2C the flow is supersonic.

If the ratio p_a/p^0 decreases with respect to the value considered in

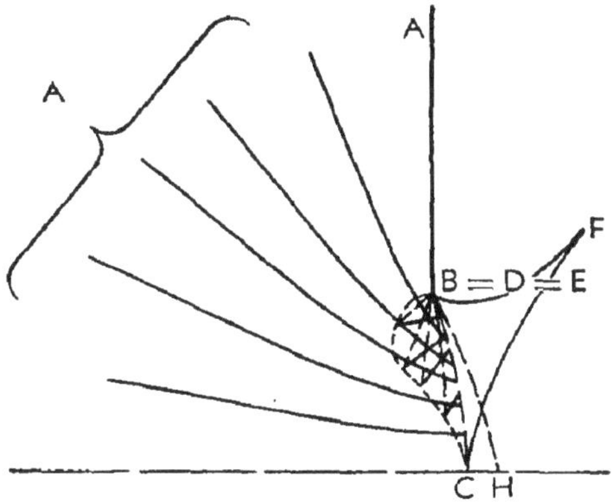

Fig. H,9e. Flow across a two-dimensional orifice. The ratio p_a/p^0 is lower than 0.0389. (The stream at D is turned 45° with respect to the direction AB.)

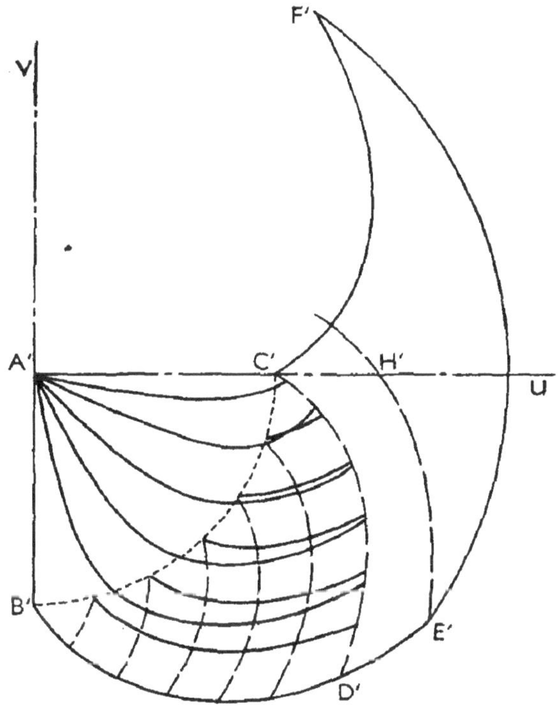

Fig. H,9f. Hodograph representation of the flow field of Fig. H,9e.

Fig. H,9c and H,9d, the point D_2 moves toward D_1 until points D_2 and D_1 become coincident with the point B. For this condition the expansion wave D_2C is a wave of the fan existing at the corner B in Fig. H,9e and H,9f.

The presence of a Prandtl-Meyer expansion around a corner, starting at $M = 1$, and adjacent to a subsonic field is general for subsonic corners in transonic flow. The sonic velocity is reached at the corner, where the sonic line and expansion fan originate.

The direction of the sonic line according to the Prandtl-Meyer expansion is at first perpendicular to the last subsonic streamline, and then bends around the corner. Indeed, if S and n are coordinates measured along and normal to the streamlines and l is the arc length along the sonic line, inclined at Λ with the direction of the streamline, then

$$\frac{dq}{dl} = \frac{\partial q}{\partial S} \cos \Lambda + \frac{\partial q}{\partial n} \sin \Lambda = 0 \tag{9-1}$$

If θ is the direction of the streamline in the hodograph plane,

$$\frac{\partial q}{\partial S} = \left(\frac{dq}{d\theta}\right)_S \frac{\partial \theta}{\partial S} \tag{9-2}$$

but for irrotational flow

$$q \frac{\partial \theta}{\partial S} = \frac{\partial q}{\partial n} \tag{9-3}$$

therefore

$$\tan \Lambda = -\frac{1}{q} \left(\frac{dq}{d\theta}\right)_S \tag{9-4}$$

In the case of the expansion around a corner, the streamline in the hodograph plane is represented by the characteristic line. For $M = 1$,

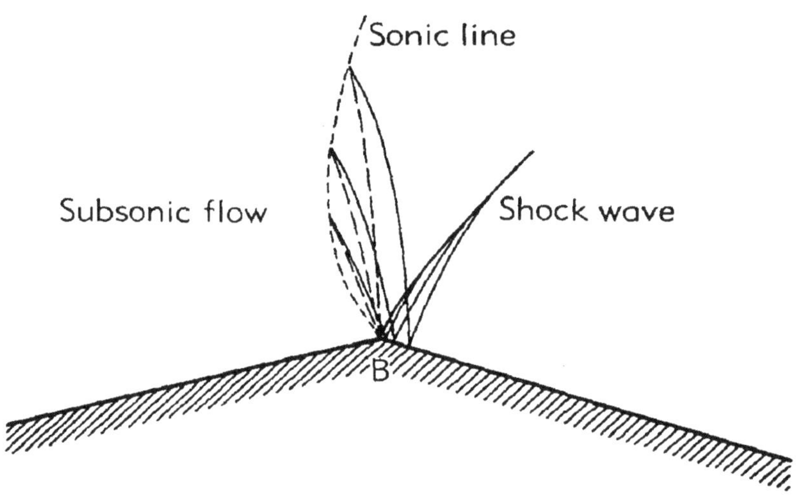

Fig. H,9g. Expansion around a sharp corner of a subsonic flow.

$dq/d\theta$ along the characteristic line at B' becomes infinite, and Λ becomes equal to $\pi/2$. The value of $(dq/d\theta)_S$ immediately becomes finite moving away from the surface AB in a direction normal to AB as shown in Fig. H,9d and H,9f; therefore, the sonic line begins normal to the surface and with infinite curvature.

The same considerations are valid when the direction is given along the boundary streamline *BE* of Fig. H,9c, H,9e, and H,9f in place of the value of the velocity. In this case also sonic velocity is reached at the corner *B*, and therefore the sonic line and expansion fan start at the corner as illustrated in Fig. H,9g. Expansion waves meet the sonic line and are reflected as compression waves, and the sonic line is normal to the subsonic surface at the corner; it has an infinite curvature at the corner

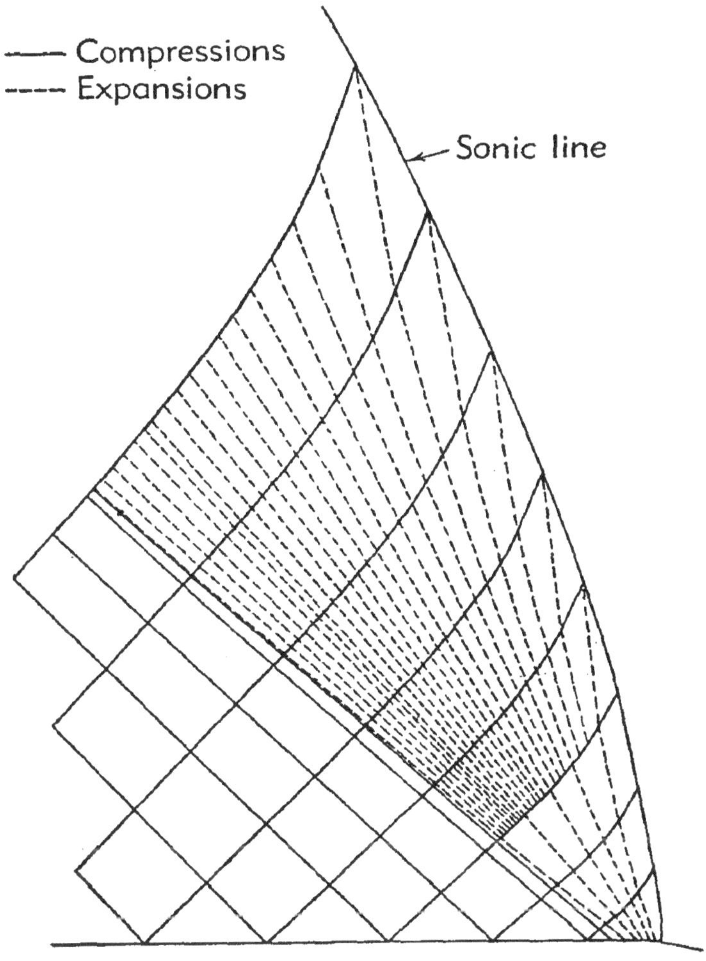

Fig. H,9h. The shape of the sonic line and the characteristic net corresponding to the flow field of the interferogram of Plate H,9c.

and is bent upstream. However, in this case the compression waves produced at the sonic line as reflections of the expansion waves are reflected once again at the solid boundary as compression waves. These compression waves coalesce and form an envelope shock wave which begins at the ridge.

Because a subsonic stream is not able to pass the slightest angle of a sharp convex deflection, this transonic pattern is general for sharp corners independent of the velocity of the undisturbed stream. However, when viscous effects are considered, other solutions with separated regions are possible, and in these cases the velocity of the undisturbed stream and the angle of the corner are important parameters of the phenomenon.

In Plates H,9a, H,9b and H,9c interferograms of flow fields around sharp corners are shown. In Plate H,9a the flow field around a sharp-nosed two-dimensional body is shown for a free stream Mach number of 0.72. The body is at a small angle of attack, and an expansion around the leading edge occurs. The picture shows that, because of the presence of viscosity, a vortex exists at the leading edge followed by a separated region. In Plate H,9b the free stream Mach number is increased to 0.81 and the expansion fan, followed by a shock wave, is found.

In Plate H,9a a transonic expansion at the ridge of a two-dimensional profile is shown. In Fig. H,9h the shape of the sonic line and the characteristic net, showing expansion and compression waves for the flow field at the ridge shown in Plate H,9c, are presented. Because of the presence of a boundary layer, the expansion fan does not originate exactly at the ridge.

H,10. Transition from Attached to Detached Shock for Sharp-nosed Two-dimensional Bodies. Consider at first a symmetrical wedge, having straight sides, at zero angle of attack and moving at supersonic speed.

For small wedge angles the shock wave is attached at the leading edge of the body, the flow is everywhere supersonic, and the shock is straight. For wedge angles larger than the angle of maximum deviation, corresponding to the stream velocity considered, a detached shock exists in front of the body. The transition from attached shock to detached shock waves, when the angle of the wedge increases, has been analyzed by Guderley with the help of the hodograph representation [*8*].

Because transonic flow is considered, the transonic coordinates α and β defined in Eq. 6-17 and 6-18 will be used for the representation.

In Fig. H,10a(1) the flow around the wedge for the condition of attached shock and supersonic flow behind the shock is presented in the hodograph and physical planes.

The flow behind the shock is uniform until expansion waves from the ridge E are produced, which expand the flow again to free stream conditions. The shock is straight and the flow along EF is represented in the hodograph plane by the point F of coordinate β corresponding to the inclination of EF. The flow in front of the expansion fan from E has constant velocity F.

When a larger wedge angle is considered, larger than the deviation corresponding to sonic velocity ($\beta_F > \beta_S$), the flow at F behind the shock becomes subsonic (Fig. H,10a(2)). The sonic velocity is reached at the ridge E; therefore, the flow along EF accelerates from the value of the

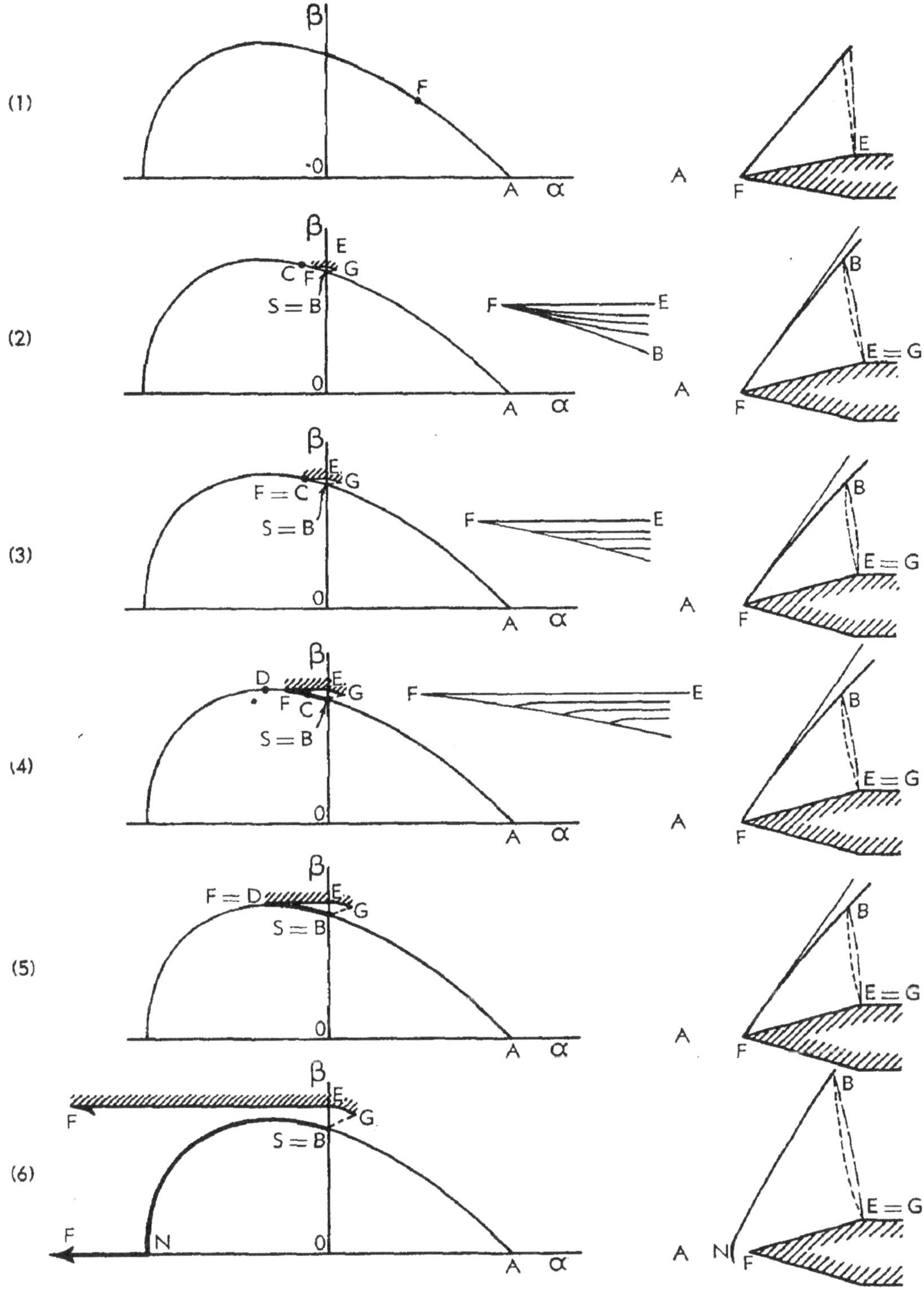

Fig. H,10a. The transition from attached to detached shock wave for wedges in the physical and hodograph planes: (1) Attached shock wave $\beta_F < \beta_S$; (2) Attached shock wave $\beta_S < \beta_F < \beta_C$; (3) Attached shock wave with subsonic flow behind the shock $\beta_F = \beta_C$; (4) Attached shock wave with subsonic flow behind the shock $\beta_C < \beta_F < \beta_{max}$; (5) Attached shock wave $\beta_F = \beta_{max}$; (6) Detached shock wave.

velocity existing at F to sonic velocity. Because FE is straight, $(\beta)_{FE}$ is constant in the hodograph plane, and FE is represented by a horizontal line.

At E an expansion around the corner occurs, and EG in the hodograph plane must be a characteristic line. GB is the last expansion wave that meets the sonic line EB. The shock FB is curved and the flow in the region FBE loses the uniform property existing for the case of Fig. H,10a(1).

The streamlines in the region of the shock BF are drawn to an enlarged scale in the hodograph plane. For the condition of point F between the sonic point S and the "Crocco point" C, the streamlines in the hodograph plane have downward slope (Eq. 7-9), while the boundary streamline runs in the direction $\beta = \text{const}$. Because of this, a singularity at the point F arises, so that the density of the streamlines increases as point F is approached. Physically the increase of density of streamlines indicates that the closer F is approached the smaller will be the differences between neighboring streamlines, and that this difference tends to zero at F. At F no difference exists between neighboring streamlines, and therefore the curvature of the shock at F tends to zero.

If the angle of the wedge increases and the point F coincides with the "Crocco point," point C of Fig. H,10a(3), the shock configuration changes, and the downward slope of the streamlines becomes zero. The density of the streamlines does not increase as F is approached, but remains constant. The change of conditions between two neighboring streamlines divided by the difference of the stream functions has a finite value near F. As a result, the shock wave has finite curvature at F even though the sides of the wedge are straight.

When point F lies between the "Crocco point" C and the point of maximum deviation D, Fig. H,10a(4), the streamlines in the hodograph plane have upward slope in the neighborhood of F. Therefore, the density of the streamlines in the immediate vicinity of F tends to zero. The differences between two neighboring streamlines as F is approached become larger and larger. Therefore, the shock begins with an infinite curvature (but the shock is still an oblique shock). Point F is again a singular point in the hodograph plane.

When the point F coincides with the point of maximum deflection D, Fig. H,10a(5), the streamlines have the same shape as for the case considered in Fig. H,10a(4). However, the singularity at the point F is of a different type, but the shock wave is still attached. (For a complete discussion of the type of singularity see [8].)

When the angle of the wedge becomes larger than the angle of maximum deviation, Fig. H,10a(6), a solution for the point F in the hodograph plane no longer exists. The boundary streamline detaches from the shock polar, and at F a stagnation point occurs. The detachment of the shock occurs very slowly, and it can be shown analytically that the transition from attached to detached shock is a continuous process [8].

From these considerations the following conclusions can be obtained: The detachment of the shock occurs at the point of maximum deviation, and the "Crocco point" has the physical significance that it represents the transition from a shock with zero initial curvature to a shock with infinite curvature.

Similar considerations are valid for a wedge with curved sides. Consider a symmetrical wedge with convex sides at zero angle of attack (Fig.

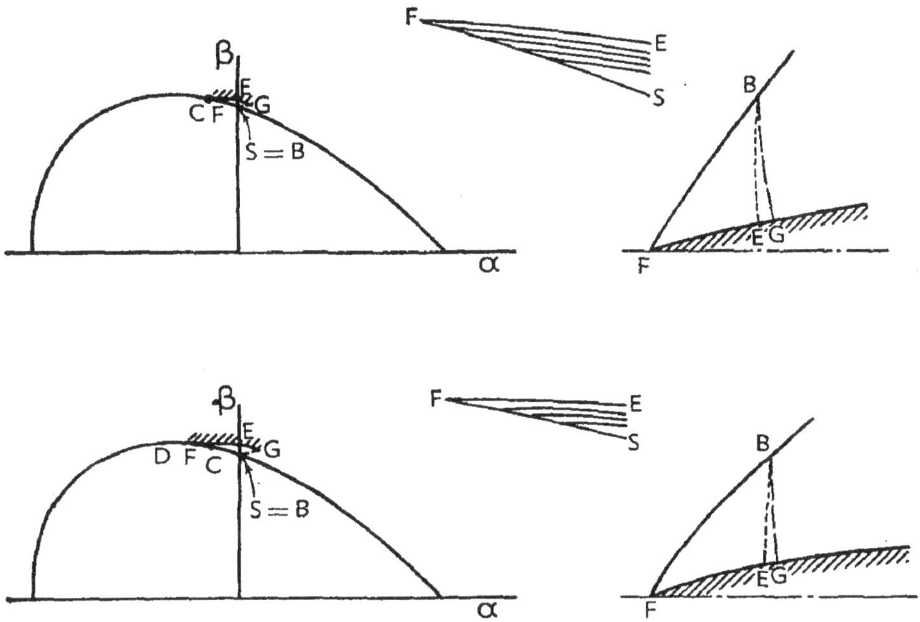

Fig. H,10b. The transition from attached to detached shock wave for wedges with curved sides. Top, $\beta_S < \beta_F < \beta_C$; bottom, $\beta_C < \beta_F < \beta_{max}$.

H,10b). The angle of the velocity vector decreases along the boundary streamline, and β on the hodograph diagram also decreases. If the angle of the wedge at F is higher than the angle of deviation corresponding to sonic velocity, as illustrated by point S of Fig. H,10b (top) but lower than the "Crocco point," the flow field is as shown in Fig. H,10b (top). Here also the boundary streamline has a downward slope; therefore, no singular point exists at F in the hodograph plane because the density of the streamlines does not increase indefinitely in moving toward F.

If the point F lies between the "Crocco point" C and the point of maximum deflection D on the shock polar as shown in Fig. H,10b (bottom), the same difficulties found for the straight wedge, Fig. H,10a(4), arise. The density of the streamlines in the neighborhood of the point F becomes very small; therefore, the distance between images of neighboring points in the physical plane becomes larger and larger as point F is approached in the hodograph plane. As a consequence, in the neighborhood of F the flow for the wedge with curved sides can be obtained from the flow for straight sides only by an infinitesimal deformation of the boundary streamline. The shock in both cases is similar and has infinite curvature.

If F lies between the "Crocco point" and the point of maximum deflection, the influence of the initial curvature of the shock on the flow pattern prevails, and the influence of the curvature of the boundary has only secondary effect; therefore, an analysis such as that presented in Art. 5 is not valid. (For the analytic considerations of the singularities at F see [*8*].)

H,11. Formation of Lambda Shocks. In III,E,2 a detailed discussion of the possible equilibrium conditions existing at the point of intersection of shock waves has been presented. Different types of lambda shocks are described by considering the conditions of equilibrium at the intersection point. These conditions are obtained from the requirement that the pressure and direction of the streamline passing through the intersection point have a single value.

In many cases this condition is sufficient to define the flow at the intersection. However, in other cases, when subsonic flow exists behind the shock waves, the equilibrium conditions at the intersection point depend upon all the subsonic region behind the shock waves; therefore, in order to find a physical solution, all the flow field in the neighborhood of the intersection point must be considered.

A physical description of these flow fields can be obtained by means of hodograph representations. However, in order to obtain hodograph representations of shock intersections, the shock polar presented in Fig. H,7 must be somewhat distorted to take into account the variation of stagnation pressure across the shock. By means of such a distortion flow pictures can be obtained if entropy gradients in the flow are assumed small and the effects of rotation are neglected.

Consider a parallel supersonic jet at static pressure p_1 which enters a region at static pressure p_a larger than p_1, and consider the flow field for increasing values of the ratio p_a/p_1 (Fig. H,11a).

For stream Mach numbers close to 1 the distortion required is small and the shock polar diagrams for incoming and reflected shocks appear as in Fig. H,11a, where corrected velocities α are plotted along the abscissa and the deviation β is plotted along the ordinate.

The shock polar I corresponds to stream flow conditions in front of the shock wave CF. The conditions behind the shock wave CF are represented by the point C, and the shock polar II corresponds to the flow conditions behind the shock wave CF.

If the pressure ratio p_a/p_1 is close to 1, point F, the intersection of the axis $\beta = 0$ and the shock polar II, which represents the point F of the axis of the jet, lies on the supersonic region of the axis α; therefore, the flow behind the reflected shock FB is supersonic and uniform.

When the pressure ratio p_a/p_1 increases, the Mach number in region II decreases and point F moves along the axis in the direction of point N.

For the condition presented in Fig. H,11a (top) point F lies in the subsonic region, and the flow behind the reflected shock FB is subsonic. At B an expansion to supersonic flow must be produced, because at B pressure p_a must be attained; therefore, at B sonic velocity must exist.

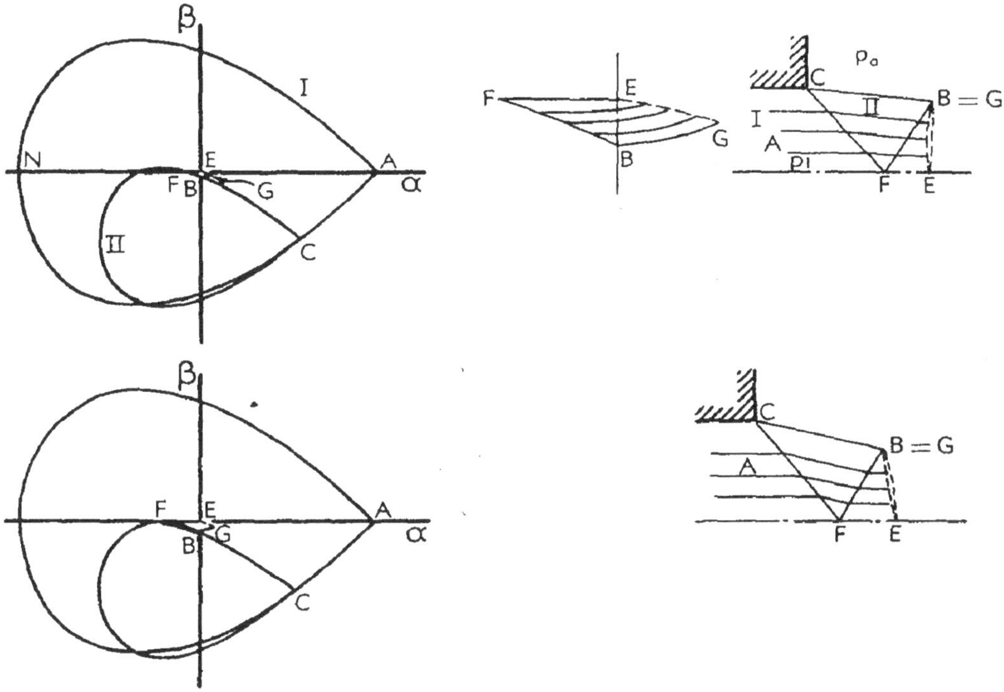

Fig. H,11a. Uniform jet entering a region of constant pressure higher than the static pressure in the jet. Reflection of the shock wave at the axis of symmetry is possible. Top, subsonic flow behind the reflected shock wave $\beta_S < \beta_F < \beta_C$; bottom, subsonic flow behind the reflected shock wave $\beta_F = \beta_{max}$.

That the flow cannot reach supersonic velocity along the shock FB can be seen from the following considerations. Assume that at B the flow is supersonic and let L be the point between F and B where sonic velocity would be reached. The shock at L would be stronger than at B; therefore, the region of the shock between L and B having supersonic flow behind the shock must be reached by expansion waves. However, the only waves that would exist in this region would be the waves reflected by the sonic line starting at L, and these waves are compression waves. Therefore, a supersonic region along the shock FB is not physically possible.

At B an expansion fan exists in accordance with the considerations of Art. 9. The stream at the axis accelerates and at E again reaches sonic velocity; *BE* is the sonic line and *GE* the last characteristic wave of the expansion fan which meets the sonic line. Details of the region *FEBG* of the hodograph are shown in larger scale in Fig. H,11a (top).

The initial curvature of the shock *FB* at *F* depends upon the position of the point *F* with respect to the "Crocco point" and to the point of maximum deviation. In Fig. H,11a (top) point *F* is between sonic velocity and the "Crocco point," and at point *F* the curvature is zero. In Fig. H,11a (bottom) point *F* is at the point of maximum deviation, and the shock has infinite curvature at *F*. The same considerations presented in Art. 10 can be applied here.

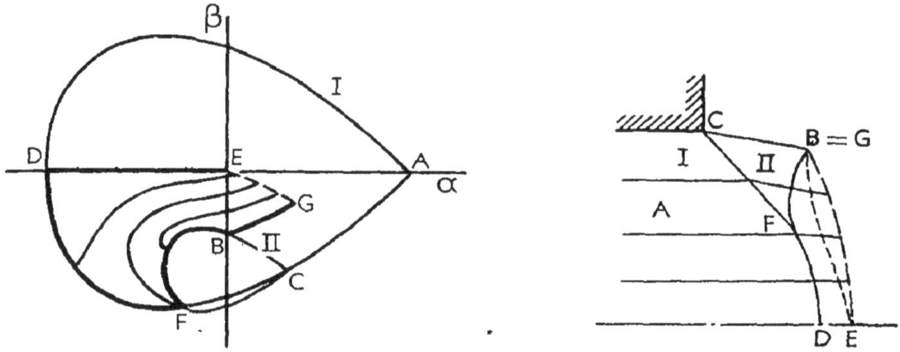

Fig. H,11b. Uniform jet entering a region of constant pressure higher than the static pressure in the jet. Formation of lambda shock.

If the pressure ratio p_a/p_1 increases with respect to the value corresponding to Fig. H,11a (bottom) the shock polar II no longer meets the axis $\beta = 0$, and the only possible solution for the shock intersection is the formation of a lambda shock, as illustrated in Fig. H,11b. Because of the flow symmetry, the shock at *D* must be a normal shock, and the shock *FD* maps to the part *FD* of the shock polar I in the hodograph. The flow behind *FD* is subsonic. The other boundary of the subsonic region is the shock *FB*, which maps into the part *FB* of the shock polar for condition II. From the hodograph diagram it is found that the inclination of the flow at *F* is larger than the inclination at point *F* upstream of the shock *FB*, and therefore the shock *FB* at *F* runs upstream. This condition is possible because the flow behind the shock is subsonic.

The flow at *B* is at sonic velocity, and at *B* the expansion fan *BG* exists. At the axis the flow accelerates to sonic velocity from *D* to *E*. The line *BE* is the sonic line, and *GE* is the last expansion wave that meets the sonic line.

The length of the shock *FD* depends upon the ratio p_a/p_1, and increases very gradually with increasing values of this ratio from the value corresponding to conditions presented in Fig. H,11a (bottom).

If the pressure p_a/p_1 still increases, conditions can be reached for which the connection point *F* between shock polar I and shock polar II

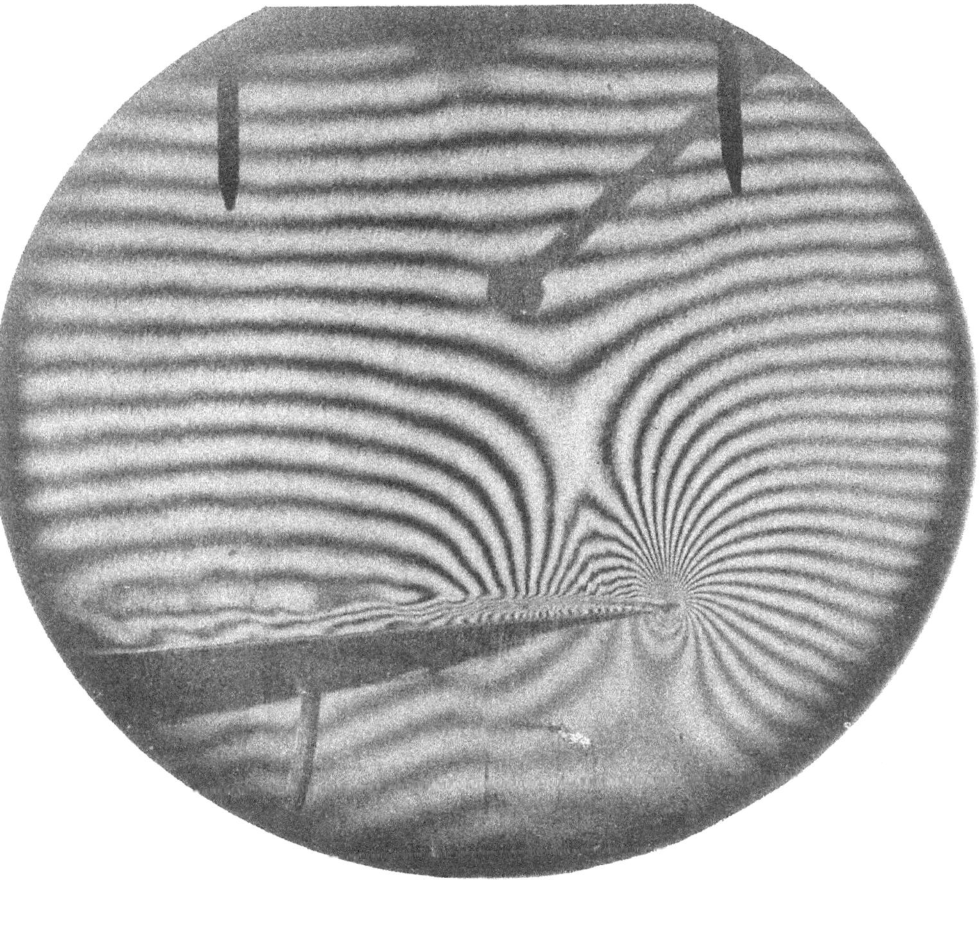

Plate H,9a. Interferogram of the flow field around a sharp corner at $M_\infty = 0.72$. Flow from right to left. (Taken at NACA Langley Field by A. Wood.)

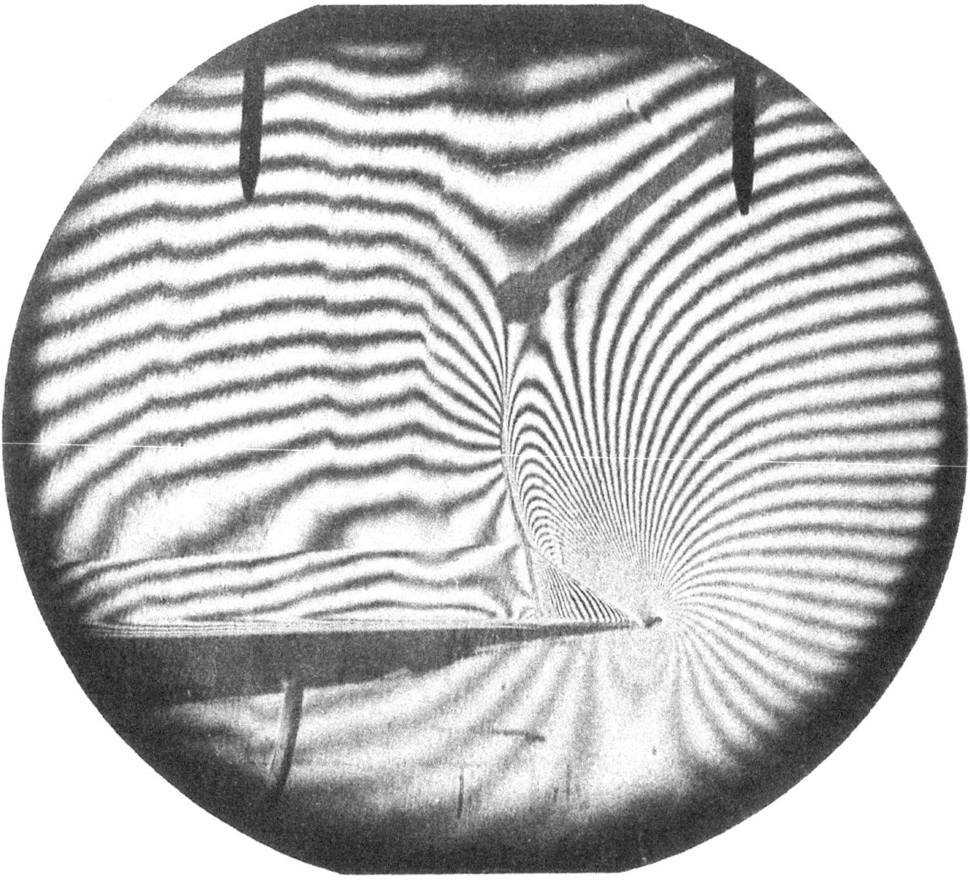

Plate H,9b. Interferogram of the flow field around a sharp corner at $M_\infty = 0.81$. Flow from right to left. (Taken at NACA Langley Field by A. Wood.)

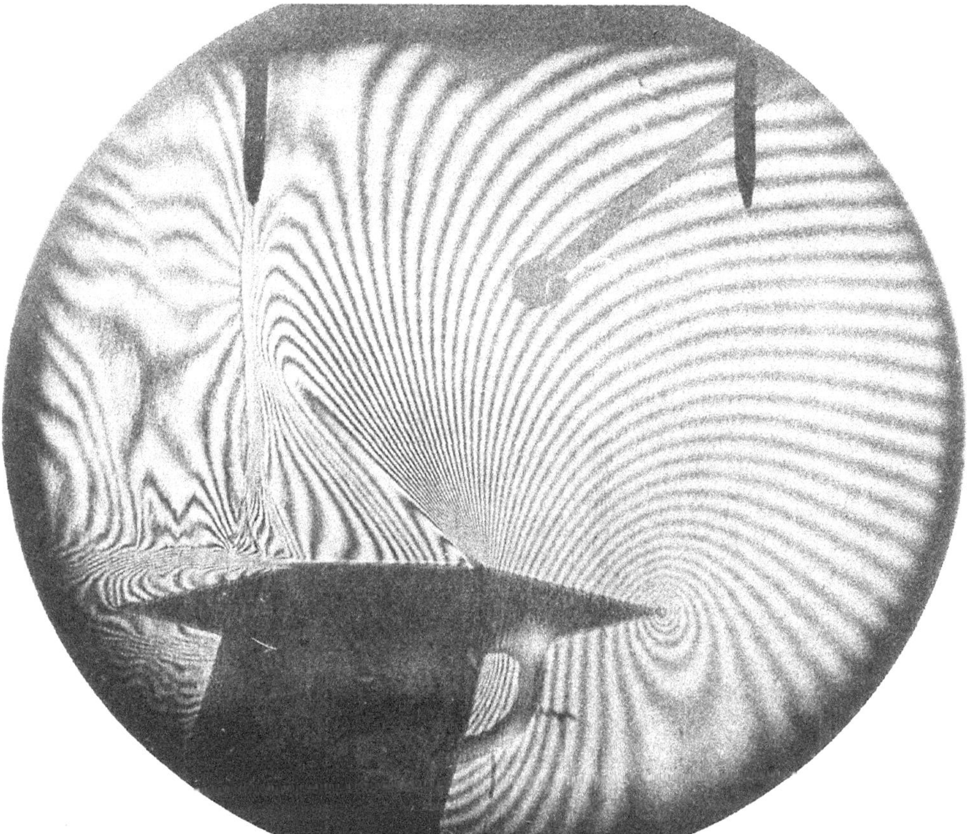

Plate H,9c. Interferogram of the flow field at the ridge of a two-dimensional profile. $M_\infty = 0.82$. Flow from right to left. (Taken at NACA Langley Field by A. Wood.)

moves into the supersonic region between point C and the axis $\alpha = 0$, or is coincident with the point C.

Conditions of this type have been investigated by Guderley [8] who found that, when the intersection point of the two shock polars is in the supersonic region or is coincident with point C of Fig. H,11b, this point cannot represent a physical solution for the connection point of the three shocks of a lambda shock.

Consider, for example, in the hodograph and physical planes the shock pattern $ACKL$ of Fig. H,11c. The only intersection point between polar I and II is the point C. Let us assume that point C represents the connection point of the lambda shock. The velocity behind the shock wave CL

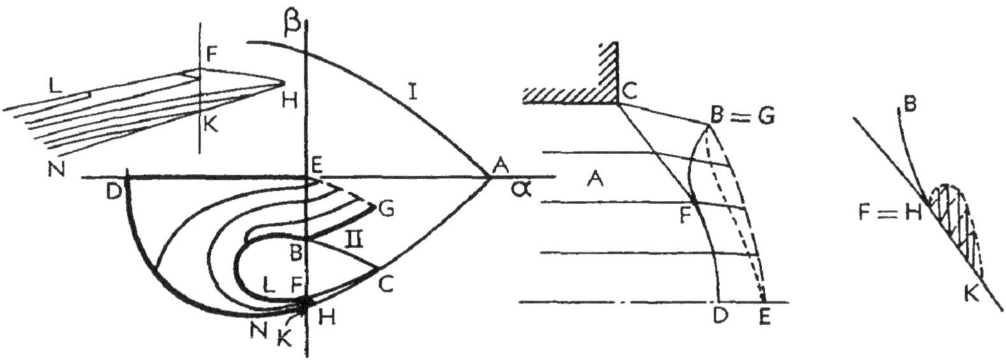

Fig. H,11c. The solution proposed by Guderley for the lambda shock when the intersection of the two shock polars is in the supersonic region. (From [12].)

decreases in moving from C to L or from C to K; therefore, expansion waves must reach the shock wave CL, because the only waves that can reach CL are the waves shown in Fig. H,11d. These waves would start from the sonic line LK. However, the sonic line can produce only compression waves; therefore, such a solution is physically impossible.

The actual flow pattern found by Guderley is shown in Fig. H,11c, where flow in the region of the point F is also shown enlarged in the physical and hodograph planes. At point F an expansion fan affecting a subsonic flow pattern exists. The shock BF reaches only up to the sonic line (point F of the hodograph). The expansion fan existing at F maps into the characteristic FH. At H the shock HKD starts and grows stronger. At K the flow behind the shock has sonic velocity, and at D it has subsonic velocity behind it since at point D the shock becomes a normal shock. The sonic line starts at F and ends at K (see enlarged sketch in Fig. H,11d) and the expansion fan from F is reflected at the sonic line as compression waves.

Point K is an exit corner of a supersonic region imbedded in a subsonic flow, and here a strong singularity arises. The compression waves starting from the sonic line coalesce before reaching HK and form a weak shock which is reflected at the shock as an expansion fan. The expansion fan is reflected at the sonic line, and the cycle can recur several times.

Point F is also singular, and as shown in the enlarged hodograph diagram, the stream function ψ increases very slowly with the distance from F; thus, the change of conditions along the sonic line is very rapid, and the curvature of the sonic line tends to infinity as F is approached.

Flow patterns as shown in Fig. H,11d are qualitatively correct also when the stream Mach number at H is large and the rotation existing in the flow can no longer be neglected.

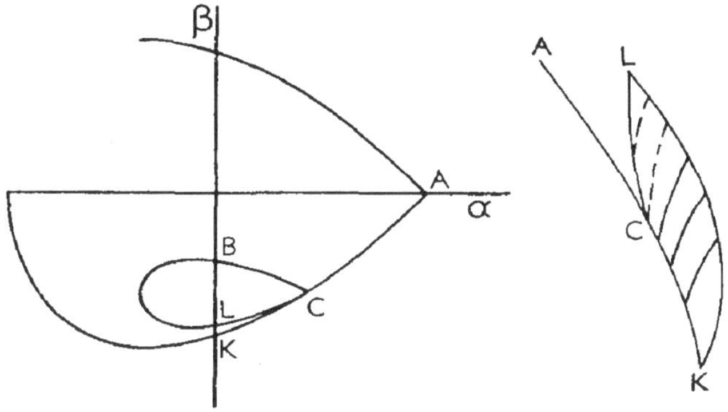

Fig. H,11d. The flow field behind a lambda shock when the intersection point of the two shock polars is in the supersonic region. Solution physically inpossible.

H,12. Transonic Properties of Two-dimensional Double Wedge Profiles. By means of considerations similar to those presented in the preceding paragraphs, a qualitative description of the flow fields having transonic regions can be obtained for many types of boundary conditions. However, if quantitative results are sought, a large amount of numerical work is required, and quantitative analytic data are available only for a few very simple boundary conditions. For more involved problems usually other approximate criteria are introduced, or information obtained from experimental investigations is used in the analysis in order to reduce the amount of numerical work.

A flow field which has been investigated in some detail experimentally and theoretically is the flow around a two-dimensional diamond-shaped profile at zero angle of attack.

Consider a symmetric double wedge profile at zero angle of attack as shown in Fig. H,12a. If the free stream Mach number is lower than 1, the flow in the front part of the wedge is subsonic and reaches sonic velocity at the ridge of the profile, where an expansion around the corner as described in Art. 9 occurs, followed by an envelope shock as illustrated in Fig. H,12a (left). If the free stream Mach number M_∞ is larger than 1, but close to 1, a detached shock occurs in front of the profile as illustrated in Fig. H,12a (right). The flow along the surface of the wedge is subsonic, and at the ridge a sonic line expansion fan and envelope shock exist as for the subsonic case. However, in this case the sonic line reaches the front shock wave, and the flow behind the sonic line is supersonic. Therefore, the envelope shock does not interfere with the subsonic region, and the flow behind the sonic line can be analyzed by means of supersonic theory. In the case of subsonic free stream conditions the flow behind the envelope shock again becomes subsonic and, therefore, the envelope shock influences the subsonic field. As a result, the analysis of the flow field must be performed by considering all the flow field in front of and behind the envelope shock.

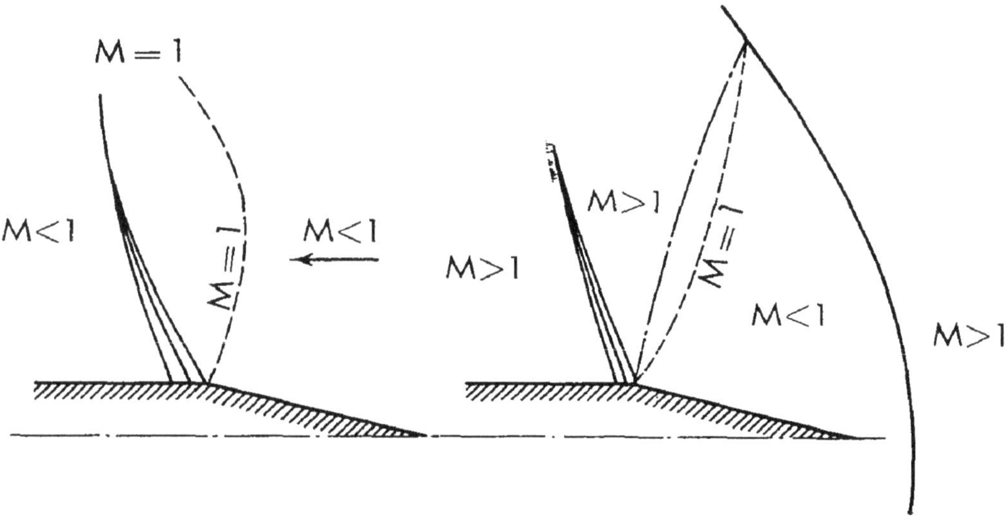

Fig. H,12a. Transonic flow field around a two-dimensional symmetric double wedge profile. Left, subsonic free stream velocity; right, supersonic free stream velocity.

The analysis can be performed in the physical or the hodograph plane. If the analysis is performed in the physical plane, an approximate shape of the sonic line and the front shock wave (when M_∞ is larger than 1) must be tentatively assumed. Satisfactory solutions of the subsonic flow region are then obtained by solving numerically the differential equations governing the motion, by the methods of squares, or by relaxation techniques [*16*].

In the hodograph plane, shock waves, characteristic lines, and the sonic lines are defined, and the solution of the boundary value problem must be obtained by means of numerical techniques, such as the relaxation technique.

Large simplifications can be introduced in the analysis by means of the transonic similarity law, which permits the application of the results of a single calculation to a family of profiles.

In [*16*] the calculations have been performed in the physical plane,

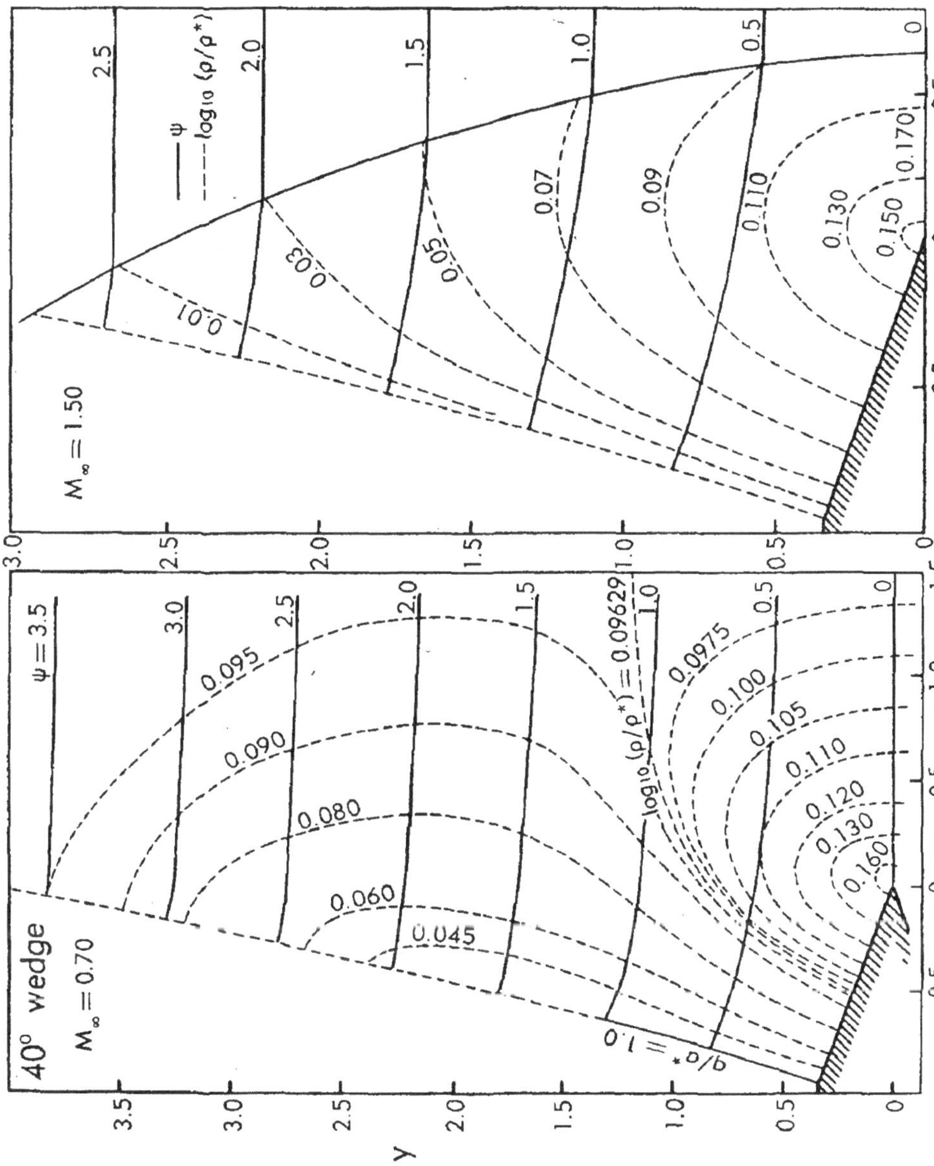

Fig. H,12b. The flow field around a 20° symmetric wedge. (From [16].) Vertical and horizontal scales are in half-chords. Left, $M_\infty = 0.70$; right, $M_\infty = 1.50$.

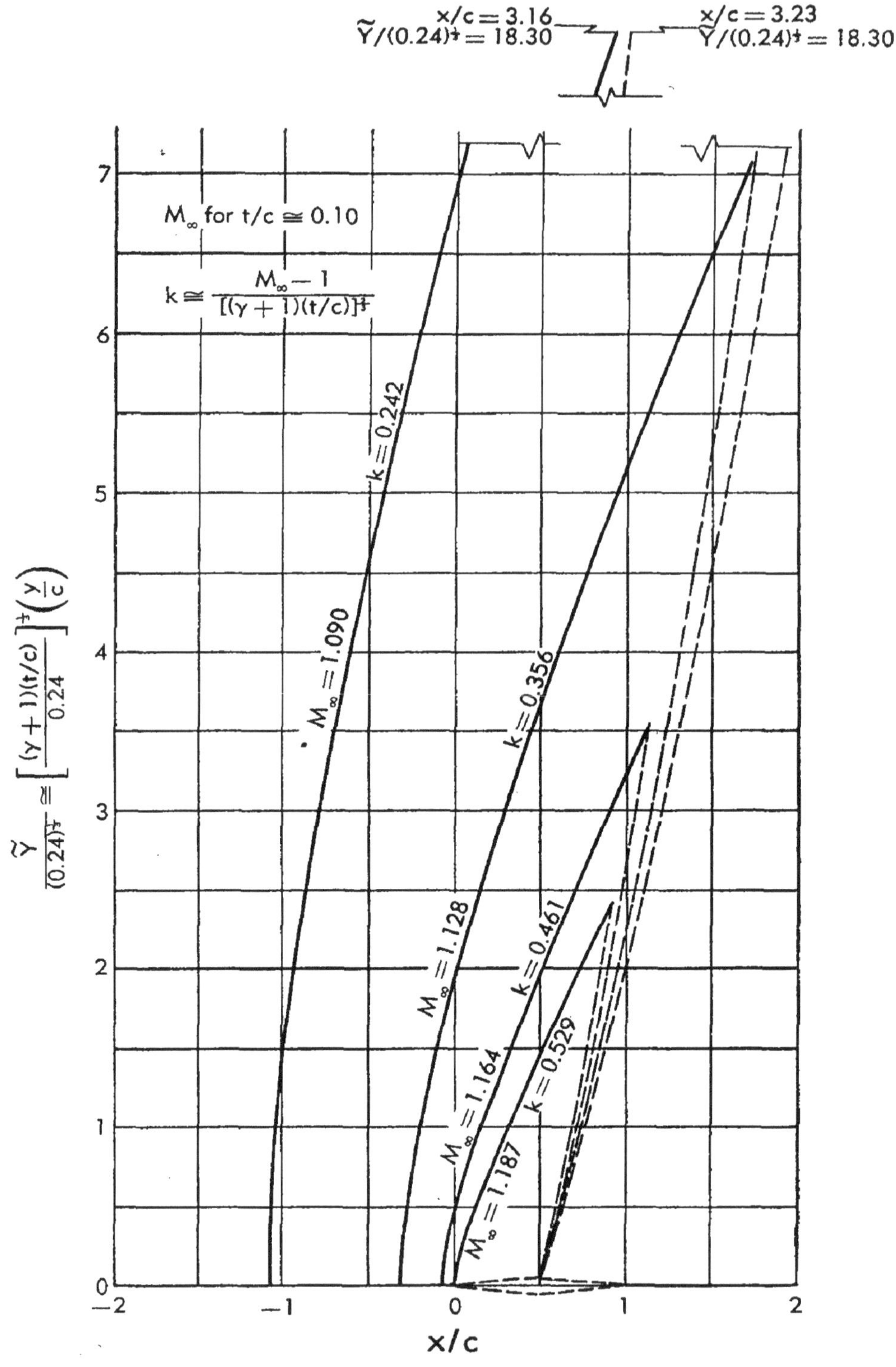

Fig. H,12c(1). The flow field around symmetric double wedge profile at zero angle of attack. (From [17].) t denotes maximum thickness and c the chord. Shape of the bow wave and sonic line.

and approximate assumptions have been made for the shape of the sonic line, while in [17] the transonic similarity law has been introduced and the calculations have been performed in the hodograph plane. Some of the results of these analyses are shown in Fig. H,12b and in the following discussion.

In Fig. H,12b streamlines and lines of constant density are shown for $M_\infty = 0.70$ and 1.50 for a 20° wedge. In Fig. H,12c(1) the shapes of the front shock wave and sonic line for symmetric wedges obtained in [17] by introducing the similarity law are shown. In Fig. H,12c(2) pressure distributions along a diamond-shaped profile for different similarity parameters k are shown (Eq. 6-32). For comparison the data of [19] for $M_\infty = 1$ are also presented. Experimental data [18] show good agreement with the theoretical results.

H,13. Two-dimensional Bodies of Arbitrary Shape. The determination of the flow field around two-dimensional bodies of arbitrary shape is of the same general nature as the determination of the flow around wedges. However, other difficulties are found because the location of the sonic line on the body is not known, and if the flow is not symmetric with respect to the stream direction, the stagnation point is also not known.

In these cases, in order to start an analytic investigation, the shape and position of the sonic line must be estimated in the first approximation. The accuracy of the first guess has a large influence on the amount of numerical work required. If the free stream has supersonic velocity and the body considered is a wing profile at a small angle of attack, the subsonic flow is restricted to the region near the leading edge of the profile (region $BEDGCA$ of Fig. H,13), and its flow properties are independent of the region of supersonic flow behind the characteristic lines BF and HC. Therefore, the analysis of the subsonic flow in the region of the nose can be used for different boundary conditions in the supersonic region.

If separation is avoided near the nose of the profile, the subsonic region, the shape of the sonic line, and the flow properties along the characteristic line BF can be determined experimentally. For example, interferograms may be used as well as characteristic calculations started at the supersonic region and at the shock wave BL, CN. The use of tunnels or shock tubes to obtain data for this subsonic region is a convenient method of attack for the analysis of such problems, because these data can be used for any shape of the boundary conditions in the supersonic region.

While an exact analytic determination of the points E and G on the sonic line and the profile of Fig. H,13 requires a complete determination of the whole subsonic flow field, the approximate position of the points E and G can be estimated by qualitative considerations, and the effects

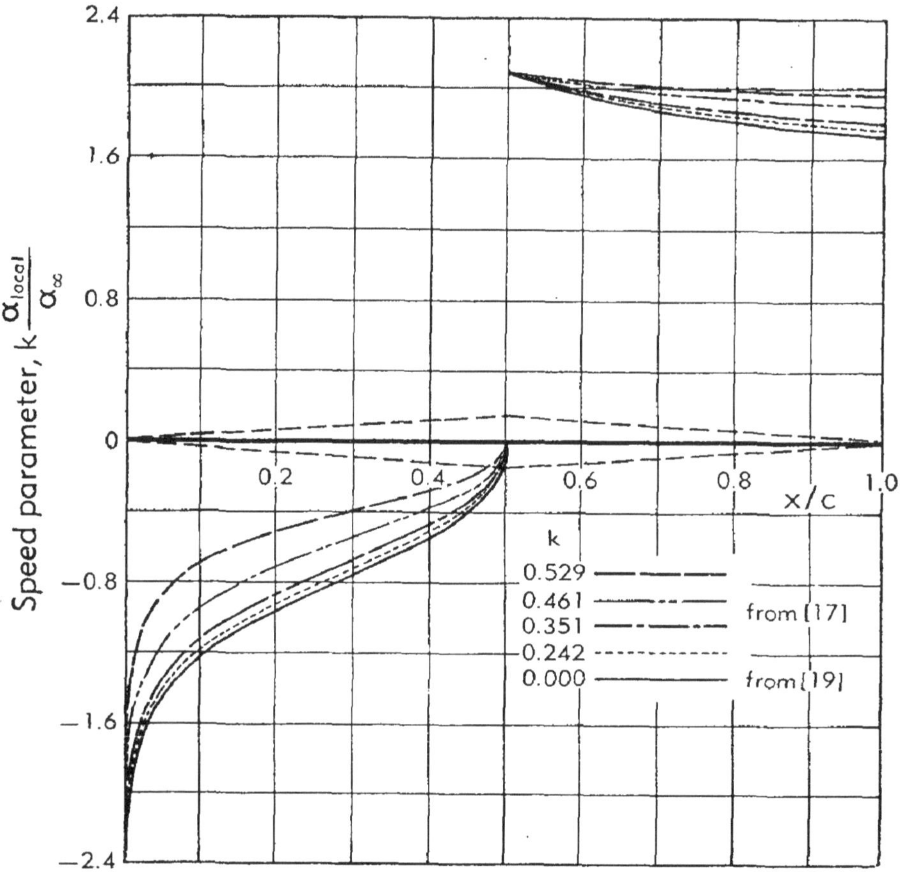

Fig. H,12c(2). Chordwise distribution of the speed parameter $k\alpha/\alpha_\infty$ at the profile surface.

on the shape and position of the shock wave due to a small change of boundary conditions can also be estimated.

Consider a two-dimensional slender profile with a round leading edge (Fig. H,13, left). At the nose a stagnation point exists; therefore, the streamline that represents the body in the hodograph plane must pass through the origin, and there a branch point must exist (Fig. H,13, right). A unique correspondence between quantities in the hodograph and physical plane exists as soon as the stagnation point D is placed in the hodograph plane at the origin.

If the free stream velocity is only slightly supersonic, the shock polar is small, and the body contour in the hodograph plane connects hodograph points which in the region of high velocity represent velocity vectors having small inclination with respect to the direction of the undisturbed stream. The result is that the entrance of the supersonic region is represented by two small channels between the shock polar diagram and the profile streamline, where the streamlines which all start at the shock polar diagram are crowded together. The lower subsonic region is contained in a long narrow bag with small density of the streamlines with almost stagnation conditions (see Fig. H,13, right). Thus, the shape of the nose is relatively unimportant. A change in the shape of the nose is stabilized by the accumulation of masses which changes the bag in the hodograph diagram. However, the small opening at the outlet of the bag, which corresponds to the two parts of the profile where the flow reaches sonic velocity, will not change, and therefore the shape of the shock wave will not change. These two parts of the profile, corresponding

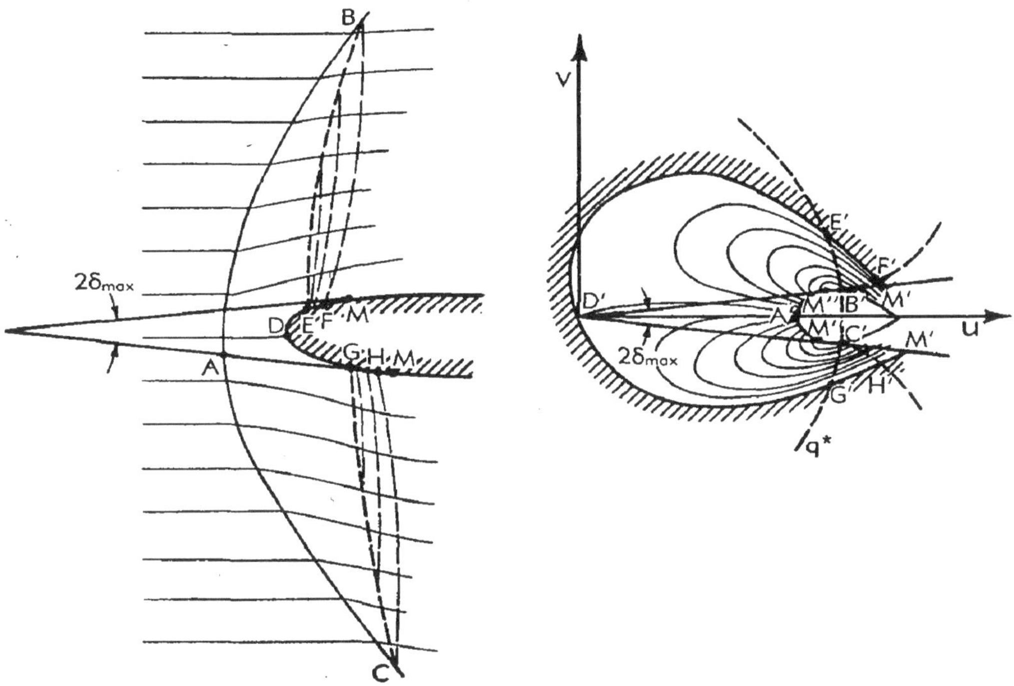

Fig. H,13. Transonic flow behind a detached bow wave, and the use of the protractor for the location of the "shoulders." Left, physical plane; right, hodograph plane.

to the small opening have the largest effect on the flow field and are called "shoulders." The regions $E'F'B'$ and $G'H'C'$ of the hodograph diagram, which define the outlet, correspond to the supersonic region which can still affect the subsonic region. Any change in the boundary of this region will change the position and shape of the shock wave a large amount.

Since the shape and position of the shock wave does not depend upon effects localized at the nose of the profile, the nose radius has little effect on the shape of the shock wave. Attempts to relate the position and shape of detached shocks to the nose radius cannot be expected to be normally successful for slender profiles or shapes different from circular.

In order to define the region of the position of the "shoulders" of the profile, a protractor can be used as proposed by Busemann [13]. Consider a protractor adjusted to twice the maximum angle of deflection δ_{max} for the free stream Mach number considered. The points M of tangency between the protractor and the profile are the "shoulders," which represent the small opening at the outlet of the bag.

If the sonic line coincided with a characteristic line no reflected wave from the sonic line would exist, point E' would coincide with point B', and the contact point M would be in the subsonic region as M'' in the hodograph diagram. However, because of the reflected waves, the inclination of the streamline at B on the shock wave is less than the inclination at E; therefore, point M' is in the supersonic region of the flow, and the transonic zone EF, GH must be slightly ahead of point M. However, the intensity of the reflected wave is usually small and point M' is only slightly downstream of the points F' or H'.

When the length of the sonic line $E'B'$ is small, the effect of the reflected waves is also small, points E', B', and F' tend to coincide, and each point can be determined directly by means of the protractor.

These considerations lead, for example, to the following result obtained experimentally in [20]. At low supersonic speed, when a detached shock exists in front of a sharp-nosed profile at an angle of attack, produced by too large a deviation of the lower surface to permit an attached shock, the pressure distribution on the upper surface can be obtained in first approximation by assuming the flow around the leading edge to be the same as it would be in the absence of a detached shock wave, and therefore the effect of the presence of the shock in the upper part of the flow may be neglected.

CHAPTER 3. CONICAL FLOW

H,14. Conical Flow Fields. A flow field is defined as conical when the physical properties such as pressure, density, velocity, energy, and entropy remain constant along every straight line starting from a given point, called the apex of the conical flow.

Conical flow is found to exist, for example, around cones which extend to infinity when viscosity and heat conductivity are neglected, because of the similarity existing between parallel cross sections of the flow field, and also around cones of finite length in supersonic flow because of the law of forbidden signals.

Point O in Fig. H,14 is the origin of a spherical coordinate system r, ψ, ω and the apex of the conical flow. The velocity components v_r, v_n, and w, along the radius, normal to the radius in the meridian plane, and normal to the meridian plane, respectively, are defined by

$$v_r = \frac{dr}{dt} \qquad v_n = \frac{rd\psi}{dt} \qquad w = \frac{rd\omega \sin \psi}{dt} \tag{14-1}$$

Because the flow is conical

$$\frac{\partial v_r}{\partial r} = 0 \qquad \frac{\partial v_n}{\partial r} = 0 \qquad \frac{\partial w}{\partial r} = 0 \tag{14-2}$$

and

$$\frac{\partial p}{\partial r} = 0 \qquad \frac{\partial \rho}{\partial r} = 0 \qquad \frac{\partial s}{\partial r} = 0 \qquad \frac{\partial h^0}{\partial r} = 0 \tag{14-3}$$

where p is the pressure, ρ the density, s the entropy, and h^0 the total

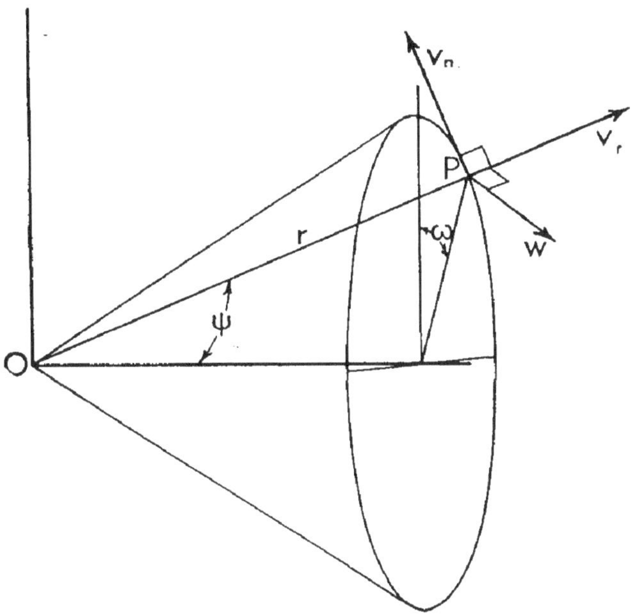

Fig. H,14. The velocity components in conical coordinates.

enthalpy. Euler's equations for conical flow in spherical coordinates are

$$v_n \frac{\partial v_r}{\partial \psi} + \frac{w}{\sin \psi} \frac{\partial v_r}{\partial \omega} - v_n^2 - w^2 = 0 \tag{14-4}$$

$$v_n \frac{\partial v_n}{\partial \psi} + \frac{w}{\sin \psi} \frac{\partial v_n}{\partial \omega} + \frac{1}{\rho} \frac{\partial p}{\partial \psi} + v_r v_n - w^2 \cot \psi = 0 \tag{14-5}$$

$$v_n \frac{\partial w}{\partial \psi} + \frac{w}{\sin \psi} \frac{\partial w}{\partial \omega} + \frac{1}{\rho \sin \psi} \frac{\partial p}{\partial \omega} + v_r w + v_n w \cot \psi = 0 \tag{14-6}$$

The continuity equation gives

$$2\rho v_r \sin \psi + v_n \sin \psi \frac{\partial \rho}{\partial \psi} + \rho \sin \psi \frac{\partial v_n}{\partial \psi} + v_n \rho \cos \psi + w \frac{\partial \rho}{\partial \omega} + \rho \frac{\partial w}{\partial \omega} = 0 \tag{14-7}$$

If the flow has constant total enthalpy, the energy equation states

$$\frac{\gamma}{\gamma - 1}\left(\frac{1}{\rho}\frac{\partial p}{\partial \omega} - \frac{p}{\rho^2}\frac{\partial \rho}{\partial \omega}\right) = -\left(v_r\frac{\partial v_r}{\partial \omega} + v_n\frac{\partial v_n}{\partial \omega} + w\frac{\partial w}{\partial \omega}\right) \quad (14\text{-}8)$$

$$\frac{\gamma}{\gamma - 1}\left(\frac{1}{\rho}\frac{\partial p}{\partial \psi} - \frac{p}{\rho^2}\frac{\partial \rho}{\partial \psi}\right) = -\left(v_r\frac{\partial v_r}{\partial \psi} + v_n\frac{\partial v_n}{\partial \psi} + w\frac{\partial w}{\partial \psi}\right) \quad (14\text{-}9)$$

Pressure and density can be eliminated from Eq. 14-7 by means of Eq. 14-4, 14-5, 14-6, 14-8, and 14-9, and there results

$$v_r\left(2 - \frac{v_n^2 + w^2}{a^2}\right) + v_n \cot\psi + \frac{\partial v_n}{\partial \psi}\left(1 - \frac{v_n^2}{a^2}\right)$$
$$+ \frac{\partial w}{\sin\psi\,\partial\omega}\left(1 - \frac{w^2}{a^2}\right) - \frac{wv_n}{a^2}\left(\frac{\partial v_n}{\sin\psi\,\partial\omega} + \frac{\partial w}{\partial\psi}\right) = 0 \quad (14\text{-}10)$$

If the flow is potential flow the potential can be expressed in the form

$$\phi = rF(\psi, \omega) \quad (14\text{-}11)$$

and

$$v_r = F(\psi, \omega) \qquad v_n = \frac{\partial F}{\partial \psi} \qquad w = \frac{\partial F}{\sin\psi\,\partial\omega} \quad (14\text{-}12)$$

Eq. 14-10 becomes

$$F\left(2 - \frac{v_n^2 + w^2}{a^2}\right) + \frac{\partial F}{\partial \psi}\cot\psi\left(1 + \frac{w^2}{a^2}\right) + \frac{\partial^2 F}{\partial \psi^2}\left(1 - \frac{v_n^2}{a^2}\right)$$
$$+ \frac{\partial^2 F}{\sin^2\psi\,\partial\omega^2}\left(1 - \frac{w^2}{a^2}\right) - \frac{2wv_n}{a^2}\frac{1}{\sin\psi}\frac{\partial^2 F}{\partial\omega\partial\psi} = 0 \quad (14\text{-}13)$$

In supersonic flow the disturbed region of the flow is bounded either by a shock wave or by an envelope of Mach cones. Therefore, for conical flow this boundary must either be a conical shock wave or a Mach cone. Inside the disturbed region the conical flow defined by Eq. 14-13 is of hyperbolic type for $(w^2 + v_n^2) > a^2$ and of elliptic type for $(w^2 + v_n^2) < a^2$.

H,15. Singular Points in Conical Flows. The entropy distribution is defined by $p\rho^{-\gamma} = \text{const}\ e^{s/c_v}$, or

$$\frac{\gamma - 1}{\mathcal{R}}\frac{\partial s}{\partial \omega} = \frac{1}{p}\frac{\partial p}{\partial \omega} - \gamma\frac{1}{\rho}\frac{\partial \rho}{\partial \omega} \quad (15\text{-}1)$$

$$\frac{\gamma - 1}{\mathcal{R}}\frac{\partial s}{\partial \psi} = \frac{1}{p}\frac{\partial p}{\partial \psi} - \gamma\frac{1}{\rho}\frac{\partial \rho}{\partial \psi} \quad (15\text{-}2)$$

therefore, because of Eq. 14-4, 14-5, 14-6, 14-8, and 14-9

$$\frac{a^2}{\gamma\mathcal{R}}\frac{\partial s}{\partial \omega} = v_n \sin\psi\frac{\partial w}{\partial \psi} - v_r\frac{\partial v_r}{\partial \omega} - v_n\frac{\partial v_n}{\partial \omega} + v_r w\sin\psi + v_n w\cos\psi \quad (15\text{-}3)$$

$$\frac{a^2}{\gamma\mathcal{R}}\frac{\partial s}{\partial \psi} = -v_r\frac{\partial v_r}{\partial \psi} - w\frac{\partial w}{\partial \psi} + \frac{w}{\sin\psi}\frac{\partial v_n}{\partial \omega} + v_r v_n - w^2\cot\psi \quad (15\text{-}4)$$

and

$$v_n \sin \psi \frac{\partial s}{\partial \psi} = -w \frac{\partial s}{\partial \omega} \tag{15-5}$$

Eq. 15-5 states that the entropy remains constant along the streamline. If t is the arc length along the projection of a streamline on the sphere

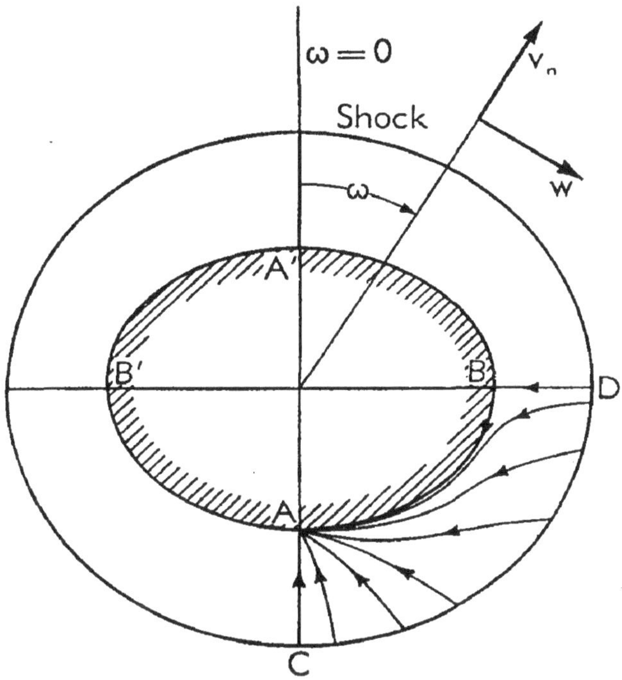

Fig. H,15a. The singular points in a conical flow field without axial symmetry.

r = const

$$\frac{ds}{dt} = \frac{\partial s}{\partial \psi}\frac{d\psi}{dt} + \frac{\partial s}{\partial \omega}\frac{d\omega}{dt} = 0 \tag{15-6}$$

and

$$\left(\frac{d\omega}{d\psi}\right)_t = \frac{w}{v_n \sin \psi} \tag{15-7}$$

Eq. 15-5 and 15-7 define the entropy at each point of the flow field with the exception of the points where w and v_n are both zero. At some of these points a singularity of the entropy can occur [21].

Consider, for example, a conical body having two planes of symmetry and analyze the streamline projections on the sphere $r = 1$ (Fig. H,15a).

In the meridian planes BB' and AA' the velocity component w is zero because of the symmetry; therefore, the entropy in these planes remains constant (Eq. 15-5). Along the surface of the body between AB, BA', $A'B'$, and $B'A$ the entropy also remains constant because the surface of the body represents a streamline. However, the entropy in the plane BD is different from the entropy in the plane AC, because the inclinations of the shock at D and C are different. Therefore, a discontinuous variation of entropy must occur at some point of the flow along the line $CABD$. At this singular point v_n and w must be zero (Eq. 15-5). Now at the points B and B' the streamline projections diverge from the plane of symmetry, and along BD

$$\frac{\partial}{\partial\psi}\left(\frac{\partial^2 s}{\partial\omega^2}\right) = -\frac{2\dfrac{\partial w}{\partial\omega}}{v_n \sin\psi}\frac{\partial^2 s}{\partial\omega^2} \tag{15-8}$$

is of the same sign as $\partial^2 s/\partial\omega^2$, because v_n is negative and $\partial w/\partial\omega$ is positive. Therefore $\partial^2 s/\partial\omega^2$ decreases as ψ decreases and at B a stagnation point

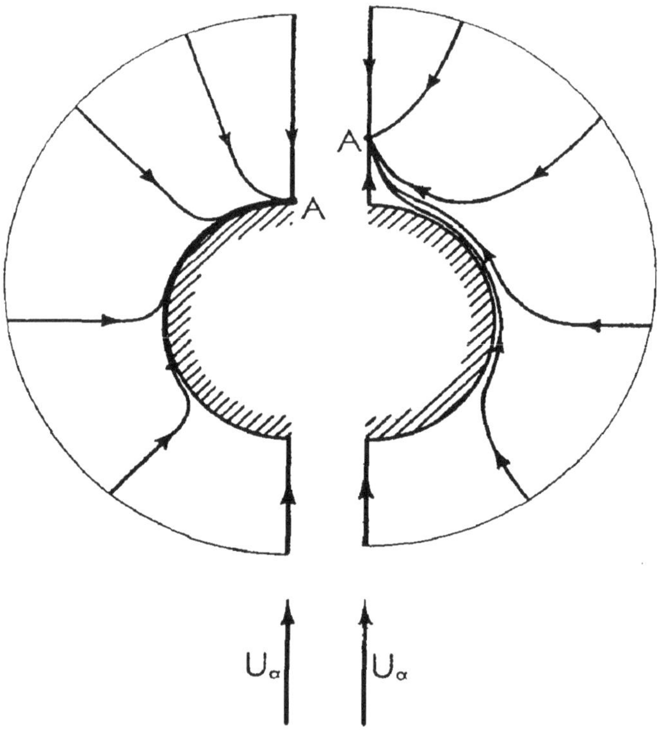

Fig. H,15b. The singular point for a circular cone at an angle of attack. Left, the singular point is at the surface of the body (small angle of attack); right, the singular point moves toward the shock (large angle of attack).

exists for the component $\sqrt{v_n^2 + w^2}$, and the entropy is constant. Along CA the expression (15-8) has the opposite sign from $\partial^2 s/\partial\omega^2$; therefore, a discontinuity occurs at the point A where v_n and w are also zero. All the neighboring streamline projections converge at A. The singular points A and A' are defined by the conditions

$$w_A = 0 \qquad (v_n)_A = 0 \qquad \left(\frac{\partial v_n}{\partial\psi}\right)_A < 0 \qquad \left(\frac{\partial w}{\partial\omega}\right)_A < 0 \tag{15-9}$$

The singular points are not necessarily on the surface of the body. For example, for a circular cone at an angle of attack a singular point exists which for small angles of attack is at the surface of the body as shown in Fig. H,15b (left) and then moves from the surface of the body toward the shock as shown in Fig. H,15b (right) as the angle is increased. The point is in the same meridian plane as the stagnation point of the cross-flow component but on the opposite side of the cone.

H,16. Conical Flow in the Hodograph Plane. Conical flow fields are important in many axially symmetric or three-dimensional problems

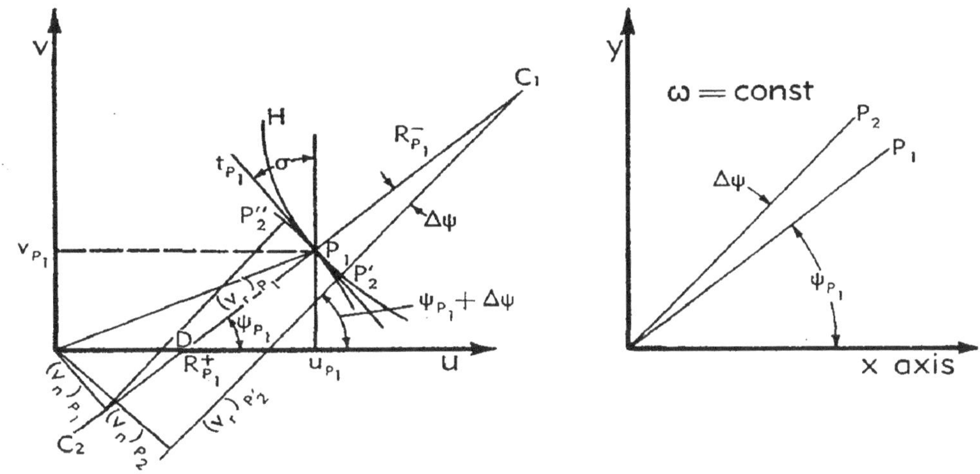

Fig. H,16. Conical flow in the hodograph plane.

in which the effects due to entropy gradients in the disturbed flow region are small. In these problems the effect due to the rotation of the flow can be neglected for practical applications, and Eq. 14-13 can be used. Eq. 14-13 can be analyzed conveniently in the hodograph plane.

Consider the variation of the velocity components v_r, v_n, w in a meridian plane ω = const. Because w is normal to the meridian plane (Fig. H,16)

$$u = v_r \cos \psi - v_n \sin \psi \tag{16-1}$$

$$v = v_r \sin \psi + v_n \cos \psi \tag{16-2}$$

But from Eq. 14-12 it appears that

$$v_n = \frac{\partial v_r}{\partial \psi} \tag{16-3}$$

or

$$\frac{\partial u}{\partial \psi} = -\left(v_r + \frac{\partial v_n}{\partial \psi}\right) \sin \psi \tag{16-4}$$

$$\frac{\partial v}{\partial \psi} = \left(v_r + \frac{\partial v_n}{\partial \psi}\right) \cos \psi \tag{16-5}$$

If the line H in the hodograph plane is the locus of the points representing the velocity vectors corresponding to each value of ψ in the plane ω = const, and l is its arc length there results

$$\left(\frac{dl}{d\psi}\right)^2 = \left(\frac{\partial u}{\partial \psi}\right)^2 + \left(\frac{\partial v}{\partial \psi}\right)^2 = \left(v_r + \frac{\partial v_n}{\partial \psi}\right)^2 \tag{16-6}$$

Eq. 16-4, 16-5, and 16-6 show that the tangent t_{P_1} to the curve H at a point P_1 is normal to the direction ψ_{P_1}. Indeed, if σ is the angle between the normal to the tangent t_{P_1} and the u axis, there results (see Fig. H,16)

$$\frac{\partial u}{\partial \psi} = -\frac{\partial l}{\partial \psi}\sin\sigma \tag{16-7}$$

$$\frac{\partial v}{\partial \psi} = +\frac{\partial l}{\partial \psi}\cos\sigma \tag{16-8}$$

Therefore σ is equal to ψ_{P_1} and the expression $[v_r + (\partial v_n/\partial\psi)]$ represents the radius of curvature of the curve H at the point P_1 considered.

If the radius of curvature is known at each point of the curve H the curve can be constructed by a step-by-step procedure, by assuming that in a small increment $\Delta\psi$ the radius of curvature does not change. From simple considerations of the sign of $\partial u/\partial\psi$ and $\partial v/\partial\psi$ it results immediately that, if the radius is positive, the point P_2 at $(\psi + \Delta\psi)$ corresponds to point P_2'' of Fig. H,16, while, if the radius is negative, the point P_2 at $(\psi + \Delta\psi)$ corresponds to P_2', where P_2' is on the circle of radius $R_{P_1}^-$ with center C_1 while P_2'' is on the circle of radius $R_{P_1}^+$ with center C_2. The radius of curvature R can be obtained from Eq. 14-10 and is

$$R = \frac{1}{1 - \frac{v_n^2}{a^2}}\left[\frac{v_n w}{a^2}\left(\frac{1}{\sin\psi}\frac{\partial v_n}{\partial\omega} + \frac{\partial w}{\partial\psi}\right) - v_n\cot\psi - \left(1 - \frac{w^2}{a^2}\right)\left(v_r + \frac{1}{\sin\psi}\frac{\partial w}{\partial\omega}\right)\right] \tag{16-9}$$

Eq. 16-9 indicates that the radius of curvature R is a function of the components v_r, v_n in the meridian plane of the component w, and of the variations $\partial w/\partial\omega$, $\partial v_n/\partial\omega$ in direction normal to the meridian plane considered. If the initial conditions at a boundary are known, Eq. 16-9 and 14-12 permit the determination of the whole flow field by a step-by-step process in several meridian planes ω = const.

H,17. Circular Conical Bodies at Zero Angle of Attack. For circular conical bodies at zero angle of attack axial symmetry exists in the flow and the shock wave that divides the disturbed from the undisturbed region is a circular conical shock. In this case the variation of entropy across the shock is constant and the flow behind the shock wave is potential flow.

Because of the axial symmetry Eq. 16-9 becomes

$$\left(1 - \frac{v_n^2}{a^2}\right) R = -v_n \cot \psi - v_r = \left(1 - \frac{v_n^2}{a^2}\right)\left(v_r + \frac{\partial v_n}{\partial \psi}\right) \quad (17\text{-}1)$$

The boundary conditions are:

At the surface of the conical body defined by $\psi = \psi_C$

$$(v_n)_{\psi_C} = 0 \quad (17\text{-}2)$$

At the shock wave surface defined by $\psi = \psi_S$ the equations of the shock waves are valid

$$(v_r)_{\psi_S} = U \cos \psi_S \quad (17\text{-}3)$$

$$-(v_n)_{\psi_S} = \frac{2}{\gamma + 1}\left(\frac{1}{M_\infty^2 \sin^2 \psi_S} + \frac{\gamma - 1}{2}\right) U \sin \psi_S \quad (17\text{-}4)$$

where the subscript ∞ indicates free stream conditions. From Eq. 17-3 and 17-4 the following relation is obtained

$$-(v_n)_{\psi_S} = \frac{\gamma - 1}{\gamma + 1} \cot \psi_S \left[\frac{q_{\max}^2}{(v_r)_{\psi_S}} - (v_r)_{\psi_S}\right] \quad (17\text{-}5)$$

Eq. 17-1 permits a step-by-step determination of the conical flow field, starting either at the shock surface or at the surface of the conical body [*22*,*23*, and *24*]. If the calculations are initiated at the shock surface the free stream Mach number and the angle ψ_S of the conical shock are assumed as data of the problem. By means of Eq. 17-3 and 17-5, v_r and v_n behind the shock can be found. The radius R is obtained from Eq. 17-1, and the values of v_r and v_n at a station $\psi_S - \Delta\psi$ are obtained from the construction of Fig. H,16. Therefore

$$(v_n)_{\psi - \Delta\psi} = (v_n)_\psi \cos \Delta\psi - (R - v_r)_\psi \sin \Delta\psi \quad (17\text{-}6)$$

$$(v_r)_{\psi - \Delta\psi} = -(v_n)_\psi \sin \Delta\psi - (R - v_r)_\psi \cos \Delta\psi + (R)_\psi \quad (17\text{-}7)$$

Then the radius $(R)_{\psi-\Delta\psi}$ is determined and the values of $(v_r)_{\psi-2\Delta\psi}$ and $(v_n)_{\psi-2\Delta\psi}$ are obtained. In this way the whole flow field can be determined. The construction can be done graphically. In this case the quantity $v_n \cot \psi + v_r$ of Eq. 17-1 can be measured in the graphical construction and is given by the segment P_1D of Fig. H,16. The value of ψ_C defining the conical body corresponding to the value of ψ_S selected is obtained at the station where $(v_n)_{\psi_C} = 0$.

If the calculation is initiated at the surface of the conical body, the values of ψ_C and $(v_r)_{\psi_C}$ are assumed as data of the problem, and the calculations are performed in a similar way for increasing values of ψ.

The calculation gives for each value of ψ a value of v_r and v_n. When the values of v_r and v_n for the same value of ψ satisfy relation (17-5), the corresponding value of ψ is equal to ψ_S. Then from Eq. 17-3 the free stream conditions can be obtained.

Because no explicit relation has been found between the angle of the conical body, the angle of the conical shock, and free stream conditions, the calculation does not permit the selection of the angle of the conical body and the free stream Mach number as data for the problem.

A large number of cone calculations for different cone angles and free stream Mach numbers have been tabulated in [*25*]. The numerical analysis can also be performed in the physical plane (see G,15 or [*24*]).

H,18. The Hodograph Diagram for an Axially Symmetric Conical Flow. Eq. 17-1 gives the radius of curvature of the hodograph diagram which describes any axially symmetric conical flow. The differential equation of the diagram in the hodograph plane can be obtained from the following considerations.

The equation of motion, for axially symmetric potential flow using cylindrical coordinates x, y with velocity components u, v, is (Sec. D, Eq. 32-14a; also III,A,9)

$$\left(1 - \frac{u^2}{a^2}\right)\frac{\partial u}{\partial x} - \frac{2uv}{a^2}\frac{\partial u}{\partial y} + \left(1 - \frac{v^2}{a^2}\right)\frac{\partial v}{\partial y} + \frac{v}{y} = 0 \tag{18-1}$$

For conical flow the velocity components u, v, and a are only functions of ψ and therefore of a parameter κ defined by

$$\cot \psi = \frac{x}{y} = \kappa \tag{18-2}$$

From Eq. 16-4 and 16-5 there results in the hodograph plane

$$\frac{\partial v}{\partial u} = -\kappa \tag{18-3}$$

and

$$\frac{\partial v}{\partial \kappa} + \kappa \frac{\partial u}{\partial \kappa} = 0 \tag{18-4}$$

while by means of Eq. 18-2, Eq. 18-1 becomes

$$\left(1 - \frac{u^2}{a^2}\right)\frac{\partial u}{\partial \kappa} - \frac{2uv}{a^2}\frac{\partial v}{\partial \kappa} - \kappa\left(1 - \frac{v^2}{a^2}\right)\frac{\partial v}{\partial \kappa} + v = 0 \tag{18-5}$$

From Eq. 18-3 by differentiation with respect to u there results

$$\frac{\partial u}{\partial \kappa} = -\frac{1}{\partial^2 v/\partial u^2} \tag{18-6}$$

And from Eq. 18-3, 18-4, and 18-6

$$\frac{\partial v}{\partial \kappa} = -\frac{\partial v/\partial u}{\partial^2 v/\partial u^2} \tag{18-7}$$

If Eq. 18-6 and 18-7 are substituted in Eq. 18-5 a differential equation which defines the hodograph diagram is obtained.

$$v\frac{\partial^2 v}{\partial u^2} = 1 + \left(\frac{\partial v}{\partial u}\right)^2 - \frac{\left(u + v\frac{\partial v}{\partial u}\right)^2}{a^2} \tag{18-8}$$

H,19. The Flow Field around Circular Cones at Zero Angle of Attack. In Fig. H,19a a typical hodograph diagram is presented, and the shape of the streamlines in the physical plane for a circular cone at zero angle of attack is shown.

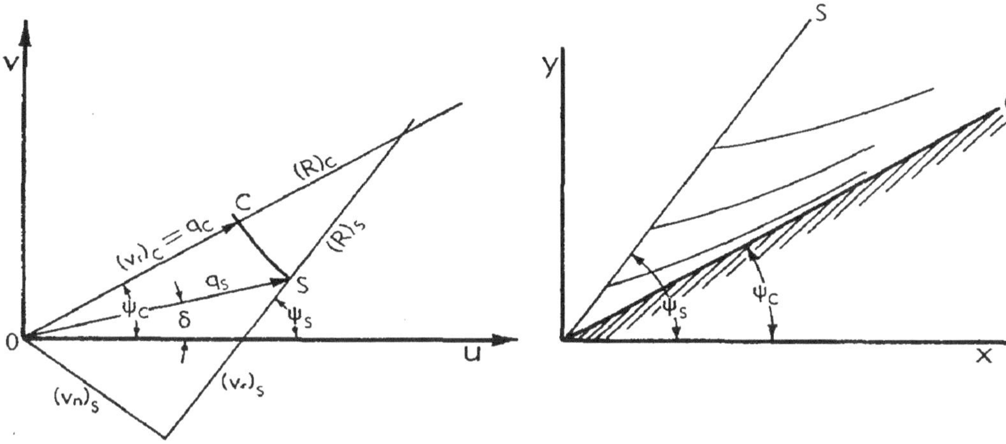

Fig. H,19a. The flow around a circular cone at zero angle of attack.

At the surface of the conical body v_n is zero; therefore, the velocity is in the radial direction and its inclination with respect to the direction of the undisturbed stream is equal to ψ_C. The radius of curvature of the hodograph diagram is negative (Eq. 17-1). Therefore, the inclination of the velocity with respect to the axis decreases as ψ increases from ψ_C to ψ_S and has the lowest value at the shock ($\psi = \psi_S$) where the inclination is equal to the deviation δ across the shock. The intensity of the velocity q increases when ψ increases and reaches its highest value at the shock wave $q = (q)_S$. Along the streamlines a gradual isentropic compression occurs and the streamlines curve and tend to become asymptotically tangent to the cone surface.

At a given Mach number, when the inclination of the shock corresponding to the maximum deviation is selected for ψ_S, a value of ψ_C is found that is the greatest possible for an attached shock for the Mach number considered. Because of the curvature of the streamlines, the value of ψ_C is larger than the deviation across the shock, and the difference is a function of Mach number. Some values of $(\psi_C)_{max}$ are plotted as functions of the free stream Mach number in Fig. H,19b. For comparison the values of δ_{max} across the shock are also presented.

Because of the compression existing along the streamlines, for conditions when the flow behind the shock is still supersonic but is very close to sonic, transition from supersonic to subsonic flow occurs along the streamlines, and a value of ψ exists between ψ_s and ψ_c where the velocity is equal to the local velocity of sound. However, it is necessary to remember that, when the flow at the surface of the body becomes subsonic, the existence of conical flow is limited to cones of infinite length, because for cones of finite length the disturbances produced at the "shoulders" travel upstream along the cone, destroying the conical properties of the flow.

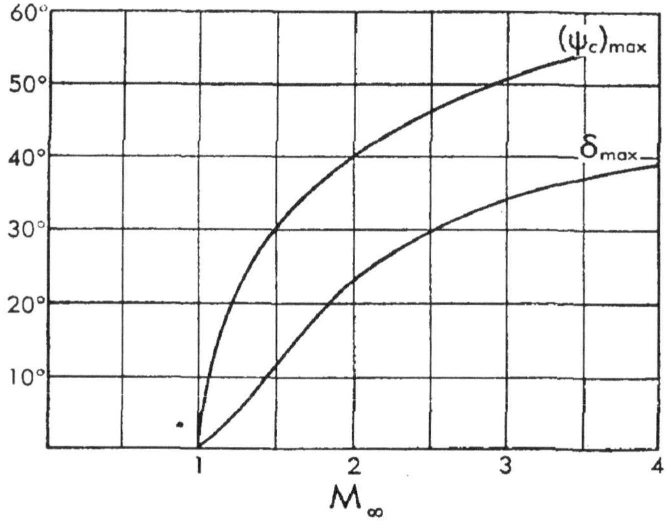

Fig. H,19b. The maximum cone angle which permits an attached shock as function of free stream Mach number.

If the value of ψ_s is larger than the value corresponding to the maximum deviation (δ_{max}), the corresponding value of ψ_c is less than $(\psi_c)_{max}$; therefore, for any conical body at a given free stream Mach number, two solutions are possible, one with a strong shock and one with a weak shock. The pressure at the surface of the body for one case is quite different from the pressure for the other case. The relation between the two solutions is determined by the downstream conditions, as both solutions are physically possible. However, the strong shock solution is exact only for infinite conical bodies, and for conical bodies of finite length it represents only the limiting value of the flow field at the apex of the body.

A graphical representation in the hodograph plane of all the possible axially symmetric conical flow solutions for a given free stream Mach number is given by the "apple curve" combined with the shock polar diagram (Fig. H,19c; [*26*]).

The "apple curve" (whose name is derived from its peculiar shape) is the locus of all the possible points representing the velocity vector at the surface of the cone. If UBq_N is the shock polar diagram for the Mach number considered, a conical flow field can be determined for each point B representing the velocity OB behind the shock which produces a deviation δ_B. The curve AB represents the hodograph diagram of the conical field. At A the velocity is in the radial direction; therefore, the point A represents the velocity at the surface of the cone. If point B moves along the curve q_NBU, the point A describes the curve q_NAU.

In Fig. H,19c several curves AB for different positions of B are presented, and the curves connecting points having constant ratio of v_n/v_r are also given. Because the ratio v_n/v_r represents the tangent of the inclination of the velocity vector with respect to the radius, the value of v_n/v_r is

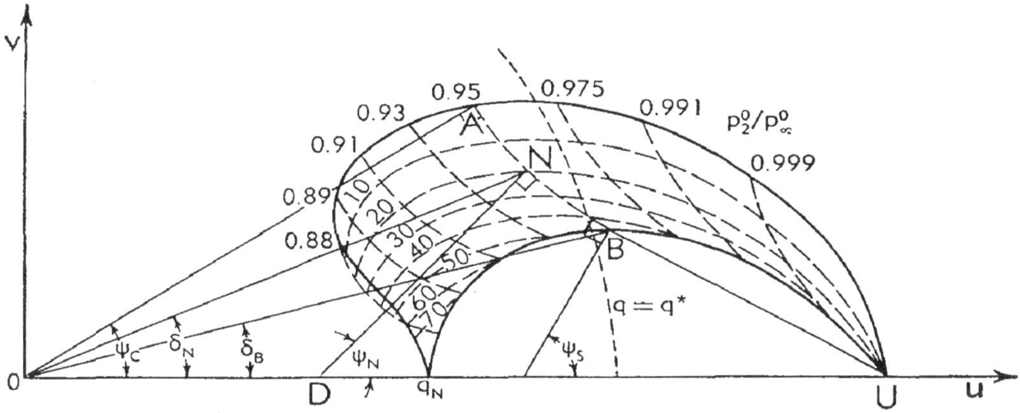

Fig. H,19c. The "apple curve" for flow around conical bodies with axial symmetry.

given as the value of the angle corresponding to the tangent. At any point N, the segment ON gives the velocity vector. Because at N $\tan^{-1} v_n/v_r = 20°$ the corresponding value of ψ, say ψ_N, in the physical plane is equal to $\delta_N + 20°$. The line DN inclined at an angle ψ_N with respect to the u axis is normal to the tangent at N to the curve AB. In the same figure the values of total pressure ratio across the shock are also given.

H,20. Other Problems Involving Axially Symmetric Conical Flow. Problems involving axially symmetric conical flow other than the flow around a conical body as described in Art. 17 can be found. The physical possibility of the existence of such flows can be discussed by means of the hodograph representations [*26*].

Consider first the possibility of transition from a uniform flow to another uniform flow by means of a conical flow field, Fig. H,20a. The flows in region 1 and region 2 are uniform and the flow in region 3 has conical properties with apex at O. In Fig. H,20a (top left), the flow in the conical region undergoes an expansion, and in Fig. H,20a (bottom, left) the flow undergoes a compression. In the hodograph plane of Fig. H,20a (right) the conical flow would be represented by the curve u_2Au_1 for the case given in Fig. H,20a (top, left) or by $u_1 B u_2$ for the case given in Fig. H,20a (bottom, left). Consider now the points u_1 and u_2 near the axis. Near the axis Eq. 18-8 becomes

$$v\frac{\partial^2 v}{\partial u^2} = 1 + \left(\frac{\partial v}{\partial u}\right)^2 - \frac{u^2}{a^2} \tag{20-1}$$

and

$$a^2 = \frac{\gamma - 1}{2}(q_{max}^2 - u^2 - v^2) \ . \tag{20-2}$$

If Eq. 20-1 is differentiated with respect to u, it appears that near the axis (where $v \to 0$) the term $(\partial v/\partial u)(\partial^2 v/\partial u^2)$ must be positive, while from

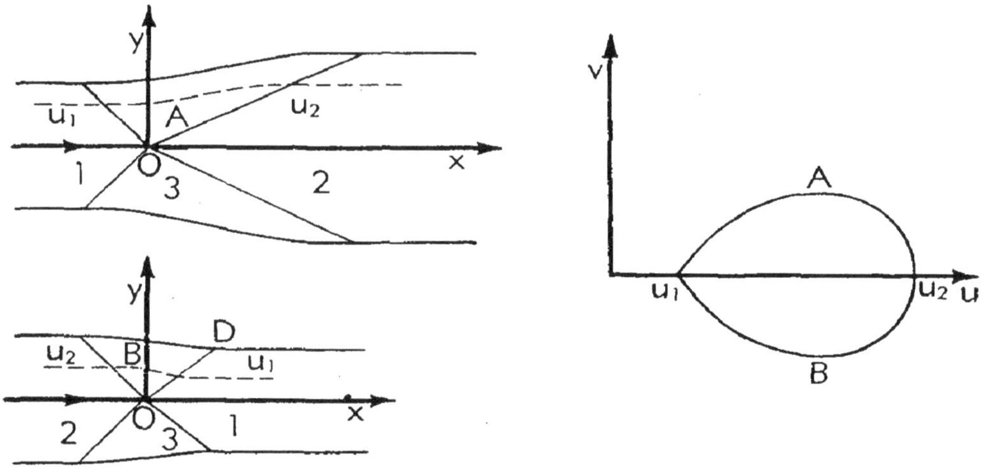

Fig. H,20a. Conical flows with axial symmetry. Top left, expansion; bottom left, compression; right, hodograph representation.

Eq. 18-6 it appears that $\partial u/\partial \kappa$ must be of opposite sign to $\partial^2 v/\partial u^2$. Now in Fig. H,20a (top left) at u_1, u increases going from u_1 to A and $\kappa = x/y$ also increases, so that $\partial u/\partial \kappa$ is positive. Therefore, from Eq. 18-6 it results that $\partial^2 v/\partial u^2$ is negative. However, at u_1, $\partial v/\partial u$ is positive; therefore, $(\partial v/\partial u)(\partial^2 v/\partial u^2)$ would be negative, in contrast to the results obtained from Eq 20-1.

Similarly the conditions at u_1 for the flow field given in Fig. H,20a (bottom left) are not possible because $\partial u/\partial \kappa$ is negative and $\partial v/\partial u$ also is negative. Therefore Eq. 18-6 requires that $\partial^2 v/\partial u^2$ be positive while Eq. 20-1 requires that $\partial^2 v/\partial u^2$ be negative. Thus, neither of the flow fields in Fig. H,20a is physically possible. A flow which is physically possible is a flow of the type given in Fig. H,20a (bottom left) but in which a shock wave produces the last part of the compression (Fig. H,20b). If OD represents a shock wave, from the conditions OD_1 behind the shock wave all the possible conditions OB_1 in front of the shock can be obtained from the shock wave equations. For each of those conditions OB_1 the components of the velocity v_r and v_n in front of the shock can be obtained and the region of conical flow in front of the shock determined by means of Eq. 17-1.

The value of $v_n \cot \psi + v_r$ of Eq. 17-1 is represented in the hodograph diagram by the segment B_1E (v_n is negative). Near the axis the segment B_1E tends to zero; therefore, the radius of curvature of the hodograph diagram becomes very small and tends to zero at the axis and near the axis the value of u changes very slowly when ψ changes. If in Eq. 20-1

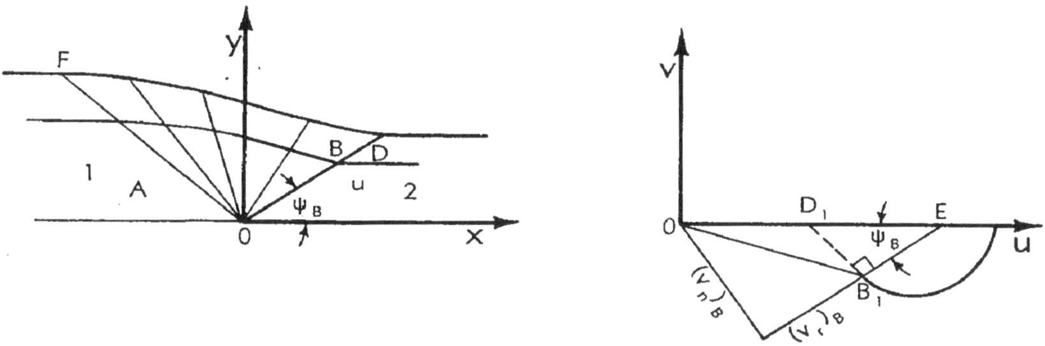

Fig. H,20b. Transition from a uniform flow to a uniform flow at lower velocity by means of conical flow. Left, physical plane; right, hodograph plane.

the value of $(u/a)^2$ near the axis is assumed to be constant and equal to the value of $(u_1/a_1)^2$ at the axis, Eq. 20-1 has solutions of the form [*26*]

$$v = A\sqrt{\frac{u_1^2}{a_1^2} - 1} \sin \frac{u_1 - u}{A} \tag{20-3}$$

$$v = A\sqrt{\frac{u_1^2}{a_1^2} - 1} \sinh \frac{u_1 - u}{A} \tag{20-4}$$

for the case of $u_1/a_1 > 1$

$$v = A\sqrt{1 - \frac{u_1^2}{a_1^2}} \cosh \frac{u_1 - u}{A} \tag{20-5}$$

when $u_1/a_1 < 1$ and

$$v = Be^{\left(\frac{u_1 - u}{A}\right)} \tag{20-6}$$

for $u_1/a_1 = 1$, where A and B are constants of integration.

Another conical flow which could be thought to exist is the conical flow behind a conical shock, reversed with respect to the shock produced by a conical body. This flow could be produced by a circular jet flowing into a higher pressure region. The shock wave would be represented in Fig. H,20b by the line FO. However, it is possible to show, by means of considerations involving the sign of $\partial^2 v/\partial u^2$, that such a conical flow is not physically possible.

H,21. Circular Cones at Small Angles of Attack. A circular cone moving at supersonic speed and at a small angle of attack α produces a conical shock that for small angles of attack (terms of the order of α^2 or higher being neglected) is a circular conical shock. This shock has the same cone angle ϵ as the shock produced by the same conical body at the same speed at zero angle of attack, but is inclined with respect to the direction of the free stream velocity at a small angle σ (of the same order as α) in the plane of the axis of symmetry of the body and the free stream velocity vector. The angle σ is, in general, different from α; therefore, the axes of symmetry of the body and the shock are different, but the angle η between the two axes is also small and of the order of α [27, 28].

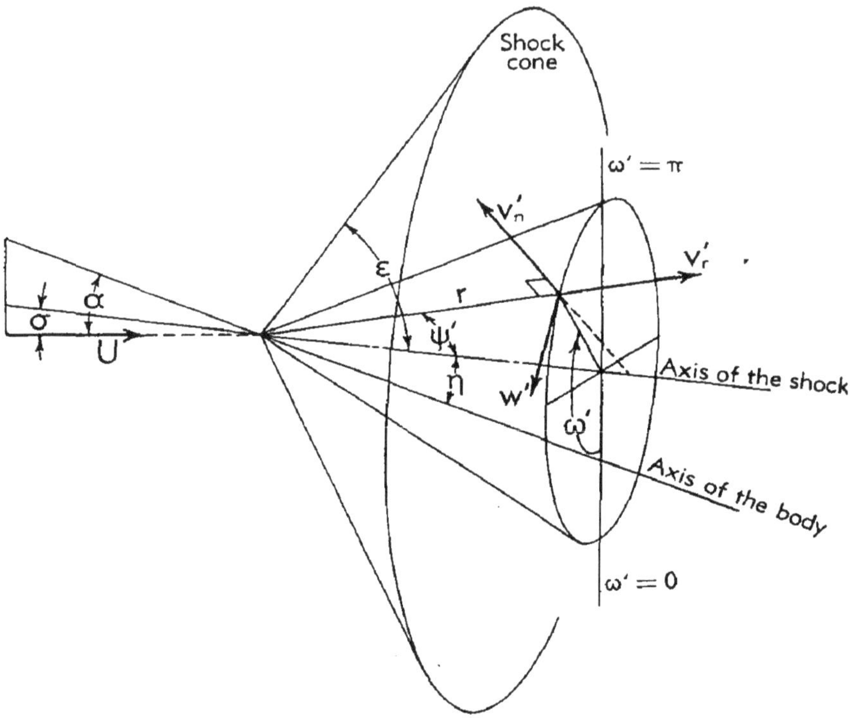

Fig. H,21. Flow around a circular cone at a small angle of attack.

In order to analyze this flow consider a circular conical shock wave with the axis of symmetry inclined at a small angle σ with respect to the direction of the undisturbed velocity. Consider first a polar coordinate system r', ψ', ω' whose axis is coincident with the axis of symmetry of the shock, and let v'_r, v'_n, w' be the velocity components in the field behind the shock wave. (The prime indicates that the quantities are referred to shock axis coordinates.) For the conditions just behind the shock, where ϵ is the angle of the conical shock (see Fig. H,21), the shock wave equations give

$$\begin{aligned} v'_{r_s} &= U\cos\epsilon\cos\sigma + U\sin\sigma\sin\epsilon\cos\omega' \\ w'_s &= -U\sin\sigma\sin\omega' \qquad (21\text{-}1) \\ v'_{n_s} &= -\frac{\gamma-1}{\gamma+1}\,\frac{q^2_{\max} - v'^2_{r_s} - w'^2_s}{U\cos\sigma\sin\epsilon - U\sin\sigma\cos\epsilon\cos\omega'} \end{aligned}$$

If the angle σ is small and terms of the order of σ^2 are neglected, Eq. 21-1 become

$$\begin{aligned} v'_{r_s} &= U \cos \epsilon + \sigma U \sin \epsilon \cos \omega' \\ w'_s &= -U\sigma \sin \omega' \\ v'_{n_s} &= -\frac{\gamma - 1}{\gamma + 1}\left(\frac{q^2_{\max} - U^2 \cos^2 \epsilon}{U \sin \epsilon}\right) \\ &\quad - \frac{\gamma - 1}{\gamma + 1}\sigma \cos \epsilon \cos \omega' \left(-2U + \frac{q^2_{\max} - U^2 \cos^2 \epsilon}{U \sin^2 \epsilon}\right) \end{aligned} \tag{21-2}$$

Therefore, a flow field defined by

$$\begin{aligned} v'_r &= v'_{r_a} + \sigma v'_{r_b} \cos \omega' \\ v'_n &= v'_{n_a} + \sigma v'_{n_b} \cos \omega' \\ w' &= \sigma w'_b \sin \omega' \end{aligned} \tag{21-3}$$

where v'_{r_a}, v'_{n_a}, v'_{r_b}, v'_{n_b}, and w'_b are functions of ψ' and are independent of ω', satisfies the boundary conditions at a circular shock wave inclined at small angle with respect to the free stream direction.

Consider now a conical surface having circular cross section, the apex coincident with the apex of the conical shock, the axis of symmetry inclined at a small angle η with respect to the axis of the conical shock, in the plane of the free stream velocity vector and the axis of symmetry of the conical shock. If ψ = const is the conical surface considered, in the coordinate system coinciding in axis with the axis of symmetry of the shock, the coordinates ψ' of a point of this surface is given by

$$\psi' = \psi + \eta \cos \omega' = \psi + \eta \cos \omega$$

where terms of the order of η^2 are neglected. The quantity ω indicates the position of the meridian plane with respect to the axis of the surface considered, and the planes $\omega' = 0$ and $\omega = 0$ are coincident (Fig. H,21). Then the velocity components v_r, along the generatrix of the conical surface v_n normal to the surface, and w perpendicular to the plane ω = const, are given, in the same approximation, by

$$\begin{aligned} (v_r)_{\psi'} &= [v'_{r_a}]_\psi + \left(\frac{\partial v'_{r_a}}{\partial \psi'}\right)_\psi \eta \cos \omega' + \sigma[v'_{r_b}]_\psi \cos \omega' = [v_{r_a}]_\psi + \alpha\,[v_{r_b}]_\psi \cos \omega \\ (v_n)_{\psi'} &= [v'_{n_a}]_\psi + \left(\frac{\partial v'_{n_a}}{\partial \psi'}\right)_\psi \eta \cos \omega' + \sigma[v'_{n_b}]_\psi \cos \omega' = [v_{n_a}]_\psi \\ &\qquad + \alpha[v_{n_b}]_\psi \cos \omega \\ (w)_{\psi'} &= \left(\sigma[w'_b]_\psi - \frac{\eta}{\sin \psi}\,[v'_{n_a}]_\psi\right) \sin \omega' = \alpha w_b \sin \omega \end{aligned} \tag{21-4}$$

where the quantity α is small and is given by

$$\alpha = \sigma + \eta \tag{21-5}$$

and v_{r_a}, v_{n_a}, v_{r_b}, v_{n_b}, and w_b are functions only of ψ and are independent of ω and α.

If the conical surface considered is the surface of the conical body ψ_C which produces the same conical shock, at the same Mach number for the conditions of the angle of attack, then at $\psi = \psi_C$ one must find that

$$(v_n)_{\psi_C} = 0$$

or, from the second of Eq. 21-4,

$$[v'_{n_a}]_{\psi_C} + \left[\frac{\partial v'_{n_a}}{\partial \psi'}\right]_{\psi_C} \eta \cos \omega' + \sigma [v'_{n_b}]_{\psi_C} \cos \omega' = [v_{n_a}]_{\psi_C} + [v_{n_b}]_{\psi_C} \cos \omega = 0 \qquad (21\text{-}6)$$

But $[v'_{n_a}]_{\psi_C}$ is the velocity component corresponding to the same conical shock for the condition of $\sigma = 0$, and therefore at the position ψ_C is zero; therefore

$$\left[\frac{\partial v'_{n_a}}{\partial \psi'}\right]_{\psi_C} + \frac{\sigma}{\eta} [v'_{n_b}]_{\psi_C} = 0 \qquad (21\text{-}7)$$

From Eq. 21-2, 21-4, and 21-6 it is seen that the components v_{r_a} and v_{n_a} are equal to the corresponding values for the conditions of zero angle of attack, and that the angle of the conical shock does not change when the conical body is at a small angle of attack. However, the axis of the shock turns with respect to the conical body through a small angle η, which is proportional to the angle of attack of the body (Eq. 21-7 and [*28*]).

Near the surface of the body v_n is small; therefore, if w is considered to be of the order of α, Eq. 14-10 becomes [*27*]

$$2v_r + v_n \cot \psi + \frac{\partial v_n}{\partial \psi} + \frac{\partial w}{\sin \psi \partial \omega} = 0 \qquad (21\text{-}8)$$

which is satisfied by an expression of the type of Eq. 21-4. Values of v_{r_b}, v_{n_b}, and w_b which satisfy Eq. 14-10 must be determined in the region between the shock wave and the conical body in order to determine the flow field.

H,22. The Entropy Field for Circular Cones at Small Angles of Attack. Because of the inclination σ of the conical shock with respect to the free stream direction, the flow between the conical body and the shock is rotational flow; therefore, singular points occur in the entropy distribution. Eq. 21-2 and 21-5 show that v_n is negative at the shock, zero at the surface of the body, and negative between the shock and the body for small angles of attack. From the second of Eq. 21-2 it appears that $\partial w/\partial \omega$ is negative at $\omega = 0$ and positive at $\omega = \pi$. Therefore, the singular point is at the surface of the body and in the plane $\omega = 0$, while in the plane $\omega = \pi$ (Fig. H,21) a stagnation point occurs for the component $U\sigma$. The entropy at the surface of the body which is equal to the entropy in the plane $\omega = \pi$ has the maximum value in the flow field considered (see Art. 15).

If expressions (21-4) are introduced in Eq. 15-3, the entropy $\partial s/\partial\omega$ is given by

$$\frac{\partial s}{\partial \omega} = f(\psi) \sin \omega \tag{22-1}$$

Therefore, the entropy in the flow field, within the approximation considered, can be expressed as [*30*]

$$s = s_a + \alpha s_b \cos \omega \tag{22-2}$$

However, expression (22-2) cannot be used at the surface of the body where the entropy is not a function of ω but is equal to $s_a - \alpha s_b$ [*21*]. In order to determine the region of validity of Eq. 22-2 consider Eq. 15-5. If Eq. 22-2 is introducd into Eq. 15-5, there results

$$\frac{\partial s}{\partial \psi} = \frac{\partial s_a}{\partial \psi} + \alpha \frac{\partial s_b}{\partial \psi} \cos \omega = \frac{\alpha^2 w_b s_b \sin^2 \omega}{(v_{n_a} + \alpha v_{n_b} \cos \omega) \sin \psi} \tag{22-3}$$

Eq. 22-3 shows that a solution as given in Eq. 22-2 implies that the entropy remains constant in every meridian plane in the region where $v_n \neq 0$ and is of zero order, since in this region $\partial s/\partial\psi$ is of the order of α^2 and can be neglected. However, near the surface of the body v_n is zero, and $\partial v_n/\partial\psi \neq 0$, therefore, in a layer contained between the surface of the cone ψ_C and a cone given by $\psi_C + \alpha$, where v_n is of the order of α, the entropy gradient $\partial s/\partial\psi$ is also of the order of α, and in a region between ψ_C and $\psi_C + \alpha^2$, $\partial s/\partial\psi$ is of zero order.

The existence of an entropy gradient $\partial s/\partial\psi$ of the order of α changes the pressure gradient by a quantity of the order of α, as can be shown by eliminating the term containing the density gradient between Eq. 14-9 and 15-2. Therefore, the effect on the pressure at the body surface due to the existence of an entropy gradient $\partial s/\partial\psi$ is of the order of α^2. Then, when terms of the order of α^2 are neglected, the effect of the entropy gradient on the flow properties at the cone surface can be obtained by assuming that a vortical layer of thickness α^2 exists at the body surface across which the pressure does not change, while velocity and density change [*21*]. Outside of the layer Eq. 22-2 can be assumed as valid and the entropy jump across the layer is given by

$$\Delta s = \alpha s_b(\cos \omega - 1) \tag{22-4}$$

Therefore, from the conditions outside of the layer the velocity and density distribution at the surface of the body can be determined.

H,23. Numerical Determination of the Flow Field. In order to obtain numerical values for cones at small angles of attack, it is convenient to perform the calculations by assuming spherical coordinates whose axis coincides with the axis of the conical shock, i.e. r, ψ', and ω'.

If Eq. 21-3 are introduced into Eq. 14-10, and terms of the order of σ^2 are neglected, and since v'_{r_a} and v'_{n_a} are defined by Eq. 17-1, the result is

$$\left(v'_{r_b} + \frac{\partial v'_{n_b}}{\partial \psi'}\right)\left[1 - \left(\frac{v'_n}{a}\right)_a^2\right]$$
$$+ v'_{n_b}\left\{\cot \psi' + \left[2\left(\frac{v'_n}{a}\right)_a + (\gamma - 1)\left(\frac{v'_n}{a}\right)_a^3\right]\frac{\left(\frac{v'_r}{a}\right)_a + \left(\frac{v'_n}{a}\right)_a \cot \psi'}{1 - (v'_n/a)_a^2}\right\}$$
$$+ v'_{r_b}\left[1 + (\gamma - 1)\frac{\left(\frac{v'_r}{a}\right)_a + \left(\frac{v'_n}{a}\right)_a \cot \psi'}{1 - (v'_n/a)_a^2}\left(\frac{v'_r}{a}\right)_a\left(\frac{v'_n}{a}\right)_a^2\right]$$
$$+ \frac{w'_b}{\sin \psi'} = 0 \qquad (23\text{-}1)$$

where the subscript a applied to ratios such as (v'_r/a) denotes the values of such ratios in the axial flow field computed by the methods of Art. 17.

Eq. 23-1 permits the determination of v'_{r_b} and v'_{n_b} by a step-by-step process starting from the shock wave, provided that the flow field $(\ )_a$ for zero angle of attack is known and w'_b is related to v'_{r_b}, v'_{n_b}, and s_b.

Since s_b remains constant in the meridian plane outside of the vortical layer when terms of the order of σ^2 are neglected, from Eq. 15-4 there results

$$\frac{\partial v'_{r_b}}{\partial \psi'} = v'_{n_b} \qquad (23\text{-}2)$$

Therefore Eq. 16-4 and 16-5 are valid also for the velocity components u'_b and v'_b, and the radius of curvature of the hodograph diagram is given by

$$R'_b = v'_{r_b} + \frac{\partial v'_{n_b}}{\partial \psi'} \qquad (23\text{-}3)$$

For given values of free stream Mach number and cone angle, the angle of the shock wave ϵ and the values of v'_{r_a} as functions of ψ' are known. From Eq. 2-9, 2-10, 2-11, and 21-2 the values of v'_{r_b}, v'_{n_b}, w'_b, and s_b at $\psi' = \epsilon$ behind the shock can be determined. The value of s_b is constant outside of the vortical layer and can be determined from the shock wave equations (Eq. 2-10). From Eq. 15-3 it follows that

$$-\frac{a_a^2}{\gamma \mathcal{R}} s_b = v'_{n_a}\frac{\partial w'_b}{\partial \psi'}\sin \psi' + w'_b(v'_{r_a}\sin \psi' + v'_{n_a}\cos \psi') + v'_{r_a}v'_{r_b}$$
$$+ v'_{n_a}v'_{n_b} \qquad (23\text{-}4)$$

Therefore, by means of Eq. 23-1 and 23-4 or the equivalent Eq. 24–19, with a step-by-step process, the whole flow field v'_{r_b}, v'_{n_b}, w'_b can be determined as functions of ψ' [27,28, 29,30]. At the surface of the conical body $(v'_n)_{\psi_c}$ must vanish; hence, from the second of Eq. 21-4,

$$[v'_{n_a}]_{\psi_C} + \left[\frac{v'_{n_a}}{\partial\psi'}\right]_{\psi_C} \eta \cos \omega' + \sigma[v'_{n_b}]_{\psi_C} \cos \omega' = 0 \tag{23-5}$$

but

$$[v'_{n_a}]_{\psi_C} = 0$$

Therefore

$$\frac{\eta}{\sigma} = -\frac{[v'_{n_b}]_{\psi_C}}{[\partial v'_{n_a}/\partial\psi']_{\psi_C}} \tag{23-6}$$

or because of Eq. 21-5 and 17-1

$$\frac{\alpha}{\sigma} = 1 + \frac{\eta}{\sigma} = 1 + \left[\frac{v'_{n_b}}{2v'_{r_a}}\right]_{\psi_C} \tag{23-7}$$

In order to determine v_r and w at the surface of the body in conical body coordinates, Eq. 21-4 can be used.

The pressure distribution at the surface of the body can be obtained from the equation (Sec. C, Eq. 11-6; also III,A,7, Eq. 7-16)

$$\frac{p}{p_1} = \left(\frac{q_{\max}^2 - q^2}{q_{\max}^2 - q_1^2}\right)^{\frac{\gamma}{\gamma-1}} e^{-\frac{s_a - \alpha s_b}{\mathcal{R}}} \tag{23-8}$$

where q^2 is defined by

$$q^2 = q_{r_a}^2 + 2\alpha v_{r_b} v_{r_a} \cos \omega + \alpha^2 w_b^2 \sin^2 \omega \tag{23-9}$$

and v_{r_b} and w_b are expressed in body coordinates.

In Eq. 23-9 some of the α^2 terms are retained while others are neglected. However, often the terms retained are much larger than the neglected terms as can be seen from the following considerations. At the surface of the body (Eq. 15-3)

$$w_b = -\frac{v_{r_b}}{\sin \psi_C} \tag{23-10}$$

therefore the term

$$\alpha^2 w_b^2 = \frac{\alpha^2 v_{r_b}^2}{\sin^2 \psi_C}$$

can be of the same order as $\alpha v_{r_b} v_{r_a}$ since $\sin^2 \psi_C$ is usually very small, and $\alpha/\sin^2 \psi_C$ is of the order of 1. Then the term $\alpha^2 w_b^2$ is large and therefore it must be retained.

In [31] values of v_{r_b}, w_b, and v_{n_b} have been tabulated for various cone angles and for various free stream Mach numbers. The coordinate system used has the axis in the direction of free stream velocity; therefore, the results must be transformed to the body axis system and then the pressure distribution can be obtained from Eq. 23-8. In [30,31] the vortical layer has been ignored and the term $\alpha^2 w_\xi^2$ has been neglected in the determination of the pressure distribution. Therefore, the tabulated values of velocity components and density at the surface of the body must be corrected for the entropy jump given by Eq. 22-4, and the pressure coefficients must be determined by means of Eq. 23-8.

H,24. Conical Flow Without Axial Symmetry. The calculation of slender pointed bodies without axial symmetry requires the determination of conical flow without axial symmetry, since this calculation starts with the conical flow field at the nose of the body. The basic problem of the determination of conical flow consists in determining the shape of the conical shock wave produced by the body. When the shape of the shock is determined by means of numerical calculations, the flow field around the body can be obtained. Because the relation between shape of the body and shape of the shock wave is not known a priori, the method of linearized analysis described in Sec. G can be particularly useful for flow determination of this kind [32,33]. An approximate shape of the shock wave is assumed as the basic solution and the flow field inside the shock is determined. Then a linearized flow field is superposed in order to satisfy the boundary conditions at the body. The calculations are simple if the basic flow can be determined analytically or numerically without a large amount of calculations. For example, for slender bodies the basic flow can be the axially symmetric flow axis used in Art. 17 and 18, which is available in tabulated form. Consider a conical shock wave which in spherical coordinates can be defined as

$$\psi_S = \psi_{0_S} + \sum_1^\infty (\psi_m)_S \cos m\omega + \sum_1^\infty (\psi_n)_S \sin n\omega \tag{24-1}$$

where all the values of ψ_n and ψ_m are small so that terms of the order of ψ_n^2 or ψ_m^2 can be neglected. Such a shock wave is approximately of circular cross section, as is found for slender conical bodies. Assume that the flow has a symmetry plane. Then the second summation of Eq. 24-1 is equal to zero.

The velocity components in radial direction v_{r_1}, in tangential direction v_{T_1}, and direction normal to the shock v_{N_1} in front of the shock wave are (see Fig. H,21)

$$\begin{aligned} v_{r_1} &= U \cos \psi_S \\ v_{N_1} &= -U \sin \psi_S \cos \beta \\ v_{T_1} &= -U \sin \psi_S \sin \beta \end{aligned} \tag{24-2}$$

where β is the angle of the dihedral between the plane normal to the shock wave and the plane containing the reference axis. Across the shock wave the following relations are valid:

$$v_{N_1}v_{N_2} = \frac{\gamma - 1}{\gamma + 1}(q_{\max}^2 - v_{r1}^2 - v_{T1}^2)$$
$$= \frac{\gamma - 1}{\gamma + 1}(q_{\max}^2 - U^2 \cos^2 \psi_S - U^2 \sin^2 \psi_S \sin^2 \beta)$$

or

$$v_{N_2} = -\frac{\gamma - 1}{\gamma + 1}\frac{(q_{\max}^2 - U^2 \cos^2 \psi_S - U^2 \sin^2 \psi_S \sin^2 \beta)}{U \sin \psi_S \cos \beta} \quad (24\text{-}3)$$

and

$$v_{T_1} = v_{T_2}, \qquad v_{r_1} = v_{r_2}$$

If the velocity components v_r, v_n, and w of the flow field in spherical coordinates are considered at each point of the shock wave $\psi_S = \psi_{0S} + \sum(\psi_m)_S \cos m\omega$, then

$$(v_{r_1})_{\psi_S} = (v_r)_{\psi_S} = U \cos \psi_S$$

$$(v_n)_{\psi_S} = -\frac{\gamma - 1}{\gamma + 1}\frac{(q_{\max}^2 - U^2 \cos^2 \psi_S - U^2 \sin^2 \psi_S \sin^2 \beta)}{U \sin \psi_S} - U \sin \psi_S \sin^2 \beta \quad (24\text{-}4)$$

$$(w)_{\psi_S} = -U \sin \psi_S \sin \beta \cos \beta + \frac{\gamma - 1}{\gamma + 1}(q_{\max}^2 - U^2 \cos^2 \psi_S - U^2 \sin^2 \psi_S \sin^2 \beta)\frac{\sin \beta}{U \sin \psi_S \cos \beta}$$

But

$$\tan \beta = \frac{d\psi_S}{\sin \psi_S \, d\omega} = \sum - \frac{m(\psi_m)_S \sin m\omega}{\sin \psi_S}$$

therefore, the velocity components behind the conical shock wave of Eq. 24-1, if terms of the order of $(\psi_m)_S^2$ are neglected, are

$$(v_r)_{\psi_S} = U \cos (\psi_0)_S - U \sin (\psi_0)_S \sum (\psi_m)_S \cos m\omega$$

$$(v_n)_{\psi_S} = -\frac{\gamma - 1}{\gamma + 1}\frac{q_{\max}^2 - U^2 \cos^2 (\psi_0)_S}{U \sin (\psi_0)_S} - \frac{\gamma - 1}{\gamma + 1}\cos (\psi_0)_S \left(2U - \frac{q_{\max}^2 - U^2 \cos^2 (\psi_0)_S}{U \sin^2 (\psi_0)_S}\right)\sum (\psi_m)_S \cos m\omega \quad (24\text{-}5)$$

$$(w)_{\psi_S} = \left[U - \frac{\gamma - 1}{\gamma + 1}\left(\frac{q_{\max}^2 - U^2 \cos^2 (\psi_0)_S}{U \sin^2 (\psi_0)_S}\right)\right]\sum m(\psi_m)_S \sin m\omega$$

Then the velocity components in the flow field behind the shock defined by Eq. 24-1 can be expressed as

$$v_r = (v_{r_0})_{(\psi_0)_S} + \sum(\psi_m)_S(v_r)_m \cos m\omega$$
$$v_n = (v_{n_0})_{(\psi_0)_S} + \sum(\psi_m)_S(v_n)_m \cos m\omega \quad (24\text{-}6)$$
$$w = \sum(\psi_n)_S w_m m \sin m\omega$$

where v_{r_0}, v_{n_0}, $(v_r)_m$, $(v_n)_m$, and $(w)_m$ are independent of ω and v_{r_0} and v_{n_0} correspond to the flow field for a circular conical shock $(\psi_0)_s$, and $(v_r)_m$, $(v_n)_m$, and $(w)_m$ at the circular cone $\psi = (\psi_0)_s$ are independent of m and are defined by

$$[(v_r)_m]_{(\psi_0)_S} = -U \sin (\psi_0)_s - (v_{n_0})_{(\psi_0)_S}$$

$$[(v_n)_m]_{(\psi_0)_S} = -\frac{\gamma - 1}{\gamma + 1} \cos (\psi_0)_S \left[2U - \frac{q_{\max}^2 - U^2 \cos^2 (\psi_0)_S}{U \sin^2 (\psi_0)_S}\right] - \left(\frac{\partial v_{n_0}}{\partial \psi}\right)_{(\psi_0)_S} \tag{24-7}$$

$$[(w_m)]_{(\psi_0)_S} = U + \left(\frac{v_{n_0}}{\sin \psi}\right)_{(\psi_0)_S}$$

and

$$\left[\frac{\partial v_{n_0}}{\partial \psi}\right]_{(\psi_0)_S} = \left[-v_{r_0} - \frac{v_{r_0} + v_{n_0} \cot \psi}{q_{\max}^2 - \dfrac{2}{\gamma - 1}\left(\dfrac{v_{n_0}^2}{1 - v_{n_0}^2 - v_{r_0}^2}\right)}\right]_{(\psi_0)_S}$$

It can be shown that the flow field defined as in Eq. 24-6 satisfies the boundary of a conical body defined by

$$\psi_C = (\psi_0)_c + \sum_1^\infty (\psi_m)_c \cos m\omega \tag{24-8}$$

if the terms $(\psi_m)_c$ are small and quantities of the order of $(\psi_m)_C^2$ or higher are neglected with respect to terms of the order of ψ_m. At the surface of the body the velocity must be tangent to the surface; thererore, at each point A of the body, the following condition must hold:

$$\left(\frac{v_n}{w}\right)_A = \frac{d\psi_C}{\sin \psi_C \, d\omega} = -\frac{\sum m(\psi_m)_C \sin m\omega}{\sin \psi_C}$$

and, therefore, the boundary conditions can be expressed, in the approximation considered, as

$$\begin{aligned}(v_n)_C = (v_{n_0})_{(\psi_0)C} &+ \left(\frac{\partial v_{n_0}}{\partial \psi}\right)_{(\psi_0)C} \sum (\psi_m)_c \cos m\omega \\ &+ \sum [(v_n)_m]_{(\psi_0)C}(\psi_m)_S \cos m\omega \\ &= -\frac{[\sum m(\psi_m)_S (w_n)_{(\psi_0)c} \sin m\omega] \sum m(\psi_m)_c \sin m\omega}{\sin \psi_C}\end{aligned}$$

where the subscript $(\psi_0)_c$ indicates quantities at the surface of the basic circular body and the subscript C indicates quantities at the surface of the conical body considered.

Because each term of the right-side summation is of the order of $(\psi_m)^2$, the boundary conditions are

$$(v_{n_0})_{(\psi_0)_C} = 0 \tag{24-9}$$

$$(\psi_m)_S[(v_n)_m]_{(\psi_0)_C} = -\left(\frac{\partial v_{n_0}}{\partial \psi}\right)_{(\psi_0)_C} (\psi_m)_C \tag{24-10}$$

Eq. 24-9 and 24-10 show that the basic flow v_{r_0}, v_{n_0} is the flow corresponding to a circular cone of angle $(\psi_0)_C$, and permits relation of the equation of the conical shock to the equation of the conical body. The coefficients $(v_r)_m$, $(v_n)_m$, and $(w)_m$ for different values of m can be determined for a given free stream Mach number and value of $(\psi_0)_C$ or $(\psi_0)_S$; therefore, from Eq. 24-1, 24-6, 24-8, and Eq. 24-9, 24-10 the flow field around any conical body of the type given by Eq. 24-8 can be obtained when the terms $(\psi_m)_C$ are small.

The determination of the quantities $(v_r)_m$, $(v_n)_m$, and $(w)_m$ as functions of ψ can be made from Eq. 14-10. Because of expressions (24-6), this equation gives

$$\left[(v_r)_m + \frac{\partial (v_n)_m}{\partial \psi}\right]\left[1 - \left(\frac{v_n}{a}\right)_0^2\right]$$
$$= -(v_n)_m \left\{\cot\psi + \left[2\left(\frac{v_n}{a}\right)_0 + (\gamma - 1)\left(\frac{v_n}{a}\right)_0^3\right] \frac{\left(\frac{v_r}{a}\right)_0 + \left(\frac{v_n}{a}\right)_0 \cot\psi}{1 - (v_n/a)_0^2}\right\}$$
$$- (v_r)_m\left[1 + (\gamma - 1)\frac{\left(\frac{v_r}{a}\right)_0 + \left(\frac{v_n}{a}\right)_0 \cot\psi}{1 - (v_n/a)_0^2}\left(\frac{v_r}{a}\right)_0\left(\frac{v_n}{a}\right)_0^2\right] - \frac{m^2 w_m}{\sin\psi} \tag{24-11}$$

where ratios such as $(v_n/a)_0$ are taken from the symmetrical flow about the basic circular cone. Because of the approximation considered, the entropy remains constant in each meridian plane outside of a vortical layer of infinitesimal thickness around the body (see Art. 15) and, in the approximation considered,

$$v_n = \frac{\partial v_r}{\partial \psi} \tag{24-12}$$

Then $v_r + \partial v_n/\partial\psi$ is the radius of curvature in the hodograph diagram in the plane ω = const and $(v_n)_m$ and $(v_r)_m$ can be obtained by a step-by-step calculation from $\psi = (\psi_0)_S$ to $\psi = (\psi_0)_C$ by means of equations

$$[(v_n)_m]_{\psi-\Delta\psi} = [(v_n)_m]_\psi \cos(-\Delta\psi) + [R_m - (v_r)_m]_\psi \sin(-\Delta\psi)$$
$$[(v_r)_m]_{\psi-\Delta\psi} = [(v_n)_m]_\psi \sin(-\Delta\psi) - [R_m - (v_r)_m]_\psi \cos(-\Delta\psi) + (R_m)_\psi \tag{24-13}$$

where R_m at station ψ is obtained from Eq. 24-11 and

$$[R_m]_\psi = \left[(v_r)_m + \frac{\partial (v_n)_m}{\partial \psi}\right]_\psi \tag{24-14}$$

can be calculated from $(v_r)_m$, $(v_n)_m$, and $(w)_m$ at ψ. The value of $(w)_{m_{\psi-\Delta\psi}}$ can be obtained from

$$(w)_{m_{\psi-\Delta\psi}} = (w)_{m_\psi} - \left(\frac{\partial w_m}{\partial \psi}\right)_\psi \Delta\psi \tag{24-15}$$

and $(\partial w_m/\partial\psi)_\psi$ can be obtained from Eq. 15-3 where

$$s = s_0 + s_1 \sum (\psi_m)_s \cos m\omega \tag{24-16}$$

Here s_1 is independent of ψ and can be determined from the equations of

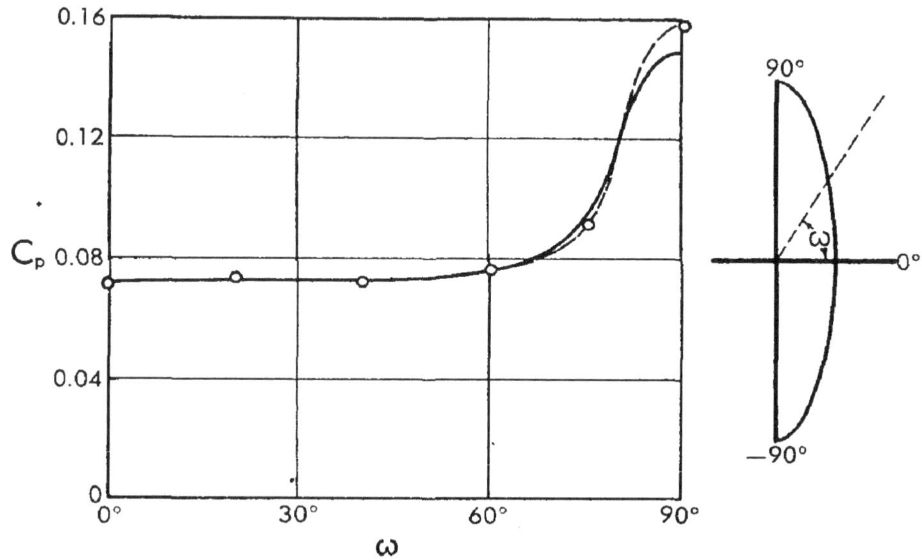

Fig. H,24. Pressure distribution around an elliptical cone at $M_\infty = 1.8$. (From [32].) The pressure coefficient is plotted against the angle ω. (See sketch at right.)

the shock from the expression

$$s_1 = \left[\frac{ds}{d\psi}\right]_{(\psi_0)_s}$$

Then, from Eq. 15-4

$$-\frac{a_0^2}{\gamma\mathcal{R}} s_1 = v_{n_0} \sin\psi \frac{\partial w_m}{\partial\psi} + v_{r_0}(v_r)_m + v_{n_0}(v_n)_m + v_{r_0}w_m \sin\psi + v_{n_0}w_m \cos\psi \tag{24-17}$$

while if the quantity s_1 is neglected and the flow is considered to be potential flow, then

$$w_m = -\frac{(v_r)_m}{\sin\psi} \tag{24-18}$$

Eq. 24-17 can be transformed and the following expression obtained

$$(v_r)_m + w_m \sin\psi = -\left(\frac{s_1}{\gamma\mathcal{R}}\right) a_0^{\frac{1}{\gamma-1}} (-v_{n_0}\sin\psi)^{\frac{1}{2}} \int_{(\psi_0)_s}^{\psi} \frac{a_0^{\frac{2\gamma-3}{\gamma-1}}}{v_{n_0}(-v_{n_0}\sin\psi)^{\frac{1}{2}}} d\psi$$

The same procedure can be used for the determination of the quantities $(v_r)_n$, $(v_n)_n$, w_n corresponding to the second summation of Eq. 24-1, when the flow does not have a symmetry plane.

As an example, the method presented has been applied to the determination of the flow field around a cone having an elliptical cross section with axes in the ratio of about 1 to 3, at zero angle of attack, at $M_\infty = 1.8$. The results are shown in Fig. H,24, where some experimental results are also presented.

When the shock shape is somewhat different from a cone having circular cross section, the basic flow field must be different from the axially symmetric field. However, if the basic conical flow components are expressed in the form

$$v_r = v_{r_a} + v_{r_b} f(\omega)$$
$$v_n = v_{n_a} + v_{n_b} f(\omega)$$
$$w = w_b f(\omega)$$

the basic flow can still be obtained by solving numerically the equations of motion in two meridian planes, and, therefore, the basic flow can be determined exactly.

H,25. Cited References.

1. Crocco, L. Singolaritá della corrente gassosa iperacustica nell'intorno di una prora a diedro. *L'Aerotecnica 17*, 519–534 (1937).
2. Guderley, G. Die Charakteristikenmethode für ebene und achsensymmetrische Überschallströmungen. *Jahrb. der deut. Luftfahrtforschung 1940*, 522–535. Also *Brown Univ. Transl. Summary A-9099.*
3. Tollmien, W. *Charakteristikentheorie.* Archiv 44/2 Technische Hochschule, Dresden, 1940.
4. Ferri, A. Application of the method of characteristics to supersonic rotational flow. *NACA Report 841.* 1946.
5. Ferrari, C. Campo aerodinamico a velocitá iperacustica attorno a un solido di rivoluzione a prora accuminata. *L'Aerotecnica 16*, 121–130, 1936. Also *Brown Univ. Transl. Summary A9-T-18.*
6. Lin, C. C., and Rubinov, S. I. On the flow behind curved shock. *J. Math. and Phys. 27*, 105–129 (1948).
7. Thomas, T. Y. On curved shock waves. *J. Math. and Phys. 26*, 62–68 (1947). Also Calculations on the curvatures of attached shock waves. *J. Math. and Phys. 27*, 280–297 (1949).
8. Guderley, G. Considerations on the structure of mixed subsonic supersonic flow patterns. *Wright Field Report F-TR-2168-ND.* 1947.
9. Munk, M. M., and Crown, C. The head shock wave. *NOL Memorandum 9773.* 1948.
10. Whitham, G. B. The behavior of supersonic flow past a body of revolution far from the axis. *Proc. Roy. Soc. 201*, 89–109 (1950).
11. von Kármán, Th. The similarity law of transonic flow. *J. Math. and Phys. 26*, 182–190 (1947).
12. Guderley, G. On the transition from transonic potential flow to a flow with shocks. *Wright Field Report F-TR-2160-ND.* 1947.
13. Busemann, A. A review of analytical methods for the treatment of flows with detached shocks. *NACA Technical Note 1858.* 1949.

14. Frankl, F. On the problems of Chaplygin for mixed sub- and supersonic flow. *Bull. Acad. Sci. URSS 9*, 1945. Also *NACA Technical Note 1155*.
15. Tricomi, F. Sulle equazioni lineari derivate parziali del secondo ordine, di tipo misto, parts 1–7. *Memorie Reale Accad. Nazl. Lincei, classe sci. fis. mat. e nat.* (*ser. 5*) *14*, 133–247 (1923). Also *Brown Univ. Transl. A9-T-26*.
16. Maccoll, J. W. Investigation of compressible flow at sonic speeds. *MOS Theoretical Research Report 7/46*. 1946.
17. Vincenti, W. G., and Wagoner, C. B. Transonic flow past a wedge profile with detached bow wave. *NACA Rept. 1095*. 1952. (Formerly *NACA Technical Note 2339*. 1950)
18. Liepmann, H. W., and Bryson, E. Transonic flow past wedge sections. *J. Aero. Sci. 17*, 745–755 (1950).
19. Guderley, G., and Yoshihara, H. The flow over a wedge profile at Mach number one. *J. Aero. Sci. 17*, 723–735 (1950).
20. Ferri, A. Alcuni risultati sperimentali ottenuti alla Galleria ultrasonora di Guidonia. *Atti di Guidonia 17*. 1939.
21. Ferri, A. Supersonic flow around circular cones at angles of attack. *NACA Technical Note 2236*. 1950.
22. Busemann, A. Drücke auf kegelförmige Spitzen bei Bewegung und Überschallgeschwindigkeit., *ZAMM 9*, 496–498 (1929).
23. Bourguard, F. Ondes ballistiques planes obliques et ondes coniques. *Comptes Rendus 194*, 846–857 (1932).
24. Taylor, G. I., and Maccoll, J. W. The air pressure on a cone moving at high speeds. *Proc. Roy. Soc. A139*, 278–311 (1933).
25. Staff of the computing section, Center of Analysis (under direction of Zdenek Kopal). Tables of supersonic flow around cones. *Mass. Inst. of Tech. Technical Report 1*. 1947.
26. Busemann, A. Conical supersonic flow with axial symmetry. *Luftfahrtforschung 19*, 134–144 (1942).
27. Ferrari, C. Determination of the pressure exerted on solid bodies of revolution with pointed noses placed obliquely in a stream of compressible fluid at supersonic velocity. *MOS R.T.P. Translation 1105*. From *Atti R. accad. sci. Torino 72*, 140–163 (1936).
28. Sauer, R. Supersonic flow about projectile heads of arbitrary shape at small incidence. *MOS R.T.P. Translation 1573*. From *Luftfahrtforschung 19*, 148–152 (1942).
29. Ferri, A. The method of characteristics for the determination of supersonic flow over bodies of revolution at small angles of attack. *NACA Technical Note 1809*. 1949.
30. Stone, A. H. On supersonic flow past a slightly yawing cone. *J. Math. and Phys. 27*, 67–81 (1948).
31. Staff of the computing section, Center of Analysis (under direction of Zdenek Kopal). Tables of supersonic flow around yawing cones. *Mass. Inst. of Tech. Technical Report 3*. 1947.
32. Ferri, A. The linearized characteristics method and its application to practical nonlinear supersonic problems. *NACA Technical Note 2515*. 1951.
33. Ferri, A., Ness, N., and Kaplita, T. Supersonic flow over conical bodies without axial symmetry. *J. Aero. Sci. 20*, 563–571 (1953).

INDEX

GPSR Authorized Representative: Easy Access System Europe - Mustamäe tee 50, 10621 Tallinn, Estonia, gpsr.requests@easproject.com

www.ingramcontent.com/pod-product-compliance
Ingram Content Group UK Ltd.
Pitfield, Milton Keynes, MK11 3LW, UK
UKHW020641110826
474212UK00003B/106

* 9 7 8 0 6 9 1 6 2 7 1 0 6 *